Advances in
Human Factors and
Ergonomics in Healthcare

Advances in Human Factors and Ergonomics Series

Series Editors

Gavriel Salvendy
Professor Emeritus
Purdue University
West Lafayette, Indiana

Chair Professor & Head
Tsinghua University
Beijing, People's Republic of China

Waldemar Karwowski
Professor & Chair
University of Central Florida
Orlando, Florida, U.S.A.

Advances in Human Factors and Ergonomics in Healthcare

Edited by
Vincent G. Duffy

CRC Press
Taylor & Francis Group
Boca Raton   London   New York

CRC Press is an imprint of the
Taylor & Francis Group, an **informa** business

CRC Press
Taylor & Francis Group
6000 Broken Sound Parkway NW, Suite 300
Boca Raton, FL 33487-2742

© 2011 by Taylor and Francis Group, LLC
CRC Press is an imprint of Taylor & Francis Group, an Informa business

No claim to original U.S. Government works

Printed in the United States of America on acid-free paper
10 9 8 7 6 5 4 3 2 1

International Standard Book Number: 978-1-4398-3497-8 (Hardback)

Visit the Taylor & Francis Web site at
http://www.taylorandfrancis.com

and the CRC Press Web site at
http://www.crcpress.com

Table of Contents

Section I: Healthcare and Service Delivery

Section II: Patient Safety

Section III: Modeling and Analytical Approaches

Section IV: Human-System Interface: Computers & Medical Devices

Section V: Organizational Aspects

Preface

This book is concerned with human factors and ergonomics in healthcare. The utility of this area of research is to aid the design of systems and devices for effective and safe healthcare delivery. Each of the chapters of the book were either reviewed by the members of Scientific Advisory and Editorial Board or germinated by them. Our sincere thanks and appreciation goes to the Board members listed below for their contribution to the high scientific standard maintained in developing this book.

Explicitly, the book contains the following subject areas:

I.   Healthcare and Service Delivery
II.   Patient Safety
III. Modeling and Analytical Approaches
IV.   Human-System Interface: Computers & Medical Devices
V.   Organizational Aspects

This book would be of special value internationally to those researchers and practitioners involved in various aspects of healthcare delivery.

April 2010

Vincent G. Duffy
School of Industrial Engineering
Purdue University
West Lafayette, Indiana, USA

Editor

Chapter 1

Challenges and Opportunities for Applying Human Factors Methods to the Development of Non-technical Skills in Healthcare Education

Michael A. Rosen[1], Moshe Feldman[2], Eduardo Salas[3], Heidi B. King[4], Joe Lopreiato[5]

[1]Booz Allen Hamilton

[2]College of Medicine
University of Central Florida (UCF)

[3]Department of Psychology,
and Institute for Simulation and Training, UCF

[4]Office of the Assistant Secretary of Defense (Health Affairs):
TRICARE Management Activity

[5]Uniformed Services University for the Health Sciences,
and National Capital Area Simulation Center

ABSTRACT

The long-term success of recent efforts to improve the quality and safety of healthcare systems will require the systematic development of non-technical skills, particularly teamwork, in care providers. More specifically, there is a need to formally integrate the non-technical skills needed for providing effective care into the full continuum of medical education, from the earliest stages of acquiring basic technical skill and knowledge to supporting professional development of practicing care providers. However, there are many challenges associated with meeting this objective. This paper proposes that Human Factors concepts and tools developed for training teamwork skills in a variety of safety-critical domains can provide partial solutions to some of the challenges faced by the medical education community. To that end, this paper 1) provides an overview of the current state of non-technical skill development for medical professionals, 2) summarizes important methods and tools in simulation-based training for teams, 3) discusses how these Human Factors strategies could be applied across the continuum of medical education as well as significant challenges to doing so, and 4) details a set of key needs for the moving forward.

Keywords: Teamwork, team training, non-technical skills, medical education, simulation-based training

INTRODUCTION

Teamwork is a fundamental component of high-reliability organizations. The provision of safe and quality care frequently requires the coordination of a diverse set of skilled professionals. However, the education and training of healthcare providers is usually conducted along disciplinary lines, and real interaction of professionals does not occur until on the job performance. Consequently, the skills needed to ensure effective *coordination* among care providers are not developed in a *systematic* way but left to develop haphazardly if at all. This represents a real problem for creating and sustaining the changes in culture and competence required to transform healthcare into a high reliability industry (To Err is Human). This paper attempts to address this issue by providing a Human Factors perspective for incorporating non-technical skills into medical education programs from the earliest stages and through life-long learning.

More specifically, simulation provides an opportunity for building, reinforcing, and refining non-technical skills throughout the course of a clinician's development. However, there are unique challenges to and differences in both the content and delivery of SBT for non-technical skill training across stages of learning (i.e. acquisition of knowledge, application of that knowledge to produce clinical competence, sustaining and continuously developing knowledge and skills) that occur over the various phases of medical education (i.e. undergraduate medical education (UME), graduate medical education (GME), continuing medical education (CME)) as physicians develop from novice to expert levels of performance. To that end, this paper seeks to accomplish several specific goals.

First, we provide an overview of non-technical skill development in medical education focusing on defining the key concepts as well as how they are implemented and evaluated in educational programs. Second, we outline the key methods of SBT for teams. The fundamentals of design and delivery of practice-based learning for teams can be used as a framework to organize the educational needs at different levels of education. Third, we discuss the key requirements, challenges, and example learning activities for each phase of development. Fourth, we highlight critical needs for widespread adoption of integrated non-technical skill training in medical education.

NON-TECHNICAL SKILLS IN MEDICAL EDUCATION

WHAT ARE NON-TECHNICAL SKILLS?

Non-technical skills are defined differently across levels of medical education, but tend to always include components of communication, interpersonal skills, and professionalism. Recently, the Accreditation Council for Graduate Medical Education (ACGME) has required that residency programs address these skills formally in preparing their residents (ACGME, 2008). While not explicitly stated as a unique competency, teamwork skills are critical for most core competencies and teamwork training continues to grow as an essential component of graduate and continuing medical education programs. Non-technical skills in undergraduate medical education are more ill defined than in graduate and continuing medical education ranging from communication between the physician and the patient to professionalism and coming in the form of come in the form of learning objectives by the Association of American Medical Colleges (AAMC) that stipulate students must have "an understanding of, and respect for, the roles of other health care professionals, and of the need to collaborate with others in caring for individual patients and in promoting the health of defined populations" (AAMC, 1998). The AAMC and Liason Committee for Medical Education (LCME) both support this objective in adopting the goal of addressing these core competencies stipulated by the ACGME to prepare students for graduate level education. A unifying framework for non-technical skills is needed.

HOW ARE NON-TECHNICAL SKILLS CURRENTLY TRAINED?

Non-technical teamwork skills are mostly addressed in graduate and continuing medical education often through on the job training or through lectures talking about teamwork in the operating environment and, more recently, using simulation based exercises. Much fewer attempts have been made to train these skills in the first few years of medical education because of a focus on medical knowledge and basic sciences. A growing trend has emerged to augment traditional curriculum by incorporating non-technical skills training into the first two years of medical education. These methods include interprofessional education programs and team based learning.

Interprofessional education programs provide opportunities for students across disciplines (e.g. medicine, nursing) to learn and practice together in an effort to increase understanding of each other's role while working together (e.g. Siassakos et al., 2009). Team Based Learning (TBL) is an educational paradigm, which groups students into learning teams (Haidet, O'Malley, & Richards, 2002). A central function of TBL is to promote student learning through various team processes. These learning teams work through application focused exercise where they can discuss and apply their knowledge on higher cognitive levels than afforded by traditional methods. The use of interprofessional education programs and TBL has been growing and research has shown that student and faculty have favorable reactions towards these formats.

FUNDAMENTALS OF SIMULATION-BASED TRAINING FOR TEAMS

The central premise of this paper is that SBT for teams can be a valuable framework for furthering the current efforts aimed at integrating non-technical skills into medical education curriculum. Consequently, this section provides an overview of SBT for teams. First, we provide a description of the core features of an SBT system for team training. These represent the critical features underlying effective practice-based learning programs for teams (Salas, Rosen, Burke, Nicholson, & Howse, 2007). Second, we highlight two aspects of programs for SBT for teams that can be used to 1) maintain continuity of teamwork concepts and skills throughout the development process of a clinician, and 2) understand the types of differences necessary to meet the needs of different phases of education and development. Specifically, we discuss the content and delivery of SBT for teams. In the following section, these concepts will be used to further explore the challenges and opportunities for non-technical skill development using SBT.

OVERVIEW OF SBT SYSTEM COMPONENTS

Simulation-based training for teams is a strategy for developing teamwork competencies using practice-based learning activities. This strategy involves the use of explicitly articulated competencies (i.e., knowledge, skills, and attitudes—KSAs) and specific learning objectives, practice activities with opportunities to perform the targeted KSAs, and diagnostic performance measurement that drives the provision of feedback and decisions about training remediation. The following sections provide an overview of each of these components.

Competencies and Learning Objectives

Training and education programs start with a specification of what is to be learned, both in terms of the program as a whole (i.e., the overall competencies) and specific learning activities (i.e., learning objectives). In the case of team training, the science of teams has produced a large body of empirical research supporting the existence of a broad variety of teamwork KSAs (Salas, Rosen, Burke, & Goodwin, 2009). This includes communication, leadership, back-up behavior, mutual performance monitoring, and shared mental models among others (Salas, Sims, & Burke, 2005).

An important distinction in the teamwork competency literature involves team specific and team generic competencies. Team generic competencies are also called *transportable* team competencies because they are 1) held at the individual level (i.e., they are knowledge, skills, and attitudes help by an individual that enable effective teamwork) and therefore move with the individual as he or she moves from team to team, and 2) are not specific to the unique characteristics of a particular set of team members or to a particular context. These transportable team competencies are generalizable across teams and situations.

Scenarios and Event Sets

In SBT, the scenario is the curriculum (Salas et al., 2005). That is, the demands placed on the learner while participating in a scenario define what can potentially be learned from that activity. Event-based methods to designing SBT for teams (Rosen et al., 2008; Lazzara et al., in press), opportunities to perform (as well as measure and learn) targeted teamwork behaviors are scripted into scenarios. These are called critical events and it is these sets of events (i.e., collections of conditions in the scenario that require team members to exhibit a targeted teamwork competency) that make SBT for teams so powerful. They provide structured opportunities to practice and measure teamwork competencies. However, these event sets are not tied to a specific type of simulation technology (e.g., mannequin patient simulators, part task trainers, role players and standardized patients, virtual environments). Events eliciting teamwork behaviors can be inserted into many different types of learning activities.

Performance Diagnosis

Measuring performance is a key component of educational and training systems as it allows for systematic performance diagnosis, that is, for determining the underlying causes of effective and ineffective performance (Salas et al., 2007).

Feedback and Remediation

Feedback is essential for learning. In general, effective feedback is timely, accurate, focused on improvement, and process oriented (i.e., descriptive of concrete behaviors involved in good performance). Ideally, team members engage in a type of guided self-learning process after an SBT scenario. Here, team members self-assess their performance and developed steps for improvement. However, this process needs scaffolding (or external support) from a skilled trainer, especially for learners new to teamwork skills (e.g., relatively novice team members will not understand the content or process they are intended to engage in).

Summary

The above sections provided a very brief overview of a relatively complex approach to training teams. However, for the present purposes, there are two core components of an SBT system for teams that are most useful to consider: content and delivery. At a high level, these are the aspects of SBT for teamwork that will vary across

First, the *content* of an SBT program is most simply the 'what' that is being targeted for training and development (i.e., competencies and learning objectives). There has been a recent emphasis on curriculum standardization in medical education. However, little has been accomplished in defining a robust and broadly accepted model of non-technical skills. While there is no widely adopted competency model for teamwork in healthcare to date, there is a large scientific literature detailing the knowledge, skills, and attitudes (KSAs) underlying effective teamwork.

Second, the *delivery* of an SBT system involves the nature of the practice activity used to provide opportunities to perform and learn. SBT is a well-articulated strategy that makes use of a variety of technologies to create simulations—ongoing representations of some aspect of a task and its environment for the purposes of learning. SBT for teams can involve a potentially broad set of learning activities. The type (or types) of learning activity which will be most effective depend on the content being trained as well as characteristics of the learner. These issues will be discussed in more detail in the following section.

IMPLEMENTING SBT FOR NON-TECHNICAL SKILL DEVELOPMENT

This section provides an initial discussion of how key non-technical skill requirements, challenges, and potential solutions for incorporating non-technical skill competencies can be mapped across the continuum of medical education. Table 1 provides a summary of this discussion focusing on differences in content and delivery across developmental phases.

NON-TECHNICAL SKILLS IN UNDERGRADUATE MEDICAL EDUCATION

The primary goal of undergraduate medical education is to prepare learners for graduate medical education. It is where future physicians acquire the fundamental scientific knowledge and general clinical skills necessary for providing care. There are several key issues with implications for non-technical skills training.

First, learners are not acquiring the domain specific technical skills at this time. Consequently, team generic teamwork competencies should be the focus of the content of SBT for teams at this phase. While learners may not be engaged in the types of tasks they will be performing on a daily basis, more general versions of effective teamwork skills can be instructed. Additionally, awareness of the importance of non-technical skills is a common component of non-technical training across domains and is appropriate for this level of learner.

Second, if teamwork involves the coordination of technical skills among team members, how is this accomplished in the absence of technical proficiency? This can be achieved by using practice activities adjusted for the level of the learner's proficiency in technical skill. In fact, certain learning activities (e.g., team-learning exercises, problem-based learning group activities) can provide a valuable opportunity for learners to practice, be assessed, and receive feedback on non-technical skills (i.e., interpersonal skills, communication, leadership, etc.). Additionally, scripted role-play activities can provide opportunities for learners to experience the targeted teamwork behaviors in meaningful ways without needing to have mastery of technical skills.

Table 1 Overview of appropriate non-technical skill education content, delivery, and challenges by phase of medical education.

	Undergraduate Medical Education	Graduate Medical Education	Continuing Medical Education
Content	Develop awareness of the importance of non-technical skills in patient safety and quality care. Develop general and transferable non-technical skills.	Build specialty specific teamwork behaviors. Develop inter-positional knowledge about the roles of other professions.	Develop mentoring and coaching skills to help develop new 'team players' and leaders. Build facilitation skills to enhance learning from experience on the job.
Delivery	Use blended learning approaches for developing awareness. Use role-play activities to demonstrate behaviors. Use TL exercises where	Use high-fidelity practice activities emphasizing teamwork. Use on the job coaching to reinforce	Use high-fidelity practice activities blending teamwork and technical performance in complex ways.

	non-technical skills are assessed. Use low-fidelity practice for generic skills.	concepts developed in SBT.	
Challenges	Learners have not yet acquired technical competence. Logistical challenges for meaningful inter-professional interaction.	Resident work hour limitations restrict the amount of time that can be spent in training activities.	Balancing the amount of Continuing Medical Education dedicated to non-technical vs. technical topics.

NON-TECHNICAL SKILLS IN GRADUATE MEDICAL EDUCATION

In Graduate Medical Education, physicians select their specialty area and engage in extensive training, both experiential on the job training as well as more structure learning. Several key issues for this phase of education are provided below.

First, the content of SBT for teams can shift from team generic to team specific. Different domains may emphasize different types of teamwork skills (e.g., teamwork in an operating room is different in some ways than teamwork in an intensive care unit) due to the nature of the work performance in that area (e.g., differences in role structures of teams, time pressure, patient volume and criticality). During Graduate Medical Education, the generic teamwork competencies developed earlier should be refined to the context of clinical practice. Additionally, learners can begin to develop inter-positional knowledge (i.e., an understanding of what other professional roles know and do in their specific domain) during this phase of learning.

Second, the delivery of SBT for teamwork can move from more abstract types of practice activities (e.g., low-fidelity simulations, role-play) to more robust types of delivery that include contextual information that may distract more novice learners.

NON-TECHNICAL SKILLS IN CONTINUING EDUCATION

Practicing physicians serve a critical role not only in providing high quality and save are, but in developing future physicians. Because of their position, they set the tone (or local culture) where they work. They serve as role-models and mentors to junior staff. Consequently, the content of Continuing Medical Education can focus on more advanced aspects of teamwork such as coaching and facilitation of team learning. High-fidelity simulations remain appropriate for advanced learners at this phase, but the scenarios used can become much more complex—integrating technical and teamwork competencies in order to provide challenging learning opportunities.

KEY NEEDS

The above discussion provides an outline of several opportunities to .. In fact, each of these has been implemented to some degree already. However, an overarching approach or method to developing and reinforcing these non-technical skills throughout a medical career is currently absent. This section discusses some of the critical needs for making this happen.

First, an important advancement would be the development of a consensus model of non-technical competencies for each level of education. Ideally, this would be an overarching framework that mapping the generic teamwork skills as well as more contextualized competency models for each clinical domain. Recently, the Accreditation Council for Graduate Medical Education (ACGME) has provided a high-level competency model for all graduate medical education programs. This model includes some aspects of non-technical skills (e.g., professionalism); however, there is a gap in terms of a robust and well-articulated model of teamwork that can be used throughout career development. This is critical to enable 'interoperability' of education programs as well as continuity across an individual's development; that is, terminology and concepts are consistent across different educational institutions and levels of education.

Second, diagnostic measurement tools capable of distinguishing between technical and non-technical performance in team scenarios is essential (Salas, Rosen, & King, 2009). This capacity allows educators to systematically and accurately identify deficiencies and provide the appropriate type of feedback. Both individual and team level feedback is important, but identifying when each is called for can be difficult without a measurement process.

Third, while some initial concepts have been outlined in this paper, systematic research is likely required to identify the most effective blend of delivery methods for a learner at a given phase of development.

CONCLUDING REMARKS

Changing the culture of healthcare requires a long-term commitment from all aspects of the system. To date, much effort has been dedicated to improving quality and safety at the point of care. This is no doubt a critical aspect of the broader change initiatives. However, a more longitudinal and holistic approach requires providing healthcare providers with access to learning activities to support the development of non-technical skills throughout their careers. This paper has outlined some of the conceptual and methodological tools available in human factors approaches to training teams with simulation. Simulation can be a valuable tool throughout career development; however, identifying the appropriate content and specific delivery methods for a given learners needs is critical. The present paper has outlined some potential solutions, but programmatic research is required to further refine and validate content and methods for different groups of learners.

ACKNOWLEDGEMENT

The views expressed in this paper belong to the authors and may not represent the University of Central Florida, Department of Defense Patient Safety Program, or the Uniformed Services University for the Health Sciences.

REFERENCES

Accreditation Council for Graduate Medical Education. (2008). *Program director guide to the common program requirements.* ACGME, Chicago.

Association of American Medical Colleges. (1998) *Report I: Learning objectives for medical student education. Guidelines for Medical School.* AAMC, Washington D.C.

Haidet, P., O'Malley, K.J., & Richards, B. An initial experience with 'team learning' in medical education. *Academy of Medicine, 77,* 40-44.

Lazzara, E. H., Weaver, S. L., DiazGranados, D., Rosen, M. A., Salas, E., & Wu, T. S. (in press). MEDS: The Medical Educators Designing Scenarios Tool A Guide to Authoring Teamwork Focused Simulation Based Training Scenarios For Patient Safety. *Ergonomics in Design.*

Rosen, M. A., Salas, E., Wilson, K. A., King, H. B., Salisbury, M., Augenstein, J. S., et al. (2008). Measuring Team Performance for Simulation-based Training: Adopting Best Practices for Healthcare. *Simulation in Healthcare, 3*(1), 33-41.

Salas, E., Rosen, M. A., & King, H. B. (2009). Integrating teamwork into the "DNA" of Graduate Medical Education: Principles for simulation-based training. Journal of Graduate Medical Education, 1(2), 243-244.

Salas, E., Rosen, M. A., Burke, C. S., & Goodwin, G. F. (2009). The Wisdom of Collectives in Organizations: An Update of the Teamwork Competencies. In E. Salas, C. S. Burke & G. F. Goodwin (Eds.), *Team Effectiveness in Complex Organizations.*

Salas, E., Rosen, M. A., Burke, C. S., Nicholson, D., & Howse, W. R. (2007). Markers for Enhancing Team Cognition in Complex Environments: The Power of Team Performance Diagnosis. *Aviation, Space, and Environmental Medicine (Special Supplement on Operational Applications of Cognitive Performance Enhancement Technologies), 78*(5), B77-B85.

Siassakos, D., Timmons, C., Hogg, F., Epee, M., Marshall, L., & Draycott, T. (2009). Evaluation of a stratedy to improve indergraduate experience in obstetrics and gynaecology. *Medical Education, 43,* 669-673.

Chapter 2

The Daily Plan®: Patients Taking Part in Patient Safety

Beth J. King, Peter Mills, Julia Neily, Amanda M. Fore, James P. Bagian

Department of Veterans Affairs
National Center for Patient Safety
Ann Arbor, MI 48106, USA

ABSTRACT

Empowerment of patients to speak-up during hospitalization is critical for the delivery of safe care. The Veterans Health Administration's (VHA) National Center for Patient Safety has long advocated a systems-based approach to health care delivery. It has often pointed out that the patient is the only component of the health care delivery system that is always present and yet the least likely to be used as a resource. The Daily Plan® is a patient-specific itinerary put into the hands of the patient to inform them of what to expect each day in the hospital. Using the VA's electronic medical record, current provider orders are extracted and displayed in a one-to-two page printed summary. The Daily Plan® includes such items as diagnostic tests, medications, scheduled appointments, diet, etc. Each day the nurse reviews this summary with the patient. The Daily Plan® can be retained by the patient and he/she can share it with family as desired. This process has been pilot tested on several nursing units at VA hospitals to measure the impact on patient safety with evaluations received by both nursing staff and patients. Of the 85 unique nurse shift evaluations, 15% identified at least one error of omission within their group of patients that shift. Additionally, 12% indicated at least one error of

12

commission was identified and corrected. From the 183 anonymous patient evaluations received, 80% reported The Daily Plan® increased their understanding of what was going to happen to them during that day of their hospital stay; 76% indicated it was easier to ask questions; and 72% reported The Daily Plan® improved their satisfaction of the hospital stay. Comments and suggestions to improve The Daily Plan® were collected and will be used to guide a more user-friendly version. The patient-centered focus of The Daily Plan® strongly demonstrates benefits in patient safety and satisfaction.

Keywords: Patient-centered, patient safety, patient involvement, daily plan

INTRODUCTION

Many authors have called for patients to participate in their care, however traditionally; patients have not been active participants in their health care (Ayana et al., 1998). The Institute of Medicine's hallmark report, To Err is Human (IOM 2000), recommended that patients be viewed as members of the health care team and be actively involved in the process of care. (Institute of Medicine Report 2000), The Institute for Healthcare Improvement (IHI) also advocates engaging patients and families as partners in care (IHI 2008). Involving patients in their care is not only ethically justified but can also improve patient satisfaction and safety (Entwistle, 2007).

A shift in thinking toward patient and family-centered care, such as involving families in daily rounds, has taken place (IHI 2008). Professional journals and books are citing data indicating patient involvement can support patient safety. Among their conclusions are that the patient can play a key role in ensuring accurate and timely diagnosis, an appropriate treatment plan, sharing side effects and preventing adverse events (Vincent, 2002); and involving patients and families in the health care experience can serve as a safeguard in the system, preventing mistakes (Spath, 2004).

The importance of communicating with patients is also gaining attention. Singh noted that communication between patients and providers is an avenue for reducing errors (Singh, 2008). Clarke found that patients and nurses were an effective defense against wrong-site surgery (Clarke, 2007). Patients should feel empowered to actively participate in their medical care. Health care professionals should encourage patients to ask questions so they have a good understanding of what is going to happen and are better equipped to increase the safety and efficacy of their care. Consequently, the time is right to develop tools to support and measure patients involvement in promoting safety in health care. (Entwistle, 2005)

Several current publications advise patients to participate in patient safety. Among them are the Joint Commission Speak Up (Joint Commission 2008) and the Agency for Healthcare Research and Quality (AHRQ) 20 Tips (AHRQ, 2008) which emphasize general actions for all patients. It goes without saying that patients and their families want to help ensure safety during their hospital stay, yet there has been little professional attention on how best to actively involve patients, or on how such involvement can affect patient safety.

The Veterans Health Administration's (VHA) National Center for Patient Safety has long advocated a systems-based approach to health care. The patient is the only component of the health care delivery system that is always present, yet is the one least likely to be used as a resource. (NCPS, 2008) With this in mind, the VHA National Center for Patient Safety initiated a pilot program to test if explicitly reviewing with the patient a written summary of what was going to happen each day could enhance patient safety during hospitalization. This program is The Daily Plan®. The pilot was approved by Research Committee at the Veterans Affairs Medical Center in White River Junction VT and the Institutional Review Board at Dartmouth College.

We wanted to determine if apprising the patient of what to anticipate daily would empower them to ask questions if something seemed different than what they had been informed to expect. We also wanted to determine if providing The Daily Plan® in writing would increase the patient's comfort in asking questions.

METHODS

The Daily Plan® is a patient-specific itinerary that lets the patient know what to expect each day in the hospital. The VHA electronic medical record allows current provider orders to be displayed in a one or two page summary. The Daily Plan® includes diagnostic tests, medications, appointments, diet, etc.

The Daily Plan® was piloted in the fall of 2007 and winter of 2007-2008 on volunteer medical-surgical units at five VA hospitals. Each pilot site tested The Daily Plan® for at least two weeks. At each site, facility and nursing leadership supported the program. Nursing, administrative staff and providers on the unit received information about The Daily Plan® pilot. Some facilities included union leadership, patient advocates, and others in the educational process surrounding The Daily Plan® pilot.

Patients who agreed to participate were given an introductory packet consisting of: an opaque folder to store The Daily Plan®, a letter explaining the program, a brochure about patient safety and the Daily Journal. The Daily Journal was a booklet of blank pages where the patient and/or family could write questions they wanted to ask.

In order to safeguard protected information, the patient's Daily Plan was put in the opaque folder and stored in their night stand or bedside drawer. Daily Plan patients were encouraged to take their information with them upon discharge. If the patient preferred not to retain the written summary, it was destroyed on the nursing unit.

The nurse reviewed The Daily Plan® with each patient daily. Patients were encouraged to ask questions if there was something they did not understand or if something seemed different than planned.

Patients were asked to voluntarily complete an evaluation. Patient evaluations were completed at the hospital prior to discharge.

Evaluations used both a 5-point Likert scale and open-ended questions. All evaluations were mailed to the National Center for Patient Safety absent patient or staff identifiers; only the name of the pilot site was known. Patients and nursing staff evaluated the program for two weeks during the four week pilot.

During the pilot period, 108 patient evaluations were returned to the National Center for Patient Safety. The rate of response is not known as an accounting of the number of evaluations given out to patients was not requested. Of the returned evaluations, the mean number of days the patient was in the hospital was 4.4; the mean number of days the patient received The Daily Plan® was 2.2. The standard deviations were 3.3 and 1.8 respectively.

Patients and/or families were invited to complete the evaluation. The majority (76%) were completed by the patient, 5% by the family and 6% by the patient and family. The remainder of the patient evaluations were incomplete, accompanied by such comments as the "patient was unwilling" or "unable to complete the evaluation."

Figure One displays patient responses to the evaluation. Note that 70% of patients responding reported that The Daily Plan® increased their understanding of what was going to happen to them that day; 65% reported that they were more comfortable because they knew what was going to happen that day; 66% reported that it made it easier to ask questions; 44% reported the patient's family asked questions after reviewing The Daily Plan®; and 65% reported that The Daily Plan® provided information that helped improve their care.

Each nurse who reviewed The Daily Plan® with at least one patient during her or his shift was asked to complete an evaluation at the end of that shift. Nurses were asked to reflect upon their experience with patients who received The Daily Plan® that shift and complete a single summarizing evaluation. This resulted in a total of 92 shift evaluations from the five pilot sites. We are not able to report a response rate for nursing evaluations because we did not capture the total number of nurses who used The Daily Plan®.

Figure Two displays the response from the nursing (RN or LPN) staff. Of particular note, 35% of shift evaluations reported at least one incident in which the nurse corrected items missing from the orders, thus avoiding an error of omission. For example, a patient identified an allergy to a medication that was not on his list of allergies on The Daily Plan®. In addition, 21% of the shift evaluations reported at least one incident in which the nurse and/or the patient noticed and corrected something that prevented a possible error of commission. One example: A patient noticed a medication in The Daily Plan® that he was not to take any more; when this was clarified with the doctor, the medication was discontinued. Sixty-six percent of the shift evaluations also reported at least one opportunity for nursing staff to provide education to patients and/or family that day, and 59% of respondents indicated at least one use of their daily journal, a booklet of several blank pages in which the patient and/or family can write questions, as a route of communication. Finally, 47% of the shift evaluations reported that The Daily Plan® helped patients understand what was going to happen to them and why.

Patients provided a variety of narrative comments regarding The Daily Plan®. One recurring comment was that The Daily Plan® should be provided earlier in the hospital stay. In future implementations of The Daily Plan®, we will examine the best way to ensure the optimal timing of its delivery. Another recurring suggestion was to simplify the information. These suggestions will help us improve the usability of the information provided.

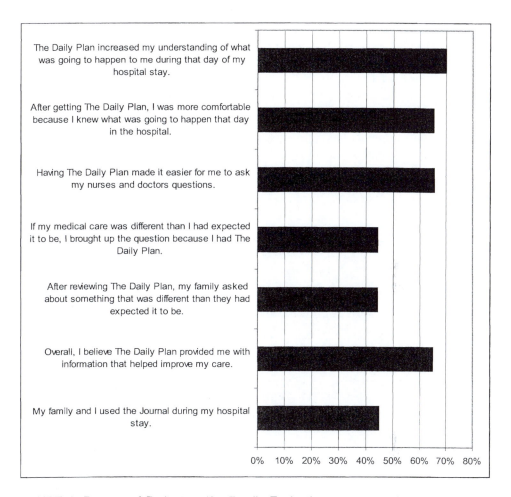

FIGURE 1. Percent of Patient and/or Family Evaluation responses that agreed or strongly agreed with the statement.

Several patients also praised The Daily Plan® with comments such as, "very good," "everything was great," "Plan is good idea, if followed through," "it was very helpful," and "I liked knowing what and when appointments were for me and testing being done." These comments support continued development of The Daily Plan®.

Nursing staff comments about The Daily Plan® frequently mirrored patient feedback, such as the need to improve the readability of the document. Nurses also made positive comments about how The Daily Plan® helped improve patient education. However, there was some concern about added workload such as "Added to a busy day by increase in my work load" and "More paperwork." This supports the need for continued discussion with nursing staff about the optimal way to implement the plan.

The majority of patient evaluations indicated The Daily Plan® increased understanding of the treatment plan, made it easier for patients to ask questions, and provided information that improved their care. Patients received The Daily Plan® an average two out of four days. We will examine the optimal timing and frequency of the delivery of The Daily Plan® in future implementation efforts.

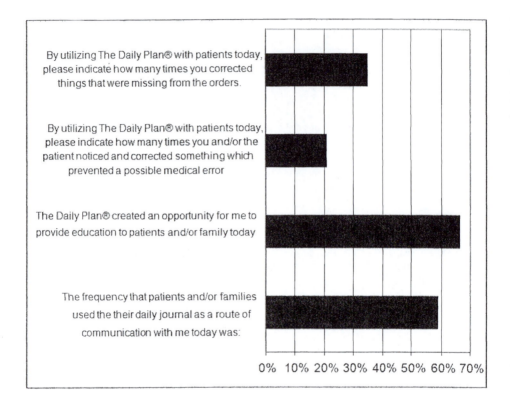

Figure 2. Percentage of nursing staff reporting this behavior or event on at least one occasion on the end-of-shift evaluation.

We also received helpful feedback from patients and nurses that we needed to improve how understandable the documents are for patients. Patients and nursing staff suggested The Daily Plan® could be enhanced by using larger print, replacing medical terminology with language more readily understood by patients. In additions patients indicated a preference for receiving the Daily Plan at specific times of the day (for example early in the morning) and some patients wanted results of their tests in addition to receiving information about what tests they would be getting on a particular day.

We were surprised and pleased that the shift evaluations indicated The Daily Plan® helped detect errors of omission (such as identifying an allergy that was not on his medication list). We were also pleased that use of The Daily Plan® help prevent possible errors of commission.

The shift evaluations showed that 47% of staff that submitted the form felt that The Daily Plan® helped their patients understand their treatment plan. However, a few respondents commented that they already accomplished this by their current approach of care delivery and that The Daily Plan® didn't result in a change. Consequently, our goal is to increase the percentage of shift evaluations in which staff perceive that The Daily Plan® helped their patients understand their treatment plan. We think improving the readability of the document is an important step in making this possible.

We will also encourage nursing leaders to continue discussions with nursing staff about any concerns of increased paperwork or workload. It is encouraging however, that 66% of the evaluations indicated The Daily Plan® provided an opportunity for patient/family education.

This pilot was initiated as a quality improvement program. It was not a controlled study with attention to controlling for diagnosis, case mix, etc. We did not have a control group for comparison purposes. In addition, this project was limited to medical-surgical units in the acute care setting. As a consequence, we have no information on how well the program may work in other settings such as long term care or mental health units. Finally, because we do not know the precise response rate of patients or staff participating in The Daily Plan® pilot, it is possible that responses were received by those who had strongly negative or positive reactions to The Daily Plan®. We suspect, however, that this is not the case because of the range of responses received.

CONCLUSIONS

Providing a patient-specific written summary of what is planned for a patient during each day of hospitalization can enhance patient safety. The evaluations indicated patients felt more empowered to ask questions when they had written information available about their planned medical care each day, and staff indicated The Daily Plan® helped detect potential errors which could be resolved, reducing the risk of possible harm.

Overall, The Daily Plan® process appeared beneficial to both patients and staff. Patients and nursing staff suggested The Daily Plan® could be enhanced by using larger print, replacing medical terminology with language more readily understood by patients. There was also an interest expressed to include laboratory and other test results for patients. Usability testing would provide valuable information for a systematic revision of the plan and how it is presented.

Further attention is warranted to evaluate this type of approach in other health care settings, including outpatient settings. Hospitals within the Veterans Healthcare Administration will participate with the National Center for Patient Safety to create the next generation of The Daily Plan® and expand its use with patients. Widespread use of The Daily Plan® may be able to improve the safety and technical quality of patient care, and can better inform patients so that they can more meaningfully participate in their own care.

18

REFERENCES

Ayana M, Pound P, Ebrahim S. (1998) The views of therapists on the use of a patient-held record in the care of stroke patients. *Clin Rehabil* 12:328-337.

Institute of Medicine (2000) *To Err is Human: Building a Safer Health System.* Washington, DC: National Academy Press.

Institute for Healthcare Improvement (2006) Delivering Great Care: Engaging Patients and Families as Partners. http://www.ihi.org/IHI/Topics/CriticalCare/IntensiveCare/ImprovementSto ries/DeliveringGreatCareEngagingPatientsandFamiliesasPartners.htm (Accessed Sept. 9, 2008).

Entwistle VA. (2007) Differing perspectives on patient involvement in patient safety. *Qual Saf Health Care.* 16:82-83.

Vincent CA, Coulter A. (2002) Patient safety: What about the patient? *Qual Saf Health Care.* 11:76-80.

Spath PL. (2004) *Partnering with Patients to Reduce Medical Errors.* Chicago: Health Forum, Inc.

Singh H, Naik AD, Rao R, Peterson LA. (2008) Reducing diagnostic errors through effective communization: harnessing the power of information technology. *J Gen Intern Med.* 23: 489-494.

Clarke J R, Johnston J, Finley ED. (2007) Getting surgery right. *Ann Surg* 246:395-403.

Entwistle VA, Mello MM, Brennan TA. (2005) Advising patients about patient safety: current initiatives risk shifting responsibility. *Jt Comm J Qual Patient Saf.* 31: 483-494.

The Joint Commission: Speak Up ™. http://www.jointcommission.org/GeneralPublic/Speak+Up/about_speakup. htm (Accessed Sept. 12, 2008).

Agency for Healthcare Research and Quality (AHRQ): 20 Tips to Prevent Medical Errors. http://www.ahrq.gov/consumer/20tips.htm (Accessed Sept. 12, 2008).

NCPS (2008) United States Department of Veterans Affairs, National Center for Patient Safety: Patient Safety for Patients. http://www.va.gov/ncps/patients.html#intro (Accessed Sept. 12, 2008).

Chapter 3

Teaching Healthcare Clinicians to Demand Safe Healthcare Delivery Systems

Linda C. Williams, James P Bagian

Veterans Health Administration
National Center for Patient Safety
Ann Arbor, Michigan, USA

ABSTRACT

Healthcare clinicians are very good at figuring out how to make things work to bring about the best for patients. Good intentions and the effort to work-around design problems can lead to unanticipated patient harm. We are teaching a faculty development workshop aimed at training physicians and other allied health professionals to observe, and report devices and systems that fall short of excellent, safe design (Aron and Headrick, 2002). Reliable manufacturers want to know when their products risk causing harm to patients. Clinicians coping with less than excellent design reduce demand for high reliability. Teaching basic human factors engineering principles provides a basis for making demands for excellence in the design of healthcare systems.

Keywords: Patient safety, device design, human factors engineering, healthcare training

INTRODUCTION

The Veterans Health Administration's National Center for Patient Safety (NCPS) is an integrated component of the VA healthcare system. NCPS is charged with promoting a safe healthcare environment and decreasing the risk of serious harm to patients. NCPS has promoted use of multiple tools for improving heath care safety including Root Cause Analysis (RCA), Healthcare Failure Mode Effect Analysis, and application of human factors engineering (HFE) principles in problem solving.

Graduate Medical Education as well as training for a variety of allied health professions is a critically important component of the VA healthcare system. Trainees are involved in providing much of the front line clinical care at VA and university hospitals (Gwande, 2002). The VA is responsible for providing a large proportion of all graduate medical education in the United States (Brotherton et al, 2004). Thus, education in the use of HFE to solve patient safety issues has potential to affect healthcare currently provided through VA, as well as the knowledge and skills of the future healthcare workforce nationally. Because of this NCPS continues to provide specialized training to faculty of healthcare training programs.

These workshops promote the use of interactive and hands-on learning methods in presenting content (Gosbee et al, 2006). Use of these teaching methods is supported by accrediting organizations which look for evidence of practice-based learning and systems-based practice. Application of HFE-based problem solving is recognized by both accrediting bodies and certifying medical boards (Kachalia et al, 2006)

USING HANDS-ON METHODS TO TEACH HUMAN FACTORS ENGINEERING

Introducing clinical faculty to a new approach to problem solving requires establishing usefulness and credibility. As clinicians are asked to consider their own experience with adverse events in which a patient was harmed or almost suffered harm, they are willing to examine these events in a different light: What happened, why did it happen, and what can be done to keep it from happening again? This is illustrated as we see seasoned faculty handling devices such as the resuscitation bag (Figure 1), finding correct assembly a challenge, and gaining insight into the unfairness of the challenge in light of the device's use in a crisis situation.

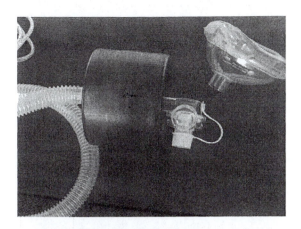

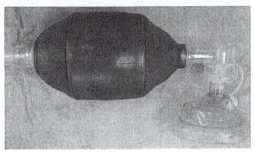

FIGURE 1. Some assembly required: the degree of opaqueness, the planes, and availability of ports provide no cues for correct assembly of mask and bag. The bag is designed for compact storage rather than efficient correct use.

Healthcare safety is best evaluated according to standards appropriate to a high reliability organization (Tamuz and Harrison, 2006). Individuals and systems must perform in a highly reliable manner consistently or risk potentially serious harm to patients. Studying adverse events and close calls allows redesign of system components to compensate for tasks and situations found to be vulnerable to human failure. As clinicians consider contributions of the environment and devices to their own experiences with adverse events, they begin to see that human factors engineering is as essential to patient safety as understanding microbiology is to infection control (Gosbee et al, 2006).

AVOIDING JARGON, UNDERSTANDING PRINCIPLES

One of the barriers to teaching HFE and patient safety is the vocabulary that rapidly becomes familiar to teachers. The dilemma becomes how best to convey what is essential without the use of terminology that may be mysterious to healthcare clinicians whose knowledge of pathophysiology includes its own complex terminology. The goal is to create understanding of concepts that are assimilated even if the terminology is not (Stone, 2008). To accomplish this, we have workshop participants conduct usability testing and heuristic evaluations of ordinary everyday items (juice drink pouches, moist wipes, and mint dispensers) to identify design issues; discuss and sketch ideas for redesign; and then examine medical devices to identify design issues (Sojourner et al, 1993 and Williams, 2005).

INVESTIGATIVE SKILLS TO AVOID PLACING BLAME ON THE HUMAN

Honing accident investigation skills is essential to learning from adverse events and close calls (Dekker, 2002). Two essential components found to improve discovery of root causes are creation of a cause and effect (or fishbone) diagram and compliance with the rules of causation (Woodcock et al, 2005). After having participants practice these two methods in just the way that we propose they do with their students – to analyze an ordinary everyday misadventure; we apply the skills to practice application to a case conference or morbidity-mortality conference.

CHANGES TO TEACHING STRATEGIES

When the faculty development workshops began in 2003, we included discussion of adult learning principles, but found this strategy took discussion away from a focus on human factors engineering and patient safety. Because those attending the workshops have expertise in both education and medical science, we resolved to demonstrate rather than discuss adult learning principles; minimizing didactic presentations, making the most of hands-on experiences, and provoking discussion. The central goal has remained to keep the principles and tools of HFE as foundational to teaching patient safety, and to train clinicians capable of effective problem solving.

Selected Learning Objectives for both Faculty and Students in Healthcare Training Programs

1. Become familiar with the basics of safety engineering and human factors engineering.
2. Understand the importance of discovering root causes and contributing factors in developing effective interventions.
3. Become familiar with human factors engineering techniques and principles related to root causes; and understand why they are crucial to effective problem solving.
4. Understand the relative strength of patient safety interventions; i.e., why the first solutions to come to mind may not produce the desired outcome. (NCPS 2005)

From the skills gained at the workshop, clinicians are able to perform a basic evaluation of devices, medications, and architecture in an operational environment. Training programs have expanded learning opportunities for their students to include patient safety electives with assigned projects, patient safety rotations during residency, and more recently an opportunity to spend a year as patient safety Chief Resident.

MEASURING CHANGE

Results of a VA-wide patient safety culture survey show favorable changes for physicians in training. In four dimensions of the survey, programs with faculty attending the workshop scored higher: non-punitive response to error, availability of education and training resources in patient safety, decreased use of shame, communication and openness, and senior management that is aware of and actively involved in promoting patient safety.

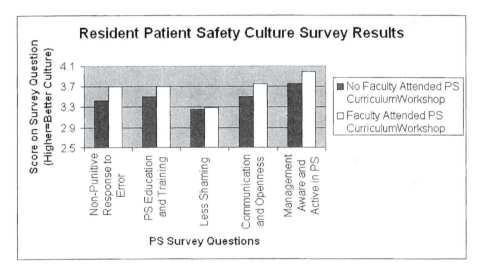

FIGURE 2. Four dimensions of culture change showing improvement in training programs affiliated with VA hospitals where faculty have attended a patient safety teaching workshop.

CONCLUSIONS AND LIMITATIONS

While it might seem ideal to dream of the results of patient safety education as a reduction in medical errors, this seems reasonable only in those few seconds of framing the answer to what you would wish for if Aladdin made the offer. Because the goal of patient safety problem solving is to protect patients from unintended harm, measuring success is rather about changes made to the way healthcare clinicians think. Success is measured in the opportunities for clinicians to become adept at identifying problematic designs (McLaughlin, 2003).

As important as recognizing when design contributes because of inherent flaws or because it fails to support human vulnerabilities; is recognizing when one's own knowledge limitations require calling for a consult. Clinicians understand that they put patients at risk when diagnosis and treatment of pathophysiology lie outside of their expertise. In the same way, we continue to teach the basics of HFE with the *caveat* that there will be problems in the design of things for which a consultation with an expert in a specialty engineering field is the best and safest choice.

The work of NCPS extends beyond education and training to patient safety investigations. In addition to physicians, nurses, and pharmacists, NCPS staff includes a variety of engineering specialties. Reports of problematic devices and equipment are evaluated (NCPS 2010). Conversations with manufacturers ensue. Responsible manufacturers are eager to correct problematic aspects of their products, and to maintain professional relationships with the VA healthcare system.

REFERENCES

Aron DC, Headrick LA. (2002) Educating physicians prepared to improve care and safety is no accident: it requires a systematic approach. *Qual Saf Health Care.* 11:168-173.

Brotherton SE, Rockey PH, Etzel SI. (2004) US Graduate medical education 2003-2004. *JAMA.*;292(9);10032-1113.

Carayon P. Editor (2007). *Handbook of human factors and ergonomics in health care and patient safety.* Lawrence Erlbaum and Associates, Inc.

Dekker S. (2002) *The Field Guide to Human Error Investigations.* Burlington, VT: Ashgate.

Gosbee JW, Anderson T. (2003) Human factors engineering design demonstrations can enlighten your RCA team. *Quality and safety in healthcare.* 12: 119-121.

Gosbee JW, Gosbee LL, Editors (2005). *Using Human Factors Engineering to Improve Patient Safety.* Joint Commission Resources.

Gosbee JW, Williams L, Dunn E. (2006) Teaching the teachers of patient safety: A progress report. *ACGME Bulletin.* Sept. http://www.acgme.org/acWebsite/bulletin/bulletin09_06.pdf (Accessed 2/28/2010).

Gwande A. (2002) Complications: A surgeon's notes on an imperfect science. Henry Hold and Co. New York

Kachalia A, Johnson JK, Miller S, Brennan T. (2006) The incorporation of patient safety into board certification examinations. Academic Medicine 81(4); 317-325.

McLaughlin R. (2003) Redesigning the Crash Cart: Usability testing improves one facility's medication drawers. *American Journal of Nursing*. 103(4):64A-64F.

Sojourner RJ, Aretz AJ, Vance KM. (1993) Teaching an introductory course in human factors engineering: A successful learning experience. Proceedings of the human factors and ergonomics society 37th annual meeting. Santa Monica, CA: Human Factors and Ergonomics Society; 456-460.

Stone NJ. (2008) Human factors and education: Evolution and contributions. *The journal of human factors and ergonomics society special 50th anniversary issue*. 50:3; pp534-539.

Tamuz M, Harrison MI. (2006) Improving patient safety in hospitals: Contributions of high-reliability theory and normal accident theory *Health Serv Res*. 41(4 Pt 2): 1654–1676.

Williams L. (2005) Changing systems the hands-on way. The VA National Center for Patient Safety's Museum. *Maryland Medicine*. 6(4):16-18.

Woodcock K, Drury CG, Smiley A, Ma J. (2005) Using simulated investigations for accident investigation studies. *Applied Ergonomics* 36(1):1-12.

NCPS (2005) - VA National Center for Patient Safety. Patient Safety Introduction Instructor's Guide.
http://www.patientsafety.gov/curriculum/Instruct_Prep.html#intro (Accessed 2/28/2010).

NCPS (2010) - VA National Center for Patient Safety. Alerts and Advisories.
http://www.patientsafety.gov/alerts.html (Accessed 2/28/2010).

Chapter 4

OR2010 – Processes Analysis Within the OR and the Consequences for Product Design

Ulrich Matern, Dirk Büchel

Experimental-OR
University Hospital of Tuebingen
Tübingen, Germany

ABSTRACT

Operating room staff experience significant problems within operating rooms (OR) in relation to functionality, working conditions and efficiency.

Two questionnaire surveys of working conditions administered to surgeons and OR nurses at the German Surgical Society's annual conferences in 2004 and 2005. Response rates of 11.7% for surgeons (n=425 respondents) and 54.3 for OR nurses (n=190) were achieved.

Real data of OR processes have been collected an put into a simulation software, which provides workflow analysis concerning ground floor and personnel.

Respondents of the questionnaires reported problems with the use of devices leading to hazards and risks for employees and patients. 70% of surgeons and 50% of nurses reported difficulties operating medical devices. Thus more than 40% were involved in situations of potential danger for staff or the patients. Recovery rooms, though available, were often not in use for this purpose due to missing equipment or staff shortages.

The analysis of the real clinical processes visualized the insufficient usage of rooms and personnel. The efficiency of the OR could be increased by a better architecture but need a more flexible employment of staff and the improvement of medical as well as building technology.

These micro- and macro-ergonomic hazards appear to relate to inadequate integration between differing aspects of OR workflow, and the lack of cost-benefit analysis.

Keywords: Micro ergonomics, macro ergonomics, operating room

INTRODUCTION

Consulting hospitals one of the most often concerns of the administrators is the inefficiency of the hospital especially of the OR-facilities as the key bottle neck of the hospital. The payment for overtime hours of the staff is one reason; the low turnover of patients another one. The surgeons themselves claim long changing times, where they lose time for the treatment of other patients at the ward or ambulance. As reason they declare the inadequate "old" architecture and missing equipment.

To evaluate the actual state in hospitals the project "OR2010" was performed which was supported by a grant of the Government of Baden Württemberg, Germany.

MATERIAL & METHOD

1. Questionnaires about working conditions in the OR were distributed among surgeons attending the 2004 annual meeting of the German Surgical Society (www.dgch.de) and among OR nurses at the nursing conference that was held during the same society's meeting in the following year (2005). [Matern et al. 2006]
2. During the visitation of several hospitals data of the OR processes were collected or provided by the administration. These data include the time of arrival of the patient at e.g. the OR entrance, the induction room, the OR suite itself, the time for the procedure, the time necessary to clean the OR after a procedure, etc. These data where fed into the computer and analysed with MedModel software regarding varies ground floors and personnel. [Matern, 2009]

RESULTS

425 of the 3 621 surgeons working in Germany who attended the 2004 meeting filled out the questionnaire (yield = 11.7%). The comparable figure among participants in the nursing conference in 2005 was 190 out of 350, or 54.3%.

Utilization of space:

86% of operating rooms had an adjacent anaesthesia induction room. In general, when such a room existed, it was used for its intended purpose (82%) – induction of anaesthesia, wiring, but not positioning of the patient. 70% of operating rooms had a separate, adjacent room for emergence from anaesthesia (a pre-PACU [post anaesthetic care unit] emergence room). Nonetheless, for about 80% of all operating rooms, the anaesthesiologists supervising patients' emergence from anaesthesia always did so in the operating room itself. The justifications that were given for this practice were that insufficient staff was available to use the emergence room, that the emergence room was architecturally unsuited to its purpose (e.g., too small), or that the necessary equipment for general anaesthesia and patient monitoring was not available there.

Climate control:

Although nearly all operating rooms had climate control equipment, the climatic conditions were still not ideal for the two professional groups' surveyed.

	Surgeons	Nurses
The air is too dry	22,0 % (n = 422)	35,1 % (n = 185)
The temperature is unpleasant	31,0 % (n = 419)	21,9 % (n = 187)
There is a draft	45,9 % (n = 422)	76,2 % (n = 189)
Smoke rises from the field during electrical cutting and coagulation	74,8 % (n = 413)	

OR apparatus:

In more than 80% of operating rooms, the electrical cables of OR apparatus lay on the floor, creating a risk of tripping, or were suspended in mid-air between a wall socket and the apparatus or the patient. More than half of all surgeons and OR nurses stated that this impaired their work. 60.5% of surgeons and 81.7% of OR nurses reported having experienced dangerous situations repeatedly because of instrument cables. Furthermore, 46.1% of surgeons and 21.1% of OR nurses reported difficulty in matching tubes and cables with their properly corresponding sockets. More than 90% of apparatuses used in the OR are free-standing and have their settings adjusted directly on the instrument console itself. Many persons working in operating rooms stated that they could not personally use all OR equipment with full intuitiveness and safety, that they felt inadequately trained to do so, and that danger might result.

	Surgeons	Nurses
Equipment could not always be intuitively used correctly	69,8 % (n = 416)	48,9 % (n = 184)
I feel inadequately trained in the use of equipment	58,8 % (n = 410)	40,3 % (n = 186)
I have read the instruction manual for all equipment in the OR	6,7 % (n = 418)	23,4 % (n = 188)
Problems with the use of equipment have repeatedly endangered persons in the OR	39,7 % (n = 388)	47,7 % (n = 176)

Efficiency of the OR-Facility

Data collected during the visitation of ORs confirmed the inefficiency claimed by the employees. Changing times between tow procedures of one up to two hours seem to be normal. Observing the actions within the OR the reasons are obvious:

- External factors:
  - Waiting for the patient
  - Waiting for colleagues to join the team
- Internal factors:
  - many processes have to be done in a relatively small area
  - Preparing the instruments
  - Conducting anaesthesia
  - Positioning, disinfection and draping of the patient
  - Anaesthesia and monitoring of the patient
  - Preparing devices
  - Documentation of nurses, surgeons and anaesthesiologists
  - Cleaning of the room

By outsourcing of the following actions in other rooms the changing time can be shortened up to 10 or 15 minutes:

- Preparation of instruments to special room
- Induction of anaesthesia, monitoring and positioning into the induction room
- Emergence of the anaesthesia into a special room

This concept was verified within the MedModel software using real data form an OR unit with two ORs performing 25 procedures in one week starting in the morning at 7:00 a.m. and finishing in the night at 11:00 p.m. The results for the new concept are:

- All procedures have been finished on Thursday at 6:00 p.m.
- 20% increase of turnover is possible
- No overtimes have to be paid
- More personal is necessary (one anaesthesiologist with a 40% employment, three staff nurses with a 5-30% employment)

CONCLUSION

The data indicate that micro and macro ergonomic deficits in the operating room, due to poor standardisation, system integration and the lack of uniformity in the use of medical devices, are a significant source of costly errors, mistakes, and potential complications.

The resulting list for ergonomic improvements in the OR is a long one. This includes the establishment of standard operation procedures (SOP), the education and training of the employees and embeds architecture as well as building and medical technology. This should be taken into account in all future decisions regarding operating rooms.

One first step would be to stipulate ergonomic design as a prerequisite for the approval of any medical product by the United States Food and Drug Administration (for the US market)

or by the corresponding bodies in the European Union (17, 18, 24).

Another important step is to explain patients and hospital personal that SOPs are not only a method to improve cost efficiency of hospitals but to improve the quality and safety of medicine and thereby, physicians and nurses will have more time to spend with their patients.

REFERENCES

Matern U., Koneczny S., Scherrer M., Gerlings T. (2006), *Arbeitsbedingungen und Sicherheit am Arbeitsplatz OP*. Dtsch Ärztebl 2006; 103(47): A 3187–92. English Version: *Working Conditions and Safety in the Operating Room.* www.aerzteblatt.de

Matern U. (2009), *Der Experimental-OP: Betriebswirtschaft, Patientensicherheit und humanitäre Patientenversorgung sind keine Gegensätze.* Der Chirurg BDC 3/2009: 149-153

Chapter 5

Benefits of a Structured User Centered Design Process For Customers – Patients and Manufacturers

Dirk Büchel, Ulrich Matern

Experimental-OR
University Hospital of Tuebingen
Tübingen, Germany

ABSTRACT

The condition precedent for usable medical devices is a user centered design process. This paper points out the actual situation in health care and the benefit of a structured user centered design process for medical products for manufacturer, customers, users and patients. Furthermore the specific requirements of such a process for medical products are shown.

Keywords: User centered design, usability engineering, patient safety, EN IEC 62366, EN IEC 60601-1-6

INTRODUCTION

Continuing modification and quality optimization affect the medical context of use. Modern and complex devices and technologies with a lot of features are catching on more and more. Koneczny shows that in the ordinary operating theater ten devices are used frequently. 75 other devices will be fetched if required. All these devices belong to 55 different device classes from different manufacturers with different user interfaces, interaction sequences and usage concepts [10]. Long ago is the time when one qualified person operated with only one device. Sufficient device training is impossible by the multiplicity of user interfaces and not usable operational concepts.

With the increasing number of complex devices, less training time per device and more stress the risk for user errors increases, too [5, 21]. The bulk of these errors are due to poor human machine communication and a lack of usability. Bleyer as well as von der Mosel call human machine misunderstanding to account for 60 percent of the incidents with medical devices [3, 20]. Nielsen shows that a lot of the problems with medical devices are well known as classic usability problems [13], including: misleading default values, new commands not checked against previous ones, poor readability, memory overload, date description errors and overly complicated workflow.

At devices that have been developed with regard to the user centered design process, these problems are reduced as far as possible. Such products are characterized by a fast learnability and an easy usage. They allow the user to achieve his goal efficiently, effectively and satisfied. Errors occur more rarely and at best work makes more fun. However usability aspects and the user centered design are not only justified by increasing user satisfaction, efficient and effective device usage, but also economic savings for manufacturers, users and clients as well as increased patients and user safety play a decisive role.

ECONOMIC REASONS FOR USABILITY

On the one hand patient treatment is combined with high costs and on the other hand health care providers have to act more and more economically. Benefit of usability quality combined with cost savings is multifarious. Positive effects due to usable devices are observable for users/clients and manufacturers.

For companies usability activities are often events that are coincided with additional effort and costs. But usability activities have to be deemed to be an investment. Generally the monetarily benefits by usability engineering and user centered design outweigh the costs [8, 12]. Wiklund visualizes the trend of expected costs savings due to usability engineering for a hypothetic product [21].

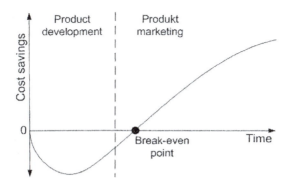

FIGURE 1: Cost savings due to usability engineering

For device manufacturer the biggest benefit of usability engineering is an intuitive product, that is adapted to the purpose, matches the user needs and requirements. Usable products succeed in the market because they come up to the clients expectations, are easy to use, will be used more often and will have a larger target audience. Furthermore for safety critical systems and devices the risk which correlates with the usability of the devices is of particular importance. In health care usability could be a matter of life or death. If device use could influence patient safety, usability problems could be critical to the success of a company.

At the user site (direct operators or companies where the product is in use) costs savings due to usable devices are on the one hand a higher productivity due to more efficient and effective task fulfillment, lower training needs and fewer errors. By decreasing the errors caused by usability problems and upgrading the usability of a user interface the performance could be increased. Shneiderman states that inconsistent colours, the size and position of buttons could reduce the working speed by 5-10%; confusing label on buttons could lead delays up to 25% [11, 18]. Arnold & Roe describes that 30 to 40% of the whole interaction time is used for error findings and corrections [1]. Lacking usability is a reason for higher risk and lower patient safety, longer working and tasks times, higher staff costs, more education and training needs and more user frustration.

Findings from Germany, USA, UK, Kanada and Australia suggest that about 10% of patients admitted to hospital may suffer some kind of adverse outcome [2, 15, 19, 22]. The US „Institute of Medicine" estimates the total national costs of preventable adverse events to be between $17 billion and $29 billion [9]. Together with Blyers assumption, that 60% of the incidents are due to usability problems, the total national costs of preventable adverse events due to bad human-machine communication is between $10.2 and $17,4 billion.

USER AND PATIENT SAFETY DUE TO USABILITY

The most important argument for usable devices in health care is user and patient safety. Safety critical situations could be caused by problems in human-machine communication, which could have effects on the user or patient health [14, 17]. Even expert users could be influenced by misleading and illogical user interfaces [16]. User errors with effect to the user or patient are often caused by the device. Possible outcome ranges between itching eyes and the death of the patient [4, 5, 16, 17]. The results of the IOM study shows for the USA that deaths due to medical errors are in the 8th-leading causes of death [9]. As mentioned before 60% of them are due to usability problems.

According to Bleyer, X-ray machines are mostly involved in use errors. A inspection of 209 incidents revealed that 88.9% of all errors are use errors for x-Ray machines, followed by radio frequency devices with 87.8% and infusion pumps with 66.7%. However for infusion pumps the outcome of the error is worse. In 33.3% of the incidents leads to the death of the patient, for radio frequency devices "only" 2%, for X-ray machines none [3]. Another study of the US Food and Drug Administration confirms that infusion pumps and radiofrequency devices are mostly involved in erroneous situations [7].

These findings demonstrate that the attention of human factor aspects and usability requirements are essential for optimizing the medical work environment and devices.

USER CENTERED DESIGN FOT MEDICAL DEVICES

User centered design process for medical devices is nearly the same process like the process for other devices, that is described in literature. This iterative process should focus on the goals, needs, characteristics and requirements of the users, too. However medical devices are generally safety critical devices. The development process had to be supplemented with safety aspects and parts of the risk management.

The fundamental process for a design process for usable devices in health care with regard to safety aspects is explained in the standards EN IEC 60601-1-6 and EN IEC 62366. These standards define specific requirements for a process to analyse, specify, design, verify and validate the usability, as it relates to safety of medical devices. They describe the usability engineering process and provide guidance on how to implement and execute the process to provide safe design [5, 6].

This design process helps to develop medical devices that raise human efficiency, decreases the risk of human errors and optimizes patient safety. The aim of usable medical devices is in the first instance a safe, fast and high quality diagnosis and therapy. Anyhow care and treatment by effective and satisfied staff is important, too. For the user the ease, flexibility and accuracy of the use of the devices is major relevant. Effective and intuitive workflows would satisfy the user. He would have more time for the patient, because he needs less time for achieving the tasks with the device. Health care provider would benefit on the one hand from increasing productiveness, efficiency and the cost savings involved, on the other hand their staff is more satisfied and lower employee turnover should be expected. Furthermore they benefit from well treated patients who enjoyed the personal time the staff had for them. Medical device manufacturer could secure a competitive advantage thanks by satisfied customers who consider their products to be safe, efficient and adapted to their needs. Furthermore development costs would increase because the engineers know early in the production cycle how the device should look like and lower support costs because intuitive devices don't need as much support.

However the process described in the standards is at a general level. Especially the assessment of usability problems and their safety relevance is a problem. Paper 524 describes the method "UseProb" that makes assessment of usability problems with regard to their risk possible.

REFERENCES

1   Arnold B., Roe R. A. (1987): User errors in human-computer interaction. In: Frese M., Ulich E., Dzida W. (Hrsg..): Psychological issues of human computer interaction in the workplace. Amsterdam. Elsevier. S.203-220

2   Baker G.R., Norton P.G., Flintoft V., Blais R., Brown A., Cox J., Etchells E., Ghali W.A., Hébert P., Majumdar S.R., O'Beirne M., Palacios-Derflingher L., Reid R.J., Sheps S., Tamblyn R. (2004): The Canadian Adverse Events Study: the incidence of adverse events among hospit. In: 2004 May 25;170(11):1678-86

3   Bleyer S. (1992): Medizinisch-technische Zwischenfälle in Krankenhäusern und ihre Verhinderung. In: Anna O., Hartung C. (Hrsg.): Mitteilungen des Instituts für Biomedizinische Technik und Krankenhaustechnik der Medizinischen Hochschule Hannover. Hannover. Fachverlag für Krankenhaustechnik

4   DGBMT-Fachausschuss Ergonomie in der Medizintechnik (2004): Ergonomie in der Medizintechnik: Potenzial zur Qualitätsverbesserung in der medizinischen Versorgung. Deutsche Gesellschaft für Biomedizinische Technik im VDE (DGBMT). Frankfurt am Main. (2004)

5   IEC 60601-1-6 (2006): Medical electrical equipment - Part 1-6: General requirements for basic safety and essential performance - Collateral Standard: Usability (IEC 60601-1-6:2006)

6   IEC 62366 (2008): Medical devices - Application of usability engineering to medical devices

36

7   Hölscher U., Laurig W. (2004): Ergonomie-Leitfaden für die Gestaltung von Medizinprodukten. Sankt Augustin. Kommission Arbeitsschutz und Normung Verein zur Förderung der Arbeitssicherheit in Europa. KAN-Bericht 31.

8   Karat C.-M. (1994): A Business Approach to Usability Cost Justification. In: Bias R. G., Mayhew D. J.(Hrsg.): Cost-Justifying Usability, San Francisco. Academic Press. S. 45-70

9   Kohn L.T., Corrigan J.M., Donaldson M.S. (1999): To err is human: building a safer health system. IOM-Reprot. Institute of Medicine. Washington DC. National Academy Press

10   Koneczny S. (2008): Erfassung und Analyse von Schwachstellen in der Funktionsstelle OP deutscher Krankenhäuser. Dissertation, Medizinische Fakultät Universität Tübingen

11   Mahajan R., Shneiderman B. (1996): Visual & Textual Consistency Checking Tools for Graphical User Interfaces. In: IEEE Transactions on Software Engineering 23, 11 (November 1997). S. 722-735

12   Mayhew D. J., Mantei M. (1994): A Basic Framework for Cost-Justifying Usability Engineering. In: Bias R. G., Mayhew D. J. (Hrsg.): Cost-Justifying Usability, San Francisco. Academic Press. S. 9-43.

13   Nielsen J. (2005): Medical Usability: How to Kill Patients Through Bad Design. In: Jakob Nielsen's Alertbox, April 11. Abruf unter: www.useit.com/alertbox/20050411.html (20.12.2007)

14   Rowland K, Pozzo D. (2007): What is Human Factors Analysis? In: Science & Technology 2007. S.36-39

15   Sauro J., Kindlund E. (2005): How long Should a Task Take? Identifying Specification Limits for Task Times in Usability Tests. In: Proceeding of the Human Computer Interaction International Conference CHI2005, Las Vegas

16   Sawyer D. (Eds.)(1996): Do it by Design – An Introduction to Human Factors in Medical Design. U.S. Department of Health and Human Services. Public Health Services. Food and Drug Administration. Center of Device and Radiological Health.

17   Shalala D., Herman A., Eisenberg J. (2000): Doing What Counts for Patient Safety. Federal Actions to Reduce Medical Errors and Their Impact. Report of the Quality Interagency Coordination Task Force (QuIC). Abruf unter: http://www.quic.gov/report/errors6.pdf (06.07.2009)

18   Shneiderman B. (1999): Das Bewusstsein wächst- Das goldene Zeitalter des Benutzers. In: SAP INFO Nr. 59 – 3/99

19   Vincent C., Neale G., Woloshynowych M. (2001): Adverse events in British hospitals: preliminary retrospective record review. In: BMJ 322 (7285), 2001, S. 517-519.

20   Von der Mosel H.A. (1971): Der klinisch – biomedizinische Ingenieur. In: Schweizerische Ärztezeitung 52 vom 29. 12. 1971. Zitiert in: Matern U, Koneczny S, Scherrer M, Gerlings T.: Arbeitsbedingungen und Sicherheit am Arbeitsplatz OP. Dtsch Arztebl 2006; 103(47): A 31

21   Wiklund M. (1995): Medical Device and Equipment Design – Usability

Engineering and Euipment Design. Boca Ration. Buffalo Grove. Interpharm Press.

22 Wilson R. M., Runciman W. B., Gibberd R. W., Harrison B.T., Newby L., Hamilton J.D. (1995): The Quality in Australian Health Care Study. In: Med J.Aust 163 (9) 1995, S. 458-471.

Chapter 6

Conflicting Benefits and Hazards of Hospital-Style Bed Rails

Sue Hignett

Healthcare Ergonomics and Patient Safety Unit
Loughborough University

ABSTRACT

This paper reviews the design and use of hospital-style bed rails. Rails were originally used as a safety feature for psychiatric patients at risk of falling in the 1800s. There are benefits, for example security in transit, facilitating repositioning, but also hazards, including deaths and injuries associated with entrapment. The developments in the technological specification of hospital beds (electric) has increased with their functionality. However, a survey in England and Wales found that patients on electric beds / pressure mattresses were three times more likely to have their rails raised. This may lead to an increase in the exposure to the risks associated with bed rails and presents a conflict for designers, staff and patients.

INTRODUCTION

Hospital beds are one of the highest volume medical devices and are found in all areas of health care. There are probably over 400,000 hospital-style beds in

England. This includes over 220,000 in hospitals (Hignett et al, 2007) as well as beds in nursing and residential homes and for care at home (Mitchell et al, 1998). The design has moved a long way since the development of the King's Fund bed specification in 1966 (Maxwell, 1997). In 1998 it was recommended that all areas (hospitals, nursing homes and private homes) should consider using powered, profiling, adjustable height beds in the UK for patients with limited or compromised mobility (Mitchell et al, 1998). There are many issues relating to the design of hospital beds, including infection control, tissue viability (mattress design), maintenance, usability (by patients and caregivers) and resuscitation support. This paper will focus on the design and use of bed (side) rails.

Bed rails (also known as side rails, bed side rails, cot sides and safety rails) are adjustable metal or rigid plastic bars that attach to the bed and are available in a variety of shapes and sizes from full, three-quarters, half, one-quarter, and one-eighth in lengths as well as one or two (split rails) for each side of the bed. Full length bed rails are available in two basic designs, trombone (telescoping) and concertina/folding (Govier and Kingdom, 2000; MDA, 2002).

WHEN ARE RAILS USED?

Healey et al (2009) carried out a survey of bed rail use at night (22.30 – 06.30) in seven randomly selected hospitals in England and Wales. They surveyed 1,092 patients in all specialities except obstetrics and paediatrics and collected data on mattress type, bed type, bed rail use, patient age, mobility, conscious state, confusion, and nurses' rationale for bed rail use. They found that 26% of patients had full rails raised and 9% of patients had partial rails raised. Patients had raised bed rails if they were described as confused (four to seven times more likely); on electric beds/alternating pressure mattresses (three times more likely). 74% of full rails were raised as a response to a perceived falls risk (by the nursing staff), with only 7% raised to be used as a turning aid and 5% raised following a patient request. For partial rails, 30% were raised to be used as a turning or rolling aid, 23% as a request from a patient, and 16% to prevent falls. The authors comment that 'although some patients may use bedrails as a movement aid, they are not designed for that purpose, and alternative equipment may be more effective.'

The use of bed rails has been discussed since the 1960s, with Fagin and Vita (1965) commenting that 'to many conscious patients, side rails are frightening and imply dangerous illness. To others, side rails are irritating and humiliating because they emphasize the confining aspects of hospitalization'. Most patients want to retain their independence, in particular with respect to elimination needs, for example 'on numerous occasions seriously ill patients climbed over the bed rails to go to the bathroom, thus averting the embarrassment of a soiled bed' (Parrish and Weil, 1958). These negative perceptions seem unchanged in 2000s. Gallinagh et al (2001) interviewed patients and elicited negative comments about the bed rail as a restraint and inappropriate use of rails by getting round the rail if they wanted to exit the bed. The patients thought the nurses were using bed rails as standard practice, firstly if the patient was restless to stop the patient and/or bed clothes slipping, and secondly as a risk averse response to staff concerns about blame or litigation if the patient fell out of bed. The use of bed rails has been identified as a component of a risk averse safety culture. Oliver et al (2008) suggest that falls could be *'cited as a failure in the duty of care (a crucial feature of successful clinical negligence claims* [in the UK])'. The increased use of rails (to prevent falls) can lead to a change in clinical practice with restriction in movement and patient autonomy which, in turn, may limit or delay rehabilitation.

WHAT ARE THE BENEFITS AND HAZARDS?

Benefits

Bed rails may serve a number of purposes (HBSW, 2003) including security in transit, facilitating turning and repositioning within the bed or transferring in or out of a bed, providing a feeling of comfort and security, facilitating access to bed controls, and providing a physical barrier to remind the patient of the bed perimeters. However no studies have been located exploring or measuring these benefits. Bed rails have been used extensively as an intervention to manage falls (McCarter-Bayer et al, 2005; Capezuti et al, 2007; Rainville, 1984; Dunn, 2001; Kilpack et al, 1991; Hanger et al, 1999; Healey et al., 2004), but there is no evidence that they prevent falls or injury (Capezuti et al, 2007; Hanger et al, 1999).

Another benefit for bed rails is as extra storage space for bed controls. Foster (2004) compared three designs of hospital bed control handsets; bed rail mounted, pendant (on cable) and control panel (figure 1). Visual (macular degeneration) and tactile impairment were simulated to evaluate any related errors with 36 participants for the task of adjusting the height of bed. Data were collected using subjective perception rating questionnaires and link analysis for error evaluation. The bed rail mounted handset was perceived to be the most complex to use but produced the fewest errors (link analysis).

Electric beds have benefits for both patients and care givers. Patient independence is enhanced through self-adjustments for both bed position and height. Staff safety has been improved by reductions in musculoskeletal demands for moving a patient in bed or transferring to/from the bed (Dhoot and Georgieva, 1996; Hampton, 1998; RCN, 1996). Milke et al (2008) suggest that the use of electric beds may contribute to a reduction in the use of bed rails as staff may feel more comfortable leaving full bed rails off electric beds since they can be lowered nearly to the floor and therefore are less likely to cause serious injury if residents happen to fall out of bed. Dhoot and Georgieva (1996) found that using an electric bed (with patient access to controls) enhanced patient independence, with 2.3 times more positional self-adjustments and 5.8 times more height adjustments (frequency) in comparison with a hydraulic (manual) bed.

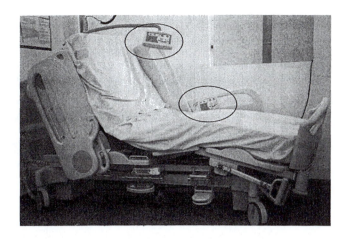

FIGURE 1 Control panel handset and bed rail hand set

Hazards

Donius and Rader (1994) suggested that the use of bed rails can have negative consequences including obstructing vision, separating the care receiver from the caregiver, creating noise, causing trauma if the patient's body strikes or becomes entangled in the side rail, dislodging tubes during raising and lowering, and creating a sense of being trapped or jailed. The patient population who have been identified as being at risk from bed rails are those who are frail, elderly and/or have conditions including agitation, delirium, confusion, pain, uncontrolled body movements, hypoxia, faecal impaction and acute urinary retention (HFCA, 2000).

There have been two papers reporting deaths, injuries and near miss adverse events associated with the use of bed rails (Todd et al, 1997; McLaughlin, 2003) using the adverse event (AE) data (JHACO, 1998) from the Food and Drug Administration (FDA) Manufacturer and User facility Device Experience database (MAUDE). The frequency of bed rail-related reports increased from 111 reports between 1985 and 1995 to 325 reports between 1998-2001 (McLaughlin, 2003). Hazards associated with rails include entrapment (1) in the mattress-rail horizontal gap (or head/foot board); (2) within the rails, including latch failure where the rail drops due to the patient struggling to free themselves; (3) with the body off the bed and the neck or chest compressed by the rail; and (4) between split side rails (Parker and Miles, 1997). Hignett and Griffiths (2005) analysed the data from MAUDE with respect to the type of rail associated with adverse events. They found that incidents involving half rails were more likely to be associated with head, neck or face entrapments and were also more likely than other bed rail types to result in death. There is very limited public domain information available about incidents involving hospital beds in the UK. Marsden (2004) reported 94 incidents in 2002 involving bed rails and 20 deaths in the UK involving bed rails since 1997.

It is possible that the use of bed rails can alter the location of a fall. Donius and Rader (1994) suggested that the use of bedrails may increase the distance of falls from the bed whereas Hignett et al (2010) found the reverse in a small pilot study, where patient falls with raised rails were clustered at the foot end of the bed.

Oliver (2002) describes the use of bed rails and covert restraints (e.g. positioning of furniture, tucking of bed clothes too tight, chair type) as a possible infringement of the autonomy and dignity of patients. Maslow's theoretical model for basic human needs consists of a hierarchy in which physiological needs and the needs for safety and belonging and love can be said to be homeostasis-related. Maslow suggested that people are '*wanting beings*', always wanting more than they already have so when one level of the hierarchy is met they will move to another level (Mullins, 1993).

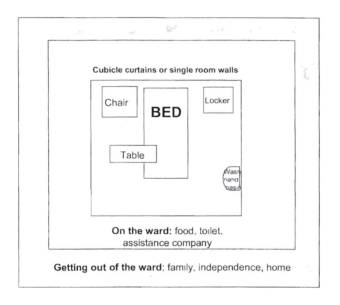

FIGURE 2 Safety and Functionality conflicts

To apply Maslow's model for falls Hignett and Masud (2006) reviewed the levels using a patient-centred model for movement and suggest specific needs for falls (figure 2). In level one (physiological needs) motivation for movement may include bladder and bowel function (to support homeostasis), hunger, thirst and activity. When these basic needs are fulfilled the motivation for movement may involve safety issues including freedom from pain, the threat of physical attack, and protection from danger or deprivation. Levels three, four and five are perhaps less likely to be activated for the 'at risk' group, although an independently mobile patient may be trying to fulfil needs at these higher levels.

CONFLICTS IN BED RAIL USE

There is a conflict in the use of bed rails. At one pole there are clear hazards associated with their use from a reduction in autonomy, movement and rehabilitation through to entrapment, injury and death. At the other pole there are potential benefits for patients with respect to access to the bed controls, turning/rolling assistance. For staff, the main benefits are based on a risk-averse behaviour relating to litigation. The question for quality in healthcare product design and clinical practice is whether the design and use of bed rails can support functionality and provision of care for patients and staff without introducing hazards. From the current evidence it would seem that balance is tipped towards a norm of bed rail use that is likely to increase rather than decrease risk of injury to patients. Can design innovate in spite of risk-averse culture which prefers to restrain patients who are perceived by caregivers to be at risk of falling. Can design lead safety behaviour or is innovation stifled by a risk-averse safety culture?

REFERENCES

Capezuti, E., Wagner, L.M., Brush, B.L., Boltz, M., Renz, S., Talerico, K.A. (2007), "Consequences of an intervention to reduce restrictive side rail use in nursing homes". *Journal of the American Geriatrics Society,* 55(3), 334-41.

Dhoot, R., Georgieva, C. (1996), *The Evolution bed in the NHS hospital environment.* Internal report. The Management School, Lancaster University.

Donius, M., Rader, J. (1994), "Use of side rails: rethinking a standard of practice". *Journal of Gerontology Nursing,* 20, 23-27.

Dunn, K.S. (2001), "The effect of physical restraints on fall rates in older adults who are institutionalized". *Journal of Gerontology Nursing,* 27(10), 40-8.

Fagin, I.D., Vita, M. (1965), "Who? Where? When? How? An Analysis of 868 Inpatient Accidents". *Hospitals,* 39, 60-65.

Foster, A.J. (2004), *A comparison of hospital bed control handsets.* Unpublished B.Sc. Dissertation. Dept of Human Sciences, Loughborough University, UK.

Gallinagh, R., Nevin, R., McAleese, L., Campbell, L. (2001), "Perceptions of older people who have experienced physical restraint" *British Journal of Nursing* 10(13), 852-859

Govier, I., Kingdom, A. (2000), "The rise and fall of cot sides", *Nursing Standard,* 14(31), 40-41.

Hampton, S. (1998), "Can electric beds aid pressure sore prevention in hospitals?" *British Journal of Nursing.* 7(17), 1010-1017.

Hanger, H.C., Ball, M.C. Wood, L.A. (1999), "An Analysis of Falls in the Hospital: Can We Do Without Bedrails?" *Journal of the American Geriatrics Society,* 47(5), 529-531.

44

HBSW. (2003), *Hospital Bed Safety Workgroup. Clinical guidance for the assessment and implementation of bed rails in hospitals, long term care facilities and home care settings.*
http://www.ute.kendal.org/learning/documents/clinicalguidance_SideRails.pdf
Accessed 25 January 2010.

HCFA. (2000), *Health Care Financing Administration guidance to surveyors in the implementation of 42 CFR part 483.13(a).* Medicare and Medicaid Programs. State Operations Manual Provider Certification Transmittal 20. Sept. 7, 45.

Healey, F., Monro, A., Cockram, A., Adams, V., Heseltine, D. (2004), "Using targeted risk factor reduction to prevent falls in older in-patients: a randomised controlled trial". *Age and Ageing,* 33, 1-5.

Healey, F.M., Cronberg, A., Oliver, D. (2009), Bedrail Use in English and Welsh Hospitals, *JAGS,* 57, 1887-1891.

Hignett, S., Griffiths, P. (2005), "Do split-side rails present an increased risk to patient safety?" *Quality and Safety in Healthcare.* 14, 113-116

Hignett, S. Masud, T. (2006), "A Review of Environmental Hazards associated with In-Patient Falls". *Ergonomics,* 49(5-6), 605-616.

Hignett, S., Fray, M., Rossi, M. A., Tamminen-Peter, L., Hermann, S., Lomi, C., Dockrell, S., Cotrim, T., Cantineau, J. B., Johnsson, C. (2007), "Implementation of the Manual Handing Directive in the Healthcare Industry in the European Union for Patient Handling tasks". *International Journal of Industrial Ergonomics,* 37, 415-423

Hignett, S., Sands, G., Youde, J., Griffiths, P. (2010), "Targeting Environmental Factors to reduce Elderly In-Patient Falls", *Proceedings of the 1st International Conference on Human Factors and Ergonomics in Healthcare / 3rd International Conference on Applied Human Factors and Ergonomics*, 17-20 July, 2010, Miami.

JCAHO. (1998), *Sentinel Event Alert: Preventing Restraint Deaths.* Issue 8, Nov. 18. www.jcaho.org. Accessed 2 April 2004

Kilpack, V., Boehm, J., Smith, N., Mudge, B. (1991), "Using research-based interventions to decrease patient falls". *Applied Nursing Research*, 4(2), 50-56.

Marsden, A. (2004), "Safe use of bed rails". Proceedings of workshop, *Are you sleeping safety?* Doncaster, 9th March 2004. Manchester: The Disabled Living Centres Council.

Maxwell, R.J. (1997), "Hospital beds and how to survive them". *Quality in Health Care.* 6, 199-200.

McCarter-Bayer, A., Bayer, F., Hall, K. (2005), "Preventing falls in acute care". *Journal of Gerontological Nursing,* 31(3), 25-33.

McLaughlin, S. (2003), *ASHE for Today's Healthcare Engineer.* 10(7), 22-25.

MDA (2002), *Bed safety equipment: An evaluation.* EL8. London: Medical Devices Agency.

Milke, D.L., Kendall, S., Neumann, I., Wark, C.F., Knopp, A. (2008), "A longitudinal evaluation of restraint reduction within a multi-site, multi-model Canadian continuing care organisation". *Canadian Journal on Aging*, 27(1), 35-43

Mitchell, J., Jones, J., McNair, B., McClenahan, J.W. (1998), *Better beds for health care: Report of the King's Fund Centenary Bed Project.* London, King's Fund

Mullins, L.J. (1993), *Management and Organisational Behaviour* (3rd Ed.) London: Pitman Publishing p452-458.

Oliver, D. (2002), "Bed Falls and Bedrails in Hospital. What should we do?" Editorial. *Age Ageing*, 31, 415 – 418.

Oliver, D., Killick, S., Even, T., Willmott, M. (2008), "Do falls and falls-injuries in hospitals indicate negligent care – and how big is the risk? A retrospective analysis of the NHS Litigation Authority Database of clinical negligence claims, resulting from falls in hospitals in England from 1995 to 2006". *Quality and Safety in Healthcare* 17, 431-436.

Parker, K., Miles, S.H. (1997), "Deaths caused by bedrails". *JAGS* 45(7), 797-802.

Parrish, H., Weil, T.P. (1958), "Patient Accidents Occurring in Hospitals: Epidemiologic study of 614 accidents". *New York State Journal of Medicine*, 58(6), 838-846.

Rainville, N.G. (1984), "Effect of an implemented fall prevention program on the frequency of patient falls". *Quality Review Bulletin*, 10(9), 287-291.

RCN. (1996), *The Hazards of Nursing: Personal Injuries at work.* Employment Information and Research Unit, Royal College of Nursing, Cavendish Square, London.

Todd, J.F., Ruhl, C.E., Gross, T.P. (1997), "Injury and Death associated with hospital bed side-rails: Reports to the US Food and Drug Administration from 1985 to 1995". *American Journal of Public Health* 87(10), 1675-1677.

Chapter 7

The Use of Human Factors to Promote Safety and Efficiency in the Operating Room

Jennifer Wagner[1,2], Jason M. Johanning[1,3], M. Susan Hallbeck[2]

[1]Nebraska-Western Iowa VA Health Care System
Omaha, NE 68105, USA

[2]University of Nebraska-Lincoln
Lincoln, NE 68588-0518, USA

[3]University of Nebraska Medical Center
Omaha, NE, 68198-3280, USA

ABSTRACT

The application of systems engineering principles, tools, and methods to health care almost coincided with the turn of the 20th century, with a focus on surgical procedures. Nearly 100 years later, the potential of systems engineering, including human factors and ergonomics, remains largely unrealized when it comes to improving the conditions in the operating room (OR).

The OR is the most complex environment in health care with an exponential increase in complexity in the past decade. With the shift from open procedures to minimally invasive (laparoscopic, natural orifice), the skills required of surgeons and surgical teams have also increased. Whereas in open procedures, surgeons view the target area directly, in minimally invasive procedures, the surgeons are required to gather, interpret, and integrate the information they need to proceed with the operation from monitors projecting images captured by cameras. This changes not only the surgeons' individual workload physically (new tools and postures) and cognitively (new interpretation and decision making requirements), but it also affects the work of the team, as additional tasks (such as manipulating the cameras or moving the monitors) and skills are now required of the team members. With all of these changes come a host of safety concerns, including those related to the physical environment and use of numerous pieces of new equipment (with cords, cables, and space requirements) as well as those related to the cognitive workload. Surgeons are expected to gather bits of information from a host of disparate sources, compile the pieces to make specific decisions, and then both communicate and act upon those decisions.

This paper reviews the history of human factors engineering in health care, focusing on the OR as the pinnacle of health care complexity. We will emphasize the changes and trends seen during the last ten years and identify different areas in which process and systems tools and methods can (and should) be applied to improve the safety and efficiency of the OR in the future.

Keywords: Human Factors, Ergonomics, Operating Room

INTRODUCTION: HISTORICAL CONTEXT

The first efforts to use systems engineering principles began early in the 1900s, focusing on efficiency and systemization/consistency in organization. The Gilbreths applied principles and tools from scientific management to the OR. In 1910, they initiated the time and motion studies in the OR, using movies to aid analysis of surgeons' movements. Their conclusions: improve the efficiency of operating room nurses and organize the instruments to be used during surgeries (Baumgart and Neuhauser, 2009). To improve conditions for staff and patients, Pool and Bancroft (1917) recommended 'systemization': weekly reviews of both the immediate and "end" results of operations; arrangement of work to minimize efforts deemed as waste and maximize time with patients; and uniformity in room set up and procedures.

Two decades later, Lawrence and Berry (1939) visited 500 ORs, looking at the efficiency of work practices and placement of instruments. They, too, utilized time and motion studies to examine the current placement of instrument tables and made recommendations for setting up the OR to minimize effort based on the bodily location of the procedure. However, another two decades passed and a similar case for the use of time and motion studies to improve health care efficiency was made once again. McKenna (1957) noted that there was opposition among surgeons to the idea that "there is always one method for doing a job which is superior to all competing methods (p. 730)." In supporting the idea of standardization, he supplied information that 40% of the time lost in delays was due to poor preoperative planning, including unavailability of desired instruments.

The types of analysis (using time and motion studies) done in the past are still necessary to promote efficient practices. However, with the complexity of the OR today, time and motion studies are not in and of themselves adequate to address the majority of issues faced by surgical teams. In an overview of the ergonomic issues to be addressed in the OR, Berguer (1999) reviewed the following considerations: visualization (including video monitor placement), manipulation, posture, mental and physical workload, and the operating room environment. Review of past studies demonstrates a focus on similar themes that have continued throughout the past 100 years without a significant change or paradigm shift in the way OR evaluations and studies occur.

EMERGING THEMES

A scan of the literature over the last ten years reveals international activities focused on a myriad of components of the operating room environment: the operating room table (Albayrak, Kazemier, Meijer, and Bonjer, 2004; Kranenburg and Gossot, 2004; Matern and Koneczny, 2007; van Veelen, Jakimowicz, and Kazemier, 2004), lights (Kaya, Moran, Ozkardes, Taskin, Seker, and Ozmen, 2008; Matern and Koneczny), floor space (Albayrak et al., 2004; Decker and Bauer, 2003; Kaya et al.; Koneczny, 2009; Kranenburg and Gossot), cables (Koneczny; Matern and Koneczny), and monitor placement (Albayrak et al, 2004; Decker et al.; Kaya et al.; Lin, Bush, Earle, and Seymour, 2007; Matern and Koneczny; van Veelen et al.). There has also been work examining the tools provided in the OR: instruments (Kaya et al.; Matern 2001; Nguyen et al., 2001; Sheikhzadeh, Gore, Zuckerman, and Nordin, 2009; Trejo, Doné, DiMartino, Oleynikov, and Hallbeck, 2006; Trejo, Jung, Oleynikov, and Hallbeck, 2007; van Veelen et al.), instrument carts (Albayrak et al. 2004; Sheikhzadeh et al.), and foot pedals (Kranenburg and Gossot; van Veleen et al.). In addition, work has continued on identifying and quantifying the negative effects surgical team members experience as a result of the postures adopted during procedures (Albayrak, van Veelen, Prins, Snijders, de Ridder, and Kazemier, 2007; Kranenburg and Gossot; Matern and Koneczny), resulting pain or work related musculoskeletal disorders (Forst, Friedman, and Shapiro, 2006; Gerbrands, Albayrak, and Kazemier, 2004; Kaya et al.; Koneczny; Nguyen et al.; Sheikhzadeh et al.).

An extensive review of these topics and findings is outside the scope of this paper, but it suffices to say that the focus has been on various components of the physical environment; studies remain fragmented, however, often neglecting to look at the larger system. Guidelines and evidence exist to improve the environmental conditions of the OR, yet it appears that adoption has been slow. A 2004 survey completed at the German Society of Surgery Congress yielded a sobering analysis of the current state of equipment in the OR (Matern and Koneczny, 2007). Findings indicated that footstands presented a problem with 49% of respondents reporting that their footswitch has fallen off the stand or that they have come near slipping off. Sixty-seven percent (67%) deemed the operating tables inadequate. OR lights presented problems for 71%. These are crucial pieces of equipment that are deemed insufficient. Clearly, an ability to use the foot pedal when needed, a surgeon slip during procedure, difficulty adjusting the operating table, and improper lighting could contribute to the success of an operation and impact both the surgical team and the patient. Overall, 97% of those surveyed found a need for ergonomic improvements.

One clear and important theme that has emerged only recently is the challenges that have accompanied the evolution of minimally invasive surgery (MIS). "The majority of current operating rooms have been designed in the second half of the 20th century to allow performance of open conventional surgery" (Albaryak et al., 2004, p. 156). These ORs are woefully inadequate to support the emersion and commonality of laparoscopic procedures, which "have altered the way surgeons interact with the surgical field" (Albayrak et al., 2007, p. 1835). Surgeons now operate by watching the procedure on a video monitor instead of direct viewing of the surgical site. This equates to new equipment, including cameras, monitors, and different requirements for lighting, as well as the assumption of new postures, which are quite often far from ergonomic. This additional equipment can cause crowding in the OR and the cables associated with new machinery contribute to some of the potential tripping hazards. The introduction of new equipment also taxes the cognitive load of the surgical team. Efforts have been made to encourage usability testing of OR devices (Hallbeck, Koneczny, Buchel, and Matern, 2008), because 70% of surgeons indicated that devices in the OR were not "intuitively usable" (Koneczny, 2009, p. 153) and 59% did not feel sufficiently trained in using them.

Many of these studies have been driven by findings that indicate that poor MIS device design is one of the leading causes of surgeon post-operation pain or numbness because of the non-neutral postures adopted during MIS procedures (Berguer, Rab, Abu-Ghaida, Alarcon, and Chung, 1997; Berguer, 1998; Berguer, Gerber and Kilpatrick, 1998; Berguer, Forkey and Smith, 1999; Berguer, Smith and Chung, 2001; Crombie and Graves,1996; Cuschieri, 2005; Emam, Frank, Hanna, and Cuschieri, 2001; Trejo et al. 2007; van Veelan and Meijer, 1999; van Veelan et al., 2004). In addition, the current laparoscopic tool design has also been shown to increase surgeon fatigue, discomfort and paresthesias in the fingers (Berguer, 1998; Trejo, et al. 2006). For all the benefits to the patient, such procedures are "fraught with disadvantages" (Matern, 2001, p. 169) for the surgeon. Therefore, recent guidelines have been offered for the design of more ergonomic tools for use in laparoscopic procedures (Matern; van Veelen et al., 2004).

A related theme to MIS and the poor ergonomics is the impact suboptimal designs have on system efficiency. "The lack of ergonomics not only leads to disturbances in the workflow, but also to physical impairment and potential hazards for all persons in the OR" (Matern and Koneczny, 2007). Much time and effort is spent training different members of the surgical team and once they start practicing, they are prone to physical pain and problems. Forst et al. (2006) surveyed 371 North American Spine Society members (surgeons and other medical professionals) and found that 107 (29%) reported carpal tunnel syndrome. For surgeons, carpal tunnel syndrome symptoms developed approximately 17 years from the beginning of their residency. Nearly one third of the surgeons reported that their work was affected in some way by the pain. Kaya et al. (2008) surveyed 82 medical professionals who participated in video endoscopic surgeries about their experience with the ergonomic conditions. Seventy-two percent (72%) reported neck pain related to head extension and 70% had back pain associated with inappropriate table height. There were also significant problems noted due to lighting and body posture. The perioperative nurses and technicians participating in focus groups with Sheikhzadeh et al. (2009) also noted back pain, with 31% reporting work absences due to the condition.

The aforementioned findings are problematic for many reasons. Members of surgical teams have acquired specialized skills, and current OR working conditions are prohibiting a fair percentage from utilizing those skills consistently in response to changes in work practice or absence that are due to musculoskeletal problems. Not only does this situation leave highly-trained workers unable to perform their jobs as planned/desired, but it is not known how the pain and fatigue affects patient safety and outcomes. Procedural modifications or workarounds to avoid painful postures or work tasks could have safety implications for patients and other members of surgical teams. Therefore, it is imperative that environmental factors be addressed. Kranenburg and Gossot (2004) summarized ergonomic problems categorically, describing the specific (postural) problems and offering suggestions to mitigate the effect on surgeons and their teams. Van Veelen et al. (2004) also outline guidelines to promote improved surgeon postures in the OR.

Efforts have begun to examine cognitive factors that impact surgical teams. "Today, surgeons' skills are no longer restricted to purely medical aspects" (van Veelen, 2004, p. 131). As previously mentioned, there is a plethora of new technology for them to learn to use. They need to be skilled in use of many devices as well as interpreting the outcome of the devices. Image guided surgery has become the standard in certain procedures, such as laparoscopic removal of the gall bladder (Matern, 2009). In this and other procedures, such as tumor resection or aortic endografting, surgeons rely on camera images (sometimes from multiple sources) to guide the course of the surgery. Digital images obtained from C-arms, colonoscopes, endoscopes, laparoscopes, and ultrasounds are presently used for guidance in virtually all surgical disciplines including cardiac surgery, ear nose and throat (ENT) surgery, neurosurgery, and orthopedics (Peters, 2001; Sindwani and Bucholz, 2005). The use of patient specific images has many benefits for the patient, but obtaining information from multiple (disparate) sources can prove taxing for the surgeon and team. As with any and all new technologies, "Without proper integration of this tool into the departmental work flow or reinvention of the work flow process in the department, the potential gains associated with the use of a PACS cannot be realized" (Siegel and Reiner, 2003, p. 165).

Based upon the historical literature and the new emerging themes, work is now expanding to examine the processes and systems in place in the OR as a whole. In a study of interruptions during cardiac surgery, Henrickson Parker, Laviana, Sundt, and Wiegmann (2009) reported that surgical interruptions due to technology, instruments, or technical factors were observed less frequently than interruptions due to other causes (such as environmental factors or communication) but were estimated to impact performance to the highest degree. Weinger and Slagle (2002) conducted cognitive task analysis interviews to develop a concept map and introduced the concept of a non-routine event as way to focus on systems, rather than individual providers. Henrickson Parker, Wadhera, Wiegmann, and Sundt (2009) developed and tested a standardized protocol system to facilitate communication between profusionists and cardiac surgeons. It is more of these types of studies, which have adopted methods utilized in other high risk industries, which are needed going forward.

SOCIOTECHNICAL SYSTEMS

The health care industry can be categorized as a complex sociotechnical system (Pasmore, 1988). It includes four elements: a social subsystem, a technical subsystem, an external environment, and an organizational design (Hendrick, 2002). When one subsystem is changed, for example by the introduction of a new technology, it necessarily affects the social system and organization and perhaps also the external environment. All four elements need to be considered jointly in design efforts. As one changes, the others are necessarily affected. Design that fails to take into account all four elements may not achieve its full potential. The military and airline and nuclear industries have looked to human factors and ergonomics to improve their safety and efficiency and it appears to be time for health care, especially those concerned with improving OR conditions, to do so as well.

The OR is especially complex both socially and technically. Many of the technical aspects have previously been discussed, but note needs to be made of the social complexity. During surgical procedures, the OR contains surgeon(s), anesthesiologist(s), surgical trainee(s), nurses, techs, and perhaps product or equipment representatives. The team's members may or may not have worked together previously. As technical changes are made, attention needs to be given to what that means for the social components: surgical teams may need additional education or skills. In addition, usability testing is crucial before devices are deployed for use in the OR.

Therefore, as new technologies are introduced into the OR, the impact on the social components (e.g., cognitive ergonomics and communication), environment (e.g, physical ergonomics and OR spacing), and organization (e.g., staffing needs and policies) needs to be considered. Patients have been one of the main driving forces behind a move towards increasingly minimally invasive surgical procedures. This change in the external environment has impacted the social, technical, and organizational components of the OR, as procedures that utilize a single larger incision require a skill set that is significantly different than that of laparoscopic procedures that are completed via multiple incisions. This interdependency also means that efforts to redesign tools, while a necessary step, may not have the intended effect of improving safety within the OR by excluding the organizational and social subsystems. Clearly there is a need to assess how new technology affects all four subsystems.

FUTURE POSSIBILITIES

"The combination of complexity, professional fragmentation, and a tradition of individualism, enhanced by a well-entrenched hierarchical authority structure and diffuse accountability" present barriers to improving safety and health care (Leape and Berwick, 2005, p. 2387), but as evidenced by other complex industries, significant improvement is achievable.

Just as van Veelen et al. (2004) noted that "to fully solve the problem of the unergonomic posture of the laparoscopic surgeon, it is important to consider all the (product) factors that influence the posture of a laparoscopic surgeon and not to focus on an isolated issue only" (p. 166), products alone are not the only factor of consideration in the OR. Armed with sociotechnical systems theory, the themes identified also represent opportunities for human factors and ergonomics to impact the safety and efficiency of the OR by considering the system as a whole. Health care adopted the use of time and motion studies because of their success in other industries and now is the time to apply human factors and ergonomics to health care to achieve success like that seen in aviation and nuclear power. ElBardissi, Wiegmann, Dearani, Daly, and Sundt (2007) have already started this migration, adopting a tool developed for the aviation industry for use in the OR.

Several recent articles have provided tools for future research. Lee, Lee, Dexter, Klein, and Park (2007) laid out research methods to be used in study of surgical ergonomics, including motion analysis, electromyography (EMG), and force plate systems. Each of these tools focuses on the members of the surgical team and how the work that they are doing (e.g., posture and muscle exertion) affects them physically. Leedal and Smith (2005) summarized the measurement of anesthesiologists' workload which has utilized objective performance, subjective workload (Subjective-Workload Assessment Technique; NASA Task Load Index), and physiological measures (such as heart rate). Koneczny and Matern (2004) also suggested methods for studying ergonomics in the OR. They list observation, checklists, time measurement, measurement of errors, questionnaires, interviews, measurements of posture, instrument motions, and EMG. All of these are tools that can be used going forward to transform the current state of ergonomics in the OR. Sociotechnical systems theory would suggest using them together for maximum efficacy.

Pool and Bancroft's (1917) rationale for systemization "With the large service of the present day... the complicated diagnosis, the large staffs, and the rapid turnover, systemization has become essential" (p. 1599) rings even more true now. What systems engineering originally provided to health care through the use of time/motion studies can now be expanded to look at the safety of the environment, evidence based practices, digital imaging, communication, and teamwork.

REFERENCES

Albayrak, A., Kazemier, G., Meijer, D.W., & Bonjer, H.J. (2004). Current state of ergonomics of operating rooms of dutch hospitals in the endoscopic era. *Minimally Invasive Therapy & Allied Technology,* 13(3), 156-160.

Albayrak, A., van Veelen, M.A., Prins, J.F., Snijders, C.J., de Ridder, H., & Kazemier, G. (2007). A newly designed ergonomic body support for surgeons. *Surgical Endoscopy*, 21(10), 1835-1840.

Baumgart, A., & Neuhauser, D. (2009). Frank and Lillian Gilbreth: Scientific management in the operating room. *Quality & Safety in Health Care*, 18(5), 413-415.

Berguer, R. (1998). Surgical technology and the ergonomics of laparoscopic instruments. *Surgical Endoscopy,* 12, 458-462.

Berguer, R. (1999). Surgery and ergonomics. *Archives of Surgery*, 134(9), 1011-1016.

Berguer, R., Forkey, D.L., & Smith, W.D. (1999). Ergonomic problems associated with laparoscopic surgery. *Surgical Endoscopy*, 13, 466-468.

Berguer, R., Gerber, S., & Kilpatrick, G. (1998). An ergonomic comparison of in-line vs pistol-grip handle configuration in a laparoscopic grasper. *Surgical Endoscopy,* 12, 805-808.

Berguer, R., Rab, G.T., Abu-Ghaida, H., Alarcon, A., & Chung, J. (1997). A comparison of surgeons' posture during laparoscopic and open surgical procedures. *Surgical Endoscopy*, 11, 139-142.

Berguer, R., Smith, W.D., & Chung, Y.H. (2001). Performing laparoscopic surgery is significantly more stressful for the surgeon than open surgery. *Surgical Endoscopy*, 15, 1204-1207.

Crombie, N.A.M., & Graves, R.J. (1996). Ergonomics of keyhole surgical instruments – patient friendly, but surgeon unfriendly. In: Robertson, S. (Ed.), Contemporary Ergonomics. Taylor & Francis, London, pp. 385–390.

Cuschieri, A. (2005). Reducing errors in the operating room. *Surgical Endoscopy*, 19, 1022-1027.

Decker K., & Bauer M. (2003). Ergonomics in the operating room - from the anesthesiologist's point of view. *Minimally Invasive Therapy & Allied Technologies*, 12(6), 268-277.

ElBardissi, A.W., Wiegmann, D.A., Dearani, J.A., Daly, R.C., & Sundt, T.M.,3rd. (2007). Application of the human factors analysis and classification system methodology to the cardiovascular surgery operating room. *The Annals of Thoracic Surgery*, 83(4), 1412-1419.

Emam, T.A., Frank, T.G., Hanna, G.B., & Cuschieri, A. (2001). Influence of handle design on the surgeon's upper limb movements, muscle recruitment, and fatigue during endoscopic suturing. *Surgical Endoscopy*, 15, 667-672.

Forst, L., Friedman, L., & Shapiro, D. (2006). Carpal tunnel syndrome in spine surgeons: A pilot study. *Archives of Environmental & Occupational Health,* 61(6), 259-262.

Gerbrands, Albayrak, & Kazemier. (2004). Ergonomic evaluation of the work area of the scrub nurse. *Minimally Invasive Therapy & Allied Technologies*, 13(3), 142-146.

Hallbeck, M.S., Koneczny, S., Buchel, D., & Matern, U. (2008). Ergonomic usability testing of operating room devices. *Studies in Health Technology and Informatics*, 132, 147-152.

Hendrick, H.W. (2002). An overview of macroergonomics. In H.W. Hendrick & B.M. Kleiner (Eds.), Macroergonomics Theory, Methods, and Applications (pp. 1-24). Mahwah, NJ: Lawrence Erlbaum Associates.

Henrickson Parker, S., Laviana, A., Sundt, T.M., & Wiegmann, D.A. (2009). Developing a tool for reliably identifying distractions and interruptions during surgery. *Proceedings of the Human Factors and Ergonomics Society 53rd Annual Meeting*, 664-668.

Henrickson Parker, S., Wadhera, R., Wiegmann, D., & Sundt, T.M. (2009). The impact of protocolized communication during cardiac surgery. *Proceedings of the Human Factors and Ergonomics Society 53rd Annual Meeting*, 684-688.

Kaya, O., Moran, M., Ozkardes, A.B., Taskin, E.Y., Seker, G.E., & Ozmen, M.M. (2008). Ergonomic problems encountered by the surgical team during video endoscopic surgery. *Surgical Laparoscopy Endoscopy Percutaneous Techniques*, 18(1), 40-44.

Koneczny, S. (2009). The operating room: Architectural conditions and potential hazards. *Work*, 33(2), 145-164.

Koneczny, S., & Matern, U. (2004). Instruments for the evaluation of ergonomics in surgery. *Minimally Invasive Therapy & Allied Technology*, 13(3), 167-177.

Kranenburg, L., & Gossot, D. (2004). Ergonomic problems encountered during video-assisted thoracic surgery. *Minimally Invasive Therapy & Allied Technology*, 13(3), 147-155.

Lawrence, W.H., & Berry, C.H. (1939). Arrangement of the operating room. *American Journal of Surgery*, XLIII(3), 669-674.

Leape, L.L., & Berwick, D.M. (2005). Five Years After To Err Is Human: What Have We Learned? *JAMA*, 293(19), 2384-2390.

Lee, G., Lee, T., Dexter, D., Klein, R., & Park, A. (2007). Methodological infrastructure in surgical ergonomics: A review of tasks, models, and measurement systems. *Surgical Innovation*, 14(3), 153-167.

Leedal, J.M., & Smith, A.F. (2005). Methodological approaches to anaesthetists' workload in the operating theatre. *British Journal of Anaesthesia*, 94(6), 702-709.

Lin, D.W., Bush, R.W., Earle, D.B., & Seymour, N.E. (2007). Performance and ergonomic characteristics of expert surgeons using a face-mounted display during virtual reality-simulated laparoscopic surgery: An electromyographically based study. *Surgical Endoscopy*, 21(7), 1135-1141.

Matern, U. (2009). Ergonomic deficiencies in the operating room: Examples from minimally invasive surgery. *Work*, 33, 165-168.

Matern, U (2001). Principles of ergonomic instrument handles. Minim Invasive Ther Allied Technol, 10(3), 169-173.

Matern, U., & Koneczny, S. (2007). Safety, hazards and ergonomics in the operating room. *Surgical Endoscopy*, 21(11), 1965-1969.

McKenna, J.V. (1957). The case for motion and time study in surgery. *American Journal of Surgery*, 94(5), 730-734.

Nguyen, N.T., Ho, H.S., Smith, W.D., Philipps, C., Lewis, C., De Vera, R.M., & Berguer, R. (2001). An ergonomic evaluation of surgeons' axial skeletal and upper extremity movements during laparoscopic and open surgery. *American Journal of Surgery,* 182(6), 720-724.

Pasmore, W.A. (1988). Designing Effective Organizations: The Sociotechnical Systems Perspective. New York: John Wiley and Sons.

Peters, T.M. (2001). Image-guided surgery and therapy: Current status and future directions. *Proceedings of SPIE*, 4319, 1-12.

Pool, E.H., & Bancroft, F.W. (1917). Systematization of a Surgical Service. *Journal of the American Medical Association*, 69(19), 1599-1603.

Sheikhzadeh, A., Gore, C., Zuckerman, J.D., & Nordin, M. (2009). Perioperating nurses and technicians' perceptions of ergonomic risk factors in the surgical environment. *Applied Ergonomics*, 40(5), 833-839.

Siegel, E.L., & Reiner, B. (2003). Work flow redesign: The key to success when using PACs. *Journal of Digital Imaging*, 16(1), 164-168.

Sindwani, R., & Bucholz, R.D. (2005). The next generation of navigational technology. *Otolaryngologic Clinics of North America*, 38, 551-562.

Trejo, A.E., Jung, M.-C. Oleynikov, D., & Hallbeck, M.S. (2007). Effect of handle design and target location on insertion and aim with a laparoscopic surgical tool. *Applied Ergonomics*, 38(6), 745-753.

Trejo, A., Doné, K.N., DiMartino, A.A., Oleynikov, D., & Hallbeck, M.S. (2006). Articulating vs. Conventional Laparoscopic Grasping Tools—Surgeons' Opinions. *International Journal of Industrial Ergonomics*, 36(1), 25-35.

van Veelen, M.A. (2004). Ergonomics in minimally invasive surgery. *Minimally Invasive Therapy & Allied Technology*, 13(3), 131-132.

van Veelen, M.A., Jakimowicz, J.J., & Kazemier, G. (2004). Improved physical ergonomics of laparoscopic surgery. *Minimally Invasive Therapy & Allied Technology*, 13(3), 161-166.

Van Veelan, M.A. and Meijer, D.W., 1999. Ergonomics and design of laparoscopic instruments: Results of a survey among laparoscopic surgeons. *Journal of Laparoendoscopic and Advanced Surgical Techniques*, Part A, 9 (6), 481–489.

Weinger, M.B., & Slagle, J. (2002). Human factors research in anesthesia patient safety. *Journal of the American Medical Informatics Association*, 9, S58-S63.

Chapter 8

Quality Improvement Projects in Patient Safety

Laurie Wolf

Barnes-Jewish Hospital
Operational Excellence Department
St. Louis, MO, 63110, USA

ABSTRACT

Lean, Six Sigma, and Human Factors Engineering were some of the tools used by a multi-disciplinary team at Barnes-Jewish Hospital to improve the entire process from patient admit to discharge on 13 Medicine divisions. A value stream analysis was conducted to reveal that the current state process includes busywork, rework and frustrations of the physicians, rework due to hand-offs and unsatisfactory patient flow. After a future state was envisioned and gaps were identified, over a dozen different projects and events were performed to improve physician satisfaction and patient safety. This paper will highlight a few of these improvement projects.

Keywords: Human Factors Engineering, Value Stream Analysis, Lean, Six Sigma

INTRODUCTION

Barnes-Jewish Hospital is a large urban teaching hospital in St. Louis, Missouri. (9,438 employees, 1,845 physicians, 803 residents/interns/fellows, 1,111 staffed beds, 83,997 Emergency Room visits and 54,733 admissions in 2009). Barnes-Jewish Hospital is part of BJC HealthCare and is affiliated with Washington University School of Medicine. Over the past year, Barnes-Jewish Hospital has worked to combine the power of Six Sigma and Lean principles and tools to reduce defects and process variability, and improve the efficiency and quality of key care delivery and support processes.

Over two years ago, Barnes-Jewish Hospital began a Lean Six Sigma transformation journey as a supplement its successful performance improvement structure under their Patient Safety and Quality Department. This effort was driven directly from the top: both the President and Chief Operating Officer endorsed and sponsored the initiative. Physician engagement is encouraged by the Chief Medical Officer. The overall objective of the journey is for Barnes-Jewish Hospital to use Lean and Six Sigma principles to reduce defects and process variability while improving efficiency and quality of care delivery and support processes.

QUALITY IMPROVEMENT APPROACH FOR PHYSICIAN SATISFACTION AND PATIENT SAFETY

The methods and strategies we use for our improvements are guided by a steering committee. The committee is a multi-disciplinary team lead by the Chief Medical Officer consisting of general medicine physicians, an Emergency Room physician, Chief Medicine Residents, Vice President of Patient Care Service/Chief Nurse Executive, Director of Patient Care, Patient Placement Manager, Director of Medicine Service, nurse managers and a Human Factors Engineer.

Physician engagement is achieved by careful preparation to ensure the most efficient use of time. Typically, improvement methods (such as FMEA, link analysis, impact and cause/effect matrix) are most successful if they are prepared and populated as much as possible before the process improvement events (with input from individual physicians). It is more time efficient to have physicians review and verify processes. Key stakeholders and sponsors are identified for each project to help prepare a first draft for steering committee review.

A tool called Value Stream Analysis was conducted to develop a plan for implementing several improvement projects. This analysis provides an understanding of the process steps required to admit, diagnose, treat, and discharge a patient. Areas for improvement were identified and provided a roadmap for future projects. Some of the improvement projects that were identified included:

Emergency Department Communication to Admitting Physicians

Reason for Action: Improve hand-off processes and communication between the emergency department (ED) and admitting services

Solution: A medicine reconciliation process was "Hard-wired" into the ED admission procedure by adding questions into the documentation system. A summary report from the emergency department physician was organized according to needs of admitting physicians to make information easier to find and decrease time to complete a history and physical on a newly admitted patient. A new admission process was also implemented that shifted the work of a medicine resident from triaging patients in the ED to admitting patients at night. A combination of all these projects reduced the length of stay in the ED by one hour, which was a statistically significant improvement.

Forms Required to be Completed by a Physician

Reason for Action: Admission forms and process are different on each of the 13 divisions making them difficult to find. Loose pages can be confused and misplaced causing incorrect patient identification.

Solution: An admission packet was developed with required forms preprinted with patient identification to increase patient safety by reducing the opportunity for patient identification errors. All forms needed by a physician are located in one place in the same arrangements on all 13 medicine floors. Time to complete admit forms decreased by >50% because wasted motion and search time was eliminated.

Transparent Plan of Care

Reason for Action: In a teaching hospital there is frequent rotation of residents and numerous specialties on the division involved with patient care. It can be difficult to identify the physician, nurse and plan of care for the patient.

Solution: A patient information board was placed inside the patient's room to show the physician, nurse, and plan of care for the patient each day. The census board in the nursing station was reorganized to show everyone which physician and nurse are responsible for each patient and their contact information. Discharge sheets that are completed by all residents at the end of each shift were made available electronically. This enabled all caregivers to review the Plan of Care tor the patients for the upcoming shift.

Discharge Process

Reason for Action: 50% of the patients were discharged after 4 pm causing a delay in admission of new patients.

 Solution: One solution was to encourage physician communication about "Anticipate Discharge Tomorrow". Communication for "Discharge home today" with the goal to discharge by 1:00 pm was done by all patient care staff. A discharge planning board with a visual reminder of tasks to be completed before discharge is located in the staff area and updated daily.

Although there was not a statistically significant increase in the percentage of patients discharged by 1:00 PM, there was a significant decrease in LOS. Staff reported improved communication and had the following comments:

- "Plans for patient disposition early in the day assists with patient assignments."
- "We can get things set up more efficiently (Home health, IV Antibiotics, SNF, Transportation, etc) when we know early about discharge".
- "This project helped get interdisciplinary care rolling more efficiently (orders for PT/OT, Diabetes/Nutrition consults, palliative/hospice care, etc)."

CONCLUSION

The value stream tool provides a useful roadmap for continuing improvement projects on the general medicine divisions. Creative implementation of tools achieves excellent physician engagement. Improvements in processes and environment have been beneficial to both patients and caregivers.

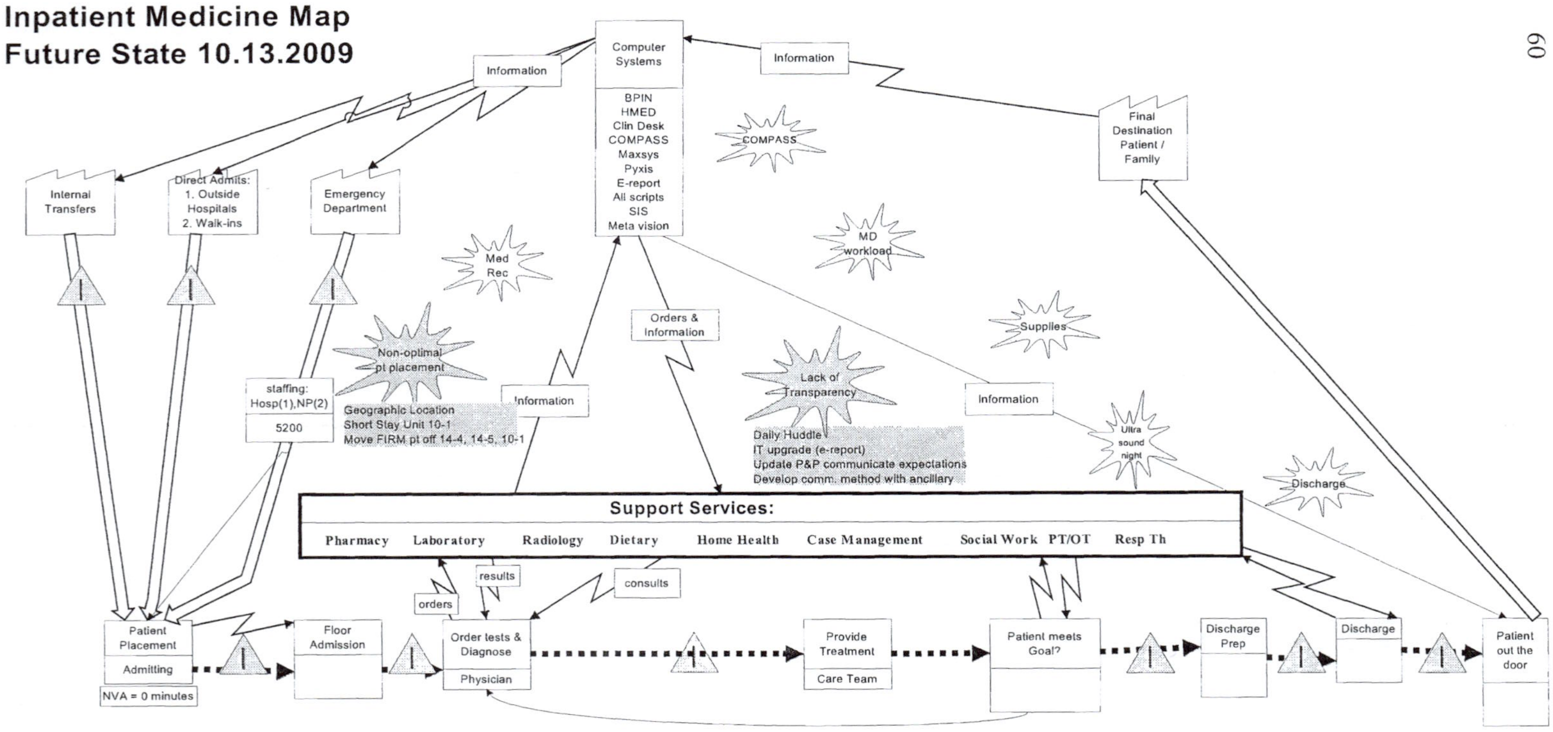

Inpatient Medicine Map
Future State 10.13.2009

FIGURE 1 Inpatient Medicine Value Stream Map

Chapter 9

Healthcare Product Design for Quality

Richard H.M.Goossens

Delft University of Technology
Faculty of Industrial Design Engineering
Landbergstraat 15
2628 CE Delft
The Netherlands

ABSTRACT

Quality in healthcare depends on several factors, varying from the skills of the medical professional and the performance of the medical team, to the quality and usability of the technical equipment. For all of these aspects new initiatives are implemented and evaluated. To study healthcare design for quality, in the research group Medisign at Delft University of Technology, a product vision was described of a system that supports and verifies quality during surgery (the Surgeons cockpit)(van Veelen 2003). In the product vision an overview is given of the factors and main issues that should be considered by designers. The aim of this paper is to give an overview of the project and discuss the results of PhD projects.

Research projects that are embedded in the design of a Surgeons Cockpit addressed the following topics in order to address quality in product design.

- A framework to address the surgeon's requirements.
- A quality control procedure for equipment.
- A method to increase awareness of ergonomic guidelines.
- An (inter)national standardized protocol, including standardized actions.
- A guideline to improve operation notes.
- A more sophisticated ranking system in order to categorize the skills of the surgeon.
- A vision for training and assessment of (IBP type) specific skills.

- Knowledge for the ergonomic design of tools that are used to manipulate objects.

Keywords: Product design, usability, guidelines

INTRODUCTION

Quality in healthcare depends on several factors, varying from the skills of the medical professional and the performance of the medical team, to the quality and usability of the technical equipment. For all of these aspects new initiatives are implemented and evaluated. For example, the effect of CRM (Crew Resource Management) Training is studied in surgical teams(Parker et al. 2007), or the use of a blackbox in the Operating Room (OR) to measure performance(Guerlain et al. 2005). It is known that although protocols for different types of surgery exist that each specialist applies these protocols in a different manner(Wauben et al. 2008). This situation differs completely from other comparable work environments where people's lives are at risk such as in aviation or nuclear/toxic industries. This lack of quality control in medicine contributes to preventable medical errors.

But not only these procedural aspects play an important role, also the quality of instruments that are used are a factor of significance. Often it is overseen that these tools need to be designed, and as a consequence the design team also plays an important role in the prevention of medical errors. For example in 2009, 34 medical devices were recalled by the FDA because there was 'a reasonable chance that they [these products] could cause serious health problems or death'. (www.fda.gov)

To study healthcare design for quality, in the research group Medisign at Delft University of Technology, a product vision was described of a system that supports and verifies quality during surgery (the Surgeons cockpit)(van Veelen 2003). In the product vision an overview is given of the factors and main issues that should be considered by designers. PhD projects for product designers were defined that are necessary to develop such a system. The aim of this paper is to give an overview of the project and discuss the results of PhD projects.

METHODOLOGY

A first step in the development of standardized quality control is to give the designers an overview of the factors that influence the quality of surgery. Base of this overview is an interaction scheme based on the factors that influence the human-product interaction during surgery.

It consists of two main factors: the human and the product, and five external factors that are of influence on the interaction between them: social, political, physical, clinical and technological. These seven factors are shortly described below.

FACTORS THAT INFLUENCE SURGICAL QUALITY

Main factors

The most important factors for designers that influence the surgical quality are discussed. These factors play a key role and must therefore be analyzed and further studied before a well functioning Surgeons Cockpit can be designed.

Product issues

The surgeon depends highly on the performance of his equipment. When the quality of a surgical instrument is poor, the quality of the surgery will also descend. In surgery there are no quality standards for the equipment that is used during surgery. Therefore the following research questions must be answered in the Surgeons Cockpit (SC) project:

What is the minimum quality standard for the equipment that is used during surgery in order to perform a high quality surgery?

How can the SC measure the quality of the equipment that is used during surgery?

How communicates the SC the quality of the equipment to the OR team?

Human factor issues

The human factor is also a very important factor of surgical quality. Especially the surgeon's knowledge and skills influence the quality of a surgery. The SC should support the surgeon in his knowledge, for instance residents need more information about protocols and anatomy and the SC should provide this information. The following research questions must be answered in the SC-project:

What kind of information about surgical procedures, anatomy etc. should be available in the SC?

How can the SC be a training tool to improve knowledge and skills?

How communicates the SC the relevant background information to the surgeon?

Social issues

A main social issue that influences the surgical quality is the fact that surgeries are always performed in a team. A general surgery needs at least 4 people: a surgeon, an assistant, an anesthesiologist and a surgery nurse. The surgeon leads the surgery and has the final responsibility of the result. Each team member has his own tasks to fulfill and has his own protocol to follow. The SC should adjust to the tasks of each surgery team member. The following research questions must be answered in the SC-project:

What kind of information is relevant for each team member?
How communicates the SC the relevant information to each team member?

Political issues

There are many political changes in healthcare the last few years. Hospitals have to publish their annual clinical results (e.g. mortality rates, complication rates and patient admissions), and insurance companies decide what budget and method is used for a specific type of surgery. Surgical quality is a very important factor in the overall quality factor of a hospital. Patients are reluctant to go to a hospital with bad annual results. The SC can facilitate the hospital in producing a surgery report of the operation. All materials that are used are recognized and stored in the SC that makes it easy to produce reports for patients, insurance companies and the management of the hospital. The following research questions must be answered in the SC-project:
What kind of information (for patients, insurance companies, the hospital, the government etc.) must be stored and recognized by the SC?
How does the SC recognizes and store this information?
What kind of output (reports) must be produced by the SC?

Environmental issues

The physical condition of the OR, such as light, temperature, laminar flow, and dimensions, determines the work environment of the OR team. The SC can measure these aspects and monitor them. The following research questions must be answered in the SC-project:
What kind of physical aspects are important to measure and monitor by the SC?
How communicates the SC the relevant information to each team member?

Clinical issues

In the Netherlands all hospitals, and sometimes even each surgeon, use their own protocol and method for a specific type of surgery. These protocols can be adjusted when there is new insight resulting from scientific studies. The SC should anticipate to these changes and must be adjustable for each surgeon. The following research questions must be answered in the SC-project:
How is a surgery protocol used/stored in the SC?
How communicates the SC this protocol to each individual surgeon?
How can the protocol be adjusted?

Technological issues

Technology is a dynamic field and the speed of development is high. The SC should therefore be very flexible, so that new technologies can be integrated easily. Furthermore, hospitals use sometimes different technologies and the SC should also be flexible to adjust to the technical systems (equipment, patient administration etc.) that hospitals use. The following research questions must be answered in the SC-project:

How can the SC be designed in a flexible way to implement new and current technologies when necessary?

PhD Projects

To answer these questions PhD projects were defined that combined design with research in the development of the Surgeon's Cockpit.

RESULTS

Framework for designers

Jalote and Badke-Schaub (Jalote and Badke-Schaub 2008) proposed a framework called the Workflow Integration Matrix (WIM). WIM uses theories of human behavior in problem solving, especially the information processing paradigm.

The proposed framework intends to provide evidence-based decision-making for the development of new surgical technologies. It addresses two needs for designers: first, the analysis of surgical requirements and processes within the surgical workspace, and second, the sharing of the requirements and processes within a multidisciplinary development team.

Quality of the equipment

Casseres and Albayrak (Cassares and Albayrak 2003) studied the safety of the equipment that is used in Dutch hospitals during minimal invasive surgery. They found in a study in 36 hospitals that 22% of all endoscopic equipment had a defect in electrical isolation and that for 36% of the light guide cables more than 30% of the fibers in the cables were defect.

Ergonomic guidelines

Wauben et al. (Wauben et al. 2006) studied whether ergonomic guidelines are applied in the operating room by sending 1,292 questionnaires to mainly European surgeons and residents. The response of 22% (284 respondents) showed that 89% of them were unaware of ergonomic guidelines, although 100% stated that they find ergonomics important. The authors concluded that the lack of ergonomic guidelines awareness is a major problem that poses a tough position for ergonomics in the operating room.

Use of protocols

Wauben et al. (Wauben et al. 2008) contacted fifteen Dutch hospitals for evaluation of their protocols for laparoscopic cholecystectomy. All evaluated protocols were divided into six steps and were compared accordingly. In total 10 protocols were usable for comparison. Concerning the trocar positions, only minor differences were found. The concept of "critical view of safety" was represented in just one protocol. Furthermore, the order of clipping and cutting the cystic artery and duct differed. Descriptions of instruments and apparatus were also inconsistent. It was concluded that the present protocols differ too much to define a universal procedure among surgeons in The Netherlands. The authors propose one (inter)national standardized protocol, including standardized actions.

Wauben et al. (Wauben et al. 2010) assessed the compliance with the Dutch guidelines for writing operative notes in nine hospitals that were asked to send 20 successive LC operative notes. All notes were compared to the Dutch guideline by two reviewers and double-checked by a third reviewer. She concluded that operative notes do not always fully comply with the standards set forth in the guidelines published in the Netherlands. This could influence adjuvant treatment and future patient treatment, and it may make operative notes less suitable background for other purposes. Therefore operative note writing should be taught as part of surgical training, definitions should be provided, and procedure-specific guidelines should be established to improve the quality of the operative notes and their use to improve patient safety.

Definition of expertise

Buzink et al. (Buzink et al. 2009) studied the expertise in manipulation of laparoscopic instruments. And found that the label 'expert', although often used as a single label for experienced surgeons that have performed over 50 procedures, cannot automatically be applied to all skills that are needed for the procedure. Therefore it can be concluded that a more sophisticated approach should be used in order to categorize the skills of the surgeons.

Because surgical techniques that draw from multiple types of image-based procedures (IBP) are increasing, such as Natural Orifice Transluminal Endoscopic Surgery, fusing laparoscopy and flexible endoscopy, and little is known about the relation between psychomotor skills for performing different types of IBP. Buzink et al. (Buzink et al. 2010) studied the relation between psychomotor skills for performing different types of image based procedures. The authors concluded that training and assessment of IBP type specific skills should focus on each type of tasks independently.

Feedback

Westebring(Westebring-van der Putten et al. 2009) et al. studied the differences in grasp control during barehanded and laparoscopic lifts. In the article the authors generated a prerequisite of knowledge for the ergonomic design of tools that are used to manipulate objects.

CONCLUSIONS

Research projects that are embedded in the design of a Surgeons Cockpit addressed the following topics in order to address quality in product design.

- A framework to address the surgeon's requirements.
- A quality control procedure for equipment.
- A method to increase awareness of ergonomic guidelines.
- An (inter)national standardized protocol, including standardized actions.
- A guideline to improve operation notes.
- A more sophisticated ranking system in order to categorize the skills of the surgeon.
- A vision for training and assessment of (IBP type) specific skills.
- Knowledge for the ergonomic design of tools that are used to manipulate objects.

REFERENCES

Buzink, S. N., et al. (2010), 'Do Basic Psychomotor Skills Transfer Between Different Image-based Procedures?', *World J Surg.*

Buzink, S. N., et al. (2009), 'Camera navigation and tissue manipulation; are these laparoscopic skills related?', *Surg Endosc,* 23 (4), 750-7.

Cassares, Y.A. and Albayrak, A. (2003), 'Keyhole operations at sight (in Dutch)', (Nederlandse Vereniging voor Endoscopische Chirurgie).

Guerlain, S., et al. (2005), 'Assessing team performance in the operating room: development and use of a "black-box" recorder and other tools for the intraoperative environment', *J Am Coll Surg*, 200 (1), 29-37.

Jalote, A. and Badke-Schaub, P. (2008), 'Workflow Integration Matrix: a framework to support the development of surgical information systems', *Design Studies*, 29, 338-68.

Parker, W. H., Johns, A., and Hellige, J. (2007), 'Avoiding complications of laparoscopic surgery: lessons from cognitive science and crew resource management', *J Minim Invasive Gynecol*, 14 (3), 379-88.

van Veelen, M. A. (2003), 'Human-Product Interaction in Minimally Invasive Surgery: a Design Vision for Innovative Products', (Delft University of Technology).

Wauben, L. S., Goossens, R. H., and Lange, J. F. (2010), 'Evaluation of Operative Notes Concerning Laparoscopic Cholecystectomy: Are Standards Being Met?', *World J Surg*.

Wauben, L. S., et al. (2006), 'Application of ergonomic guidelines during minimally invasive surgery: a questionnaire survey of 284 surgeons', *Surg Endosc*, 20 (8), 1268-74.

Wauben, L. S., et al. (2008), 'Evaluation of protocol uniformity concerning laparoscopic cholecystectomy in the Netherlands', *World J Surg*, 32 (4), 613-20.

Westebring-van der Putten, E. P., et al. (2009), 'Force feedback requirements for efficient laparoscopic grasp control', *Ergonomics*, 52 (9), 1055-66.

Chapter 10

Effect of Emotional Intelligence on Healthcare IT Acceptance

Renran Tian[1], Xiaosong Zhao[1,2], Vincent G. Duffy[1,3]

1 School of Industrial Engineering
Purdue University
West Lafayette, IN 47906, USA

3 School of Agricultural and Biological Engineering
Purdue University
West Lafayette, IN 47906, USA

2 Department of Industrial Engineering
Tianjin University
Tianjin, P. R. China, 300072

ABSTRACT

Nursing IT systems, as various other computerized systems, have been implemented in health care providers to aid daily nursing work. As with the implementation of other computerized systems in different industries, it is critical to understand and improve nurses' acceptance of these IT innovations to optimize utility. However, due to the unique features of nursing work system, general IT acceptance models need to be modified and extended to fit this complex environment. Also, individual factors in most IT acceptance studies are limited to basic demographic factors which may not be strong enough to directly reflect individual differences. In this study, we adopt the concept of emotional intelligence into classic IT acceptance model in

For more information, please contact corresponding author: Xiaosong Zhao,
Email: zhaoxs_tju@tju.edu.cn

individual factor, is proposed to reflect nursing work features on IT acceptance.

Keywords: Emotional Intelligence, Healthcare IT Acceptance, Nursing Informatics

INTRODUCTION

New technology acceptance has been one important topic since the rise of automation in all industries. During the past two decades when various computerized systems experienced fast progress, researchers paid much attention on the acceptance of new IT innovation. Models were proposed based from two directions to study users' attitudes when facing new technology or system:

1. Social psychology based studies try to analyze user's decision making process based on the theory of reasoned action (Fishbein and Ajzen, 1975) which emphasizes the relationship of attitude->intention->behavior. TAM (Technology Acceptance Model) proposed by Davis (1989) and planed behavior theory proposed by Ajzen (1991) are the two most influential models. Based on these models and other related confirmatory and extended studies, Venkatesh and associates (2003) proposed UTAUT (Unified Theory of Acceptance and Use of Technology) model summarizing all important achievements in this field.

2. Technology/User feature based studies try to predict users' acceptance based on the feature and characteristics of the targeted system. One representative is IDT (Innovation Diffusion Theory) proposed by Moore and Benbasat (1991), which studied eight most important features of technology and environment for IT acceptance. Another representative is the construct of perceived innovativeness proposed by Agarwal and Prasad (1998) which distinguish users from others based on their internal ability of accepting innovations.

After Institute of Medicine report (Kohn and Corrigan, 2000), healthcare IT systems have been more and more widely used to aid different tasks and improve patient safety. Due to the complex and unique features of healthcare environment, IT acceptance models meet some problems and limitations to analyze healthcare staffs' attitudes (Hu, et al., 1999; Chau and Hu, 2002; Chismar and Wiley-Patton, 2002, 2003). One of the most important limitations is the difference in individual users between healthcare staffs and those in other model validation environment like intelligence or competence.

In this study, we propose one new construct of emotional intelligence (EI) as individual factor to analyze IT acceptance in healthcare environment. As one part of general intelligence, emotional intelligence refers to "the ability to monitor one's own and others' emotions, to discriminate between them, and to use the information to guide one's thinking and actions" (Salovey and Mayer, 1990). The research purpose is to (1) adopt EI as one powerful individual factor to study IT acceptance so that users' basic status can be better analyzed, and (2) prove IT acceptance in healthcare can be better predicted because EI level of healthcare staffs reflect their ability of decision making within the highly dynamic, uncertain and critical working process.

THEORETICAL BASIS

INTELLIGENCE AND EI

According to Mayer and Geher (1996), a person's general intelligence represents that individual's overall level of intellectual attainment and ability, and has often been used to successfully predict a person's academic and occupational achievement, which includes: (1) abstract, analytic, and/or verbal intelligences; (2) mechanical, performance, visual-spatial, and/or synthetic intelligences; and (3) social and/or practical intelligences. EI belongs to the social intelligence, and is much less studied among different kinds of intelligence.

Based on the definition of EI cited in previous section, EI contains four aspects of monitoring one's own emotion, monitoring others' emotion, understanding emotion, and control and use emotion to serve the working purpose. Also, as one specified intelligence, EI is correlated to other types of intelligence like cognitive and verbal intelligence (Cote and Miners, 2006).

EFFECT OF EI ON JOB SATISFACTION

Based on our best knowledge, there is no literature about EI's effect on intention in IT acceptance literature. However, some researchers have studied the effect of EI on job satisfaction. Abraham (2000) investigated several factors to job satisfaction including job control, organizational commitment, and EI. The author found that EI and job control jointly explained a significant 26% of the variance of job satisfaction, and further analysis showed that EI is only associated with job satisfaction significantly with the moderator effect of job control. Based on Abraham's study, Chiva and Alegre (2008) studied the effect of EI on job satisfaction with the consideration of organizational learning capacity (OLC) as mediator. They conclude that EI has significant effects on job satisfaction only through OLC construct.

Thus, although the number of literatures addressing EI and attitude/intention is very limited, we can conclude based on current studies that EI has directly or indirectly significant effect on attitude. Based on the theory of reasoned action, EI is proposed to be one factor for IT acceptance in this study.

HEALTHCARE WORK SYSTEM FEATURES

Healthcare work system, as one complex sociotechnical system, has many prominent features that may affect the IT system design and IT diffusion process. According to Wears and Perry (2007), the features and complexity of healthcare can be summarized in following aspects: homeostatic activity, equifinality, remoteness, unstoppability, mortality. To summarize, healthcare work process is highly dynamic, uncertain, and critical.

EI AS AN INDIVIDUAL FEATURE

As one specified type of intelligence, EI can reflect user's ability and features directly. Since EI can be measured continuously, quantitative analysis can also be performed for EI-controlled IT acceptance analysis.

Since EI is about the ability to interpret and control emotion, it may not be always true that EI will affect user intention in some circumstances. However, in healthcare environment, especially for some major professionals like nurses, emotion plays an important role in daily work and among cooperation. Along the fast changeable, unpredictable, and interruptive process, it is essential to control a good emotion and atmosphere to confront high time-pressure and critical tasks.

Thus, using EI as one individual variable can better explain healthcare staffs' attitude towards their job, and also their intention to use IT systems.

CONCEPTUAL MODEL

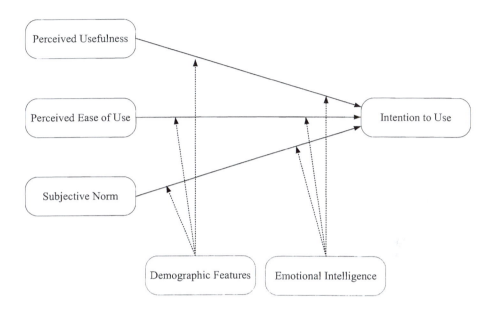

Figure 1. Emotional Intelligence controlled Healthcare IT Acceptance Model

Conceptual model of EI controlled healthcare IT acceptance is shown in figure 1. The three main factors of "perceived usefulness", "perceived ease of use" and "subjective norm" are based on TAM and TPB models. One group of control variables is demographic features of individual, which is based on the UTAUT, including age, gender, and experience. Another control variable is EI. Definitions of all constructs are based on literatures:

- **Perceived Usefulness:** "the degree to which a person believes that using a particular system would enhance his or her job performance." (Davis, 1989)
- **Perceived Ease of Use:** "the degree to which a person believes that using a particular system would be free of effort." (Davis, 1989)
- **Subjective Norm:** "the degree to which an individual perceives that important others believe he or she should use the new system." (Venkatesh, et. al., 2003)
- **Emotional Intelligence:** "the ability to monitor one's own and others' emotions, to discriminate between them, and to use the information to guide one's thinking and actions." (Salovey and Mayer, 1990)

The main hypothesis is that <u>EI of individual user will moderate the effect of his perceived usefulness, perceived ease of use, and subjective norm on his intention to use targeted healthcare IT system.</u>

CONSTRUCT MEASUREMENTS

The constructs of "perceived usefulness" and "perceived ease of use" will be measures using the survey proposed by Davis (1989), which includes 6 items for each construct. The construct of "subjective norm" will be measured using the 4-item survey proposed by Venkatesh and others (2003) for "social influence" since these are actually two identical constructs.

The control variable of EI will be measured using WLEIS survey proposed by Wong and Law (2002). This survey tool includes 4 sections with 4 items in each section, and all these four sections are corresponding to the 4 aspects of EI definition. Authors of this questionnaire have used several independent datasets to prove the reliability and validity of this tool.

FUTURE WORK

The overall questionnaire contains 32 items for all constructs and demographic questions. Data are planned to be collected in urban hospitals in both USA and China to also consider culture influences. Nurses and pharmacists will be mainly surveyed in first stage because they are using more IT systems directly than physicians.

REFERENCES

Abraham, R., (2000), "The Role of Job Control as a Moderator of Emotional Dissonance and Emotional Intelligence-Outcome Relationships." *The Journal of Psychology*, 134(2), 169 – 184.

Agarwal, R., and Prasad, J., (1998), "A Conceptual and Operational Definition of Personal Innovativeness in the Domain of Information Technology." *Information Systems Research*. 9(2), 204 – 215.

Ajzen, I., (1991), "The Theory of Planned Behavior." *Organizational Behavior and Human Decision Processes*, 50, 179 – 211.

Chau, P. Y. K., and Hu, P. J. H., (2002), "Examining a Model of Information Technology Acceptance by Individual Professionals: An Exploratory Study." *Journal of Management Information Systems*, 18(4), 191-229.

Chiva, R., and Alegre, J., (2007), "Emotional Intelligence and Job Satisfaction: the Role of Organizational Learning Capability." *Journal of Managerial Psychology*, 19(2), 88 – 110.

Chrismar, W. G., and Wiley-Patton, S., (2002). "Test of the Technology Acceptance Model for the Internet in Pediatrics." *AMIA 2002 Annual Symposium Proceedings.*

Chrismar, W. G., and Wiley-Patton, S., (2003). "Does the Extended Technology Acceptance Model Apply to Physicians." *Proceedings of the 36th Hawaii International Conference on System Sciences.*

Cote, S., and Miners, C. T. H., (2006), "Emotional intelligence, cognitive intelligence, and job performance." *Administrative Science Quarterly*, 51(1), 1 – 28.

Davis, F. D., (1989), "Perceived Usefulness, Perceived Ease of Use, and User Acceptance of Information Technology." *MIS Quarterly*, 13(3), 319 – 340.

Fishbein, M., and Ajzen, I., (1975), *Belief, Attitude, Intention, and Behavior: An Introduction to Theory and Research. Reading.* MA: Addison-Wesley.

Hu, P. J. H., Chau, P. Y. K., Sheng, O. R. L., and Tam, K. Y., (1999), "Examining the Technology Acceptance Model using Physician Acceptance of Telemedicine Technology." *Journal of Management Information Systems*, 16(2), 91 – 112.

Kohn, L. T. and Corrigan, J. M., (2000), *To Err is Human: Building a Safer Health System.* From Committee on Quality of Health Care in America, Institute of Medicine. National Academy Press, Washington, D.C.

Mayer, J. D., and Geher, G., (1996), "Emotional Intelligence and the Identification of Emotion." *Intelligence*, 22, 89 – 113.

Moore, G. C., and Benbasat, I., (1991), "Development of an Instrument to Measure the Perceptions of Adopting and Information Technology Innovation." *Information Systems Research*, 2(3), 192 – 222.

Salovey, P., and Mayer, J. D., (1990), "Emotional Intelligence." *Imagination, Cognition, and Personality*, 9(3), 185 – 211.

Venkatesh, V., Morris, M. G., Davis, G. B., and Davis, F. D., (2003), "User Acceptance of Information Technology: Toward a Unified View." *MIS Quarterly*, 27(3), 425 – 478.

Wears, R. L., and Perry, S., J., (2007), "Human Factors and Ergonomics in the Emergency Department." *Handbook of Human Factors and Ergonomics in Health Care and Patient Safety by Pascale Carayon (Ed.),* 851 – 864.

Wong, C. S., and Law, K. S., (2002), "The Effects of Leader and Follower Emotional Intelligence on Performance and Attitude: An Exploratory Study." *The Leadership Quarterly*, 13, 243 – 274.

Chapter 11

Targeting Environmental Factors to Reduce Elderly In-Patient Falls

Sue Hignett[1], Gina Sands[1], Jane Youde[2], Paula Griffiths[1]

[1]Loughborough University

[2]Derby Hospitals NHS Foundation Trust

ABSTRACT

This paper will describe the results from two exploratory studies on un-witnessed elderly in-patient falls in acute facilities. The first study analyzed incident reports from England and Wales between 1 September 2006 and 31 August 2007 (n=215,784). We found a difference in the location of falls for patients described as frail and those described as confused. This was further explored in a pilot study to collect detailed information about contributory factors and the location of falls through staff interviews. We found that the use of bedrails seemed to alter the location of the fall, with falls from beds with raised rails clustered around the foot end of the bed.

INTRODUCTION

In-patient falls have consistently been the biggest single category of reported incidents from since the 1940s; they are a significant cause of morbidity and mortality and have a high prevalence after admission to hospital (Morgan et al, 1985; Oliver et al, 2004; Mahoney, 1998). The risk factors have been identified and reported since the 1950s (Parrish and Weil, 1958; Fine, 1959; Fagin and Vita, 1965). Although only a small percentage of patient falls result in death and serious injury they represent a serious financial, governance and resource burden in terms of on-going healthcare costs and litigation (Boushon et al., 2008). The incident rate for falls is approximately three times higher in hospitals and nursing homes than in community-dwelling older people (American Geriatrics Society, 2001). It has been suggested that this may be due to a combination of extrinsic risk factors (relating to the environment), for example, unfamiliar environment and wheeled furniture, combined with intrinsic risk factors (relating to the patient) such as confusion, acute illness and balance-affecting medication (Tinker, 1979; Tinetti, 2003; Salgado et al, 2004; Kannus et al, 2006). Many papers have reported that the majority (over 70%) of in-patient falls are un-witnessed with the patient found on the floor and little information in the incident report. (Fagin and Vita, 1965; Hitcho et al, 2004; Healey et al, 2008).

This paper offers an exploration of contributory factors with a detailed analysis of falls risks from reported incidents and a pilot case study of unwitnessed in-patient falls on Care of the Elderly wards in an acute hospital.

METHOD

215,784 reports were retrieved from the UK National Reporting and Learning System (NRLS) database for slips, trips and falls between 1 September 2006 and 31 August 2007, with 44,202 reports from Care of the Elderly wards in acute and community hospitals in England and Wales. A random 15% sample was taken (6,577 reports), of which 4,571 were un-witnessed. The free text narratives on the incident reports were coded into intrinsic and extrinsic contributory factors using an initial conceptual framework (Hignett and Masud, 2006). As coding progressed more factors emerged from the data and were added to the coding framework. All the reports were re-coded with the final set of factors to ensure inclusivity of the coding process. Reports were coded as frail if the patient was described as weak, frail or needing a walking aid, and as confused if described as having any type of dementia, confusion or lack of awareness. As most falls were the result of a combination of factors, few were coded to a single code. The contributory risk factors for the frail and confused groups were compared with the whole sample and explored with the Chi-squared and Fisher's exact tests. The effect size was calculated to determine the strength of the relationships using the Phi statistic and statistical significance was assessed with a two-tailed P-value <0.05.

The pilot case study reviewed 26 reported incidents for un-witnessed patient falls from March to September 2009 in 4 Care of the Elderly wards (n=112) in a large acute hospital (1,150 beds). The nurse reporting the incident was interviewed with a structured proforma (figure 1) to add factual information, for example the exact location of the fall, whether the bed rails were raised (3/4 length rails) and the type of footwear worn by the patient at the time of the fall. The study was granted Ethical Approval from Nottingham Research Ethics Committee 1 (08/H0403/149) and Research Governance by Royal Derby Hospital (DHRD/2008/071).

78

Where was the patient before they fell?
Where was the patient found?
Were the bed rails up or down?
Were they using a mobility aid (e.g. stick, frame, wheelchair)?
Were they wearing slippers or shoes?:
What was the flooring?
What was the lighting level?
Does the patient use glasses? Were they wearing glasses?
Was the patient attached to anything e.g. catheter, drip?
What happened after the fall?
How did the patient get up from the floor?
Were any injuries sustained?

FIGURE 1 Interview Proforma

RESULTS

Analysis of the NRLS revealed that most falls occurred at the bedside (n=1,726), with 416 patients reported to have fallen in the bathroom. 356 patients were coded as frail, 481 as confused, and 3,814 reports had insufficient information to code these factors. The highest number of reported intrinsic and extrinsic risks were toilet-related (n= 508; e.g. incontinence), bed rails up (n=230) and slippery/wet floor (n=121). The location of falls (figure 2) indicated that more than the expected number of frail patients fell in the toilet/bathroom (17%, n=62, expected n=32.4). Less than the expected number of frail patients fell by the bed (23%, n=81, expected. n=134.4) and less than the expected number of confused patients fell in the toilet/bathroom (5%, n=24, expected. n=43.8).

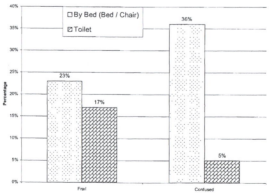

FIGURE 2 Incident data analysis

The case study also found that most patients fell by the bed (n=18), with 5 falling from an adjacent chair or commode, and 3 falling in the bay or bathroom. 18 patients were found on the floor by the bed, 4 were found by their chair, 3 in the middle of the bay and 1 in the bathroom (figure 3). 10 of the 18 patients falling from bed had raised bed rails; for 7 patients bed rails were either not applicable as they were not in bed or there was no information. Most patients had bare feet (n=17) at the time of the fall, with 8 wearing shoes, socks or slippers. 14 falls occurred under 'good lighting' (day light or artificial light), with 12 falls in poorly lit conditions

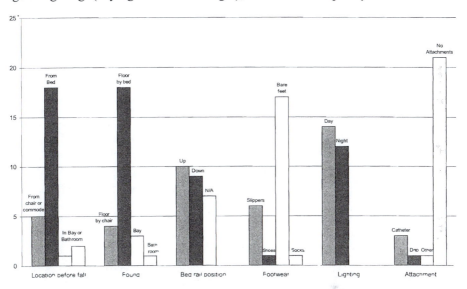

FIGURE 3 Case study results

The data were plotted on three location maps (figures 4, 5 and 6). In the 10 cases the falls occurred from the bed when the bedrails were raised (figure 4), with the patient found on the floor at the lower end of the bed, having 'wriggled to the bottom' of the foot end of the bed.

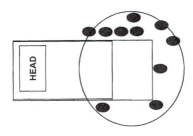

FIGURE 4 Location map of falls from the bed with raised bed rails

80

In the 8 cases the falls occurred from the bed when the bedrails were not raised (figure 5). The location of the falls is less clustered than figure 4 (with raised bedrails).

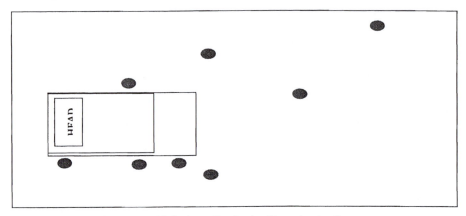

FIGURE 5 Location map of falls from the bed with no bed rails

In 5 cases where the patient fell from a chair or a commode they were found on the floor by the head of the bed (n=4) or at the end of the bed (n=1). Three other patients were found in the middle of the bay and in the bathroom (figure 6).

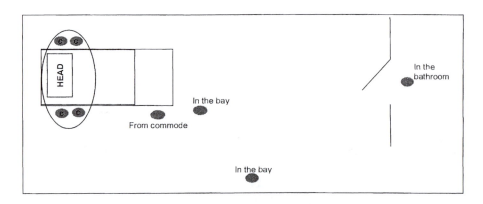

FIGURE 6 Location map of falls from chair/commode or in the bay/bathroom

DISCUSSION AND CONCLUSION

Although the statistical effect size for the NRLS associations was small and this analysis does not allow us to show causal relationships, some interesting issues have emerged that require further investigation, for example bed rails and location of the fall. The use of bed rails has been discussed since the 1960s, with Fagin & Vita (1965) commenting that *'to many conscious patients, side rails are frightening and imply dangerous illness. To others, side rails are irritating and humiliating because they emphasize the confining aspects of hospitalization.'* Bed rails have been used extensively as an intervention to manage falls (McCarter-Bayer et al., 2005; Capezuti et al., 2007; Hanger et al, 1999; Healey et al., 2004), but there is no evidence that they prevent falls or injury (Capezuti et al., 2007; Hanger et al., 1999).

The location of falls suggest that patients described as frail were able to achieve a greater distance from the bed (possibly with a walking aid), with a higher than expected number reported to have fallen in the toilet/bathroom. Interventions to address elimination needs have included scheduled toileting (Kilpack et al, 1991; Krauss et al, 2008). This has reported limited success for two-hourly schedules due to poor patient compliance (Krauss et al, 2008), although a pilot trial indicated that hourly rounding may reduce the number of falls (Meade et al, 2006). In 1987 Morse et al. reported a study where patients were interviewed about contributory factors for falls. Patients identified difficulties with distance perception (frequently underestimating distances between objects) due to the greater size and distance between hospital fixtures compared to those at home. There have been several interventions to facilitate the route from bed-to-bathroom. These include bringing the toilet to the bed by placing the commode adjacent to the bed (Krauss et al., 2008; Chung, 2009), locating the patient in a bed near to toilet in multi-bed bays (Parker, 2000; Krauss et al., 2008; Janken et al., 1986), removing obstacles from the bed-toilet pathway (Krauss et al., 2008; Becker et al., 2003; Ray et al., 1997; Semin-Goossens et al., 2003), and marking a path from the bed to the bathroom (West, 2009 personal communication).

It has been suggested that changes in hospital design may affect the risk of falls (Gulwadi & Calkins, 2008; Janken et al., 1986; Feldbauer et al, 2008) but research studies have failed to systematically evaluate environmental design interventions. The lack of high quality research on physical environment interventions might be, as Oliver et al. (2007) suggest due to the *'inherent logistic difficulties in performing or interpreting studies in care homes or hospitals associated with population, setting, design, and outcome measurement. Getting consent from or randomizing frail, confused, unwell elderly people, who are often in the institution for only a short stay, is challenging'.*

In the US, falls resulting in patient death or serious disability while being cared for in a healthcare facility are included in the 28 'never event' categories by the National Quality Forum (2007) and Institute for Healthcare Improvement (2008). This is likely to raise the priority for finding effective interventions as it is an emerging belief that hospitals may not be reimbursed for events that should never occur, this would include falls (Odom-Forren, 2008).

Making environmental changes can be very expensive. If the research evidence is not available to show that different layouts and technology can reduce both the incidents and injuries associated with elderly in-patient falls at the time of construction then retro-fitting is unlikely to happen.

REFERENCES

American Geriatrics Society. (2001), "Guideline for the Prevention of Falls in Older Persons." *Journal of the American Geriatrics Society*, 49(5), 664-672.

Becker, C., Kron, M., Lindemann, U., Sturm, E., Eichner, B., Walter-Jung, B. et al. (2003), "Effectiveness of a multi-faceted intervention on falls in nursing home residents." *Journal of the American Geriatrics Society*, 51, 306-13.

Boushon, B., Nielsen, G., Quigley, P., Rutherford, P., Taylor, J., Shannon, D. (2008), *Transforming Care at the Bedside How-to-Guide: Reducing Patient Injuries from Falls*. Cambridge, MA:Institute for Healthcare Improvement.

Capezuti, E., Wagner, L.M., Brush, B.L., Boltz, M., Renz, S., Talerico, K.A. (2007), "Consequences of an intervention to reduce restrictive side rail use in nursing homes." *Journal of the American Geriatrics Society,* 55(3), 334-41.

Chung, H. (2009), *The lived experience of older adults who fall during hospitalization*. PhD Dissertation. College of Nursing, Graduate School of the Texas Woman's University

Fagin, I.D., Vita, M. (1965) "Who? Where? When? How? An Analysis Of 868 Inpatient Accidents." *Hospitals* 39:60-5.

Feldbauer, R., Boan, D., Nadzam, D., Finis, N., Nadzam, B. (2008), "Design of a patient-safe environment: The Joint Commission Position." *HERD*, 1(2), 65-68.

Fine, W. (1959), "An analysis of 277 falls in hospitals." *Gerontological Clinics*, 1, 292-300.

Gulwadi, G.B., Calkins, M.P. (2008), "The Impact of Healthcare Environment Design on Patient Falls." *The Center for Health Design*. www.healthdesign.org

Hanger, H.C., Ball, M.C. Wood, L.A. (1999), "An Analysis of Falls in the Hospital: Can We Do Without Bedrails?" *Journal of the American Geriatrics Society,* 47(5), 529-531.

Healey, F., Monro, A., Cockram, A., Adams, V., Heseltine, D. (2004). "Using targeted risk factor reduction to prevent falls in older in-patients: a randomised controlled trial." *Age and Ageing,* 33, 1-5

Healey F, Scobie S, Oliver D, Pryce A, Thompson R, Glampson B. (2008), "Falls in English and Welsh hospitals: a national observational study based on retrospective analysis of 12 months of patient safety incident reports." *Quality and Safety in Healthcare* 17:424-430.

Hignett, S. Masud, T. (2006), A Review of Environmental Hazards associated with In-Patient Falls. *Ergonomics,* 49(5-6), 605-616.

Hitcho E, Krauss M, Birge S, Dunagan W, Fischer I, Johnson S. et al (2004) "Characteristics and circumstances of falls in a hospital setting." *Journal of General Internal Medicine* 19:732-739.

Institute for Healthcare Improvement (2008), *IHI Global Trigger Tool for Measuring Adverse Events (UK Version).* Cambridge MA: Institute for Healthcare Improvement

Janken, J.K., Reynolds, B.A., Swiech, K. (1986), "Patient falls in the acute care setting: identifying risk factors." *Nursing Research,* 35(4), 215-9.

Kannus, P., Khan, K.M., Lord, S.R. (2006), "Preventing falls among elderly people in the hospital environment." *Medical Journal of Australia,* 184(8), 372-3.

Kilpack V, Boehm J, Smith N, Mudge B. (1991), "Using research-based interventions to decrease patient falls." *Applied Nursing Research,* 4(2), 50-56.

Krauss MJ, Tutlam N, Costantinou E, Johnson S, Jackson D, Fraser VJ. (2008), "Intervention to prevent falls on the medical service in a teaching hospital." *Infect Control Hosp Epidemiol,* 29(6), 539-45.

Mahoney, J.E. (1998), Immobility and Falls. *Clin Geriatr Med,* 14(4), 699-727.

McCarter-Bayer, A., Bayer, F., Hall, K. (2005), "Preventing falls in acute care." *Journal of Gerontological Nursing,* 31(3), 25-33.

Meade CM, Bursell AL, Ketelsen L. (2006), "Effects of Nursing Rounds on patients' call light use, satisfaction and safety." *American Journal of Nursing,* 106(9), 58-70.

Morgan, V.R., Mathison, J.H., Rice, J.C., Clemmer, D.I. (1985), "Hospital falls: a persistent problem." *Am J Public Health,* 75(7), 775–777.

Morse, J.M., Tylko, S.J., Dixon, H.A. (1987), "Characteristics of the Fall-Prone Patient." *The Gerontologist,* 27(4), 516-522.

National Quality Forum (2007), *Serious Reportable Events in Healthcare 2006 Update. A Consensus Report.* Washington DC: National Quality Forum

Odom-Forren, J. (2008), "Never Events: A Patient Safety Imperative." *Journal of PeriAnaesthesia Nursing*, 23(4), 223-225.

Oliver, D., Daly, F., Martin, M., McMurdo, M. (2004), "Risk Factors and risk assessment tools for falls in hospital inpatients: a systematic review." *Age Ageing*, 22, 122-130.

Oliver, D., Connelly, J.B., Victor, C.R., Shaw, F.E., Whitehead, A., Genc, Y. et al. (2007), "Strategies to prevent falls and fractures in hospitals and care homes and effect of cognitive impairment: systematic review and meta-analyses." *BMJ*, 334, 82-7.

Parker, R. (2000), "Assessing the risk of falls among older patients." *Professional Nurse*, 15(8), 511-514

Parrish, H., Weil, T.P. (1958), "Patient Accidents Occurring in Hospitals: Epidemiologic study of 614 accidents." *New York State Journal of Medicine*, 58(6), 838-846.

Ray, W.A., Taylor, J.A., Meador, K.G., Thapa, P.B., Brown, A.K., Kajihara, H.K. et al. (1997), "A randomized trial of a consultation service to reduce falls in nursing homes." *JAMA*, 278(7), 557-62.

Salgado, R.I., Lord, S.R., Ehrlich, F., Janji, N., Rahman, A. (2004), "Predictors of falling in elderly hospital patients." *Archives of Gerontology and Geriatrics*, 38, 213-219.

Semin-Goossens, A., van der Helm, J.M.J., Bossuyt, P.M.M. (2003), "A failed model-based attempt to implement an evidence-based nursing guideline for fall prevention." *J Nurs Care Qual*, 18(3), 217-225.

Tinetti, M. (2003), "Clinical practice; Preventing falls in elerly persons." *New England Journal of Medicine*, 348(1), 42-49.

Tinker, G.M. (1979), "Accidents in a Geriatric Department." *Age Ageing*, 8(3), 196-198.

Chapter 12

Design for Patient and Staff Safety

Laurie Wolf

Barnes-Jewish Hospital
Operational Excellence Department
St. Louis, MO, 63110, USA

ABSTRACT

Barnes-Jewish Hospital is a large urban teaching hospital in St. Louis, Missouri. (9,438 employees, 1,845 physicians, 803 residents/interns/fellows, 1,111 staffed beds, 83,997 Emergency Room visits and 54,733 admissions in 2009). Barnes-Jewish Hospital is part of BJC HealthCare and is affiliated with Washington University School of Medicine. Over the past few years, Barnes-Jewish Hospital has worked to combine the power of Six Sigma and Lean principles and tools to reduce defects and process variability, to enhance patient and staff safety.

This presentation will be a collage several different projects using Lean, Six Sigma, and Human Factors techniques to improve processes and design that will achieve efficiency and safety. Projects reviewed will include equipment design (such as retrofitting operating room connectors to prevent incorrect hook up), patient room layout (to prevent patient falls) and solutions for preventing employee slip and falls. Projects will highlight Human Factors solutions that were used to improve the process or make it "mistake proof" if possible.

This article will discuss one project that designed a process to decentralize equipment cleaning to improve patient and staff safety.

Keywords: Human Factors Engineering, Lean, Six Sigma, Decentralized Equipment Cleaning, Patient Safety, Hospital Worker Safety, Patient Safety, IV pump availability

INTRODUCTION

As one nurse described, "our process of providing our patients with necessary equipment was broken. We lacked a mechanism for tracking equipment, no feel for if our inventories are adequate, no follow-up for broken equipment, and most importantly, no way of ensuring that equipment got to patients in a timely manner."

This broken process resulted in a classic "work-around" by nursing staff. In the case of IV pumps, nurses would be fast to claim a pump from a discharged patient's room, do a quick cleaning off the record, and use it for their just-admitted patient. Or, they would go to a neighboring unit to "borrow" one— without really intending to return it. Needless to say, nurses were extremely frustrated at the lack of readily available equipment.

Barnes-Jewish Hospital's equipment supply partner, had frustrations as well. Staff had to retrieve, clean and deliver patient equipment, moving it back and forth from the centralized cleaning area in the basement up to the nursing divisions—as much as ½ mile away! There were long elevator waits going up and coming back down. They awkwardly maneuvered lots of equipment at one time, resulting in the use of poor body mechanics. In addition, everyone had concerns about transporting used patient equipment through public areas, risking potential exposure to other visitors and staff.

DESIGN NEW PROCESS TO IMPROVE EQUIPMENT AVAILABILITY

To remedy the situation, a Lean/Sigma project was initiated to design a new equipment cleaning process. The primary goal was to enhance patient safety by reducing equipment delivery time and the time nurses spent locating equipment. The scope of the project was initially to improve the availability of IV pumps only. Additional equipment was added after the process was proven to be a success.

Sponsorship for the project came from the top, and further support was found throughout many various disciplines and departments. The medicine patient care director served as the project's executive sponsor. A clinical nurse specialist, conducted the project as part of her Six Sigma training, and a management engineer, was the project facilitator. The project team also included nurses, clinical nurse managers, unit secretaries and representatives from IS, receiving, central sterile processing and materials and supply. An important partner in this project was management and staff from Barnes-Jewish Hospital's equipment supplier.

CURRENT STATE

With a combination of Lean and Six Sigma tools, the team analyzed the current process for getting equipment to a patient, cleaning it after use and getting it ready for the next patient. The current state was mapped out in a value stream so all participants could understand the required tasks and roles of everyone involved. The process began with the nurse ordering an IV pump, then the next steps were listed such as; receiving order, delivering pump to requesting nurse, using pump to deliver medication, discontinuing use, returning to soiled utility room, and transporting to decontamination. The process ended with cleaning the pump and preparing for next use. The group then took a "field trip" to walk the entire process to see where each activity was performed. Each process step was reviewed to see if it was value or non-value added. Time estimates were made for each process step.

Next an affinity diagram was developed to understand the reasons why equipment (such as an IV pump) was not available in a timely manner. Numerous reasons were grouped into the following categories: communication, equipment availability and nursing process.

FUTURE STATE

The team identified the "preferred future state" for the patients, that is, the ideal end result of the equipment delivery process. The future state required the equipment to be cleaned on the nursing division in the soiled utility room and then placed in the clean equipment storage area ready for next use. This brought up the challenge to determine how many IV pumps must be stored on each division.

This critical element of the new process involved setting up and maintaining "par levels," of equipment on specific divisions based on usage. To do this, the team approached unit leadership with an empowering request: "how many clean IV pumps would you need on-hand to avoid ever calling downstairs for pump delivery?" Initially, the par levels were set to this requested amount, and then adjusted as usage patterns were observed by the equipment supplier. One challenge was to balance the number of available IV pumps with the timing of the cleaning process. A 2-hour rounding schedule was established. Each division was checked at least every 2 hours to see if there was any equipment that needed to be cleaned. Typically they would find that the original par level estimate was too high and the number could be reduced once the nurses trusted that they would always have a pump available when it was needed.

The process begins when the nurse puts the used equipment in the soiled utility room (they are still required to remove the tubes from the IV pumps).
- Inventory is cleaned and stored on the floors rather than the department.
- Par levels of equipment are maintained in designated clean rooms throughout the facility.

- Charge slips are placed on each equipment item when it is ready for use. When needed, nursing staff retrieve item from the nearest clean room, complete the charge slip and turn it in to the unit secretary.
- After use, the items are brought to a soiled utility room.
- Throughout the day, equipment jess decontaminated in the soiled utility rooms and par levels are restocked in each area.
- Equipment inventory and rounds verification are performed each day.

DESIGN OF NEW SOILED AND CLEAN UTILITY: VISUAL CUES

Visual cues such as tape on the floor to indicate proper space for equipment location were placed in clean and dirty storage areas. A designated cleaning area (approximately 3'x 2') was labeled and stocked with the appropriate supplies. There is also tape on the floor in the Clean Equipment room to designate the proper location and number of clean equipment. The best place for storage in the clean and dirty utility rooms was determined by the Lead Charge Nurse. The configuration varies according to the building layout of each division. The cleaning supplies are maintained by the unit secretary and housekeeping.

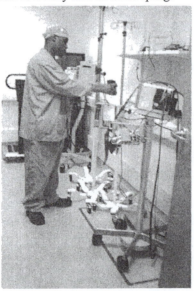

FIGURE 1 Visual indicators show area for clean IV pumps

RESULTS

Since introducing the decentralized equipment cleaning process throughout Barnes-Jewish Hospital, the wait for equipment has improved considerably. The range of time it took to get an IV pump was 40 minutes to 4 hours, 36 minutes (with a mean of 2 hours, 7 minutes). With the new process, the wait time is completely eliminated! In addition, the rate of lost equipment dropped almost in half (from 12% to 6.9%).

To make decentralizing equipment cleaning a success required building trust. The new process was implemented on one volunteer division as a 2-week trial. It was so successful the division did not want the trial to end and there were several divisions that had heard positive results and wanted to be the next division to be selected.

The new process has now been in place for the entire hospital for almost 2 years. The program has been expanded to equipment including: bedside commodes, sequential compressive devices, walkers, bed checks, portable suction and bariatric equipment.

CONCLUSION

In conclusion, this project empowered multiple disciplines to work together and to partner with an outside supplier to create the best possible system for all. The results demonstrate how even a "low-tech" solution with minimal capital expense can generate sustainable improvements in service delivery and tremendous gains in staff satisfaction. As one nurse described it, the project was a true gift to nurses; the best process improvement she had seen in 30 years!

REFERENCES

Norman, Don. (1988), *Design of Everyday Things*.
Wickens, C.D., Holland, J. G. (3rd edition), "Engineering Psychology and Human Performance."
The Wrong Patient http://www.annals.org/cgi/content/full/136/11/826

Chapter 13

Design Considerations in the Provision of Safe Patient Handling Environments

Mary W. Matz

Patient Care Ergonomic Solutions, LLC
17702 Esprit Drive
Tampa, FL 33647, USA

ABSTRACT

The use of patient handling technology positively influences the quality of patient care, mobilization, rehabilitation, quality of life, and the risk to staff and patients from moving and handling patients. Architecture and design that take into account factors that impact patient handling and the use of patient handling equipment will thus foster improved patient care and outcomes as well as safer and more professionally satisfying work environments for staff. Design/architectural features that must be addressed in this context include: Flooring Materials and Finishes, Space Constraints, Storage Space, Door Openings, Hallway Widths, Floor/Walkway Slopes and Thresholds, Elevator Dimensions, Headwalls/Service Utility Columns, and Weight Capacities of Toilets and Mounted Objects.

Keywords: Patient Care Ergonomics, Safe Patient Handling, Patient Handling Technology, Hospital Design

INTRODUCTION

Patent care environments are fraught with hazards for both caregivers and patients. The focus of this paper is on the manner in which building design specifications can work to decrease ergonomic hazards and thus provide a safer environment of care for both patients and caregivers. Although patients are not directly affected by ergonomic hazards in a health care environment, they are impacted by care from injured staff and from incidents that occur during use of manual patient handling techniques rather than the use of technology. The ramifications of providing an ergonomically safe environment of care for staff and a safer clinical environment for patients are significant. Caregivers will benefit from reductions in pain and injury, patients from improvement in clinical outcomes, and organizations will realize cost savings associated with caregiver injuries and less lengthy patient stays.

The primary ergonomic control measures in a patient care environment are varying types of patient handling equipment, including overhead/ceiling lifts, floor-based lifts, lateral transfer devices, repositioning devices and others. With the inclusion of these in a patient care environment come increased space and often storage requirements, floor surfaces and designs that facilitate use of rolling equipment, and adequate weight capacities for use with larger patients. And, these design considerations must be included in not only nursing units, but in all areas where patient handling and movement occur: acute care, critical care, therapy clinics, radiology/diagnostics, treatment areas, procedure rooms, and others, even the morgue.

The inclusion of patient handling technology into health care environments is relatively new in the United States, however, increasing interest is seen by the adoption of safe patient handling legislation in ten states and proposed national legislation. And, because patient handling equipment and its design parameters are new to many design professionals in the United States, they have had no consensus standards or master specifications to follow and depended on the word and expertise of manufacturers and the limited design recommendations currently available. Proactively, the *2010 Guidelines for Design and Construction in Health Care Facilities,* includes completion of a patient handling and movement assessment to generate recommendations to decrease risk from patient handling tasks and relays design considerations when introducing fixed equipment such as ceiling lifts. As well, other design considerations related to a safe patient care environment are included and are discussed below.

DESIGN CONSIDERATIONS IN THE PROVISION OF SAFE PATIENT HANDLING ENVIRONMENTS

FLOORING MATERIALS AND FINISHES:

There has been growing concern about musculoskeletal injuries associated with the movement of patients and health care-related equipment on flooring surfaces with high coefficients of friction and the possible resultant excessive forces on the spine. Increases in the shear forces to the spine are attributable to the difficulty in overcoming inertia when initially pushing or pulling a wheeled object, surface resistance of the flooring material, wheel design and condition, and the weight being pushed/pulled. From a safe patient handling perspective, rolling lifts over carpeting as compared to less resilient flooring materials is a factor to consider when specifying flooring materials.

SPACE CONSTRAINTS:

"Working" space is a critical design aspect for safe patient handling. For instance, moving rolling equipment in tight spaces compounds already difficult patient handling tasks and when caregivers must continually move items in order to provide proper patient handling, their risk of injury is compounded. Additionally, awkward postures resulting from lifting and moving patients in small spaces increase the risk of injury. Adequate space will enhance the quality of nursing by facilitating mobilization of patients, reducing strain-related injuries to staff, and increasing staff productivity.

STORAGE SPACE:

Inadequate storage space is universally problematic in health care facilities. Limited and inaccessible storage space for mobile patient handling equipment significantly impacts staff compliance in the use of safe patient handling techniques. Similarly, other non-portable patient handling equipment is less likely used when less available. If a storage room is quite a distance away from patient rooms or if a storage room is filled to the brim with other equipment and caregivers must move that equipment to access the patient handling equipment, caregivers will often decide to transfer patients manually, increasing their risk of injury.

DOOR OPENINGS:

Insufficient doorway dimensions prevent the use of mobile patient handling lifts and other rolling equipment. Entry and exit, especially in emergency situations involving bariatric beds, are problematic in many health care facilities. And, when doorways are too narrow for bariatric beds, bariatric patients must receive treatments and procedures in their rooms rather than in a designated treatment or procedure area.

HALLWAY WIDTHS

Narrow hallways can add a level of difficulty to moving patients and equipment. An inadequate turning radius in a hallway creates an unsafe situation when staff must push a heavy bed sideways in order to turn sharply around a corner or into a patient room.

FLOOR/WALKWAY SLOPES AND THRESHOLDS

Hospitals are filled with rolling equipment, yet high to medium thresholds abound, making it difficult for staff to use rolling equipment and unsafe for the patients moving themselves or being moved. Pushing patients up and down inclines in beds or wheelchairs has the potential for causing serious injury to both the patient and the caregiver.

ELEVATOR DIMENSIONS

Interior dimensions of elevators may prevent the use of certain types of high-tech and bariatric beds.

HEADWALLS/SERVICE UTILITY COLUMNS

Headwall and service utility column/system designs can promote or interfere with the installation and use of overhead lifts—especially traverse track systems. If lacking good ergonomic design, these structures can also limit easy access to patients and items required for care.

WEIGHT CAPACITIES OF TOILETS AND MOUNTED OBJECTS

Serious injuries may occur if the weight capacities of toilets, chairs, hand rails, and other mounted objects in patient rooms, bath and shower rooms, hall ways, waiting rooms, and elsewhere are not taken into consideration when there is potential for use by morbidly obese patients and visitors.

CONCLUSIONS

The time has come for the inclusion of design specifications to ensure an ergonomically safe health care workplace to protect our caregivers as well as provide a safer environment of care of our patients in the United States. The guidelines found within the *2010 Guidelines for Design and Construction in Health Care Facilities* and the accompanying White Paper provide this information for the first time.

REFERENCES

AORN Workplace Safety Taskforce (2007). "Safe Patient Handling and Movement in the Perioperative Setting". Denver, CO: Association of periOperative Registered Nurses [AORN].

ARJO (2005). *Guidebook for Architects and Planners*, 2nd ed. ARJO Hospital Equipment AB.

Facility Guidelines Institute (2010). *Guidelines for Design and Construction of Health Care Facilities.* Chicago:American Society for Healthcare Engineering.

Hignett, S. (2005). "Determining the space needed to operate a mobile and an overhead patient hoist". *Professional Nurse,* 20:7,39–42.

Marras, W.S., Knapik, G.G., & Ferguson, S. (2009). "Lumbar spine forces during manoeuvring of ceiling-based and floor-based patient transfer devices". *Ergonomics*, 52:3,384-397.

Muir, M. & Haney, L. (2004). "Designing space for the bariatric resident". *Nursing Homes/Long Term Care Management*, November:25–28.

Rice, M.S., Woolley, S.M., & Waters, T.R. (2009). "Comparison of required operating forces between floor-based and overhead-mounted patient lifting devices". *Ergonomics*, 52:1,112-120.

Villeneuve, J. (2006). "Physical Environment for Provision of Nursing Care: Design for Safe Patient Handling". In: Nelson, A.L. (Ed), *Handle with Care: Safe Patient Handling and Movement.* New York: Springer Publishing Company.

CHAPTER 14

Use of Effective Patient-Handling Techniques within Healthcare

Andrea Simone Baptiste,
Shawn Paul Applegarth

HSR&D/RR&D Center of Excellence:
Maximizing Rehabilitation Outcomes
James A. Haley VA Medical Center
8900 Grand Oak Circle
Tampa, FL 33637-1022

ABSTRACT

This paper will recommend use of technological solutions for patient handling tasks in an effort to reduce the risk of injury to caregivers and patients. In addition, it aims to offer more detailed information regarding patient transfers and provides safe solutions to commonly performed nursing tasks.

Keywords: technological solutions, fatigue, high risk tasks, patient handling tasks

INTRODUCTION

In 2006, registered nurses had the fifth highest number of musculoskeletal disorders in the US, exceeding truck drivers and construction workers. 56% of injuries and illnesses involved healthcare patients of which 86% were due to overexertion (BLS, 2007). The average cumulative weight that a nurse lifts in an 8 hour day is 1.8 tons (Tuohy-Main, 1997). In an effort to better understand why injury rates are high in the nursing profession, the ergonomic risk factors have been examined more carefully in the recent years (Baptiste et al, 2006; Nelson and Baptiste, 2006, Lloyd & Baptiste, 2006; Marras et al, 1999; Jang et al, 2007; Waters et al, 2007; Nelson et al, 2003; Fujishiro et al., 2005; McCoskey, 2007; Milholland et al, 2007; Nelson, Motacki & Menzel, 2009).

The goal of this paper is to objectively provide data that ranks ten high risk tasks and to offer direct technological solutions which attempts to drastically diminish risks to users. Such solutions have been implemented in healthcare facilities and have already proven to decrease the risk of injury amongst nursing staff. Furthermore, this study has also considered fatigue factors when performing patient handling transfer tasks. By measuring impulse, the effect of fatigue, which is used as a measure of cumulative trauma of a muscle, can clearly be seen. The research goals of this project were to: 1) Rank the patient handling tasks that required the highest forces 2) Rank the tasks that cause the most fatigue 3) Identify potential solutions.

METHODS

This laboratory based study was designed to simulate common tasks that are performed amongst nurses and caregivers but aimed to capture the forces specifically required in patient handling tasks. The tasks chosen for this study were selected based on a comprehensive review of the literature (Nelson and Baptiste, 2006). The essential transfer tasks selected were: transferring from a bed to wheelchair, lateral transfers from bed to stretcher, seated to standing transfers, seated and supine repositioning, lifting and holding a patient's leg, making an occupied bed, bathing a patient in bed, changing an absorbent pad, diapering a patient, changing a patient's gown, dressing a patient, and lifting a patient from floor level.

VARIABLES

The biomechanical stressors of each transfer task (peak force and impulse), are considered to be the dependent variables. The independent variables were all the subtasks which were sometimes the same within different transfer tasks. For example log rolling a patient needs to be performed to change a diaper but also is required in making an occupied bed. These variables are the subtasks that were manipulated or tested in this study. The last variable is the mediating variables, which can be defined as those variables that serve to clarify the nature of the relationship between the independent and dependent variable.

To better understand the outcome variables, an operational definition of the terms is presented. The measured force represents the total force taken for the specific task. Impulse can be described as change in momentum and it considers the direction of consequent changes in velocity whilst work does not. Work (force x distance) was not used since work does not capture true internal forces and fails to reflect the difficulty of the task whereby impulse does.

Peak forces may be a better indicator of acute trauma and impulse may be interpreted as a good indicator of cumulative trauma and fatigue. Length of the subtask performed is an important variable in addition to amount of force required and posture adopted. All of these factors contribute to muscular fatigue, and trauma, predisposing the caregiver to risk of injury.

MATERIALS

Real time forces of each transfer task were measured with 2,200 N capacity tri-axial load cells from Advanced Mechanical Technology Inc (AMTI). This type of load cell has the ability to measure forces in all three planes x, y and z. Several mechanical attachments were custom fabricated in order to interface with the load cells and the technologies/caregiver. These special attachments insured a proper fit between the assistive device and the load cell, and allowed the caregiver to perform the transfer task as normally as possible thus decreasing measurement error in capturing push and pull forces.

Assistive technologies used in this study included: a floor based lift, a sit to stand lift, a ceiling lift, friction reducing devices, a slide board and a gait belt. A variety of slings were also used including a standing sling (used with sit to stand lift), a universal or seated sling (used with floor based lift), a supine sling (used with an overhead ceiling lift), a limb support sling and a bathing sling.

Mannequins that represent the 67th, 77th, and 90th percentile U.S adult male population in weight, were used as the primary patient surrogate for the transfers. The caregiver was represented by a male who was approximately six feet tall and 900 N in weight.

DATA COLLECTION PROTOCOL

Each transfer task was performed three times on each of the three mannequins. However, some of the tasks were not possible to accomplish with a mannequin, due to the inability to hold posture and weight bear. To overcome this limitation in the study, tasks using the gait belt were tested with a live male of weight 860 N, and 1.85 m in height.

A trial run was performed prior to any real time data collection to assure the data acquisition system and load cells were working properly and calibrated. Load cells were zeroed prior to each task to ensure accurate data acquisition. An excel spreadsheet was created to organize and record all the data for the three mannequins, tasks and trials.

Once the transfer task was completed, all force data files were saved individually and converted into text files. These force files were then opened in Matlab and the corresponding moments for each subtask was then calculated with a custom written algorithm. Objective findings were compiled and analyzed across mannequins to identify those components of the transfer tasks that impose the greatest peak forces and total impulse.

RESULTS

Results show that some patient handling tasks are more physically demanding than others. Analysis of the final data indicates that there are subtasks with can be rated as high risk sub-tasks, based on very high force demand and total impulse.

The advantage of objectively measuring these subtasks is that data analysis can show exactly what about the transfer task makes it high risk and where the peaks occur in terms of highest force demands. Any sub task or task resulting in high force demand will place the musculoskeletal system at risk of injury.

Data is separated into high risk force and high risk fatigue tasks (based on impulse measurements). The highest risk tasks, based on collected data, are listed with potential solutions described to minimize the risk.

HIGH RISK TASKS - FORCE

The tasks that required the most force to perform are depicted in Figure 1 and are further broken down below.

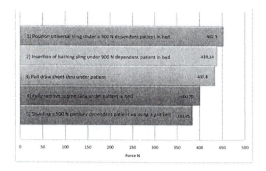

FIGURE 1: The top 5 high risk tasks according to force demands

The positioning of a universal sling under a 900 N dependent patient in bed required 452.5 N of force. This force can be eliminated completely by use of a repositioning sheet that stays under the patient, thus eliminating the repeated sling insertion. The sling should be breathable so there is little risk of increasing pressure ulcers.

The insertion of bathing sling under 900 N dependent patient in bed required 439.14 N of force. A potential solution for this task is to use a ceiling lift to raise the patient, and then insert the sling. Moving the ceiling lift along the track with the same patient requires 15.5 N of force which is a 96% reduction in force. An alternative solution would be to use a floor based lift to raise the patient out of the bed, insert the bathing sling, then lower the patient.

Pulling a draw sheet under the patient requires 433.80 N. The recommended solution would be to use a ceiling lift then remove the sheet or use a friction reducing device (FRD) to turn the patient half way, remove half the sheet then repeat the process and remove the other half. Both of these solutions would significantly reduce the force imposed by manually performing this task.

Fully removing a supine sling under patient in bed requires 400.9 N of force. This task can be corrected by using a ceiling lift or FRD as mentioned above. Moving the ceiling lift on the track takes 15.5 N.

Performing a sit to stand transfer of a 900 N partially dependent patient using a gait belt manually required 383.85 N of force. Use of a powered sit to stand lift reduces the effort as the force to move the lift to the patient was calculated at 60 N. This will vary based on distance the lift will have to be moved.

HIGH RISK TASKS - FATIGUE

A sub task or task that also has a high total impulse represents a task that should be done over a sustained period of time as research has shown that static postures impose high risk of injury (Iowa State University, 2005; National Institute for Occupational Safety and Health [NIOSH], 1997; Pope, Goh, & Magnusson, 2002; Jang et al, 2007).

Figure 2 offers the top five subtasks that will cause the most fatigue based on impulse. Tasks causing fatigue can be an indicator of cumulative trauma. The key findings with potential technological solutions are described below:

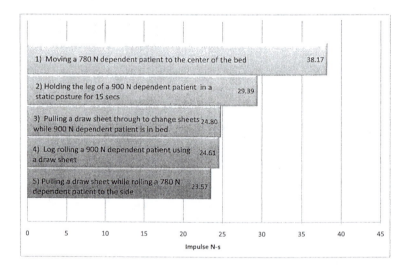

FIGURE 2: The top 5 high risk tasks associated with fatigue

Moving a 780 N dependent patient in the center of the bed takes a long time causing the caregiver to be in a forward flexed and awkward posture. This can be prevented by using either a ceiling lift with repositioning sling or FRD to reposition the patient. The FRD significantly reduces friction against the sheets and makes bed mobility tasks much easier. It also eliminates shear forces against the patient skin, because one surface slides over the other (Lloyd and Baptiste, 2006, Baptiste et al, 2006). In addition it reduces the awkward sustained posture adopted by the caregiver if this task were performed manually.

Holding the leg of a 900 N patient for 15 seconds restricts the blood flow, and ultimately results in fatigue, predisposing the caregiver to risk of injury. A solution for this is a limb-support strap/sling attached to a floor based or ceiling lift. In this case, the weight of the leg is off set by the lift, freeing the caregiver to perform wound care or cleaning as needed safely.

Pulling a draw sheet through and log rolling a patient in bed are both tasks that take some time and place caregivers in awkward postures over the bed, increasing their risk for injury due to the static posture required. Solutions for these tasks are use of a ceiling lift to turn the patient to one side, or use of a FRD sheet. Both of these solutions take less time, are easier to perform and are safer for both caregiver and patient.

DISCUSSION

Traditionally, healthcare professionals, including nurses have been taught use of body mechanics when moving and transferring patients. This has proven unsuccessful (Engkvist et al., 2001; Fanello et al., 1999; Harper et al. 1994; Lagerstrom & Hagberg 1997) as evidenced by the rising number of work related patient handling injuries amongst nurses and nursing assistants (BLS, 2007). The standard of care needs to change as healthcare technology becomes more readily available and evidenced based studies exist, and have proven the benefits of such technologies (Nelson and Baptiste, 2006, Lloyd & Baptiste, 2006; Marras et al, 1999; Jang et al, 2007; Waters et al, 2007; Nelson et al, 2003, Daynard et al, 2001; Milholland et al, 2007; Nelson, Motacki & Menzel, 2009).

Evidence has shown that the mobile devices do decrease risk of injury and keep healthcare professionals safer during patient transfers and nursing tasks (Daynard et al., 2001; Evanoff, Wolf, Alton, Canos, & Collins, 2003; Garg, Owen, Beller, & Banaag, 1991a; 1991b). There is further documented evidence at the national level that states the safe working limit is 156 Newtons (35 lbs) for patient handling. It should be noted that all of the tasks in Figure 1 exceed 156 Newtons or 35 lbs and that using a ceiling lift and sit to stand lift requires 15.5 Newtons and 60 Newtons respectively, both of which fall below the national limit for patient handling.

There is also legislature that is supportive of using technology in seven U.S states. Globally, in the UK, Canada, and Australia, healthcare facilities have already instituted strict minimal lift policies and have made it mandatory that staff use mobile lifts and other patient handling devices when applicable.

This study adds to the evidenced based research in the support of using technological solutions for performing typical nursing tasks, such as bed mobility (repositioning up or turning in bed), bed to chair transfers, insertion and removal of slings (which is commonly done), log rolling, and static limb holding.

CONCLUSIONS

There are effective patient handling techniques which will allow caregivers to safely move and transfer patients. These techniques include use of lifts and friction reducing devices. This study has identified ten high risk tasks that can be done differently with use of technology, thus reducing the risk of injury to both caregiver and patient. The standard of care is changing with regards to use of mechanical lifts, overhead ceiling lifts, friction reducing devices and repositioning slings for patient care. All of these technologies currently exist on the market and should be implemented to keep patients and workers safer, reduce risk of patient falls, and skin tears.

REFERENCES

Baptiste A, Boda SV, Nelson AL, Lloyd JD, Lee WE 3rd. (2006). Friction-reducing devices for lateral patient transfers: a clinical evaluation. American Association of Occupational Health Nurses, Apr; 54(4):173-80.

Daynard, D., Yassi, A., Cooper, J. E., Tate, R., Norman, R., & Wells, R. (2001). Biomechanical analysis of peak and cumulative spinal loads during patient handling activities: a sub-study of a randomized controlled trial to prevent lift and transfer injury health care workers. Applied Ergonomics, 32, 199-214.

Engkvist, I-L., Kjellberg, A., Wigaeus, H.E., Hagberg, M., Menckel, E., & Ekenvall, L. (2001). Back injuries among nursing personnel – identification of work conditions with cluster analysis. Safety Science, 37, 1-18.

Evanoff, B., Wolf, L., Aton, E., Canos, J., & Collins, J. (2003). Reduction in injury rates in nursing personnel through introduction of mechanical lifts in the workplace. American Journal of Industrial Medicine, 44, 451-457.

Fanello, S., Frampas-Chotard, V., Roquelaure, Y., Jousset, N., Delbos, V., Jarmy, J., et al. (1999). Evaluation of an educational low back pain prevention program for hospital employees. Revue Du Rhumatisme (Eng. Ed.), 66(12), 711-716.

Fujushiro, K., Weaver, J. L., Heaney, C. A., Hamrick, C. A., & Marras, W. S. (2005). The effect of ergonomic interventions in healthcare facilities on musculoskeletal disorders. Am J Ind Med, 48(5), 338–347.

Garg, A., Owen, B., Beller, D., & Banaag, J. (1991a) A biomechanical and ergonomic evaluation of patient transferring tasks: Bed to wheelchair and wheelchair to bed. Ergonomics, 34, 289-312.

Garg, A., Owen, B., Beller, D., & Banaag, J. (1991b). A biomechanical and ergonomic evaluation of patient transferring tasks: wheelchair to shower chair and shower chair to wheelchair. Ergonomics, 34, 407-419.

Harper P., Pena, L., Hsu, P., Billet, E., Greer, D., & Kim, K. (1994). Personal history, training and worksite as predictors of back pain of nurses. American Journal of Industrial Medicine, 25, 519-526.

Iowa State University, Department of Environmental Health and Safety. (2005) Health & Safety. Retrieved January 21, 2010 from http://www.ehs.iastate.edu/cms/default.asp?action=article&ID=88

Jang R, Karwowski W, Quesada PM, Rodrick D, Sherehiy B, Cronin SN, Layer JK. (2007). Biomechanical Evaluation of Nursing Tasks in a Hospital setting. Ergonomics, Nov;50(11):1835-55.

Lagerstrom M., & Hagberg, M. (1997). Evaluation of a 3-year education and training program for nursing personnel at a Swedish hospital. AAOHN Journal, 45, 83-92.

Lloyd, J. D., & Baptiste, A. (2006). Friction-reducing devices for lateral patient transfers: a biomechanical evaluation. American Association of Occupational Health Nursing, 43(3), 113-119.

Marras, W. S., Davis, K. G., Kirking, B. C., & Bertsche, P. K. (1999). A comprehensive analysis of low-back disorder risk and spinal loading during the transferring and repositioning of patients using different techniques. Ergonomics, 42(7), 904-926.

McCoskey, K. L. (2007). Ergonomics and patient handling. AAOHN J, 55(11), 454-462.

Milholland, G. A., Deckow-Schaefer, G., & Kapellusch, J. M. (2007). Justification for a minimal lift program in critical care. Crit Care Nurs Clin North Am, 19(2), 187–196.

National Institute of Occupational Safety and Health (NIOSH). (1997). Low Back Musculoskeletal Disorders: Evidence for Work-Relatedness. Musculoskeletal Disorders and Workplace Factors, NIOSH Publication Number 97-141. Retrieved January 21, 2010 from http://www.cdc.gov/niosh/docs/97-141/ergotxt6.html.

Nelson, A., & Baptiste, A. S. (2006). Update on evidence-based practices for safe patient handling and movement. Orthopaedic Nursing, 25(6), 366–368.

Nelson, A., Lloyd, J. D., Menzel, N., & Gross, C. (2003). Preventing nursing back injuries: redesigning patient handling tasks. AAOHN J, 51(3), 126-134.

Nelson, A. L., Motacki, K., & Menzel, N. (eds.). (2009). The Illustrated Guide to Safe Patient Handling and Movement. New York: Springer Publishing

Pope, H., Goh, K., & Magnusson, M. (2002). Spine ergonomics. Annual Review of Biomedical Engineering, 4, 49-68

The Future of the Nursing Workforce in the United States: Data, Trends and Implications. Retrieved January 21, 2010 from http://www.jbpub.com/catalog/9780763756840/

Tuohy-Main, K. (1997). Why manual handling should be eliminated for resident and career safety. Geriatrician, 15, 10-14.

U.S. Department of Labor Bureau of Labor Statistics (2007). Nonfatal occupational injuries and illnesses requiring days away from work, 2006. Retrieved January 21, 2010, from http://www.bls.gov/iif/oshwc/osh/case/osnr0029.pdf

Waters, T. R. (2007). When is it safe to manually lift a patient. American Journal of Nursing, 107(8), 53–58, quiz 59.

Chapter 15

A Tool to Compare All Patient Handling Interventions

Mike Fray, Sue Hignett

Healthcare Ergonomics and Patient Safety Unit
Loughborough University, UK

ABSTRACT

Patient handling intervention strategies are many and varied. The focus of interventions has primarily been on the health, safety and welfare of care givers. Data from 4 EU focus groups and 2 world-wide expert panels were used to evaluate whether other types of outcomes were perceived as having relative importance. Qualitative and quantitative analysis showed that organisational and patient outcomes were also highly rated by the participants. The data showed 12 outcomes as being of the highest priority with good agreement between the 4 EU sources (Kendall's Concordance significant at 0.005). In parallel, a systematic analysis of patient handling intervention literature was considered to evaluate the qualities of each study. Using the 12 most important outcomes from the initial study and the most appropriate and accessible measurement tools from the literature analysis, the Intervention Evaluation Tool (IET) is proposed. The IET is a single set of measurements that can be used for evaluating all organisational and individual patient handling interventions. The IET has been trialled at 2 sites in 4 EU countries.

Keywords: Patient handling, risk management, safety performance, interventions, healthcare workers, MSD.

INTRODUCTION

Patient handling is a known cause of musculoskeletal risk for healthcare staff. A range of ergonomic and other approaches have been used to reduce the effects of these tasks, e.g. risk assessment and management, training, equipment provision, culture change. Comparing the effectiveness of these interventions has been difficult due to the different outcome measures used to evaluate success. Fray and Hignett (2006) found that published patient handling studies used staff outcomes in 77% and represented patient outcomes in less than 8% of the investigations. Recent systematic reviews have concentrated on the specific measures of musculoskeletal disorders (MSD) in healthcare staff using the highest level of scientific data (Bos *et al*, 2006, Amick *et al.*, 2006, Martimo *et al.*, 2008) and deduce there is little high quality evidence available and little proven benefit on the rate of MSDs. In comparison Hignett et al (2003) used an inclusive methodology and a quality assessment system to allow a wider range of information to be accessed and included.

The literature does include many different methods for measuring outcomes from patient handling interventions. At present it is very difficult to conduct a meaningful comparison between different styles of interventions or different methods of measuring outcomes. This paper will discuss the relative values of the wider range of outcome measures used to evaluate patient handling interventions and describe the development of an 'inclusive' evaluation tool. If a wide application evaluation tool can be developed then it would prove useful to report successful patient handling interventions and guide organisations to a more directed and streamlined approach of future investment to improve their services.

Experts and practitioners from four European Union (EU) countries participated to add to the content validity and strengthen the evaluation tool for the use across all members of the European Panel of Patient Handling Ergonomics group (EPPHE).

The overall aim of this study is to develop an intervention evaluation tool (IET) that allows the comparison of different types of interventions on a single score system using a range of outcomes. The three stages of this process are a) identify which outcomes are preferred by patient handling practitioners, b) develop a tool that measures all the preferred outcomes in a single calculation, c) use the tool in 4 EU locations. This study is part of a longer term research partnership investigating patient handling interventions sponsored by Arjo-Huntleigh ab and the EPPHE group.

METHODS

DEVELOPMENT OF THE INTERVENTION EVALUATION TOOL (IET)

The first aim, to create identify the most important outcomes was achieved using focus groups across 4 EU countries. The 4 countries included in the study were selected using the following criteria:

- Ability to access a range of suitable participants.
- Support delivered by a key facilitator to recruit and organise the focus group.
- Achieve a geographical and demographic spread across the EU.
- A mix of levels within the actions taken to implement/answer the EC directive on manual handling in healthcare (Hignett et al, 2007).

The countries selected for the focus groups were UK, Finland, Italy and Portugal.

The structure for the focus group was based on a model by Higgins (1994), known as the 'Nominal Group Technique'. This allows participants to individually record their own thoughts based on a question set. The method was tested at 2 UK and 2 international pilot sessions. Recruitment for the focus group facilitators took place through the EPPHE group network. Several key stages were used to improve the between-groups validity. Advice was circulated to the focus group facilitators. The following items were included in the guidance:

- The documentation was translated into Finnish / Italian / Portuguese (in the UK) and sent to the facilitator to checking for translation and content errors.
- Standards for recruiting participants for the group.
- Guidance for the focus group room, facilities and timetable.

As it was essential to standardise the translation and cross checking of feedback from the focus groups, the following process was used:

- A translator was supplied to translate the participant feedback information
- A whispering interpreter was provided to report on the group in real time.
- The discussion groups were transcribed in Finnish / Italian / Portuguese. The transcriptions were translated into English.
- The whispering interpreter was taped and transcribed in English
- The transcriptions from the discussion groups (Finnish / Italian / Portuguese) and whispering interpreter (English) were compared for differences.

Specific instructions were developed to assist the EU facilitators. The EU facilitators all participated in the international pilot studies to learn the process and format. The Principal Investigator (MF) was present at all the EU groups to assist with the standardisation of the process and with the development of the discussion group check list of topics for discussion in collaboration with the EU facilitator and the interpreter. Each focus group was centred around a scenario describing a patient care centre. The participants were invited to give advice to the centre and they were asked which outcomes would they like to measure.

The relative importance of the outcomes was considered within each of the individual and homogenous groups. The following simple analytical style was used:

a) The initial recordings of the preferred outcomes were scored on the content.
b) A computerised qualitative analysis package (NVivo) was used to identify themes and content from the focus group discussion. This will be reported in a future paper.

c)   The ranked priority lists created by the participants at the close of the focus
group interview were scored and ranked on a 5 point scale.

Corrections were made for unequal group sizes before comparisons were made
between groups. The ranking scores were added to give a group preference list. All
group lists were compared to check for similarities and differences. All the groups
were accumulated to give an overall list of the preferred outcomes.

LITERATURE ANALYSIS

To achieve the second aim of selecting methods for measuring each of the preferred
outcomes a detailed analysis of published patient handling studies was completed.
Studies were collected using the search strategy used for Hignett et al. (2003) was
extended to December 2008. 752 additional papers were assessed against the
inclusion criteria and 328 included in the analysis. Each paper was analysed by two
independent researchers and the following data were recorded:
- Design of the study
- Characteristics of the intervention
- Quality Rating (QR; Downs and Black, 1998)
- Level of outcome measure (Robson, 2007)
- Ranking of outcome (12 factors from EU study)
- Practitioner rating (from Hignett et al, 2003)

The full comparative data found in this analysis will be presented in a future paper.

EU TRIAL

Two pilot trials were conducted to assist with the development of the tool. The tool
was evaluated in two ward areas in 4 EU countries. The full tool was independently
translated and checked by the EU facilitators prior to distribution. EU facilitators
collected the data on each site. The primary researcher (MF) was present to observe
and record the process. An expert review panel (EPPHE) was conducted after the
EU trials to discuss the tool and the methods used.

RESULTS

Four EU focus groups and 2 worldwide expert panels were completed (n=44, 9
countries were represented).

DEVELOPMENT OF THE INTERVENTION EVALUATION TOOL (IET)

The results from the focus groups and the individual scores were analysed for
content and theme (table 1). 210 outcome qualities were recorded in the focus group
discussions. The outcome qualities were grouped and compared to give a complex

definition for each theme. The translated material was returned to the Finnish / Italian / Portuguese focus group facilitator to check for errors in language and translation. The three sets of qualitative and quantitative scores were combined to identify the most highly valued outcomes. The 12 most highly rated outcomes were worthy of further inclusion.

Table 1. Number of different recorded outcomes

Beneficiary	Outcomes included	Outcomes included in rankings
Organisational	65	13
Staff	57	14
Patient	40	7
Task	30	3
Others	18	1
Totals	**210**	**38**

The ranked scores recorded at the end of the focus groups were then combined against the thematic definitions to give the ranked list for each country and in total (table 2). Themes that scored less than 5 in any countries combined scores were removed. The same 12 outcomes were seen as most important in each of the 4 countries.

Table 2. Ranked themes for individual and combined EU countries

	Outcome theme	Italy	Portugal	Finland	UK	Total
Organis-ation	Accident numbers	8	3	11	6	6
	Absence or staff health	3	8	2	4	4
	Financial	12	12	7	10	12
	Safety Culture	2	1	1	2	1
Staff	MS health	1	5	8	1	2
	MSD Exposure measures	12	9	5	12	10
	Competence, compliance	4	2	4	6	3
	Psychological well-being	10	7	9	4	7
Patient	Patient injuries	8	12	11	9	11
	Patient perception	8	10	11	8	9
	Patient condition	6	7	6	11	8
	Quality of care	5	4	3	7	5

Statistical Analysis

There is similarity between the 4 EU sets of rankings as the same highly ranked outcomes (safety culture, compliance and MSD measures) are seen in all countries, as are the lower ranked outcomes of finance and patient related measures. It was therefore more appropriate to conduct an analysis for association rather than difference. Kendall's Measure of Concordance was performed using the correction factor for tied ranks and W=27.66 (N=12, df 11, k=4) is significant at the 0.005 level and indicates close agreement between the four EU groups.

The literature analysis examined all the methods used to measure outcomes in the included studies (n=343). All papers with a QR of >50% were included. Table 3 shows the number of methods used for each outcome.

Table 3. Number of methods used to measure outcomes.

Preferred outcome	No. methods included
1 Safety Culture	5
2 MS Health	45
3 Competence Compliance	21
4 Absence or staff health	19
5 Quality of care	1
6 Accident numbers	2
7 Psychological well being	8
8 Patient condition	1
9 Patient perception	26
10 MSD exposure measures	170
11 Patient injuries	0
12 Financial	10

The outcome measurement methods were assessed using the following inclusion criteria:
- Level of academic quality of the study (QR rating >50%).
- Evidence of peer reviewed validation studies for the method.
- Previously used to score a peer reviewed intervention trial.
- Most frequently used measurement devices.
- Complexity of the data collection in healthcare.

The IET incorporated the 12 most preferred outcomes (Table 2) and the most suitable method for measuring each of those outcomes (above). Some of the methods chosen (Table 4) were closely related to known peer reviewed tools and studies (1,2,3,4,7,9). But others required careful consideration of a range of tools (5, 6, 10,12). The patient outcomes were poorly represented in the literature review and needed new methods of measurement to be devised.

Table 4. The measure and sources of the IET

Preferred outcome	Method for collection	Source paper
1 Safety Culture	Organisational audit of safety systems (PHOQS)	Hignett (2005)
2 MS health	MSD level in staff (Nordic Questionnaire)	Dickinson(1992)
3 Competence/ Compliance	Observational checklist (DiNO)	Johnsson (2004)
4 Absence or staff health	Standard absence per work population (OSHA)	Charney (1997)
5 Quality of care	Ward and patient survey to evaluate care quality	Nelson (2008)
6 Accident numbers	Accident numbers and non-reporting ratios	Menckel 1997)
7 Psychological well being	3 part worker for satisfaction and well being (Bigos)	Evanoff (1999)
8 Patient condition	Patient survey to evaluate clinical needs	Nelson (2008)
9 Patient perception	Survey for comfort, security, fear etc	Kjellberg (2004)
10 MSD exposure measures	Workload based on patient handling tasks	Knibbe (1999)
11 Patient injuries	Measure for detrimental effects of poor handling	No source
12 Financial	Calculation of costs versus investment	Chokar (2005)

EU TRIALS

Two wards were selected by the local facilitators to allow the IET to be used to assess the performance of the patient handling management systems. It was not possible to use the IET as a pre-post intervention assessment. The range of scores in each section was clarified with the results from the trial to give best differentiation across the sample. Table 5 shows the percentage scores in each section and the total score for the IET (%). During the trial some data were not available and the appropriate maximum or minimum score was inserted (*italics*). The cost benefit analysis was not conducted in this trial but no ward area had access to the costs of sickness absence.

The IET scores differentiated between performance levels, UK 2 had one staff on reduced capacity for 12 months, safety culture scores had weakness for all countries, Portugal scored poorly for compliance and MSD exposure due to poor equipment provision and high risk tasks. The emergency medicine ward in Italy scored the highest overall score, which matched with the EU facilitator's assessment of the ward. Low injury rates and low levels of physical handling reduced risks and improved the IET.

Table 5 EU trials - % scores for each IET section and total IET

	UK 1	UK 2	Po 1	Po 2	Fi 1	Fi 2	It 1	It 2
Safety Culture	55.6	46.7	13.8	23.3	30.7	39.8	15.6	25.2
MS health measures	40.0	50.0	55.0	51.5	22.6	21.6	38.5	100
Compliance/competence	29.2	47.9	3.5	11.5	59.6	29.3	56.9	29.6
Absence or staff health	0.0	10.7	95.9	64.6	71.2	0.0	100	99.5
Quality of care	75.0	80.0	100	69.0	64.2	86.7	88.8	79.5
Incidents and accidents	0.0	97.3	89.5	69.8	82.5	72.0	89.8	88.5
Psychology well-being	76.2	82.4	77.7	70.7	75.0	70.3	71.7	81.2
Patient condition	64.5	79.9	45.0	65.9	64.2	62.5	69.1	84.4
Patient perception	68.7	100	100	66.7	100	52.1	93.3	90.0
MS exposure measures	64.0	70.8	52.1	55.2	79.4	75.8	71.6	97.1
Patient injury	0.0	0.0	91.8	66.8	100	100	100	100
Financial	100	100	100	100	100	100	100	100
IET SCORE	38.5	53.0	53.2	46.0	53.5	42.3	58.4	65.6

DISCUSSION

This study has reviewed a wide variety of data from literature and empirical sources. The process of measuring and comparing different types of patient handling interventions has been addressed with the development of the Intervention Evaluation Tool. Every effort has been made to draw the content from studies and measurement methods that have either a good academic score or have proven validation. Some outcome areas were poorly represented in patient handling studies, in particular those relating to patient conditions and quality of care. The IET has undergone several peer review evaluations and has been translated into a further 3 EU languages to allow for further evaluation.

The IET is created to collect a comprehensive set of data from a ward or unit and calculate 12 individual section scores and an overall score to show the effectiveness of the management processes for patient handling. It can be used as a before and after intervention comparison or to compare different types of interventions in similar settings. The IET has been developed to include two distinct forms. Firstly there is a guidebook for managers which outlines the structure of the IET and clearly shows the calculation and scoring process. Secondly there is a data collection format and a series of data collection forms to aid the process. The data collection consists of only 4 sections; a management survey for workload and staff structure, a safety culture audit, transfer observations for

25% of patients, and a questionnaire survey for 50% of staff and 25% of patients. The calculation of the IET scores for the separate and combined scores has been developed into an excel spreadsheet.

The detailed investigation of the different intervention studies and the focus groups has developed a clearer picture of the outcomes that are valued among the patient handling specialists in healthcare. There has been a move towards a more organisational and behavioural focus. The measures of safety culture and competence/compliance have featured highly alongside the traditionally high ranking MSD and sickness absence. This shift of perspective may suggest that most patient handling specialists consider the physical risks are manageable with the equipment/engineering solutions that are available in the marketplace and that the future developments are to be focussed on delivering a more compliant organisation.

The IET scores in Table 5 show differentiation between the wards. A simple 1-13 score was been assigned to the 12 sections (IET total, 87). The weightings and calculation structure will need to be part of any future evaluation and validation. The initial results are encouraging and allow for differentiation between different management systems. If a local facilitator can distribute the staff survey and prepare the access for the transfer observations, the time on the ward is approximately 3 hours. Complications of missing data and lack of observation do increase data collection time.

Much work needs to be completed to develop and validate this proposed tool. If the IET proves to be a usable and efficient measurement tool then it will be possible to identify the strengths and weaknesses in an organisation from the individual rating scores in the 12 sections and an overall performance score for patient handling interventions. This will allow future interventions to be designed with specific outcomes and gains for the participating organisation, giving the opportunity for more directed interventions to enable best return on financial investment.

REFERENCES

Amick B., Tullar J., Brweer S., Irvine E., Mahood Q., Pompeii L., Wang A., Van Eerd D., Gimeno D., Evanoff B. (2006). Interventions in health-care settings to protect musculoskeletal health: a systematic review. Toronto: Institute for Work and Health, 2006

Bos E.H., Krol B., Van Der Star A., Groothof J.W., (2006), The effect of occupational interventions on reduction of musculoskeletal symptoms in the nursing profession. *Ergonomics* 49, 7, 706-723.

Charney W., (1997). The lift team method for reducing back injuries. *AAOHN*. Vol. 45, No. 6, 300-304.

Chhokar R; Engst C; Miller A, Robinson D, Tate R, Yassi A, (2005). The three-year economic benefits of a ceiling lift intervention aimed to reduce healthcare worker injuries. *Applied ergonomics*. 2005 Mar; 36(2): 223-9.

Dickinson C. et al, (1992). Questionnaire development: an examination of the Nordic Musculoskeletal Questionnaire. *Appl Ergon*. Vol 23 No 3 June 1992.

Downs S.H. Black N. (1998). The feasibility of creating a checklist for the assessment of methodological quality of both randomised and non-randomised studies of healthcare interventions. *Journal of Epidemiological Community Health* 52: 377-84.

Evanoff B., Bohr P., Wolf L., (1999). Effects of a participatory ergonomics team among hospital orderlies. *Am. J. Industrial Medicine* 35:358-365.

Fray, M. Hignett, S, (2006) An Evaluation of Outcome Measures in Manual Handling Interventions in Healthcare. In Pikaar, R.N., Konigsveld, E.A.P., Settels, P.J.M. (Eds.) *Proceedings of the XVth Triennial Congress of the International Ergonomics Association, Meeting Diversity in Ergonomics* 11-14 July 2006, Maastricht, Netherlands.

Higgins, J.M. (1994) *101 Creative Problem Solving Techniques*. Florida: New Management Publishing Co.

Hignett S, (2003). Intervention strategies to reduce musculoskeletal injuries associated with handling patients: a systematic review. *Occup Environ Med;*60:e6.

Hignett S, Crumpton, E., Alexander, P., Ruszala, S., Fray, M., Fletcher, B (2003) *Evidence based patient handling- Interventions, tasks and equipment*. London Routledge.

Hignett S. and Crumpton E. (2005). Development of a patient handling assessment tool. *Int. JTR* April 2005, Vol 12, No 4 178-181.

Hignett, S., Fray, M., Rossi, M. A., Tamminen-Peter, L., Hermann, S., Lomi, C., Dockrell, S., Cotrim, T., Cantineau, J. B., Johnsson, C. (2007). Implementation of the Manual Handing Directive in the Healthcare Industry in the European Union for Patient Handling tasks. *International Journal of Industrial Ergonomics* 37, 415-423.

Johhnsson C., Kjellberg K., Kjellberg A., Lagerstrom M., (2004). A direct observation instrument for assessment of nurses' patient transfer technique (DINO). *Appl Ergon.* 35 (2004) 591-601.

Kjellberg K., Lagerstrom M., Hagberg M., (2004). Patient safety and comfort during transfers in relation to nurses' work technique. *J. of Adv. Nursing,* 47(3), 251-259

Knibbe J., Friele R., (1999). The use of logs to assess exposure to manual handling of patients, illustrated in an intervention study in home care nursing. *IJIE,* 24 (1999) 445-454.

Martimo K.P., Verbeek J., Karppinen J., Furlan A.D., Takala E.P., Kuijer P., Jauhianen M., Viikari-Juntura E. (2008). Effect of training and lifting equipment for preventing back pain in lifting and handling: systematic review. *BMJ* doi: 10.1136/bmj.39463.418380.BE

Nelson A., Collins J., Siddarthan K., Matz M., Waters T., (2008). Link between safe patient handling and patient outcomes in long term care. *Rehabilitation Nursing,* Vol 33, No 1 Jan 2008.

Robson L.S., Clarke J.A., Cullen K., Bielecky A., Severin C., Bigelow P.L., Irvin E., Culyer A., Mahood Q. (2007) The effectiveness of occupational safety management system interventions: A systematic review. *Safety Science 45 (2007) 329-353.*

CHAPTER 16

Evaluation of an Intervention to Reduce Upper Extremity Pain in Ultrasound

Sharon Joines[1], Tamara James[2], Gisela Suarez[2]

[1]School of Design
North Carolina State University
Raleigh, NC 27607 USA

[2]Ergonomics Division
Duke University and Health System
Durham, NC 27710 USA

ABSTRACT

Work-related musculoskeletal disorders (WMSDs) are increasing around the world among health care workers who routinely perform ultrasounds (sonographers). This is due to the upper extremity-intensive nature of their work and the high grip force required to scan increasing numbers of obese patients. The purpose of this study was to evaluate a product (transducer rings) to determine if it could potentially reduce required grip force and upper extremity pain among sonographers. The results of this product evaluation indicate there may be a benefit to installing rings on ultrasound transducers, particularly for those who already have high levels of upper extremity pain.

Keywords: Ergonomics, Ultrasound, Upper Extremity Pain, Musculoskeletal Disorders, Intervention, Patient Obesity

INTRODUCTION

Ultrasound images or sonograms use the reflections of high-frequency sound waves to construct an image of a fetus, body organs, blood flow, or soft tissues inside the body. Diagnostic medical sonographers hold transducers with their hands to direct the sound waves into areas of a patient's body in a cone-shaped or rectangle-shaped beam. They sometimes grip transducers tightly and often hold them in static and awkward postures for prolonged periods of time.

The nature of this occupation requires intense use of the upper extremity putting sonographers at risk for developing work-related musculoskeletal disorders (WMSDs), including pain or discomfort of the upper limbs and shoulders, as well as Carpal Tunnel Syndrome (Lamar, 2004). According to the Bureau of Labor Statistics (BLS), Diagnostic Medical Sonographers held more than 50,000 jobs in 2008 in the United States and are projected to grow more than 18% by 2018 (BLS, 2010). One study in Canada found the direct costs of WMSDs can be estimated at $1218 for each full time equivalent (FTE) sonographer (Muir, et al., 2004). That equates to over $60 million in the US for direct costs of sonographer WMSDs, using the 2008 BLS employment figures. Internationally, reports of musculoskeletal pain or discomfort for sonographers range from 80 – 95.4% with a point prevalence of 80% and lifetime incidence of 91% with an average duration of 5 years (Russo, et al., 2002). A recent survey of about 3,000 sonographers shows that nearly 90% are scanning in pain. That is an increase of 9% from the previous large survey of WMSDs among sonographers, which was published in 1997 (Orenstein, 2009). Past research has also shown that one in five will eventually experience a career-ending injury (Brown and Baker, 2004).

Studies examining risk factors associated with sonographers' upper extremity musculoskeletal disorders have identified gripping the transducer, applying sustained pressure, and scanning with a flexed or hyperextended wrist were significantly correlated with increasing severity of symptoms in the hand, wrist and forearm area (Mirk, et al., 1999). Investigations into the relationship between Carpal Tunnel Syndrome (CTS) and ultrasound have shown that 65% of sonographers have experienced symptoms of CTS at some point in their career and high force to grip the transducer has a positive correlation with CTS (Schoendfeld, et al., 1999).

MOTIVATION

During the past 20 years there has been a significant increase in obesity in the United States and in those individuals requiring ultrasound examinations. According to the Center for Disease Control (CDC, 2010) in 2008 only one state (Colorado) had a prevalence of obesity less than 20%. With this increase in obesity among patients, sonographers are more often required to use excessive force to grip transducers and to push against layers of tissue to obtain quality scans for these patients. Sonographers at the host site frequently visited Employee Health with reports of pain due to these increasingly demanding physical requirements.

Researchers worked together with the host site's sonographer ergonomics committee and industrial designers to develop interventions to reduce some of these awkward postures and excessive forces. These prototype interventions were received with mixed success. One intervention that seemed promising was a small, flexible ring that could be attached to transducers to help reduce required grip force and to give sonographers greater leverage when pushing against many layers of patients' tissue.

OBJECTIVE

The purpose of this study was to evaluate the flexible transducer rings to determine the feasibility of pursuing this as a possible solution for reducing the required grip force and resulting upper extremity pain among sonographers performing ultrasound scans.

METHODS

INTERVENTION DESCRIPTION

The ergonomic ultrasound transducer ring (shown in Figure 1) is made of solid plastic tubing formed into a ring that fits over the ultrasound probe. The ring provides the sonographer with an area that their fingers can push against during exams. This reduces the force required to grip the transducer, allows for better load distribution, and a non-rigid surface to push against. Varying sizes can be easily produced to accommodate different transducer styles. Different diameter tubing can be used to provide varied widths to push against and multiple rings can be stacked to accommodate different size hands and user preferences.

The material provides a small, flexible ridge against which sonographers can comfortably push during exams. The extremely low profile allows for a more secure grip without increasing the surface contact area between the transducer and the patient. This ring would have applications for any ultrasound procedure (fetal diagnostic, general adult, echo, vascular or with an obese patient) requiring high force or where gel may be transferred from the patient to the transducer.

DEPENDENT AND INDEPENDENT VARIABLES

There was one independent variable evaluated in this field test – intervention. There were three dependent measures used to assess the performance of the interventions: discomfort, productivity, and employee acceptance (perception of benefit, usability and interest in permanent installation).

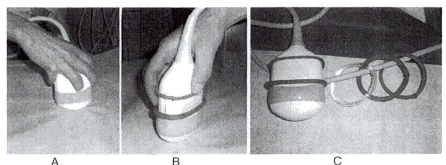

A   B   C

Figure 1: Ergonomic Intervention for Ultrasound Transducers — Small, Flexible Rings

A: 3-D transducer gripped to apply force without ring
B: 3-D transducer held using two rings affording a ledge for the finger tips (lower ring) and thumb (upper ring) to push against when applying force with the transducer
C: 3-D transducer and set of rings with varied thicknesses, textures, and stiffness

PARTICIPANTS

A total of 18 sonographers (14 females) participated in the product evaluation. Participants experience ranged from novice (1 year) to expert (28 years) with an average of 10.7 years and an average age of 39.2 years old (see Table 1). All participants were right-handed and 60% reported having had 'significant job related hand/UE injuries'. The groups' demographics were similar with some expected anthropometric (height, weight and hand length) differences between the sexes (see Tables 1 and 2). The highest average rating of discomfort was reported in the upper back followed by the trapezius then deltoid for all participants (see Table 3). The highest rating reported discomfort was 9 of 10 in the upper back, 8 in the trapezius and 2 in the deltoid reported by a 38 year old female weighting 95 lbs with 9.5 years of experience.

Table 1: Participant Demographics

	AGE	HEIGHT (IN)	WEIGHT (LBS)	EXPERIENCE (YRS)
All (n=18)	39.2[1] (23.0-61.0)	65.2 (60.8-73.0)	164.6 (95.0-280.0)	10.7 (1.0-28.0)
Males (n=4)	42.3 (24.0-61.0)	68.3 (66.0-73.0)	171.0 (150.0-190.0)	6.0 (2.0-10.0)
Females (n=14)	38.3 (23.0-57.0)	64.8 (61.0-69.0)	159.6 (95.0-280.0)	11.1 (1.0-28.0)

Note: [1] Average
(Min – Max)

Table 2: Participant Hand Strength and Length

| | STRENGTH (LB) | | |
	GRIP	PINCH	HAND LENGTH[1] (CM)
All (n=18)	77.3[2] (55.5-124.5)	18.8 (12.5-28.5)	18.0 (16.5-20.3)
Males (n=3)	114.7 (98.0-124.5)	23.3 (20.5-28.5)	19.3 (19.0-20.0)
Females (n=15)	68.7 (55.5-79.5)	17.6 (12.5-25.0)	17.7 (16.5-20.3)

Note: [1] measured from the distal wrist crease to tip of long finger
[2] Average
(Min – Max)

Table 3: Participant Initial Comfort Ratings

| | INITIAL COMFORT RATING[1] | | |
	DELTOID	TRAPEZIUS	UPPER BACK
All (n=18)	2.5 (0.0-7.0)	3.3 (0.0-8.0)	3.9 (0.0-9.0)
Males (n=3)	3.9 (0.5-7.0)	4.0 (2.0-7.0)	4.3 (2.0-7.0)
Females (n=15)	2.1 (0.0-5.0)	3.1 (0.0-8.0)	3.9 (0.0-9.0)

[1] assessed using a modified CR-10 Borg scale

SEQUENCE OF EVENTS

Sonographers who work in various ultrasound clinics at the host site (University affiliated Medical Center) and a University hospital in an adjacent state were invited to participate in this product evaluation. Participation was optional and the purpose as explained to participants was to determine whether or not the transducer rings were usable and beneficial. All sonographers were told they could discontinue using the rings at any time. Field testing of the rings was performed in four different clinical settings.

Baseline and Follow-up Data Collection Session

Each sonographer reported demographic information and completed a baseline survey comprised of upper body discomfort questions (assessed using a modified CR-10 Borg Scale). Participants performed pinch and power grip strength tasks grasping a hand dynamometer (Baseline Hydraulic) and pinch grip dynamometer with their dominant hand. The standard dynamometer grip testing position, as recommended by the American Society of Hand Therapists was used. Participants' hand length was measured from the distal wrist crease to the tip of long finger. Six different rings of various width and material type were then attached to transducers.

Approximately six weeks later, a follow-up survey was distributed to participants. The follow-up survey asked questions about upper body discomfort, usability of the rings, and impact on productivity. Participants responded using 6-point strength of agreement scale to statements regarding benefits, usability, productivity and desire for installation. On the 6-point strength of agreement scale, 1 was defined as 'Strongly Disagree' (2 - Slightly Disagree, 3 - Disagree, 4 – Agree, 5 - Slightly Agree, 6 - Strongly Agree).

RESULTS

Pre and post-intervention survey data were evaluated as a potential intervention to reduce WMSDs and to determine the feasibility of conducting further research. Pre and post-survey analysis found the following results.

PAIN LEVEL

Participants were separated into three groups based on their peak discomfort rating at the beginning of the study (see Table 4). Only one participant had a peak discomfort rating of 0, .3, .5, .7, 1 or 1.5 and was placed in the low discomfort group. Six participants with peak discomfort rating of 2, 2.5, 3, or 4 were placed in the moderate discomfort group. The remaining eleven participants with peak discomfort ratings of 5 or greater were placed in the high discomfort group.

Table 4: Grouping of Participants and Average Discomfort Ratings

Pain Level Group	Timing	Average Discomfort		
		Deltoid	Trapezius	Upper back
Low (rating of 0-1.5, n=1)	Pre	1.0	1.0	1.0
	Post	5.0	3.0	3.0
Moderate (rating of 2-4, n=6)	Pre	3.8	1.6	1.7
	Post	1.8	2.8	1.5
High (rating of 5+, n=11)	Pre	4.5	3.2	4.4
	Post	3.2	2.8	3.0

For the individual in the low group, the discomfort at the end of the study was reported as moderate or high increasing for each of the areas of interest (see Figure 2). For those in the moderate group, deltoid pain levels decreased between the pre and post-intervention. For those in the moderate group, upper back pain levels slightly decreased and increased in the trapezius. For those in the high pain group, their pain levels decreased for each of the areas of interest. Due to the small sample size it is unknown if these changes in pain level were significant. Those with the greatest pain appeared to benefit from the probe ring intervention.

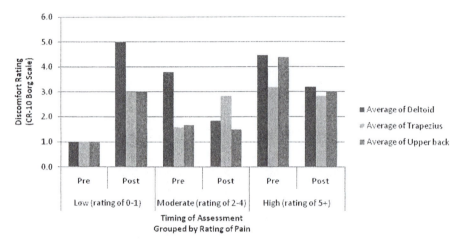

Figure 2: Participant Discomfort Ratings

In one case, the individual reporting the highest pain levels in the baseline survey (8 of 10 in the upper back, 2 in the trapezius and 2 in the deltoid) reported discomfort levels of 1, 2, and 3 respectively in the post-intervention survey. This 70% reduction in peak pain was the most dramatic impact in the study, yet the individual indicated no interest in having the rings permanently installed.

ACCEPTANCE OF RING FOR PERMANENT INSTALLATION

When asked if they would like the rings permanently installed, 41% of participants responded positively (see Figure 2). A positive response refers to a rating of 4, 5, or 6. Similarly when asked if they would use the rings if they were available, 44% of participants responded positively.

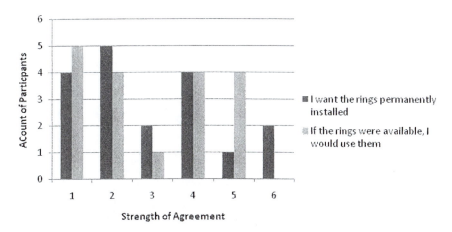

Figure 2: Participant Rating of Ring Acceptability

Sonographers who reported high pain levels overall, found the rings to be helpful — 64% reported the rings as beneficial. Those who indicated a more moderate level of pain indicated they did not want the rings permanently installed on their equipment.

Over half of the participants (53%) indicated the rings did not impact their productivity (see Figure 3). Of the participants reporting an impact on their productivity 62% of those reported only slight (agreement) impact.

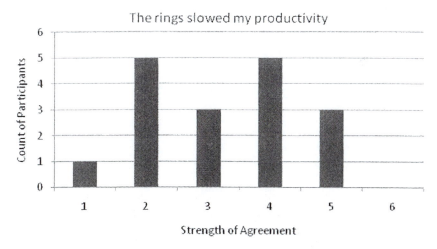

Figure 3: Participant Support of Permanent of Rings

RING USABILITY

Most participants regardless of pain level found the rings were easy to position (83%), easy to clean (83%), and were not in the way when not being used (88%). See Figure 4 (a and b).

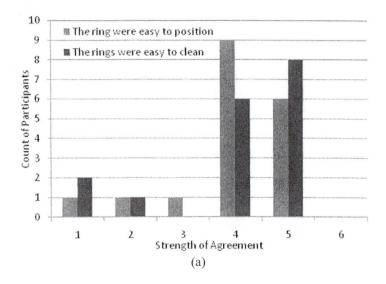

(a)

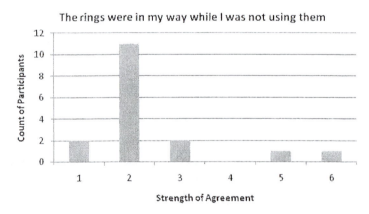

(b)

Figure 4: Participant Ring Usability Ratings

SUMMARY

The results of this product evaluation demonstrate there may be a benefit to having rings installed on ultrasound transducers. The rings are inexpensive, versatile, have no effect on patients, are easy to clean, have minimal to no impact on users' productivity, can be quickly added to existing equipment, are readily available, yet can be moved out of the way if desired,

The evaluation revealed a possible connection between individuals with high levels of pain and use of the rings which should be examined more closely. Further research on the rings as an intervention to reduce WMSDs is needed and the authors are currently communicating with an equipment manufacturer to assess interest in a larger study and possible use of this product in the manufacture of new equipment.

REFERENCES

Bureau of Labor Statistics (February 25, 2010), National Employment Matrix for Diagnostic Medical Sonographers, Bureau of Labor Statistics Website: http://data.bls.gov:8080/oep/nioem/empiohm.jsp.

Brown, G. Baker, J. (2004), "Work-related musculoskeletal disorders in sonographers." *Journal of Diagnostic Medical Sonography,* Volume 20, pp. 85–93.

Center for Disease Control (February 27, 2010), US Obesity Trends, Center for Disease Control Website: http://www.cdc.gov/obesity/data/trends.html

Lamar, S. (2004), *"Investigation of factors associated with prevalence and severity of musculoskeletal symptoms among the workers in clinical specialties of radiologic technology: An ergonomic and epidemiological approach."* Department of Industrial Engineering. North Carolina State University, Raleigh, 2004.

Mirk, P. Magnavita, N. Masini, L. Bazzocchi, M. and Fileni, A. (1999), "Frequency of Musculoskeletal Symptoms in Diagnostic Medical Sonographers. Results of a Pilot Survey". *Radiol Med*, October 1998, Volume 4, pp. 236-241.

Muir, M. Hrynkow, P. Chase, R. Boyce, D. Mclean, D. (2004), "The Nature, Cause, and Extent of Occupational Musculoskeletal Injuries among Sonographers: Recommendations for Treatment and Prevention", *Journal of Diagnostic Medical Sonography*, Volume 20, pp. 317-325.

Orenstein, Beth. (2009),"Scanning in Pain - Sonographers Seek Relief from Job-Related Hazard". *Radiology Today*, Volume 10, Number 18, p. 24.

Russo A, Murphy C, Lessoway V, Berkowitz J, (2002), "The Prevalence of Musculoskeletal Symptoms among British Columbia Sonographers*", Applied Ergonomics*, Volume 33, Issue 5, pp. 385-393.

Schoenfeld, A. Goverman, J. Weiss, D. Meizner, I. (1999). "Transducer User Syndrome: An Occupational Hazard of the Ultrasonographer", *European Journal of Ultrasound*, Volume 10, Issue 1, pp. 41-45.

Chapter 17

A Systems Engineering Approach to Improve Healthcare Quality

Kevin Taaffe, Scott A. Shappell, Paris Stringfellow, Joel S.Greenstein, Sandra K. Garrett, Kapil Chalil Madathil, Anand K. Gramopadhye

Department of Industrial Engineering
Clemson University
Clemson, SC 29634-0920, USA

ABSTRACT

The healthcare crisis in the United States has prompted the Institute of Medicine [IOM] and National Academy of Engineering [NAE] to collaborate to address how systems engineering could be used to redesign processes and delivery systems in healthcare. A collaborative effort funded by Robert Wood Johnson Foundation, the National Science Foundation, NIH, and NAE, brought together engineers and healthcare professionals to examine engineering applications and systems engineering tools as they apply to the healthcare system. The full report, Building a Better Delivery System: A New Engineering/Health Care Partnership, outlines several recommendations to close the growing gap. In response the Clemson University-GHS partnership demonstrates how systems engineering principles can be used. In this research, the authors present several concurrent research efforts being pursued as a partnership between Greenville Hospital System and Clemson University, and variety of tools that can be used to study and solve complex issues faced in healthcare. In particular, the authors suggest where both human factors and quantitative systems modeling can be used to provide insight and recommendations for changing existing processes.

Keywords: Root Cause Analysis, Human Systems Integration, Simulation, Scheduling

DESIGNING A HEALTHCARE IMPROVEMENT FRAMEWORK AT GHS

Greenville Hospital System (GHS) is a leader in providing healthcare for the Upstate of South Carolina. As with any healthcare organization, GHS is striving for continuous improvement, and there seems to always be room for improvement in this field. Whether the performance measures are based on patient satisfaction, overall healthcare delivery, adverse outcomes, errors, etc., significant improvements can be realized.

In partnering with Clemson University, GHS can combine the ability to deliver state-of-the-art modeling and data analysis while grounding the results and outcomes in research. One main focus of GHS is to standardize healthcare delivery. While this may sound contradictory to providing personal care on an individual-to-individual basis, it actually is directly in line with this concept. In fact, GHS would like to standardize the method by which personal care is delivered. Providing a more predictable experience for their patients will likely lead too many benefits, including improved patient satisfaction.

DESIGN FOR SIX SIGMA PROJECT APPROACH

While no two process improvement projects are the same, the methodology by which a project is executed can be standardized. GHS and Clemson have adopted the Design for Six Sigma (DFSS) approach across all projects. Not only is a proven approach being used to ensure each project has accountability and continues to be pursued only based on its likelihood of success, but DFSS provides a template for each project that helps all persons involved understand and be aware of the current status and progress within each effort.

DFSS employs an improvement framework using DMAIC or Define-Measure-Analyze-Improve-Control (See Figure 1). We discuss each step in the improvement process in more detail next. It is a very adaptable framework that allows for improving existing processes and products. A second framework that focuses on the development of new processes and products is called DMADV or Define-Measure-Analyze-Design-Verify. While the focus will be on introducing DMAIC, an example of using DMADV is shown in Section 3 on one of the project efforts.

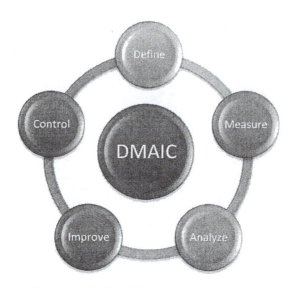

Figure 1. The DMAIC Project Approach.

DEFINE

In this first phase, the project stakeholders and team members must decide if a business case exists for the project. The project should clearly state the potential impact of the project on customers, profits, and its relationship to business strategies. Another key factor is the problem statement – detailing (when) the problem has been seen, (what) is the problem, what is the (magnitude) of the problem and what is the (impact or consequence) of the problem.

Without a clear goals statement, with achievable results and measurable targets, a project can begin to flounder immediately. Often a project has immediate buy-in from several parties, but then great ideas only seem to get discussed in meetings, with little or no action being taken. With DFSS, the team members must provide these details in order to successfully complete the Design Phase.

It is also important to state any barriers to remove or lessons learned to date so that project risk can be assessed, and the likelihood for success can be continually updated and hopefully increased.

MEASURE

It is at this key juncture that extensive data analysis is necessary to understand what the current state of the process truly looks like. For example, has a value stream or process map been completed to better understand the process and problem, and where in the process the root causes might reside? Has the measurement system been checked for repeatability and reproducibility, potentially including the training of data collectors? Data collection is a real issue on each and every project. Knowing when to use historical data or to acquire new/updated information from the existing system will influence the data analysis. All team members must believe in the analytical approach, and this is the opportunity for members of the team to demonstrate that the analysis being carried out is based on a valid representation of the underlying process.

It is important to understand the satisfaction level of the customer (whether it is staff, doctors, or patients), and to be aware of any large gaps between current performance and the customer requirements.

ANALYZE

In the Analyze project phase, this is typically where potential root causes to the problem are identified. As individual root causes are discussed and studied, they can be eliminated from consideration, thus reducing our potential root cause list.

There must be a mechanism for identifying the root cause to output relationship. By collecting the correct data in the Measure phase, the team can be better prepared to conduct the right data mining analysis in hopes of estimating the impact of root causes on key outputs. This will ultimately lead to the prioritizing of specific root causes.

IMPROVE (OR DESIGN)

The Improve phase is dedicated to carrying out the testing of one or more improvement opportunities in the process. Such improvements can be implemented in the real system or in a simulated system, with the latter option offering the opportunity to test several competing scenarios and to select only the best for implementation in the real system.

CONTROL (OR VERIFY)

Lastly, the Control phase provides a framework for measuring how well the process improvements are being maintained. It also opens the door to continuous improvement, since the project stakeholders periodically revisit the current state and are more aware of the possibility for further system changes. It is important to understand the satisfaction level of the customer (whether it is staff, doctors, or patients), and to be aware of any large gaps between current performance and the customer requirements.

ROOT CAUSES, ERROR REPORTING, AND COMMUNICATION – A HUMAN FACTORS APPROACH

In this section, we review several concurrent research efforts between GHS and Clemson that all have one common thread – applying human factors principles in process improvement. Each research effort is presented below.

ROOT CAUSE ANALYSIS PROCESS DEVELOPMENT

The driving issue for this work was that GHS, like many other healthcare providers, has non-systemic and sometimes ineffective means for understanding the true "root causes" of sentinel events (the Joint Commission defines a sentinel event as "an unexpected occurrence involving death or serious physical or psychological injury, or the risk thereof", Joint Commission, 2010). This means that many of their intervention efforts to improve patient safety are local and often only graze the surface, and thereby do not address the root causes of the problems.

HOW TO ADDRESS THIS?

Since we know that the majority of unsafe events involve human error on some level, we wanted to introduce a method for root cause analysis which focuses on the human operator and those factors that influence his/her behavior. The Human Factors Analysis and Classification System (HFACS) is a framework that allows investigators to evaluate accidents and incidents across all levels of the event including, the unsafe acts themselves, the preconditions for unsafe acts, unsafe supervision and organizational influences (Wiegmann and Shappell 2003). This framework has shown tremendous success in helping to manage safety in a number of different domains (i.e. Wiegmann and Shappell 2001, Broach and Dollar 2002, Reinach and Viale 2006), but only preliminary work had been done to explore the use of this approach in the healthcare industry (ElBardissi et al. 2007). This work aimed to bridge the gap by using HFACS as a foundation on which to improve the root cause analysis process within GHS.

RESEARCH METHODOLOGY AND RESULTS THUS FAR

The methodology used to conduct this work started by defining the problem at hand. Next, archival data (e.g. documentation of past sentinel events) was evaluated for contributing and leading causal factors using the HFACS framework. To do this, expert HFACS analysts reviewed each incident case together and reached consensus as to the causal factors present. The data was compiled and analyzed to determine leading causal factors and trends in human error.

Thus far, preliminary analysis has been conducted and high-level trends have been identified. The next steps will involve using these findings to refine the process for conducting root cause analyses of sentinel events.

MEDICATION SAFETY REPORTING SYSTEM

GHS maintains a number of data bases that can potentially provide information about the causes, frequency, location, and severity of adverse drug events. An adverse drug event is defined as an injury resulting from the use of a drug. These data bases, however, are not coordinated. This makes it difficult to detect trends in the data or opportunities for improvement of patient safety. The goal of this project is to identify, collect, and organize useful medication event information so that it may be used to systemically reduce the frequency of adverse drug events.

This project is currently in the Define phase. We have interviewed stakeholders and observed hospital processes to elicit stakeholder needs. We have observed general agreement among stakeholders that we should place initial emphasis on adverse drug events that involve high-risk medications. A detailed process map of the use of warfarin in anticoagulation therapy has been developed and reviewed by GHS. It is through this process map that Clemson and GHS can collectively assess what is right and what is wrong with the current method of administering warfarin.

Anticoagulation therapy is most commonly involved in the treatment and care of atrial fibrillation, deep vein thrombosis, pulmonary embolism, and mechanical heart valve implants. Warfarin, an orally administered anticoagulant, requires complex dosing, careful monitoring, and consistent patient compliance. We will use the process map to identify metrics to measure that are relevant to adverse warfarin events. After process map evaluation, the following key project milestones will be addressed. First, a process for collecting the data necessary to calculate these metrics will be developed. At the same time, this methodology will also include a procedure for organizing the data and the continual tracking of these metrics. The resulting information will be used by the hospital to support decision making to improve patient safety. This is a closed loop process in that in addition to making improvements to administering high-risk medication, improvements to the design of the data collection methodology can be implemented. This DFSS approach allows the process for reviewing data and information to improve and adjust to the actual system.

LABOR AND DELIVERY TEAM COMMUNICATION AND COORDINATION

Critical to the performance of high reliability organizations are a series of processes and initiatives to optimize teamwork, communication, and care coordination in critical settings. The Joint Commission has identified communication and teamwork as significant root causes contributing to nationally reported sentinel events. Previous analysis within GHS has identified concerns related to teamwork and communication within the Greenville Memorial Hospital (GMH) Labor and Delivery setting placing patient safety potentially at risk. Nationally, in a number of other industries, the importance of effective team coordination manifested by optimal communication and teamwork has been demonstrated to substantially improve organizational performance and safety.

KEY OBJECTIVES

There is strong desire on the part of GHS and Labor and Delivery leadership to optimize teamwork and communication within this setting. A secondary desire is to learn how to optimize teamwork and communication in a critical setting and then transfer this knowledge to other critical settings within the hospital system. Key objectives to achieve from this effort include:

1. Identify the root causes of less than optimal communication and teamwork within the GMH Labor and Delivery Unit.
2. Identify appropriate metrics to monitor / measure quality labor and delivery processes and outcomes.
3. Identify, develop, and implement corrective actions to culture and processes within the GMH Labor and Delivery Unit to optimize communication and teamwork.
4. Develop a methodology to transfer learning from this pilot study to all other critical settings within GHS to optimize communication and teamwork within and across units.

This work will address open-ended questions about the process of team communication and coordination in dynamic, time-critical, high-risk settings with the aim of ultimately reducing medical errors and improving patient safety within healthcare. As in the other efforts in Section 3, the researchers here are utilizing engineering and human factors techniques and drawing upon principles for team coordination in other similar settings (NASA, aviation and air traffic control, etc.) Subsequently, healthcare-specific advances to this body of research are anticipated through this work.

RESEARCH / EDUCATIONAL OUTPUTS

This research will focus on all functions that occur within GMH's Labor and Delivery Department, including all care handoffs with other units, but will exclude the tracking of patient and baby once inside the other unit. The project will involve all clinicians, staff, and support personnel that are responsible for delivery of care to patients on the unit. In keeping with the DMAIC process outlined in Section 2, noteworthy tasks that will be addressed in this research include:

- Development and delivery of metrics to monitor process and outcome quality.
- Development of a high level process flow map for activities of care within the unit.
- Measurement and analysis of the current culture and process of communication and teamwork on the unit.
- Analysis of potential and actual failure points and root causes for less than optimal communication and teamwork on the unit.
- Review and recommendation of potential solutions to address failure points and optimize communication and teamwork on the unit.
- Development and implementation of a plan to improve communication and teamwork on the unit.
- Assessment of the plan's effectiveness and the development of monitoring processes.
- Development of a methodology to transfer learning to other critical settings within GHS.

Outputs from this research will yield contributions to the local and national healthcare community as well as to the body of research that makes up systems engineering and quality improvement.

HEALTHCARE DELIVERY IN SURGICAL UNITS – A SYSTEMS MODELING APPROACH

INCREASING FIRST CASE, ON-TIME STARTS

First cases of the day are elective surgeries that are scheduled to start at 8:00 am, with patient arrivals between 5:30 am and 6:30 am. At Greenville Memorial Hospital (GMH), cases are considered on time if the actual incision time is no more than 10 minutes after the scheduled incision time. Very few hospitals track on time starts based on such a strict measure. First case delay has been identified as a contributor to surgical cancellations, overtime, and dissatisfaction for patients, staff and surgeons. GHS and Clemson University are tasked with identifying the root causes of late starts and finding mitigating strategies. As inputs to these models, data has been collected by shadowing patients and hospital staff. After undergoing reliability testing with Minitab, a hospital database has been mined for historical data and then passed to the simulation model, permitting extensive testing of current and alternative policies. Considering all cases throughout the day, there is a significant increase in late starts from the first to the second case of the day,

132

indicating that an improvement in first case on-time starts would have a direct benefit to reducing subsequent late starts.

PATIENT PROCESS IN PRE-OP

Figure 2 represents the process flow that each outpatient encounters. Similar to most perioperative settings, the flow is not sequential for all tasks that need to be completed. Prior to requiring a Pre-op RN for completing the operation prep process, the patient may require lab work, a visit to the business office, or a separate triage by a triage nurse. Once the Pre-op RN has completed her/his tasks, additional resources interact with the patient before being ready to move to the OR.

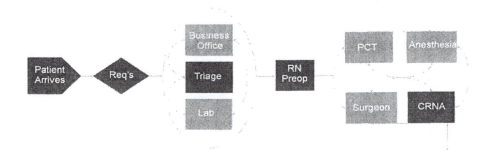

Figure 2. Outpatient pre-op flow (patient flow).

The boxes in dark blue represent critical tasks or resources that are required for every patient. Note that triage can always be performed by the Pre-op RN. Based on the data collected during the project, patients are not leaving Pre-op early enough to ensure 85% on-time starts. There are three components of the Pre-op process that can have a significant effect on delivering the patient to the OR. Clemson and GHS considered specific improvements within these three components as follows:

1. Suggest appropriate scheduled patient arrival times to Pre-op.
2. Experiment with various process changes in Pre-op (focusing on the triage and RN prep time functions)
3. Predict when specific OR resources (certified registered nurse anesthetist or CRNA, anesthesiologist, and surgeon) must arrive in order for patients to leave Pre-op on time.

Historical data indicates that higher acuity level patients require longer in the OR prior to incision. GHS currently plans for this by appropriately staggering the arrivals of patients to Pre-op, with higher acuity patients arriving earlier than lower acuity patients (see Figure 3).

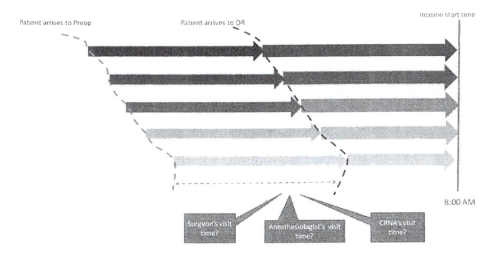

Figure 3. Patient arrivals to pre-op and OR to attain on-time starts.

In order to evaluate the various alternatives, a detailed discrete event simulation model was developed using Rockwell Software's Arena simulation tool. Clemson used historical data, as well as actual patient and resource observations, to represent the actual process and validate the model for accuracy.

SOLUTIONS AND RECOMMENDATIONS

Surprisingly, the patient arrival time had little effect on the efficiency with which patients were prepped and ready to move to the OR. There is simply a natural spreading of patient arrivals based on personal choice of each patient on when to arrive. This prevented the crowding or bunching of arrivals at any one time.

While several improvements were attempted within Pre-op, the limiting factor appears to be how the resources external to Pre-op are notified and respond to patients being ready in Pre-op. The project helped GHS realize the importance of having a well-defined trigger mechanism to alert the surgeon or anesthesiologist that they are now needed. By improving this key line of communication, GHS stands to gain 20-30 minutes in patient waiting time and, ultimately, reduce the number of procedures that are starting late.

This project is in the final phase, with the improvements in Pre-op and to the communication method between resources still being tested.

IMPROVING OPERATING ROOM AND SCHEDULED BLOCK UTILIZATION

Of all projects currently underway between Clemson and GHS, the OR and block utilization project has the potential for the most significant revenue generation for the hospital, as the issue is one of using available resources as efficiently as possible. Through this effort, GHS could not only improve access to OR time for surgeons and surgical groups, but also increase its profitability through the ability to schedule more cases and not leave ORs staffed but idle.

The fundamental question in OR and block scheduling is whether or not a surgeon should be given a reserved time on specific days (i.e., a block) for performing surgeries.

BLOCK ASSIGNMENTS AND ALLOCATION

Block times are assigned to individual surgeons, surgery groups or services, and surgeons can assign a case to their block if they can finish the surgery within the block. Typically, facilities determine a surgical group's share of available block time using formulas based on OR utilization, a contribution margin, or some other performance metric (Dexter *et al.*, 2002). Once each group's share of time has been calculated, a method must be found for fitting each group's allocated OR time into the surgical master schedule. This involves assigning specific ORs on specific days (Blake *et al.*, 2002). Blake describes a hospital's experience using integer programming to solve the problem of developing a consistent schedule that minimizes the shortfall between each group's target and actual assignment of OR time (Blake *et al.*, 2002).

While block booking solved this issue by reducing variability and allocating more predictable demand to specific surgeons, it does not allow block times to be filled by other surgeons or services which can increase the idle OR times (Charnetski, 1984; Ozkarahan, 2000). That is why most of the hospitals use a combination of open booking and block booking approaches in which the block is reserved for the block owner(s) until a certain time prior to operation or *release time* which depends on the service types and hospital policies (In some type of surgeries like plastic surgery, blocks are released ten days prior to the date where some others are released two days prior to the date). At this point the released times could be used by every surgeon and service on a first come first serve basis or under some other policies (Dexter *et al.*, 1999). Some other hospitals schedule a proportion of their ORs according to open booking and use block booking to schedule the balance of their ORs (Gupta, 2007). Cardoen *et al.* (2009) provides an excellent review of how mathematical programming models have been used to solve operating room planning and scheduling problems.

BEST PRACTICES IN BLOCK AND OPEN SCHEDULING

Dexter and Traub (2002) describes a method of assigning time to the groups/services/surgeons that maximizes efficiency. Using past data that tracks the number of surgery hours used on the same day of week over several months, the user can assess how many blocks would be most appropriate if this group continued to provide a similar level of business in the future.

In addition to the above approach, Blake *et al.* (2002) proposes using a more equitable method among all case specialists. Blake *et al.* (2002) and Santibáñez *et al.* (2007) developed mixed integer programming models for assigning blocks to the surgeons/services, using several specific criteria. In particular, they examined how to assign ORs by day of week to specific surgical groups in order to attain pre-specified group time allocations.

Industry experience suggests anywhere between 75%-85% utilization is an appropriate target, considering that room turnover and idle time at the end of the day are common components of OR time (Dexter, 2003). Each facility has a different approach, but one common theme is that blocks are assigned quickly when a surgeon or surgical group continues to have significantly higher case hours than block time assigned.

BLOCK RELEASE STRATEGIES

After decisions have been made to assign blocks to surgeons and surgical groups, equally important policy decisions must be made to determine when unused block time should be released. Dexter *et al.* (2003) discusses several methods that could be used to release blocks when there are requests that could not be satisfied. In a system that wants to maximize efficiency, releasing block times should only be done if there are cases waiting to be added to the schedule. This tends to be a little harder to manage with multiple staff manipulating the schedule, and in practice many hospitals have specific time for releasing blocks (Shaneberger, 2003). Some hospitals won't count the released times toward unutilized times of the surgeon/service/group, if they have been released early enough.

NEXT STEPS

The next step in this project is to compile historical data that can indicate how far in advance surgeons requested case time, along with the details of each case. This will help Clemson develop a decision support tool (using simulation) that can provide utilization, and perhaps even profitability, estimates based on various block allocations. Clemson is also investigating an integer programming-based approach that would use GHS-driven objectives for improved efficiency to set a block (and release) schedule for perioperative Services.

CONCLUSIONS

It is no surprise that healthcare research spans many disciplines, as has been documented in this discussion. From human factors to communication to simulation to mathematical modeling, an abundance of engineering-based tools can be directly applied to the needs of healthcare professionals. Still, for the project sponsor (in this case GHS), navigating the tools unique to each project can be a challenge. This is why GHS has chosen to use the DMAIC project approach that provides a standard method for evaluating and assessing progress across all efforts. Without the combination of introducing skills unique to each healthcare issue and managing the complete set of research projects under a proven accountability tracking system, GHS would not be able to achieve the successes already noticed within the departments involved in the various process improvement efforts already.

REFERENCES

Blake, J. T., Dexter, F., and Donald, J. (2002), "Operating Room Managers' Use of Integer Programming for Assigning Block Time to Surgical Groups: A Case Study." *Anesth Analg*, 94, 143-148.

Broach, D. M., and Dollar, C. S. (2002). *Relationship of Employee Attitudes and Supervisor-Controller Ratio to En Route Operational Error Rates*. Research Report No. DOT/FAA/AM-02/9, Federal Aviation Administration, Civil Aeromedical Institute, Oklahoma City, OK.

Cardoen, B., Demeulemeester, E., and Belien, J. (2009), "Operating room planning and scheduling: A literature review." *European Journal of Operational Research*.

Charnetski, J. R. (1984), "Scheduling operating room surgical procedures with early and late completion penalty costs." *Journal of Operations Management,* 5, 91-102.

Dexter, F., Macario, A., Traub, R. D., Hopwood, M., and Lubarsky, D. A. (1999), "An Operating Room Scheduling Strategy to Maximize the Use of Operating Room Block Time: Computer Simulation of Patient Scheduling and Survey of Patients' Preferences for Surgical Waiting Time." *Anesth Analg*, 89, 7-20.

Dexter, F., and Macario, A. (2002), "Changing Allocations of Operating Room Time From a System Based on Historical Utilization to One Where the Aim is to Schedule as Many Surgical Cases as Possible." *Anesth Analg,* 94, 1272-1279.

Dexter, F., and Traub, R. D. (2002), "How to Schedule Elective Surgical Cases into Specific Operating Rooms to Maximize the Efficiency of Use of Operating Room Time." *Anesth Analg*, 94, 933-942.

Dexter, F. (2003), "How can ORs best manage block time for scheduling surgical cases?" *OR Manager*.

Dexter, F., Traub, R. D., and Macario, A. (2003), "How to Release Allocated Operating Room Time to Increase Efficiency: Predicting Which Surgical Service Will Have the Most Underutilized Operating Room Time." *Anesth Analg*, 96, 507-512.

ElBardissi, A.W., Wiegmann, D. A., Dearani, J. A., and Sundt, T.S. (2007). "Application of the Human Factors Analysis and Classification System methodology to the cardiovascular surgery operating room." *Annals of Thoracic Surgery*, 83, 1412-1419.

Gupta, D. (2007), "Surgical Suites' Operations Management." *Production and Operations* Management, 16, 689-700.

Joint Commission, (2009). Sentinel Event. Available at: http://www.jointcommission.org/SentinelEvents/

Ozkarahan, I. (2000), "Allocation of Surgeries to Operating Rooms by Goal Programming." *J. Med. Syst.,* 24, 339-378.

Reinach, S. and Viale, A. (2006). "Application of a human error framework to conduct train accident/incident investigations." *Accident Analysis and Prevention*, 38, 396-406.

Santibáñez, P., Begen, M., and Atkins, D. (2007), "Surgical block scheduling in a system of hospitals: an application to resource and wait list management in a British Columbia health authority." *Health Care Management Science*, 10, 269-282.

Shaneberger K. (2003), "What are the best ways to handle block-time release?" *OR Manager*.

Wiegmann, D. A., and Shappell, S. A. (2003). *A Human Error Approach to Aviation Accident Analysis: The Human Factors Analysis and Classification System*. Burlington, VT, Ashgate Publishing Company.

Wiegmann, D. A., and Shappell, S. A. (2001). "Human Error Analysis of Commercial Aviation Accidents: Application of the Human Factors Analysis and Classification System (HFACS)." Aviation, Space, and Environmental Medicine, 72(11), 1006-1016.

Chapter 18

Patient Satisfaction in Emergency Departments – First Results of a Survey

B. Podtschaske, U. Schmuntzsch, A. Volkmer, D. Fuchs., W. Friesdorf

Department Human Factors Engineering and Product Ergonomics
Berlin Institute of Technology
Fasanenstr. 1, 10623 Berlin, Germany

ABSTRACT

Emergency departments become more and more a success factor of a patient-oriented, safe, efficient and tasked-tailored health care delivery system; thus it moves increasingly in the focus of optimization. Aim of the presented survey was to ask patients/affiliated for (critical) incidents" while their hospital stays as an indicator for perceived health care. Two questionnaires were developed and pre-tested. In a pilot project focusing the metropolitan region Berlin, experiences gained from the study design and questionnaires. First results show that the developed questionnaires are consistent, valid and reliable (Cronbach's alpha ranges from 0.944 to 0.638). The feedback of the survey in accordance to the promoting effort is rather disappointing (N=346 patients, N=160 affiliated). Nonetheless a first conclusion can be derived to create improvement in EDs: to arouse interest for feedback, dialogue between patients and medical staff must be improved urgently.

Keywords: Patient Satisfaction, Emergency Department, Quality, Survey, Questionnaire, Evaluation

INTRODUCTION

Continuous increase of medical knowledge in combination with a high number of medical and technological innovations has led to an enormous quality improvement in healthcare. But in accordance health care costs are continuously rising. To assure quality of health care delivery systems and access to health care goods and services under limited budgetary conditions, optimization is necessary.

The human factors and ergonomic approach defines outcome quality of a working system as the degree of fulfillment of a given task (effectiveness); in consideration of the used resources (= structural quality, e.g. infrastructure, staff, equipment) the degree of efficiency is obtained as a result. If the correct medical treatment is chosen, the question arises regarding the correct completion of the medical procedure (= procedural quality). The ideal accomplishment process should possibly be without any potential danger or waste of resources (Friesdorf and Marsolek, 2009; Carayon and Friesdorf, 2006).

In the medical field tasks and processes are characterized by a higher complexity. Therefore the examination level needs to be varied corresponding to the investigation focus. Depending on which medical procedure is considered as a work task, a more or less distinguished picture can be detected by decomposition or aggregation of workflow. Thus, "patient care" can be decomposed into different subsystems, so called "fractals". Focusing on these fractals helps to identify optimization potential for the overall working system without struggling with system complexity (Marsolek and Friesdorf, 2007). An example for such an organizational unit is the emergency department (ED).

Beside the clinical emergency care as a primary task other new or enlarged tasks can be identified, such as

- Ambulatory emergency care and home-care in rural regions;
- Navigation of clinical treatment processes;
- Controlling of the hospital capacity utilization;
- Establishing reputation for the hospital.

Significance of EDs as a key success factor for patient-oriented, safe, efficient and tasked-tailored health care delivery systems is increasing. Although patient satisfaction may not be statistically associated with the quality of care, it is associated with the concept of overall quality as perceived by the patient and allows conclusions on problems in working processes and structures to be drawn. Consequently, it is necessary to discover variables and interrelations gaining valuable information for optimization potential.

OBJECTIVE OF THE SURVEY

Objective of the survey was to develop and evaluate an adequate instrument for measuring patient/affiliated satisfaction of perceived quality in an ED. First impressions of overall patient satisfaction and sample composition were of main interest. The following questions were addressed:

- Which expectations does this group associate with "high quality" in EDs?
- Which experiences (positive/negative) has this group collected in EDs?
- How do these expectations/experiences influence satisfaction with EDs?

METHODS

Our survey approach is an (critical) incident-oriented-concept. Keys of this problem-oriented approach are concrete or critical incidents, patients experience during their hospital stays. (Critical) incidents are situations, which indicate the quality of the service (Klein, 2004).

An advanced comprehensive process analyses in different EDs as well as a patient satisfaction survey for an ED in Potsdam, Germany were made. On this extensive database the research field "clinical urgent care" (stakeholders, tasks, processes and structures) was comprehensively analysed and described.

For the study two standardized questionnaires were developed: one for the patients and the other for accompanying persons (family members, friends, or significant others). The questionnaires were developed and evaluated in the following steps:

1. Developing: on the extensive database of preliminary studies in different ED the causal model and questionnaires were designed;
2. Pre-test: both questionnaires (patient/affiliated) were pre-tested (online and paper) by several focus groups (survey experts, senior research group). In this process content consistency, usability and comprehensibility were checked.

Because of the dependent situation and a more limited view during their hospital stay, patients/affiliated were anonymously asked about their experiences after their hospital stay to assure more representative results (Yellen et al., 2002; Müller-Staub et al., 2008). Inclusion criteria for convenience sample were:

- Urgent care was an ambulant treatment;
- Urgent care was no longer than a year ago.

In order to avoid a bad response rate the survey was highly promoted, e.g. all involved (20 out of 37) Berlin EDs got material (in total 40.000 post cards) in order to inform patients/affiliated. Additionally information was placed in „Berliner Fenster" (TV of the Berlin Public Transport Services), and in high-circulation regional daily newspapers. Approximately 7.7 million contacts were made. The survey processed between the middle of October 2009 and the middle of February 2010 primary as an online-survey. Interested people also could get a paper based questionnaire, return develops included.

RESULTS

QUESTIONNAIRES

The questionnaires were process-oriented and comprised of six parts: firstly socio-demographic background, secondly incident-related data (emergency treatment) and thirdly arrival and registration. The final three sections were waiting period, treatment and departure, and satisfaction with overall performance.

To avoid stereotyped response behavior alternate questions were created using yes-or-no-answers and 5-points-likert-scaled answers. Two are open questions for personal comments. Control questions secure the content consistency of the affected answers for central aspects. To cover the influence on patient/affiliated satisfaction five distinct categories were established (see Figure 1):

1. Person-related issues (5 resp. 4 items), such as sex, age, level of education achieved etc.
2. Institution-related issues (3 items), such as location of the department, last consultation in the department, selection process etc.
3. Incident-related issues (8 items), such as weekday, physical injury etc.
4. Perceived and expected (service) quality (16 resp. 10 items), such as tangibles, empathy, perceived performance etc.
5. Overall satisfaction (4 items), such as overall satisfaction, willingness to revisit and a willingness to recommend to friends (these items measure what we termed customer loyalty).

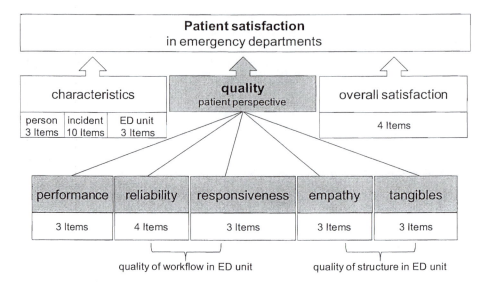

Figure 1. Dimensions and Items of Patient Satisfaction in ED.

SURVEY: "BERLIN AND ITS EMERGENCY DEPARTMENTS"

In a pilot project focusing on the metropolitan region of Berlin, first experiences gained from the study design and evaluated questionnaires are displayed below. In addition to a descriptive analysis of all questionnaire items, a statistical review of the five satisfaction dimensions was undertaken.

Survey Participants

A total of 346 persons responded to the patient questionnaires and 160 to the affiliated questionnaire. Target group of this survey are ambulatory emergency patients and affiliated, whose last stay was no longer than 1 year ago. Consequently, all persons were excluded from the data analysis that did not meet these demands or did not specify this. The number of patients that falls within the sample group is 224, the number of affiliated persons 99. For all other collected variables, in the case of missing data all valid cases were analyzed. This may lead to varying sample values.

The patient sample consists of 41.3 % men and 58.7 % women. The average age is 37.24 years (median = 35 years and SD = 18.47 years). More than one third (36.3 %) have a final degree, one quarter of the participants have completed vocational training (26 %), 15.7 % have an A-level, 11.7 % a secondary school level and 3.6 % an elementary school level degree.

In the affiliated sample, the ratio is 49.5 % men to 50.5 % women. The average age is 42.06 years (median = 42 years and SD = 18.73 years). Almost half of respondents (44.3 %) has a final degree and more than one quarter (27.5 %) completed vocational training. 16.5 % have an A-level, 6.2 % a secondary school level and 2.1 % an elementary school level.

Reliability and Validity of Questionnaires

One aim of this study is to view the quality of the rated satisfaction dimensions. As a measure of internal consistency Cronbach's alpha was used, considering Nunally (1978) values above 0.7 to be satisfactory. In addition, the corrected item-total correlation and the Cronbach's alpha, if each item was to be removed from the scale, were evaluated. Regarding the corrected item-total correlation values, a value of approximately 0.5 is regarded as high (Bortz and Döring, 1995, p. 200).

To control the criterion-related validity the item ("How was your overall stay in this ED?") was used. The consistency between this overall satisfaction and the values of the five satisfaction dimensions was calculated using the product moment correlation (Bortz and Döring, 1995, p. 200). As level of significance an alpha error probability of 0.05 was set. Additionally the criterion validity was recorded by means of product moment correlation (Bortz and Döring, 1995, p. 200).

As shown in Table 1, all quality scales have satisfactory Cronbach's alpha values above 0.7, except to "responsiveness". The corrected item-total correlation of all items in the scales is above 0.5. The Cronbach's alpha values if the corresponding item was to be removed, show that almost all items contribute to an increased internal consistency of their scale.

Table 1. Cronbach's Alpha (C.A.) Values of Quality Scales.

Dimension	No. of items	No. of answers (N)	Mean	Corrected Item-scale-correlation	C.A.	C.A. with missing item
Performance	3	136	2.85	0.715/ 0.676/ 0.642	0.821	0.781/ 0.752/ 0.792
Reliability	3	212	2.81	0.522/ 0.765/ 0.751	0.818	0.904/ 0.662/ 0.666
Responsiveness	2	221	2.98	0.472 /0.472	0.638	
Empathy	3	175	2.53	0.631/ 0.518/ 0.523	0.724	0.534/ 0.700/ 0.673
Overall Satisfaction	2	210	2.49	0.834/ 0.834	0.897	
Service Quality	15	107	2.55	0.911- 0.502	0.944	0.938-0.945

Regarding the review of criterion validity the four dimensions "performance" ($r = 0.721$), "reliability" ($r = 0.828$), "responsiveness" ($r = 0.713$) and "empathy" ($r = 0.782$) correlated significantly with the overall assessment ("How was your overall stay in this ED?") ($p = 0.000$).

First Results of Patients' Satisfaction

Looking at the distribution in the question of overall satisfaction, it must be stated that in both samples there is no normal distribution (patient: p = 0.000 / affiliated: p = 0.003) (see Figure 2).

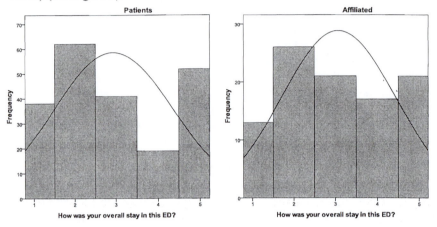

Figure 2. Frequency Distribution of Overall Satisfaction.

At a scale from 1 "very good" to 5 "poor" is the overall patient satisfaction 2.92 (SD = 1.442, statistical spread 1-5 and median = 3) and the affiliated satisfaction 3.07 (SD = 1.357, statistical spread 1-5 and median = 3). Overall, the mean numbers of the affiliated are in the range 2.17 (Item 25) to 3.78 (item 21). According the patients, it ranges from 2.18 (Item 25) to 3.55 (Item 23 and 24).

DISCUSSION

Social scientists investigated that the response-rate for online-surveys is comparable to those of offline-surveys. Weaknesses such cognitive stress by reading on a screen, missing access or affinity to the internet are balanced by strengths such as easy to create and to spread, without loss of time, or very high data quality due to the reduction of transcription errors (Kahnwald and Köhler, 2009, p. 292). Indeed, the feedback of the survey considering the promotion effort made is rather disappointing. The relatively low participation in questioning admits interpretations in two directions: firstly patients/affiliated are not highly motivated to give their opinion and impressions. Secondly, the staff in ED is not in a position to or does not wish to motivate their patients/affiliated to give feedback. The commonly discussed "mature" patient appears not (yet) to be present, at least in Berlin's EDs.

CONCLUSIONS

The motivation of this survey was to identify aspects of perceived health care quality by patients/affiliated. First results show that the developed questionnaires are consistent, valid and reliable. In spite of the rather low response rate, first conclusions can be derived from the data base to create improvement in EDs: to arouse interest for feedback, dialogue between patients and medical staff must be improved urgently. It is hoped, that the evaluation of the data base will deliver more concrete findings of how structural and procedural quality in EDs from the patient/affiliated point of view could be improved. The approach to ask for more or less "critical incidents" seems a reliable method to discover optimization potentials. However, to examine perspective of the patients/affiliated only is rather not enough – it must be complemented by other stakeholder views (physicians/ nurses, practitioner, rescue services and the hospital management). A comprehensive understanding of how to create an "ergonomic" ED, which considers its complexity, then becomes possible.

REFERENCES

Bortz, J., Döring, N. (1995). Forschungsmethoden und Evaluation, 2.vollst. überarbeitete und akt. Auflage, Springer-Verlag, Berlin-Heidelberg-New York.

Carayon, P. and Friesdorf, W. (2006). Human Factors and Ergonomics in Medicine. In Salvendy, G. (Ed.), Handbook of the Human Factors and Ergonomics, New Jersey: John Wiley and Son. pp. 1517–1537.

Friesdorf, W. and Marsolek, I. (2009). Medicoergonomics – A Human Factors Engineering Approach for the Healthcare Sector. In Schlick, C. M. (Ed.), Industrial Engineering and Ergonomics. Visions, Concepts, Methods and Tools. Festschrift in Honor of Professor Holger Luczak. Berlin, Heidelberg: Springer Verlag. pp. 165-176

Kahnwald, N. and Köhler, T. (2009). Die Sifa-Langzeitstudie: Design und Umsetzung einer Online-Erhebung zur Tätigkeit von Fachkräften für Arbeitssicherheit. in: Jackob, N., Schoen, H., Zerback, T., (Hrsg.) (2009). Sozialforschung im Internet – Methodologie und Praxis der Online-Befragung, Verlag f. Sozialwissenschaften, Wiesbaden. S. 292.

Klein, K. (2004). Ereignisorientierte Patientenbefragung – Entwicklung und Validierung eines ereignisorientierten Fragebogens zur Bewertung der stationären medizinischen Rehabilitation (Diss.), Albert-Ludwig-Universität, Freiburg i.Br.

Marsolek, I. and Friesdorf, W. (2007). Arbeitswissenschaft im Gesundheitswesen – Balancierte Rationalisierung statt banaler Rationierung. In Gesundheitsstadt Berlin e.v. (Ed.), Handbuch Gesundheitswirtschaft – Kompetenzen und Perspektiven der Hauptstadtregion. Berlin: Medizinisch Wissenschaftliche Verlagsgesellschaft. pp. 88-94.

Müller-Staub, M., Meer, R., Briner. G., Probst, M.-T., Needham, I. (2008). Erhebung der Patientenzufriedenheit im Notfallzentrum eines Schweizer Universitätsspitals: Konzept und Ergebnisse (Teil 1), in: Pflege, S. 172-179, Bd. 21, Heft 3, Verlag Hans Huber, Bern.

Nunnally, J. C. (1978). Psychometric theory, 2nd ed., New York: McGraw-Hill.

Yellen, E., Davis, G.C., and Ricard, R. (2002). The Measurement of Patient Satisfaction, in: Joun Nurs Care Qual 2002, 16(4). pp. 23-29.

Chapter 19

Medical Team Training in the Veterans Health Administration: Checklist-Guided Preoperative Briefings and Postoperative Debriefings

Douglas E. Paull, Lisa M. Mazzia

Veterans Health Administration
National Center for Patient Safety
Ann Arbor, MI 48106-0486

ABSTRACT

The Veterans Healthcare Administration (VHA) National Center for Patient Safety (NCPS) established the Medical Team Training (MTT) Program to improve patient outcomes and staff morale through communication and teamwork. The cornerstone of MTT is the preoperative briefing and postoperative debriefing guided by a checklist in the operating room (OR). One hundred twenty nine facilities conducting surgical care of the Veteran participated. The Program included preparation and planning, a learning session, and follow-up coaching interviews.

Ninety-eight percent of facilities were routinely briefing operative cases 12 months following the learning session. Positive changes following checklist-guided briefings included higher patient safety and teamwork scores on surveys, better compliance with antibiotic prophylaxis, improved morale as evidenced by decreased nursing turnover, and improved efficiency. Preliminary analysis also suggests decreased surgical morbidity and mortality.

Keywords: Briefing, Checklist, Operating Room, Teamwork, Training

INTRODUCTION

Communication and teamwork failure are leading causes of adverse events in healthcare including those errors that occur in the OR (Gawande et al., 2003). Medical team training and implementation of aviation-based *Crew Resource Management* (CRM) communication principles and techniques have been associated with improved teamwork and fewer errors in the OR (McCulloch et al., 2009). The modern OR is a dynamic environment requiring collaboration among surgeon, anesthesiologist and/or certified nurse anesthetist, circulating nurse, scrub nurse and/or technician, and others. A checklist is one solution addressing the increasing complexity of operations and the limits of human memory. A well-designed checklist, utilized just prior to the operation, would ensure team members were introduced, critical information shared, and the necessary equipment available (Gawande, 2009). VHA NCPS piloted and then initiated a nationwide MTT Program focusing on checklist-driven preoperative briefings and postoperative debriefings in the OR. The goals of the MTT Program are to improve patient outcomes and staff morale. By June, 2009 the MTT Program had trained 12,425 OR staff. Checklist-guided preoperative briefings and postoperative debriefings have positively influenced the culture of patient safety in the VHA.

MEDICAL TEAM TRAINING PROGRAM

PREPARATION AND PLANNING

The MTT Program consists of several distinct phases (National Center for Patient Safety, 2010a): preparation and planning, learning session, and follow-up coaching interviews. The preparation and planning phase starts 3 months prior to the learning session with a leadership conference call. Leadership including the Director, Chief of Staff, Nurse Executive, Chief of Surgery, Nurse Manager, and Patient Safety Manager discuss their facility and the MTT Program with NCPS faculty. Facility leadership support is critical to the success of the checklist-guided preoperative briefing, postoperative debriefing project (Paull et al., 2009).

Following the leadership call, a series of three preparation and planning conference calls between the facility's implementation team and NCPS faculty discussed potential areas for improvement. Projects that addressed these areas were chosen for implementation. Preoperative briefings and postoperative debriefings was a mandatory project, but facilities could also choose among structured hand-offs, code debriefings, multidisciplinary patient-centered rounds in the intensive care unit, and interdisciplinary administrative briefings as additional patient safety initiatives. Physician and nurse champions were identified and outcome measures established. The work product of the three calls was the facility's implementation table, which served as a guide for the project (s).

LEARNING SESSION

The one day learning session consisted of didactic teaching, films, interactive exercises, role playing and doc-U-dramas. Prior to the learning session, staff completed a voluntary, anonymous *Safety Attitudes Questionnaire* (Sexton et al., 2007), establishing a baseline measure of patient safety. Following the learning session, the implementation team met with NCPS faculty to finalize their implementation table. Teams were instructed to begin preoperative briefings and postoperative debriefings within 72 hours of the learning session.

FOLLOW UP INTERVIEWS

Semi-structured follow-up coaching interviews were conducted between NCPS faculty and facility implementation teams. Interviews occurred at one month and then quarterly for one year following the learning session. The semi-structured interview process was focused on helping teams overcome obstacles. The *Safety Attitudes Questionnaire* was repeated at 9-12 months.

THE CHECKLIST

During the preparation and planning calls, implementation teams were encouraged to review checklists developed by other facilities posted on the NCPS MTT website. Teams were challenged to develop an OR checklist. The goal of the checklist-driven preoperative briefing was to ensure that the right personnel, equipment and information were available for the operation. The checklist-guided postoperative debriefing captured patient safety issues, reasons for delays, and equipment problems. The learning session included opportunities to practice utilizing checklists to conduct preoperative briefings and postoperative debriefings. In 2009, a checklist pilot study was carried out. Checklists were solicited from participating facilities. Seventy four facilities shared their checklists. A spreadsheet was created. Aggregate data was analyzed. If a given element appeared in over 60% of all checklists, it was included in the first draft of a generic VHA pilot checklist. Ten sites were then chosen to use the VHA pilot checklist over a two month period. Pilot sites purposely included large and small, urban and rural, inpatient and outpatient surgical facilities. Every two weeks, a conference call between NCPS faculty and facility OR teams was conducted. Checklist elements were added or removed. The resulting checklist draft was sent to VHA Leadership (Surgery, Anesthesiology, Nursing) for final editing. The VHA checklist is now available on the NCPS MTT website (see Figure 1).

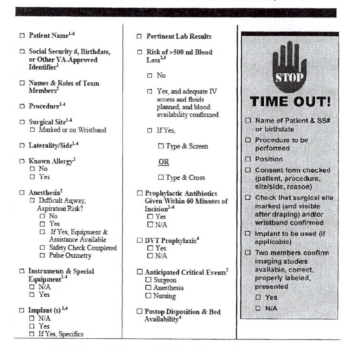

FIGURE 1. VHA Preoperative Briefing Checklist.

OUTCOMES

In a study of the first 110 facilities to complete at least one follow up interview, there was continuous improvement in the degree of implementation of checklist-guided preoperative briefings and postoperative debriefings. By the fourth interview 98% of facilities were conducting preoperative briefings and postoperative debriefings. There was a statistically significant increase in *Safety Attitudes Questionnaire* scores for patient safety and teamwork among OR providers participating in the MTT Program (see Figure 2). Nursing turnover in the OR, a surrogate marker for staff morale, decreased by 33%, falling from 9% the year prior to implementation to 6% the following year, $p = 0.02$. Compliance rates for antibiotic prophylaxis improved from 88% to 97%, $p \leq 0.05$. Ninety-one percent of facilities reported at least one undesirable event "caught" by the checklist-guided preoperative briefing during the one year follow-up period. Similar to other studies, there was an improvement in OR efficiency associated with checklist-guided briefings (Nundy et al., 2008). Fifty-four percent of facilities reported an increase in "on-time" starts for the first case of the day, and 63 % of facilities noted better use of equipment. Preliminary data analysis suggests a decrease in surgical morbidity and mortality with implementation of the checklist-guided preoperative briefing and postoperative debriefing.

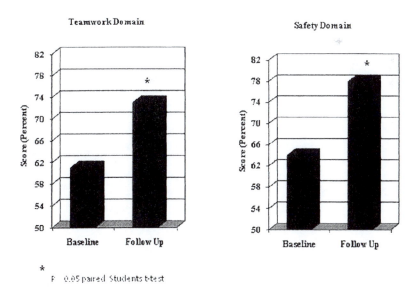

FIGURE 2.   *Safety Attitudes Questionnaire* scores before and after implementation of checklist-guided preoperative briefings and postoperative debriefings.

SUMMARY

Teamwork and communication failure are a leading cause of patient adverse events in the OR. The implementation of checklist-guided preoperative briefings and postoperative debriefings was associated with improved patient outcomes, enhanced staff morale, and a fundamental change in the culture of patient safety in the OR throughout the VHA organization. We believe these results are substantiated by other studies, and the process would be reproducible throughout healthcare.

REFERENCES

Gawande, A.A., Zinner, M.J., Studdert, D.M. et al. (2003), "Analysis of errors reported by surgeons at three teaching hospitals." *Surgery,* 133, 614-621.

Gawande, A. (2009). *Checklist Manifesto.* Metropolitan Books, Henry Holt and Co., New York.

McCulloch, P., Mishra, A., Handa, A., Dale, T., Hirst, G., Catchpole, K.. (2009). "The effects of aviation-style non-technical skills training on technical performance and outcome in the operating theatre." *Qual Saf Health Care*, 18, 109-115.

National Center for Patient Safety (Feb. 7, 2010a). Medical Team Training White Paper, NCPS MTT Website: http://www.patientsafety.gov/MTT/WhitePaper.pdf

National Center for Patient Safety (Feb. 7, 2010b). Medical Team Training Briefing Guides, NCPS MTT Website: http://vaww.ncps.med.va.gov/Education/MTT/guides/index.html

Nundy, S,, Mukherjee, A., Sexton, J.B., Pronovost, P.J., Knight A., Rowen, L.C., Duncan, M., Syin, D., Makary, M.A. (2008). "Impact of preoperative briefings on operating room delays." *Archives of Surgery* 143:1068-1072.

Paull, D.E., Mazzia, L.M., Izu, B.S., Neily, J., Mills, P.D., Bagian, J.P. (2009). "Predictors of successful implementation of preoperative briefings and postoperative debriefings after medical team training." *The American Journal of Surgery*, 198, 675-678.

Sexton, J.B., Paine, L.A., Manfuso, J., Holzmueller, C.G., Martinez, E.A., Moore, D., Hunt ,D.G., Pronovost, P.J. (2007). "A check-up for safety culture in my patient care area." *The Joint Commission Journal on Quality and Patient Safety*, 33, 699-703.

Chapter 20

VA Health Information Technology: Computerized Patient Record System (CPRS), Bar Code Medication Administration, (BCMA) and My HealtheVet

Danielle Hoover, Max Theis

Veterans Health Administration
National Center for Patient Safety
Ann Arbor, Michigan, USA

ABSTRACT

The Veterans Health Administration has been a national leader in developing and implementing health information technology (HIT) with the award winning Veterans Health Information Systems and Technology Architecture (VistA). VistA has several components including: the Computerized Patient Record System (CPRS), computerized physician order entry (CPOE), and Bar Code Medication Administration (BCMA). CPRS allows access to a single medical record from multiple clinical settings. BCMA is an application which aids in assuring the right patient gets the right medication at the right time.

154

VA HIT goes beyond an internal electronic medical record through data sharing with the Department of Defense and Kaiser Permanente to create an interoperable electronic health record; and in 2003, VA launched My HealtheVet, a web-based personal health record. This paper will give an overview of these systems, examples of increased patient safety, and instances of unanticipated consequences.

Keywords: Bar Code Medication Administration, Electronic Health Record, Personal Health Record, Computerized Physician Order Entry

INTRODUCTION

The human body is a complicated and intricate system. The systems designed to care for the health of the human body, are also complicated and intricate. In healthcare as in all systems, there are inputs and outputs. Sometimes a piece of the patient needs to be removed (blood sample, biopsy, etc) and other times something needs to be added (medication, implant, etc) in order to diagnose or cure. Adding to the complexity of treating human disease, patients present to multiple care settings such as a physician's office, an emergency room or even a local pharmacy. Although it is rare to hear it spoken in public, physicians, as humans, are fallible and prone to error, adding further complication to providing good care.

As noted in the Institute of Medicine's groundbreaking report "To Err is Human", almost 100,000 deaths occur in hospitals every year due to preventable medical errors (more than cancer or automobile accidents) (IOM, 2001). Seemingly physicians are aware of the issue, as 95% report having witnessed a serious medical mistake (Ball, 2002). While the problems which lead to errors in healthcare are interrelated and difficult to fix, many can be prevented through the use of well designed health information technology (HIT). One of the most common tasks in healthcare is medication administration, which requires making sure that patients get the right medication at the right time, in the right dose (Agrawal, 2009 & Koppel, 2008). Unfortunately errors in the medication administration process are all too common, with illegible handwritten orders, drug packages which look alike or drug names that sound alike. It is estimated that almost 2% of patients admitted to a major teaching hospital will experience a preventable adverse drug event during their stay, at an average increased cost of $4700 or an annual cost of $2.8 million for a 700-bed teaching hospital (Ball, 2002). However HIT systems exist which have been shown to reduce upwards of 90% of adverse drug events (Ball, 2002).

The technology and software for health information systems have been developing at exponential rates in recent years (Ball, 2002), as there is a demonstrated need for them, however they have not yet been universally adopted (Jha, 2006). A 2005 survey by Jha et al. estimated that approximately 24% of physicians used an electronic medical record (EMR) system in the outpatient setting, and only 5% percent of hospitals used a computerized physician order entry (CPOE) system for medications. More recently it has been estimated that 43.9% of physicians are now using some sort of EMR in an outpatient setting, however only 20.5% have a basic EMR as defined by Hsiao et al. (system includes patient demographic information, patient problem lists, clinical notes, orders for prescriptions, and viewing capability for laboratory and imaging results). Only 6.3% have a fully functional EMR (basic features plus medical history and follow-up, ability to send orders for tests and prescriptions electronically, provision of warnings of drug interactions or contraindications, notifications for out-of-range test levels, and reminders for evidence-based interventions) (Hsiao, 2009).

The United States Department of Veterans Affairs Veterans Health Administration has been a worldwide a leader in transforming healthcare through the use of its award winning HIT system: Veterans Health Information and Technology Architecture (VistA) (VA Innovations Website 2010). Created over 20 years ago, the system is now in place at over 1,300 sites including 156 medical centers, 875 community based outpatient clinics, 136 skilled nursing facilities, and 44 resident rehabilitation programs (VistA FAQ 2010). VistA is the technology backbone for multiple programs that support the day-to-day clinical and administrative functions of the Veterans Health Administration. In this paper, we will illustrate the use of VA HIT with a patient Mr. X, presenting for care to a VA emergency room, proceeding through admission to the hospital, and returning home to outpatient care. For each healthcare setting we will present issues posed by the traditional paper-based record keeping in a typical US hospital – the characteristics that could lead to errors in care, how the HIT in place at the VA has improved patient care (Amarasingham, 2009), and some of the unanticipated issues that arise from the use of HIT.

COMPUTERIZED

PATIENT RECORD SYSTEM

CLINICAL DOCUMENTATION

To look at how information technology is used in healthcare, let us imagine that it is a Sunday morning and Mr. X, a 50 year old male veteran, arrives at the emergency room (ER) for care. When he presents for emergency care at the VA, he tells his healthcare providers that he has been coughing for two days and this morning he vomited. He also tells them that he was at his primary care provider's office last week and had some medications changed, but he can't remember what they were. In a typical non-VA ER, providers (doctors, nurses, pharmacists, social workers, therapists, etc) do not have access to the patient's outpatient medical record. A clerk could try contacting the office with a request to fax information, but Mr. X has arrived at the ER on a Sunday morning when most physician offices are closed. This problem of real time access to records from different locations is solved at the VA through the Computerized Patient Record System (CPRS). CPRS creates a single electronic chart that allows access to notes from both outpatient and inpatient settings. Providers in the VA ER are able to retrieve notes from the office visit, find what medications were changed, and access any other relevant health information from the visit. All providers on the team put their clinical notes into CPRS and no matter where the patient is: a specialist's office, in the operating room, at a rehab facility, or any other location in the VA system, all the notes are in CPRS rather than in separate files at each location. This improves care because critical data such as allergies and medication lists are uniformly displayed at each patient interface. A newly entered allergy is immediately visible to all other providers in the system.

If Mr. X arrives in a typical non-VA ER, his medical record is a paper chart on a clipboard. All of the medical information for the patient including medical history, medications, orders, and test results are attached to the paper chart. It isn't unusual for a team member to spend a significant amount of time looking for a patient's chart. The patient's physical chart can only be in one place at a time yet multiple providers need to use it at the same time. It is easy to imagine Mr. X's condition and treatment changing even while providers wait for access to the clipboard. Waiting for the chart reduces the healthcare team's ability to work on tasks in parallel; they must work on tasks in series, which can lead to delays in care.

This problem is addressed in the VA through CPRS. The entire healthcare team can have simultaneous access to Mr. X's electronic chart. Healthcare providers can also easily switch between patient charts to manage multiple patients from the same computer terminal. Although this solves the problem of access to the chart, a provider can still inadvertently "pick up the chart" of the wrong patient by selecting the wrong patient record from a drop-down list, similar to a provider in a paper-based system physically picking up the wrong chart. This is a common problem in both electronic and paper-based systems when patients have similar names. Some VA facilities have added a visual cue in the form of a photo of the patient as a strategy to reduce such errors. The patient's image appears when their record is accessed in CPRS. Although photographic images are not entirely reliable and may not reflect a sick patient's current appearance, the connection of image and information in CPRS reduces the risk of selecting the wrong record.

Although having access to notes from all members of the healthcare team across the spectrum of care (inpatient, outpatient, physicians, nurses, pharmacists, etc) can be helpful, this can also create information overload. With so much information available it is sometimes difficult for healthcare providers to easily find information relevant to the current clinical situation. Ideally, an EMR is configurable to the tasks each healthcare provider performs. A system designed to support a nurse performing daily tasks may not support a physician's tasks (Russ, 2009).

ALERTS, NOTIFICATIONS, AND CLINICAL REMINDERS

Mr. X has now been assessed and examined, his records reviewed, and the healthcare team has decided to order blood tests. In many non-VA ERs, the orders are handwritten or checked off on a paper chart, the blood is drawn and sent to the lab, and the results printed back to the clerk's station. This process may cause delays in care while waiting for the clerk to physically transfer the laboratory results from the printer to the paper chart. There is also a risk that Mr. X's results will be put into another patient's chart. To prevent these potential errors, the VA system automatically posts lab results directly to the electronic patient chart and electronically alerts the provider that labs results are ready. The VA system also notifies the provider about results for radiology, procedures, consult reports, admissions, and other timely medical information.

In addition to notifying or alerting the provider that results are available, the VA system can also prompt providers when a clinical task is due. The system will remind Mr. X's team that he does not have an advance directive on file. This will prompt the team to discuss the patient's wishes regarding cardiopulmonary resuscitation. These clinical reminders are often used in the outpatient setting to prompt providers to screen patients for age appropriate diseases such as colon cancer and prostate cancer. They are similar in function to a paper checklist or other cognitive aids used in some office charts to remind providers when preventive care such as an influenza vaccination or mammogram is due. However in the VA, clinical reminders are very robust and used in a variety of situations to prompt providers to follow clinical guidelines. A 2004 RAND study showed that VA patients received recommended clinical care 67% of the time compared to 51% in a national sample of non-VA patients (Asch, 2009). Multiple studies have shown VA care has better outcomes than other national hospitals, possibly due to the use of clinical reminders (Asch, 2009).

Although there are benefits to directly alerting the provider to lab results and reminding the provider about clinical care recommendations, there have been reported problems with "alert fatigue" (Hysong, 2010). This occurs when providers receive so many alerts and notifications that they become background noise creating risk that a critical alert will be overlooked. There is a need therefore to increase the signal to noise ratio (less noise per signal) in order to reduce alert fatigue.

COMPUTERIZED PHYSICIAN ORDER ENTRY

For Mr. X, diagnostic test results indicate that he is to be admitted to the hospital for treatment of pneumonia. His provider has decided to order an antibiotic. In a typical hospital without computerized physician order entry (CPOE), medication orders are handwritten, scanned and faxed to the pharmacy. This can lead to adverse medication events with sound-alike medications (cetirizine vs sertraline, cefoxitin vs cefixime) and potential over dosages (10mg vs 10 μg) due to verbal orders and poor handwriting (ISMP, 2010). Also, the systems in place to check for contraindications or dangerous medication interactions can be paper-based and potentially out of date or difficult to access (Russ, 2009).

These errors are addressed in the VA with the use of a CPOE system. Using CPOE, a provider types or selects the desired medication from an electronic list, then also enters the dosage, route, and frequency. Computerization of order entries allows for automatic alerts of drug interactions (based on a drug database) rather than relying on human memory (Norman, 2002). Computers can verify that drug doses are appropriate for the given patient and symptoms, and can warn providers about patient specific drug interactions (ISMP, 2010 & Russ, 2009). Sending pharmacy orders electronically reduces the time it takes to dispense the medication, and reduces the possibility of miscommunication due to a verbal or handwritten order. Use of a CPOE system for medications has been shown to reduce medication errors by over 80% (Bates, 1999). As is the case with any technology, CPOE is dependent on reliable software and hardware. CPOE can be potentially more dangerous than hand written orders (Nebecker, 2005). Less than intuitive electronic interfaces may result in a greater need for training of new staff. Facilities should also have contingency plans in place to ensure that safe effective care can still be delivered if the electronic system goes down.

BAR CODE MEDICATION ADMINISTRATION

Mr. X is now in an inpatient hospital unit with intravenous fluids infusing, the orders for an antibiotic are in the computer system, pharmacy has dispensed the medication, and it is time for the nurse to administer the medication. Mr. X is in a double room and there is another patient with same name a few rooms down. In a typical non-VA hospital, the nurse pushes a medication cart (or carries a tray) between patient rooms with medications for multiple patients. This can lead to potential errors with patients receiving medications intended for another patient (Bargren, 2009). Safe medication administration practices include assuring the right medication, in the right dose, is administered to the right patient, at the right time, via the right route. Approximately 10 years ago, the VA implemented Bar Code Medication Administration (BCMA) to prevent errors during medication administration (Mims, 2009). The BCMA system requires scanning a barcode on the medication packaging and a bar coded wristband on the patient when administering the medication. This helps assure the right medication is administered to the right patient, at right time (Mims 2009).

The VA experienced some problems with the initial roll-out of BCMA. The physical size of the medication carts and scanning equipment made it difficult for nurses to maneuver in small patient rooms. New wireless technology is making BCMA more user-friendly and the groundwork has been laid to expand bar code technology into the labeling of laboratory and pathology specimens in VA. Currently specimen labels are often handwritten and sometimes placed on the specimen before or after collection, leading to potential specimen misidentification errors (Dunn, 2010). Technology, such as handheld bar code label printers commonly used in retail, is being investigated for positively identifying patients and matching specimen labels to them at the time of collection.

MY HEALTHEVET

Mr. X is now ready to be discharged from the hospital. In a non-VA healthcare system, all the records remain at the hospital, even those that are electronic stay at the hospital and are often not shared with other healthcare systems. This can lead to fragmented care and increased costs. The VA electronic health record can be viewed from any VA facility in the country and also shares information with the Department of Defense and Kaiser Permanente (VA Press Release, 2010). In addition to sharing health data with other facilities, the VA also has an online personal health record called My Healthevet (MyHealth.va.gov). Patients can renew medications and view upcoming appointments online. Giving Mr. X easy access to his own medical information makes him an active member of his own healthcare team.

CONCLUSION

The widespread use of health information technology in the VA has helped remove some of the cognitive burden and limited access to information common in typical paper-based medical facilities. The VA is nationally recognized for efforts to prevent medical errors and for providing excellent healthcare (VA Innovations Website, 2010). As with all innovative technology, HIT has gone through, and will continue to go through growing pains. As the VA and private sector healthcare organizations work to address the problems created by HIT systems, such as work stoppage due to system downtime and alert fatigue from information overload, new HIT will be developed to improve the quality and safety of healthcare.

Acknowledgements

The authors would like thank Linda Williams, RN, MSI, and Scott Wood, PhD, VA National Center for Patient Safety for their contributions to this paper.

REFERENCES

Agrawal A. Glasser AR. (2009) Barcode medication. Administration implementation in an acute care hospital and lessons learned. *Journal of Healthcare Information Management.* 23(4):24-9.

Amarasingham R, Plantinga L, Eiener-Wesi M, Gaskin DJ, Powe NR. (2009) Clinical information technologies and inpatient outcomes: A multiple hospital study. *Arch Intern Med.* 169(2):108-114.

Asch S, McGlynn E, Hogan M, Hayward R, Shekelle P, , et al. (2009) Improving quality of care: How the VA outpaces other systems in delivering care. Rand Publication. http://www.rand.org/pubs/research_briefs/2005/RAND_RB9100.pdf. (Accessed 3/1/2010.)

Ball M, Douglas J. (2002) IT, patient safety, and quality care. *Journal of Healthcare Information Management* Vol. 16, No. 1, 28-32.

Bargren M, Lu D-F. (2009) An evaluation process for an electronic bar code medication administration information system in an acute care unit. *Urologic Nursing.* Vol. 29(5) 355-391.

Dunn EJ, Moga PJ. (2010) Patient misidentification in laboratory medicine: a qualitative analysis of 227 root cause analysis reports in the Veterans Health Administration. *Arch Pathol Lab Med.*134(2):244-55.

Hsiao CJ, Beatty P, Hing E, Woodbell D, Rechtsteiner E, Sisk J. (2009). Electronic medical record/electronic health record use by office based-based physicians: United States, 2008 and preliminary 2009. http://www.cdc.gov/nchs/data/hestat/emr_ehr/emr_ehr.pdf (Accessed 2/21/2010)

Hysong S, Sawhney M, Wilson L, Sittig D, Espadas D, et al. (2010) Provider management strategies of abnormal test result alerts: a cognitive task analysis. *Journal of the American Medical Information Association.* Vol 17. 71-77.

Institute of Medicine. (2001) *To Err is Human: Building a Safer Health System.* Washington, DC: National Academy Press, 26.

Jha A, Ferris T, Donelan K, DesRoches C, Sheilds A, et al. (2006) How common are electronic health records in the United States? A summary of the evidence. *Health Affairs.*, Oct. 11, 496-507.

Koppel R. Wetterneck T. Telles JL. Karsh BT. (2008) Workarounds to barcode medication administration systems: their occurrences, causes, and threats to patient safety. *Journal of the American Medical Informatics Association.* 15(4):408-23.

Mims E, , Tucker C, Carlson R, Schneider R, Bagby J. (2009) Quality-monitoring program for bar-code-assisted medication administration.

American Journal of Health-System Pharmacists. 66, 1125-1131.

Nebeker J, Hoffman J, Weir C, Bennett C, Hurdle J. (2005) High Rates of Adverse Drug Events in a Highly Computerized Hospital. *The Archives of Internal Medicine.* Vol. 165, 1111-1116.

Norman D. (2002) Knowledge in the head and in the world. *The design of everyday things.* Basic Books. 54-80.

Russ AL, Saleem JJ, Justice CF, Hagg H, Woodbridge PA, Doebbeling BN. (2009) Healthcare workers' perceptions of information in the electronic health record. *Proceedings of the Human Factors and Ergonomics Society 53rd Annual Meetings-2009.* 53:11: 635-639.

Delivering Patient Care. Rand Health Research Highlights. Rand.org

Russ AL, Saleem JJ, McManus MS, Zillich A, Doebbeling B. (2009) Computerized Medication Alerts and Presciber Mental Models: Observing Routine Patient Care. *Proceedings of the Human Factors and Ergonomics Society 53rd Annual Meetings-2009.* 53:11: 655-659.

HealtheVet: http://www.myhealth.va.gov/
VA Innovations Website Reference: http://www.innovations.va.gov/ (Accessed 2/23/2010).

Unknown Authors. A Call to Action: Eliminate Handwritten Prescriptions Within 3 Years! The Institute for Safe Medication Practices. http://www.ismp.org/newsletters/acutecare/articles/whitepaper.asp (Accessed 2/23/2010)

VistA FAQ:
http://www.innovations.va.gov/innovations/docs/InnovationsVistAFAQPublic .pdf (Accessed 2/23/2010).

VA Press Release: http://www1.va.gov/opa/pressrel/pressrelease.cfm?id=1824 (Accessed 2/23/2010).

Chapter 21

Model-Based Usability and Error Analysis of an Electronic Health Record Interface

Scott D. Wood[1], Dean F. Sittig[2], Adol Esquivel[2], Daniel Murphy[2], Brian Reis[2], Hardeep Singh[2]

[1]Department of Veterans Affairs National Center for Patient Safety
Ann Arbor, MI

[2]Houston VA HSR&D Center of Excellence, and The Center of Inquiry to Improve Outpatient Safety Through Effective Electronic Communication, both at the Michael E. DeBakey Veterans Affairs Medical Center and the Section of Health Services Research, Department of Medicine, Baylor College of Medicine, Houston, TX

ABSTRACT

Electronic health records, such as the Veterans Affairs Computerized Patient Record System (CPRS), have the potential to fundamentally improve many aspects of health care. Evaluating usability for such systems, however, remains challenging due to inherent system and task complexity. Similarly, testing prototypes for system augmentation or redesign is perhaps more challenging not only due to interdependence and tight coupling between system components, but also due to factors such as patient safety and privacy. The notification system within CPRS is one such tightly-coupled component that permits health care providers to receive patient-specific alerts and reports from other members of a patient's health care team. We propose the use of human performance modeling to help quantify usability issues with notification processing, to better understand how the current user interface affects patient safety, and to help develop and test alternative designs.

Keywords: Electronic Health Records, EHRs, Usability, Cognitive Modeling

INTRODUCTION

Electronic health record systems (EHRs) have the potential to fundamentally improve the quality and safety of the health care delivery system while simultaneously reducing its costs (Chaudhry et al. 2006). While these benefits have been realized in many organizations, simply implementing existing commercially available EHRs has not lead to universal improvements in cost, quality or safety (Linder et al. 2007). In addition, several investigators have identified new kinds of errors resulting from the use of these systems (Koppel et al. 2005; Han et al. 2005; Campbell et al. 2006). Finally, even with these mixed results, the U.S. federal government is placing intense pressure on health care organizations of all sizes and types to implement these EHR systems as rapidly as possible.

The Computerized Patient Record System (CPRS) is the main application with which health care providers (HCPs) in the Veterans Health Administration interact with the electronic health records (EHRs) of their patients. CPRS has been in routine use throughout the VHA since the late 1990's. Although CPRS remains innovative in many ways, problematic aspects of the CPRS user interface continue to contribute to increased physician workload and dissatisfaction, user errors, and potentially, suboptimal care for patients (Nebeker et al. 2005). Specifically, the design of the CPRS View Alert window, where notifications to HCPs are displayed, does not support efficient and error-free operation under normal conditions (although current usage may be very different from original design assumptions). While there are numerous anecdotal reports by physicians of frustration, wasted time, and data loss, and emerging empirical evidence to support this, the extent and source of the design problems have not been rigorously analyzed. Recent research on physician follow-up of test results (Singh et al. 2009a) suggests that usability problems with the View Alert window may have a direct adverse impact on patient safety. This paper describes ongoing research to apply human-performance modeling to the notification-processing tasks performed by primary care providers. The described work will analyze the actions and knowledge required by physicians to properly process CPRS notifications communicated through the View Alert window. We propose to develop a computational cognitive model to simulate provider usage of the View Alert window and test this model using a simulated interface and a realistic set of clinical scenarios. We expect this work to lead to a better understanding of how the current design of the View Alert window affects usability and increases HCP error, how the error recovery procedures affect normal workflow, and how the user interface may be redesigned to improve patient safety.

NOTIFICATIONS IN CPRS

Notifications are used in CPRS to transmit information to a patient's care team, and include non-critical information, such as appointment scheduling issues, as well as critical information that needs formal acknowledgement of receipt and subsequent follow-up actions (e.g., ordering additional tests, procedures, or initiating a referral to specialist in response to a chest x-ray that is suspicious of cancer of the lung). HCPs can potentially receive hundreds of such notifications per day depending on the number of patients they are responsible for and their role within the care team. Figure 1 shows the Patient Selection Screen in CPRS, the bottom portion of which is called the View Alert window. The View Alert window is a scrollable table view containing notifications (also called alerts), where each notification occupies a row in the table and each column displays standard information describing the notification. Notifications are processed by double-clicking on them in the table and performing an appropriate action, after which the notification disappears from the window. Although this functionality is minimally sufficient for performing the task, the interface does not scale well for the actual number of notifications received on a daily basis. Anecdotal evidence suggests that the behavior of the user interface has a strong potential to elicit human error. One reported problem is that HCPs have to sift through hundreds of purely informational notifications to find the few critical notifications that must be attended to immediately. Another problem is that notifications can disappear before the HCP is done with the processing task.

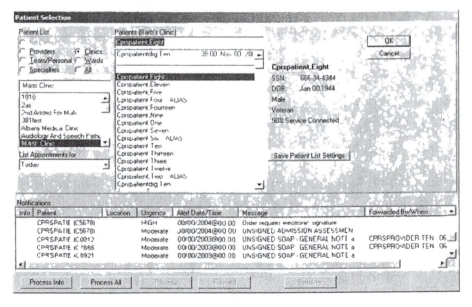

FIGURE 1. View Alert processing window in CPRS.

Due to the combination of high notification rate, interface usability and variable degree of user proficiency, some notifications are inadvertently lost, intentionally discarded, or are not processed in a timely fashion. As reported by Singh and colleagues (2009a), almost 7 percent of all abnormal test results are potentially missed. Missing abnormal imaging and lab test results means that patient care may be delayed and is considered to be an adverse event with the potential for harm to the patient. Using the system as designed requires a high level of vigilance by HCPs to correctly perform the notification processing tasks. If errors are made while processing, the system design may not allow for error recovery or it may require significant additional time for recovery. For example, if an HCP is interrupted while processing a notification, the notification may disappear prematurely because the system automatically deletes notifications after they are first seen. After such events, if the HCP is also unable to recall the name of the associated patient, there is no easy way for the HCP to recover the information lost from the deleted notification.

Since 2007, there has been a focused effort within the VA's Houston Patient Safety Center of Inquiry to understand the nature of communications breakdowns that lead to untimely follow-up of abnormal test results (Hysong, et al. 2009). In addition to quantifying the frequency of such errors, Singh and colleagues utilized focus groups, cognitive task analysis, structured interviews, chart reviews, and other qualitative techniques to understand reasons, in depth, why such errors are so prevalent. One such study suggests that there are other strong factors in poor performance with test result follow-up, such as lack of user proficiency with CPRS, use of external cognitive artifacts, and ability to schedule dedicated time to processing notifications (Hysong et al, 2010). Related work has led to several types of improvements, including technological and procedural changes (Singh, et al. 2009b; Singh et al, 2009c), as well as policy recommendations (Singh & Vij, 2010). Despite these successes, there are inherent limitations in qualitative observational approaches that motivate the need for a mixed-methods approach (Creswell, 2009).

LIMITATIONS OF CURRENT APPROACHES

Although qualitative observational research is best for identifying and understanding usability issues from the end-user perspective, quantitative techniques to evaluate system usability are also necessary for the broader challenge of system redesign. The "gold standard" for usability evaluation is testing the system with real users, in their working environment, performing actual tasks (Kieras, 2007), however user interface evaluation of complex systems is notoriously difficult in real-world contexts (e.g. Kieras and Santoro, 2004). In such cases there is often a trade-off between scenario simplicity and test validity – the testing scenario either accurately reflects reality or it is simple enough that the resulting data can be analyzed. In health care, such analyses are further complicated by the wide variety of case types, user roles, and individual differences in how HCPs use technology to perform their jobs. Evaluating the notification-processing task in CPRS is similarly complicated because different providers have developed their own strategies for using CPRS to both accomplish their health care goals, as well as to work around limitations in the CPRS interface. Furthermore, other factors such as privacy laws (i.e. HIPAA), limited availability of representative system users, and lack of an instrumented test environment make it difficult to obtain good evaluation data even when other complexities are mitigated.

HUMAN PERFORMANCE MODELING

Human performance modeling is a general technique for characterizing and analyzing the knowledge and actions required to perform structured tasks (Kieras, 2003). GOMS modeling is a specific modeling technique for analyzing human-computer interaction tasks (Card et al, 1983). GOMS is also an acronym for the main representational components of the technique: Goals, Operators, Methods, and Selection Rules. Goals represent the goals a user is trying to accomplish such as "Send an email message" and are usually represented as hierarchy with a top-level goal. Operators are the atomic-level actions that a user performs to accomplish a goal such as "Press the Send Button." Methods are sequence of operators grouped to accomplish a single goal, like "Attach a File to the Message." Thus, methods are subroutines or procedures that are composed of operators, and which can invoke other methods. Selection Rules are decisions used to choose between methods when more than one may apply. For example, a decision to send an encrypted email message might be represented by the following selection rule: "If message is private, then Send-Encrypted-Email."

GOMS modeling allows analysts and designers to represent a hierarchical task analysis in a standardized notation that is both human readable and machine executable (Kieras, 1988). Methods are developed to reflect the way people think about their work and can be segmented into one of three broad categories. *High-level* methods are those that reflect user goals irrespective of any particular technology, such as "Communicate with a colleague." *Low-level* methods describe actions that are specific to a particular interface platform and which would apply to any application developed on that platform. Methods for goals such as "Select Print from the File menu," for example, are platform-specific and are considered to be low-level. *Mid-level* methods are design-specific and determine the workflow the user must follow to accomplish high-level goals using lower-level methods. This is important in user interface design because mid-level methods are those over which designers have the most control.

GOMS modeling has been successfully used in other mission- and safety-critical domains such as military and nuclear energy, but GOMS modeling by hand can be very tedious and error-prone. GLEAN (GOMS Language Evaluation and ANalysis) is a software tool that automates the execution of GOMS models and allows the analyst to create an environment in which to simulate how a target system is used by a simulated user utilizing the knowledge and actions that an actual person would require (Wood 1993, Kieras et al, 1995). At the heart of GLEAN is the simulated user, which is represented primarily through a simplified cognitive architecture, and contains standard information processing components such as cognitive, perceptual and motor processors (Kieras, 1999). GLEAN's simulated human interacts with a simulated (or actual) device using the procedural knowledge contained in the corresponding GOMS models. As output, GLEAN produces estimated task execution and learning times, as well as workload estimates such as the number of items stored in short-term (working) memory.

To obtain reasonable estimates, GLEAN requires that specific task instances be defined in addition to the procedural knowledge and the simulated device. These task instances are specific, concrete descriptions of the goals and tasks that are representative of actual user tasks. Example task instances might include, "Acknowledge receipt of CBC lab results for patient XYZ and annotate in the patient's health record" or "Consult lung specialist for an abnormal chest-x ray". ·These task instances represent the types of benchmark tasks used in human-subject usability tests and contain much the same information.

GLEAN also supports the analysis of human error (Wood, 2000). This allows the analyst or designer to model error recovery routines alongside normal system procedures. For example, GLEAN could be used to model the steps an HCP would perform to cancel an incorrect lab test was mistakenly ordered. GLEAN also allows analysts to better understand why a particular interface element is more prone to error than another (Wood & Kieras, 2002).

MODELING NOTIFICATION PROCESSING TASKS

The top-level method of the notification processing task, as shown in Figure 2, is deceptively simple: Using the View Alert window from Figure 1, the HCP searches for a notification in the Notification table, double-clicks on it, and processes the notification as necessary. After processing an individual notification, it then disappears from the list; then the HCP selects the next one to process. This continues until there are no more notifications to process.

> **Method for Goal: Process List of Notifications**
> Step 1. Accomplish Goal: Sort notifications
> Step 2. Accomplish Goal: Search for next notification to process
> Step 3. Point to next notification
> Step 4. Double-click on notification row
> Step 5. Accomplish Goal: Process a notification
> Step 6. Decide: If more notifications to process, go to Step 1
> Step 7. Return with goal accomplished

FIGURE 2. Notification processing task structure.

The Process List of Notifications method is similar in structure to what one might use for processing email messages, and in fact, shares many similarities regarding usage. One similarity is that, while certain types of notifications will be handled similarly, all notifications are unique and most of these include some degree of creative problem-solving. While processing email, for example, one message may require finding an open meeting slot on the calendar, while another may require solving a design problem and writing the explanation in as a reply. In CPRS, the HCP may order a specialized blood test based on abnormal results from an earlier test, or the HCP may forward the notification to a specialist asking for advice. The point here is that processing alerts requires performing varying actions and takes an unpredictable amount of time.

Also like email, some notifications are part of a formal workflow and require HCP sign-off to legally acknowledge or approve certain actions, such as prescription requests. Similarly, there is also a class of notifications that is considered "information only", not requiring any specific action. An email example might just be a reminder that hand-washing kills germs or to drive safely during the holidays. Figure 3 shows a simplified task hierarchy for the notification processing task. The top-level method, *Process List of Notifications*, invokes three other methods to sort the list, search for a single notification, and then process it. The *Process a Notification* method describes the actions to process a single notification and can, in turn, invoke several other procedures, depending on the type of notification being processed.

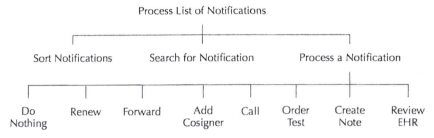

FIGURE 3. Notification processing task structure.

Another similarity between notification processing and email is that HCPs differ in how and when they process notifications - some deal with notifications first thing in the morning, some at the end of the day, and others deal with them a few at a time between patients. Likewise there are individual differences in how HCPs prioritize the order in which they process alerts – some strictly process alerts in the order in which they appear in the Notification table (by default, the most recent notifications are listed first), while others scan the list to process the most urgent notifications first.

Although there are many similarities between processing notifications and email messages, there are also many subtle differences that make the notification task much more difficult and prone to error.

PRELIMINARY WORK AND RATIONALE FOR MODELING

An old adage among military leaders states that "Plans are useless, yet planning is essential". Similarly, even if we were to do nothing else with a model, the task analysis necessary for modeling often yields essential knowledge about the usability of an interface. Even considering just our description of the top-level method (Process List of Notifications), for example, suggests several areas where human performance modeling can inform evaluation of the system and task. To illustrate this further, we will focus more on the structure of the top-level method and how GOMS modeling can inform both future research and applied work.

The structure of the top-level method makes clear that visual search plays a key role in the notification task. Although CPRS training documentation makes no mention of notification processing strategies, it is clear that some HCPs have well-defined strategies. Hysong et al (2010) found that HCPs utilizing such strategies were significantly more likely to follow-up on abnormal test results than HCPs who relied only on superficial visual scans. One such strategy might include sorting notifications by patient first, so that the HCP can complete all necessary tasks for one patient before moving on to another, but relying on their memory as a way to protect against premature deletion of notifications. Part of this strategy would require the HCP to first remember all of the notifications received for a patient, then perform all tasks for that patient within the patient record (as recalled from the notification list), and finally to manually remove the processed notifications from the list. Without external cognitive aids, though, this type of strategy puts an additional cognitive burden on HCPs, making it more likely that they will confuse unprocessed notifications with those that actually were processed. This type of problem shows up in GLEAN because the virtual human must store the list items in short-term, working memory throughout the processing task. The longer we need to retain such things, of course, the more likely we are to forget them. From a design perspective, modeling can thus provide valuable quantitative information to compare redesign options. Designs that minimize cognitive workload (including memory load) are less likely to cause errors than those that do not.

The GOMS model can also be used to help discover optimal search and performance strategies for processing notifications. By making small changes to the top-level method structure, such as by altering or eliminating the sorting step, we can directly compare performance by testing the competing strategies against a common set of benchmark tasks. This type of information can be used in design, for instance, to modify interface layout or content so that it better supports optimal strategies. Even if no redesign were possible, analyzing strategies might to develop improved training materials – perhaps not surprising, but GOMS models contain just the kind of "how to do it" information required for good task-based documentation and help design (Elkerton & Palmiter, 1991).

In some cases, however, the actions required of users are just too complex. This shows up in GOMS models as long, convoluted methods, or containing placeholders for complex cognitive tasks (e.g., developing a diagnosis). In these cases, the modeling process can help quantify the value of new technology (e.g.

how would a new notification-prioritizer affect performance?) – clarifying design alternatives in this way was key to the approach taken by Butler and colleagues (2007) to radically improve an aircraft-maintenance scheduling-system.

From a research perspective, the working memory load example also makes an empirically testable prediction. If we need a deeper understanding of how working memory affects performance and errors, the GOMS model provides a strong basis for designing an experiment, including a basic experimental paradigm, and key variables to manipulate (such as number of notifications and duration of the necessary processing actions). The model can also be used to determine minimal instrumentation requirements for measuring outcomes (e.g. recording the time a particular stimulus is presented) and necessary properties of secondary tasks (e.g. that they stress a particular modality) (Wood, 2000).

CONCLUSIONS

Human performance modeling can be a useful technique by providing both a theoretical foundation on which to research human-system interaction issues and as an engineering tool for designing new systems. In health care research human-performance modeling can provide additional structure and quantitative measures to qualitative methods. It can also be used as a basis for experimental design and for interpreting results. As such, human-performance modeling can help shape future health care research by adding yet another mixed-method bridge between qualitative and quantitative approaches.

From a practitioner standpoint, human-performance modeling represents a practical, engineering-level tool for identifying usability problems and evaluating design alternatives. Modeling error and error recovery should be particularly useful in the context of EHR use because of the high frequency of certain information-intensive tasks (such as processing notifications) and the high consequence of failure. Understanding error-inducing aspects of the interface will allow us to develop systems that are more tolerant to human limitations and which can better capitalize on human strengths. In health care system design, the ability to clearly consider error recovery alongside normal task procedures could help us improve key patient safety aspects of our systems, along with reducing costs and increasing the effectiveness of health care providers. The ability to capture the user's procedural knowledge in a simple, human-readable format means that human performance models such as GOMS can facilitate the entire system development process, including design, evaluation, and training.

A final point is that human performance modeling can help us determine whether we are solving the right problem. One measure of this is how well our technological solutions match the non-technological goals of our users. Without a clear connection, it is hard to determine through modeling whether what we are doing is having a positive effect on the situation. Perhaps a revealing indicator in our notification-processing example is that "caring for patients" is not explicitly in the top-level method. Instead the work process centers on notifications. Although this may just be a matter of semantics, the danger is that while we may optimize the notification processing task, we may miss larger opportunities to improve patient care. It may be useful to expand our modeling scope to include a top-level method for providers along the lines of "Care for my Current Patients." Although we could just end up at the same place, but we might also discover more fundamental improvements to dramatically improve patient care. For those of us working at the intersection of research and practice, human performance modeling can provide the right structure to more easily explore and develop game-changing ideas.

REFERENCES

Butler, K. A., Zhang, J., Esposito, C., Bahrami, A., Hebron, R., and Kieras, D. 2007. Work-centered design: a case study of a mixed-initiative scheduler. In *Proceedings of the SIGCHI Conf on Human Factors in Computing Systems.*

Campbell E, Sittig DF, Ash JS, Guappone K, Dykstra R.(2006). Types of unintended consequences related to computerized provider order entry. Journal of American Medical Informatics Association. Sep-Oct;13(5):547-56.

Card, S. K., Moran, T. P., & Newell, A. (1983). *The Psychology of Human-Computer Interaction.* Hillsdale, NJ: Lawrence Erlbaum.

Chaudhry B, Wang J, Wu S, et al. (2006). Systematic review: impact of health information tech-nology on quality, efficiency, and costs of medical care. *Ann Intern Med.* May 16;144(10):742-52.

Elkerton, J., Palmiter, S.L. (1991). Designing help using a GOMS model: an information retrieval evaluation, *Human Factors*, v.33 n.2, p.185-204.

Han YY, et al. (2005). Unexpected increased mortality after implementation of a commercially sold CPOE system. *Pediatrics*;116:1506 –12.; 63.

Hysong SJ, Sawhney MK, Wilson L, Sittig DF, Esquivel A, Watford M, Davis T, Espadas D, Singh H. (2009). Improving outpatient safety through effective electronic communication: A study protocol. *Implement Sci.* Sep 25;4(1):62.

Hysong SJ, Sawhney MK, Wilson L, Sittig DF, Espadas D, Davis TL, Singh H. (2010). Provider Management Strategies of Abnormal Test Result Alerts: A Cognitive Task Analysis. *J Am Med Inform Assoc* 17:71-77.

Kieras, D. E. (1999). *A Guide to GOMS Model Usability Evaluation Using GOMSL and GLEAN3* (No. 38). Ann Arbor, Michigan: University of Michigan.

Kieras, D. E., Wood, S. D., Abotel, K., & Hornof, A. (1995). GLEAN: A Computer-Based Tool for Rapid GOMS Model Usability Evaluation of User Interface Designs. Proceedings of *User Interface Software and Technology.*

Koppel R, Metlay JP, Cohen A, et al. (2005). Role of computerized physician order entry systems in facilitating medication errors. *JAMA*;293:1197–203.

Linder JA, Ma J, Bates DW, Middleton B, Stafford RS. (2007). Electronic health record use and the quality of ambulatory care in the United States. *Arch Intern Med.* Jul 9;167(13):1400-5.

Nebeker JR, Hoffman JM, Weir CR, et al. (2005). High rates of adverse drug events in a highly computerized hospital. *Arch Intern Med.* 165:1111– 6.

Singh, H., Thomas, E. J., Mani, S., Sittig, D. et al. (2009a). Timely follow-up of abnormal diagnostic imaging test results in an outpatient setting: are electronic medical records achieving their potential? *Arch Intern Med*, 169(17).

Singh H, Wilson L, Petersen LA, Sawhney M, Reis B, Espadas D, Sittig DF. (2009b). Improving Follow-up of Abnormal Cancer Screens Using Electronic Health Records: Trust but Verify Test Result Communication. *BMC Med Inform Decis Mak*. Dec 9;9(1):49.

Singh H, Vij M. (2010). Eight Recommendations for Policies for Communication of Abnormal Test Results. In press. *Jt Comm J on Quality and Patient Safety*.

Wood, S. D. (2000). *Extending GOMS to Human Error and Applying it to Error-Tolerant Design*. Doctoral dissertation, U of Michigan, Ann Arbor, MI.

Chapter 22

Team Coordination and Performance During Simulated and Real Anesthesia Inductions

Gudela Grote[1], Michaela Kolbe[1], Johannes Wacker[2], Donath Spahn[2]

[1]Department of Management, Technology, and Economics, ETH Zurich
Kreuzplatz 5, 8032 Zurich, Switzerland

[2]Institute of Anaesthesiology, University Hospital Zurich
Raemistrasse 100, 8091 Zurich, Switzerland

ABSTRACT

Two studies are presented that investigated coordination in anesthesia teams. In the first study, 15 teams were observed while performing a simulated induction of general anesthesia with a cardiac arrest occurring during intubation. In the second study, 27 teams were observed during real anesthesia inductions. In both cases behavior was coded based on a taxonomy of coordination behaviors which distinguishes action and information coordination and within each of these categories explicit versus implicit coordination. In the simlutor study, coordination behavior of residents and nurses were compared and it was found that good teams were characterized by more explicit action coordination by residents, more implicit action coordination by nurses, and less explicit information coordination by both residents and nurses. During real inductions, team coordination patterns were compared for good and bad teams. In particular, effects of team monitoring and talking to the room were analyzed. In the good teams, team monitoring led to speaking up and unsolicited assistenace, while talking to the room seemed to be effective in reducing requirements for explicit coordination. Overall, the results demonstrate the intricate nature of effective coordination in anesthesia teams. They also indicate the importance of detailed behavioral analyses in order to provide substantiated recommendations for team training.

Keywords: Coordination, high-risk teams, patient safety

INTRODUCTION

Anesthesia is a complex work environment involving a variety of risks and potentially life-threatening consequences of failures (e.g. Catchpole, Bell, & Johnson, 2008). There is growing evidence that the success of medical crews in dealing with their demanding tasks depends on coordination which is adaptive to situational demands (Künzle et al., in press; Manser et al., 2008; Salas, Rosen, & King, 2007; Zala-Mezö et al., 2009). There is little specific empirical evidence, however, on what kind of coordination behavior is effective for maintaining good clinical performance in the transition from routine situations to managing non-routine events. This is especially true for "daily hassles" – non-routine events which are not critical per se but might be early heralds of later incidents.

We attempt to fill this gap by analyzing coordination behavior and performance in a simulated major non-routine event and in the course of real-life intubations with whatever non-routine events happened to occur. Our interest was focused on three modes of coordination that have been much discussed in recent literature, namely explicit versus implicit coordination (Serfaty & Entin, 2002) and "talking to the room" (Tschan et al., 2009; Waller & Uitdewilligen, 2008).

METHOD

Two studies were carried out with different samples and oberservational settings. In the first study, we examined coordination behavior of 15 anesthetic crews (each consisting of one resident and one nurse) in simulated inductions to general anesthesia involving a cardiac arrest during laryngoscopy as non-routine event. Cardiac arrest during anesthesia induction represents a rare but potentially dangerous situation, where a timely response is crucial. Thus, the clinical crew performance was assessed using a time-based measure (execution time). The teams were video-taped and coordination behavior was coded of the video-tapes by trained observers using a taxonomy distinguishing action and information coordination and within each of these between explicit and implicit coordination (Kolbe et al., 2009).

The second study investigated anesthesia crews performing inductions to general anesthesia in a clinical setting. The 25 three to four person crews consisted of anesthetists and registered nurses. The patients had ASA 1 (normal or healthy) or ASA 2 status (mild systemic disease or condition). Coordination behavior was analyzed using the same taxonomy as in the first study (Kolbe et al., 2009). Additionally, talking to the room was coded if a crew member appeared to address a communication not to a specific crew member, but to the whole crew. Clinical crew performance was assessed by an anesthetist using a checklist of required tasks during anesthesia induction.

RESULTS

STUDY 1

In the simulated setting, non-parametric comparisons between different phases of the anesthesia inducation, different coordination mechanisms and the two team members revealed that (1) overall anesthesia crews relied heavily on implicit coordination, (2) they coordinated actions (however not information exchange) with higher levels of explicitness in non-routine than in routine situations, (3) residents' explicit action coordination, nurses' implicit action coordination, and little explicit information coordination by both residents and nurses during non-routine situations was positively related to crew performance.

Additionally, we analyzed micro-coordination patterns by means of lag sequential analysis (Bakeman & Gottman, 1986). In particular, we investigated immediate reactions of the other team member to those coordination mechanisms that were either positively (residents' explicit action coordination; nurses' implicit action coordination) or negatively (residents' and nurses' explicit information coordination) related to team performance. As such, we integrated individual behavior into dynamic team patterns which allowed us to draw conclusions on the team level (see Uhlig, 2009 for a critical comment on focusing only on individual behavior in health care team analysis). In doing so we found a plausible explanation for the effectiveness of residents' explicit action coordination: it frequently led the nurses to perform the requested behavior accordingly or to pay special attention to the correctness of the current procedure or information. Thus, by being explicit, the residents were able to effectively elicit valuable task-relevant behavior from the nurses who could then contribute to task performance based on their clinical experience. This pattern resembles what Shelly (1997) in his research on social interaction called "fundamental interaction sequence", where 'action opportunities' (clear instruction) are followed by 'performance' (providing action upon request).

The findings of the lag sequential analysis are less clear regarding a possible explanation for the ineffectiveness of nurses' explicit action coordination. The reactions of residents to explicit action coordination from nurses resembled those of nurses to residents' explicit action coordination described above: residents provided action or information upon request. We can only speculate that these behaviors were not task adaptive because they consumed too much time and thus prevented the residents from keeping ahead of the situation and giving necessary instructions. There may be a discrepancy in how literal an explicit action request is received by a nurse from a resident vs. a resident from a nurse. In other words, the explicit action coordination from a resident may initiate a cascade of tasks performed by the nurse, whereas an explicit action coordination from a nurse may initiate the performance of the specified task only and therefore cause the need for even more (rather than less) time-consuming explicit action coordination from the nurse. Testing this assumption would have required an investigation of lag 2 and more transitions, which was not possible due to the small amount of event sequences during the non-routine phase.

Concerning the negative relation between explicit coordination of the information exchange and team performance, the lag sequential analysis revealed that explicit information coordination initiated further information exchange and produced time-consuming patterns of explicit information coordination. Particularly the requests for information in response to information provided upon request seemed to represent an inefficient way of coordination method in a situation where quick response is vital.

Finally, the time-saving character of implicit coordination is represented in the effectiveness of nurses' implicit action coordination. We found that when nurses provided unsolicited action, residents frequently reacted by also providing unsolicited action. Seeing that nurses were executing the correct procedures saved the residents from giving instructions and allowed them to spend their time and mental concentration on task-relevant behavior as well. This pattern might presuppose not only shared knowledge (Michinov, Olivier-Chiron, Rusch, & Chiron, 2008) but also mutual trust (Rico, et al., 2008).

STUDY 2

In the 25 real inductions, 162 non-routine events occurred with an average of about seven events per crew. Anesthesia crews spent about 7% of induction time with talking to the room. No direct relation between talking to the room, the occurrence of non-routine events and crew performance was found. However, talking to the room was positively related to system monitoring, which in turn was positively related to crew performance and negatively to the occurrence of non-routine events caused by the crew.

Again, lag sequential analyses were used to study micro-coordination patterns, in particular regarding two behaviors which have been discussed as crucial for good team performance, namely team monitoring (Marks, Mathieu & Zaccaro, 2001) and talking to the room (Tschan et al., 2009; Waller & Uitdewilligen, 2008). As expected, we found that immediate consequences of team monitoring were speaking-up and providing unsolicited assistance – both patterns occurred only in high-performing teams. Team monitoring also led to giving instructions – in high as well as in low-performing teams. These results highlight the important functions of team monitoring for further team coordination. Monitoring enables team members to recognize when their teammates need guidance or assistance and immediately react with either providing assistance or speaking up and giving instructions. It seems to be a characteristic of well-performing teams that monitoring led not only to giving instructions but also to speaking up and providing assistance, indicating that team members are more engaged in helping each other. These behaviors have been described as backing-up behavior in terms of the "discretionary provision of resources and task-related effort to another member of one's team that is intended to help that team member obtain the goals as defined by his or her role when it is apparent that the team member is failing to reach those goals" (Porter, Hollenbeck, Ilgen, et al., 2003, pp. 391-392). Our results indicate that backing-up following team monitoring is indeed valuable for team performance, as suggested elsewhere (Salas, Sims, & Burke, 2005).

We differentiated two distinct facets of talking to the room. Information-related talking to the room refers to interpreting and sharing information, while action-related talking to the room includes commenting the own current behavior. As expected, we found that these communications invited other team members to actively participate in effective coordination - action-related talking to the room led to further action-related talking to the room, the same was found for information-related talking to the room. Furthermore, by constantly updating each other about own behavior, action-related talking to the room seemed to make it unnecessary to give instructions. Thus, it substituted for more direct forms of team coordination. The overall results indicate that in high-performing teams, talking to the room initiates task-relevant communication, which could be interpreted as a facilitator for the development of shared situation awareness and an easy-to-use coordination tool.

CONCLUSION

While in the first study concerning handling a major non-routine event the importance of explicit coordination was evident, the analyses in the second study focusing on routine situations and minor non-routine events indicated that talking to the room might substitute for explicit forms of crew coordination by constantly updating each other about own behavior and thoughts. Also, in the second study the importance of team monitoring was evident as a mechanism to ensure team members to tailor their behavior to the specific needs of other team members. These might entail forms of explicit coordination like speaking up or of implicit coordination like providing unsolicited assistance.

Overall, the results demonstrate the intricate nature of effective coordination in anesthesia teams. They also indicate the importance of detailed behavioral analyses in order to provide substantiated recommendations for team training.

REFERENCES

Bakeman, R., & Gottman, J. M. (1986). Observing interaction: An introduction to sequential analysis. Cambridge: Cambridge University Press.

Catchpole, K., Bell, M. D. D., & Johnson, S. (2008). Safety in anaesthesia: A study of 12606 reported incidents from the UK National Reporting and Learning System. Anaesthesia, 63, 340-346.

Kolbe, M., Kuenzle, B., Zala-Mezö, E., Wacker, J., & Grote, G. (2009). Measuring coordination behaviour in anaesthesia teams during induction of general anaesthetics. In R. Flin & L. Mitchell (Eds.), Safer surgery. Analysing behaviour in the operating theatre, pp. 203-221. Aldershot: Ashgate.

Künzle, B., Zala-Mezö, E., Kolbe, M., Wacker, J., & Grote, G. (in press). Substitutes for leadership in anaesthesia teams and their impact on leadership effectiveness. European Journal of Work and Organizational Psychology.

Manser, T., Howard, S. K., & Gaba, D. M. (2008). Adaptive coordination in cardiac anaesthesia: A study of situational changes in coordination patterns using a new observation system. Ergonomics, 51, 1153-1178.

Marks, M. A., Mathieu, J. E., & Zaccaro, S. J. (2001). A temporally based framework and taxonomy of team processes. Academy of Management Review, 26, 356-376.

Michinov, E., Olivier-Chiron, E., Rusch, E., & Chiron, B. (2008). Influence of transactive memory on perceived performance, job satisfaction and identification in anaesthesia teams. British Journal of Anaesthesia, 100, 327–332.

Porter, C. O. L. H., Hollenbeck, J. R., Ilgen, D. R., Ellis, A. P. J., & West, B. J. (2003). Backing up behaviors in teams: The role of personality and legitimacy of need. Journal of Applied Psychology, 88, 391-403.

Rico, R., Sánchez-Manzanares, M., Gil, F., & Gibson, C. (2008). Team implicit coordination processes: A team knowledge-based approach. Academy of Management Review, 33, 163–184.

Salas, E., Sims, D. E., & Burke, C. S. (2005). Is there a "big five" in teamwork? Small Group Research, 36, 555-599.

Salas, E., Rosen, M. A., & King, H. (2007). Managing teams managing crisis: Principles of teamwork to improve patient safety in the Emergency room and beyond. Theoretical Issues in Ergonomics Science, 8, 381-394.

Serfaty, D., & Entin, E. E. (2002). Team adaptation and coordination training. In R. Flin, E. Salas, M. Strub & C. Martin (Eds.), Decision making under stress. Emerging themes and applications (pp. 170-184). Alderhot: Ashgate.

Shelly, R. K. (1997). Sequences and cycles in social interaction. Small Group Research, 28, 333–356.

Tschan, F., Semmer, N. K., Gurtner, A., Bizzari, L., Spychiger, M., Breuer, M., et al. (2009). Explicit reasoning, confirmation bias, and illusory transactive memory. A simulation study of group medical decision making. Small Group Research, 40, 271-300.

Uhlig, P. (2009). Commentary and clinical perspective. In R. Flin & L. Mitchell (Eds.), Safer Surgery. Analysing behavior in the Operating Theatre (pp. 437–443). Aldershot: Ashgate.

Waller, M. J., & Uitdewilligen, S. (2008). Talking to the room. Collective sensemaking during crisis situations. In R. A. Roe, M. J. Waller & S. R. Clegg (Eds.), Time in organizational research (pp. 186 – 203). Oxford UK Routledge.

Zala-Mezö, E., Wacker, J., Künzle, B., Brüesch, M., & Grote, G. (2009). The influence of standardisation and task load on team coordination patterns during anaesthesia inductions. Quality and Safety in Health Care, 18, 127-130.

Chapter 23

Leadership and Minimally Invasive Training Enhance Performance in Medical Emergency Driven Teams: Simulator Studies

Norbert K. Semmer[1], Franziska Tschan[2], Sabina Hunziker[3], Stephan Marsch[3]

[1]University of Bern
Bern, Switzerland

[2]University of Neuchâtel
Neuchâtel, Switzerland

[3]University Hospital Basel
Basel, Switzerland

ABSTRACT

We report four studies with participants being confronted with a cardiac arrest of a "patient" in a high-fidelity simulation. Based on video-tapes, performance was assessed in terms of the time until meaningful treatment was started and "hands-on" time; leadership behavior was coded on the basis of video-tapes. We could show (1) that directive leadership of first responding nurses and later arriving residents was associated with higher performance, (2) that there was more directive leadership and better performance if groups of three physicians were all present when a cardiac arrest occurred, as compared to a condition where they entered sequentially, indicating that deficiencies in leadership behavior are not simply due to a lack of skills, (3) that medical students who received a leadership-oriented debriefing after a cardiac arrest scenario showed more leadership, and better performance during a second visit four months later, and (4) that a brief instruction to first responding physicians encouraging them to lead improved leadership and performance. Team-building is important for the performance of medical emergency teams should receive more attention in guidelines and training.

Keywords: Training; minimally invasive training; leadership; medical emergency; cardiac arrest; simulation

INTRODUCTION

Many emergency situations are time-sensitive; they require quick and decisive action, and they do not permit extended planning and deliberation. A cardiac arrest is an example for such a situation; it requires immediate attempts to assure the flow of oxygen to the brain, for instance by cardiac massage and ventilation. It has been estimated that any minute that a patient does not receive adequate treatment after a cardiac arrest reduces chances of survival by 8 to 10 per cent (Larsen, Eisenberg, Cummins & Hallstrom, 1993. Medical teams dealing with a cardiac arrest do not have much time to establish team coordination; they have to function as a team instantly.

The actions required in a cardiac arrest situation are well defined in pertinent algorithms (Nolan, Deakin, Soar, Bottiger, & Smith, 2005). These algorithms specify medical acts, such as cardiac massage, ventilation, defibrillation, and medication; they do not, however, refer to processes of team coordination. Team coordination involves (a) decisions, such as who will do what (e.g., who ventilates, who performs cardiac massage, who organizes a defibrillator), and (b) communication, for instance in terms of informing other members about the condition of the patient, previous acts performed, etc.

Based on these considerations, the present paper examines (a) if there are, indeed, differences in leadership behavior, (b) if these behaviors are related to medical performance, (c) if leadership behavior is more or less adequate because of differences in leadership skills or in differences in thresholds for exercising leadership behaviors, and (d) if short interventions that do not teach leadership skills but encourage their mobilization will enhance leadership and team performance.

THE SETTING: HIGH FIDELITY SIMULATION

For our studies we used a high fidelity manikin with the possibility of remote control of vital signs (Human Patient Simulator, (Medical Education Technologies, Inc)). The full body simulator is a computer-based manikin with human physiology emulation capability; it allows very realistic interactions. For the studies presented, the manikin was lying on a bed. A connected monitor displayed a continuous electrocardiogram and the non-invasively measured blood pressure at selected time intervals. An intra-venous catheter was placed in a peripheral vein to allow direct administration of medication. A commercially available manual defibrillator was placed next to the bed.

Participants were taking care of a patient whose condition at first was not too serious (e.g., the patient complained of dizziness after a bicycle stress test). Two minutes after participants had started to take the medical history, the patient fainted, his pulse and respiration stopped and the monitor displayed ventricular tachycardia. This situation calls for immediate resuscitation attempts.

After the session, participants received a debriefing with feedback by a medical expert on their performance. Participants rated the scenario as highly realistic (typically between 8 and 9 on a scale from 1 to 10; cf. Hunziker. Tschan, Semmer, Zobrist, Spychiger, Breuer, Hunziker, and March, 2009)

STUDY 1: DIRECTIVE LEADERSHIP AND MEDICAL PERFORMANCE

The considerations presented above suggest that leadership is important for medical performance and that teams differ in the extent that a clear leadership is established. Our first study tested if leadership behavior does, indeed, predict medical performance (Tschan, Semmer, Gautschi, Hunziker, Spychiger, & Marsch, 2006).

THE STUDY

The study involved a cardiac arrest situation as described above, and it consisted of three phases. In phase one, three nurses were present, one of whom witnessed the cardiac arrest (first responder) and called the two others. In phase two, a resident joined, and in phase three a senior physician entered. All participants acted in the roles they actually had in their daily professional life.

We hypothesized (1) that directive leadership of the first responding nurses would enhance performance in phase one, (2) that directive leadership of the incoming resident would enhance performance in phase 2; however, we predicted that the resident should establish coordination rapidly; therefore, we hypothesized more specifically that the residents' leadership behavior during the first 30 seconds would predict performance in phase 2; (3) that directive leadership of the senior physician should predict performance in phase 3. We further assumed that the senior physician should act in accordance with his or her role as a mentor, and therefore predicted (4) that the senior physician would enhance performance by asking structured questions ("structuring inquiry").

After eliminating one group who did not diagnose the cardiac arrest correctly, data of 21 groups were analyzed. Analyses are based on video-tapes. The following parameters were coded: (1) state of the patient: circulation vs. no circulation (coded every 5 seconds; Cohen's $\kappa > .9$); (2) actions (e.g., cardiac massage; preparing the defibrillator; coded every 5 seconds; $\kappa > .75$); (3) verbal behavior (all utterances; $\kappa > .8$). Directive leadership was coded if a person gave specific directions for an action (e.g., please hand me the tube; you start cardiac massage), including corrections (you are massaging too slowly). Questions were coded as structuring inquiry if they referred to the distribution of tasks in the group, to technical acts performed, or to the condition of the patient – but only if they were specific (e.g., are you preparing epinephrine); whereas general questions (what have you already done?) did not count as structuring inquiry.

Performance was assessed as the percentage of the time that cardiovascular support was given (i.e., any of the behaviors specified in the algorithm, such as cardiac massage, defibrillation, etc.).

RESULTS

As expected, directive leadership behavior of the first responding nurse predicted performance in phase 1 ($r = .45$, $p < .05$), and directive leadership behavior of the resident during the first 30 seconds after his/her arrival predicted performance in phase 2 ($r = .52$, $p < .05$). These associations also held when controlling for potentially confounding variables, such as the length of the phase, or (for phase 2) performance of the preceding phase. In phase 3, only structuring inquiry of the senior physician emerged as predictor of performance.

These results show a) that directive leadership is important for medical performance in these situations, b) that more indirect forms of leadership (structuring inquiry) by senior physicians is optimal once the basic leadership structure is established. The results also show that leadership cannot be taken for granted; had leadership been optimal in all groups, leadership would have a very small variance and could not predict performance.

STUDY 2: DIRECTIVE LEADERSHIP: BASIC SKILLS VS. SITUATIONAL TRIGGERS

If leadership is important for medical performance, and if medical professionals differ in their leadership behavior, the question arises where these differences come from. Basically, there are two possibilities. First, people may differ in basic leadership skills, due to their leadership experience and training throughout their career. Second, people may differ in their tendency to exercise leadership; some may recognize more than others that leadership is required in a given situation, and some may be more or less reluctant than others to actually take the lead. In this latter case, people may have the necessary skills but do not focus their attention on situational cues signaling that they should exercise their skills, or do not interpret such cues appropriately. Competing cues, such as cues that signal task affordances, might catch their attention more than cues about coordination requirements; the latter would become sufficient priority only if there is ample opportunity, without competing cues. Put differently, people often are confronted with (at least) two different tasks. There is the primary task, which in our case is resuscitation. But there also is a secondary task, which in our case is establishing a coordinative structure (as part of team building). Optimal performance would imply switching between the primary and the secondary task, giving just enough attention to the secondary tasks that performance in the primary task is enhanced, but not so much that the secondary tasks becomes the primary one.

There is research in group psychology showing that group members often tend to attend to the primary task only. For instance, they neglect planning (Weingart, 1992) and (re-) consideration of strategy (reflexivity; cf. West, 1996). More generally speaking, they tend to attend to the primary task and to neglect issues of group process (cf. Gurtner, Tschan, Semmer, and Nägeli, 2007; Hackman, Brousseau & Weiss, 1976). Matters are worse under stress, which narrows the attentional focus and may induce a loss of team perspective (Driskell, Salas & Johnaston 1999). On an individual level, similar phenomena are known in research on learning, as learners often forego systematic learning, preferring to go for immediate results (often by trial and error); this has been called the "production paradox" (Carroll, 1998).

Of course, these two basic reasons are not mutually exclusive. However, to the extent than one or the other is dominant, they have quite different implications. If differences in leadership behavior are mainly due to leadership skills, training in leadership skills is required. Such training is rather time consuming, requiring discussion and demonstration of relevant behavior, opportunities for practice, and feedback (Yukl, 2010). If differences in leadership are mainly due to attending to pertinent cues and drawing appropriate conclusions, it may suffice either to draw their attention to such information, or to provide an opportunity for establishing a coordination structure without competing cues being present.

We assumed that people with medical training, being of high status and being highly educated, would have exceptionally good chances to acquire leadership skills throughout their education and their career, and to be exposed to role models demonstrating such behavior. We therefore assumed that not taking leadership would most likely not be due to a lack of basic skills. Possessing basic skills does, however, not necessarily imply that one is able to mobilize these skills, unless one's attention is drawn to the necessity for their use, or unless the situation provided an opportunity for exercising them without strong competing affordances.

THE STUDY

We tested the assumption that opportunity for team building would enhance leadership behavior and performance by varying situational conditions that should make it easier vs. more difficult to exercise leadership (Hunziker et al, , 2009). Ninety-nine groups were confronted with the cardiac arrest scenario as described above. The groups consisted of three physicians (general practitioners or hospital physicians) plus a nurse who was introduced as technically proficient (e.g., in using the defibrillator) but instructed to act on command only.

In one condition, all three physicians were present when the cardiac arrest occurred. In the second condition, only one was present, and the two others entered the scene sequentially after the onset of the cardiac arrest; they arrived quickly, groups were complete in less than 10 seconds after the cardiac arrest started. The "all present" condition implies a short period where the three participants are together without urgent requirements for emergency action, as the state of the patient requires only routine action. This situation should yield an opportunity for establishing a coordinative structure in the absence of urgent competing cues. Participants could rely on this structure once the emergency starts. By contrast, if others join sequentially, opportunities to establish a coordination structure are limited, as the situation requires urgent medical action; with regard to communicating with colleagues, such aspects as informing them about what happened may be more salient than establishing a coordinative structure.

We therefore hypothesized (1) that participants in the all-present condition would show more leadership behavior, which (2) should go along with improved medical performance. Performance was assessed as "hands-on time" (i.e. treatment that corresponded to the algorithm for cardiac arrest treatment) during the first three minutes. Utterances were coded in terms of decisions what should be done (e.g., we should defibrillate), decisions how something should be done (e.g., we should defibrillate with 360 Joule), direction/command (e.g., you should perform massage quicker), task assignment (e.g., will you ventilate?), reflection (e.g., what should we do next?), and other. With regard to directive leadership, direction/command and task assignment are most pertinent.

RESULTS

Preformed groups showed more hands-on time during the first three minutes than ad-hoc groups (124 seconds [SD = 33] vs. 93 seconds [SD = 37]; this difference was highly significant ($p < .0001$). Preformed groups also took less time until the first appropriate meaningful intervention (24 [16] vs. 43 [28]; $p < .0001$); until the first defibrillation (67 [42] vs. 107 [46]; $p < .0001$), and until administering epinephrine (157 [55] vs. 210 [70] $p < .0001$).

Thus, preformed groups performed significantly better, and the differences are clinically relevant (see Hunziker et al., 2009). Of importance for the issue of leadership and coordinative structure are the results concerning leadership utterances. Preformed groups showed significantly more direction/command (6.8 [3.4] vs. 3.8 [2.4]), decision how (4.0 [2.2] vs. 2.8 [1.7]), and task assignment (1.7 [1.5] vs. 0.7 [0.8]) than ad-hoc groups. When these variables were entered into a regression analysis, controlling for type of team and type of physician, direction/command emerged as a significant predictor for hands-on time, the parameter estimate being an additional 4.6 seconds of hands-on time per direction/command utterance (see Hunziker et al., 2009).

Thus, (1) preformed groups performed better in terms of hands-on time; (2)

preformed groups showed more direction/command utterances, and (3) direction/command utterances predicted performance. These results suggest that the condition "preformed group" induced more leadership utterances in terms of direction/command, which, in turn, improved performance. If it is, indeed, the greater number of leadership utterances that is responsible for the difference between the conditions, type of group should become nonsignificant once direction/command is controlled for. This is exactly the result obtained in the multiple regression analysis.

Therefore, these results suggest (a) that our participants did have basic leadership skills; (b) a leadership structure is more likely to be established in a condition that offers the opportunity for establishing it in the absence of competing action requirements, and (c) that this leadership structure does enhance medical performance.

STUDIES 3 AND 4: CAN LEADERSHIP BEHAVIOR BE ENHANCED BY MINIMALLY INVASIVE INSTRUCTIONS?

If leadership behavior can be enhanced by situational changes as introduced in Study 3, the question arises if it also can be enhanced by training that is minimally invasive in the sense that it consists of directing people's attention to the necessity for leadership rather than of intensive training. If participants focus on the necessity to lead, they might mobilize their skills (which basically are present) even in situations that require emergency action (i.e., treating the cardiac arrest) and thus contain cues that compete with cues indicating the necessity to establish a coordinative structure.

Okhuyzen (e.g., Okhuyzen and Eisenhardt, 2002) have advocated, and tested, such an approach. Arguing that minimal instructions (called "formal interventions by these authors) would direct people's attention to a secondary task related to group process, they could show that short instructions like asking people to question other group members or managing time carefully resulted in greater knowledge integration in problem-solving groups.

STUDY 3

The Study

Building on the work by Okhuyzen, we designed a minimally invasive intervention. It involved 79 groups composed of three fourth-year medical students who were exposed to the cardiac arrest scenario twice (Hunziker, Bühlmann, Tschan, Legeret, Schumacher, Balestra, Semmer, Hunziker, & Marsch, 2010). As always, the scenario was followed by a debriefing, during which an experienced physician gave feedback to the participants about their performance. In this study, the debriefing after the first session (baseline) was supplemented by an additional instruction, which focused either on technical issues (e.g., correct position of the arms during cardiac massage) or on team leadership; the latter emphasized four points: (1) decide what to do; (2) tell your colleagues what they should do (task assignment); (3) make short and clear statements; and (4) ensure adherence to algorithm. At baseline, all three members of a group were present from the beginning.

The follow-up session took place four months after the baseline session. This time, only group member was present when the cardiac arrest occurred. Students were assigned to groups randomly, but with three constraints. First, all members of a group had received the same type of instruction (technical or leadership oriented), second, no two team members had worked together in the same group at baseline, and third, an equal number of males and females were assigned the role of the first responder. At baseline, 79 groups of three students each took part. At follow-up, there were 32 groups in the technical instruction condition, and 31 in the leadership instruction condition. The main performance measure again was hands-on time during the first three minutes. Leadership utterances were coded as in Study 2, yielding a κ of .9.

Results

There were no differences in performance between the two conditions at baseline, except that groups assigned to the technical condition showed more (!) leadership utterances. Groups in the leadership-instruction condition had more hands-on time (median = 120 seconds, Interquartilrange (IQR) = 98-135) than groups in the technical-instruction condition (87 sec, IOR = 61-108). The difference is highly significant, and so are differences with regard to time to first meaningful measure, time to chest compressions, and time to ventilation; only time to defibrillation did not show a significant difference.

Thus, groups in the leadership condition performed significantly better in time-based performance measures. They also showed more leadership utterances (median = 7; IQR = 4-10) than groups in the technical condition (median = 5, IQR = 2-8); the difference is significant, and so are differences in task assignments and commands; decisions what to do were marginally significant ($p = .08$). By contrast, groups in the technical condition performed significantly better in terms of the correct arm and shoulder position; this result makes supports the conclusion that the differences were, indeed, due to the different instructions.

This study suggests that a brief leadership instruction was sufficient to induce significant changes in leadership behavior and performance four months later. In line with the results of Study 2, these results support the notion that participants did not lack basic leadership skills but that the skills they had needed to be activated either by opportunity (lack of competing demands in Study 2) or by instruction (Study 3).

In addition to the results reported in Hunziker et al. (2010), we performed a mediator analysis (see above). However, we could not confirm that the performance improvement in the leadership condition was mediated by leadership behavior. Leadership behavior did predict performance, but the coefficient for condition was not substantially reduced when leadership behavior was controlled. This result may be due to what Okhuyzen and Eisenhardt (2002) called a "spillover effect. These authors found that a brief intervention had effects beyond the behavior specified in the instruction; for instance, an instruction to attend to time management also improved questioning others, and the instruction to question others also improved attending to time management. It is therefore possible that the leadership instruction induced processes that go beyond leadership behavior; candidates might be a greater willingness to accept instructions, or more attendance to coordination requirements by the group as a whole, resulting, for instance, in implicit coordination. We are currently analyzing the data along these lines to determine what might explain the fact that leadership behavior predicted better performance but did not mediate the effects of condition.

STUDY 4

The fourth study is also refers to minimally invasive instructions, but involved physicians rather than medical students, and a brief instruction to the first responding physician just before the scenario started. A total of 34 teams consisting of three physicians each were randomly assigned a) to a standard handover of the "patient", or (b) to a standard handover followed by a brief leadership instruction (e.g., "If necessary, lead the group"). This study is not completed yet, but first analyses indicate that leadership utterances of the first responder were more frequent, cardiovascular support was given during a greater percentage of the available time, and time until the first meaningful intervention was shorter in the leadership instruction group.

OUTLOOK

The results obtained clearly indicate that team building in medical emergency groups constitutes a task in its own right, in addition to the primary task of medical treatment. In the specific scenario investigated in these studies, the diagnosis poses no special problems, and treatment is clearly prescribed by internationally accepted algorithms (Nolan et al., 2005). Therefore, quick and decisive action is required, and such behavior can be supported by directive leadership. Medical professionals typically have the social skills necessary for effective leadership, but they do not necessarily exercise these skills unless either given an opportunity that involves an absence of competing action requirements, or being specifically reminded of the need to lead by way of minimally invasive training.

Further research will have to investigate in more detail the exact mechanisms that are responsible for the effects of minimally invasive training. However, it is clear at this point that team building skills need to be included into medical training and into treatment algorithms.

REFERENCES

Carroll, J.M., 1998. *Minimalism beyond the Nurnberg Funnel*. Cambridge, MA, MIT Press.

Driskell, J. E., Salas, E., & Johnston, J. (1999). Does Stress Lead to a Loss xof Team Perspective? *Group Dynamics: Theory, Research, and Practice, 3(4)*, 291-302.

Gurtner, A., Tschan, F., Semmer, N. K., & Nägele, C. (2007). Getting groups to develop good strategies: Effects of reflexivity interventions on team process, team performance, and shared mental models. *Organizational Behavior and Human Decision Processes, 102(2)*, 127-142.

Hackman, R. J., Brousseau, K. R., & Weiss, J. A. (1976). The interaction of task design and group performance strategies in determining group effectiveness. *Organizational Behavior and Human Performance, 16*, 350-365.

Hunziker, S, Bühlmann, C, Tschan, F, Legeret, C., Schumacher, C., Balestra, G., Semmer N. K., Hunziker P, & Marsch S..(2010). Brief leadership instructions improve cardiopulmonary resuscitation in a high fidelity simulation: a randomised controlled trial. *Critical Care Medicine, 38 (4)*. 1-4

Hunziker, S., Tschan, F., Semmer, N., Zobrist, R., Spychiger, M., Breuer, M., et al. (2009). Hands-on time during cardiopulmonary resuscitation is affected by the process of teambuilding: a prospective randomised simulator-based trial. *BMC Emergency Medicine, 9(1), 3*.

Larsen, M. P., Eisenberg, M. S., Cummins, R. O., & Hallstrom, A. P. (1993). Predicting survival from out-of-hospital cardiac arrest: A graphic model. *Annals of Emergency Medicine, 22*, 1652-1658.

Nolan, J. P., Deakin, C. D., Soar, J., Bottiger, B. W., & Smith, G. (2005). European resuscitation council guidelines for resuscitation 2005: Section 4. Adult advanced life support. *Resuscitation, 67(Supplement 1)*, S39-S86.

Okhuysen, G. A., & Eisenhardt, K. (2002). Integrating knowledge in groups: How formal interventions enable flexibility. *Organization Science, 13(4)*, 370-386.

Tschan, F., Semmer, N. K., Gautschi, D., Hunziker, P., Spychiger, M., & Marsch, S. C. U. (2006). Leading to recovery: Group performance and coordinative activities in medical emergency driven groups. *Human Performance, 19(3)*, 277-304.

Weingart, L. R. (1992). Impact of group goals, task domponent domplexity, effort, and planning on group performance. *Journal of Applied Psychology, 77(5)*, 682-693.

West, M. A. (1996). Reflexivity and work group effectiveness: a conceptual integration. In M. A. West (Ed.), *Handbook of Work Group Psychology* (pp. 555-579). Chichester: Wiley.

Yukl, G. (2010). Leadership in organizations. Upper Saddle River, NJ: Prentice Hall.

Chapter 24

Decisive Action vs. Joint Deliberation: Different Medical Tasks Imply Different Coordination Requirements

Franziska Tschan[1], Norbert K. Semmer[2,]
Sabina Hunziker[3], Stephan Marsch[3]

[1]University of Neuchâtel
Neuchâtel, Switzerland

[2]University of Bern
Bern, Switzerland

[3]University Hospital Basel
Basel, Switzerland

ABSTRACT

Medical emergency teams have to act in a coordinated manner. One might assume that the behaviors necessary for good team coordination are the same across many different situations. In contrast, we hypothesized that coordination behaviors have to match task requirements, which may be different across tasks. In a medical simulator we studied performance of medical professionals in two different scenarios. In the cardiac arrest scenario requiring resuscitation, the diagnosis is clear, and task requirements are prescribed by guidelines. This situation calls for decisive action with minimal delays; it should profit from directive leadership. In the diagnostic task scenario the diagnosis was difficult, as conflicting cues were present. This calls for joint reflection and weighing all available evidence; it should therefore profit from explicit reasoning and from involving the whole group by "talking to the room". Micro-analysis of coordination behaviors revealed that clear leadership was related to performance in the resuscitation, but not the diagnostic task. In contrast, "talking to the room" was related to performance in the diagnostic, but not the resuscitation scenario.

Keywords: Medical teams; coordination requirements; group performance

INTRODUCTION

Many medical emergencies are treated by ad-hoc formed temporary action teams (Faraj & Xiao, 2006); their members may or may not have collaborated before (Fiedor, Hunt, & DeVita, 2006; S. Marsch et al., 2003; S. C. Marsch et al., 2001). Such ad-hoc teams often cannot count on previous common experiences, nor do the team members have prior knowledge about the level of expertise of the other group members; which may well influence performance (Sabina Hunziker et al., 2009; S. U. Marsch et al., 2004).

Different medical emergencies require very different skills from ad-hoc emergency teams. For example, treating a sudden cardiac arrest may not be very problematic to diagnose, and treatment is prescribed by a clear algorithm. This situation requires a team to perform highly coordinated actions under high time pressure (Nolan, Deakin, Soar, Bottiger, & Smith, 2005). In contrast, if a patient presents ambiguous symptoms in an emergency situation, the primary task of a team may be collective decision making in order to reach the correct diagnosis. This latter situation requires coordinating 'thoughts', and integrating information and expertise, whereas less emphasis is given on coordinating actions as long as the diagnosis is not clear.

TASK COORDINATION REQUIREMENTS

Given that emergency teams are confronted with different task types, different coordination mechanisms may be crucial for high performance in these different situations (McGrath, 1984; F. Tschan & von Cranach, 1996; M. J. Waller, 1999; Xiao, Seagull, Mackenzie, & Klein, 2004). Indeed, a recent study on cockpit crews (Grote, Kolbe, Zala-Mezö, Bienefeld-Seall, & Künzle, 2010) not only showed that coordination behaviors varied between phases of the flight in accordance with changing task requirements. It also revealed that the same coordination behavior (leadership) was positively related to performance in some, but negatively related to performance in other phases of the flight. This finding underlines the importance of a contingency approach in the study of group processes and group performance that relates task coordination requirements to specific coordination behavior.

Coordination-requirement analysis of group tasks (F. Tschan et al., in press) is a method based on hierarchical task analysis (Annett, Cunningham, & Mathias-Jones, 2000), which permits developing hypotheses concerning which coordination behavior may be especially important for high performance for a given task.

Coordination requirements of a cardiac resuscitation task

Coordination requirements of a cardiac resuscitation task including ventricular fibrillation (pulseless uncoordinated rapid electrical activity of the heart) can be derived from resuscitation algorithms (Nolan et al., 2005; F. Tschan et al., in press). In terms of a hierarchical task analysis, resuscitation guidelines consist of three general goals, each with several subgoals: Goal (a) refers to the diagnosis of the cardiac arrest and includes the subgoals of "checking pulse", "checking breathing" and "checking consciousness". Goal (b) is to ensure oxygenation of the brain despite the fact that the patient has no own circulation and includes the subgoals of opening airways, performing ventilation, and performing cardiac massage. Goal (c) is related to reestablishing spontaneous circulation and includes the subgoals defibrillation and administering drugs (epinephrine).

Coordination requirements for this task include rapid transitions from goal (a) to goals (b) and (c), which are performed in alteration. Further coordination requirements are a gapless alternation between (b) and (c), as well as assuring smooth coordination between the two group members who ventilate and perform cardiac massage (subgoals of b), and the coordination of all group members present during defibrillation (subgoal of c). Thus, coordination requirements of this task include coordinated and rapid changes between subgoals as well as coordinated pacing of the different activities of three or more people. It can therefore be hypothesized that direct leadership and clear task distribution are crucial for high performance of a resuscitation task.

Coordination requirements for a group diagnostic task

For a diagnostic task in an ambiguous situation, coordination requirements are very different. Coordination requirements can be derived from a task analysis of the diagnostic process. In the medical literature (Bowen, 2006; Patel, 2002; Swartz, 2006), the diagnostic task is described as consisting of three main steps that can also be seen as goals: Goal (a) is data collection, with subgoals of collecting data based on patient history, physical exams, and other information sources. Goal (b) is to integrate this information in order to generate a hypothesis about a possible diagnosis. Subgoals are integrating information, and comparing information with illness scripts. Goal (c) is to confirm the initial diagnosis; subgoals are considering and testing plausible alternative hypotheses.

Thus coordination requirements for a diagnostic task performed in a group imply explicitly sharing information and knowledge in the group. This is a prerequisite for an informed discussion aimed at integrating this information. Thus, a further coordination requirement relates to integrating information, impyling explicitly relating pieces of information to each other. Based on this sharing and integrating information, a preliminary, and then a final, diagnosis can be reached.

Previous research shows that these coordination requirements are not easy to fulfill in groups. For example, groups often do not collect expert information from individual group members and thus rely too heavily on information initially known by all group members (Stasser & Stewart, 1992; Mary J. Waller, Gupta, & Giambatista, 2004). Groups also are reluctant to explicitly plan and develop strategies (Hackman, Brousseau, & Weiss, 1976). Furthermore, in discussions, group members often simply state facts but do not relate different pieces of information to one another, thus not integrating information (cf. Kerr, 2004). This latter finding may not be restricted to diagnoses made in groups. Indeed, studies examining physicians who individually diagnosed patients found important and frequent shortcomings in the individual reasoning process (Elstein, 2002; Graber, Franklin, & Gordon, 2005), and particularly an underuse of explicit reasoning (Elstein, 2002; Eva, 2005).

A behavior that may help overcome these shortcomings in groups is "talking to the room", which has been investigated by Artman and Waller (cf. Artman & Waern, 1999; M. Waller & Uitdewilligen, 2009). In their studies of emergency situations, group members spoke in a louder voice, and did not address a specific colleague but the whole room when explaining their assessment of the situation. This behavior increases the chance of getting the attention of the whole group and also invites group members to participate in a mutual problem solving process (M. Waller & Uitdewilligen, 2009). Talking to the room has been shown to correlate with explicit reasoning, as indicated by the occurrence of words such as "therefore, because" and the like (F. Tschan et al., 2009). It can therefore be hypothesized that talking to the room is related to higher performance in groups confronted with ambiguous diagnostic tasks in emergency situations.

COORDINATION AND PERFORMANCE IN A RESUSCITATION AND A DIAGNOSTIC TASK

We present results of two studies investigating groups of medical professionals who were confronted with either a resuscitation task or an ambiguous diagnostic situation. As shown above, coordination requirements are very different for these tasks; we therefore expect that clearer directive leadership is related to higher performance in the resuscitation, but not in the diagnostic tasks. Conversely, we expect that behavior aimed at fostering the integration of information (talking to the room) is related to performance in the diagnostic, but not the resuscitation tasks. We draw on results of earlier studies as well as on additional analyses.

Resuscitation task: leadership, talking to the room and performance

In several studies, we confronted groups of medical students (S. Hunziker et al., 2010), general practitioners, nurses and hospital physicians (Sabina Hunziker et al., 2009; S. C. U. Marsch et al., 2004; F. Tschan et al., 2006) with a resuscitation task. The scenario involved a witnessed cardiac arrest: At the beginning, the patient was alert, but after two minutes, he complained of dizziness and suffered a cardiac arrest requiring an advanced resuscitation intervention. The scenario presented had the same basic structure but was adapted to the circumstances students or physicians may encounter in their daily life. Performance was measured as percentage of time the patient had no pulse and received appropriate care (cardiac massage, defibrillation, and intubation). Leadership behavior was assessed as the number or percentage of leadership utterances.

In accordance with the contentions developed above, we hypothesized that directive leadership behavior would be related to performance of groups confronted with a resuscitation task. Indeed, in all studies involving the resuscitation task, more directive leadership was related to higher performance. These results also corroborate findings from other studies (Cooper & Wakelam, 1999; Künzle, Kolbe, & Grote, 2010).

We hypothesized that behavior that aims at sharing and integrating knowledge would be less important for high performance in the resuscitation task. To test this hypothesis, we analyzed the relationship between talking to the room and resuscitation performance for twenty groups of three physicians each (general practitioners and hospital physicians) who were confronted with the resuscitation task described above.

Talking to the room was coded as all utterances that were expressed in a loud voice, did not address a specific group member but "the room", and referred to general aspects of the patient state or the procedure rather than specific aspects such as clear commands. Performance was assessed as the percentage of time the pulseless patient received adequate treatment (cardiac massage, intubation, defibrillation).

Mean group performance was 54.7% (SD = 15.5), and in 9 of the 20 groups (45%), at least one group member engaged in talking to the room. Performance of groups that did engage in talking to the room was not higher (52.5%) than performance of groups that did not engage in talking to the room (57.4%; t (18) = .692, p = .498). This result supports the hypothesis that talking to the room does not enhance performance in the resuscitation task.

Diagnostic task: talking to the room, leadership and performance.

Using the same high fidelity patient simulator, we confronted groups of physicians with an ambiguous diagnostic task (F. Tschan et al., 2009). The main goal was to relate the reasoning process to diagnostic performance. For the current paper, we ran additional analyses to test the hypothesis that in this situation, leadership would not be related to performance.

Procedure and simulation scenario

A total of 53 experienced physicians participated, working in groups of two (7 groups) or three (13 groups). In this scenario, the physicians were handed over a patient suffering from left sided pneumonia by an emergency physician who was a research confederate. Participants were given basic patient information and history, were informed that medication treatment had already started (penicillin), and were handed the patient file containing additional information. The physician handing over the patient also informed the group about a failed attempt to insert a subclavian catheter (a vein access below the collarbone).

The scenario unfolded as follows. First, the patient complained about the pain caused by the failed attempt to insert the subclavian catheter; he then started complaining about increasing difficulties to breathe, and about dizziness. The surveillance monitor showed a gradual increase in heart and respiratory rate, and a gradual decrease in blood pressure and blood oxygen saturation. Breathing sounds, which could be auscultated with a stethoscope, became more 'obstructive' (indicating possible fluid in the lung) over time, but were present on both sides. These symptoms match the typical symptoms of an anaphylactic shock (a severe allergic reaction), which was the correct diagnosis.

This situation is ambiguous because some of the symptoms (increased heart rate, low oxygen saturation and difficulties breathing) are also typical symptoms of a tension pneumothorax, a collapsed lung. As a tension pneumothorax is a possible complication after an attempt to insert a subclavian catheter (Mansfield, Hohn, Fornage, Gregurich, & Ota, 1994), and as the symptoms developed gradually, both diagnoses should be considered. However, the patient also presented symptoms that make a tension pneumothorax unlikely. First, breathing sounds were present on both sides; in case of a tension pneumothorax, they would be absent in the affected lung. Second, the patient declined being in pain, and the decrease of the patient's state was very rapid; both aspects are not typical for a tension pneumothorax. Finally, the fact that the patient was allergic was stated in the patient file.

Diagnostic performance

The physicians collaborating in a group had never worked together before participating in this study. A scripted nurse was also in the room. He answered questions, but only participated in the medical treatment when explicitly instructed by the physicians. If groups did not correctly diagnose the anaphylactic shock, two levels of help were provided. First, the confederate nurse was instructed to provide cues pointing to the anaphylactic shock diagnosis (e.g. by mentioning that the patient's skin turned red, or by drawing attention to the fact that penicillin was administered). Groups that did not find the correct diagnosis even with the help of the nurse received a phone call from the 'emergency department' informing them that the patient's wife had alerted them to the allergy.

Based on this procedure, we found that six groups (30%) declared the correct diagnosis without help from the confederate nurse; 14 groups (60%) did not find the correct diagnosis without help; of these 8 diagnosed the anaphylactic shock after having received additional information from the confederate and 6 groups only after having received an outside phone call).

Reasoning, talking to the room, leadership and performance

Based on the video-tapes and word-by word transcripts of all communication, we coded *explicit reasoning* each time two or more diagnostic information were related using conjunctions such as "therefore", "because", etc., thus explicitly relating pieces of diagnostic information to each other. *Talking to the room* was coded as a dummy variable based on whether at least once a physician communicated in a loud voice, and did not address another group member directly. Leadership was coded as a dummy variable (present / absent) based on a coding of whether one physician was taking a more dominant leadership position. All variables were double coded by two independent researchers. Cohen's kappa for each category was higher than .75.

Groups that engaged in talking to the room were significantly more likely to find the diagnosis on their own, and so were groups that engaged in more explicit reasoning (F. Tschan et al., 2009). When explicit reasoning and talking to the room were entered into a regression analysis simultaneously, talking to the room remained a significant predictor (β = .392, p = .026, one-tailed), whereas explicit reasoning was no longer significant (β = .308, p = .092, one-tailed). The two were correlated substantially (r = .51; p = .029); an index combining the two was significantly associated with performance.

We had predicted that directive leadership would not be crucial for performance in the diagnostic task. Indeed, a Chi2 test was not significant (p = .826); parametric and nonparametric correlations yielded the same result. Our hypothesis that directive leadership is not crucial for this task is supported

DISCUSSION

Based on the assumption that coordination requirements in medical emergency groups are related to task type, we assessed the relationship of (a) directive leadership and (b) talking to the room (indicating information integration) to performance in two different tasks. One task (cardiopulmonary resuscitation) was well structured and required the "coordination of acts"; the second taskinvolved an ambiguous diagnostic task requiring the "coordination of thoughts". In accordance with our hypotheses, leadership was related to performance in the resuscitation but not the diagnostic task, and talking to the room was related to performance in the diagnostic but not the resuscitation task.

Our results have implications for theory, research approaches, and application and training. Regarding theory, our results suggest that it is important for the group to match the specific coordination requirements of the task. Identifying a limited set of variables assumed to be important for all groups and all tasks will therefore be possible only on a very general level. For example, Salas and colleagues (2005) suggested a set of five basic coordination mechanisms for teams (team leadership, mutual performance monitoring, backup behavior, adaptability, and team orientation), proposing "that the coordinating mechanisms will be needed in all cases and will have little variance across the team type or task" (p. 564). We feel that this position can be upheld only if the five concepts are defined in a very general manner; for instance, by regarding talking to the room as a form of indirect leadership. Thus, we would go further than Salas et al., who state that task type, group development stage, and maturity may render some of the aspects more important; we would emphasize the specifics (i.e. directive vs. indirect leadership) and their match to task requirements more strongly.

From a research point of view, we feel it is necessary to develop better methods for defining and assessing coordination requirements of different tasks. Well-known known task typologies (e.g. McGrath, 1984; Steiner, 1972) or broad distinctions between coordination and collaborative tasks (Salas et al., 2005) may not be specific enough to derive specific coordination requirements. We believe that a task analysis that specifies coordination requirements (Annett et al., 2000; F. Tschan et al., in press) can be useful in guiding researchers to assess the pertinent coordination variables, and may also help to integrate conflicting results (Künzle et al., 2010).

Finally, if tasks indeed require different coordination behaviors, this has implications for group training. Specifically, training should go beyond teaching general skills (e.g., in terms of the big five of teamwork and include a focus on identifying conditions under which specific coordination behaviors may be useful, irrelevant, or even damaging.

REFERENCES

Annett, J., Cunningham, D., & Mathias-Jones, P. (2000). A method for measuring team skills. *Ergonomics, 43*(8), 1076-1094.

Artman, H., & Waern, Y. (1999). Distributed cognition in an emergency co-ordination center. *Cognition, Technology & Work*(1), 237-246.

Bowen, J. L. (2006). Educational strategies to promote clinical diagnostic reasoning. *The New England Journal of Medicine*(355), 2217-2225.

Cooper, S., & Wakelam, A. (1999). Leadership of resuscitation teams: 'Lighthouse Leadership'. *Resuscitation, 42*, 27-45.

Elstein, A. S., & Schwarz, A. (2002). Evidence base of clinical diagnosis: Clinical problem solving and diagnostic decision making: selective review of the cognitive literature. *BMJ, 324*, 729-732.

Eva, K. W. (2005). What every teacher needs to know about clinical reasoning. *Medical Education, 39*(1), 98-106.

Faraj, S., & Xiao, Y. (2006). Coordination in fast-response organizations. *Management Science, 52*(8), 1155-1169.

Fiedor, M. L., Hunt, E. A., & DeVita, M. A. (2006). Teaching organized crisis team functioning using human simulators. In M. A. DeVita, K. Hillman & R. Bellomo (Eds.), *Medical Emergency Teams. Implementation and Outcome Measurement* (pp. 232-257). New York, NJ: Sptinger.

Graber, M. L., Franklin, N., & Gordon, R. (2005). Diagnostic error in internal medicine. *Archives of Internal Medicine, 165*(13), 1493-1499.

Grote, G., Kolbe, M., Zala-Mezö, E., Bienefeld-Seall, N., & Künzle, B. (2010). Adaptive coordination and heedfulness make better cockpit crews. *Ergonomics, 53*, 211-228.

Hackman, R. J., Brousseau, K. R., & Weiss, J. A. (1976). The interaction of task design and group performance strategies in determining group effectiveness. *Organizational Behavior and Human Performance, 16*, 350-365.

Hunziker, S., Bühlmann, C., Tschan, F., Balestra, G., Legret, C., Schumacher, C., et al. (2010). Brief leadership instructions improve cardiopulmonary resuscitation in a high fidelity simulation: a randomized controlled trial. *Critical Care Medicine, 38*, 1-4.

Hunziker, S., Tschan, F., Semmer, N., Zobrist, R., Spychiger, M., Breuer, M., et al. (2009). Hands-on time during cardiopulmonary resuscitation is affected by the process of teambuilding: a prospective randomised simulator-based trial. *BMC Emergency Medicine, 9*(1), 3.

Kerr, N. L., & Tindale, R. S. (2004). Group performance and decision making. *Annual Reviews of Psychology, 55*, 623–655.

Künzle, B., Kolbe, M., & Grote, G. (2010). Ensuring patient safety through effective leadership vehaviour: A literature eeview. *Safety Science, 48*, 1-17.

Mansfield, P. F., Hohn, D. C., Fornage, B. D., Gregurich, M. A., & Ota, D. M. (1994). Complications and failures of subclavian-vein catheterization. *The New England Journal of Medicine, 26*, 1735-1738.

Marsch, S. C., Marquardt, K., Conrad, G., Spychiger, M., Eriksson, U., & Hunziker, P. R. (2001). The success rate of cardiopulmonary resuscitation depends on the quality of team building. *Intensive Care Medicine*, abstract.

Marsch, S. C. U., Müller, C., Marquardt, K., Conrad, G., Tschan, F., & Hunziker, P. R. (2004). Human factors affetc quality of cardiopulmonary resuscitation in simulated cardiac arrests. *Resuscitation, 60*(1), 51-56.

Marsch, S. U., Hunziker, P., Spychiger, M., Breuer, N., Semmer, N., & Tschan, F. (2004). *Teambuilding delays crucial measures in simulated cardiac arrests.* Paper presented at the Schweizerische Gesellschaft für Intensiv Medizin; 12.-15. Mai 2004, Interlaken.

McGrath, J. E. (1984). *Groups, interaction and performance.* Englewood Cliffs, NJ: Prentice-Hall.

Nolan, J. P., Deakin, C. D., Soar, J., Bottiger, B. W., & Smith, G. (2005). European resuscitation council guidelines for resuscitation 2005: Section 4. Adult advanced life support. *Resuscitation, 67*(Supplement 1), S39-S86.

Patel, V. L., Kaufman, D. R., & Arocha, J. F. (2002). Emerging paradigms of cognition in medical decision making. *Journal of Biomedical Informatics*.

Salas, E., Sims, D. E., & Burke, C. S. (2005). Is there a "Big Five" in teamwork? *Small Group Research, 36*(5), 555-599.

Stasser, G., & Stewart, D. (1992). Discovery of hidden profiles by decision-making groups: Solving a problem versus making a judgement. *Journal of Personality and Social Psychology, 93*(3), 426-434.

Steiner, I. D. (1972). *Group processes and productivity*. New York: Academic Press.

Swartz, M. H. (2006). *Textbook of Physical Diagnosis, 5th edition*. Oxford: Elsevier.

Tschan, F., Semmer, N. K., Gautschi, D., Spychiger, M., Hunziker, P. R., & Marsch, S. (2006). Leading to recovery: Group performance and coordinating activities in medical emergency driven groups. *Human Performance, 19*, 277-304.

Tschan, F., Semmer, N. K., Gurtner, A., Bizzari, L., Spychiger, M., Breuer, M., et al. (2009). Explicit reasoning, confirmation Bias, and illusory transactive Memory: A simulation study of group medical decision making. *Small Group Research, 40*(3), 271-300.

Tschan, F., Semmer, N. K., Vetterli, M., Gurtner, A., Hunziker, S., & Marsch, S. U. (in press). Developing observational categories for group process research based on task analysis: Examples from research on medical emergency driven teams. In M. Boos, M. Kolbe, P. Kappeler & T. Ellwart (Eds.), *Coordination in human and primate groups*. Berlin: Springer.

Tschan, F., & von Cranach, M. (1996). Group task structure, processes and outcome. In M. West (Ed.), *Handbook of Work Group Psychology* (pp. 95 - 121). Chichester: Wiley.

Waller, M., & Uitdewilligen, S. (2009). Talking to the room. Collective sensemaking during crisis situations. In R. A. Roe, M. J. Waller & S. R. Clegg (Eds.), *Time in Organizational Research* (pp. 186-203). London: Roudledge.

Waller, M. J. (1999). The timing of group adaptive behaviors during non-routine events. *Academy of Management Journal, 42*, 127-137.

Waller, M. J., Gupta, N., & Giambatista, R. C. (2004). Effects of adaptive behaviors and shared mental model creation on control crew performance. *Management Science, 50*(11), 1534–1544.

Xiao, Y., Seagull, F. J., Mackenzie, C. F., & Klein, K. (2004). Adaptive leadership in trauma resuscitation teams: a grounded theory approach to video analysis. *Cognition, Technology & Work, 6*, 158-164.

CHAPTER 25

Proactive Design of a Hospital Environment with Multifaceted Ergonomic Considerations

Hyojin Nam, Fiammetta Costa, Maximiliano Romero, Giuseppe Andreoni

Department of INDACO, Politecnico di Milano
Via Durando 38/a 20258 Milano Italy

ABSTRACT

This paper presents a research project in the field of Design for Healthcare. The research has been focused in improvimg User Experience in relationship with hospital services, more specific with pediatric department in one of the most important Italian hospitals. The research process has been finalized through an iterative cycle of project-evaluation work performed by students of the Industrial Design Faculty in Politecnico di Milano.

Keywords: Design for health and wellbeing, User centred design, Action Research, Healthcare

INTRODUCTION

Ergonomics intervenes in the different levels of healthcare between users and facilities, and influences its planning, designing and managing [1]. Following the new vision of health services for quality of care, ergonomic issues are becoming more and more important. Especially, pediatrics needs discreet considerations for various users like children, adolescents and family, and diverse multidisciplinary issues such as sensorial perception, visual communication, physical interaction, usability, etc.

A hospital environment includes continuous and dynamic interaction between people and their surroundings that produces physiological and psychological strain on the person [6]. Children in a hospital have anxiety and fear which sometimes are based on false expectations [2]. The children also have their own needs and preferences such as activities, a familiar and safe environment, and their parents nearby [9]. Furthermore, in a hospital, there are various environmental and systematic factors that have an impact on patient care and outcomes such as distractions, interruptions and workload to caregivers [3]. Therefore, the quality of pediatric care is affected by multifaceted factors beyond one single point of view.

The research unit 'Ergonomics and Design' of the Dept. of INDACO in Politecnico di Milano designed a built environment solution for the pediatric care of a general hospital in Milan. The pediatric department is a specialized ward for children with diabetes, obesity and other endocrinologic diseases. The goal of the project was to improve healthcare experiences and outcomes with a built environment solutions including new products. This paper describes the methodological pathway to analyze and develop new concepts of products for a new pediatric environment which is dedicated to diabetes and obesity treatment. The objective is to investigate the diverse ergonomic issues to make the pediatric experiences different.

METHODOLOGY AND PROCESS

The whole process was composed of 2 parts and was developed according to User-Centred Design guidelines; user involvement, iterative process with design/evaluation/modification cycles, and interdisciplinarity of the design team [5]. In the first part, the directions of positive changes were defined through ethnographic research and multidisciplinary discussions. In the second part, each product concept was analyzed and developed by undergraduate students of Politecnico di Milano, who attended at the courses 'Biodesign' and 'Ergonomics for Design' of the academic year 2009/2010.

From basic research and ethnographic study to design practice and ergonomic analysis, the project introduced the application of an integrated set of tools to provide users' information on explicit and latent needs [10]. To improve pediatric experiences and environments, the reality and possibility of each product were explored with multifaceted ergonomic considerations. This mixed methods approach created reliable explanation of a new pediatric environment through triangulation [7].

[1] User-Centred Design

The environment such as the pediatric ward, the ambulatory space and the emergency room were observed and analyzed. To reveal different aspects of empirical reality, patients and families, doctors and nurses were interviewed on how they work, what they need, what they have around them, how their relationships are, and so on. With these ethnographic studies, the context of the possible future solutions were identified, and the requirements and goals were defined.

Several meetings with designers, ergonomists, bioengineers, pediatricians and people in charge or purchasing and reception at the hospital helped to figure out right problems for right solutions. Several assumptions were explored such as 'What if kids were motivated with joy, not fear?', 'What if kids had active responsibility for their health?', 'What if parents were encouraged and strengthened fully?', 'What if parents had active responsibility for kids' health?', 'What if caregivers could use the tools efficiently with pleasure? and 'What if the pediatric ward was a living lab for innovation?'. Finally, about 30 new concepts were decided under 5 new groups which are 1) 'From painful moments to joyful moments', 2) 'Active responsibility for eating', 3) 'Kid-centred space', 4) 'Family-centred space', and 5) 'Efficient & cheerful caregiving'.

Students were involved to create concrete design solutions and refine the solutions with diverse ergonomic considerations. Before doing deep dives to each product, they were introduced to the objectives and context of the assigned concepts. Visiting the pediatric ward which was researched during the ethnographic studies helped the students to keep balance between the reality and possible changes. A set of research templates were provided for guiding the whole research and analysis process.

[2] From Design to Analysis

Differently from the standard process which advances from Analysis to Design, the researchers adopted an unusual process from Design to Analysis while designing products. This experiment was intended to maximize design possibilities in medical and healthcare products.

It has been often observed that many requirements and stereotypes in the healthcare field limit and restrain innovation itself. Like the Hippocratic Oath, any new changes and innovation must do no harm to users of the new design. However, it does not mean that the regulation can hinder the potentials of design. Therefore, the inversed process from Design to Analysis allowed designers to fully explore positive opportunities and find a way to consolidate promising opportunities in safe and sound ways.

[3] Involvement of students with different roles

For the design process, students from 2 different classes were involved with different roles. The students involved in the design process were majoring in Industrial Design, Interior Design, Furniture Design, Communication Design and

Fashion Design. Each student from the 'Biodesign' and 'Ergonomics for Design' courses chose a product concept from the same product pool considering his/her specialty. The 'Biodesign' students who were introduced to human anatomy and physiology, physiological signals and related clinical applications applied the knowledge and techniques into design practice. Afterwards, the 'Ergonoimcs for Design' students who learnt ergonomic tools and contents to integrate human factors in different stages of product design took part in research and analysis of the new products. The final outcomes were evaluated by various human factors and considerations for real pediatric services.

These divided roles to take the responsibility for a same product encouraged students to explore possible directions with various factors. At the same time, it also became the opportunity to investigate the relationship of research and didactics and to put in practice an action research approach [8].

In the beginning, the 'Biodesign' students were emphasized to propose design solutions exploring the concept and context in details such as 'Experience process', 'Mind & Needs, 'Relationship' and 'Body boundaries' of children and families and caregivers, and 'Market analysis'. Subsequently, the ergonomic evaluation of the proposed solutions by the 'Ergonomics for Design' students were executed following the topics below with different objectives.

- Anthropometrics: Verify dimensions according to users' anthropometric characteristics.
- Physiology & Biomechanics: Verify postures and movements according to physiologic and biomechanical characteristics of users.
- Perception & HMI: Verify commands and controls according to perceptual characteristics of users.
- Cognition: Verify principles of usability and understandings according to cognitive characteristics of users.
- Questionnaire or Focus group: Discover explicit needs of users.
- Contextual interview or Ethnographic observation: Discover latent needs of users.
- Design suggestion(Concept): Propose design revisions for improving the product.
- Design suggestion(Storyboard): Illustrate desirable scenarios for using the product.

RESULTS

As the pediatric department was a specialized for diabetes and obesity, the main target users were decided to children with the chronic diseases and their families as well as caregivers. Moreover, students were encouraged to develop products and surrounding service systems together beyond pieces of new products. This is because every new product for healthcare services should be used and operated as a component of the healthcare service system.

The final design outcomes show the proactive collaboration of ergonomics and design for healthcare. The authors present 'Table ware', one of the projects developed in the product group 'Active responsibility for eating', as an example to show the process and result. The target age of users was down to 6 -10 considering the local circumstances.

Table ware for 'Active responsibility for eating'

Table wares in hospitals are very optimized for maintenance and delivery. Meals are distributed to directly to rooms or beds. For the efficiency, physical ergonomic factors are often appreciated more than any others. In this process, patients cannot have active roles for choosing and eating foods.

Photo 1. A table ware for breakfast in a hospital

While patients with diabetes are admitted in the ward, they learn how to adjust meals and activities and how to manage blood glucose and insulin. Also patients with obesity hold down drug treatment and education for behavioral change at the same time. Considering the importance of proper eating in those diseases, the researchers devised new experiences to make fully use of eating as training and to give active responsibility to children.

Design proposal

Three students from Product design (Claudia Gomes, Daniele Greco, Rodrigo Brito) and one student from Communication design (Barbara Agostinelli) were assigned to the 'Table ware' group. The original design proposal of the group consists of a plate, a set of cutlery and a self-heating tray. The proposal aims to improve the awareness of good nutrition and healthy food consumption. This goal is to make a change in habits of everyday life, more than treatments for diseases.

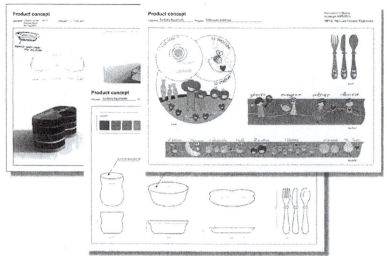

Photo 2. Some concepts from the Biodesign group proposal

Ergonomic evaluation and considerations

The evaluation was carried out by three students (Federico Biraghi, Gianni Rehkopf, Lorenzo Rossetti). It includes;

- Anthropometrics: The dimensions of the products, as the maximum diameter of the glass and the minimum dimension of the cutlery have been defined on the basis of literature data for the target user (from 5%til of 6 years old kids to 95%til of 10 years old).
- Physiology & Biomechanics: The evaluation group advised biodesigners to offer a handle point on the plate as well as on the bowl. The handle point is important to improve safe transportation in case of hot foods.
- Perception & HMI: As explained before, the aim of this project was to improve the awareness about well nutrition and consumption of healthy food. Therefore, the evaluation was focused principally on the perception of graphical signals printed on the product. The evaluators gave an opinion that "humanization" of vegetables and fruits could give a negative sense to the child consumers therefore causing the opposite result. The evaluators propose to use healthy cartoon characters to represent the advantages of correct nutrition. In addition, to divide the plate in three food zone was accepted as a good idea to represent the necessity of balanced diet.
- Cognition: Obesity is one of the most popular diseases in the pediatric department. Considering this, the goal of the product was not only to improve the quality of the food but also to reduce the quantity. The evaluators judged the proposal positively because 1) the cutlery is small, and then reduces the quantity of food for each mouthful; 2) the glass is large in proportion with highness, and then it seems more capacious; and

3) the plate, as divided in three zones and full of printed colors, seems full of food.

- Questionnaire or Focus group: as required, students performed a sort of focus group with 4 kids between 6 and 10 years old. It began with some questions about eating habits and proceeded to free discussion about their desires. Some interesting points were found in relation with eating habits of some cartoon characters. Those knowledge were useful to reinforce the idea to apply a cartoon character. Though children knew the importance of eating vegetables, most of them (4 of 5) prefer not eating them.

- Contextual interview or Ethnographic observation: To observe users in a real context allowed evaluators to find unaware needs of kids as well as caregivers. Contextual interview permited students to define the importance of cleaning easily. Other important conclusions were related to the necessity to manage the plates and bowls with only one hand (normally, the second hand is occupied with cutlery or glass).

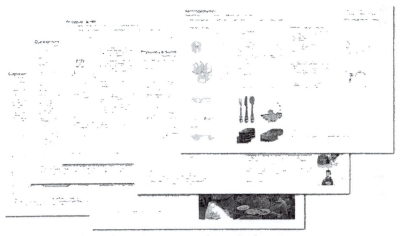

Photo 3. Templates from the ergonomic analysis

- Design suggestion: As the conclusion of the ergonomic evaluation, students were invited to suggest a new concept based on the original concept but applying ergonomics improvements. The suggestion was represented in 2 templates. One template was for the products (photo 4) and the other was for the scenario of use as storyboards (Photo 5).

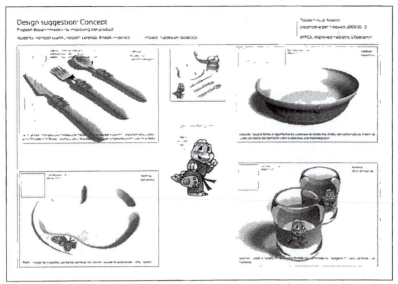

Photo 4. The concept design suggestion as a result of ergonomic analysis

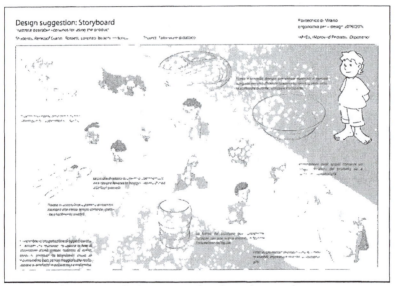

Photo 5. The concept design suggestion for using products

The final design

As is possible to see in the final design, most of the suggestions given by the ergonomics evaluators were accepted and applied. It is interesting to note that, despite changes, the essence of the products were not modified. For this reason, from methodological point of view, this process could be considered as positive.

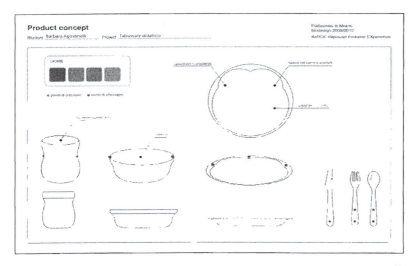

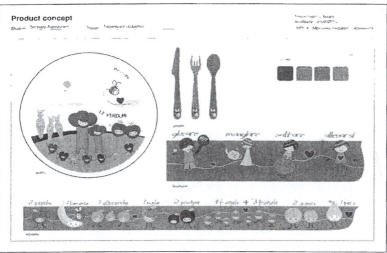

Photo 6. The final project presentation

DISCUSSION

As conclusion, it is possible to affirm that an iterative model: research-design-evaluation is recommended. This experience, performed with more than 20 groups, has given a result of improvement in final design results on the 80% of the cases. The students of "ergonomic evaluation group" have constrained the "design group" to rethink the project for paying more attention on real users' need and desire.

Both groups of students, Biodesign as well as Ergonomics for Design, have received a set of graphical templates(A3 sheets) with the entire features to analyze and design. As result of the experience, we can asseverate that the use of templates has been a good choice to arrive a wide ergonomic research.

The experimental process highlighted also critical issues which can be analyzed in further research. The first issue concerns the quality of the ergonomic evaluation which is related to the completeness of the design proposals. In this research, at the beginning of analysis, a project is often full of 'lack of information'. In many cases, the evaluators could not received enough information. In those cases, the analyses became advices. Regarding the User-Centred Design approach, a further development of the presented research could be the direct involvement of users in the design process. [4].

ACKNOWLEDGEMENT

The authors would like to express their gratitude to the students who participated in the concept design and ergonomic evaluation.

REFERENCES

[1] Baglioni, A. and Tartaglia, R. (2002), Ergonomia e ospedale. Il Sole 24 Ore S.p.A., Milano.
[2] Gorfinkle, K. (1997), Soothing your Child's Pain. Contemporary Books, Illinois.
[3] Hall, L.M, Hubley, P. and Watson, C. (2008), Interruptions and Pediatric Patient Safety. Journal of Pediatric Nursing
[4] Martin , P. and Schmidt, K. (2001), "Beyond ethnography: redefining the role of the user in the design process", Inca, 1.
[5] Norman, D.A., and Draper, S.W. (1986), User centered system design: New perspectives on human-computer interaction. Lawrence Erlbaum Associates, New Jersey.
[6] Parsons, K.C. (2000), Environmental ergonomics: a review of principles, methods and models. Applied Ergonomics, 31, 581-594.
[7] Patton, M.Q. (1990), Qualitative Evaluation and Research Methods (2nd ed.). Sage Publications, California.
[8] Reason, P. and Bradbury, H. (2001), Handbook of action research. Sage Publications, London.
[9] Runeson, I., Hallström, I., Elander, G. and Hermerén, G. (2002), Children's needs during hospitalization: An observational study of hospitalized boys. International Journal of Nursing Practice, 8, 158–166.
[10] Sanders, E.B. and Dandavate, U. (1999), Design for Experiencing: New Tools. In the Proceedings of the First International Conference on Design and Emotion, TU Delft, the Netherlands.

Chapter 26

First Do No Harm: The Unintended Consequences of Embedding Technology in Healthcare

Atif Zafar[1], Mark Lehto[2]

[1]Indiana University School of Medicine
Regenstrief Institute Inc.

[2]Purdue University
Department of Industrial Engineering

ABSTRACT

In 2000 the Institute of Medicine reported that medical errors and adverse drug events may be causing up to 98,000 deaths per year in the United States at a cost of $38 billion! They cited health information technology (HIT) as one potential solution that could enhance the safety, quality and efficiency of healthcare while reducing cost. Since that report the Federal Government has provided more than $1 billion in funding to support the implementation of HIT. However, recent evidence suggests that without consideration of human factors issues, HIT has the potential to actually *introduce new medical errors* and increase harm. We report in this paper on our experiences with implementing HIT in real-world clinical environments and the reality of unintended and often unpredictable consequences.

Keywords: Health Information Technology, Medical Errors, Unintended Consequences, Adverse Drug Events, Clinical Workflow, Human Factors.

INTRODUCTION

The Institute of Medicine (IOM) published a seminal report in 2000 entitled "To Err is Human: Building a Safer Health System" (Kohn et al., 2000). In this report they concluded that adverse drug events and medical errors may be responsible for up to 98,000 unnecessary deaths per year in the United States. The key conclusion from this was that there is no one responsible entity but that this is a *systems-wide problem*. They cited lack of public attention to the issue, failure of appropriate licensure and accreditation, the perceived threat of litigation that lets errors go unreported, the decentralized and fragmented nature of the healthcare system and the lack of incentives for improved safety as key contributing factors to this problem. They concluded that there is no "magic bullet" that could reduce the harm done but that a "comprehensive approach" is needed.

One potential solution that could help mitigate some of these issues is health information technology. Health information technology (HIT) is an umbrella term that encapsulates many technology tools, processes and policies present in healthcare today. Some of these include electronic medical records, clinical decision support systems, health information exchange networks, computerized provider order entry systems, security and privacy policies and processes for patients' "protected health information" (HIPAA), personal health records, and ePrescribing systems. One of the key advantages of HIT is that it provides for *better communication and coordination of care* in an increasingly specialized and fragmented healthcare delivery environment. The features and functions of HIT systems can lead to a reduction in duplicate testing, better handoffs between care sites and care providers, point-of-care advice about drug-drug, drug-diagnosis and drug-lab interactions, reminders about clinical practice guidelines using evidence-based medicine, checking for corollary orders using order-sets and guidance on selection of cost-effective medications and management strategies.

Despite the perceived advantages of HIT systems, implementation in real-world environments is still hampered by technical, cultural and financial barriers. Integration with legacy systems, managing workflow change and keeping costs affordable are real-world implementation issues faced by purchasers of HIT systems. Furthermore, recent evidence (Harrison et al., 2007; Ash et al. 2004) suggests that unless careful attention is also paid to human factors issues, implementation may lead to faulty systems that could cause additional harm which these systems are designed to prevent! Indeed the US Federal Government has and continues to spend significant dollars on the implementation of HIT systems as shown in the graph below:

Figure-1: Federal Health IT Expenditures in Recent Years

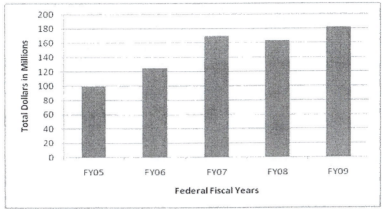

At Indiana University and the Regenstrief Institute we have studied health information technology systems for over 40 years. We have also played a pivotal role in the recent incarnation of the AHRQ National Resource Center for Health IT (http://healthit.ahrq.gov) NRC and have provided technical assistance to implementers of HIT systems around the country. The NRC was instantiated in 2004 to support the $100 million in grants awarded to front-line HIT implementers throughout the country. The goal of the NRC was to provide technical assistance to these grantees and assimilate and disseminate the best practices and lessons learned from these projects. In the discussion that ensues we share our experiences from this effort and illustrate how HIT systems may lead to unintended effects and provide case examples from real-world settings.

UNINTENDED CONSEQUENCES OF HIT

The types of unintended consequences have been categorized previously into several classes (Harrison et al., 2007; Ash, et al., 2004). Broadly speaking, these classes include (1) changes in existing workflow patterns that undermine vital interactions among care providers, (2) user-interface issues that lead to faulty or inefficient interaction with the systems for data entry and data access, and (3) changes in power-structures that reduce provider autonomy and impose practice restrictions, and which can lead to confusion among less able personnel making higher level decisions. We next describe our experiences with each of these.

First, there are interactions within an existing system that are often vital to the successful completion of an intended task. If the task network is *perturbed* by the HIT implementation then these interactions may disappear with the result that the intended task cannot be successfully completed. For example, in one large academic medical center, providers were used to looking in a basket for paper-based cardiac echo and EKG reports for their primary care patients. One day the basket disappeared and the providers were poorly trained to access the same results online. This resulted in critical echo and EKG results being ignored for a period of time until the requisite training was completed asynchronously.

In another medical setting, the nurses would alert providers to critically abnormal lab results returned by fax on paper and then would schedule the patients for urgent return visits for result adjudication. When the same critical results were re-routed for computer display, the *provider* was required to login and note the critical results. This *task-shifting* then caused delays in adjudication of abnormal results and the nurses had to be approached asynchronously for rescheduling these urgent appointments.

Second, user-interface problems abound with the current generation of HIT systems. In one esteemed medical center, a new inpatient provider order-entry system was implemented that had a major flaw. It permitted orders to be entered without first verifying the currently selected patient. In one case, a blood-thinner called Warfarin was improperly administered to a patient who was not on it. In another case a blood transfusion was given to the wrong patient resulting in an adverse reaction. Use of bar-codes on medications and strict verification systems can reduce these types of problems.

We have also previously studied the nature of clinical workflows using time-motion studies and task-analysis (Louthan et al., 2006; Overhage et al., 2001; Zafar and Lehto, 2009). Clinical workflows can be depicted as complex, intercalating networks of directed graphs of task sets. They are highly interruptive and require frequent signing-in and signing-out of a computer system for simple tasks such as results lookup or order entry. Furthermore, these workflows occur in highly mobile settings and systems that incorporate fixed, *stationary HIT systems* can impede the workflows. For example, in one study we found that "walk-time" in a busy emergency room cost the system $70,000 per year per provider (Zafar and Lehto, 2009)! In another study, the rate of patient check-ins into a primary care clinic resulted in nurses aides' sitting idle for 45% of their time due to bottlenecks generated elsewhere in the care pipeline (Zafar and Lehto, 2009). Having *mobile devices* for data access, data entry and inter-personnel communication could potentially alleviate these problems. And use of push-technologies for result notification can reduce the need to constantly log-in or log-out to check if results are available.

In another case, a specialist consult request and lab and radiology test order-entry system was in effect an *order-communication* system where orders entered electronically were not "sent anywhere" but just stored. This was because the intended receivers of the orders had strict control over schedules and would not let a "computer system change schedules". A human operator on the other end had to physically "extract the orders" and make telephone calls to set up appointments for tests and consults. This resulted in "ordered tests and consults" never occurring with the patient coming back to their primary care provider several months later without the requisite tests or consult visits completed.

Third, power structure preservation is important for maintaining workflows. In another example, an older physician shied away from entering orders in a computer and had an inpatient ward clerk enter these orders. The physician would write orders on paper and have the ward clerk "translate" them into the computer system. In many cases, these orders would not translate uniformly and slightly different orders were entered. This resulted in delays in care and missed opportunities for treatment. This is an example of a shift in power-structure where a less competent person is making decisions about how the orders translated into the order types the order entry system could understand.

These examples highlight the need to perform a comprehensive human factors analysis prior to implementing HIT systems. This entails understanding the entire task pipeline or network, understanding roles and responsibilities and then mapping out how the new HIT system would be overlaid on the task networks and role maps. Replacing a human operator by a computer equivalent often equates to a shift in the power structure and unless the newly assigned roles in this shift can be adjudicated, accepted and understood errors will undoubtedly occur. In the next section we look at some guiding principles for HIT system implementation that can help mitigate some of these problems.

IMPLEMENTATION STRATGEIES FOR HIT

There are some simple guiding principles that can be applied during the technology needs assessment and implementation processes that can help mitigate problems down the road. These principles are time tested in the medical informatics literature and can prove useful when dealing with the complex cultural, financial and technical challenges encountered during HIT implementation.

First, it is helpful to view the implementation process as a cycle of needs assessment, implementation and evaluation. Technology implementation should be driven by a need to solve real-world problems and not as an end in itself. The key types of problems for which HIT solutions exist today include the following:

1. Inadequate communication between providers – lack of relevant clinical data at the point of care in a fragmented healthcare delivery system – leading to duplicate testing, increased cost of care, missed diagnoses and poor patient satisfaction.
2. Inadequate knowledge of current clinical practice guidelines and research results leading to poor quality and geographic disparities in care.
3. Lack of knowledge about drug-drug, drug-lab and drug-diagnosis interactions and inadequate monitoring of treatments leading to adverse drug events.

Second, formative and summative evaluation is a critical ingredient for success, both for the technology selected and for the implementation process itself. The evaluation drives the quality and process improvement in a continuous manner. Evaluation strategies exist for many types of HIT implementations (Cusack et al. 2007-2009), ranging from simple electronic medical record systems to full regional health information exchange. Evaluation provides the answer to the question of need assessment and whether the technology solution is actually doing what it was intended for. Evaluation need not be rigorous (such as a randomized controlled trial of interventions) but in most cases a simple before-after observational study will suffice.   And evaluation should encompass both quantitative *and qualitative* measures of performance. Satisfaction by users is as important as cost-reduction or medical errors prevented.

Third, understanding the current workflows using a human factors approach and predicting how the HIT system would interact with, augment or replace elements of the workflow is of critical importance. An HIT system should be well integrated into the existing workflow or it will not be accepted (Poon et al, 2004). Furthermore, the degree to which the information provided by an HIT system is well integrated into the workflow can influence the quality of provider decision making (Wu, et al., 2009). Understanding workflows requires tracking of data elements, task lists and roles and responsibilities over time along with the relevant technology pieces that are interfaced with at each step. Task analysis is an important part of this. Inefficient processes that are currently done manually should *not* be automated as they will likely result in bottlenecks in the workflow that could prove difficult to circumvent. Rather these processes should be *replaced* by more efficient automatic processes. An example of this is when handwritten medication prescriptions are replaced with electronic versions. In one environment, the new HIT system allowed providers to type prescriptions into the system that would then print out on paper. The clinic staff still had to hand-carry these prescriptions to the pharmacy, and in many cases would lose the scripts. By replacing the printed prescriptions with an ePrescribing system that would send electronic messages to the pharmacy this problem disappeared.

Fourth, involve the users early! This cannot be stressed enough. There are numerous cases of spectacular systems failures because users were "surprised" by how the systems were implemented – in one case a $20 million HIT system had to be turned off after 2 months of use because the users complained so much. The stakeholders include not only the healthcare providers but all of the other support staff that may interface with the system at one time or another. Bringing all stakeholders together at the outset helps offset cultural issues down the line.

Fifth, start small and pilot test implementations before going live all over your care environment. A small set of dedicated users and promoters can go a long way towards enticing less enthusiastic users to try the system. In one case, a senior physician used to work with a nurse practitioner who was an enthusiastic user. The senior physician had very little computer experience so the nurse practitioner was the on-the-ground trainer for the physician.

Sixth, understand the data flows. You should account for how every piece of data entered into a system (or retrieved from a system) gets from point A to point B. Data quality issues abound in electronic systems and as the adage suggests "garbage-in, garbage-out". Alert fatigue is a very real issue, where too many "warnings" are displayed by a clinical decision support system, that are based on incorrect or out-of-date information and this irritates and frustrates users so they ignore all warnings, including the critical ones. As an example, at one hospital, the system would alert providers to provide flu shots for many of the eligible patients in the system, despite the fact that these patients already had their flu shots. What had happened was that these shots were being administered at another site and the information about these "events" was not uploaded into the database. As a result of this incomplete data, a faulty alert was being generated that irritated users. In another case, medications and diagnoses from several years ago still persisted in the system so incorrect alerts were being generated regarding drug-drug interactions for drugs that had long been discontinued.

CONCLUSION

Health Information Technology has the potentially to dramatically reduce the rates of medical errors and adverse drug events in this country and help reduce the cost and improve the quality of care. However, unless properly implemented, these systems have the potential to cause harm. A rigorous human factors analysis is a critical process needed during the implementation phase to ensure that such harm does not occur. Some commonsense principles discussed in this paper can be applied during planning and implementation to mitigate problems and improve success rates.

REFERENCES

Ash J.S., Berg M., Coiera E., (2004), "Some Unintended Consequences of Information Technology in Healthcare: The Nature of Patient Care and Information System-related Errors", *J Am Med Inform Assoc*, 2004;11:104-112.

Cusack, C., Poon E. G.. (2007-2009), "Health IT Evaluation Toolkit", "HIE Evaluation Toolkit", *AHRQ National Resource Center for Health IT Website* (http://healthit.ahrq.gov)

Harrison, M., Koppel, R., Bar-Lev (2007), "Unintended Consequences of Information Technologies in Health Care – An Interactive Sociotechnical Analysis", *J Am Med Inform Assoc.* 2007;14:542-549.

Kohn, L.T., Corrigan, J.M., Donaldson, M.S., (2000), "To Err is Human: Building a Safer Health System", *Committee on Quality of Health Care in America*, National Academy Press, Washington D.C., 2000.

Louthan M, Carrington S, Bahamon N, Bauer J, Zafar A, Lehto M.(2006) , "Workflow Characterization in a Busy Urban Primary Care Clinic", *Proc AMIA Symposium,* 2006:1015, Washington DC.

Overhage, J., Perkins S., Tierney, W., McDonald, C., (2001), "Controlled Trial of Direct Physician Order Entry: Effects on Physicians' Time Utilization in Ambulatory Primary Care Internal Medicine Practices", *J Am Med Inform Assoc,* 2001;8(4):361-71

Poon, E.G., Blumenthal, D., Tonushree, J., Honour, M.M., Bates, D.W., Kaushal, R., (2004). "Overcoming Barriers to Adopting and Implementing Computerized Provider Order Entry Systems in U.S. Hospitals", *Health Affairs,* 23(4):2004;184-190

Wu, S., Lehto, M.R., Yih, Y., Saleem, J,J., Doebbeling, B., (2009), "On Improving Provider Decision Making with Enhanced Computerized Clinical Reminders", chapter in *Lecture Notes in Computer Science*, Springer: Berlin / Heidelberg, 569-577.

Zafar, A., Lehto, M., (2009 – in review), "Workflow Characterization in Busy Urban Primary Care and Emergency Room Settings: Implications for Clinical Information System Design", *International Journal of Industrial Ergonomics* – In Review.

Chapter 27

How to Prevent Medication Errors

Fumito Tsuchiya

Tokyo Medical and Dental University
Dental Hospital Pharmacy

ABSTRACT

Many accidents and incidents are occurred in the medical treatment. The medication is related to many of these errors. How to prevent medication errors are big problems in the world. But in JAPAN, not only government and pharmaceutical companies, but also researchers made grate efforts for preventing medication errors. We developed the drug name similarity checking systems, and pharmaceutical companies used this system before approval. Other efforts for improvement of Labels of drugs reduced medication errors. This paper introduces these efforts and results of ergonomic and other field studies.

Keywords: Patient Safety, Medication errors, drug-name, safety-usage

INTRODUCTION

Operational accidents due to mixing up of patients happened in 1999 in Japan have immediately raised national interest about medical accidents. Coincidentally in the U.S., the book titled, "To Error is Human" was published and the president Clinton (at the time) had put up the goal to reduce medication errors into half in five years. These incidents indicate that people all over the world has started to show higher interest in medical accidents. Also what's common throughout the world is that enough researches on human errors in medical field have not been made due to the closed nature of medical world though human errors must be occurring in this field as well as in other fields.

Not only in Japan, but there have been a very large number of medicinal product-related medical accidents and incidents cases in various countries throughout the world. This is entirely because the present-day medical services center around drug therapy; and because during the process from prescription of ethical drugs to taking of the drugs, a very large number of people are involved, such as doctors, pharmacists, nurses, and patients, and hence various types of interactions occur between these "people" and medications or "substances."

In the past, concerning the "safety of medications," only the safety of medications themselves has been demanded, laying stress on the prevention of adverse events; and for many years, a great deal of effort has been exerted for ensuring this kind of safety. However, with regard to the safety for the prevention of human errors when using medications in terms of names, packaging, etc. of medications, almost no interest has been shown worldwide, and it was not until ten years ago that various kinds of investigations in this area had begun to be conducted. Currently, the safety evaluation in this aspect is still in a fundamental study stage.

In Japan, in a report titled "Comprehensive Measures for Promotion of Medical Safety" submitted by the "Committee on Medical Safety Measures" which was set up in the Ministry of Health, Labour and Welfare(MHLW) in 2002, two concepts were defined: the safety of medications themselves as the "safety of substances", and the safety when using medications as the "safety of use"; and for particularly the latter, it was pointed out that governmental agencies, medical institutions, pharmaceutical companies, etc. should take a variety of measures in their respective positions.

In order to ensure the "safety of use" of medications, MHLW set up the "Sub-committee on Pharmaceuticals and Medical Device" and the "Sub-committee on Error by Human Factors" as subordinate organizations of the Committee on Medical Safety Measures. The purpose of the former subordinate organization is to investigate regarding creation of concrete measures for preventing occurrence of accidents by changing names, packages, etc. of medications; and the latter organization conducts investigations of schemes for preventing human errors including organizational structures.

Additionally, in 2002, with the aim of preventing medical accidents, a network project for medical safety measures to collect information on Incidents cases that occurred at medical institutions was initiated by MHLW. As this system enabled collection of various Incidents cases from medical institutions, measures for medications and preventive measures against human errors have begun to be investigated by analyzing these cases.

Particularly with regard to medications, the "Working Group on the Similarity of Pharmaceuticals" was set up under the "Sub-committee on Pharmaceuticals and Medical Device" in 2003, and concrete measures were investigated in five areas: "Specification," "Similarity of names," "Similarity in appearance of injection drugs," "Transfusion," and "Ophthalmic drugs." At this Working Groups (WGs), in order to investigate preventive measures against accidents not only from the clinical viewpoint but also from the standpoint of human engineering, members were organized in a way that specialists in human engineering must be included in each WG.

In 2004, a report by WGs was submitted, and in response to it, schemes for preventing accidents were shown by the MHLW to pharmaceutical companies and medical institutions.

In 2005, to supplement the "Comprehensive Measures for Promoting Medical Safety," a WG was set up and a report titled "Regarding Future Medical Safety Measures," and then in response to it, the Medical Service Law was revised. The revision will be put into effect in April 2007. In this way, the governmental agencies, medical institutions, pharmaceutical companies, etc. came together and made efforts to ensure medical safety, and as a result, we have begun to see some changes in the details of medical accidents in Japan. In the beginning, markedly crude accidents and the like continuously appeared, however in recent years, as seen in accidents caused by very poor working conditions or a shortage of the staff in medical services, or accidents occurring due to insufficient acquisition of skills, instances due to complications, and the like, it can be said that there have been qualitative changes to accidents due to environments or reasons peculiar to medical services. Many types of sensors or alarming devices have begun to be installed in various medical equipment including order entry systems with the aim of preventing medical accidents.

However, this mechanization has led to the occurrence of new accidents. Based on medical accidents that occurred via order entry systems, this paper examines the measures taken to prevent the occurrence of such accidents in terms of both the system and "safety of use" of medications.

MEASURES

LOOK-ALIKE AND SOUND-ALIKE DRUG NAME

"Drugs Name Similarities" are big problems. These similarities are classified into two types: Look-alike and Sound-alike. Table 1 shows the case of similarity related problems in Japan and Table 2 shows the ones in the U.S.

Table 1 Case example of similarity problem occurrence
(Japan:Two-bites characters)

Stem1	Stem2
サクシゾン	サクシン
ウテメリン	メテナリン
アルマール	アマリール
タキソール	タキソテール

Table 2 Case example of similarity problem occurrence
(U.S.:One-bite character)

Stem 1	Stem 2
Lamisil	Lamicel
Accutane	Accolate
Sinequan	Singulair

Traditionally prescriptions have been written by hand. But with the spread of order entry system at hospitals or introduction of receipt issuance system at medical clinics, prescriptions now have been issued through such systems; in other words, issuance of prescriptions in printed letters have begun to gain popularity. However on the other hand, at hospital wards most of the instructions are given in a hand writing form, thus the problem of illegibility still remains. In Japan in the past, normally English letters were used for drug names, but today the use of Katakana is on the increase. Katakana is a phonogram, therefore look-alike and sound-alike, attributed similarity problems related to drug names, are almost the same in its characteristic. This is the difference from the U.S. where they have to come up with the measures for each one of them.

The relationship between drug names and human errors was pointed out in the presentations by Tsuchiya at The International Ergonomics Association (IEA 1982). Also in Japan, the studies have been continued led by Tsuchiya and colleagues since then. However, in-depth studies about the "similarities" done today are very few in number, though the cases of "errors" are reported.

In Japan, accidents due to order entry systems, "miss-selecting of Drug-Name" are occurred. I conducted research regarding the relationship between the number of entered letters and the number where the brand name of drug is identified; and published that "the probability of one brand name being obtained through an entry of 2 letters is 11%, through an entry of 3 letters is 67%, and through an entry of 4 letters is 91%", and proposed that at least 3 letters should be entered when considering both convenience and safety. In response to this proposal, national university hospitals are now obliged to use the 3- letter method as the entry method; and the Japan Council for Quality Health Care (JCQHC) began to check whether the entry of 3 letters or more is used as an evaluation item for hospital functions.

Development of the system to objectively evaluate similarities of drug names

In Japan, the number of ethical drugs which can be used at medical institutions is more than 20,000. The similarity of names of ethical drugs is the cause of errors in selecting ethical drugs on computers as stated above, or errors made by pharmacists when preparing drugs.

Tsuchiya, et al. (2001) developed a technique to quantify the similarity of a pair of drug names(Fig1). The purpose of this study was to test whether the probability of confusion errors is related to the similarity indices calculated by their formulas: "cos1", "htco", and "edit". The developed system was decided to be practically used through the pilot study at the Japan Pharmaceutical Information Center (JAPIC).

There is now a mechanism that when a pharmaceutical company submits an application for approval of a new drug, the name can be checked by using this system to avoid the similarity of names. Pharmaceutical companies are not under any obligation to check the name by using this system, however, most pharmaceutical companies are currently using the system to avoid a post-marketing problem of the similarity of names in the market. Also, with regard to the decision concerning the similarity, the flowchart approved at the above-mentioned WG on the similarity of names is open to the public, and pharmaceutical companies can judge whether or not a name change is necessary by applying values for each item shown in the search system to this flowchart. This system is now available by website (https://www.ruijimeisyo.jp).

Fig 1 system to objectively evaluate similarities of drug names (examples)

224

Fig 2 Website of Similar name search system

PRESCRIPTION AMOUNTS OF DRUGS AND THE PREVENTIVE MEASURES

Change the design of Drug label

In Western countries, with regard to oral drugs, tablets or capsules are used in most cases, but in Japan, we often use the powdered drugs. Regarding powdered drugs, there are multiple specifications such as those for active pharmaceutical ingredients (the volume of content is 100% drug substance), some-fold powder (thinned drug substance), and hence it may be uncertain as to which dosage form was used for the amount entered in the prescription. As a result, this may cause a quantitative error. A 100-fold powder has a lower concentration than a 10-fold powder (the figure before-fold represents the degree of thinning drug substance; a 10-fold powder means a powder thinned by 10-fold). This is common knowledge in the area of pharmacy, but not necessary common knowledge among doctors. At a particular university, when medical trainees were given a question about which one contains a higher concentration of drug substance, a little more than 10% of the trainees answered that a 100-fold powder contains a higher concentration. In order to avoid such misunderstanding, MHLW changed the way of labeling the concentration from "-fold" to "%."

However, with the labeling of %, it is quite difficult to instantly and accurately judge the amount of drug substance contained in 1 g. There have been quite a few instances where entry errors were prevented by a measure where the labeling for specification has been changed to "XX mg/g" (Fig.2), rather than using "%" to prevent entry errors for prescription amounts of powdered drugs. Because such a calculation mistake is frequently seen when pharmacists prepare drugs, MHLW, in the light of the report by the above-mentioned WGs, gave a notification to pharmaceutical companies that "XX mg/g" should be shown near the trade name even when "%" is included in the specification for the trade name. Also, at many medical institutions, a note such as "As the amount of preparation" is added after the amount of prescription in order to make clear whether the prescription amount entered is for the amount of drug substance or the amount of preparation.

 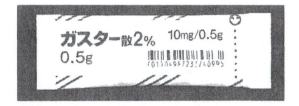

Fig 2 Label of powder drugs

MHLW required the same rule for the injection drug labels, so drug name contains strength as "%" have to display "XXmg/YYmL" (Fig3).

Fig.3 Label of Injection drugs

Change the design of Drug label

Dug name printed in label is very small , so we cannot recognize the drug name immediately. In some study, pharmacists and nurses looked the drug labels only on/two seconds, so the design of label is required "emphasis of items". Recently Japanese pharmaceutical company made a effort as Fig4.

Fig.4   Example of changed Design of Injection drugs labels
(Left: Old design,   Right: New design)

CHANGE THE PRESCRIPTION DOSAGE WRITING RULE

In Jan 2009, WHLW changes the rule of prescription dosage writing rule. As the former chapters mentioned, we have powder drugs, so, Japanese style dosage written in prescription was daily dosage. Almost countries the medical doctors write the dosage as unit dose. But in Japan we use the daily dose not only prescription but also most of package inserts. But developments of new drugs are globalized, and clinical trials are same conditions, so recently new drugs dosages are written by unit doses. And communication errors due to dosage confusion ware occurred, so "Committee on Medical Safety Measures" required to study for the purpose of avoiding communication errors. After four years study, they proposed to change the rule of dosage, from daily-dose to unit-dose. This new rules starts new school year of 2010, and MHLW thinks a dual circulation periods are about five years.

CONCLUSIONS

In Japan, not only government and pharmaceutical companies, but also researchers made grate efforts for preventing medication errors. It is important to point out that as good results have been produced from WGs and the like set up by the MHLW for the preventive measures against accidents, various approaches have begun to be made to ensure the "safety of use" for the areas of human engineering, psychology, etc. Also, data analyses have begun to be performed by using the method of data mining or text mining, and though a little too late, the basic research in this field is beginning to be conducted.

There has been practical research that conducted eye-tracking analysis using eye mark cameras to change labels or PTP package designs, etc. In addition, a transfusion manufacturer has conducted an investigation from the viewpoint of human engineering to change designs for transfusion products of the said company, and prepared documents for medical institutions which included the results of the investigation. It has been drawing attention as a way to show how a pharmaceutical company makes approaches to ensure the "safety of use."

Medical accidents related to medications occur not only in Japan but also in many countries throughout the world. It can be said that "ensuring the safety of use" should be a common theme worldwide, but it is no exaggeration to say that the research system in Japan is at the forefront. Therefore, there may be a high possibility that a "proposal for ensuring the safety of use" can be presented from Japan to the world in the future.

As an actual situation, medications may have been designed with an idea that they be handled by doctors, pharmacists, nurses, and other specialists, and hence, amateurs would not handle them. As mentioned at the beginning of the paper, in Japan, safety has been discussed as the "safety of substances," but over the past 10 years, the "safety of use" has begun to be pursued. In this sense, we are still at a stage where people have begun to understand fundamental research. It is greatly expected that research in this field will develop prosperously in the future

REFERENCES

[1] Tsuchiya, F., Tsuchiya, I.: A Study of Human-error Causing by Drug Naming. Proceeding of IEA'82, Tokyo (1982) 64-65

[2] Tsuchiya, F., Kawamura, N.:Development of system that objectively judges similarity of medicine name, Proceeding of 15th Triennial Congress of the International Ergonomics Association, Vol.5. Seoul (2003) 506-509

[3] Yamade, Y., Haga, S., Shin, H., Tsuchiya, F.: Experimental Study on Similarity of Drug Names and Confusion Errors, Proceeding of 15th Triennial Congress of the International Ergonomics Association, Vol.5. Seoul (2003) 693-695

[4]F. Tsuchiya: "Malpractice prevention and ideal way of packaging of medical products and display." PHARM TECH JAPAN, Jiho, Vol.19, No.11, pp.27–37, 2003 (in Japanese)

Chapter 28

Analysis of Questionnaire on Appearance Similarity of PTP Sheets

*Akira Izumiya[1], Yoshitaka Ootsuki[1], Masaomi Kimura[1],
Michiko Ohkura[1], Toshio Takaike[2], Fumito Tsuchiya[3]*

[1]Shibaura Institute of Technology

[2]AstraZeneca

[3]Hospital, Faculty of Dentistry, Tokyo Medical and Dental University

ABSTRACT

In recent years in Japan, an increasing number of malpractice cases have been reported, more than half of which are related to drug labels or packages [1]. The Japanese Ministry of Health, Labor and Welfare is now using a similar name search engine to ban new names for medicines that resemble existing names. However, no such action has addressed the issue of appearance similarity. We carried out a questionnaire evaluation with 412 pharmacists. They looked at images of imaginary PTP sheets and then replied with the names of real medicines having PTP sheets that they thought resembled the displayed sheets in appearance. This article presents our analysis of the results.

Keywords: appearance similarity, evaluation method, medical accident, PTP sheet

INTRODUCTION

In recent years, many medical accidents have been reported, including mistakes made by medical personnel who were confused by pharmaceutical packages and displays [2].

The Japanese Ministry of Health, Labor and Welfare is now using a similar-name search engine to ban new names for medicines that resemble existing names. However, no such action has addressed appearance similarity. We previously concentrated on the appearance similarity of the PTP sheets wrapped around tablets, and we are continuing this research [3].

We conducted a questionnaire study with pharmacists, where they gave the names of other medicines whose appearances resembled the imaginary PTP sheets shown in the questionnaires. We analyzed the front and back sides of the PTP sheets separately.

QUESTIONNAIRES

Figure 1 shows an example of the imaginary PTP sheets. Pharmacists looked at these images and provided the names of medicines whose appearances resembled the sheets they were shown. Nine imaginary PTP sheet images were used. Table 1 shows the questionnaire used for each imaginary PTP sheet. A lot of 412 pharmacists answered the questionnaires.

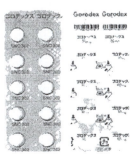

Fig.1 Example images of an imaginary PTP sheet

Table 1 Questionnaire items

No.	Question
1	Do you know of a medicine whose PTP sheet resembles this PTP sheet?
	If so, what medicine?
2	If they are similar how?

ANALYSIS OF FRONT SIDE OF PTP SHEETS

Separate analyses were made for the front and the back sides of each PTP sheet. This section reports the front side results.

METHOD OF ANALYSIS

We compared the shown images and the images of the PTP sheets of medicines named by pharmacists to examine their similarities according to the attributes shown in Table 2. These attributes are described in detail as follows:

The details of these attributes are explained as follows:

(1) Color of sheet
   The color of the front side of the PTP sheet.

(2) Color of tablets
   The color of the tablets in the PTP sheets.

(3) Shape of tablets
   The shape of the tablets in the PTP sheets.

(4) Printing positions and text contents
   Printing positions and text contents in four parts on the front side of a PTP sheet as shown in Fig. 2.

(5) Colors of printing
   Colors of printed on the four parts of the front side of a PTP sheet (Fig. 2).

Table 2 Attributes(front side)

No.	Attribute
1	Color of sheet
2	Color of tablets
3	Shape of tablets
4	Printing positions and text contents
5	Colors of printing

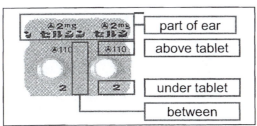

Fig.2 Four parts of front side of PTP sheet
for examining contents and colors.

RESULTS

Tables 3 and 4 show the questionnaire results. "Number of medicines named" in Table 3 shows the number of medicines named by the pharmacists as having PTP sheets with a similar appearance as those shown in the questionnaire. Figure 3 shows an example of the front sides of a pair of a displayed imaginary PTP sheet and that of a medicine named for its similarity in appearance.

These two example sheets have the following factors of similarity:

-Color of sheet: gold
-Colors of tablets: white
-Shape of tablets: circular
-Color of printing: blue

After comparing each imaginary PTP sheet in the questionnaire with the PTP sheets of all real medicines named by pharmacists as having similar appearance to that imaginary PTP sheet, we obtained the features of similarity. Table 5 shows the results for the imaginary PTP sheet of imaginary medicine A.

For each factor of similarity in Table 5, "number of similar sheets" shows the number of medicines whose PTP sheet was named as similar by the respondents. "Number of pharmacists" shows the number of respondents who named any medicine with similar-appearing PTP sheets for each factor. "Ratio" shows the ratio of the number of the pharmacists who named similar medicines to the number of responding pharmacists. From Table 5, as well as the tables for the other imaginary medicines, the following results were obtained:

-The ratios of tablet color (2) and tablet shape (3) are highest.
-The ratio of the combination of factors (2) and (3) is higher than any other combinations of two factors.
-The ratio of the combination of factors (1), (2), and (3) is higher than any other combination of three factors.

Table 3 Number of medicines.

Imaginary medicine name on imaginary PTP sheets	Number of medicines
medicine A	10
medicine B	36
medicine C	15
medicine D	15
medicine E	22
medicine F	8
medicine G	1
medicine H	4
medicine I	8

Table 4 Partial results of questionnaire.

Imaginary medicine name on imaginary PTP sheet shown in questionnaire	Name of medicine pointed out	Number of pharmacists who pointed it out
medicine A	medicine A1	103
	medicine A2	15
medicine B	medicine B1	46
	medicine B2	13
medicine C	medicine C1	65
	medicine C2	63
medicine D	medicine D1	21
	medicine D2	10
medicine E	medicine E1	124
	medicine E2	6
medicine F	medicine F1	8
	medicine F2	2
medicine G	medicine G1	19
medicine H	medicine H1	24
	medicine H2	1
medicine I	medicine I1	11
	medicine I2	6

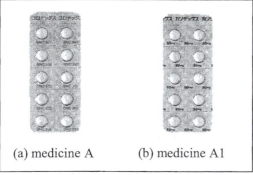

(a) medicine A   (b) medicine A1

Fig.3 Example for pair A.

Table 5 Example of analysis results form questionnaire (Medicine A).

Similarity factor	Number of similar sheets	Number of pharmacists	Ratio
1. Color of sheet	31	106	95%
2. Color of tablets	35	111	100%
3. Shape of tablets	35	111	100%
4. Color of printing	15	83	75%
1*2	31	106	95%
1*3	31	106	95%
1*4	13	80	72%
2*3	35	111	100%
2*4	15	83	75%
3*4	15	83	75%
1*2*3	31	106	95%
1*2*4	13	80	72%
1*3*4	13	80	72%
2*3*4	15	83	75%
1*2*3*4	13	80	72%

DISCUSSION

From the analysis results for all nine imaginary PTP sheets, three important factors of similarity emerged:

-Color of sheet
-Color of tablets
-Shape of tablets

We also confirmed that the printing positions and text contents were not such important similarity factors. The factors of appearance similarity were thus clarified by the analysis results of this questionnaire.

ANALYSIS OF BACK SIDE OF PTP SHEETS

METHOD OF ANALYSIS

We compared the images of the PTP sheets for the imaginary medicines shown in the questionnaire and those for the real medicines named by pharmacists to examine the similarities based on the attributes shown in Table 6. The details of these attributes are explained as follows:

(1) Color of sheets
   The color of the back side of the PTP sheet.
(2) Color of printing
   Color printed on the back side of the PTP sheet. In the case of two colors, it is the color other than black. (Fig. 4)
(3) Number of lines in repeated unit
   The number of lines in a repeated unit of printing on the back side of the PTP sheet. Figure 5 shows an example. In this case, because one repeated unit of printing contains two lines, this value is 2.
(4) Total number of printed items
   The total number of printed items on the back side of the PTP sheet.
(5) Printed contents of part of ear
   The printed contents in part of ear.
(6) End line"
   The line or figure at the bottom of the back side of the PTP sheet.

Table 6 Attribute(back side)

No.	Attributes
1	Color of sheet
2	Color of printing
3	Number of lines in repeated unit
4	Total number of each printed items
5	Printed contents of part of ear
6	End line

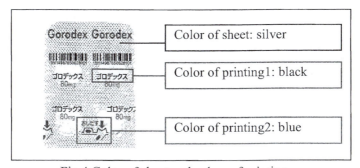

Fig.4 Color of sheet and colors of printing.

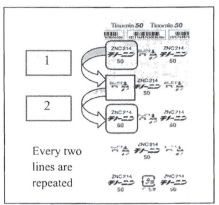

Fig.5 Number of lines in repeated unit.

RESULTS

For the back side of PTP sheets, analysis procedures similar to those for the front side were carried out. Table 7 shows the ratio of each attribute, where we took the weighted average of the number of respondents. From the table, we can find two highly significant factors of similarity between the imaginary PTP sheets shown and the similar-appearing PTP sheets of real medicines named by pharmacists:

-Color of sheet
-Color of printing (color other than black)

As for the color of printing, we defined the "dominant color of printing" for the case of two printed colors, one of which is black. If the ratio of the printing area of black is higher than that of the other color, the "dominant color of printing" is black. On the other hand, if the ratio of the other color is higher than that of black, that color is the "dominant color of printing."

Table 7 Analysis results from questionnaire.

Attribute	Ratio
Color of sheets	100.00%
Color of printing (black)	25.15%
Color of printing (other color)	76.25%
Number of lines in repeated unit	13.27%
Printed contents of part of ear	31.27%
Number of medicine's name	8.70%
Number of contents	7.67%
Number of identification code	0.00%
Number of company name	0.00%
Number of care-mark	21.09%
Number of PT-mark	6.49%
End line	61.46%

DISCUSSION

The color of the back side of all PTP sheets we showed this time was silver, which is the case for most of the PTP sheets used in Japan. Therefore, we could not determine whether the back-side color is actually an important factor of appearance similarity.

As for color of printing, we obtained the following results:

-In the case of one printed color, many medicines whose PTP sheets have the same printed color were pointed out by the pharmacists.
-In the case of two printed colors, one of which is black, many medicines whose PTP sheets have the same "dominant color of printing" were named.

Since such attributes as the number of lines in repeated units and the number of contents showed small ratios, the printing position was not an important factor of appearance similarity. Consequently, for the back side of PTP sheet, the results suggested that "the dominant color of printing" was an important factor of appearance similarity. In addition, our evaluation method using a questionnaire is considered useful because it could work well to clarify the factors of appearance similarity.

CONCLUSIONS

We administered a questionnaire evaluation on appearance similarity of PTP sheets with pharmacists as respondents. The pharmacists gave the names of medicines whose PTP sheets resembled the imaginary PTP sheets shown in the questionnaire. From our analysis of the results, we made two key conclusions regarding the appearance similarity factors on the front side of the PTP sheets:

-The color of the sheet and the color and shape of the tablet are important factors of appearance similarity.
-The printed contents and their positions are not important factors of appearance similarity.

Furthermore, we made two key conclusions regarding the appearance similarity factors on the back side of the PTP sheets:

- The "dominant color of printing" is an important factor of appearance similarity.
- The position of printing is not an important factor of appearance similarity.

In the future, we will continue to conduct research toward establishing evaluation indexes of appearance similarity.

ACKNOWLEDGMENT

We thank all of the pharmacists who answered our questionnaire.

REFERENCES

[1] Linda T. Kohn, Janet M. Corrigan, and Molla S. Donaldson: TO ERR IS HUMAN, Committee on Quality of Health Care in America, Institute of Medicine, 2000
[2] F. Tsuchiya: "Malpractice prevention and ideal way of packaging of medical products and display." PHARM TECH JAPAN, Jiho, Vol.19, No.11, pp.27–37, 2003 (in Japanese)
[3] Y. Ohtsuki, A. Izumiya, M. Ohkura, F. Tsuchiya: "Examination of method of display medicine to prevent human error (X) –Examination of an appearance similarity evaluation method of PTP sheet-." Japan Ergonomics Society, Vol.44, No. extra, pp.76–66, 2008 (in Japanese)

Chapter 29

Information System for Health: a Proposal of an Animated Instruction Guide to Design Medicine Inserts in Brazil

Carla Galvão Spinillo, José Marconi B. de Souza, Larissa Lívia R. Barbosa

Department of Design
The Federal University of Paraná
Curitiba, PR, 80060-150, Brazil

ABSTRACT

A digital guide with animated instructions to aid the design of medicine inserts was produced and tested in Brazil with five developers, in a qualitative manner. The results showed that participants improved the design of inserts by using the guide. They considered it a user-friendly system, although pointing out some drawbacks. Participants' suggestions to improve the guide were considered for its final version, contributing to enhance health information quality in Brazil.

Keywords: medicine inserts, animation, instructions

INTRODUCTION

The medicine insert is the main document to inform patients and health professionals about the composition, usage and storage of medicines. Several researchers have demonstrated that graphic presentation of these inserts affects their comprehension and consequently influences on task performance of taking medicines (e.g. Waarde, 2004; Fujita, 2004). A study carried out by Spinillo et al. (2009) in Brazil found that drawbacks in the visual organization of text and pictures in medicine inserts had a negative impact on using medicines by participants. The authors investigate the effectiveness of inserts regarding five medicines, varying in their pharmaceutical formula and ways of use, with 60 participants in a simulated manner. A total of 352 errors occurred during task performance, most regarding action errors (N=179) and information processing errors (N= 121). The former errors were in general due to deficiencies in the design of some medicines containers (e.g. insulin applicator, dosage lid of antibiotic bottle), which lead to difficulties in performing the steps. The information processing errors regard the comprehension of the inserts by participants whether for pitfalls in typography or in pictorial instructions. As a result, most participants did not perform the tasks satisfactorily. These outcomes ratify the need of improvements in the design of medicine inserts as for their impact on patient's welfare and on safety of medicine usage. In this sense, governments of several countries establish regulations for producing medicine inserts, such as Holland, United States and Brazil.

In general, the medicine inserts regulations focus on the information content (e.g., composition, warnings), giving little attention to or even neglecting aspects of their graphic presentation (e.g. page layout, legibility, graphic hierarchy, pictorial instructions). Most pharmaceutical manufacturers restrict the design of inserts to what is imposed by the regulations, not going further in seeking communicational efficacy of visual organization of information. This is, perhaps, due to lack of interest on graphic language matters, or lack of expertise in the fields of ergonomics and information design by the developers of inserts of the pharmaceutical industry.

The large amount of information presented in medicine insert makes it a dense document, mainly employing text and eventually pictorial instructions to aid in the comprehension of medicine usage. From information design and information ergonomics perspectives, medicine inserts urge to become a user-friendly document with good legibility and effective visual organization of information. This is particularly relevant when considering the elderly population, which increases yearly in the world. They consume far more medication and face more constraints in reading texts than young populations. In Brazil, the number of elder people reaches 16 million – approximately 9.3% of the country population (IDEC 2010)–, disserving the attention of health governmental institutions and of organizations for consumer rights. The Brazilian Institute for Consumer Protection (IDEC, 2010) recently proposed a set of recommendations to the Ministry of Health towards increasing the rigor in observing manufacturers on their compliance with regulation on production of medicines and their inserts. The importance that patients read and understand medicine inserts was also highlighted in the recommendations proposed.

By considering the lack of recommendations on the graphic presentation of information in the Brazilian legislation, and the need for aiding developers in the design of medicine inserts, an interactive system was proposed presenting animated graphic design guidelines. Before introducing the systems proposed, it seems necessary to briefly explain how medicine inserts are currently produced in Brazil, and to bring in some aspects regarding the design of interactive systems and the use of animation with instructional purpose.

THE PRODUCTION OF MEDICINE INSERTS IN BRAZIL

The production of medicine inserts in Brazil is part of an information system available to health professionals and patients, which is controlled by the Ministry of Health. In recent revision of the legislation, the Brazilian government recommends to medicine manufacturers that the inserts should be more straightforward and concise, with adequate and distinct contents to patients and professionals. The insert sections should come with titles as questions to patients (i.e., in moderate conversational style), and should be also available in electronic format to ease users' access to the document. Similarly, the Brazilian government makes a medicine inserts database – The Electronic *Bulario* (Insert) - available for consultation by health professionals and general public. It presents the insert texts of all medicines registered and commercialized in Brazil. The registration of medicines in Brazil is also an electronic process. This is mandatory for manufacturers to obtain authorization to produce and to commercialize medicines. The registration is done through the Medicine Insert Electronic Management System – the *E-Bula* (E-Insert). This is a tool of restrict access to manufacturers in order to submit their medicine insert contents to ANVISA – National Agency for Sanitary Surveillance/ The Brazilian Ministry of Health. Only after obtaining the registration and permission to produce the medicine, the manufacturer is allowed to design the medicine inserts.

The design process of medicine inserts requires developers not only to specify the graphic aspects of the content, but also to comply with the regulation in force. Thus, developers of inserts should be provided with pharmaceutical information on the medicine and with legal information about the document. Besides, they should also possess the necessary knowledge to visually organize the content provided. However, very often the information design skills of developers are not sufficient to make proper decisions on the graphic presentation of medicine inserts. As a result, it is common to find weaknesses in medicine inserts regarding information hierarchy, use of typographic resources (e.g., upper case and bold), emphasis in warnings, and pictorial instructions (e.g., illustration style, observer's point of view). Figure 1 shows a diagram of procedural task analysis of designing medicine inserts by developers of the pharmaceutical industry in Brazil. It presents the steps (rectangles) and the decision points (diamond shapes) of the design process. This process considers that developers of inserts should have information design skills that could assure the effectiveness of their graphic-oriented decisions (gray areas), which includes among other things, layout, typography and pictorial instructions issues. As pictorial instructions may not always be employed in the inserts, their referring step is then represented by a dotted outline in the diagram.

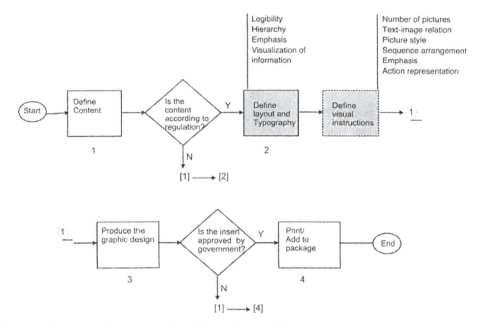

Figure 1. Procedural task analysis of designing medicine inserts in Brazil.

INSTRUCTIONAL INTERACTIVE SYSTEMS

Filatro (2004) considers the communication of instructional contents a dialog between the instruction producer and its interpreter. He states that the use of interactive technology in instructional materials may optimize learning processes, and makes flexibility in time and place for learning possible. In this sense, an instructional interactive system in health field may allow, for instance, developers of medicine inserts to be trained on ergonomic and graphic aspects of the document, according to their availability in time and place to access the instructional system.

In addition, Padovani and Moura (2008) claim that a well-designed navigation system prevents users from spatial disorientation, which is in general caused by deficiencies in the organization of information in the interface. The authors highlight that the design of interactive systems should seek intuitive navigation, allowing users' cognitive resources to be fully employed in the tasks required by the system. In this sense, they recommend that a navigation system should: (a) be ease of learning by making the position, the function and the use of navigation tools noticeable to users; (b) be distinguishable by differing the main content from tools and links in the interface; (c) be properly labeled to inform the functions and uses of the tools, and the location of the links to users; (d) be consistent by showing the navigation tools/devices always in the same place; (e) provide feedback to users to inform on the outputs of ongoing and completed tasks; (f) provide information about the system context to guide users in their path regarding information nodes and their connection to other nodes; and finally, it should (g) find a balance between the optimization of usability with the minimization of task-time, number of actions, complex structure and number of hierarchical levels and sequence of links.

In relation to instructional interactive systems, the learning of their contents can be also facilitated by other visual resources, such as animation. This has been largely employed in the fields of digital entertainment and education, as for instance, e-learning games and digital material for training professional in several areas.

ANIMATION IN INSTRUCTIONAL MATERIAL

Besides other instructionally relevant principles (see Mayer 2005a; Mayer 2005b for a complete list), animation should be designed in accordance with the "attention guiding principle" (Betrancourt, 2005). This guideline suggests that animation should direct learner's attention to thematically relevant instructional contents; otherwise learners might just miss them (Lowe, 2003). This is specially important if, like in our case, learners are novice who are not trained to differentiate between significant and insignificant graphic design decisions. In order to gain attention and guide learners, instructional designers should locate "signaling animation" (e.g. flash highlights, moving arrows etc.) close to the element under focus (Betrancourt 2005). Animated transitions, such as slow fade in, can be used to highlight "connectors" (Engelhardt 2002) linking verbal commentary (e.g. a typographic principle) and corresponding graphic features (e.g. application of the principle). In other occasions, animation can be used to demonstrate how reading direction could be affected by controlling the small distances between graphic elements (e.g., gestalt proximity law). In this case, animated arrows with directional fade in transition can be used to indicate how changes on reading trajectories were affected by subtle changes on the proximity between visual elements (e.g. pictures and text). Without such signaling feature beginners might not realize how subtle graphic design decisions can have an impact on reading fluency (a process that in itself is invisible).

Animation, thus, plays an important part in digital instructional systems by promoting content learning. According to Filatro (2004) the learning success of a digital system mainly relies on the fulfillment of users' information needs. This regards the degree of users' knowledge of a system, and of the context in which the system will be used. Hence, validating interactive instructional systems with potential users is of prime importance to verify their effectiveness as learning tools. By considering this, the interactive instructional system proposed for the design of medicine inserts in Brazil was tested with developers of such inserts, employing a user-centered approach. This allowed accounting for the developers' information needs on learning the guidelines, presented through animation. The system design and its validation with users are briefly presented next.

ANIMATED INSTRUCTION GUIDE TO DESIGN MEDICINE INSERTS

The proposed interactive instructional system is referred to as *A Practical Guide: How to improve Medicine Inserts*. It is addressed to developers of such inserts in Brazil to improve their skills in information design and ergonomics, as previous mentioned. Its animated guidelines are based upon outcomes of previous studies conducted on the effectiveness of medicine inserts and on graphic presentation of information (e.g. Waarde, 2004; Spinillo et al., 2009)

The design process for the Guide prototype occurred in three phases: (a) planning, (b) production, and (c) validation. In the former, a context study was carried out looking at how medicine inserts are produced in Brazil, the legislation in force, and the developers' information needs in the scope of graphic design. In general, the outcomes showed developers were acquainted with the regulation on medicine inserts. They also acknowledged the lack of information on graphic aspects relevant to design the inserts, mainly on typography and on pictorial instructions. By considering these outcomes together with the literature on interactive instructional systems, the Guide was conceived: (a) to be easy and quick to access (technical requirements for the system); (b) to have the content organized into sequential topics; (c) to provide structured navigation to allow following the content in sequence; (d) to present the guidelines in plain and concise language; (d) to use images, animation and interactivity to support guidelines learning; (e) to provide additional information to support developers in the design process; and (f) to provide a printed version of the content to allow consultation of guidelines in alternative medium.

In the following phase (the Guide production), the information architecture and the graphic interface of the system were developed. The content topics were numbered in step-by-step approach to the instructional system, but flexible in navigation, i.e., users decide on the paths to access information. As additional contents to developers, check lists on information design principles for text and visual instructions were made available, as well as suggestions for testing procedures on the insert's comprehension and patient's task performance. These intend to emphasize the importance of validating the medicine inserts with potential users. Regarding interactivity of the Guide, pop-up windows, navigation buttons (forward and backward) and animation in pictures were made available to developers. Pop-up windows were employed to explain certain concepts and terminologies in the main text, and could be accessed by clicking on underlined words. This prevented navigation digression of the principal content (Figure 2b). To demonstrate the applicability of the guidelines, animation was used as magnifying lens effect (zoom out) to show details in pictures of simulated medicine inserts

(Figure 2a). They come out whether by clicking on highlighted areas of a picture, or by clicking on a button for showing all animations in sequence. The Guide also presents two menus for navigation. The main menu (column in the left) displays all section headings of the content at once. User's navigation across the menu is shown by highlighting the section heading s/he is in, and by an indentation placed near the heading. The secondary menu is displayed in a line on the top of the screen, and shows links to information on the Guide design team and contact, among others.

 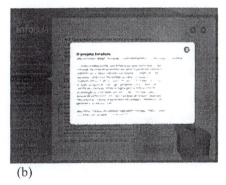

(a)   (b)

Figure 2. Screen capture of the Guide interface (a) and of pop-up windows (b).

After the design of the interface, the Guide was then validated with medicine inserts developers. Since this phase required experimental approach, it seems necessary to explain its procedures and outcomes in more detail in this paper, as follows.

VALIDATING THE GUIDE

In order to verify the effectiveness and the suitability of the Guide, it was tested with five developers of the pharmaceutical industry in Curitiba, Brazil through questionnaires and a workshop/hands-on activity. The former were to measure participants' acquaintance with the graphic aspects regarding the design of inserts and the importance they append to such aspects before and after using the Guide (pre-test and post-test). This allows measuring the effect of the Guide information content on participants' knowledge and attitude towards graphic presentation of medicine inserts. A questionnaire was also employed to measure their level of satisfaction with the Guide. The workshop was conducted to assess participants' task performance of designing medicine insert based upon the recommendations available in the Guide. Participants' responses on the questionnaires and the outcomes of the workshop were analysed qualitatively. In this sense, it seems important to highlight that a small number of participants was considered appropriate to validate the instructional system, as this allowed a closer contact with participants in the design process, and specially in the discussion session.

PROCEDURES AND MATERIAL

Initially, each participant received a questionnaire presenting a list of information design principles to produce medicine inserts. To each item of the list participants could mark a value ranging from 1 (minor importance) to 5 (major importance). When finished, they returned the questionnaires to the researchers. Then, they were asked to take part in the workshop, which was carried out in two phases: production of a medicine insert and discussion of the inserts designed. For the former, a desktop computer was provided to each participant, presenting the Guide and a text file with simulated content of a medicine insert. Participants were given 15 minutes to navigate in the instructional system to be acquainted with the Guide. Afterwards, they were asked to design a medicine insert according to the proposed guidelines. Two hours were given to this task, and they could interact with the researchers in case of doubts about the use of the equipment or of the instructional system. Observation technique was employed, allowing taking notes on participants' attitudes towards the task and/or the instructional system.

When participants finished the design of the medicine insert, they were asked to answer again the previous questionnaire with design principles (post-test); and another questionnaire regarding their satisfaction with the Guide. It presented positive statements about the Guide, to participants mark a value ranging from -2 (minor degree) to +2 (major degree) according to their degree of agreement. Afterwards, a discussion was conducted about the medicine inserts designed with the aid of the Guide. They were printed out and affixed in a panel to be compared across their design solutions. The outcomes were registered in writing.

RESULTS AND DISCUSSION

The responses of the questionnaire on design principles presented before using the Guide (pre-test) indicate participants considered most listed items important to the production of medicine inserts. After using the Guide, the degree of importance regarding the items, in general, increased (post-test). It is worth pointing out that among the design principles, those on illustration and text alignment were considered of less importance by participants in the first questionnaire. However, after using the Guide, participants reviewed their responses on illustration, but not on text alignment. This seems to indicate the guidelines on this topic were not clear or convincing as much as necessary to make participants aware of the relevance of this typographic issue for designing the insert. Thus, adjusts in the guidelines on text alignment were needed. Regarding participants acquaintance with graphic terms, the responses showed three out of five participants ignored simple typographic terms, such as, font size and condensed font. This indicated that revision in the Guide content was also necessary for typographic terms. Despite these, the majority of responses on satisfaction was positive, suggesting that the Guide was approved by the participants as an instructional tool for designing medicine inserts.

Regarding the outcomes of the workshop, in the development phase the inserts were designed according to the given time, and no relevant attitude or doubt was registered during participants' task performance. In the discussion session, the relationship between the decisions made by participants and respective support provided by the tutorial guidelines was the main focus. Participants' difficulties and facilities in designing the inserts, and suggestions to improve the instructional system also emerged from the discussion. As for example, participants stressed their need for guidelines on how to emphasize the medicine name and the patient care hotline. Despite this, they considered the graphic presentation of the inserts produced with the aid of the Guide better than those they used to produce for the pharmaceutical industry. They also considered that their knowledge and skills on graphic presentation of information in medicine inserts were improved. Such feedback from participants seems to indicate that the proposed animated guide was effective on supporting learning about the graphic presentation of medicine inserts.

The results on the validation of the Guide with developers of medicine inserts ratify the findings of previous studies on instructional interactive systems and animation. This latter was employed to direct developers' attention to the guidelines through examples provided in the pictures of inserts. This is in agreement with Betrancourt (2005) and Lowe (2003). The interface layout and the resource used for interactivity - such as navigation buttons, highlighted menu headings, pop-ups windows - seem to prevent developers from spatial disorientation, in accordance with Padovani and Moura (2008). Finally, by considering developers' information needs and suggestions for improving the Guide, the proposed instructional system may reach learning success, as said by Filatro (2004).

CONCLUSIONS AND FINAL REMARKS

The outcomes indicate that, in general, the Guide has achieved success in terms of learning support and interface usability. In other words, pictorial and animated examples helped developed to apply learned guidelines effectively. The system information architecture seems to have promoted participants' intuitive navigation. Flexibility allied to structured navigation (linear step-by-step) had a positive effect on participants' opinions on the Guide as an instructional system. Moreover, the guidelines seem to be apprehended promptly, as the task of producing a medicine insert in accordance with the Guide was conducted in a given (limited) time.

However, adjustments were required to the Guide content based upon participants' suggestions to improve its potential as an instructional system. Thus, further guidelines and explanations on typographic aspects of the medicine inserts were considered. Accordingly, an enhanced version of the Guide was developed by the research team to be available in Brazil in a near future.

Finally, we feel confident that the *Practical Guide: How to improve Medicine Inserts* as an interactive instructional system may contribute to patient health safety by enhancing the quality of information produced by the Brazilian pharmaceutical industry.

248

ACKNOWLEDGEMENT

Thanks are due to the participants who volunteered to this study for making it possible, and to CNPq – National Council for Scientific and Technological Development, Brazil for funding this research.

REFERENCES

Betrancourt, M. (2005). The animation and interactivity principles in multimedia learning. The Cambridge Handbook of Multimedia Learning. R. E. Mayer. Cambridge, USA, Cambridge University Press: 287-296.

Engelhardt, Y. (2002), *The language of graphics.* Faculteit der Natuurwetenschappen, Wiskunde en Informatica, University of Amsterdam. PhD: 197.

Filatro, A. (2004), *Design instrucional contextualizado: educação e tecnologia.* São Paulo: Editora Senac.

Fujita, P. T. L. (2004), A comunicação visual de bulas de remédios: análise ergonômica da diagramação e forma tipográfica com pessoas de terceira idade. Disponível em: www.infodesign.org.br, *Infodesign: Revista Brasileira de Design da Informação,* n° 1/1, ISSN 1808-5477.

IDEC- Instituto Brasileiro de Defesa do Consumidor (2010). Available at www.idec.org.br/rev_idec. Retrieved in 21/02/2010

Lowe, R. K. (2003), Animation and learning: selective processing of information in dynamic graphics. *Learning and Instruction* (13): 157-176.

Mayer, R. E. (2005a), Principles for managing essential processing in multimedia learning: segmenting, pretraining, and modality principles. *The Cambridge handbook multimedia learning.* R. E. Mayer. New York, USA, Cambridge University Press: 169-182.

Mayer, R. E. (2005b), Principles for reducing extraneous processing in multimedia learning: coherence, signaling, redundancy, spatial contiguity, and temporal contiguity principles. *The Cambridge handbook multimedia learning.* R. E. Mayer. New York, USA, Cambridge University Press: 169-182.

Padovani, S.; Moura, D. (2008), *Navegação em hipermídia: uma abordagem centrada no usuário.* 1. ed. Rio de Janeiro: Ciência Moderna.

Spinillo, C.G.; Padovani, S.; Lanzoni. C. (2009), Patient Safety: Contributions from a Task Analysis Study on Medicine Usage by Brazilians.. In: Smith, M.J.; Salvendy, G. (Eds.). (Org.). *Human Interface and the Management of Information. Designing Information Environments.* New York: Springer, v. 5617, p.

Waarde, van der K. (2004), Visual information about medicines. Providing patients with relevant information. In: Spinillo, Carla G.; Coutinho, Solange G. (Eds). *Selected Readings of the Information Design International Conference 2003.* Recife, SBDI | Sociedade Brasileira de Design da Informação. p. 81-89.

Chapter 30

Task Analysis for Reading Strategies in Medicine Inserts: A Methodological Proposal

Patricia Lopes Fujita, Carla Galvão Spinillo

Postgraduate Program in Design, Department of Design
Federal University of Paraná
Curitiba, Paraná, Brazil

ABSTRACT

This paper discusses a task analysis method developed to visually organize qualitative data collected using the 'Think Aloud' technique to investigate reading strategies on medicine inserts textual content, in order to improve results visualization and interpretation. Information content, graphic presentation and the text structure of medicine inserts in Brazil are investigated, considering patients' information reading strategies to verify difficulties and metacognitive process through a qualitative approach. A study was conducted with six highly educated adult participants using the 'Think Aloud' technique, which is a cognitive approach. Each participant was interviewed individually and their answers were registered by digital audio recorder and in the presence of the researcher, which made a transcription of the recorded data for analysis. Thus, to interpret and obtain results through the collected data, a task analysis methodology was developed. The method proposes to visually organize 'Think Aloud' qualitative data into four categories: textual structure, graphic presentation, difficulties in order to improve the data visualization, examination and mainly to compare different participants' outcomes. According to the results, it is possible to say that the 'Think Aloud' and the task analysis method contribute to data collection regarding reading strategies on medicine inserts.

Keywords: Task analysis, reading strategies, medicine inserts

INTRODUCTION

The medicine insert is a printed instruction document, essential in the use of any kind of medication, providing specific information, such as: indication for medical treatment, pharmaceutical composition, warnings, and instructions of use. Although, this document contains information directed to different kinds of readers/users, in this case are, health professionals and patients, which make the medicine insert into a considerable complex document in both linguistic and graphic aspects. The medicine insert content may present different kinds of information, such as, warnings, tables, diagrams and pharmaceutical terms. For patients/users with little or no acquaintance with this kind of document, understanding information can be a difficult task, to say the least.

The medicine inserts content in Brazil is regulated by the ANVISA – National Agency of Sanitary Vigilance of the Ministry of Health. It consists mainly of medicine identification, pharmaceutical forms, patient, technical, storage information. Even though the regulation regards the quality and type of content in medicine inserts, the graphic presentation and text structure of information seems to be ignored by developers. Aspects such as text legibility and visual configuration of the medicine insert are not considered, despite their influence on reading and comprehension of written information (e.g. Wright, 1999; Van der Waarde, 2004, 2006).

Information transmitted through visual instructions play an important role on the use of medicines and good information is part of efficient communication between healthcare providers and patients (Van der Waarde, 2004, 2006). The graphic presentation of information in medicine inserts may affect reading and comprehension, consequently the use of medicines (Fujita & Spinillo, 2006, 2008). Therefore, the success on taking a medicine with safety depends on the task of reading and understanding the information of medicine inserts.

Considering the reading task of medicine inserts a very important aspect on taking medicines safely by patients and by taking into account the above mentioned aspects, this paper approaches a study regarding reading strategies on medicine inserts (Fujita, 2009) focusing on a task analysis method developed by the authors. The study was conducted with six highly educated adult participants using 'Think Aloud' technique, which is a qualitative data collection cognitive approach. Thus, to interpret and obtain results through the collected data, a task analysis methodology was developed. The method proposes to visually organize 'Think Aloud' qualitative data in order to improve the data visualization, examination and mainly to compare different participants' outcomes.

READING STRATEGIES IN MEDICINE INSERTS

The act of reading consists of creating meaning, in a process where the comprehension of written language occurs while the reader interacts with a text. Once reading, previous knowledge of the theme is activated, allowing to attribute meaning to words, sentences and paragraphs. Reading is an activity that demands a deeper look at the text and what is retained from it, and at the same time it relates knowledge stored in the reader's memory. Therefore, reading is defined as an interactive activity involving "thinking and creativity" through the comprehension of a text. Giasson (1993) considers reading as an interactive process between: reader, text and context. The reader is able to comprehend what is read from her/his cognitive structure. This concerns the knowledge stored in the memory, together with the cognitive activities and strategies.

The cognitive studies in reading demonstrate that text processing is developed through the use of strategies by readers, which are based upon their cognitive activities while interacting with a text.

Kato (1985) identified two types of information processing in a text: cognitive and metacognitive process. The former regards subconscious actions, whereas the latter regards the conscious actions taken in front of a problem. Brown (1980) considered the involvement of consciousness as the main criterion to distinguish cognitive and metacognitive activities. In this sense, Cavalcanti (1989) referred to cognitive actions as automatic actions and to metacognitve actions as controlled actions. By considering strategy any deliberated control and planned activity that leads to comprehension (Brown, 1980), it is pertinent to state that they occur only in the metacognitive process, since readers are aware of their reading activities. In the metacognitive process readers may accompany and assess their understanding of the message while reading a text, and even take measures when comprehension fails.

In view of that, it is possible to assert that cognitive activities are performed spontaneously, without readers' control of the information process, allowing text interaction simultaneously with readers' previous knowledge stored in the LTM and recovered in the WM. Metacognitive activities, on the other hand, allow monitoring those actions, and making conscious. According to Wright (1999) cognitive and metacognitive processes include the capacity of understanding and interpreting the visual structure and the content of written instructions.

In reading strategies metacognition is a process that monitors comprehension of a text, making use of introspection to assess readers' interaction with written information. By introspection is meant one's reflection about him/herself, in this case, regards readers' self-perception when interacting with medicine inserts. Among the techniques for examining reading strategies, the Think Aloud stands out as it allows the verbalization of readers' introspection while reading a text.

In order to examine readers' metacognitive process when interacting with medicine inserts, the present study employed the Think Aloud technique to collect qualitative data. The aim of this study was to identify patients difficulties and information needs during medicine inserts reading process through the observation of their reading strategies. The collected data, consisted on the verbalization of participants thoughts accompanied by reading, providing information on strategies, difficulties and procedures during the execution of a task (in this case, reading a medicine insert), while the sequence of processed information from the verbal manifestation of the mental processes of the reader.

With the purpose to improve the outcomes from the analysis of the collected data and its comprehension, a task analysis method was developed to analyze qualitative and introspective data generated by participants performing the Think Aloud technique.

A STUDY ON READING STRATEGIES FOR MEDICINE INSERTS IN BRAZIL

The study was conducted with six highly educated adult participants (over 21 year old) using Think Aloud technique to assess their metacognitive strategies when reading a medicine insert produced in Brazil.

The material tested was a medicine insert of an anti-inflammatory drug. The criteria for decide on this is medicine were: to comply with Brazilian regulation for medicine inserts established by ANVISA and to be purchase with medical prescription only.

THINK ALOUD PROCEDURE

The experiment was conducted with each of the six participants individually and in isolation. Initially the Think Aloud technique was explained to participants, and a text was handed to them to offer some training. On this regard, participants were asked to read a medicine insert leaflet and to think aloud their thoughts concerning its information and instructions. Each individual interview was registered by digital audio recorder and in the presence of the researcher, which afterwards made a written transcription of the recorded data for analysis.

TASK ANALYSIS (DATA ANALYSIS PARAMETERS)

The outcomes from the introspective reading task performed by the participants were organized using a task analysis methodology developed by the authors (Fujita, 2009) in order to interpret and obtain results through the collected data; a task analysis methodology was developed (Figure 2). The method proposes to visually organize 'Think Aloud' qualitative data and provide a better look at participants reading performance, metacognitive comments (reading strategies actions) and difficulties; as well as improving the data visualization, examination and mainly to compare different participants' outcomes.

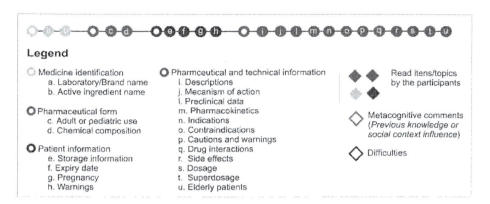

FIGURE 2 Task analysis model

According to Diagram 1.1, the above line with coloured spheres marked with letters inside, represents a medicine insert structure based on ANVISA'S regulation (Resolution RDC Nº 140, 2003), detailed by the Legend below. The Green, red yellow and purple diamond forms are used to define the read topics and the blue and black outline signalize when the participant verbalized metacognitive comments (concerning previous knowledge or social context influence) or difficulties while reading a particular topic. Figure 3, presents an example of application using this parameters for task analysis:

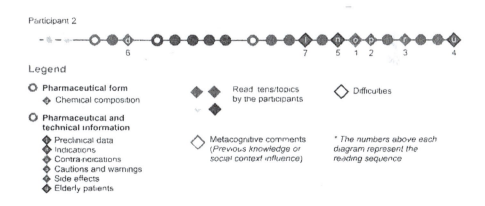

FIGURE 3 Task analysis example (participant 2)

The Figure 3 presents Partipant's 2 task analysis, according with the Legend below, the read content is defined by the diamond form with the letter that represents the topic in the reading sequence. The items with black and blue outline determine the moment when difficulties and metacognitive comments occurred about that particular topic. The number below the diamond forms (read topics) spheres denote the reading sequence, and the spheres are the topics that are part of the medicine insert text structure, but weren't read by that participant.

RESULTS AND DISCUSSION

The results are discussed based on the task analysis of each one of the six participants verbal recorded data, organized using the developed task analysis method. The results are presented by Figure 4:

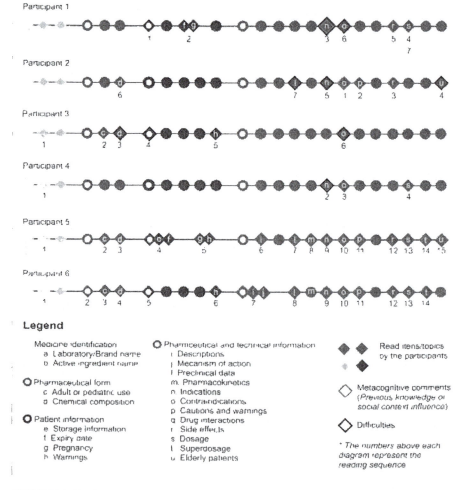

FIGURE 4 Task analysis final results (six participants)

In reference to Figure 4, it is observed that Participants 1 and 2 performed a non-linear reading sequence in relation to the medicine insert structure and quite different from the others. The Participants 4, 5 and 6 presented a linear reading sequence based on the medicine insert's structure. Although, Participant 3 executed the reading task in a similar way as the last ones, it is important to notify that this reader gave up to keep on reading the medicine insert reporting that he felt discouraged by the difficulties caused by deficiencies on the graphic presentation. This participant also explained that those difficulties forbade him to access or find the information he needed. Therefore, it is possible to infer that this fact interfered on reader's reading strategies performance, because he couldn't visualize the content.

It was possible to notice that Participants 1, 2, 3 and 4, expressed difficulties concerning technical terms and graphic presentation. The majority of this difficulties refers to the content of "Pharmaceutical and technical information", such as: (n) Indications; (o) Contraindications and (l) Preclinical data. However, while reading the (n) Indications and the (o) Contraindications, all the participants reported to have a lot of interest in these two topics. In this sense, it is also important to outline that Participants 1 and 2, presented metacognitive comments (concerning previous experiences/knowledge and context information) while reading (n) Indications, (o) Contraindications and (s) Dosage.

As for Participants 5 and 6, according to Figure 4 it is possible to notice that both read almost all the topics of the medicine insert structure and didn't revealed any difficulties or metacognitive actions during the reading task. Although those participants covered a many topics of the structure, sometimes they read only the titles, and the first sentences, until they realize the topic didn't interested them. It is possible to infer that the reading sequence performed by Participants 5 and 6 was guided by the titles (in bold or uppercase) contained in the medicine insert's structure.

By taking into account the analysis of all the results it is also possible to affirm that graphic presentation aspects in medicines inserts information content can influence users reading process.

FINAL CONSIDERATIONS

This paper approached a study regarding reading strategies on medicine inserts (Fujita, 2009) focusing on a task analysis method developed by the authors. The study was conducted with six highly educated adult participants using the 'Think Aloud' technique. To interpret and obtain results through the collected data, a task analysis methodology was developed. The method proposes to visually organize Think Aloud's qualitative data in order to improve the data visualization, examination and mainly to compare different participants' outcomes.

According to the results, it is possible to say that the 'Think Aloud' technique and the task analysis methodology contribute on qualitative data collection, organization and observation of reading strategies for medicine inserts. Although some improvements are still necessary in this proposed methodology, it was possible to have a closer and general look at the same time into the results, and to improve qualitative data visualization in task analysis.

REFERENCES

Brasil. Ministério da Saúde. (2003). *Resolução RDC N° 140*
http://e-legis.anvisa.gov.br/leisref/public/showAct.php?id=6311

Brown, M. (1980). Metacognitive development and reading. In: Spiro, R. J., Bruce, B. C., Brewer, W. F. *Theoretical issues in readingcomprehension: perspectives from cognitive psychology, linguistics,artificial intelligence, and education*. Hillsdale, NJ: L. Erlbaum Associates, pp. 453-481.

Cavalcanti, M. C. (1989). *I-N-T-E-R-A-Ç-Ã-O leitor texto: aspectos de interação pragmática*. Campinas: Unicamp.

Ericsson, K. A., Simon, H. A. (1987). Verbal reports on thinking. In: Faerch, C., Kasper, G. (Eds) *Introspection in second language research*. Clevedon: Multilingual Matters, pp. 24-53

Fujita, P. T. L.; Spinillo, C. G. (2006). A apresentação gráfica de bula de medicamentos: um estudo sob a perspectiva da ergonomia informacional. In: *Congresso Internacional de Ergonomia e Usabilidade ' Ergodesign', 2006, Bauru*. Anais. Bauru: Unesp, p. 1-6, ISBN: 85-99679-02-3.

Fujita, P.T.L., Spinillo, C.G. (2008) Verbal Protocol as an information ergonomics tool to verify reading strategies in medicine inserts. In: AEI 2008: 2nd International Conference on Applied Ergonomics, *Proceedings of the AHFE International Conference 2008*, Las Vegas, Nevada, vol. 1. USA Publishing | AHFE International, Louisville.

Fujita, P. T. L. (2009) *Graphic presentation analysis of medicine inserts leaflets textual content from patients reading perspective regarding the context of use*. Dissertation (Master) – Federal University of Paraná, Postgraduate Design Program. 160p.

Giasson, J. (1993). *A compreensão na leitura*. Lisboa: Asa, 317p.

Kato, M. (1986). *No mundo da escrita: uma perspectiva psicolingüística*. São Paulo: Ática, 144p.

Van der Waarde, K. (2004). Visual information about medicines. Providing patients with relevant information. Spinillo, Carla G.; Coutinho, Solange G. (Eds). *Selected Readings of the Information Design International Conference 2003*. Recife, Sbdi | Sociedade Brasileira de Design da Informação, p. 81-89.

Vand der Waarde, K. (2006). Visual information about medicines for patients. Jorge Frascara. (Eds). *Designing Effective Communications: Creating contexts for clarity and meaning*. New York: Allworth Press, p. 38-50.

Wright, P. (1999). Comprehension of printed instructions: examples from health materials. In D. Wagner, R. Venezky, & B. Street (Eds.) *Literacy: an international handbook*. Boulder, CO: Westview Press, p. 192-198.

Chapter 31

Analysis on Descriptions of Precautionary Statements in Package Inserts of Medicines

Keita Nabeta[1], Masaomi Kimura[1], Michiko Ohkura[1], Fumito Tsuchiya[2]

[1]Shibaura Institute of Technology
3-7-5 Toyosu, Koto City
Tokyo 135-8548 JAPAN

[2]Tokyo Med. & Den. University
1-5-45 Yushima, Bunkyo City
Tokyo 113-8549 JAPAN

ABSTRACT

To prevent medical accidents, users must be informed of the cautions written in medical package inserts. To realize countermeasures by utilizing information systems, we must also implement a drug information database. However, this is not easy to develop, since the descriptions in package inserts are too complex and their information poorly structured. In this paper, we report on the result of analysis concerning the descriptions of 'precautions for application' in package inserts via text mining methods, and propose a data structure for the same, in order to organize the information in the descriptions.

Keywords: Text Mining, Medical Safety, Drug Information

INTRODUCTION

To prevent medical accidents, users must be informed of the cautions written on the package inserts of medicines. These are exclusive legal documents that describe detailed information for each drug, such as the composition, efficacy, dosage and cautions. However, since the description in package inserts is excessive and there are over 20,000 medicines in Japan, it is not easy for healthcare workers to read such documents.

As a countermeasure, the utilization of information technology has been focused on in recent years. To develop an information system that alerts healthcare workers at the user interface of an ordering system and an electronic health record system, we need to implement a drug information database as a source of information alerts. Actually, the Pharmaceutical and Medical Devices Agency (PMDA) has provided SGML (Standard General Markup Language) formatted data of package insert information since the year 2003 in order to utilize the same in information systems. However, this is not easy since the descriptions in package inserts are too complex and their information poorly structured.

In this paper, we report on the results of analysis concerning the descriptions of 'precautions for application' in package inserts in terms of meaning via text mining methods. The reason for analyzing 'precautions for application' data is that it contains information concerning the usage situation that must be known at the timing of prescription. We also propose their data structure to organize the information in the descriptions.

TARGET DATA

We obtained the SGML data from the PMDA Web site, utilizing the 'YJ code' as a key, which is a drug identification code and included in the 'standard drug master (9/30/2007)' provided by MEDIS-DC (The Medical Information System Development Center). The number of SGML files was 11685.

PREPROCESSING

In our target SGML data, since precautionary statements are described in the 'precautionsforapplication' elements as shown in Figure 1, we extracted text data from the 'item' and 'detail' elements in the same. After this extraction, we split each sentence and converted 1byte character to 2 bytes. In this study, we removed the sentences described in parentheses, since these can be regarded as additional information. As a result, we obtained 43639 statements from 9130 SGML data.

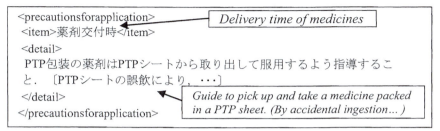

FIGURE 1 Sample Data of Precautionary Statement

Additionally, we applied dependency analysis to the extracted statements via software named 'Cabocha'. In this study, we investigated them by using information such as morphemes, segments, dependency relations and parts of speech provided by the software.

ANALYSIS

EXTRACTION OF PREDICATE WORDS

Since predicate words are often described at the end of the sentence in Japanese, we extracted only headwords in the final segment from the results of dependency analysis. Table 1 shows the top 10 extracted words and their frequencies.

This result shows that 'こと' (must) was the most commonly occurring word, with a frequency of 83%. It emerges with meanings of 'order', if we place it after verbs, e.g. '使用すること' (must use). We therefore focused on verbs that modify 'こと' (must), although there are other order expressions, e.g. weaker '望ましい' (desirable) and that of possibility 'できる' (can).

Table 1 Extracted Predicate Words (Top 10)

Keywords in segment	Frequency	Ratio (%)
こと (must)	24929	82.86
望ましい (desirable)	1578	5.24
ある (exist)	812	2.70
洗い流すこと (must wash)	250	0.83
注意(caution)	194	0.64
場合 (case)	105	0.35
できる (can)	73	0.24
粘膜 (mucous membrane)	68	0.23
行う (do)	51	0.17
開瞼する (keep the eyelids apart)	47	0.16

EXTRACTION OF ORDER STATEMENT

To extract summarized information concerning the order description, we collected verbs modifying 'こと' (must) in the previous subsection and obtained 415 kinds of headwords in Table 2. In this table, we can see that there are verbs expressing direct orders such as '使用する' (use), causative orders such as '使用させる' (make someone use), direct interdictions such as '使用しない' (do not use) and causative interdictions such as '使用させない' (do not make user use). These do not, however, provide enough information to fully understand the statements.

To extract such information, we focused on the case markers (particles) appearing at the end of segments modifying the verbs. These are words that indicate the dependency relationship of a certain word to a predicate verb. Although we obtained three types of case markers in the result of Cabocha, which are 'case markers - general', 'case markers - collocation' and 'case markers - citation', we used only 'case markers – general', since it accounted for 90.65% and there were fewer case markers in that type than others. In this study, we focus on the case markers, 'を' (wo) , 'に' (ni), 'が' (ga), 'から' (kara), 'で' (de), 'と' (to), 'へ' (e), 'の' (no), 'より' (yori) and 'にて' (nite).

We aggregated ternaries, a verb, a case marker and the verb modifier, which also appears in the segment containing the case marker. After aggregation, we visualized the ternaries as nodes of a network to find major information in the order statement concerning the four cases, which are direct order, interdiction, causative order and causative interdiction. In the network, the nodes are collected verbs, case markers and keywords, and the edges express the connections among them, while their frequencies are expressed as their width. The threshold of appearance frequencies in the data limits the nodes and edges in Figs. 2, 3, 4 and 5.

Table 2 Extracted Order Words (Top 10)

Keyword in segment	Frequency	Ratio (%)
指導する (guide)	4771	18.27
注意する (be careful)	3696	14.15
使用する (use)	3417	13.08
使用するない (do not use)	2440	9.34
注射する (inject)	986	3.78
避ける (avoid)	968	3.71
投与する (administer)	918	3.52
行う (perform)	760	2.91
行うない (do not perform)	436	1.67
する (do)	420	1.61

Direct Order Statements

This figure shows that major 'direct order' verbs are '使用する' (use), '注意する' (be careful) and '中止する' (stop), and we can see statements such as '点眼用に使用する' (use for eye drops), '静脈内に注射する' (inject into vein) and '洗い落としてから使用する' (use after washing).

 It should be noted that the statement '神経走行部位を避けること' (avoid part of nerves) appeared in this network. Though it is an order to avoid, it can be interpreted as an interdiction, since the avoidance is implied in the concept of interdiction.

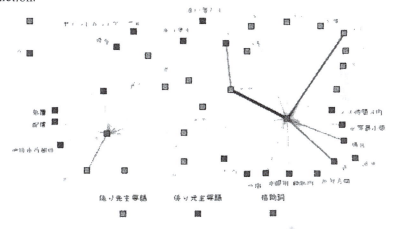

FIGURE 2 Network of Words related to Direct Orders (Threshold: 90)

Direct Interdiction Statements

Figure 3 shows that the major expressions concern the interdiction of usage. In particular, we can see an interdiction to apply medicines to the affected areas, such as '炎症部位に使用しない' (do not apply to inflammatory part). Moreover, we obtained statements that indicate an interdiction of mixing drugs, e.g. '製剤と混合しない' (do not mix with drugs).

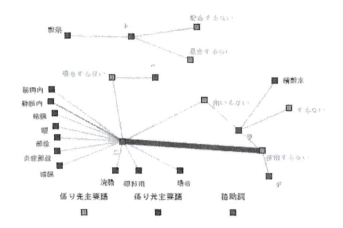

FIGURE 3 Network of Words related to Direct Interdiction (Threshold: 50)

Causative Order Statements

In Figure 4, we see two types of causative verbs. The first is for medical experts and patients, and another type is for medicines.

The former gives examples such as '噛まずに服用させる' (make patients take the medicine without chewing), '食後に服用させる' (make patients take the drug after a meal) and '水で服用させる' (make patients take the medicine with water). These are statements that indicate orders to patients via pharmacists and nurses.

The latter gives examples such as '中和させる' (neutralize) and '乾燥させる' (dry). They express methods for the preparation and storage of medicines.

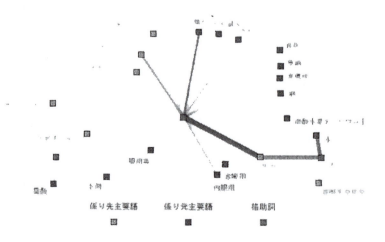

FIGURE 4 Network of Words related to Indirect Orders (Threshold: 5)

Causative Interdiction Statements

As causative order statements already mentioned, the subjects of the former include both humans (pharmacists, nurses and patients) and medicines. In the figure, statements such as '水なしで服用させない' (do not make patients avoid water) and '注射筒から逆流させない' (do not regurgitate from syringe) can be found.

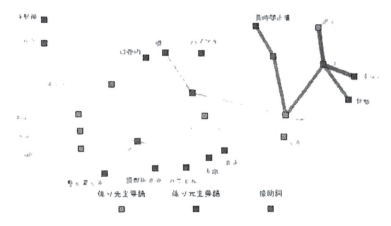

FIGURE 5 Network of Words related to Indirect Interdiction (Threshold: 0)

Based on these results, we found two points concerning precautionary statements that should be considered. The first is that both order statements and interdiction statements chaotically coexist. Since this may lead to the misreading of these statements, they should be described separately. The second point is that we should distinguish two kinds of subjects of statements, namely patients and medical experts. This is important for system checking because the system must know the subjects to whom warnings should be given.

EXTRACTION OF WORDS REPRESENTING TIMING

In the previous section, we found words that express the time to perform an action, e.g. '投与前' (before administration) and '使用時' (in use). Since it is important for us to know the timing to execute the order, we extracted these from the target data.

Initially, we compiled keywords where the part of speech is an 'adverbial noun' (Table 3). We defined '場合' (case), '前' (before), '後' (after), '時'/'とき' (time) and '際' (when) as feature words that express the concept of timing, since most segments expressing timings contain these words.

Table 3 Extracted Adverbial Nouns (Top 10)

Keywords of segment	Frequency	Ratio (%)
場合 (case)	2336	25.13
刺したとき (time of stinging)	599	6.44
ため (because)	536	5.77
後 (after)	357	3.84
開封後 (after opening)	346	3.72
とき (time)	303	3.26
長期間 (long-term)	235	2.53
長時間 (extended period)	218	2.34
アンプルカット時 (time of ampoule cutting)	160	1.72
溶解後 (after dissolving)	153	1.65

Next, we extracted expressions of timing via the characters defined above. The precautionary statements contain mainly two types of descriptions, one of which is that the order is described after the timing expressions in statements e.g. '使用時に注意すること' (be careful at the time of use), and another of which where the timing expression appears singly e.g. '薬剤交付時' (delivery time of medicines) in the figure. In the former case, we extracted keywords that modify the verb expressing the order and whose segments contain the feature words defined above. In the latter case, we extracted keywords that are at the tail of sentences and have feature words. We regarded the keywords obtained as timings to execute the order.

The resultant keywords contained those with similar meanings. To reduce the variety of expressions, we identified such synonymous expressions. We investigated the top 50 extracted words and found six groups: 'at the time of delivery', 'at the time of preparation', 'in use', 'after use', 'at the time of storage'. We show the results of this classification in Table 4.

Table 4 Classification of Words Meaning Time

Class	Example of extracted words	Frequency	Ratio (%)
Time of delivery	薬剤交付時 (delivery time of medicines)	3791	13.77
Time of preparation	調製時 (time of preparation), アンプルカット時 (time of cutting ampoule), etc.	6067	22.04
In use	投与時 (in administration), 注射時 (time of injection), etc.	12,599	45.77
After use	投与後 (after administration), 使用後 (after use)	279	1.01
Time of storage	保存時 (time of storage)	406	1.48
Others	やむを得ない場合 (in the inevitable case), 反復使用する場合 (for the repetitive use), etc.	1438	5.23

We found that the caution related to 'in use' were the largest group. Utilizing this classification, we obtained characteristic words that express an order and appear only at specific instances, e.g. '廃棄する' (dispose of) after use and '遮光保存する' (store under light interception) at times of storage. The obtained timings for alerts are vital to prevent accidents in order for the system to specify the time at which to send an alert.

CONCLUSION AND DISCUSSION

In this study, we analyzed precautionary statements in the package inserts of medicines using a text-mining method to develop a drug information database. Consequently, we found the major statements used to express orders and interdictions in use. Moreover, we obtained points of consideration concerning the subjects of orders in the statements and timings of their execution.

Utilizing the knowledge obtained, we can define the information structure used to describe the precautionary statement. It should contain information such as the actions described in the statement, the flag to express an 'order' or 'interdiction', the subject to be ordered and the timing.

In future work, we will analyze other information in the package inserts and develop a drug information database based on the results.

REFERENCES

Pharmaceuticals and Medical Devices Agency, http://www.pmda.go.jp

K. Nabeta et al. (2008), "Investigation into problems of utilization of drug information in package inserts to ensure the safety of drug usage." *Proceedings of AHFEI2008*

M. Kimura et al. (2009), "Analysis on Descriptions of Dosage Regimens in Package Inserts of Medicines" *HCI* (9), pp. 539-548.

The Medical Information System Development Center, http://www.medis.or.jp

CHAPTER 32

The Standardization of Medicine Name Structures Suitable for a Prescription Entry System

Masaomi Kimura[1], Keita Nabeta[1], Michiko Ohkura[1], Fumito Tsuchiya[2]

[1]Shibaura Institute of Technology
3-7-5 Toyosu, Koto City, Tokyo 135-8548 JAPAN

[2]Tokyo Med. & Den. University
1-5-45 Yushima, Bunkyo City, Tokyo 113-8549 JAPAN

ABSTRACT

Ensuring safe medical usage is one of the keys to preventing medicine-related accidents. With this in mind, there have been efforts to avoid medicines whose name or visual packaging appearance is similar to others: namely in terms of the naming, design and/or adoption of medicines. However, such countermeasures remain insufficient, since medicine-related accidents are still occurring.

Because of this, we conducted a study to create a database containing information to ensure safe medical usage and construct a prescription checking system based on the same.

In this paper, we report on the result of a study concerning the method used to extract information contained in medicine names, which can help verify users' input of the same into a prescription input system.

Keywords: Medical safety, medicine name structure, prescription entry system

INTRODUCTION

Safe medical usage is one of the keys to preventing medicine-related accidents. To ensure such safety, there have been many efforts to avoid medicines whose name or visual packaging appearance is similar to others, including avoiding similar names, designs and adoption among different medicines. Unfortunately, these are insufficient, since medicine-related accidents persist. In fact, in a hospital in Japan, the muscle relaxant, Succin, was wrongly injected into a patient instead of the sound-alike steroidal anti-inflammatory medicine, Saxison. Though Saxison had not been adopted in the hospital as a means of avoiding confusion, the doctor confused Succin with the former, prescribed Succin, and caused a fatal accident.

In order to prevent such wrong prescriptions, we need a means of verifying whether the selection of a medicine is appropriate for the purpose of treatment. In Japan, computerized prescriber order entry (CPOE) systems are widely used to prescribe medicines, and there have been many efforts to prevent the wrong input of medicine names: improved order in the list of medicines, highlighting frequently confused medicine names by adding certain symbol characters, and so on (Furukawa et. al., 2001, Tsuchiya et. al., 2007). Besides such efforts, we consider it important to provide suitable medicine-related information for doctors in order to recognize the selection errors of medicines. With this in mind, we conducted a study to create a database containing the information required for a CPOE prescription checking system based on such database.

Though the name of a medicine is obviously a significant part of the prescription information, it is difficult to identify the information included in the names. This is because the information structures they contain are not fully standardized. Previous study has shown that product names are composed of 23 kinds of elements, e.g. premodifiers, stems, postmodifiers, standard units, dosage forms and so on (Tsuchiya et. al., 2001). Remember that the usage of medicines with the same stem differs depending on the standard units or dosage forms. It is obvious that wrong usage can cause an accident. Because of this, in these elements, the stem of names, standard units and dosage forms are essential as a minimum to identify a medicine. Unfortunately, not all medicines sold in Japan have such information in their names. Even for the names of medicines containing all this information, there is no guarantee that they will be arranged in the determined order. For instance, although the order whereby a brand name comes first, the dosage form second and the standard unit last, such as 'トリアラム 錠 0.25 mg' (Trialam Tablets 0.25 mg), is typical, other orders also exist such as 'カムリトン 0.25 mg 錠' (Camriton 0.25mg Tablets).

For this reason, in order to treat the information included in medicine names in the CPOE, the medicine names must be broken down and the parts identified based on master data for each kind of information. Since such master data do not, unfortunately, exist, we first create master databases of information including standard units and dosage forms. We subsequently introduce the method to identify the information in medicine names by matching substrings to those obtained by utilizing N-grams and then extract the stems in medicine names by subtracting the matched substrings from the same. We evaluate this method by measuring the extraction precision.

TARGET DATA

In this study, we identified each medicinal product in Japan by a HOT9 code, which is identical if and only if the medicines have the same brand name, the same standard unit, the same dosage form and are produced by the same pharmaceutical company. In the rest of this paper, we count the number of products based on this code. As the master data of original medicine names, we used official name data in the 'MEDIS standard medicine master', the version dated July 31, 2009. This master data includes HOT9 code information that has a one-to-one relationship with each of 24,261 medicines.

We used dosage form master data contained in the dictionary data of single ingredient medicines, as compiled by the Health and Labour Sciences Research project, 'The study on medicine dictionary data items and standards (ICH M5)' in 2008.

MAKING OF MASTER DICTIONARIES

Besides master dictionaries of the original medicine name and dosage form respectively, we made additional dictionaries of symbol master data, standard unit master data, parenthetic expression master data and application master data.

For the symbol master dictionary, we broken down the original medicine name data into single characters and gathered those other than Kana (Katakana, Hiragana), Kanji, alphabetical, and numeric characters, e.g. 「, 」 , (,), and ⟨, ⟩ .

For the standard unit master dictionary, we split the standard unit data contained in the MEDIS standard medicine master by characters in the symbol master dictionary, and extracted letter strings and numbers (including the decimal separator characters, namely, commas and periods) followed by Kana, Kanji and letters, e.g. 25 mg, 10000 国際単位 (International Unit).

For the application master dictionary, we extracted letter strings with a certain pattern 'X用'. The Kanji character '用' denotes that 'X' is the application objective, e.g. 手術用 (surgical purpose), 注射用 (injection purpose). We assumed that 'X' consists of the same sort of letters (only Kanji, only Kana and so on).

For the parenthetic expression master dictionary, we extracted letter strings in parentheses and brackets included in the symbol master dictionary. Following extraction, we removed the duplicate entries in this and other dictionaries.

All data in these dictionaries are labeled with two alphabetical letters denoting the dictionary to which the data belongs and a sequential number in each dictionary

270

(Fig. 1).

Standard unit master dictionary		Dosage form master dictionary		Application master dictionary		Parenthetic expression master dictionary	
NewCode	data	NewCode	data	NewCode	data	NewCode	data

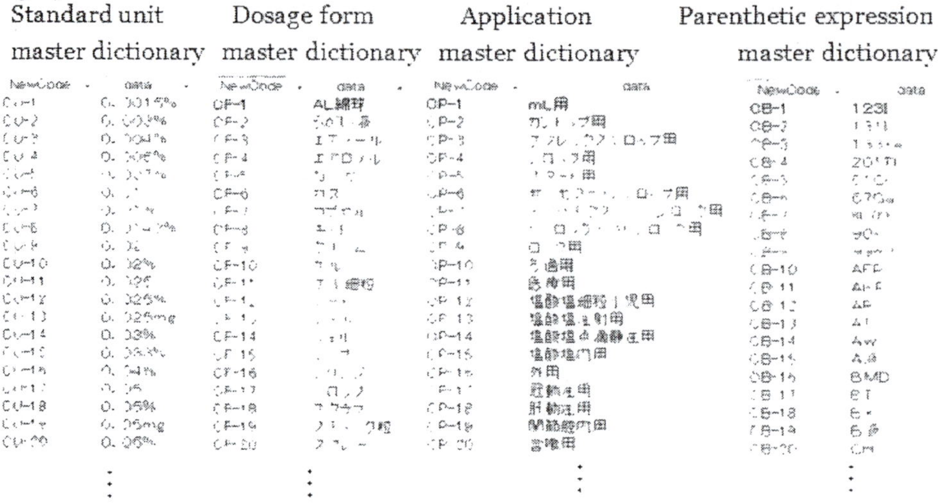

FIGURE 1   The master dictionaries (parts). The left column in each table shows the code corresponding to each data.

METHODS

In this study, we propose a method to extract a stem part in a medicine name by removing the matched parts to the dictionary data.

MATCHING PARTS OF MEDICINE NAMES TO MASTER DICTIONARIES

Though we may naively expect to find character substrings in medicine names that match data in the dictionaries by applying full-text searches, it is obviously computationally expensive. We can evaluate its cost as O(NLM), where N is the number of medicine names, L is the (typical) length of medicine names and M is the number of registered data in the dictionaries.

There is another problem in the naïve application of full-text search for cases where the abbreviated string of the registered data should also be registered in the dictionaries. For example, there are two words '注射液' and '注' that mean 'injection drug' ('注' is an abbreviation of '注射液'), which appear in names such as 'サルソニン注射液' (Salsonin injection) and 'ロヒプノール注' (Rohypnol injection). In such cases, both these words should be registered in the dictionaries. The difficulty is that 'サルソニン注射液' can be regarded as matching both '注射液' and '注', though we expect that only '注射液' to match.

To overcome such difficulties, we propose a means of comparison based on an N-gram, which is popular in the area of natural language processing (NLP).

First, we extract all contiguous substrings of definite length from 1 to L (the length of the medicine name) from a medicine name and compare them to the data in the dictionaries. We note that the calculation cost is $O(NL\log L \cdot \log M)$, under assumption that the dictionaries are indexed and their searching cost is in the order of $O(\log M)$. We can therefore anticipate the cost of our method to be much smaller than that for naïve full-text searches, since M is usually far more than $\log L \cdot \log M$.

Then, assuming that the original words of abbreviations are included in the dictionaries, we neglect the matched substring extracted in the first step, if it is included elsewhere as a substring. Consequently, we can expect to identify the substrings in each medicine name that correspond to the data listed in the dictionaries. Following this identification, by finding the dictionary in which the data corresponds to the substring, we can reorganize the information to find whether a medicine name lacks information on the dosage form, standard unit and so on.

Finally, we extract the stem part in a medicine name by removing the matched substrings. This is based on the assumption that no extra information is needed other than the data in the dictionaries for which we prepared.

A schematic figure of the method mentioned above is shown in Fig. 2.

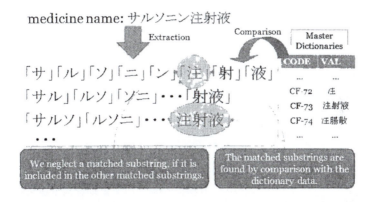

FIGURE 2 The schematic figure of the proposed method. The medicine name is broken down into string segments, the length of which ranges from one to that of the original medicine name. Each segment is compared to each dictionary and the matched substrings are compared to the other matched substrings to see whether they are part of the latter.

FINDING PAIRS OF MEDICINE NAMES WHOSE KANA STEMS ARE SIMILAR

As we mentioned in the Introduction, the problem is the confusion associated with

272

similar-sounding medicines. As for prescription input, this can be embodied as the wrongful input of medicines with names similar to the intended medicine. Since a name stem is usually used to specify a medicine, in this study, we extracted name pairs whose stems are mutually similar by calculating some similarity indices. (Since the name stems of Japanese medicines are composed of Katakana characters except for Chinese herb medicines, we focus on the Katakana part of the name stems.) As similarity indices, we use HTCO (the number of coincident characters in the first and final two characters respectively), H3COS1 (the number of coincident characters in the first three characters) and the edit distance that measures the number of operations required (addition, deletion and replacement) to transform one stem to another.

Taking the similarity of Katakana characters themselves into account, we identified the characters in the groups, 'アァヤカ', 'ツッシミ', 'ンソリ', 'クケフワ', 'エコユ', 'スヌ', 'ナメ', 'テラ', 'イィ', 'ウゥ', 'エェ', 'オォ' before calculating HTCO, H3COS1 and the edit distance.

EVALUATION

MATCHING PARTS OF MEDICINE NAMES TO MASTER DICTIONARIES

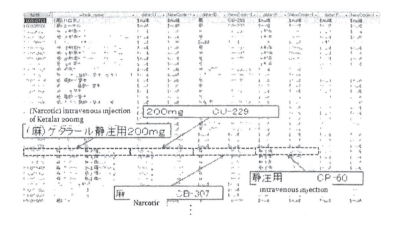

FIGURE 3 Matching result of medicine names (part). This shows that for the medicine named '(Narcotic) Ketalar intravenous injection 200 mg', the parts '200 mg', 'Narcotic' and 'intravenous injection', which are listed in the dictionaries, have been successfully broken down.

As shown in Fig. 3, we can see that our method breaks down the medicine name into certain parts corresponding to the dictionary data. For example, the original name, '(Narcotic) Ketalar intravenous injection 200 mg' is successfully broken down into parts in dictionaries, namely, 'Narcotic' in the parenthetic expression master dictionary, 'intravenous injection' in the dosage form master dictionary, and '200 mg' in the standard unit master dictionary respectively.

To evaluate the effectiveness of matching medicine names to the data in the dictionaries, we measure the accuracy of finding the stems in names by removing the matched data from the original names. (The symbols registered in the symbol master dictionary such as parentheses are removed from the original names prior to the extraction of the stems.) For 500 randomly selected medicine names, we successfully obtained 474 stems (94.8%). Fig. 4 shows some examples of extracted stems. The wrong extractions for 26 medicine names were caused by the multiplicity of meanings of the parts of names. For example, we obtained 'ター' (ter) as the 'stem' of the name, 'ガスター 散 2%' (Gaster powder 2%). The substring 'ガス' (gas), although only part of the brand name 'ガスター' (Gaster), has its own meaning as the state of a substance. Of cause, for human eyes, it is clear that the part does not refer to the physical state of the medicine. However this is implicitly based on the assumption that the string 'ガスター' (Gaster) is known as the brand name of a medicine. Since our objective is to identify the semantics of the parts of names, we cannot assume that their meanings are already known in advance.

In future study, we may solve this problem by taking account of the co-occurrence relationships of substrings appearing side-by-side in medicine names. If some substring invariably appears in another substring, we may presume that they suggest the existence of the stem containing those substrings as parts.

	hot9	whole_name	name_stem
1	100310901	（局）注射用チアミラールナトリウム	チアミラールナトリウム
2	100311601	（局）笑気ガス（ショウワ）	笑気
3	100311604	（局）亜酸化窒素	亜酸化窒素
4	100311608	（局）※液化亜酸化窒素（日産）	化亜酸化窒素
5	100313001	アネソキシン－５０	アネソキシン－
6	100314701	（麻）ケタラール静注用２００ｍｇ	ケタラール
7	100323912	（局）※ブロムワレリル尿素（山善）	ブロムワレリル尿素
8	100328401	ユーロジン１ｍｇ錠	ユーロジン
9	100333802	ネルガート１５	ネルガート
10	100343701	ニトラゼパム錠５ｍｇ「トーワ」	ニトラゼパム

FIGURE 4 The extracted name stems (part). The column name 'hot9', 'whole_name' and 'name_stem' respectively denote HOT9 codes, the original medicine names and the stem parts.

FINDING PAIRS OF MEDICINE NAMES WHOSE KANA STEMS ARE SIMILAR

Fig. 5 shows pairs of medicine names extracted under the condition HTCO=1, H3COS1=1 and edit distance equal to 1. Most are combinations of a brand name and its generic name, e.g. 'カルテオロール' (Carteolol) is the generic name of 'カ

ルテロール' (Cartelol). However, Fig. 5 also contains pairs of different efficacies. For example, the pair of 'タキソテール' (Taxotere) and 'タキソール' (Taxol) and that of 'アルマトール' (Almatol) and 'アルマール' (Almarl) are well-known as confusing names that cause severe accidents. It is striking that 'クインスロン' (Kuinsron) and 'クインロン' (Kuinlon) are confusing, have different efficacies and are produced by the same pharmaceutical company. It is preferable to discuss countermeasures against the confusion of such medicines within the company concerned.

カルテオロール	カルテロール
トリアゾラム	トリアラム
デキストセラン	デキストラン
テストビロンデポー	テストロンデポー
ラキソロン	ラキソロン
タキソテール	タキソール
フカノーゼリン	フカノーリン
ニカルジピン	ニカルピン
クィンスロン	クィンロン
ダイアコート	ダイアート
カプトプリル	カプトリル
オメプラゾール	オノフラール
オフロキサシン	オフロキシン
エフトリオールデポー	エストリールデポー
エフトリオール	エストリール
アルマトール	アルマール
アラセプリル	アラセリル
グリセリン	グリセロリン
フラボキサート	フラボサート
ベルマゾロン	ベルゾロン
フルコナゾール	フルコアール
プロビルアルコール	プロビールアルコール
フェノバール	フェノール
フェニタレン	フェニレン

FIGURE 5 The pairs of medicine names (Japanese) for whichh HTCO=4, H3COS1=3 and edit distance=1.

SUMMARY AND CONCLUSION

Though medicine names are a significant part of prescription information, their information structures are not fully standardized, making it difficult to identify the information therein. Not all medicines sold in Japan have sufficient information in their names and even for medicine names containing all the relevant information, there is no guarantee that they will be arranged in the determined order. For this reason, in order for the CPOE to treat the information in medicine names, the latter must be broken down into the parts matched to master dictionaries for each kind of information.

In this study, we proposed master dictionaries: the original medicine name master dictionary, the dosage form master dictionary, the symbol master dictionary, the standard unit master dictionary, the parenthetic expression master dictionary and the application master dictionary. We next introduced a method to identify the information in medicine names by matching pieces of strings obtained by N-grams to the data in the dictionaries. Subsequently, we extracted the stems in medicine names by subtracting the matched substrings from the same.

REFERENCES

Kimura, M., Tatsuno, K., Hayasaka, T., Takahashi, Y., Aoto, T., Ohkura, M., Tsuchiya, F. (2007), "Application of Data Mining Techniques to Medical Near-Miss Cases." *Proceedings of the 12th International Conference on Human-Computer Interaction.*

Tsuchiya, F. (2007), "Medication Errors Caused by Order Entry System and Prevention Measures." *Proceedings of the 12th International Conference on Human-Computer Interaction.*

Furukawa, H., Tsuchiya, F., Onishi, H., Masue, T., Bunko, H., Miyamoto, K. (2001), "Investigation of Possibility for Physician Order Entry System on Risk Management Strategy Related To Medication Errors." *Japan Journal of Medical Informatics*, 41, pp. 909–996.

Tsuchiya, F., Kawamura, N., Wang C., Hara, A.(2001), "Standardization and similarity deliberation of Drug-names." *Japan Journal of Medical Informatics*, 21(1), pp. 60–68.

The Medical Information System Development Center (1996), *http://www.medis.or.jp*

Chapter 33

Patient Safety in Education (Emergency Dept., OR, ICU)

D. Fuchs., B. Podtschaske, W. Friesdorf

Department Human Factors Engineering and Product Ergonomics
Berlin Institute of Technology
Fasanenstr. 1, 10623 Berlin, Germany

ABSTRACT

The main objective of clinical efforts is an increasing patient safety in treatment processes. Therefore well education is one of the most important issues in clinical work systems worldwide. In the past 20 years simulators has been established, particularly in Anesthesia Education Programs. Simulation training improves not only technical-skills of the medical experts; it also aims for better quality and safer treatment processes. But it is not sufficient to train only the medical staff members on the personal level. Patient safety can only be improved in a safely designed overall system. In this work a connection between education programs and patient safety is used and expanded to a system ergonomic education concept. All stakeholders of clinical environment are considered in this new systematic approach to guarantee a safe clinical work system, therefore a better patient safety.

Keywords: Patient Safety, Education Concept, Simulation, Medico Ergonomics

INTRODUCTION

Education is one of the most important issues in every work system. Particularly in the health care sector well educated personnel is needed, because every human or system error can lead to a fatal harm of the patient. The clinical work system is characterized by a system of medical experts working with high developed technique and inhomogeneous patient groups in dynamic processes. For reaching a higher patient safety or rather a secure system all involved stakeholders have to be considered in a systematic education concept.

SITUATION

The quality of medical treatment processes is often associated with the evaluation of the Patient safety. Safe processes are defined by the handling confidence of the clinical experts (cf. Rall et al, 2002). For this reason especially Anesthesiologists have implemented different education programs since more than 20 years. Particularly medical staff members of anesthesia are working in areas like Intensive Care Unit (ICU), Operation Room (OR) or Emergency Department (ER), where they are concerned with maximum pressure, high workload and critically ill patients.

Therefore more than 100 full-scale patient simulators are in use for Anesthesia qualification programs worldwide (cf. Rall et al (2002)). Most of them are used in university hospitals within the medical training. According to the aviation sector, Gaba et al. developed a "Simulated-Based Training in Anesthesia Crisis Resource Management (ACRM)". Residences in Anesthesia achieve practical experience and theoretical background (specific technical skills) as well as teamwork and communication skills (non-technical skills) within simulator training. These computer based programs can be adjusted on individual training status and skills of the participants and their know-how. Training concepts include different approaches for individual training and interdisciplinary groups; it is called "combined team training" especially to improve decision-making and teamwork skills.

Simulation training and other education programs are main parts of the medical education since years. Using simulators in combination with different safety concepts also from other areas e.g. aviation "crew resource management" considering the following targets:

- Specific technical skills
- Attitude towards mistakes
- Individual and team behavior, teamwork
- Decision making
- Risk management
- Communication management

But, is this the right way to get really safety health care systems?

PROBLEM

Simulator training is still not evaluated and their effect on mortality/mobility or patient safety is rarely proofed (Draycott T et al., 2006). More over individual training of medical staff members is only a single part of the overall system. For really improving treatment processes and patient safety it is necessary to analyze and understand the whole system (cf. Manser T et al., 2000). Only if all stakeholders get involved and concentrated on system safety, they can achieve real improvements for the system and finally the patient.

Stakeholders are not only medical staff members within the patient treatment process but the hospital administration, e.g. hospital management as well as building planer and medical technical industry. Especially medical technical industry often produces high developed technical devices without integration in the clinical work system. Can the complex interaction of different elements be improved by using only the simulated-modular training of the medical staff members (Users)?

All stakeholders together are responsible for the clinical environment and structural basics. Ergonomics (or Human Factors) discipline concerning with the understanding of interactions among humans and other elements of a system. What do they say?

Medico Ergonomic Education Concept

Following the system ergonomic HTO approach, which aims at a well balanced design of all system elements (Human, Technology and Organization) not only the Human (medical stuff) should be considered in an education concept also the organization and technology (e.g. medical technical industry) need to be considered for a safe and efficient work system. The design priority for a systematic education concept is structured according to the TOP-Approach: "first selecting safe technological solutions (T), before investigating organizational alternatives (O) or initiating personal behavior changes (P)" (Friesdorf W & Marsolek I, 2008). But what does it mean being implemented in a system ergonomic education concept?

An adapted TOP approach for a medico ergonomic education concept focuses not only on the clinical staff qualification but also on qualifying the involved organizational institutions and suppliers of the medical technical devices. For involving all stakeholders we need an interdisciplinary cooperation, which is directed to all specific experts. Sub processes of Communication, Coordination and Shared Knowledge according to the ergonomic cooperation model from Steinheider & Legrady (2000) should be optimized to reach a successful work system. The question remains: who is appropriated to coordinate this concept?

Ergonomics (or Human Factors) are highly recommended to coordinate interdisciplinary collaboration and thus to develop a systematic qualification concept – the Medico Ergonomic Education Concept (as illustrated in Figure 1).

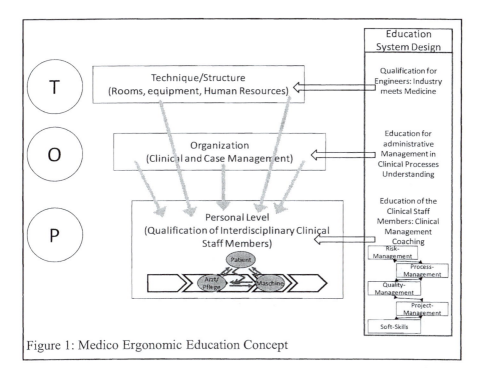

Figure 1: Medico Ergonomic Education Concept

It consists of three different levels of qualification:

1) Qualification of environmental institutions (engineers of the medical technical industry, hospital planers, hospital administration etc.)
   "Industry meets Medicine"
2) Qualification of organizational members and leading medical staff members within the clinic
   "Clinical Management Coaching"
3) Further and advanced Training of the Medical Staff Members in technical skills, non-technical skills and Management methods
   "Simulation-based Training"

On the personal level (P) there are already well established training methods and concepts, as described above. However there are hardly any approaches addressing the organizational or technical level. For this reason we developed and implemented additional training courses for all stakeholders of the health care system in our Medico Ergonomic Education Concept. The following detailed description focuses on the innovative parts of the concept. Two different qualification programs, that address the T(echnical) and O(rganizational) level, will presented below.

Medical technical equipment, which is suitable for the task, error tolerant and user-/patient friendly, is an essential basis for safe and efficient treatment processes. In order to develop those products it is necessary for the engineers to be familiar with the clinical work system and its treatment processes. For this reason we developed a custom-built three day qualification program for the biomedical industry. The program is called "Industry meets Medicine" and consists of the following components:

- system ergonomic model to convey a view on the whole system
- comprehensive overview about the entire medical rescue and treatment chain for poly-traumatized patients
- basic treatment components from diagnostics and therapy
- clinical information systems and resource management processes
- interaction among "human", "technology" and "organization" in clinical work system
- simulations of selected poly-traumatized treatment scenarios
- visitation program of different clinical work places ER, ICU,OR

The qualification for biomedical engineers refers to the technical level (T) of the TOP Approach within the Education Concept. The same applies to hospital building designers, architects or the hospital administration management. They also have to gain an understanding of the treatment processes and the medical way of thinking.

In order to address the organizational level (O) within the hospital, we designed a "Clinical Management Coaching". It is designed particularly for the head physicians and leading medical staff members The program considers the following requirements:

- Integration of hospital strategies
- Support of the collaboration of nurses and physicians as a team, theoretical basis on process- and project-management
- Adaption to the scope of working activities of the clinical staff members

The participants gain knowledge in management methods in the fields of process and project management, risk management, quality management as well as controlling. This approach facilitates interdisciplinary collaboration of the team and is adjusted to *intra*departmental characteristics as well as *inter*departmental processes. Thus it contributes to a sustainable organizational development of the whole clinic.

Education programs are essential for maintaining medical standards and high quality. But only by integrating all three levels (TOP) it is possible to systematically improve treatment processes and achieve high patient safety all over the system.

REFERENCES

Draycott, T., Sibanda, T., Owen, L. (2006), *Does training in obstetric emergencies improve neonatal outcome?* BJOG 113:177-182.

Manser T., Wehner, T., Rall, M. (2000), *Analysing action sequences in anaesthesia.* Europ Journal Anaest 17: 526

Friesdorf, W. & Marsolek, I. (2008), *Fehlerhafter Umgang mit Medizingeräten.* Zeitschrift für Evidenz, Fortbildung und Qualität im Gesundheitswesen 102 (9), 563-567, Germany.

Gaba, D.M., Howard, S.K., Fish, K.J., Smith, B.E., Sowb, Y.A. (2001), Simulated-Based Training in Anesthesia Crises Resource Management (ACRM): A Decade of Experience, Simulated Gaming: 32; 175-193, USA.

Lussi, C., Grapengeter M., Schüttler J. (1999), *Simulatortraining in der Anästhesie*, Anaesthesist 48: 433-438, Springer-Verlag, Germany.

Rall, M., Schaedle, B., Zieger, J., Naef, W., Weinlich, M. (2002), Neue Trainingsformen und Erhöhung der Patinetensicherheit, Unfallchirirg 105: 1033-1042, Germany.

Steinheider, B. (2000). Cooperation in interdisciplinary R&D Teams, Proceedings of the ISATA 2000, International Symposium on Automotive Technology and Automation (33rd: 2000 Sep : Dublin) (pp. 125-130). Epsom: ISATA-Düsseldorf Trade Fair, Germany.

Chapter 34

Patient Safety? Usability? Experience in Real Clinical Life

Ulrich Matern, Dirk Büchel

Experimental-OR
University Hospital of Tuebingen
Tübingen, Germany

ABSTRACT

This study analyzes the 1,330 occurrence messages for the sector OR received between the years 2000 and 2006 by the German Federal Institute for Drugs and Medical Devices (BfArM). The connection between the usability errors of the medical devices and the damage of patients and co-workers is shown, as well as thereupon taken measures of the manufacturers.

Keywords: Usability, medical devices, surgery

INTRODUCTION

The operating room is one of the most expensive and most complex units within a hospital. To safely and efficiently operate an OR all components (building and equipment technology) should be harmonized with each other. Own experience being a surgeon, showed that this is not the case. To analyze this hypothesis and to accordingly draw the resultant conclusions various investigations were performed using custom-built tools.

Via a checklist ORs were evaluated, surgeons, OR nurses and anesthetists were surveyed using standardized questionnaires, internet based critical incident reporting system were researched.

The hypothesis was corroborated. Thus, among other things 70% of the surgeons and 49% of the OR nurses are not able to correctly handle the equipment. There is a nasty draft in the OR. Rooms are unsuitable as there is a lack of supply connectors, equipment or staff. The staff is overstrained from the nowadays operating ORs, leading to a high potential of hazards of all persons within the OR and making the operation uneconomic [3].

Ergonomics & equipment design, workflow processes, architecture, climate & hygiene, navigation & positioning, system-interfaces as well as supply are the substantial topics to improve safety and economic efficiency of the ORs.

In co-operation with the BfArM (German Federal Institute for Drugs and Medical Devices, Bonn, Germany), that is similar to the FDA, and our group occurrences within the OR from critical incident archive of the BfArM were evaluated. The study is dedicated to the aim to ascertain, how far ergonomic deficiencies and an unsatisfactory usability were involved in these occurrences.

MATERIAL AND METHOD

Altogether 1,330 cases from the archive of the BfArM were examined, which had been registered and worked on in the years from 2000 to 2006. Firstly, the notifications were listed by the class of the device and the announced error (loss of function, electrical error, mechanical problems...) was documented. Next, the descriptions of the occurrence made by the first reporter, the statement of the manufacturer and the final evaluation of the BfArM were reviewed.

RESULTS

1.330 critical incidents had been reported between 2000 and 2006 to the BfArM.

From these 42% (643 application and use errors) could be led back to a problem in communication between the user and the medical device. In 21% of all cases no reason could be determined. This was the matter because of the fact that the involved medical device was disposed by the hospital personnel directly after the event and was not available for further investigation.

In 17% (48 occurrences) the unsatisfactory description of the product leads to misunderstandings. The reasons for these problems can be found within the users manual or an unsatisfactory feedback represented by the product itself, e.g. numeric codes, leads to faulty operations.

In 63% (229 occurrences) the medical device was used for the wrong application, including, the use of the medical device for something it had not been intended for.

95 % of the reported occurrences are connected to an endangerment of the persons (1.264), in the OR.

Consequences for products in the field

Constructional changes were made in approximately a quarter of all cases.
Regarding the cases of suitable usability the recall rate lies by 14%.
In 52% of the cases the manufacturers decided to optimize the information for the users (e.g. user manual) or the training of the users.
A Selling -, production- and application stop, did not play any substantial role.

DISCUSSION

This study shows only the peak of the iceberg, because most users in German hospitals have little knowledge of the legal obligation to register occurrences to the medical-device-operator-regulation [4], respectively the safety-plan-regulation [5]. The BfArM is responsible for the central collection, analysis and evaluation of possible upcoming risks concerning the use of medical devices. The processes of reporting often are not sufficiently known in hospitals and/or in individual cases are not known at all.
It has to be considered that those users, who reported to BfArM already assumed a technical error of the medical device. Incidents due to ergonomic causes by the users were recognized only rarely and were also not thought of as compulsorily modifiable. Keeping this aspect in mind it is amazing that numerous ergonomic problems are still hidden behind these reports. Through current changes of the MDD [2] especially these occurrences should be seized in future.
This study describes the connection between usability errors, the endangerment of the patients and co-workers in the OR, as well as the taken measures. Thereby, it is made clear that this matter of fact concerns a relevant medicine-technical problem, with which human suffering and also costs of the treating hospital, the manufacturers, the insurance companies and the entire health system is connected. Additionally, it is defined in the fundamental requirements of the MDD that primarily the equipment design is responsible for security.
Therefore it is amazing that in only 31% of the occurrences caused by unsatisfactory usability the design of the criticized devices was changed; in 52% however the product information / operating instruction was improved. These measures have a very limited effect, since many users consider neither having read nor taken notice of the operating instruction [3] avoiding the operator regulation.
It seems to be necessary to seize more suitable measures in the future. Directions are given by the new international standard EN60601-1-6, which demands the consideration of the usability as an independent process during the product development [1].
But also on the part of the users and operators everything should be done, in order to ensure the security of the patients and the personnel. Therefore error messages should be considered and function tests, which are demanded by the equipment before the application should be accomplished. In future users and operators should

consider ergonomic aspects in the selection of new medical devices more than it is usual today. The current ways of procurement including the sample order have to be tested and possibly replaced by specific usability tests if necessary.

REFERENCES

1. DIN/EN 60601-1-6: *Medical electrical equipment – Part 1-6: General requirements for safety – Collateral standard: Usability 1. Edition.* Berlin, Germany, Beuth 2008
2. European „Medical Device Directive" 2. August 1994. Bundesgesetzblatt Nr. 52, vom 9. August 1994, S. 1963ff.
3. Matern U., Koneczny S., Scherrer M., Gerlings T.: *Working Conditions and Safety in the Operating Room.* Dtsch Arztebl 2006; 103(47): A 3187–92
4. "Medical-Device-Operator-Regulation" *Verordnung über das Errichten, Betreiben und Anwenden von Medizinprodukten (Medizinprodukte-Betreiberverordnung MPBetreibV).* Edition from 21.08.2002. BGBl. I S. 3396
5. „Safety-Plan-Regulation" *Verordnung über die Erfassung, Bewertung und Abwehr von Risiken bei Medizinprodukten (Medizinprodukte-Sicherheitsplanverordnung - MPSV).* Edition from 24.06.2002

Chapter 35

How to Develop Usable Surgical Devices – The View From a U.S. Research University

M. Susan Hallbeck

University of Nebraska-Lincoln
Lincoln, NE 68588-0518, USA

ABSTRACT

User-centered design (UCD) is a broad term used to describe design processes in which the end-users influence how a design takes shape (Abras, et al., 2004). It is both a philosophy and a process. It is a philosophy in that it puts the user in the center of the design process, and it is a process in that it focuses on the cognitive factors that come into play during the interaction of people with objects (U.S. Department of Health & Human Services, 2007). UCD is a comprehensive approach to product development that is based on two fundamental premises: that the design should be focused on users and that a scientific, data-driven or engineering approach to design must be taken.

Most medical devices, including surgical and laparoscopic tools, have not been designed using UCD principles; in fact some appear not to have considered there was a user. In an effort to reduce the discomfort that surgeons experience, UCD principles were developed and employed in University design projects. These methodologies involve an integration of users' needs and expectations in combination with the tasks performed. The methodologies followed include the Quality Function Deployment (Akao, 1999; Marsot, 2005; Kuijt-Evers et al., 2009), Bobjer and Jansson's (1997) "11-point program to design ergonomic tools, an adaption of Nielsen's (1993) usability principles (Hallbeck and Oleynikov, 2006), and a practical ergonomic product design process (Jung et al., 2008). In addition, ISO, ANSI, and IEC standards as well as AAMI, AHRQ, FDA and IHI guidelines must be utilized. Teaching University students to combine these techniques with standard and guidelines for medical device design allows the students to approach the development process from a number of different angles and challenges them to think about the user and understand the socio-technical., cognitive and physical parameters, as well as the interactions of these parameters, with respect to the (re)design effort for task performance.

Keywords: User-centered design, usability, medical device design, pedagogy

THE PROBLEM

Biomedical., mechanical and even industrial engineering students at US Universities can and do follow standards from ISO and ANSI, etc. for surgical and medical device design in general. However, those standard requirements (shall) and guidelines (should) are not enough to create a good and usable medical device. Usability requires human factors /ergonomics input following a user-centered design methodology. Better design pedagogy can change patient safety by reducing medical errors and also by enhancing safety for the clinical staff.

FOLLOWING THE RULES

The ISO and ANSI standards are a great first place to begin the journey to better device design. Many of these standards work well to specify device characteristics. These standards include:

- IEC (2004) 60601-1-6: Medical electrical equipment – Part 1-6: General requirements for safety- Collateral standard: Usability
- IEC (2000). EN 894-3: Safety of Machinery – Ergonomics Requirements for the Design of Display and Control Actuators Part 3: Control Actuators.
- IEC (2002). IEC 60073: Basic and Safety Principles for Man-Machine Interface, Marking and Identification.
- ISO (2006). 9241-110: Ergonomics of human-system interaction
- ISO/IEC 25062 (2006). Software engineering — Software product Quality Requirements and Evaluation (SQuaRE) — Common Industry Format (CIF) for usability test reports

However, simply following the rules not enough to ensure a usable, intuitive, great device design. Other standards and guidelines from the Association of Medical Instrumentation (AAMI), Agency for Healthcare Research and Quality (AHRQ) and the Food and Drug Administration (FDA) will go the next step:

- AAMI HE 48(1993): Human factors engineering guidelines and preferred practices for the design of medical devices
- AAMI HE 75 (2009). Human factors engineering - Design of medical devices
- ANSI/AAMI HE 74 (2001): Human factors design process for medical devices

- FDA (1996) Good Manufacturing Practices / Quality Systems Regulations
- FDA (2000). Medical Device Use-Safety: Incorporating Human Factors Engineering in to Risk Management.
- FDA (1997). Do it by Design-An intro to human factors in medical devices.
- AHRQ Quality Improvement and Patient Safety guidelines (AHRQ.gov)

These latter guidelines and standards include human factors/ergonomics methodology. The FDA defines Human Factors as a science devoted to the interaction of people and equipment. "Human Factors," "human engineering," "usability engineering," and "ergonomics" are often used interchangeably. In the field of medicine, the objective of Human Factors is to improve human performance and reduce the likelihood of use error and patient injury (FDA, 2003).

While these rules are a great start for developing individual surgical devices, it isn't enough, the system in which the device will be used and maintained needs to be considered.

WHAT IS MISSING?

There needs to be a systems approach and there needs to be a focus on user-centered design – for the whole system. Surgical suite *systems* errors arise from two main factors: 1) inadequate training regarding device handling and use, including poorly organized routines and 2) poorly designed user interfaces inducing errors or not facilitating recovery from error during operations (Hallbeck et al., 2008).

Medicine traditionally has treated errors in the operating room as failings of the medical staff, implying inadequate knowledge or skill performance. Engineering or a systems safety approach focuses on the process of hazard, safety and risk analysis. One major tenet of the systems approach is that most errors are the product of predictable human failings in the context of poorly designed systems. Thus, the systems approach seeks to identify situations or factors likely to give rise to human error and implement "systems changes" that reduce their occurrence or minimize patient impact. Full understanding of what users want or need and how they perform tasks permit a systems approach to reducing error and designing better medical devices (Hallbeck et al., 2008). This important information can be obtained by using various evaluation methods (e.g. observations, questionnaires, interviews, checklists, expert appraisals, accident or injury analyses, task analyses, safety analyses, and so on) (Cushieri, 1995; Hedge, 2001; Koneczny and Matern, 2004; Patkin, 1997, Rall et al., 2002; Thomas et al., 2003). A good way to elicit the information follows user-centered design principles, in which a design or process is focused on users and a scientific, data-driven engineering approach to design/use is taken (Chapanis, 1995; Hallbeck and Oleynikov, 2006; Hallbeck, et al., 2008).

The poorly-designed system-provoked errors do not inevitably lead to hazardous situations for patients or operating room teams; often they merely lead to a time-consuming and costly interruption of a surgical procedure. They may also not provide for near miss recovery – later leading to an error. The sum of these errors, however, may induce a combination of causes that might result in hazardous situations for all persons in the operating room (patient and surgical team). Reason

(1990; 1997; 2000) referred to this as the Swiss cheese model, where all the holes (minor provoked error and hazardous situations) align to allow a serious event or incident to occur in the operating room. A major source of these bad-design provoked errors is related to device operation. Norman (2002) summed this up well when he stated, "If it needs a sign or explanation, it is not well-designed".

Approximately two-thirds of the incidents related to the devices can be ascribed to incorrect operation by the user and not to technical faults (Backhaus, 2004; Bleyer, 1992; Runciman et al., 1993). In addition, cleaning and maintenance must be taken into account for medical devices or life-threatening error can occur (Poovey, 2009).

Patient-centeredness and patient safety were identified as two of the six specific aims for improvement by the Institute of Medicine (IOM) in its 2001 report *Crossing the Quality Chasm* (Corrigan, et al., 2001). The IOM defines patient safety as "freedom from accidental injury" (2001). Nielsen (1993, as cited by Sonderston and Rauch, 1996) states that usability engineering can double a system's learnability, give a 25% increase in its efficiency, reduce the error rates by 500%, and double the user's satisfaction – this refers to software, but there is no reason this cannot happen for medical devices as well. This patient-centeredness and safety can come from user-centered design.

USER-CENTERED DESIGN

User-centered system design starts with the needs of the user (Soderston and Rauch, 1996). One way to design and evaluate the medical devices is to look at user-centered design (in this case surgeon- or nurse-centered design) principles for the tool design and operation. These user-centered principles include: 1) *Ease of learning and use* – the tool needs to allow users who have never seen it before to learn to use it quickly to accomplish basic laparoscopic surgical tasks, 2) *Efficiency of use* – the tool needs to be designed to allow rapid accomplishment of tasks for more experienced nurses or laparoscopic surgeons, 3) *Error minimization* – the tool should be designed to minimize the number and severity of errors, 4) *Subjective satisfaction* – the experience of using a tool should be a pleasant and comfortable one, and 5) *Accommodation* – the tool needs to be fit (or adjust) to as wide a group of surgeons as possible and minimize potential damage to the surgeon (Nielsen, 1993; Hallbeck and Oleynikov, 2006).

Developing successful, safe and useful medical products relies on the product developers understanding of the target group for whom they are designing (Gould, 1995; Janhager, 2005; Margolin, 1997; Preece, 2002). The user-centered design (UCD) model incorporates the needs, abilities, skills and performance of the user into the redesign process. UCD is an approach for employing usability. Usability has been defined by the International Standards Organization (ISO) as, "The effectiveness, efficiency and satisfaction with which specified users achieve specified goals in particular environments" (ISO DIS 9241-11). Simply put, usability designates the ease with which people can use a tool / product in order to achieve a goal. If a product has good usability, the user can employ the product without much thought and/ or effort. However, if a product has poor usability the safety of the user can be at risk.

As usability becomes more and more common, customers such as medical staff are starting to expect products to be easy to use. This means usability has moved from being what marketing professionals call a "satisfier" to being a "dissatisfier" (Jordan , 2000). People are no longer pleasantly surprised when a product is usable, but they are unpleasantly surprised when it is difficult to use. This shows that usability is something that is absolutely necessary for product acceptance and success. If it takes the user a long time to operate the product and it is difficult to use, the product has poor usability. So how do we know if a product is "quick and easy" to use? First, it is important to understand that usability is not a one-dimensional property; it is a combination of factors including: error frequency, subjective satisfaction, ease of learning, memorability and efficiency of use (Liljegren, 2006, Herring, 2008). Usability and user-centered design both follow the same principles. Liljegren (2006) found that in medical technology, users determined that the component "difficult to make errors" was 30% of overall usability, while "easy to learn", "efficient to use" and "easy to remember" were each 20% of the overall usability and satisfaction was only 10%.

Usability has also been directly linked to safety concerns in products. In fact, a lack of usability has been shown to be a casual factor in many industrial., domestic and transport accidents (Jordan, 1998). If a product has poor usability, the repercussions can range anywhere from simply annnoying the user to actually putting the users' lives at risk (Green and Jordan, 1999). This can be attributed to designers not understanding both the physical and psychological human characteristics of the population for whom they are designing. This shows the importance, and implications, of not designing products for proper usability of the intended population. Teaching usability and UCD are important for the future of surgical device design.

TEACHING USER-CENTERED DESIGN

Design must address a set of needs and clearly-defined constraints through what Charles Eames describes as, "a plan for arranging elements to accomplish a particular purpose". The identification of these needs, purposes or elements, requires systems engineering or *problem-framing* skills prior to any activities of the problem-solving kind (Eames, 1989, as cited by Comazzi, 2008). Essential components of this understanding include the actual device users, (e.g., patient, family member, physician, nurse, professional caregiver), typical and atypical device use, device characteristics (domain most easily understood via standards and guidelines, etc.), characteristics of the environments in which the device will be used, and the interaction between users, devices, and use environments.

Over the years, companies have realized the importance of ergonomics in product design because ergonomically designed products have a competitive advantage in the marketplace (Cayol and Bonhoure, 2004). Therefore, ergonomists are often requested to develop a new product or improve an existing product (Marmaras and Zarboutis, 1997; Cai and You, 1998; Das et al., 2002).

The nature of a product and the functional requirements may be simple or complex; however, the product design process involves a series of events that would involve identifying user's needs, defining design concepts, making a prototype, testing usability, and releasing a product to the market (Hedge, 2001). Iterative application of the most relevant current knowledge and experience throughout this process will yield an ergonomically sound product (Jung et al., 2008). Based on this general process, a rather practical design process for ergonomic products was developed. These 9 iterative steps are assessing design needs, examining standards and guidelines, getting user information (e.g. anthropometry), creating an idea sketch, making a preliminary model, making a working prototype, getting assessments from the end-users and finally design validation (after iteration of the latter steps) – which are similar to the 11-points of Bobjer and Jansson (1997).

These methodologies involve an integration of users' needs and expectations in combination with the tasks performed – hence a systems approach. Full understanding of who (target population) wants what kind of products or customer needs is the first and most important factor (Chapanis, 1995). The user's needs can be specified by using various evaluation techniques, e.g. observations, questionnaires, interviews, checklists, expert appraisals, accident or injury analyses, task analyses, safety analyses, and so on (Stanton and Young, 1998; Page, 1998; Hedge, 2001). We can also incorporate 5S or lean methodology such as "Simplification" by including process mapping, visual systems (a.k.a. 5S), designing benign failures to avoid costly errors as well as facilitating correct action to reduce the errors and increase patient safety.

Other design and analysis techniques can aid the engineering student in the medical/surgical tool design process. These techniques include QFD- whose basic premise is "what does the customer want?". These are translated into customer needs with the next question aiding that "How can the device/system fulfill those needs?". We can then link the "Whats" and "Hows" by quantifying how they are related (Akao, 1993). Since these relationships must be quantified, this leads to use of other methods such as systematic user evaluation (Kuijt-Evers et al., 2009). Contextual inquiry, cognitive task analyses, usability tests, heuristics, cognitive walkthrough, focus groups and Delphi techniques are all recommended, for various development phases and needs by Martin et al. (2008). Liljegren (2006) found that usability testing was the primary methodology to evaluate designs, but that hierarchical task analysis and cognitive walkthrough are also useful. Others (Zhang et al., 2003) found that heuristic evaluation was an efficient, useful and low-cost method for evaluation and identification of usability problems and their severities. We can teach students to query the users about task, challenges, benefits and ask the pivotal question "why?" they do it, as users perform their jobs or visualize a future job.

The FDA (2003) asserts that adding human factors considerations into medical devices can result in intuitive operation and low reliance on manuals; easy-to-read displays and controls; positive and safe connections; effective alarms and easy repair and maintenance. Overall, human factors engineers and students can make life easier by producing better, customer-focused products. We should strive to produce products that are simple, yet value-adding; elegant, pragmatic and intuitive.

REFERENCES

Abras, C., Maloney-Krichmar, D. and Preece, J. (2004) User-Centered Design. *In*: Bainbridge, W., ED. *Birkshire Encyclopedia of Human-Computer Interaction.* Thousand Oaks, CA: Sage Publications, 2004.

Akao, Y. (1993). *Quality Function Deployment.* Productivity Press, Cambridge, UK.

American National Standards Institute AAMI. (1993). *Human factors engineering guidelines and preferred practices for the design of medical devices* (ANSI/AAMI HE-48). Arlington, VA: Association for the Advancement of Medical Instrumentation, 1993.

American National Standards Institute AAMI. (2001). *Human factors design process for medical devices (ANSI/AAMI HE-74). Arlington, VA: Association for the Advancement of Medical Instrumentation,* 2001.

American National Standards Institute AAMI. (2009). *Human factors engineering – Design of medical devices* (ANSI/AAMI HE-75/3rd Edition). Arlington, VA: Association for the Advancement of Medical Instrumentation, 1993.

American National Standards Institute. (2001). *Common industry format for usability test reports. ANSI-NCITS 354-2001, Washington, DC, USA*

Backhaus, C. (2004) Einige Zahlen zur Sicherheit von Medizintechnik. mt-Medizintechnik 124: 202-203

Bleyer, S. (1992). Medizinisch-technische Zwischenfälle in Krankenhäusern und ihre Verhinderung. In: Anna O, Hartung C (Eds.), Mitteilungen des Instituts für Biomedizinische Technik und Krankenhaustechnik der Medizinischen Hochschule Hannover. MH Hannover, Germany

Bobjer, O. and Jansson, C. (1997). A research approach to the design of ergonomic hand tools. The 11-point programme. Triennial Congress of the International Ergonomics Association, Volume 2, 193-195.

Cai, D. and You, M., 1998. An ergonomic approach to public squatting-type toilet design. Applied Ergonomics 29, 147-153.

Chapanis A (1995) Ergonomics in product development: a personal view. Ergonomics 38: 1625-1638

Corrigan, J.M., Donaldson, M.S., Kohn, L.T. et al. (2001). Crossing the Quality Chasm. A New Health System for the 21st Century. Washington, DC: Institute of Medicine, National Academy of Sciences, National Academies Press.

Cuschieri, A. (1995) Whither minimal access surgery: Tribulations and expectations. American Journal of Surgery 169: 9-19.

Das, B., Wimpee, J., Das, B., 2002. Ergonomics evaluation and redesign of a hospital meal cart. Applied Ergonomics 33, 309-318.

Eames, C. (1989)."Design Q&A: Questions by Mme. L. Amic & Answers by Charles Eames," in *The Films of Charles & Ray Eames*, vol. 4 - DVD, 1989. As cited by Making Material Matter: Design in Education *John Comazzi (2008) 4th* Creative Engagements conference in Oxford, UK

EN (IEC) (2000) Safety of Machinery – Ergonomics Requirements for the Design of Display and Control Actuators Part 3: Control Actuators. EN 894-3. European Standard Adopted 2000

EN (IEC) (2004) Medical Electrical Equipment; Part 1-6. General Requirements for safety. EN (IEC) 60601-1-6. European Standard Adopted 2004

FDA (2003) OHIP Annual Report - Reducing Use Error http://www.fda.gov/ AboutFDA/CentersOffices/CDRH/CDRHReports/ucm130329.htm

Gould, J. (1995). How to Design Usable Systems. I*n:* Baecker, Grudin, Buxton and Greenberg (eds.)*, Readings in Human-Computer Interation: Toward the Year 2000,* San Francisco, CA:Morgan Kaufmann Publishers, 93-121.

Green, W.S., and Jordan, P.W., 1999. *Human Factors in Product Design.* London: Taylor & Francis.

Hallbeck, M.S. and Oleynikov, D. (2006). Hands on... Surgeon-Centered Laparoscopic Tool Design. Triennial Congress of the International Ergonomics Association.(Eds. R.N. Pikaar, E.A.P. Koningsveld and P.J.M. Settels), ISSN 0003-6870, Elsevier Ltd. 2006, 1777-1781

Hallbeck, M.S., Koneczny, S., Büchel, D. and Matern, U. (2008). Ergonomic Usability Testing of Operating Room Devices. *Studies in Health Technology and Informatics*, 132, 147-152.

Hedge, A (2001) Consumer product design, in: W. Karwowski (Eds.), International Encyclopedia of Ergonomics and Human Factors. Volume II, pp. 888-891, Taylor and Francis, London

Herring, S.R. (2008) Implications of not using user-centered design: A three part investigation. Unpublished Master's thesis University of Nebraska – Lincoln

IEC (2002) Basic and Safety Principles for Man-Machine Interface, Marking and Identification. IEC 60073. International Standard Adopted 2002

IEC (2005) Medical devices – General requirements for safety and essential performance – Usability. IEC 62366

ISO (1998) Ergonomic requirements for office work with visual display terminals (VDTs) – Part 11: Guidance on usability. ISO 9241-11

ISO/IEC (2003) Medical electrical equipment -- Parts 1-8: General requirements for safety - Collateral standard: General requirements, tests and guidance for alarm systems in medical electrical equipment and medical electrical systems. ISO/IEC 60601: 1-8

Jordan, P. W., 1998. *An Introduction to Usability.* London: Taylor and Francis.

Jordan, P., 2000. *Designing Pleasureable Products: An introdution to the new human factors.* London: Taylor & Francis.

Jung, M.C., Hallbeck, M.S., Kim, J.-Y. and Haight, J.M. (2008). Practical design process for ergonomic hand tool handle and desk and chair development. *International Journal of Occupational Safety and Ergonomics*, 14(2), 247–252.

Kaye, R. and Crowley, J. (2000). Medical Device Use-Safety: Incorporating Human Factors Engineering into Risk Management Identifying, Understanding, and Addressing Use-Related Hazards, FDA. http://www.fda.gov/cdrh/humfac/1497.pdf

Keyes, M.A., Ortiz, E., Queenan, D., Huges, R., Chesley, F., Hogan, E.M. (2005) A Strategic Approach for Funding Research: The Agency for Healthcare Research and Quality's Patient Safety Initiative 2000-2004. In: Advances in Patient Safety: From Research to Implementation, http://www.ahrq.gov/qual/advances/

Kohn, L.T., Corrigan, J.M., Donaldson, M.S. (1999). To Err is Human - Building a Safer Health System. National Academy Press, Washington

Koneczny, S. and Matern, U. (2004). Instruments for the evaluation of ergonomics in surgery. Minimally Invasive Therapy & Allied Technologies, 13, 167-177

Kuijt-Evers, L.F.M.,Morel,K.P.N., Eikelenberg, N.L.W. and Vink,P. (2009). Application of QFD as a design approach to ensure comfort using hand tools: Can the design team complete the House of Quality? *Applied Ergonomics*, 40(3), 519-526.

Leape, L.L. (1994). Error in medicine. *JAMA,* 272, 1851-1857.

Leape, L.L., Berwick, D.M. (2005). Five years after "To Err is Human". What have we learned? The Journal of the American Medical Association, 293, 2384-2390

Lewis, J.R. (1992). Psychometric evaluation of the Post-Study System Usability Questionnaire: The PSSUQ. Proceedings of the Human Factors Society 36th Annual Meeting: 1259-1263

Lewis, W.G., Narayan, C.V. (1993). Design and sizing of ergonomic handles for hand tools. Applied Ergonomics, 24, 351-356.

Liljegren, E. (2006). Usability in a medical technology context assessment of methods for usability evaluation of medical equipment. International Journal of Industrial Ergonomics, 36, 345–352

Margolin, V. (1997). Getting to know the user. *Design Studies*, 18(3), 227-236.

Marmaras, N. and Zarboutis, N. (1997). Ergonomic redesign of the electric guitar. Applied Ergonomics 28, 59-67.

Marsot, J. (2005). QFD: A methodological tools for integration of ergonomics at the design stage. *Applied Ergonomics*, 36, 185-192.

Martin, J.L., Norris, B.J., Murphy, E. and Crowe, J.A. (2008). Medical device development: The challenge for ergonomics. *Applied Ergonomics*, 39, 271-283.

National Patient Safety Agency (2005) National Incident Reporting System. NPSA, National Reporting and Learning System, http://www.npsa.nhs.uk/health

Nielsen, J. (1993). Usability Engineering. Academic Press, San Diego, CA.

Norman DA (2002) The design of everyday things. Basic books, New York

Page M (1998) Consumer products – more by accident than design? In: Stanton N (Ed.) Human Factors in Consumer Products: 127-146, Taylor & Francis

Patkin M (1997). A checklist for handle design. Ergonomics Australia On-Line 11(2). http://www.uq.edu.au/eaol/apr97/handle.html

Poovey, W. (2009). Doctor: HIV Infections Will Never Be Traced to VA. May 8, 2009. http://abcnews.go.com/US/wireStory?id=7542087

Porter, S., Porter, J.M., 1999. Designing for usability; Input of ergonomics information at an appropriate point, and appropriate form, in the design process, In: Green, W.S. and Jordan, P.W. (Ed.), Human Factors in Product Design: Current Practice and Future Trends. Taylor & Francis, London, pp. 15-25.

Preece, J., 2002. *Interaction Design - beyond human-computer interaction.* New York: John Wiley and Sons, Inc.

Rall, M., Zieger, J., Schaedle, B., Haible, T., Dieckmann, P. (2002). The Critical Incident Analysis Tool: Facilitating to Find Underlying Causes of Critical Incidents in Anaesthesiology for Novices in Human Error. Workshop on Investigating and Reporting of Incidents and Accidents (IRIA), Glasgow 2002. http://www.dcs.gla.ac.uk/~johnson/iria2002/IRIA_2002.pdf

Reason J (1990) Human Error. Cambridge University Press, Cambridge (UK)

Reason J (1997) Managing the Risks of Organizational Accidents. Ashgate Publishing, London

Reason J (2000) Human error: models and management. British Medical Journal 320: 768-770

Runciman WB, Webb RK, Lee R, Holland R (1993) The Australian Incident Monitoring Study. System failure: an analysis of 2000 incident reports. Anaesthesia and Intens. Care 21: 684-695

Sawyer D. (1997). Do It By Design: An Introduction to Human Factors in Medical Devices. FDA. http://www.fda.gov/MedicalDevices/ DeviceRegulationandGuidance/GuidanceDocuments/ucm094957.htm

Sawyer D. (2000). Medical device requirements, human factors, and the food and drug administration. In: *Proceedings of the Human Factors and Ergonomics Society 44th Annual Meeting.* Santa Barbara: Human Factors and Ergonomics Society, 526-527.

Soderston, C. and Rauch T. (1996). The case for user-centered design. In: *Proceedings of the STC 44th Annual Conference. Society for Technical Communication.*

Stanton N, Young M (1998) Is utility in the mind of the beholder? A study of ergonomics methods. Applied Ergonomics 29: 41-54.

Thomas AN, Pilkington CE, Greer R (2003) Critical incident reporting in UK intensive care units: a mail survey. J. of Evaluation in Clinical Practice. 9, 59-68

U.S. Department of Health & Human Services (2007). *What is Usability?* Available from: www.usability.gov. [Accessed 7 Sept. 2007].

Zhang, J., Johnson, T.R., Patel, V.L., Paige, D.L., and Kubose T. (2003). Using usability heuristics to evaluate patient safety of medical devices, Journal of Biomedical Informatics, 36, 23-30.

Chapter 36

Standardized Evaluation Process of Usability Properties

Dirk Büchel, Ulrich Matern

Experimental-OR, University Hospital of Tuebingen
Tübingen, Germany

ABSTRACT

Newly usability aspects become an issue in health care. Of particular importance for usable products is the evaluation. In this context it is necessary to attend among the defined usability criteria to safety, too. In this chapter a standard procedure for usability evaluation and a usability assessment method regarding to safety aspects is presented. It is shown that these methods helps to avoid the evaluator effect and that it enhances objectivity, reliability and validity of the usability evaluation in health care.

Keywords: Usability evaluation, evaluator effect, reliability, objectivity, UseProb, heuristic evaluation, user test

INTRODUCTION

Clinical errors resulting from usability problems are not a seldom occurrence. The steadily growing increase in the use of technology in medicine, as well as equipment which is becoming progressively more complex and more complicated, puts an enormous strain on clinical staff. This can be observed particularly in intensive care units and in the operating theatre, where it is common to find 15 pieces of medical apparatus which are in constant use and which, if the need arises, can be supplemented by possibly 75 more [19].

This situation is provoked by inadequate usability of the equipment and appliances. Very many pieces of equipment are not constructed for intuitive use and their handling is difficult to learn [6, 21]. Operator errors are inevitable. A large proportion of the clinical errors that occur is associated with the use of equipment and approximately 60% of these errors can be ascribed to usability problems [16, 21]. In a survey with 425 surgeons and 190 OR nurses 70% of the surgeons and 50% of the nurses reported difficulties operating medical devices (Tab. 1). Thus more than 40% were involved in situations of potential danger for staff or the patients [21].

Thus, ergonomic considerations and the observance of usability requirements are absolutely essential for optimizing the clinical working environment.

Table 1: Comments on the use of equipment by surgeons and nurses

Comment	Surgeons	Nurses
Equipment could not always be intuitively used correctly	69,8 % (n = 416)	48,9 % (n = 184)
I feel inadequately trained in the use of equipment	58,8 % (n = 410)	40,3 % (n = 186)
I have read the instruction manual for all equipment in the OR	6,7 % (n = 418)	23,4 % (n = 188)
Problems with the use of equipment have repeatedly endangered persons in the OR	39,7 % (n = 388)	47,7 % (n = 176)

Recently, equipment manufacturers started to raise demands for the inclusion of a usability inspection in obtaining CE certification conforming to the standards EN IEC 60601-1-6 and EN IEC 62366. Medical equipment now has to be designed with the help of a usability engineering process. However, the procedure has not yet been defined, nor do special test criteria exist. Nor are there standardized guidelines for the evaluation and assessment of usability.

Because of this, usability tests are carried out in extremely diverse ways and produce different sets of results. Independent investigations [18, 23, 25] have examined the extent to which international usability laboratories discover the same problems of a system and have reported on substantial differences both in the results obtained and in the method used.

Hertzum and Jacobsen describe this finding as the "evaluator effect". They conclude that the effect is due to vague goal analysis in determining the task scenarios for each test, an inadequate evaluation procedure and the unspecific selection of criteria for problem classification [13].

Further difficulties lie in the evaluation of usability problems. Various suggestions have been put forward for the classification of usability problems but because they are designed for consumer products they do not take the safety aspect into consideration. The current evaluation methods of usability engineering are not intended specifically for safety-critical equipment and therefore neglect the safety aspect [5, 10, 14, 15, 26]. Risk management methods do take usability into consideration, but only to a minimal extent [1, 7, 8, 9, 20, 22].

THE STANDARD PROCEDURE FOR EVALUATING USABILITY IN HEALTH CARE

There is still no method that, in conforming to the international standards EN IEC 60601-1-6 and EN IEC 62366, enables the usability evaluation of medical equipment under consideration of the safety aspect. To avoid the evaluator effect and to improve the quality of the evaluation, a standard procedure and an evaluation method has now been designed in the "experimental operating room" (Experimental OR). This provides the basis for evaluating a wide range of usability problems which can then be assessed according to the requirements both of usability engineering and risk consideration.

The usability evaluation for medical equipment will thus be standardized and the test quality criteria of objectivity, reliability and validity will be fulfilled as far as possible. Thus, usability and risk considerations are combined in one standard procedure [4]:

1. Requirement analysis for medical products
The standard procedure starts with an analysis of the context of use of the product. User needs, user tasks, work environment and the used devices are in this analysis of particular importance. With the results organizational, functional and user requirements on the product will be established.

2. Test criteria to measure usability
The next step is to deduce goals to evaluate them in the following usability evaluation. These goals include among others:
- features and performance of the device
- relevant legal stipulations and standards
- safety requirements
- user tasks and goals
- workflows and work organization

With the evaluated functional and organizational requirements plus the requirements of the users on the product, qualitative and quantitative usability goals for the product had to be extracted, with regard to the results of the context of use analysis. These goals provide a basis for evaluating and assessing the usability of the product. Usability evaluation in healthcare is based on gathering the usability measures effectiveness, efficiency and user satisfaction with regard to safety aspects, extracted from user interaction data observed during the use of the product.

3. Expert review
To get first impressions of the device usability and collect potential usability problems a heuristic evaluation [26] should be carried out. Several studies show that this usability method together with a user test provides best results [3, 10, 11, 15, 26]. If possible a review of existing critical incidents reports related to the device

should be conducted to explore possible problems, incidents and near-incidents based on the device use.

4. User test
With the user test, representative test persons will be observed in a realistic environment performing realistic tasks. Potential problems evaluated in the expert review and identified tasks will be checked. Objective and subjective data concerning the usability measures will be collected during the observance of the users, who are fulfilling tasks.

5. Determination of the random sample for the user test
For the user test a representative random sample according to the user profile had to be determined. The following points are important:
- If indirect users (e.g. cleaning and maintenance staff) are important for the usability of the device, they should be considered for the usability evaluation, too.
- Different user groups (e.g. surgeons, scrub nurses, circulators) and experiences have to be considered.
- For a usability assessment a random sample of minimum 9 participants is necessary [3]. The size depends on how many different user groups use the system.

6. Test set-up and test environment
For high quality test results a realistic test environment according to the context analysis and relevant observation technique should be used (Fig. 1). If necessary stress situations should be prepared and special hardware for eye-tracking, EMG measurement etc. should be integrated.

FIGURE 1: Typical usability test at the experimental-OR in Tuebingen

7. Scenarios and tasks of the test
The tasks for the user test should represent real tasks from the context of use. Mostly a complete evaluation of all tasks with the product is not possible. In such cases the choice of the tasks for the test is of particular importance because it has a

considerable impact on the test quality [13]. With respect to the reliability of the test attention should be paid to the following aspects, when choosing tasks for the test:

- if possible choose all primary tasks and the important and main used secondary tasks.
- choose the tasks that have been identified as potentially problematic at the expert review
- if safety critical situations and tasks have been identified during the context analysis, choose tasks that evaluate these situations.

For a more realistic user test create scenarios to evaluate the tasks [13]. Furthermore role-playing with an observer that takes part in the situation and verbal allocation of the tasks make the test situation more realistic.

8. Usability assessment

For usability assessment the method "UseProb" was designed to fit the requirements of the medical work system. This method combines usability evaluation and risk management. In a first step the errors and problems found by the user test have to be categorized. Errors due to usability problems have to be extract from human errors and machine errors. Furthermore for the usability problems the influencing factors for efficiency, effectiveness and satisfaction have to be detected and documented.

For the assessment an interdisciplinary team of experts (medical and human factors) is responsible for judging the factors delineating the problem: frequency of occurrence, severity of the problem (task conclusion) and effects concerning safety. By doing so can assign usability problems to a particular 10-points rating level with a three-dimensional matrix (Fig. 2). The mean value of the experts' assessment is the result. If the standard deviation is too high further experts have to be asked.

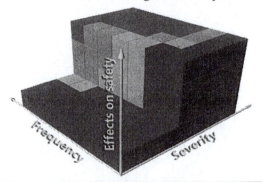

X	Medical device is outstanding, user-friendliness positively convince.
IX	Medical device is good and provides satisfaction; expectations regarding usability and user-friendliness are mainly met.
VIII	Medical device is good. Usability problems are marginal and rarely appear. Bodily injury and material damage due to usability problems are highly unlikely.
VII	Medical device is sufficient and in some cases confusing; slight usability problems are detectable. Bodily injury and material damage due to usability problems are highly unlikely.
VI	Medical device does not meet all expectations and is barely adequate. Bodily injury and material damage due to usability problems are highly unlikely.

V	Medical device generates dissatisfaction. Usability problems occur frequently. Bodily injury and material damage due to usability problems are unlikely.
IV	Medical device is insufficient and causes frustration. Bodily injury and material damage due to usability problems are unlikely.
III	Medical device unsettles the user and causes severe frustration. Usability problems may lead to hazardous situations. Bodily injury and material damage may be the result if problem is not detected and eliminated.
II	Medical device lacks safety due to usability problems. Hazardous situations with bodily injury or material damage due to usability problems can not be excluded.
I	Medical device lacks safety due to usability problems. It is a safety hazard risk for patient and/or the environment, if problem is not detected and eliminated.

FIGURE 2: Usability assessment method "UseProb"

After assessing the problem, the next step is to examine whether it is possible to by-pass the risk. To do so, solutions for avoiding the usability problem must be found. Should the solutions for avoidance or reduction of the risk prove to be technically or economically impractical, it is necessary to determine whether the risk arising from a usability problem is so low as to be reasonably acceptable; if it is low enough, it must not be included in the assessment. Justification is possible if the risk is acceptable on account of the benefit gained.

9. Report

A lot of usability reports are not usable. Frequent errors are shown by Dumas & Redish, as well as Molich et al. [10, 24].

To avoid errors and to meet more reliability, a standardized report is mandatory [24]. Since 2006 the ISO 25062 standardizes usability reports [17]. This international standard defines the information that has to be measured and documented within a usability evaluation for software products. It allows the involved organization and each other a rerun. Hence this standard redounds to more reliability of a usability evaluation.

This standard could be used for a usability evaluation for medical products, too. Additionally the specific requirements from the medical work system and safety aspects must be considered.

EVALUATION OF THE STANDARD PROCEDURE

In various studies it could be demonstrated that the standard procedure fulfills the test quality criteria sufficiently, detects more problems than in comparable studies and allows assessment [3].

32 experts (100%) in the fields of medicine and man-machine communication in medicine have declared the method to be an efficient tool in usability evaluation (Fig. 3).

302

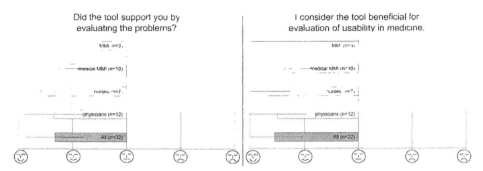

FIGURE 3: Satisfaction measurement of 32 experts

In independent, repeated studies involving a different random sample, different test supervisors and different evaluation experts, as well as in a study repeated after seven months with the same evaluation experts, it was demonstrated that the standard procedure and the evaluation method fulfil the test quality criteria of objectivity, reliability and validity.

Objectivity was measured by two autonomous tests using the standard procedure. In booth tests the primary tasks selected for the test matched to 100%. From all tasks identified by tasks analysis 72% were evaluated in booth studies.

Together booth studies could find 42 usability problems. 22 of them could be observed with booth tests, which is a match of 52%. After choosing the 20 and the 15 important problems by the respective test supervisor the match was 68% for the 20 and 73% for the 15 most important problems.

The objectivity of the assessment was measured by a comparison of the problem assessment. Two workshops with a different expert group of five persons and different instructors have been carried out to assess the usability problems of the two tests. The first workshop assessed 17 and the second 16 usability problems. 13 problems of them agreed for both workshops. Fig. 4 shows the results which have been generated in both workshops.

Beside one value, the assessment results were nearby. The average deviance was 0.9 assessment marks. Most of the results (11 from 13) don't deviance more than one mark. The correlation (Pearson) of the two tests was 0.734 with a significance of 0.01.

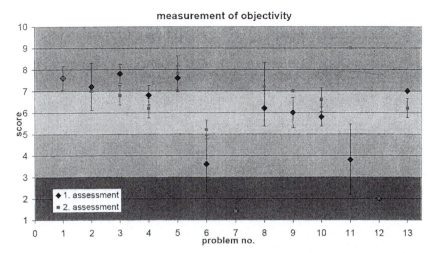

FIGURE 4: Measurement of the objectivity by two autonomous assessments

Reliability was measured by another study which repeats a usability assessment with the equal expert group after seven month (Fig. 5).
Within this study the maximum deviation between two assessments was 1.75 marks (problem 21). For 70% of the problems the deviation between two assessments was lower than one mark. The average deviation was 0.6 marks. The correlation of the first assessment and the repeated was 0.81 with a significance lower 0.001.

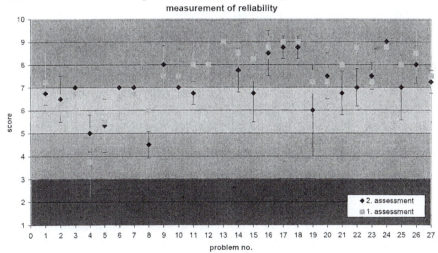

FIGURE 5: Measurement of the reliability by repeating a assessment after seven month

DISCUSSION

With three studies the relevance and the quality of the standard procedure was evaluated. For this an expert survey and evaluation of the quality criteria validity, reliability and objectivity was carried out. A point that influences the quality of a usability evaluation is the evaluator effect. Hertzum & Jacobsen point out, that the systematic choice of suitable tasks for a user test is an appropriate action to avoid this effect and to increase the test quality [13]. Only by choosing relevant tasks it is possible that different evaluators get comparable results. By using the standard procedure in two independent studies all primary tasks has been chosen by two different test supervisors. Studies that didn't use such a standardized procedure show drastically deviations. Molich et al. demonstrate in a comparison of nine different evaluators, who tested the same system, that there was only a low match of the chosen tasks for the test. Nearly half of the evaluation tasks (49%) have been selected by only one evaluator and only 6% have been evaluated by more than five evaluators [24]. In the study shown in this paper ten of 13 tasks have been chosen by both evaluators for a user test. This is a match of 77% and even a match of 100% for the primary tasks.

However, if the evaluating device is very complex the results could differ more. In such cases, according to the standard procedure the most relevant primary tasks have to bee selected. As a basis for decision-making results of the expert review have to be used. If the task analysis shows that in a user test not all primary tasks of the device could be evaluated, the participants of the expert review should select the tasks for the test. Selection criteria should be potential problems, frequency and importance of the task and safety aspects.

To verify the objectivity, the objectivity of execution and the objectivity of results have been measured. Objectivity of the execution was measured by comparing the problems found in the two independent tests. Totally the percentage match of detected problems was 52%, even 68% for the 20 and 73% for the 15 important problems. At the already explained study by Molich et al. only 25% of the detected problems could be found by two ore more evaluators. Further studies show similar results [13, 18, 23, 24, 25]. This approves the advantage of the standardized approach. However differences will occur independent from the standardization. This is due to different test users with no standardized level of education, experience and attitudes as well as different test conditions

The objectivity of the results was measured by comparing the assessment results of the two tests and the correlation. The aim was a correlation between the two tests of 0.8. With the result of 0.734 the aim couldn't be achieved to total, but it was close by. Most of the potential problems have been assessed similarly, but one outlier value affects the correlation. For this potential problem the expert groups in the different studies were at odds with each other. In the first workshop the experts ranked this potential problem as safety critical. In the follow up workshop the other experts thought that the situation in that the potential problem occur won't be relevant in praxis, so that the problem was classified as no problem. Therefore this deviation is due to disagreement respectively missing information of the experts, not due to problems of the method. If this problem is excluded from the correlation calculation the correlation is 0.933 with significance less than 0.001.

For measuring retest reliability the usability assessment was repeated with the same evaluators. The correlation was 0.81 with a significant less than 0.001. This means reliability [2, 12].

Validity was achieved by adhere to validity criteria. These include amongst others the use of standardized, scientific valid methods, representative sample, representative and appropriate tasks for the test [3]. Furthermore the context of use has to be simulated and if possible tasks could be verbalized by a role play instead of a sheet of paper. If this aspects will be considered validity is given.

Overall the developed and evaluated standard procedure reveals itself to be a suitable instrument to evaluate and assess usability of medical products while considering safety aspects.

REFERENCES

1   Börcsök J. (2006): Funktionale Sicherheit- Grundzüge sicherheitstechnischer Systeme. Hüthig GmbH & Co. KG, Heidelberg
2   Bortz J., Döring N.(2006): Forschungsmethoden und Evaluation für Human- und Sozialwissenschaftler. Heidelberg. Springer Medizin Verl.
3   Büchel D. (2009): Entwicklung einer allgemeingültigen Standardprozedur zur Überprüfung der Gebrauchstauglichkeit medizintechnischer Produkte. Dissertation an der Universität Tübingen
4   Büchel D., Scherer M., Matern U. (2009): Validierung medizinischer Geräte auch unter Berücksichtigung des Menschlichen Faktors. In: Hermeneit A., Steffen A., Stockhardt J. (Hrsg.): Der CE-Routenplaner: Medizinprodukte planen, entwickeln, realisieren. Köln, TÜV Media GmbH, TÜV Rheinland Group
5   Desurvire H. (1994): Faster, Cheaper: Are Usability Inspection Methods as Effective as Empirical Testing? In: Nielsen J., Mack R.L.: (Eds.): Usability Inspection Methods. New York: John Wiley & Sons, S.173-202
6   DIN EN 60601-1-6 (2005): Medizinische elektrische Geräte Teil 1-6: Allgemeine Festlegungen für die Sicherheit - Ergänzungsnorm: Gebrauchstauglichkeit. Berlin. Beuth
7   DIN EN 61508 (2006): Funktionale Sicherheit elektrischer/elektronischer/programmierbar elektronischer sicherheitsbezogener Systeme, Berlin. Beuth
8   DIN EN 61511 (2005): Funktionale Sicherheit - Sicherheitstechnische Systeme für die Prozessindustrie. Berlin. Beuth
9   DIN EN ISO 13849-1 (2009): Sicherheit von Maschinen. Sicherheitsbezogene Teile von Steuerungen. Teil 1. Allgemeine Gestaltungsleitsätze. Berlin. Beuth
10   Dumas J. S., Redish J. C. (1999): A Practical Guide to Usability Testing. Bristol UK. Intellect Books
11   Fu L., Salvendy G., Turley L. (2002): Effectiveness of user testing and heuristic evaluation as a function of performance classification. Behaviour & Information Technology, 21(2). S.137-143
12   Hamborg K.C., Hoemske T., Ollermann F. (2006): Qualitätssicherung im Usability Testing: Heinecke A., Paul H. (Hrsg.): Mensch & Computer Mensch und Computer im Strukturwandel. München, Oldenbourg. S. 115-124
13   Hertzum M., Jacobsen N.E. (2001): The Evaluator Effect: A Chilling Fact about Usability Evaluation Methods. In: International Journal of Human-Computer Interaction, Vol. 13, No. 4 (2001). S. 421-443

14 Heuer J. (2003): Expertenevaluation. In: Heinsen S., Vogt P. (Hrsg.): Usability praktisch umsetzen – Handbuch für Software, Web, Mobile Devices und interaktive Produkte, Carl Hanser Verlag, München, Wien. S. 115-135

15 Hollingsed T., Novick D.G. (2007): Usability Inspection Methods after 15 Years of Research and Practice, In.: SIGDOC'07, October 22–24, 2007, El Paso. S. 249 – 255

16 Hölscher U., Laurig W., Müller-Arnecke H. W. (2008):Prinziplösungen zur ergonomischen Gestaltung von Medizingeräten. Forschung Projekt F 1902. 2. Aufl.. Dortmund, Berlin. Bundesanst. für Arbeitsschutz und Arbeitsmedizin.

17 ISO/IEC 25062 (2006): Software-Engineering - Qualitätskriterien und Bewertung von Softwareprodukten (SQuaRE) - Gemeinsames Industrieformat (CIF) für Berichte über Gebrauchstauglichkeitsprüfungen. Berlin. Beuth

18 Kessner R. (2001): On the reliability of Usability testing. In: CHI 2001 Extended Abstracts. New York, NY: ACM, S. 97-98

19 Koneczny S. (2008): Erfassung und Analyse von Schwachstellen in der Funktionsstelle OP deutscher Krankenhäuser. Dissertation, Medizinische Fakultät Universität Tübingen

20 Leveson N.G.(2001): Safeware–System safety and computers. Amsterdam. Addison-Wesley

21 Matern U., Koneczny S., Scherrer M., Gerlings T. (2006): Arbeitsbedingungen und Sicherheit am Arbeitsplatz OP. In: Dtsch Arztebl 2006; 103(47): A 3187–92

22 MIL-STD-1629A (1980): Procedures for performing a Failure Mode, Effects, and Criticality. Military Standard, US-Department of Defense, Washington DC, 1980

23 Molich R., Dumas J. (2008): Comparative usability evaluation (CUE-4). In: Behaviour and Information Technologie Volume 27, Issue 3, May 2008, S. 263-281

24 Molich R., Ede M., Kaasgraad K., Karyukin B. (2004): Comparative usability evaluation. In: Behaviour & Information Technology, vol. 23, no. 1. S. 65-74

25 Molich R., Thomsen A.D., Karyukina B., Schmidt L., Ede M., van Oel W., Arcuri M. (1999): Comparative evaluation of usability tests. In: Extended Abstracts of ACM CHI 99 Conference. New York: ACM Press.

26 Nielsen J. (1993): Usability Engineering. San Diego. Academic Press

CHAPTER 37

Analysis of Cross-Professional Communication in Thoracic Operating Rooms

Chelsea Kramer[1], Avi Parush[1],
Kathryn Momtahan[2], Seneca Brandigampola[1]

[1] Carleton University
1125 Colonel By Drive
Ottawa, Ontario
K1S 5B6 Canada

[2] The Ottawa Hospital
General Campus
501 Smyth Rd
Ottawa ON
K1H 8L6 Canada

ABSTRACT

Healthcare teams defined by interdependent activities rely on coordination through timely information sharing. Surgical teamwork is especially vital to patient safety and overall quality of care. This study analyzed communication patterns in the operating room as a means of exploring teamwork processes in complex, cross-disciplinary environments. Using ethnographic methods of data collection, we observed 25 healthcare workers throughout 15 thoracic procedures at a large hospital in eastern Ontario. We classified 1506 utterances by speaker and receiver, and analyzed them as a function various surgical contexts. Results showed that the staff surgeon was the 'hub' of most information exchanges. High complexity procedures (i.e., more invasive, longer duration, and rated more difficult by the staff) tended to exhibit more team communication, while lower complexity procedures were surgeon dominated. This research offers a pragmatic method to investigate teamwork in complex environments. Specifically, communication analysis can be used to understand team activity through patterns of information flow.

Keywords: Teamwork, communication, cross-professional, patient safety, operating room, human factors

INTRODUCTION

Healthcare teams are characterized largely by the need for interdependence, yet patient focus is partitioned into distinct perspectives (Pugh, Santacaterina, DaRosa & Clark, 2010), responsibilities, and incentives that drive individual actions and attentions (Reddy, Dourish, & Pratt, 2001; Reddy & Spence, 2008). Information exchange facilitates teamwork by enabling coordination of unique perspectives on the same information through interaction (Gorman, Cooke, & Winner, 2006), thus making communication a distinguishing factor of effective teams (Mickan & Rodger, 2005).

Operating teams face contingencies and evolving circumstances that entangle nursing, anesthesia, and surgical responsibilities (Ash, Berg, & Coiera, 2004). In the operating room (OR), the way in which information is shared can reflect procedure complexity, team performance, as well as individual skill (Blom, Verdaasdonk, Stassen, Stassen, Wiering & Dankelman, 2007). As such, accurate information transfer among people and technology is essential to effectively coordinate interdependent activities as part of the teamwork process (Reddy, Pratt, Dourish & Shabot, 2003).

TEAMWORK IN THORACIC SURGERY

The field of Thoracics requires cross-professional teamwork; surgical site real-estate competes with anesthesia's maintenance of patient homeostasis, while simultaneously nurses oversee the needs of both. In addition, the adoption of Video Assisted Thoracic Surgery (VATS) for most procedures increases dependence on technology and coordination compared to many other areas (e.g., orthopedics). VATS constrains physical mobility and attentional resources by separating visual and tactile feedback. Shifting attention from mentally integrating data from disparate displays, as well as other members of the operating team, may inhibit the ability to formulate a complete and coherent picture of the situation (Patel, Zhang, Yoskowitz, Green, & Sayan, 2008). As a result, information shared among team members may be incomplete -or worse, inaccurate- and consequently lead to degradation in teamwork and quality of care.

To better understand this type of information loss, there is a need to study the specifics of communication within context. Communication issues are a frequent cause of inadvertent patient harm (Leonard, Graham, & Bonacum, 2004), and cause and estimated 70% of the adverse events (The Joint Commission, 2010). The current study aimed to evaluate how patterns of verbal communication in the operating room may yield insight into teamwork processes, procedure complexity and the potential impact on patient safety.

METHOD

Participants

Two graduate students with experience in cardiac surgery (Parush et al., 2008) observations attended 15 thoracic procedures over the course of 14 non-consecutive days between February and April of 2009. All procedures were performed at a large Ontario hospital consisting of Esophagoscopies, Bronchoscopies, Mediastinoscopies Lobectomy/Thoracotomies, an Esophagectomy, and Bullectomy/Pleurectomies. One seldom performed chest lavage was considered an outlier and was not analyzed. The fourteen remaining procedures included four staff surgeons, six staff anesthesiologists, two surgical fellows, three anesthesia residents, six circulating nurses, and four scrub nurses. Patient consent was obtained two hours prior to each case, and the OR teams were informed that two observers would be monitoring the case for the purpose of understanding surgical team communication.

The data presented here reflect approximately 24 hours of observation. The Procedure durations were based on the patient's noted entry and departure from the OR. The five non-surgical procedures ranged from approximately 28 to 46 minutes ($M = 34.23$ minutes; $SD = 7.93$), and the nine surgical procedures ranged from 68 to 241 minutes ($M = 137.29$; $SD = 62.40$).

Study Approach

This research was qualitative in nature such that variability (e.g., case complexity, duration, and number of team members involved) within and across cases was influenced by the needs of the patient and OR scheduling, and not by experimental design. Ethnographic methods of data collection included: 1) background education and pre-observation interviews with senior surgical staff formed the researcher's foundation of knowledge for thoracic procedures; 2) real-time observations and field notes of thoracic procedures to capture team communication; and 3) post-observation interviews with observed healthcare workers.

Materials

Observational data was collected in part using a hand-held computer tablet (Fujitsu Lifebook USeries, U820) installed with Remote Analysis of Team Environments (RATE, Guerlain et al., 2005). RATE is an open-source program that facilitates the capture of communication descriptors using customized lists of the people and actions/events that re-occur in the OR. Here, RATE recorded the speaker, responder, type of event, content of communication, and free-form comments made by the researcher. Each data entry was time-stamped and then imported to Excel spreadsheets, which was later combined with hand notes also taken during observations.

Analysis

Communication as a teamwork process was derived from verbal information exchange between team members in the OR. The unit of analysis was each utterance produced by an OR member. The sequence of utterances shown here contains two surgeon (S) utterances and one anesthesiologist (A) utterance. S to A: "Okay time to go." S to A: "Art[erial] line?" A to S: "Yeah, then good to." Utterances could also be from one person to the team where the response is implied by the action. The following is an example of a one-sided communication sequence. S to Team: "Let's get scrubbed"

Non-verbal communication was documented when it was relevant to other verbal communication. For example, researchers noted that a question was asked, and acknowledged with an action (e.g., a head nod). However, it should be emphasized that there may have been other non-verbal communication, such as body language, that was beyond the scope of this study. Identifiers were removed from the data to ensure patient and OR team's anonymity.

RESULTS

Procedure Characteristics

In order to analyze OR communication as a function of transferable contextual categories and not as a function of diverse individual procedures and cases, the following categories were defined: 1) Common surgery structure and 2) Procedure complexity.

Common surgery structure. We coded a common surgery structure based on team activity that provided chronological context and facilitated comparison of communication among diverse procedures. The five chronological phases typical of thoracic surgeries were marked by the following actions: 1) the insertion of intravenous lines and patient intubation (Lines); 2) the scoping of the surgical site and patient positioning on the table (Scope); 3) draping and sterilization of the patient's surgical site (Prep); 4) the actual surgical procedure (Surgery); and 5) the closing of the surgical site (Close). The content of each category was reviewed and validated by a staff surgeon to ensure accuracy.

Procedure complexity. Complexity was derived from ratings of 11 OR team members during post-observation interviews (i.e., scale of 1-5, with 1 being the least complex and 5 being the most complex). Mean ratings fell into two complexity groups that also corresponded to a general degree of surgical invasiveness: surgical procedures all scored above three (M = 3.80, SD = 1.21) and non-surgical procedures scored below three (M = 1.70, SD = .80), with the exception of mediastinoscopies which involve minor surgery. Procedures were thus classified as either high complexity (i.e., lobectomy/thoracotomy, esophagectomy, and bullectomy/pleurectomy), or low complexity (i.e.,esophagoscopy, broncoscopy and mediastinoscopy). Due to the small sample size, non-parametric analysis assessed the difference between complexity ratings of high and low groups. Significant results from the Friedman's Test (X^2 = 56.8, df = 6, p = .000), Wilcoxon's signed ranks (z = -2.95, p = .003) and the sign test (p = .001), suggest that high complexity cases had significantly higher ratings than low complexity cases.

To gain insight into the specific patterns of verbal communication processes, the following communication analyses were conducted as a function of the two complexity levels.

Communication Characteristics

Utterances were chosen as the unit of analysis because they best reflected the volume of communication. 1506 utterances were documented over the course of the 15 observations (Min = 20; Max = 189; M = 108.97, Mdn = 101.5). 1270 utterances were from high complexity cases, and 236 communications were from low complexity cases. Verbal utterances were then analyzed as a function of 1) procedure duration, 2) distribution by phase, 3) speaker, and 4) direction of communication.

Communication and procedure duration. The number of utterances was positively correlated with the total duration of procedure (Pearson's r = .91, p < .01), that is, longer procedures were associated with more utterances (see FIGURE 1. Relationship between procedure duration and number of utterances (r = .91, p < .01).

1). As duration was also associated with procedure complexity, it can be inferred that high complexity cases were associated with more utterances than low complexity.

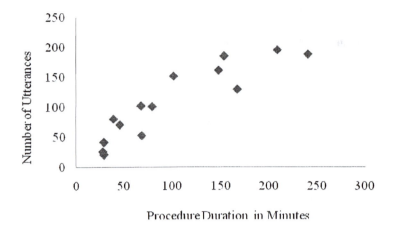

FIGURE 1. Relationship between procedure duration and number of utterances (r = .91, p < .01).

Communication distribution by phase. The number of utterances within each phase was used to calculate the proportion relative to the total number of utterances. FIGURE 2 shows that communication was primarily concentrated within the surgery phase. Surgery contained almost half of all utterances in both high complexity (45.3%) and low complexity (50.4%) procedures. Lines and intubation, scoping and patient positioning, and surgical close ranged from 13% - 20% of total utterances across both groups. The exception was for the low complexity procedures where patient prep phases were either extremely brief (i.e., during a mediastinoscopy where surgery is minimal) or did not exist for non-surgical procedures (e.g., scoping only).

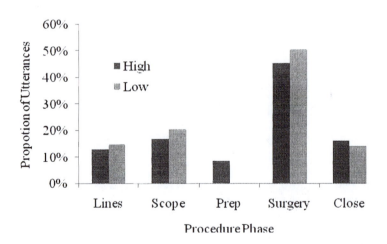

FIGURE 2. Proportion of communication utterances as function of phase for high and low complexity procedures.

Analysis of speaker as a function of profession. The proportion of speaker-specific communication was analyzed relative to total communications to assess the distribution of utterances among the team. In both high and low complexity procedures, the staff surgeon participated in the largest proportion of communication (49% and 61% for high and low complexity, respectively). Circulating nurses participated in a combined percentage of 16% in high, and 21% of low complexity surgeries. Staff anesthetists participated in 17% of high complexity, but only 3% of low complexity surgeries. Scrub nurses, surgical fellows and anesthesia residents appeared underrepresented with low proportions of about 7% and below in both complexity groups.

Cross-professional information flow. Sequential analysis (Siegel & Castellan, 1988) mapped the flow of teamwork-related communication by assessing the probability of utterances between each profession. Using zero-order probabilities, the proportion of times that each team member engaged in conversation with any other member was compared to the total number of utterances. As the thoracic teams often contained six or more people at one time, this data was more clearly illustrated using the three general professions (i.e., surgery, anesthesia, nursing), instead of individual professions (i.e., resident, staff). The entire OR team was also represented as a fourth receiver of information, since a large proportion of communications were broadcasted to the team as a whole versus specific people.

The results from the sequential analysis (Figure 3) show the surgery team as the main information hub in both high and low complexity surgeries, with most of the communication either originating from surgeons or directed to them. The previous analysis of speaker confirms that the staff surgeon and not fellow can be considered the main source of communication. In high complexity surgeries, the most likely paths of communication involved members of surgery, specifically from surgery to anesthesia (.16), followed by surgery to nursing (.15), nursing to surgery (.12), surgery to surgery (.11). In low complexity surgeries, the most likely paths also involved surgery: surgery to surgery (.21), surgery to nursing (.21), surgery to team (.14), and nursing to surgery (.11). Nurses had similar proportions of inter-team communication (.08 and .09 for high and low respectively), while anesthesia had .09 in high and only .03 in low. In both groups, there was a very low proportion of interaction among the nursing and anesthesia teams, (.04 and below). The low complexity group also had a higher proportion of communications directed to the team from surgery (.14), compared to the high complexity group (.09).

314

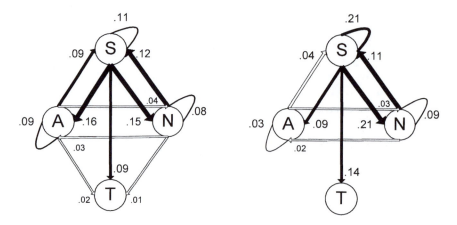

FIGURE 3. Sequential analysis diagram showing probabilities of information paths; left diagram - high complexity, right diagram – low complexity.

CONCLUSIONS

The aim of this research was to analyze cross-professional communication in thoracic OR's. Our study showed that verbal communication was primarily concentrated within the surgery phase, which suggests a higher level of inter-dependency in terms of teamwork: the surgeons who perform the surgery, the anesthetists who maintain the patient's homeostasis, and the nurses who assist with tools and tasks outside of the sterile environment. The other phases do not necessarily require the physical presence of all individuals, however, inconsistent team presence could lead to information loss during absences or interruptions. For example, at times the surgeon would not appear until the scoping process, only to discover the wrong sized tube was being used. As such, smaller proportions of overall communication may reflect times where independent activities prevent cross-professional communication, potentially degrading team situation awareness later on.

While verbal communication was indicated certain roles and responsibilities within the team, little communication occurred between the nursing and anesthesia teams. Nevertheless, the nurses were still repeatedly observed assisting the anesthesiologists. In the absence of communication, these members may rely on visual and auditory cues to build and maintain their situation awareness. Future studies should consider how non-verbal communication is used above and beyond verbal communication in order to capture whole team behaviour.

The staff surgeon appeared to take on a leadership position, acting as the hub of information sharing, although many self-described as equal team participants. Surgeons consistently initiated and received most of the communications, and were the most likely to address the team as a whole. In turn, this orchestrated the team's behaviour and set the overall "mood" of the OR. This finding has implication for leadership training with regards to the influence of verbal behaviour in shaping team performance. Parush et al. (2010) also defined surgeons as the communication hub based on having the least consistent access to information sources, yet needing to know most of the information. This need for information sharing has implications for future research such as the development of augmented display technology to facilitate information sharing (Parush et al., 2010).

Of note to healthcare research is a common obstacle of obtaining permission to record surgery. When this is not possible or needed, this study offers a pragmatic and cost-effective way to capture team communication in OR settings. Similar research (Blom et al., 2007; Elbardissi et al., 2008) has used advanced data-capturing technology such as multiple video cameras and audio recording to find consistent communication patterns related to procedure complexity, team situation awareness and teaching behaviour within cardiac surgery and laparoscopic cholecystectomies. Our approach (see also Parush et al., 2009) yielded comparable results using only two observers and note taking.

Although this data may not transfer to all healthcare contexts, similarities among communication data between thoracic and cardiac provide more substantial implications, which can then be tested in other contexts in the future. Generalization across several surgical procedures is the first step towards extending findings to other domains of healthcare. Ultimately, exploring the multifaceted underpinnings of team communication will promote further understanding of the fundamental relationship between team communication and team performance.

Acknowledgements

This research was supported by the Ontario Graduate Scholarship for Science and Technology (OGSST) partnered with The Ottawa Hospital. We would especially like to thank Jim Worthington for his on-going interest and support. Finally, thank you to all the healthcare staff who made this research possible.

REFERENCES

Ash, J., Berg, M. & Coiera, E. (2004). Some unintended consequences of information technology in health care: the nature of patient care information system-related errors. *Journal of American Medical Informatics Assoc, 11*, 104–112.

Blom, E.M., Verdaasdonk, E.G.G., Stassen, L.P.S., Stassen, H.G., Wiering, P.A., & Dankelman, J. (2007). Analysis of verbal communication during teaching in the operating room and the potential for surgical training. *Surgical Endoscopy, 21*: 1560–1566.

Gorman, J.C., Cooke, N.J., & Winner, J.L. (2006). Measuring team situation awareness in decentralized command and control systems. *Ergonomics, 49*, 1312-1325.

Guerlain, S., Adams, R., Turrentine, T. S., Guo, H., Collins, S. R. & Calland, F. (2005). Assessing team performance in the operating room: development and use of a "black box" recorder and other tools for the intraoperative environment. *Journal of American College of Surgery, 200*(1), 29-37.

The Joint Commission (2010). Sentinel events. Retrieved February 5, 2010 from: http://www.jointcommission.org/SentinelEvents/Statistics/.

Leonard, M., Graham, S. & Bonacum, D. (2004). The human factor: the critical importance of effective teamwork and communication in providing safe care. *Qual Saf Health Care, 13*, 85-90.

Mickan, S. M., & Rodger, S. A. (2005). Effective health care teams: A model of six characteristics developed from shared perceptions. *Journal of Interprofessional Care, 19*(4), 358-370.

Patel, V.L, Zhang, J., Yoskowitz, N.A, Green, R. & Sayan, O.R. (2008). Translational cognition for decision support in critical care environments: a review. *Journal of Biomedical Informatics,41*(3), 413-31.

Parush, A.,. Kramer, C., Foster-Hunt, T., Momtahan, K., Hunter, A., Sohmer, B. (2010). Communication and Team Situation Awareness in the OR: Implications for augmentative information display. *Revised for publication in the Journal of Biomedical Informatics.*

Parush, A., Momtahan, K., Foster-Hunt, T., Kramer, C., Hunter, A., Nathan, H. (2009). A Communication Analysis methodology for developing cardiac operating room team-oriented displays. *In the proceedings of the Annual Human Factors and Ergonomics Society Conference*, October 2009, pp. 728-732.

Parush, A., Momtahan, K., Foster-Hunt, T., Kramer, C., Holm, C., and Nathan, H. (2008). Critical information flow analysis in the cardiac operating room. In the *Proceedings of the International System Safety Conference.* Vancouver, BC, Canada.

Pugh, C.M., Santacaterina, S., DaRosa, D.A. & Clark, R.E. (2010). Intra-operative decision making: More than meets the eye. *Journal of Biomedical Informatics Article in Press, Corrected Proof.*

Reason, J. (1995). Understanding adverse events: Human factors. *Qual. Health Care 4*(2): 80-89.

Reddy, M., Dourish, P., & Pratt, W. (2001). Coordinating heterogeneous work: Information and representation in medical Care. *Proceedings of European Conference on Computer Supported Cooperative Work* (pp. 239-258). September 16-20, Bonn, Germany.

Reddy, M., Pratt, W., Dourish, P., &, Shabot, M.M. (2003). Sociotechnical requirements analysis for clinical systems. Methods Inf Med, 42, 437-44.

Reddy, M. C., & Spence, P.R. (2008). Collaborative information seeking: A field study of a multidisciplinary patient care system. *Information Processing and Management 44*, 244-255.

Shields, T. (2004). General Thoracic Surgery. Lippincott Williams & Wilkins. pp. 524.

Siegal, S. & Castellan, N.J. (Eds). (1988). Nonparametric statistics for the behavioral sciences (2nd ed.). United States: McGraw-Hill.

CHAPTER 38

Airline Safety Practices in Clinical Nursing

Gary Laurence Sculli

Veterans Health Administration National
Center for Patient Safety
Ann Arbor, Michigan, USA

ABSTRACT

Nurses are the largest group of healthcare workers providing patient care and play a crucial role in patient safety. Crew Resource Management (CRM) training offers tools and strategies that nurses can readily apply to reduce the risk of patient harm. Examples include the use of standardized communication tools, the application of specific mechanisms to reduce interruptions during safety sensitive tasks, the use of checklists, and the adoption of operational paradigms that foster the appropriate management of automation. A CRM program modified for the clinical nursing environment has great promise in the promotion of patient safety in multiple areas of practice. This chapter will discuss those elements of airline CRM that have a direct application to professional nursing and can be applied to clinical processes to improve patient safety.

Keywords: Crew Resource Management, Nursing, Distractions, Interruptions, Assertive Communication, Intimidation, Checklists, Automation, Technology

INTIMIDATION, CULTURE AND COMMUNICATION

Crew Resource Management (CRM) was initiated in the early nineteen eighties by airlines in the United States in response to several highly publicized accidents that demonstrated poor communication among the flight crew. Investigations of these accidents revealed a consistent theme was the failure of subordinate crew members to effectively assert their deepening concerns about the course of action the captain of the aircraft was taking. In 1978 while attempting to land in Portland Oregon, United Airlines Flight 173 ran out of fuel, crashing just short of the airport (NTSB, 1979). Cockpit transcripts clearly demonstrated that the First Officer and Second Officer were concerned about the aircraft's fuel status, but were timid and ineffective in attempts to raise these concerns. Airline industry culture at the time emphasized a captain's authority, not teamwork. "Speaking up" proved to be difficult for subordinate crew members, despite signs of an impending disaster.

An element of CRM that has utility in both the cockpit and on the nursing unit is the concept of followership. When airline crew members practice followership, they are engaged, critically thinking, and able to effectively mount assertive challenges to preserve operational safety in all circumstances. Pilots learn to continually practice inquiry, respectfully questioning and clarifying team decisions in the interest of safety. With the aid of standardized communication tools, pilots gain skill in the art of assertive advocacy, learning to provide timely feedback that is specific, direct, and concise. One such tool is the Four Step Assertive Communication Tool (see Figure 1). The tool provides a template to construct an assertive statement thus increasing the likelihood that concerns are heard and resolution gained. The use of such a tool by non-airline industry teams is especially important when team members experience intimidation in the face of authoritative and/or dictatorial leadership styles: a work condition often found on the typical nursing unit.

Assertive Communication Tool
1) Get Attention
2) State Concern (Preface with "I'm concerned with" or "I'm uncomfortable with")
3) Offer Alternative
4) Get Resolution (End with "Do you agree?")

FIGURE 1. A Four Step Assertive Communication Tool (Adapted from Kern, 2001)

High Power Distance cultures are described as cultures stressing the absolute authority of leaders (Helmreich et al, 1999). In such cultures subordinates are disinclined to question or challenge those in charge. This has direct implications for the nurse-physician dyad as it exists within the professional nursing culture. Nurses often experience hostile, even aggressive behavior when questioning medical regimes. In 2006 The American Association of Critical Care Nurses (AACN) surveyed four thousand nurses with an average experience range of 17.5 years. Sixty-five percent said that they experienced verbal abuse in their practice, most often from physicians. An Institute for Safe Medication Practice (ISMP) survey in 2003, which queried over two thousand nurses and pharmacists, found that 75% of respondents attested to using avoidance techniques to clarify orders rather than directly questioning an intimidating prescriber. The intimidation ranged from voice intonation to rare cases of physical abuse. Healthcare has by tradition tolerated such behavior based on one's stature in an organization (Physician Executive, 2004). Nurses often possess critical information about a patient's status, yet if they are reluctant, fearful, or ineffective when transferring that information to medical decision makers, the outcome can be perilous for those patients under the care of healthcare teams.

Exposing nurses and physicians to CRM training, emphasizing teamwork and the use of standardized tools to ensure effective communication, will reduce the risk of inadvertent patient harm. CRM curriculum introduced at the medical and nursing student level will provide the most promise for lasting cultural change (Musson et al, 2004)

DISTRACTION AND CRITICAL TASKS

Medication variances are a frequent source of patient harm. The National Center for Patient Safety (NCPS) has received over 3,700 individual Root Cause Analyses (RCAs) and Aggregate Reviews involving some phase of the medication delivery process, and has received over 121,000 individual reports involving medications since 2000. Distractions and interruptions are contributing factors in 45% of medication errors in hospitals and health systems (USP, 2008). Medication administration is a frequent and critical task for nurses; it can be viewed as analogous to flight regimes experienced by pilots that demand concentration due to high task and mental loads. A mechanism in place to protect pilots during such regimes is the so called "Sterile Cockpit Rule". The rule is derived from Federal Regulation and mandates that pilots will refrain from extraneous conversation and activities during all operations below 10,000 feet. The rule also prohibits employers from requiring pilots to engage in non flight related activities during this time (Sumwalt, 1993).

Nurses frequently administer medications in environments replete with interruptions, and cite distractions as a leading causal factor of medication errors (Mayo and Duncan, 2004). In addition to multiple environmental distractions, nurses are often interrupted by the requirement to complete tangential functions not directly related to patient assessment (IOM, 2004). The sterile cockpit concept applied to medication administration can reduce the frequency and number of interruptions experienced by nurses, and has been demonstrated in clinical practice. Implementations of this concept have included the use of "Do Not Disturb" signs strategically placed in the clinical area, and / or the use of vests with similar wording worn by nursing staff during the medication delivery process. Additional measures include establishing quiet zones around medication dispensing areas. A 2003 study entitled Applying Airline Safety Practices to Medication Administration by Dr. Theresa Pape RN Ph. D found a significant reduction in the number of interruptions per medication pass on a medical surgical nursing unit through the use of staff education, a "Do Not Disturb" vest, and a checklist (Pape, 2003). Pape demonstrated similar results in a 2005 study using only conspicuous signage and a protocol checklist (Pape, 2005). The corollary to reducing distractions through the application of the sterile cockpit concept is twofold. In addition to greater efficiency in medication delivery, a reduction in medication error and in patient harm will be realized.

CHECKLISTS

Checklists are a staple in the airline industry. They reduce reliance on memory especially in emergency situations. Checklists aid in directing attention to important items or steps in a process that if omitted, can adversely affect safety. In 2009, the World Health Organization published a global study which demonstrated that the use of a Surgical Safety Checklist at key points in the peri-operative process significantly reduced post operative complications and mortality (Haynes et al., 2009). Checklists applied in the Intensive Care Unit (ICU) setting guiding adherence to infection control practices during central venous catheter insertions, yielded a sustained and significant decrease in central catheter associated blood stream infections (Pronovost et al., 2006)

Checklists have been applied to nursing processes in the past. For example nurses have used checklists before sending a patient to surgery, or just prior to discharge. While it is fitting to use a checklist in these situations, it is important that checklist tools are developed appropriately to match the task at hand. Often checklists, though developed with good intent, miss the mark and are viewed as cumbersome or ineffective by the user. In one study, nurses used a checklist to support the medication administration process. Although nurses read and agreed with the process steps outlined in the checklist, only one nurse in the study actually held and used the checklist throughout the entire medication delivery process. One explanation for this result may be a mismatch between the required task and the philosophy used for checklist development. Medication administration is a routine practice, yet the checklist tool used in the study was laden with instructions and detail. This can lead to the unintended consequence of increasing task load rather than supporting memory. Poorly designed checklists create inefficiencies in care delivery, and may pose increased risk to patients (Winters et al., 2009).

Pilots use *read and verify* checklists for normal procedures such as starting the aircraft engines, inspecting the cockpit prior to the first flight of the day, and preparing the aircraft for takeoff or landing. All of these are normal actions that a pilot completes repeatedly. For such procedures the expectation is to accomplish all of the steps in the process from memory, then verify completion as a final step using the checklist. *Read and verify* checklists are not loaded with narrative or expanded verbiage. Below is an example of a generic read and verify checklist in aviation (see Figure 2). Items and their required actions are straightforward, clear and concise, allowing ease of use.

Before Takeoff

```
Window Heat.................................ON  HIGH
Anti-Ice ...    ..    ..    ... ..    .. ..ON
Flight Instruments & Radios    ..    . .SET
Yaw Damper    ...    .. ........ . ...........ON  & CHECKED
Flight Controls    .    .    .    .CHECKED
Stabilizer Trim    . ..    .  .SET
Flaps / Slats ..    .. ..... .    . .. ..... .. ..15  & GREEN
Electrical... ..... ............ ......  ...........NO  LIGHTS
Fuel Pumps... .  .. ......... ... ..........SET  FOR TAKEOFF
Fuel Heat.. .    ..    .  ..    .....OFF
Hydraulics......................... ...  ........PRESS  & QTY NORMAL
Elevator & Rudder Lights .. ..  ..  ...... OFF
Air Cond & Press.............. ... ..........SET  FOR TAKEOFF
EPR & Airspeed Bugs .  .. ... .. .. .. SET
Transponder    .    .. ..    .  .ALTITUDE
Take Off Briefing............................COMPLETE
```

FIGURE 2. Generic example of an airline *read and verify* checklist

There are many routine processes and procedures in clinical nursing analogous to normal operations in the cockpit that can benefit from the use of a well constructed *read and verify* checklist. Consider the task of establishing intravenous access on a hospitalized patient. Nurses accomplish this multiple times over the course of a week, or even a day; the steps in the procedure are familiar. Yet, there are sometimes iatrogenic consequences associated with this process. Clinicians forget or fail to remove tourniquets from extremities or catheters are placed in limbs where venipuncture is contraindicated. These errors can be prevented with the consistent use of a *read and verify* checklist. Below is an example of a normal checklist that may be used for IV insertion, based on the read and verify philosophy (see Figure 3). The nurse completes items from memory, then references the checklist prior to inserting the intravenous catheter (before insertion), and prior to leaving the patient's room (after insertion).

322

Before Insertion

Patient Identification.................................CONFIRMED
Correct Side...CONFIRMED
Catheter Size...CONFIRMED
Equipment..AT BEDSIDE
Patient...BRIEFED

After Insertion

Tourniquet..REMOVED
Line...FLUSHED
Pump...SET (with fluids)
Sharps...DISPOSED
Site...LABELED
Documentation...COMPLETE

FIGURE 3. Example of proposed nursing *read and verify* checklist for intravenous catheter insertion.

Read and verify checklists are often carried out by using a challenge - response methodology. In this case, one individual reads the item to be checked or step to be completed in the process (challenge), and another individual says aloud the appropriate action dictated on the checklist (response). Both individuals are expected to confirm that the response reflects the true state of the system. For example, if an aircraft checklist calls for a response that the hydraulic pressure is "checked and normal," both pilots must confirm that this is truly the case. A similar example in nursing practice occurs when two nurses use the challenge and response method to verify patient identification.

Read and do checklists are used in situations that are more critical such as abnormal or emergent scenarios (Gaffney et al, 2005). In emergent situations, individuals can experience strong physiologic reactions such as fear or anxiety, which must be managed while performing required tasks. In these circumstances human memory may be less dependable, the ability to assess and critically think is negatively affected, and situational awareness is reduced (Winters et al, 2009). Often emergencies present a complex set of variables; there may be more than one appropriate response depending on the presence or absence of specific circumstances. Analysis is required to determine the correct course of action. *Read and do* checklists allow humans to move through this process with guidance and direction so that assessments are correctly focused, and that actions taken are ultimately suitable for the conditions experienced. With *read and do* checklists, beyond a few initial items that may be completed from memory, all other items are read directly from the checklist, and then completed one by one. The checklist may present questions and continue in an "If yes than complete action A, if no complete action B" format. Whatever the format, the overriding point is that *read and do* checklists reduce the burden on human memory in critical scenarios by directing attention and assessment, then spelling out specific actions to be taken in response.

Nurses have available a *read and do* checklist when implementing Advanced Cardiac Life Support Protocol (ACLS) as prescribed by the American Heart Association for patients experiencing life threatening arrhythmias. The checklist is presented in a decision tree format providing a series of questions, actions and sequences that guide the user through the protocol. On acute care nursing units, life threatening scenarios can emerge quickly, requiring that interventions be both timely and accurate. Considering this fact, reducing the probability that patient outcomes will rely solely on memory or practitioner experience through the use of a well developed checklist is appropriate. *Read and do* checklists use can and should be expanded within the clinical environment to reduce risk and increase the level of safety for patients.

MANAGING AUTOMATION

In 1987, an airliner departing from Detroit's Metropolitan Airport crashed shortly after takeoff killing all but one passenger (NTSB, 1989). It was determined that the crash was caused by the crew's failure to configure the wing flaps and slats for takeoff. Wing configuration is critical to the development of lift on takeoff; this item appears multiple times in the taxi and before takeoff checklists. It was also determined that the crew did not execute the required before takeoff checklists as prescribed by standard operating procedure. Modern airline aircraft are highly automated and comprise multiple systems which support pilots in their efforts to maintain safe flight. For example if the aircraft wing is not configured properly, and a pilot applies takeoff thrust, the cockpit warning system will provide an immediate visual and aural warning to the crew. With the accident aircraft previously described, the circuit to the warning system failed, rendering it useless. Such warning systems exist to back up, not replace human vigilance. Knowing that automated surveillance systems exist in the background may breed complacency over time, inverting the relationship between vigilance and automated systems. Since automated systems fail, such an inversion can be hazardous to operational safety.

Managing automation and technology presents great challenges for pilots. With the proliferation of highly sophisticated flight management and auto pilot systems, pilots spend more time monitoring system states than physically flying the aircraft. Automation complacency can take hold as pilots experience what is known as the "along for the ride" syndrome (Kern, 2001; Hawkins, 1993). An overriding paradigm for pilots is to avoid implicit trust in technology; this can be accomplished by specific behaviors. Verification of computer inputs with a second crew member prior to execution is one method that can trap error at an early stage. Continual monitoring and cross checking of computer navigation systems with other navigational sources can also detect faults (Helmreich et al, 1999). Adhering to standard operating procedures with checklist verification can also mitigate the consequences of automation failure. It is also paramount for pilots to remain ready to disengage automation and assume manual control of the aircraft when automated systems malfunction or create prolonged confusion. Automation should never supplant basic flying skills.

Nurses continually interface with technology in practice and would benefit from similar approaches to managing the proliferation of automation and technology in patient care. Actual and potential harm to patients resulting from the use of Health Information Technology (HIT) and clinical monitoring systems is well documented. Computerized Physician Order Entry Systems (CPOE) in many cases have induced medical errors rather than reduce them. In one case, patient mortality increased after CPOE implementation (Han et al., 2005; Koppel et al., 2005; Zhang et al. 2005). Patient deaths have been attributed to false readings displayed by blood glucose monitoring devices, physiologic monitoring systems in ICU settings, and unreliable low pressure alarms in dialysis units (NCPS, 2008). CRM training, in addition to teamwork and communication, also focuses on human factors and human performance topics. Training for nurses that emphasizes an appropriate model for managing technological resources at the bedside should be provided and reinforced at intervals. Key areas of focus for nursing staff include: cross checking readouts from monitoring systems with a patient's actual condition, avoiding complacency and reduced vigilance when physiologic monitoring systems are in use, maintaining basic clinical assessment skills when automation creates prolonged confusion, and verifying programming inputs with another staff member for infusion pumps or other automated systems.

CONCLUSION

Airline pilots and nurses perform work that has similar characteristics. Both professional groups work in teams, perform safety sensitive tasks, and endure periods of high mental load. A CRM program modified for the clinical nursing environment has great promise in the promotion of patient safety in multiple areas of practice. Standardized communication tools that promote assertive advocacy can facilitate the flow of important patient information, especially in *high power distance* cultures. Applying mechanisms to mitigate environmental distractions during peak medication administration periods can reduce the probability that patients will be harmed as a result. Checklists for use in nursing practice will be effective provided the correct philosophy is used in their development. Distinctions must be made between normal procedures and emergent scenarios to maximize checklist efficiency. Nurses can adopt a paradigm that resists an implicit trust in technology, allowing better management of automation at the bedside.

REFERENCES

Gaffney F.A., Harden S.W., Seddon R. (2005). *Crew resource management: The flight plan for lasting change in patient safety.* Marblehead, MA: HCPro, Inc.

Han Y.Y., Carcillo J.A., Venkataraman S.T., Clark R.S.B, Watson R.S., Nguyen T.C., Bayir H., Orr R.A. (2005). Unexpected increased mortality after implementation of a commercially sold computerized physician order entry system. *Pediatrics.* 116(6):1506-1512.

Haynes A.B., Weiser T.G., Berry W.R., Lipsitz S.R., Breizat A-H.S., et al. (2009). A surgical safety checklist to reduce morbidity and mortality in the global population. *New England Journal of Medicine, 360,* 491-499.

Helmreich, R.L., Merritt, A.C., & Wilhelm, J.A. (1999). The evolution of crew resource management training in commercial aviation. *International Journal of Aviation Psychology, 9*(1), 19-32.

Institute of Medicine. (2004). *Keeping patients safe: Transforming the work environment for nurses.* Washington, DC: The National Academies Press.

Kern, T. (2001). *Controlling pilot error: Culture, environment, & CRM.* New York: McGraw-Hill.

Hawkins, F.H. (1987). *Human Factors in Flight.* Vermont: Ashgate.

Koppel R, Metlay JP, Cohen A, et al. (2005). Role of computerized physician order entry systems in facilitating medication errors. *JAMA.* 293:1197-1203.

Mayo M, Duncan, D. (2004). Nurse perceptions of medication errors: What we need to know for patient safety. *Journal of Nursing Care Quality, 19,* 209-217.

Musson D.M., Sandal G.M., Helmreich R.L. (2004). Personality characteristics and trait clusters in final stage astronaut selection. *Aviation, Space, and Environmental Medicine, 75*(4), 342-349.

NCPS   (2008).   Bleeding   episodes   during   dialysis. http://www.patientsafety.gov/alerts/BleedingEpisodesDuringDialysisAD09 -02.pdf (Accessed 3/1/2010)

NTSB   (1979)   http://www.ntsb.gov/ntsb/GenPDF.asp?id=DCA87MA046&rpt=fi (Accessed 2/28/2010)

NTSB (1989). http://www.ntsb.gov/ntsb/Response2.asp (Accessed 2/28/2010)

Pape TM, Guerra DM, Muzquiz M, Bryant JB, Ingram M, Schranner B, Alcala A, Sharp J, Bishop D, Carreno E, Welker J. (2005).Innovative Approaches to Reducing Nurses' Distraction During Medication Administration. *The Journal of Continuing Education in Nursing, 36* (3) 108-116.

Pape, T. M. (2003). Applying airline safety practices to medication administration. *Medsurg Nursing,* 12, 77-93.

Pronovost P, Needham D, Berenholtz S, Sinopoli D, Chu H, Cosgrove S, Sexton B, Hyzy R, Welsh R, Roth G, Bander J, Kepros J, Goeschel C. (2006) An Intervention to Decrease Catheter-Related Bloodstream Infections in the ICU. *New England Journal of Medicine.* 355:26:2725-2732

326

Sumwait R.L., (1993). The Sterile Cockpit. *ASRS Directline*. No. 4. http://asrs.arc.nasa.gov/directline_issues/dl4_sterile.htm#anchor524636 (Accessed 3/1/2010).

USP (2008) *Pharmacopeial Forum* Vol. 34(6). *http://www.usp.org* (Accessed 1/7/2010)

Winters B.D., Gurses A.P., Lehmann H., Sexton J.B., Rampersad C., Pronovost P.J. (2009) Clinical review: Checklists - translating evidence into practice. *Critical Care* 13:210.

Zhang J, Patel VL, Turley JP, Johnson TR. (2005) Health Information and Medical Errors. *Business Briefing: North American Pharmacotherapy* http://www.touchbriefings.com/download.cfm?fileID=6341 (Accessed 3/1/2010)

Chapter 39

Improving the Quality of Patient Care in an Emergency Department Through Modeling and Simulation of its Process

Adam Langdon, Praveen Chawla

Edaptive Computing, Inc.
Dayton, OH, 45458, United States

ABSTRACT

Maximizing the effectiveness of a process such as in an emergency department while minimizing the cost of the resources needed to do so requires complex analysis. The unpredictable nature of these types of processes makes such a task even more difficult. Primary methods to improve critical metrics such as length of stay include reducing the time needed to perform individual tasks, reorganizing the process to eliminate unneeded tasks, or improving resource allocations to reduce wait times. All of these steps first require an intimate understanding of the process as it currently operates. This alone can be difficult without the proper tools to precisely and intuitively capture all aspects of the process. However, once an understanding of the process is achieved, deciding what steps to take to impact performance metrics is even more challenging. A manager may attempt to implement certain changes on a trial basis, but this often proves costly in both time and staff confidence. Furthermore, many times these types of decisions must be made in response to rapidly changing conditions. An unexpected surge in patient arrivals or acuity levels may require a rapid reassignment of resources to different areas within the department. Without the proper tools for analysis and real-time situation awareness, methods for process improvement are ad-hoc at best.

We present an intuitive solution based on our modeling and simulation framework, EDAptive® Syscape™. This tool provides an intuitive environment for creating, executing, and visualizing emergency department process models. The result is the type of analysis and situation awareness that can significantly improve performance. Modeling and simulation software offers hospitals and staff the unique ability to perform accurate, highly detailed predictive analysis of the specific and systemic impacts of operational, process, and layout changes before decisions are made. Through modeling and simulation, one can examine the complex and numerous effects of proposed changes as they will impact: patient flow analysis; staff utilization and efficiencies; resource, bed, and spatial demand patterns; ancillary departments; and throughput and wait times. When connected with real-time data, the capture model becomes a dashboard for real-time situational awareness. In summary, modeling and simulation tools offer a level of detail, accuracy, and quantitative analysis that is simply unavailable through spreadsheets, flowcharts, and traditional consulting methodologies.

Keywords: Process modeling, emergency planning, education and training

INTRODUCTION

Typically, staffing requirements for emergency departments are developed in an experienced-based and sometimes ad-hoc fashion. While this method may result in sufficient staffing, the process will likely be sub-optimal. This can lead to unnecessary wait times, inefficient patient care, and higher operating costs. However, attempting to improve this process usually requires trial and error. In addition, introducing new processes and ideas cannot be done with any high level of confidence. Potential improvements cannot be tested without significant risk to patients. Attempting to manually analyze and improve processes and resources deployed in emergency departments can be challenging, especially since they can involve hundreds of inter-related key performance parameters. Therefore, a more methodical approach is required. Process modeling and simulation can provide such an approach and is worthy of consideration as a tool for process improvement.

Specific benefits of such a technique include the ability to analyze staffing requirements and utilization, test new processes, and simulate current processes for training and educational purposes (Patvivatsiri et al., 2006; T Takakuwa and Wijewickrama, 2008; Van Oostrum et al., 2008). Aside from process improvement, there may be a general lack of understanding of current processes. Simulation is well suited for capturing and modeling uncertainty, which is a major component of the health care domain (Lowery, 1998). Without a methodical planning approach, it is nearly impossible to foresee and prepare for the different scenarios that may unfold.

This paper discusses how a modeling and simulation methodology was used to analyze an Emergency Department Process with the objective of determining cost-effective methods to reduce the average Length of Stay (LOS) to 180 minutes or less. First, we discuss the general methodology used to construct a process model, and instantiate the model with specific data, and validate its performance using historical data. Next we describe the task of identifying potential process improvements and capturing them as To-be models. Finally, we present strategies for improving the LOS as derived from analysis of a specific Emergency Department process.

AN EMERGENCY DEPARTMENT AS AN EXECUTABLE MODEL

Modeling and simulation tools offer a level of detail, accuracy, and quantitative analysis that is simply unavailable through spreadsheets, flowcharts, and traditional consulting methodologies. Because simulation accounts for the variability, interdependencies and complexities of healthcare's working environments, simulation tools are capable of precise predictive analysis. Via "what if" scenario analysis, options for addressing various problems and issues can be examined, compared, tweaked, and fully understood. And because simulations are objective and quantitative, they are invaluable tools for solving hotly debated change issues through precise analytics.

Process modeling and simulation offers a way to better understand and develop emergency department processes (Jacobson et al., 2006; Jurishica, 2005). Many such approaches exist, including mathematical modeling and structured decision charts (Aguilar, 2006). A more general approach worthy of attention is the Business Process Modeling Notation (BPMN) (Object Management Group, 2006). This modeling language was designed to provide a common graphical notation for business analysts, managers and engineers to design and evaluate business process workflows. By unifying the various modeling notations and methodologies that currently exist, BPMN provides a common notation for expressing both basic and advanced business modeling concepts. Furthermore, the BPMN standard is equally suitable for modeling complex processes within the medical field (Sanchez et al., 2000).

Using the primary elements of workflow and connecting objects, BPMN depicts workflows among specific activities and the data exchange between them. By providing robust, graphical editing, users can quickly create reusable libraries of BPMN elements, which can then be quickly combined to create process diagrams. For example, models could be used to assess the workload capacity of a network of medical and triage stations and associated personnel to handle a postulated flow of casualties, as well as assess optional configurations of stations, people, and processes to determine which is best. Table 1 describes some of the BPMN elements utilized for modeling an Emergency Department process.

Table 1: Standard BPMN Elements

Element	Purpose	Graphical Notation
Events	Represent the start, intermediate, and end states	Start Intermediate End
Activities	Represent tasks and sub-processes	Task Process
Gateways	Represent decision points within the workflow	Gateway Fork/Join Inclusive Decision/Merge
Sequence flow	Represents the order in which activities will be performed	Sequence Flow
Message Flow	Represents the flow of data between participants	Message Flow
Association	Associates a data artifact to a flow object	Association
Pool	Represents a participant in the process and serves to partition related activities	Name
Data Object	Represents the data required or produced by an activity	Data

A computer-sensible model of an Emergency Department provides an intuitive view of the various steps within the process and the critical decision points. In addition to using BPMN to capture the structure of the process, a critical part of creating an executable model is defining the semantics of the process elements. As patients enter the system, they are received and triaged to determine the severity of their condition. Based on this first decision, patients are either given urgent care or sent to a specific treatment area. Each patient is evaluated, treated, and observed for a short period of time as needed. Each task within the process requires specific amounts and types of resources, as well as varying amounts of time to complete. All of these business rules and operating constraints must be captured in a precise way suitable for execution.

To do this, we employed Rosetta (Alexander, 2006), a system-level design language, to formally specify the behavior of the Emergency Department. Attached to each task, gateway, and event is Rosetta source code that captures how that particular BPMN element operates in relation to the process. For example, when a patient arrives at a particular task, that task cannot begin unless enough resources are available. In the meantime, the patient is placed in a queue. The rules needed to manage the queues, determine resource availability, and calculate performance metrics are captured using Rosetta. The following code fragment (Figure 1) describes the logic of updating the current patient queue based on available resources.

```
getNewQ: newQ =
     if (t==0) then
             updateQState(inputList, tick, averagetime,
                             resourceMap, resourcegroup)
     else
             updateQState(queue@previous_t & inputList,
                             tick, averagetime, resourceMap,
                             resourcegroup)
     end if;
```

Figure 1: Example model semantics expressed using Rosetta

Rosetta provides a means of capturing the semantics of a process in an unambiguous way. By defining the rules of operation and interaction for each component separately, we can then combine them using the BPMN notation to generate complex system behavior. As multiple rules are applied, their combined effects generate realistic and often unexpected results. Such behavior is difficult if not impossible to predict in the absence of a formally specified, executable model. With the aid of an intuitive modeling framework, users can capture all of this information in a methodical and reusable manner.

MODEL CREATION AND VALIDATION

Next we discuss how we applied this methodology to address a critical issue facing a large Emergency Department. The objective was to determine cost-effective methods to reduce the average length of stay for the department to 180 minutes or less. To accomplish this, we first worked with Emergency Department staff and utilized existing data from a previous Lean Process exercise to construct a computer- sensible model using Syscape. We first captured the structure of the process, which includes the process tasks, decision points, and task interdependencies. Next, we documented resource needs and key performance parameters of each process component. Finally, we modeled process-specific behavior using Rosetta. An example of such a model is shown in Figure 2.

With the Syscape modeling tool, the Emergency Department model could be configured for different types of scenarios by modifying specific properties. Each task can be assigned an average time and a deviation, which represents the expected time to complete the task. The staffing levels could be modified to simulate morning shifts versus afternoon shifts. The type of resource allocation can also be modified. Resources can be allocated to serve a specific area in the Emergency Department, or a global allocation scheme can be applied.

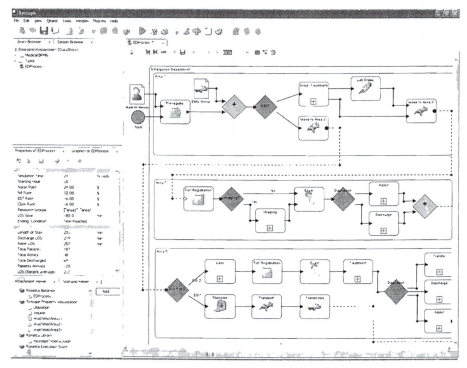

FIGURE 2: An example representation of an Emergency Department Process

Next, the probabilities for each decision point could be modified to simulate various scenarios in which the severity or volume of patients changes rapidly. Finally, the simulation time could be varied to simulate various time periods. Each of these simulations could be executed quickly and the results could be visualized as they changed over time using standard charts and graphs (Figure 3). The results could also be exported to a spreadsheet format for further analysis.

Once we had completed the general Emergency Department model, we then sought to capture and validate the As-Is process. Using previously collected data on task times, patient volumes, and LOS, we configured the model parameters to replicate the performance of the Emergency Department over an average 24-hour period. With this historical data, we determined that the model produced simulated results within 5% of what was expected for LOS and patient throughput.

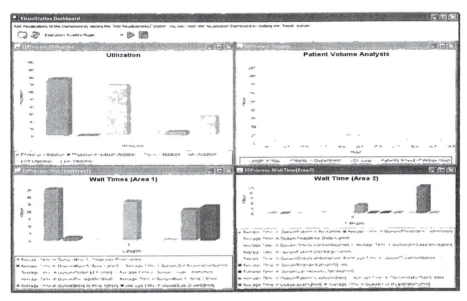

FIGURE 3: Visualizing Simulation Results

TO-BE MODELING AND ANALYSIS

With a validated As-Is model as the foundation, we next began to develop several To-be models. We first simulated the model for various sets of parameters to determine which had the most impact on overall LOS. This sensitivity analysis process helped identify the most relevant simulation parameters when attempting to optimize a specific performance metric. As a result of this effort, we determined that the number and type of resource allocation had the most significant affect on LOS. We first modeled resource allocation based on 'pods', in which resources are assigned to specific patient areas. This reflected the current practice of the Emergency Department. We next modeled a global allocation scenario in which resources could be assigned to any appropriate task in the Emergency Department.

In addition to resource allocation strategies, we also created several To-be scenarios to reflect proposed expansion plans in the Emergency Department. This included the addition of a new patient area, additional nurses, and additional bed capacity. By simulating these scenarios, we could determine 1) if the proposed expansion would lead to a reduction in LOS and 2) would this reduction provide a reasonable return on investment.

Through the simulation and analysis of several To-be scenarios, we developed several strategies that could be employed to reduce the LOS to 180 minutes or less. While we cannot report the specific results due to their proprietary nature, the significant findings we established through simulation included 1) static resource allocation creates inefficient resource utilization and 2) real-time resource movement between process areas will minimize cost and still achieve the LOS goal.

We also determined through collaboration with Emergency Department staff that a key benefit to this approach is the ability to visualize process performance over time. Using a dashboard, we were able to present the results of a simulation scenario through intuitive animations of key performance metrics. However, since patient arrival rates will likely vary over time compared to the average rates used for simulated analysis, we recommended that the executable model be connected to real-time data. The Emergency Department model could then be updated automatically on a periodic basis to serve as a real-time visual dashboard. Staff could use the dashboard to assess status and make informed resource allocation decisions in real-time. This has significant potential to reduce LOS, lead to better utilization of resources, and ultimately allow doctors to see more patients. Such a dashboard will provide a better view of bottlenecks as they develop, permit effective decisions regarding movement of floating resources, trigger alarms (sent through email/pager/text message) when certain thresholds are reached, and provide cost effective ways to mitigate bottlenecks. Furthermore, the model can be used to provide a "futurecast" capability to predict the near-term status of the Emergency Department based on current conditions.

CONCLUSIONS

By applying a methodical modeling and simulation approach to a critical problem facing Emergency Departments, we have demonstrated the benefits of such an approach to process improvement and patient care. Specifically, we showed that we can graphically model a business process i.e., patient flow, in an Emergency Department using BPMN. We can then simulate the resulting model and accurately reproduce its behavior in terms of key performance metrics. Through the analysis of several To-be scenarios, we determined that one of the key causes of inefficiency in an Emergency Department process is resource imbalance created due to a lack of real-time situational awareness. We further determined and confirmed with the Emergency Department staff that substantial savings can result from the ability to move human resources in real-time to tasks within the Emergency Department that need them the most.

In addition to the improvements we recommended for the specific Emergency Department, the use of process modeling and simulation can be beneficial in many ways. The first is the analysis of current processes. By modeling process workflows, valuable information can be obtained quickly on the efficiency of the processes and the resources deployed within it. Key staff can potentially make informed decisions rather than ad-hoc ones. The model can first be configured to represent current practices within the system. Using the model, various patient volumes can be simulated to predict performance of the system. Other metrics, such as resource utilization and average patient wait times can be quickly examined. These activities can provide insight into subtle nuances of the process that would otherwise be difficult to detect. For example, a process model may reveal that adding additional staff to patient registration may cause other resources to be better utilized. Furthermore, by applying real-time data to these models, stakeholders can apply predictive analytics to significantly improve decision-making.

Another benefit results from the ability to easily model many different process scenarios. Instead of incurring the high cost of live exercises, potential process improvements can be simulated and quickly compared. This is especially beneficial for disaster response, in which it is extremely difficult to perform live exercises for the myriad of emergency scenarios that may occur. Ultimately, the use of process modeling will likely encourage continuous process improvement by significantly reducing cost and effort of testing new ideas. In addition, simulating a process before it is implemented will provide increased confidence. A final benefit of process modeling relates to training and education. An accurate and intuitive picture of a current process is a powerful educational tool.

REFERENCES

Aguilar, E. R., Ruiz, F., García, F., and Piattini, M. (2006), "Evaluation measures for business process models." In *Proceedings of the 2006 ACM Symposium on Applied Computing* (Dijon, France, April 23 - 27, 2006). SAC '06. ACM, New York, NY, 1567-1568.

Alexander, P. (2006), *System Level Design with Rosetta (Systems on Silicon)*. Morgan Kaufmann Publishers Inc., San Francisco, CA

Object Management Group (2006), Business Process Modeling Notation (BPMN) Version 1.0. OMG Final Adopted Specification. 2006

Jacobson, S. H., Hall, S. N., and Swisher, J. R. (2006), "Discrete-event simulation of health care systems," Patient Flow: Reducing Delay in Healthcare Delivery (Ch. 8), International Series in Operations Research and Management Science, Springer, New York, USA

Jurishica, C. J. (2005), Emergency department simulations: medicine for building effective models. In *Proceedings of the 37th Conference on Winter Simulation* (Orlando, Florida, December 04 - 07, 2005). Winter Simulation Conference. Winter Simulation Conference, 2674-2680

Lowery, J.C, (1998), "Getting started in simulation in healthcare," *Proceedings of the 1998 Winter Simulation Conference,* 31–35, Institute of Electrical and Electronics Engineers, Piscataway

Patvivatsiri, L., Fraticelli, B.M.P., and Koelling, C.P. (2006), "A simulation-based approach for optimal nurse scheduling in an emergency department," *Proceedings of the 2006 Industrial Engineering Research Conference*, May 20-24, Orlando, FL

Sanchez, S. M., Ferrin, D. M., Ogazon, T., Sepúlveda, J. A., and Ward, T. J. (2000), "Emerging issues in healthcare simulation." In *Proceedings of the 32nd Conference on Winter Simulation* (Orlando, Florida, December 10 - 13, 2000). Winter Simulation Conference. Society for Computer Simulation International, San Diego, CA

Takakuwa, S. and Wijewickrama, A. (2008), "Optimizing staffing schedule in light of patient satisfaction for the whole outpatient hospital ward," *Proceedings of the 2008 Winter Simulation Conference*, 1500-1508, Miami, FL

Van Oostrum, J. M., Van Houdenhoven, M., Vrielink, M. M. J., Klein, J., Hans, E. W., Klimek, M., Wullink, G., Steyerberg, E. W., Kazemier, G. (2008), "A simulation model for determining the optimal size of emergency teams on call in the operating room at night," *Anesthesia & Analgesia*, Vol. 107, No. 5, 1655-1662

Chapter 40

Virtual Reality: From Training to Rehabilitation

Maurissa D'Angelo[1], Susan Kotowski[2], David B. Reynolds[1],
S. Narayanan[1], Jennie Gallimore[1]

[1]Wright State University
Department of Biomedical Industrial
and Human Factors Engineering
Dayton, OH 45435-0001, USA

[2]University of Cincinnati
College of Allied Health Sciences
Department of Rehabilitation Sciences
Cincinnati, OH 45267-0394, USA

ABSTRACT

Few rehabilitation programs and techniques have shown promise in standardizing long term amputee rehabilitation. Providing real-time, objective feedback will help to continuously improve the quality of life of amputee individuals. Existing systems and techniques are subjective and vary significantly from location to location and currently there is no established long term rehabilitation plan for amputees. The objective of this research was to determine the impact of a virtual reality rehabilitation (VRR) training program. This program provided real-time feedback and quantitative metrics to the amputee user. The focus of this study involved a male, five years post amputation, who exhibited hip adduction and fatigued quickly when walking. The individual completed a baseline session and four training visits. Following intervention, the subject exhibited an increased walking speed and a more narrow base of support. The subject was also able to walk a longer distance during the 2-minute walk test and fatigued less quickly. The individual stated that he could stand and walk for longer periods of time following the training and felt that his activities of daily living were easier and his overall quality of life had improved. VRR, incorporating real-time feedback and objective performance metrics, appears to be very promising. Five additional amputee patients completed VRR training and all reported positive outcomes. To our knowledge, this was the first study documenting a successful VRR gait-based strategy in patients with lower limb amputations.

Keywords: Virtual Reality, Rehabilitation, Amputee Training, Gait Techniques, Lower Limb Amputation

INTRODUCTION

There is a need for improved and more efficient rehabilitation for disabled individuals, specifically the amputee population. Despite a growing incidence of amputations and the evolution of rehabilitation, therapies have yet to be standardized for lower limb amputees. The goal of rehabilitation is to help individuals reach their highest level of potential and return to an activity level as close to possible as that prior to injury or disease.

There are approximately two million people living the United States with limb loss, and this number is growing by approximately 185,000 each year (Ziegler-Graham, 2008; Amputee Coalition, 2008). Centers of excellence for amputee rehabilitation are growing and further documenting the need for universal and superior amputee rehabilitation through the interaction of multiple specialties (Gauthier-Gagnon, 2006). Amputees are benefiting from continued improvements in prosthetics; however, a gap exists for integrating these prosthetic improvements with integrated rehabilitation. There is a lack of standards and individualized training, additionally training and rehabilitation vary significantly between rehabilitation locations.

Through the combination of expertise and customized programming in a virtual reality (VR) rehabilitation training environment, rehabilitation potential is unlimited. Using VR, newly developed prosthetics can be integrated with dynamic and improved rehabilitation techniques in a universal setting. With the advancement of computer capabilities, it is now possible to combine the expertise of an entire rehabilitation team into an individualized program specific to a disabled individual's needs. A consistent training program using VR will help to relieve rehabilitation staff and allow them to spend more quality, individual time with patients. Although VR has been used for over thirty years, progress in the rehabilitation field has only recently been documented (Keshner, 2004). Through the combination of computational and sensory technologies, an immersive environment can be created, allowing for a participatory rehabilitation program.

There is a well-accepted notion throughout the rehabilitation community that teamwork is essential for optimal rehabilitation. Interdisciplinary teamwork has been shown to improve both long term and short term outcomes of rehabilitation (Pasquina, 2006). This teamwork must be both across and within disciplines and it is essential to ensure that the patient and family members recognize the importance of their roles on the team and along the rehabilitation path. Together as a team, individual's needs should be identified and functional goals created to facilitate rehabilitation.

This research will address one of the many challenges amputees face during the course of rehabilitation and training – gait abnormalities. It is hypothesized that through appropriately designed visualization methods, amputees will be able to more effectively and efficiently ambulate. The three most common gait deviations amputees exhibit are uneven stride length, unequal weight distribution between limbs and a wide base of support (R. Gailey, personal communications, March 2007). These parameters will be addressed. Figure 1 details the overall design scheme for the research. A case study from the full evaluation will be discussed.

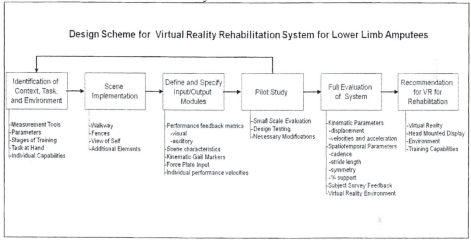

FIGURE 1. Virtual Reality Rehabilitation System Design and Test

METHOD

BACKGROUND

Rehabilitation specialists across many disciplines must work together in order to form an integrated, coordinated and effective treatment plan for the amputee. Through this rehabilitation effort the rehabilitation team will assist and encourage the amputee to once again reintegrate into the community (Pasquina, 2006).

Research has determined that muscle strength, balance and natural gait parameters all decline following lower limb amputation (VanVeltzen, 2006). Early and continued rehabilitation is necessary in order to restore walking capabilities with the use of a prosthesis. Standardization of both rehabilitation techniques and measures is necessary in order to improve the outcome and quality of life of amputee individuals. Conventional lower limb rehabilitation needs to become just that, an agreed upon, standardized, baseline rehabilitation program that can be individualized, as necessary, for each amputee.

In order to optimize lower limb amputee rehabilitation the patient, family, and medical and rehabilitation team, must form a partnership of tolerance, trust and understanding (Figure 2). Once this partnership is developed, the team as a whole must work together to implement rehabilitation. VR is an excellent assistive rehabilitation technology to assess both motor and cognitive abilities and to help plan and execute rehabilitation training. Lower limb amputation can require a hospital stay of up to six weeks (Braddom, 2000), therefore reducing environmental interaction and enrichment. This reduction of natural environmental interaction is counterproductive to rehabilitation and restoration of daily living functions (Optale, 2001).

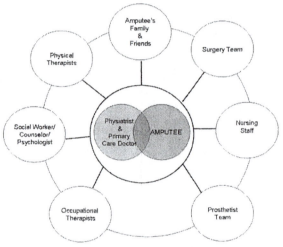

FIGURE 2. Radial Diagram of Amputee Rehabilitation (D'Angelo, 2010)

VR is a broad field that allows for a novel way to interact with information. It is different from typical graphical user interfaces in that it allows 3D stereoscopic rendering of the world (Figuera, 2005). VR attempts to replace some or all of the user's experiences in the physical world with 3D information such as graphics and sound and enables the user to experience a world that does not exist or exists in another time or place (Feiner, 1993). Virtual Reality (VR) environments present the capabilities to provide all individuals, regardless of their mobility level or cognitive capabilities, the ability to participate in rehabilitation tasks in an enriched environment. VR environments can exploit normal everyday experiences and immerse the individual, allowing them to focus on real life tasks. These environments have been shown to reduce the consequences of disabilities (e.g. Traumatic Brain Injury (TBI) and Stroke (Wang, 2004)).

Standardizing amputee rehabilitation is not only possible, but also convenient through the use of advanced computing technologies currently available. Computers can not only provide the capabilities to store and process data, but with technologies such as virtual reality, computers can also display information in real-time to the user and rehabilitation staff. Virtual reality allows for proven rehabilitation techniques to be used respectively in a safe, effective and engaging environment.

SUBJECT

Inclusion criteria included (1) lower limb amputation at least 12 months prior to study enrollment; (2) discharged from all other rehabilitation for prosthesis/gait; (3) have had current prosthesis for at least a month.

Exclusion criteria included: (1) < 18 years old or > 75 years old; (2) excessive or serious pain; (3) currently participating in any experimental rehabilitation or drug studies; (4) history of serious cardiopulmonary abnormalities or severe hypertension; (5) inability to walk for at least ten minutes at a time.

Using these criteria, a volunteer who responded to an advertisement was screened. Prior to screening and participation, the subject signed an informed consent form approved by the local institutional review board. The case study involved a 55 year-old male who had experienced an above knee, lower limb amputation on his left side 5 years prior to study participation resulting from diabetes complications. He was right hand dominant. He has been using his prosthetic, an OttoBock™ C-Leg, for approximately a year and a half.

The individual participated in five sessions and completed a 2-minute walk test prior to each session. Following the 2-minte walk test he completed a series of walking exercises at self selected slow, fast and normal paces. The subject was baselined for a period of two weeks and then following the baseline period began training. This training involved self visualization in a virtual reality environment (Figure 3). Training occurred over four weeks with one session each week lasting approximately one and a half hours.

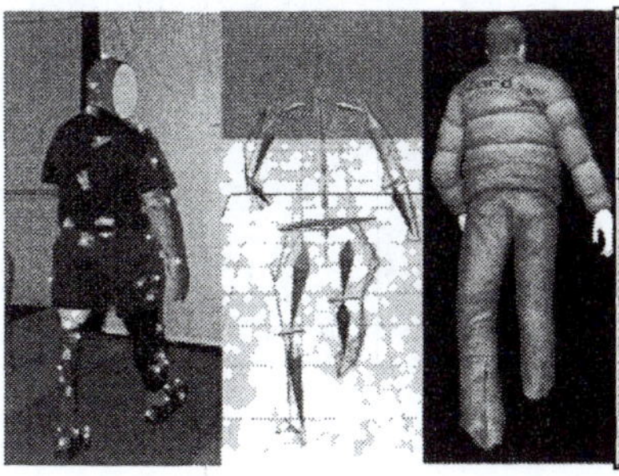

Visit 1	Visit 2	Visit 3	Visit 4	Visit 5
Initial vist and gait analysis	VR Gait	VR Gait	VR Gait	Final vist and Gait Analysis

FIGURE 3. Training Session and Schedule

APPARATUS/INSTRUMENTS

This training involved the use of a six camera motion capture system (Motion Analysis Corporation®) located around a 20-foot walkway. Two projectors were used to display images of the amputee, real time, as he walked on the walkway. The virtual reality software, WorldViz®, was used to create the avatar or character that was displayed on the screen in front of the subject. A general avatar, a male with jeans and a coat in this case, was used to remove any potential psychological effects of self-image. The male avatar moved in unison with the subject. 55 Retro-reflective markers were placed on the subject using a modified combined Helen-Hayes and Cleveland-Clinic cluster system.

The primary outcome measurements were taken using the Motion Analysis system data collection system, Cortex. This system was used to collect spatiotemporal and kinematic data. In addition to the data collected during gait analysis, subjects filled out questionnaires at the end of each session. These questionnaires helped to qualify the subject's overall satisfaction with the virtual environment and training sessions. They also served to assess the subject's general ability to complete activities of daily living and determine if there was any improvement. The questionnaires (Figure 4) were modified based on a review and studies of the Functional Independence Measures (FIM), Amputee Activity Scale (AAS), Prosthetic Profile of an Amputee (PPA), and Locomotor Capabilities Index (LCI) (Leung, 1996).

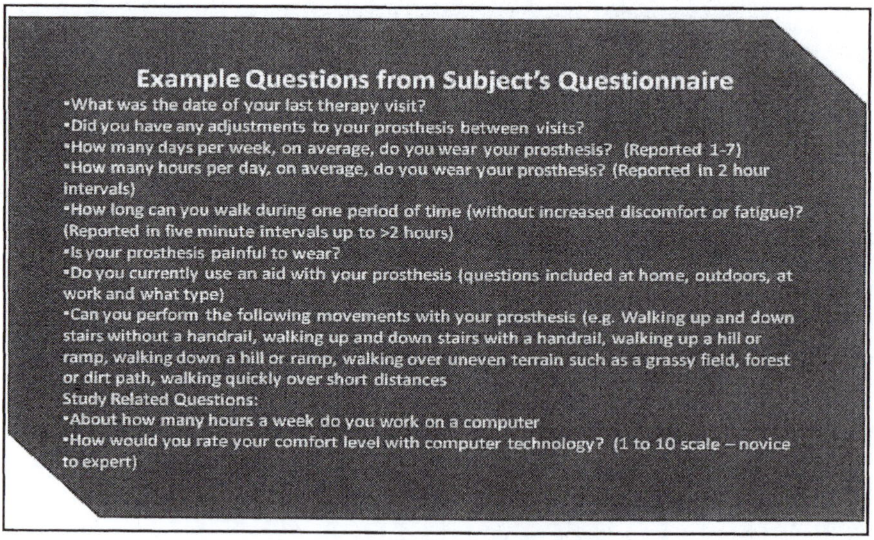

Example Questions from Subject's Questionnaire
•What was the date of your last therapy visit?
•Did you have any adjustments to your prosthesis between visits?
•How many days per week, on average, do you wear your prosthesis? (Reported 1-7)
•How many hours per day, on average, do you wear your prosthesis? (Reported in 2 hour intervals)
•How long can you walk during one period of time (without increased discomfort or fatigue)? (Reported in five minute intervals up to >2 hours)
•Is your prosthesis painful to wear?
•Do you currently use an aid with your prosthesis (questions included at home, outdoors, at work and what type)
•Can you perform the following movements with your prosthesis (e.g. Walking up and down stairs without a handrail, walking up and down stairs with a handrail, walking up a hill or ramp, walking down a hill or ramp, walking over uneven terrain such as a grassy field, forest or dirt path, walking quickly over short distances
Study Related Questions:
•About how many hours a week do you work on a computer
•How would you rate your comfort level with computer technology? (1 to 10 scale – novice to expert)

FIGURE 4. Sample Questions from Questionnaire

TESTING AND INTERVENTION

The subject was screened and, after completing an approved consent form, performed a 2-minute walk test and series of walking exercises (self-selected slow, fast and normal speeds). It took approximately an hour and a half to administer all measures to the participant. Following the two week baseline period (during which time there were no study requirements), the subject returned to the laboratory to begin virtual reality training. The baseline period was used to show no significant changes in gait patterns and walking activities were occurring without training. During the training sessions the subject repeated the 2-mintue walking test and series of walking exercises. The subject then began virtual reality training. This training involved placing the 55 retro-reflective marker set on the individual. The individual then walked up and back on the 20-foot walkway at self selected normal, fast and slow speeds. The projector screens at either end of the walkway displayed the backside view of a male avatar depicted as if the individual was walking approximately five feet behind himself. Each subject completed three trials at the three self selected velocities. Spatiotemporal (velocity, stride length, cadence, percent double support, time symmetry) and kinematic (displacement, velocities, acceleration) gait variables were collected. A force plate mounted in the middle of the walk way recorded 3D ground reaction forces.

DATA ANALYSIS

Five parameters related to gait quality were calculated from the kinematic data collected. These parameters included walking speed, step length, stride length, width of base of support, and length walked from the 2 minute walk test. All calculations were made based on the center of the walking volume; the length of the volume used was 20 feet and all calculations were based off the center ten feet (five feet on either side discarded to account for ramp up and slow down of walking). Reported calculations are based on the individuals self-selected normal walking speed. Walking speed was determined as feet/minute and was calculated using the center of the volume. The walking speed was calculated as a product of cadence and step length. This was calculated for self selected normal, fast and slow speeds (normal reported). Cadence (steps/min) was determined by the total number of steps divided by the time needed to complete these steps. Step length (feet) was calculated as the distance between two consecutive contralateral heel strikes. Step lengths were not normalized to leg length as the subject served as his own baseline and the data analysis focused on individual improvement ratios rather than subject to subject data values. The stride length (feet) was based on the distance between two consecutive heel strikes of the same foot.

RESULTS

The subject completed a baseline visit, baseline period of no study requirements, three training sessions and a final training and gait assessment session. Compliance was 100% and he attended all clinical sessions.

Prior to intervention, the subject exhibited significant fatigue from walking and had difficulty completing the 2-minute walk test without stopping to take a break. As a result of his fatigue and difficulty walking, he was very unsatisfied with his ability to perform activities of daily living.

Following intervention, the subject was able to complete the 2 minute walk test without stopping (final visit and gait assessment). The two minute walk test has been shown to be an adequate correlation to measures of physical function and prosthetic use within the lower limb amputee population (Brooks, 2001). The two main parameters assessed were stride length and width of base of support. The results are shown in Table 1. The self selected normal walking speeds improved over the training and are equivalent to very slow, slow and moderate walking speeds, respectively from baseline to final (Gailey 2002; R. Gailey, personal communications, March 2007). The individual's base of support narrowed slightly but this improvement was minimal as compared to the greater walking distance the subject could complete and the improved walking speeds.

Table 1. Calculated Gait Assessment Parameters

(*avg)		Baseline	Intervention	Final
Walking Speed (feet/sec)	N	15.5	19.7	27.7
Stride Length (in)	L	43.4	30.2	32.2
	R	43.3	38.2	34.0
Step Length (in)		24.5	23.9	24.10
Base of Support (in)		12	10	9
2 Minute Walk Test		211	272.5*	298

Qualitatively the subject was very pleased with his outcomes as the result of this study. He hopes to continue in future clinical trials and continue to improve his gait. The subject exhibited reduced fatigue while walking.

CONCLUSION

Chronic and repetitive exercises have been shown to create permanent structural changes in the brain and a reorganization of the nervous system (Miles, 2003). An environment that consists of functional real world demands and repetitive procedures can aid in the restoration and rebuilding of an individual's cognitive processes (Optale, 2001). VR has the capability to immerse a disabled individual into a real world setting where he/she can participate in realistic scenarios as part of his/her rehabilitation program.

This research presented a case study to test the hypothesis that through appropriate visualization methods, amputees will be able to more effectively and efficiently ambulate with a more symmetrical gait. Our subject exhibited more uniform gait at all speeds (self selected normal, fast and slow) at the end of his training. He was able to walk a greater distance during his two minute walk test and reported that "this visualization has dramatically improved my gait."

This research shows great promise for future research using VRR to improve gait techniques and enable more efficient and effective rehabilitation for amputees and other disabled individuals. Opportunities for using VRR to improve rehabilitation programs include expanding the currently limited opportunities for rehabilitation scenarios, enhancing primitive spatial and temporal training scenarios, addressing staff and facility limitations, creating user friendly interfaces improving individual motivation through real-time feedback, and integrating an interactive environment. VR rehabilitation systems are emerging as valuable tools in the reestablishment of functionality and improved quality of life for individuals suffering from disabilities. In the case of lower limb amputees, virtual environments can help individuals to understand where their limb is in space and in relationship to other objects (proprioception). By creating an environment in which the user's limbs (both prosthesis and natural) are graphically shown in the "real world" environment, the user can begin to understand the necessary motions for walking. This system provides a platform for multidisciplinary effort to bridge technology gaps and create a successful VRR system.

REFERENCES

Braddom, R. (2000). *Physical medicine and rehabilitation* (Second ed.). Philadelphia, Pennsylvania: W.B.   Saunders Company.

Brooks, D., Parsons, J., Hunter, J. P., Devlin, M., & Walker, J. (2001). The 2-minute walk test as a measure of functional improvement in persons with lower limb amputation. *Archives of Physical Medicine and Rehabilitation, 82*(10), 1478-1483.

D'Angelo, M., Kotowski, S., Reynolds, D., Narayanan, S. (2010). Application of Virtual Reality to the Rehabilitation Field to Aid Amputee Rehabilitation: Findings from a Systematic Review. *Disability and Rehabilitation: Assistive Technology.* 5(2), 136-142.

Feiner, S., Macintyre, B., & Seligmann, D. (1993). Knowledge-based augmented reality. *Communications of the ACM, 36*(7), 53-62.

Figueroa, P., Bischof, W. F., Boulanger, P., & James Hoover, H. (2005). Efficient comparison of platform alternatives in interactive virtual reality applications. *International Journal of Human – Computer Studies, 62*(1), 73-103.

Gailey, R. S., Roach, K. E., Applegate, E. B., Cho, B., Cunniffe, B., Licht, S., et al. (2002). The amputee mobility predictor: An instrument to assess determinants of the lower-limb amputee's ability to ambulate. *Archives of Physical Medicine and Rehabilitation, 83*(5), 613-627.

Gauthier-Gagnon, C., & Grisé, M. (2006). Tools to measure outcome of people with a lower limb amputation: Update on the PPA and LCI. *Journal of Prosthstics and Orthotics, 18*(1S), 61.

Keshner, E. (2004). Virtual reality and physical rehabilitation: A new toy or a new research and rehabilitation tool? *Journal of NeuroEngineering and Rehabilitation, 1*(1), 8.

Leung, E. C., Rush, P. J., & Devlin, M. (1996). Predicting prosthetic rehabilitation outcome in lower limb amputee patients with the functional independence measure. *Archives of Physical Medicine and Rehabilitation, 77*(6), 605-608.

Miles, T. Reorganization of the Human Motor Cortex by Sensory Signals: A Selective Review. *Clin Exp Pharmacol P.*, vol. 32, pp. 128-131, 2005.

Optale, G. (2001). Music-enhanced immersive virtual reality in the rehabilitation of MemoryRelated cognitive processes and functional abilities: A case report. *Presence: Teleoperators, 10*(4), 450.

Pasquina, P. F., Bryant, P. R., Huang, M. E., Roberts, T. L., Nelson, V. S., & Flood, K. M. (2006). Advances in amputee care. *Archives of Physical Medicine and Rehabilitation, 87*(3), 34-43.

United Amputee Services Association, Inc. (UASA) (2003). *A survivor's guide for the revent amputee.* Winter Park, Florida: United Amputee Services Association, Inc.

van Velzen, J., van Bennekom, C., Polomski, W., Slootman, J., van der Woude, LHV, & Houdijk, H. (2006). Physical capacity and walking ability after lower limb amputation: A systematic review. *Clinical rehabilitation, 20*(11), 999-1016.

Wang, P., Kreutzer, I. A., Bjärnemo, R., & Davies, R. C. (2004). A web-based cost-effective training tool with possible application to brain injury rehabilitation. *Computer methods and programs in biomedicine, 74*(3), 235-243.

Ziegler-Graham, E., MacKenzie, E., Ephriam, P., Travison, T., Brookmeyer, R. (2008). Estimating the Prevalance of Limb Loss in the United States: 2005 to 2050. *Archives of Physical Medicine and Rehabilitation.* 89(3), 422-429.

CHAPTER 41

Designing a Virtual Patient for Communication

April Barnes, Jennie Gallimore, Rosalyn Scott

Ohio Center of Excellence for
Human-Centered Innovation
Wright State University
USA

ABSTRACT

Communication is a core clinical skill for healthcare professionals, but training can be inconsistent. The amount of time focused on training communication, the quality and quantity of feedback, and the diversity of actual patient interactions may vary widely among students. Virtual patients (VPs) are an innovative method to provide communication training in a safe environment. This paper discusses formative research on the development of an interactive VP designed with a communication framework. A communication model is being developed based on input from expert clinicians, extensive literature analysis, and theories of communication. A prototype VP has been developed that uses speech recognition and a script-mapping mechanism to allow the learner to interact in a natural verbal conversation flow. To assess affective communication, content of the spoken words and voice characteristics are evaluated for emotional qualities. The integration of the communication model, speech recognition and emotion detection will result in a high fidelity VP. The VP will have more human-like conversation skills that are representative of actual physician-patient interactions. Metrics for communication effectiveness will be developed so that performance can be objectively analyzed. It is expected that the improved fidelity of the VP interaction will significantly improve the training of communication and interpersonal skills for health care providers. This research is the first phase of a long-term goal to develop a completely immersive, longitudinal surgical scenario using different forms of medical simulation including computer-based VPs, surgical simulators with haptic controls, high-fidelity mannequins and virtual environments.

Keywords: Virtual patient, communication, medical education

INTRODUCTION

The importance of communication in the physician-patient relationship cannot be understated. Lipkin, Putnam & Lazare (1995) describe the significance of the medical interview, "The medical interview is a core clinical skill. It is the medium of doctor-patient communication and relationship, the most important single source of diagnostic data, the means through which the physician elicits the patient's partnership and participation in the process of care"(Lipkin, Putnam, & Lazare, 1995). Despite its importance, communication training for physicians can be inconsistent. The amount of time focused on training communication, the quality and quantity of feedback and the diversity of patient cases may vary widely among students. Standardized patients (SPs), people trained to portray real patients, are often used to provide consistent patient scenarios to students. However, SPs introduce some elements of human error as well. Virtual patients are an innovative method to provide communication training and objective feedback. This paper will discuss formative research on the development of an interactive VP built with a communication framework. A communication model is being developed based on input from expert clinicians, extensive literature analysis, and on theories of communication. The integration of the communication model, speech recognition and emotion detection will result in a higher fidelity VP. The VP will have more human-like conversation skills that are more representative of actual physician-patient interactions. Metrics for communication effectiveness will be developed so that performance can be objectively analyzed. It is expected that the improved fidelity of the VP interaction will significantly improve the training of communication and interpersonal skills for health care providers.

BACKGROUND

Virtual patients are defined by the Association of American Medical Colleges (AAMC) as "interactive computer programs that simulate real-life clinical scenarios in which the learner acts as a healthcare professional obtaining a history and physical exam and making diagnostic and therapeutic decisions"(AAMC). VPs have different capabilities and levels of fidelity but most have similar formats. A case scenario is presented to the user with the patient represented by still images, video clips and/or audio clips. The learner interacts with the patient by selecting pre-scripted questions from a menu, typing free-form text, or speaking to the VP. The user navigates systematically through the patient history, physical examination and investigation including laboratory tests and diagnostic imaging. Many VPs follow a "string of pearls" format, popular in video games, where one level must be completed before proceeding to the next (Ellaway, Poulton, Fors, McGee, &

Albright, 2008).

BENEFITS OF VIRTUAL PATIENTS

VPs provide an alternative to the traditional "apprentice-type" model of clinical instruction and patient-based training (Kneebone, 2003; Zary, Johnson, Boberg, & Fors, 2006). They provide opportunities for repeated practice in a safe environment, present a standard performance to every learner and can be modified to represent a variety of patients and outcomes (Kneebone, 2003).VPs also give objective, individualized feedback which can be integrated with updated medical evidence and links to relevant literature (Kneebone, 2003; Triola et al., 2006). In addition, assessment with VPs can evaluate knowledge, data interpretation and management skills with one tool (Round, Conradi, & Poulton, 2009). Web-based VPs provide the added benefits of lower production costs, improved accessibility, easier sharing and the capability for multiple users to use a single VP simultaneously (Triola et al., 2006; Zary et al., 2006). Also, changes made in web-based cases are immediately accessible and users do not have to carry out extensive installations, maintenance, or upgrades of software (Zary et al., 2006).

DRAWBACKS OF VIRTUAL PATIENTS

VPs can be very costly in time, staff and other resources needed to develop, maintain and integrate VPs into the curriculum (Cook & Triola, 2009; Huang, Reynolds, & Candler, 2007; Zary et al., 2006). For the student, another potential drawback is a lack of interactivity. Many VPs have videos or static photos of SPs and do not have voice recognition or natural speaking capabilities. Those that are capable of conversing usually have a narrow range of responses and limited expressiveness. This can be a distraction for learners, decrease realism and may raise cognitive load (Cook & Triola, 2009; Raij et al., 2006; Triola et al., 2006).When used for assessment, VPs are often more objective than SPs but there can be difficulty evaluating the learner on nonverbal communication, eye contact, and use of jargon (Triola et al., 2006).

ADVANCES IN VP DEVELOPMENT

Advances in technology are allowing the development of higher fidelity VPs that address many of these issues. Improvements in the usability of software and the reduced amount of computing power required to develop and maintain VPs facilitate the development of VPs by non-technical staff and sharing among institutions. High-fidelity VPs are more appropriate for training interpersonal skills that are currently trained via interactions with SPs or real patients (ACGME). For example, DIANA (DIgital Animated Avatar) a life-sized, 3D VP with high-quality rendering, animation with speech and gesture recognition that responds to natural speech was found to be effective in teaching communication skills to medical students(Johnsen et al., 2006; Stevens et al., 2006). Similar results have been shown by other VPs that facilitate interaction (Kenny, Parsons, Gratch, & Rizzo, 2008; Zary et al., 2006; Alverson et al., 2005; Conradi et al., 2009; Parsons, Kenny, & Rizzo, 2008). Learners training with VPs have similar results in education measures and interview skill performance compared to learners using SPs (Johnsen et al., 2005; Raij et al., 2006; Johnsen, Raij, Stevens, Lind, & Lok, 2007). Although there is some debate over the effect of the artificiality of VPs, most students are able to overcome this and become immersed in the simulation. Studies have shown that medical students training with VPs demonstrate nonverbal communication behavior, respond empathetically to VPs, and experience substantial emotional effects (Bearman, 2003; Deladisma et al., 2007). Overall, learners are receptive to the use of VPs and find them to be realistic, appropriately challenging and valuable as educational and assessment tools (Gesundheit et al., 2009; Kenny et al., 2008; Stevens et al., 2006). Speech emotion detection is an emerging technology for VP applications. Speech analyses have found distinct measurable vocal changes that correlate to certain emotions (Bänziger & Scherer, 2005; Yu, Chang, Xu, & Shum, 2001).Rodriguez (2008) found changes in vocal frequency and temporal metrics in Human-Virtual Human (H-VH) interactions were similar to those in Human-Human (H-H) interactions (Rodriguez, Beck, Lind, & Lok, 2008). This suggests that VPs can evoke similar emotions in the user as an actual patient.

COMMUNICATION

Physician-patient communication can be very complex and dynamic. Each party brings their own cognitive abilities, experiences, emotions, and biases to the interaction. Evaluating communication is also complex. Many times, checklists are used by evaluators to quantify the subjective nature of communication skills. For example, the Kalamazoo consensus statement (2001) lists the seven essential elements of communication in medical encounters (Makoul, 2001):

Build a relationship
Open the discussion
Gather information
Understand the patient's perspective
Share information
Reach agreement on problems and plans
Provide closure

Another approach is to evaluate the presence of empathy in physicians' communication. One frequently used measure is the Jefferson Scale of Physician Empathy (JSPE)(Hojat et al., 2001). The JSPE is a 20-item scale that measures physician's self-reported empathy.

In an effort to move to objective measures of training and evaluating communication, we propose that VPs should be created based on theories and models of communication.

COMMUNICATION MODEL APPROACH

Current research at the Ohio Center of Excellence for Human-Centered Innovation is focusing on development of a VP built upon a communication and personality models to enhance training and objective evaluation. A model-based approach allows one to replace open-ended, trial and error analysis of the data with guided exploration. One communication model being evaluated is the Cognitive-Affective Model of Organizational Communication Systems (CAMOCS) developed by Te'eni (Te'eni, 2001).

The CAMOCS model is based primarily on research in organizational communication in business management and is based on 301 research articles. The model includes three main factors: **1) inputs** to the process including task attributes, physical and cognitive distances between senders and receivers, and values and norms of communication, **2) cognitive-affective processes** including communication strategies, message form, and the communication medium, and **3) impact** of the communication including levels of understanding, and relationships between the sender and receiver. The model also takes into account communication complexity. Examples of **cognitive complexity** are intensity of information exchange; multiple views held by communicators; incompatibility between representation technique and information communicated. Impacts for these are misunderstandings, understanding messages in a different context than intended, or cognitive work increase to recode information. Examples of **Dynamic Complexity** are time constraints, deficient feedback, and asynchronous communication leading to impacts on communication strategies and forgetting. **Affective complexity** examples are complex emotional feelings or sensitivity to changes in disposition toward team members leading to mistrust, lack of disclosure, or misunderstanding. Te'eni's model provides us with the ability to evaluate physician-patient communication via a model based on extensive experimental research. This model can help to guide the research and specifically allows us to code communications into relevant categories for analysis.

EXAMPLES OF INPUTS

Task attributes include defining specific procedures required to complete the task, potential variations of the task and time demands. For example, referring to the Kalamazoo statement on essential communication elements, specific phrases and affects used by expert physicians to build a relationship with a patient would be identified. Examples of variations are the setting (office visit, ER, hospital inpatient, etc), or whether this is the initial meeting with the patient or an established patient. Time demands are also a factor in physician-patient communication. The time allotted for a routine office visit or the patient's condition that may require immediate treatment.

Sender-receiver distance is also applicable to physician-patient communication. The physician has to present the information to the patient at a level that the patient can comprehend. This will vary from patient to patient based on education level, cultural differences and language barriers. The physician also has to be cognizant of the emotional nature of the communication. For example, a physician smiling when delivering somber news may be confusing to the patient. Cultural differences may also create discrepancies between the values and norms of the physician and patient. For example, a patient may reject a blood transfusion based on religious beliefs.

EXAMPLES OF PROCESS

Examples of communication **goals** in the medical domain include obtain current health status and health history, understanding of surgical outcomes and risk, building trust, etc.

Examples of communication **strategies** are provided in Table 1 along with definitions of strategies.

Table 1. Definitions of Communication Strategies

Strategy	Definition	Example
Contextualization	Provision of explicit context in messages	"I had a CT scan last week."
Affectivity	Provision of affective components (emotions, moods) in messages.	"I am feeling very overwhelmed."
Control - adjusting	Testing and adjusting communication according to feedback during the process	"I don't think you understood what I meant, let me put this another way"
Control - Planning	Planning the pattern of communication and contingencies ahead of the process	First I will discuss ...., followed by ....
Perspective taking	Considering the receiver's view and attitude.	"I understand why in your case you might be hesitant to try this option."
Attention focusing	Directing or manipulating the receiver's information	"I want you to focus on ..."

EXAMPLES OF IMPACT

The impact of the communication depends on achieving mutual understanding. An example would be a physician explaining potential side effects of a treatment. The patient conveys understanding of the risks. Building a relationship with a patient is vital to effective communication. The physician may be presenting the best available treatment to the patient but if the patient does not trust the physician's judgment or feel they are working for the patient's best interest they will be less likely to adhere and actively participate. Studies have also shown patients are less likely to sue physicians they feel have established a rapport (Vicente, 2004).

VP CONCEPT

Using the CAMOCS model, communication can be coded and evaluated to measure performance, and the model can be used to inform design to provide the type of learning examples that are needed for communication education.

Figure 1 illustrates the basic concept of the VP. The VP is developed based on a learning objective and scenario. The full body VP is a 3D model that can be displayed in an immersive virtual environment with stereoscopic viewing, on computer screens, or 3D televisions. The animated VP can move as well as make facial expressions and blink. The 3D model is programmed using software by Haptek. A speech recognition algorithm analyzes the learner's spoken words. Signal processing is used to analyze speech tone to pick up speaker emotion. Keywords are processed followed by coding of the information into the communication model. Based on the users input, the VP will provide a response including voice output and non-verbal communication via movement and facial expressions. The selection of outputs will vary based on probabilities so that the VP is not providing a canned answer each time. Coding the communication between the learner and the VP will allow for a communication analysis to provide feedback to the user. For example, if the VP portrays emotion and the learner does not respond as might be expected, feedback on lack of affectation can be provided to the learner. In the future, gesture and facial emotion recognition, and eye tracking of the learner can be added to the system similar to DIANA (Johnsen et al., 2005; Stevens et al., 2006).

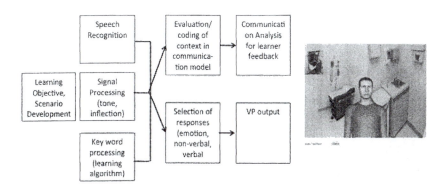

Figure 1. Block diagram of VP Concept

With assistance from Richard Bell, MD, from the American Board of Surgery we are currently creating a colon cancer scenario. The communication model is under development based on expert clinician input, literature analysis, and theories and models of communication.

FUTURE PLANS

A goal of this research is to develop a completely immersive, longitudinal surgical scenario utilizing multiple forms of medical simulation. The integration of different modalities will allow the user to practice clinical skills, decision-making, and communication skills. The user will experience the complete range of interactions that a surgeon would have with a patient over the course of treatment. The scenario will begin with the surgical consultation with a VP. The patient interview and physical exam will be conducted with a computer-based VP integrated with a high-fidelity mannequin simulator. Next, hands-on procedures would be performed using a part task or full body VR simulator. The student would then visit with the VP to provide follow-up care. The scenario would include built-in time schedules so that training does not take place all in one day but follows more realistic time periods between visits.

CONCLUSION

Virtual patients are becoming more prevalent for medical education. Although often focused on clinical reasoning and decision-making skills, they can also be used for training communication and interpersonal skills. Advances in technology have made high fidelity VPs more accessible. This research proposes the development of a VP that is able to communicate in a natural, conversational manner including the appropriate emotion and non-verbal feedback. In addition, the VP will also utilize a communication model to objectively evaluate the communication skills of the user.

REFERENCES

ACGME. *ACGME Outcome Project.* Retrieved 10/1/2009, 2009, from http://www.acgme.org/outcome/assess/toolbox.asp

Alverson, D. C., Saiki Jr, S. M., Caudell, T. P., Summers, K., Sherstyuk, A., Nickles, D., et al. (2005). Distributed immersive virtual reality simulation development for medical education. *Journal of International Association of Medical Science Educators, 15*(1), 19-30.

Bänziger, T., & Scherer, K. R. (2005). The role of intonation in emotional expressions. *Speech Communication, 46*(3-4), 252-267.

Bearman, M. (2003). Is virtual the same as real? Medical students' experiences of a virtual patient. *Academic Medicine : Journal of the Association of American Medical Colleges, 78*(5), 538-545.

Conradi, E., Kavia, S., Burden, D., Rice, A., Woodham, L., Beaumont, C., et al. (2009). Virtual patients in a virtual world: Training paramedic students for practice. *Medical Teacher, 31*(8), 713-720.

Cook, D. A., & Triola, M. M. (2009). Virtual patients: A critical literature review and proposed next steps. *Medical Education, 43*(4), 303.

Deladisma, A. M., Cohen, M., Stevens, A., Wagner, P., Lok, B., Bernard, T., et al. (2007). Do medical students respond empathetically to a virtual patient? *The American Journal of Surgery, 193*(6), 756-760.

Ellaway, R., Poulton, T., Fors, U., McGee, J. B., & Albright, S. (2008). Building a virtual patient commons. *Medical Teacher, 30*(2), 170-174.

Gesundheit, N., Brutlag, P., Youngblood, P., Gunning, W. T., Zary, N., & Fors, U. (2009). The use of virtual patients to assess the clinical skills and reasoning of medical students: Initial insights on student acceptance. *Medical Teacher, 31*(8), 739-742.

Hojat, M., Mangione, S., Nasca, T. J., Cohen, M. J. M., Gonnella, J. S., Erdmann, J. B., et al. (2001). The Jefferson scale of physician empathy: Development and preliminary psychometric data. *Educational and Psychological Measurement, 61*(2), 349-365.

Huang, G., Reynolds, R., & Candler, C. (2007). Virtual patient simulation at US and Canadian medical schools. *Academic Medicine : Journal of the Association of American Medical Colleges, 82*(5), 446-451.

Johnsen, K., Dickerson, R., Raij, A., Harrison, C., Lok, B., Stevens, A., et al. (2006). Evolving an immersive medical communication skills trainer. *Presence: Teleoperators & Virtual Environments, 15*(1), 33-46.

Johnsen, K., Dickerson, R., Raij, A., Lok, B., Jackson, J., Shin, M., et al. (2005). Experiences in using immersive virtual characters to educate medical communication skills. *Proc. of IEEE Virtual Reality,* 179-186.

Johnsen, K., Raij, A., Stevens, A., Lind, D. S., & Lok, B. (2007). The validity of a virtual human experience for interpersonal skills education. *Proceedings of the SIGCHI Conference on Human Factors in Computing Systems,* 1058-1069.

Kenny, P., Parsons, T. D., Gratch, J., & Rizzo, A. A. (2008). Evaluation of Justina: A virtual patient with PTSD. *Lecture Notes in Computer Science, 5208,* 394-408.

Kneebone, R. (2003). Simulation in surgical training: Educational issues and practical implications. *Medical Education, 37*(3), 267-277.

Lipkin, M., Putnam, S. M., & Lazare, A. (1995). *The medical interview: Clinical care, education, and research.* New York: Springer-Verlag.

Makoul, G. (2001). Essential elements of communication in medical encounters: The Kalamazoo consensus statement. *Academic Medicine : Journal of the Association of American Medical Colleges, 76*(4), 390-393.

Parsons, T. D., Kenny, P., & Rizzo, A. A. (2008). Virtual human patients for training of clinical interview and communication skills. *International Conference on Disability, Virtual Reality and Associated Technology,* Porto, Portugal.

Raij, A., Johnsen, K., Dickerson, R., Lok, B., Cohen, M., Bernard, T., et al. (2006). *Interpersonal scenarios: Virtual ≈ real?*

Rodriguez, H., Beck, D., Lind, D., & Lok, B. (2008). Audio analysis of Human/Virtual-human interaction. *Lecture Notes in Computer Science, 5208,* 154-161.

Round, J., Conradi, E., & Poulton, T. (2009). Training staff to create simple interactive virtual patients: The impact on a medical and healthcare institution. *Medical Teacher, 31*(8), 764-769.

Stevens, A., Hernandez, J., Johnsen, K., Dickerson, R., Raij, A., Harrison, C., et al. (2006). The use of virtual patients to teach medical students history taking and communication skills. *The American Journal of Surgery, 191*(6), 806-811.

Te'eni, D. (2001). Review: A cognitive-affective model of organizational communication for designing IT. *MIS Quarterly, 25*(2), 251-312.

Triola, M., Feldman, H., Kalet, A. L., Zabar, S., Kachur, E. K., Gillespie, C., et al. (2006). A randomized trial of teaching clinical skills using virtual and live standardized patients. *Journal of General Internal Medicine : Official Journal of the Society for Research and Education in Primary Care Internal Medicine, 21*(5), 424-429.

Vicente, K. J. (2004). *The human factor: Revolutionizing the way people live with technology.* New York, NY: Routledge.

Yu, F., Chang, E., Xu, Y. Q., & Shum, H. Y. (2001). Emotion detection from speech to enrich multimedia content. *Lecture Notes in Computer Science, ,* 550-557.

Zary, N., Johnson, G., Boberg, J., & Fors, U. G. (2006). Development, implementation and pilot evaluation of a web-based virtual patient case simulation environment--Web-SP. *BMC Medical Education, 6,* 10.

CHAPTER 42

Mobile Devices as Virtual Blackboards for M - Learning

Danco Davcev, Vladimir Trajkvik

Faculty of Electrical Engineering and Information Technologies
"Ss. Cyril and Methodius" University
Skopje, REPUBLIC OF MACEDONIA

ABSTRACT

Mobile learning (m-Learning) offers solutions that address the shortcomings of the traditional classroom especially for healthcare students and workers. In this paper, we propose small mobile devices as virtual blackboards for collaborative learning.

The feasibility and usability of mobile devices as virtual blackboards for educational purposes was investigated by a survey done with the medical students and workers. We found that functions of the mobile virtual blackboard were effective in cooperative learning. From the ergonomics point of view connected with the usage of the mobile devices, some improvements are recommended.

Keywords: Mobile devices, Mobile Learning, Collaborative Learning, Ergonomics in Healthcare

INTRODUCTION

Mobile Learning (M-Learning) is commonly used to increase effectiveness of educational process by offering solutions that address some of the shortcomings of the traditional classroom (Adewunmi, A., et al., 2003), (Fulp, C.D., and Fulp, E.W.,

2002), (Kool, S., et al., 2003), (Shotsberger, P., and Vetter, R., 2001).

Collaborative learning is a term drawn from educational psychology to describe guided support that helps students mindfully engage in unfamiliar new work. Mindful engagement, in the context of Learning Environment, means actively participating in learning by finding out about the new concepts by asking within certain learning community (Tucker, T. G., Winchester, W. W., 2009).

Wireless LAN Solutions provide deployment of high-speed Internet and LAN connectivity not only in the classroom but throughout the campus as well. Accessibility of information, regardless of the location, is available for both, the students, healthcare workers and instructors.

M-Learning is learning with the help of a mobile device (typically PDA, or mobile phone). In our opinion, mobile devices should also provide wireless communication between professor(s), student(s) and other healthcare workers (HCW), (McAlister, M.J., Xie, P.H., 2005), (Yoshino, et al, 2003), (Wang, X., 2009).

The use of mobile devices as virtual blackboards will be elaborated in this paper. It serves for consultations purposes and it is especially suitable for acquiring small and exact pieces of knowledge that can be immediately used. The possibility of instant help from another colleague creates an ideal environment for collaborative learning.

Our application delivers communication via UDP protocol, which enables transmission of data packets among participants with use of IP addresses, so UMTS - IMS networks can also be used.

MOBILE BLACKBOARD ARCHITECTURE AND DESIGN

The mobile virtual blackboard is made for attending consultations on mobile devices, precisely on pocket PCs. When compared with other kinds of distance learning methods, mobile learning should give fast and exact answers to needed questions. Consultations in a group create implicit collaboration among students and/or HCW which gives more quality knowledge, because they can find answers to certain questions, that they may never think of. General equipment Support for Mobile Learning System is given in (Dongsheng, D., 2009), (Haitao, P., et al., 2009), (Yi, J., 2009).

Our Wireless mobile virtual blackboard, consultation system implements a client - server architecture that requires less communication load on the client applications, since mobile devices have limited memory and processor power. The server part is implemented as desktop PC application, since servicing clients might demand system with high performances and data throughput. The entire communication among the clients is carried out through the server (see Figure 1).

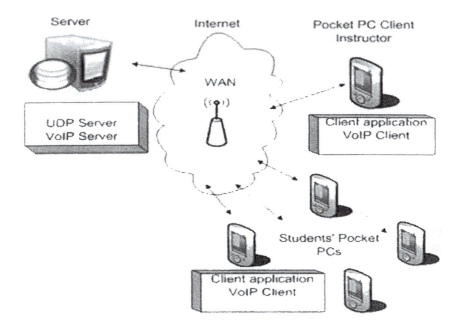

Figure 1. Architecture of the mobile blackboard system.

There are two kinds of clients, both implemented on a pocket PC platform. The first client is instructor application, which controls the resources in the group, and is the source of relevant knowledge. The other client is a student and/or HCW which uses the knowledge provided from the instructor. The client user's interface has controls that enable chat, drawing, feature for file upload/download, authority control (who should use the blackboard). The virtual blackboard consultation system also contains features for audio conversation with means of VoIP. The significance of this feature is obvious. Talking enables more precise discussion to be led in less time.

Using the mobile blackboard, the student and/or HCW can pose a question by using several ways of communication.

Interactivity of this system is very important issue. Speaking, combined with drawing, where also the colleagues largely participate, should be the most appreciated way of conducting conversation and explaining learning topics. All details about the collaboration and interaction possibilities among colleagues are given on the mobile virtual blackboard UML sequence diagram (Figure 2).

360

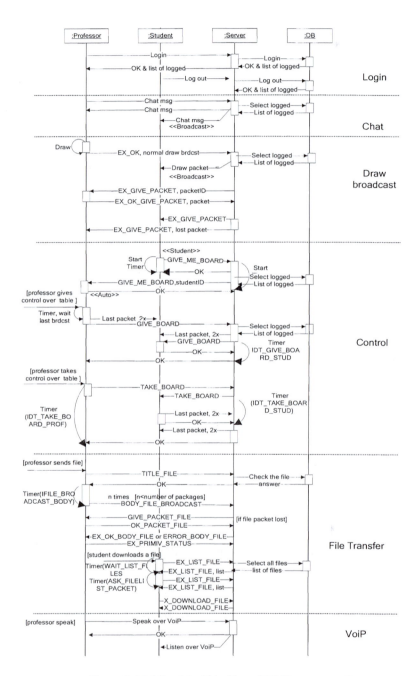

Figure 2. Mobile virtual blackboard UML sequence diagram.

MOBILE BLACKBOARD IMPLEMENTATION

The wireless virtual blackboard system is realized and functional. Since all of the students and/or HCW are still not equipped with pocket PCs, the version of the virtual blackboard for desktop PCs is also implemented.

Applications for Wireless mobile virtual blackboard are made in C++ Visual Studio .NET 2005 development framework.

The instructor's interface is shown on the Figure 3. The control of the blackboard is represented with the button named "get control". The chat module consists of the input text box the display text box, where all chat messages are written, and the send button. The button named "Chat" enables seeing the chat on the full screen. The buttons named "fUp" and "fDown" serve for upload and download.

The place where the graphics is shown is called the virtual blackboard. This is the place where the downloaded files are opened and where notes can be added using colours and pens with different thicknesses. The "Clean" button serves for clearing the blackboard. With clicking on the "Save" button, the contents shown on the virtual blackboard can be saved as a bitmap file. This is the view of an instructor.

Figure 3. Mobile blackboard interface.

The student's and/or HCW's user interface is very similar to the instructor's except that it has a button to send a drawing request, and does not have the authority control, and does not have the ability for upload.

Below the chat boxes, a combo box with logged students is shown in line with buttons for file upload. The rest of client's window is drawing area where a picture can be downloaded and opened. Pens with several colours and pen thicknesses can be used for drawing on the picture. There is also an eraser, for the entire picture or for parts of it. All these controls are set on the right side of the drawing area. Mobile virtual blackboard at work showing discussion about ECG diagram is given on Figure 4.

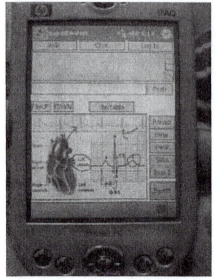

Figure 4. Mobile Virtual blackboard at work.

MOBILE BLACKBOARD EVALUATION

The mobile blackboard was evaluated by several groups of healthcare workers. The questionnaire includes three types of questions regarding: interface usability, participants experience in technology usage for medical purposes and specific features of mobile virtual blackboard. The following set of questions was given to the group of 14 medical students.

The questions set concerning the interface usability are:

    1)   Are the colours from the drawing pens enough?

    2)   Are you satisfied with the way of presentation on the blackboard?

    3)   What do you think of giving drawing controls to more than one user?

4)   What is your opinion on the logging concept on the system?
5)   What is your opinion on the controls position on the user interface and do you have any suggestions about it?

The X axis in Figure 3 represents questions with the same number from questionnaire above.

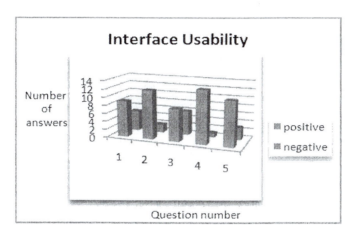

Figure 5. Results from interface usability part of questionnaire.

General opinion among participans for the interface usability is positive. Interface is functional and suggestions for controls position will be included in the future project. Option for more than one users to have drawing controls divided participants. The current system design left this option for the instructor who can decide about number of drawing control holders.

The questions list that examines participants experience in technology usage for medical purposes are:
1)   Do you use personal organizers for private or professional purposes?
2)   Do you use collaborative software of any kind?
3)   Have you used medical collaborative applications with similar purpose before?
4)   Do you find the technology helpful in your scope of work?

The Y axis in figure 4 represents questions with the same number from questionnaire above.

Results show that most of the examinees have experience with technology application in medical cases. All of them agreed that technology can be useful in their work.

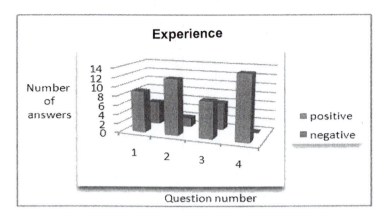

Figure 6. Results from technology experience part of questionnaire.

The questions used to provide information about features of mobile blackboard and future directions are:

1) Do you consider the option for sending files (e.g. ECG) to be useful for your work?
2) Do you use an option for VOIP communication?
3) Do you use an option that saves current image with additional drawings on the blackboard for further analyses?
4) Do you think that an option to save chat history should be provided?
5) What is your opinion on providing communication between mobile phones and virtual blackboard using MMS and SMS messages?

The Y axis in figure 6 represents questions with the same number from questionnaire above.

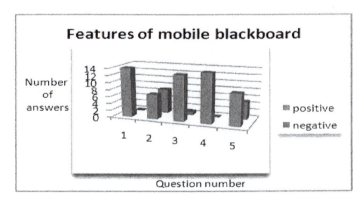

Figure 7. Results from the features of the mobile blackboard part of questionnaire.

The research on students' interest on adopting mobile devices in educational context shows that most of the students have positive thinking about the introduction of mobile virtual blackboard into the educational process. General opinion of all examinees was that virtual blackboard could increase effectiveness of the learning process. Similar results could be found in (Luchini, K., et al, 2003) and (Luchini, K., et al, 2004). On the other hand, an assessment of the usability of mobile information and communication technology (ICT) for hospitals is done in (Alsos, O.A., Dahl, Y., 2008). They found that the usability of mobile ICT used at the point of care is highly contextual and is closely dependent on physical and social aspects of the care situation.

CONCLUSIONS

Mobile services will provide means for building new learning environments. In our opinion, most needed services are communication and collaboration services among healthcare colleagues and workers.

General opinion among the examinees was that the mobile virtual blackboard is effective and useful tool for efficient consultations among colleagues.

From the ergonomics point of view connected with the usage of the small mobile devices, some improvements are recommended and it is certainly one of the topics for our future research. The potential for aches, pains and ultimately acquiring a cumulative trauma disorders are much higher when mobile devices with smaller size and portability are used. Similar results for notebooks are presented in (Wyatt, P., et al., 2006).

REFERENCES

Adewunmi, A., et al, 2003, "Enhancing The In- Classroom Teaching/Learning Experience Using Wireless Technology", *In Proc. of 33rd ASEE/IEEE Frontiers in Education Conference,* Boulder, CO, USA, pp. S4E -20-5

Alsos, O.A., Dahl, Y., 2008, "Toward a Best Practice for Laboratory-Based Usability Evaluations of Mobile ICT for Hospitals", *NordiCHI 2008: Using Bridges,* 18-22 October, Lund, Sweden, pp.3-12

Dongsheng, D., 2009, "Research on Equipment Support Mobile Learning System", *IFITA 09, Volume 3,* 15-17 May, pp. 139 – 141

Fulp, C.D., and Fulp, E.W., 2002, "A Wireless Handheld System For Interactive Multimedia-Enhanced Instruction", *In Proc. of 32nd ASEE/IEEE Frontiers in Education Conference,* Boston, MA, USA, pp. T3F6-T3F9

Haitao, P., et al., 2009, "The Approach of Device Independence for Mobile Learning System", *IEEE Int. Symposium ITIME 09, Volume 1,* 14-16 Aug, pp. 575 - 578

Kool, S., et al., 2003, "Graduate-Undergraduate Interaction In Wireless Applications Research And Development", *In Proc. of 33rd ASEE/IEEE Frontiers in Education Conference*, Boulder, CO, USA, pp. S2D - 9-12

Luchini, K., et al, 2003, "Pocket PiCoMap: A Case Study in Designing and Assessing a Handheld Concept Mapping Tool for Learners". *In Proceedings of CHI 2003 International Conference*, pp. 321-328, Ft. Lauderdale, Florida, USA

Luchini, K., et al, 2004, "Design Guidelines for Learner-Centered Handheld Tools" *In Proc. of CHI2004 International Conference*, Vienna, Austria, pp. 135-142

McAlister, M.J., Xie, P.H., 2005, "Using PDA for Mobile Learning", *In Proceedings of the IEEE International Workshop on Wireless and Mobile Technologies in Education (WMTE'05)*, Tokushima, Japan

Park, K.S., et al., 2005, "PDA based Point-of-care Personal Diabetes Management System", *Proceedings of the IEEE Engineering in Medicine and Biology 27th Annual Conference*, Shanghai, China, September 1-4, pp. 3749-3752

Shotsberger, P., and Vetter, R., 2001, "Teaching and learning in the wireless classroom," *IEEE Computer, vol. 34, no. 3*, pp. 110 – 111

Tucker, T. G., Winchester, W. W., 2009, "Mobile Learning for Just-In-Time Applications", *ACMSE '09*, March 19-21, Clemson, SC, USA

Yi, J., 2009, "Research of One Mobile Learning System", *International Conference on Wireless Networks and Information Systems, WNIS '09*, 28-29 Dec., pp. 162-165

Yoshino, et al, 2003, "Group Digital Assistant: Shared or Combined PDA Screen" I*n IEEE Transactions on Consumer Electronics, Vol. 49, No. 3*, pp. 524-529.

Wang, X., 2009, "The Key Issues Research of Apple based Courseware Center of Mobile Learning System", *4th Int. Conf on Computer Science & Education*, 25-28 July, pp. 1565 – 1568

Wyatt, P., et al., 2006, "Oh, My Aching Laptop: Expanding the Boundaries of Campus Computing Ergonomics", *SIGGUCCS'06*, November 5–8, Edmonton, Alberta, Canada, pp. 431-439

CHAPTER 43

The Central Totem for Hospital Stay: An Example of Applied Patient-Centred Design

Cinzia Dinardo

Consorzio CETMA
Brindisi, ITALY

ABSTRACT

The central totem for hospital stay has been developed in accordance with patients' needs, which were determined through preliminary analysis, interviews and surveys in some hospitals and ALFs (Assisted Living Facilities) throughout Regione Puglia. As for current trends, the totem also aims to meet the requirements deriving from the changes that are taking place in hospitals: at most two patients per room, more room and less furniture, technology present but not invasive, suspended furniture, ever growing demand for connection to electric appliances.

Planning has lead to the integration of different functions in a single, compact and easy to install element, that guarantees the organization of a two-beds hospital room. The preliminary ergonomic analysis have been carried out on the basis of three aspects that are considered to be top priority to improve comfort: a room with a minimum net surface of 9 square metres per bed and equipped with the totem; easy access to bedside table; the need to define and separate the patient's personal from the social space during hospitalization.

This space-saving structure - in compliance the planning guidelines - reduces the overall hindrance resulting from the two bed units, nightstands and partition, thus allowing for the circulation of a wheelchair.

The totem integrates the functionality of the nightstands, which are suspended by means of articulated arms, optimizes ease use and access for the patient and simplifies room cleaning. The overall accessibility has been achieved through hinges for the rotational translation of the nightstand, which, thanks to a 360° rotation, has become a prosthic element for the patients who are confined to the bed. The nightstands are equipped with angular handles boasting an innovative design and soft anti-shock material designed to guarantee an easy grip and rotation of the nightstand straight from the bed.

The totem is equipped with an innovative system to separate the area between the two beds. It consists of a completely foldaway curtain, that can be pulled out through a horizontally pivoted and well-balanced movement by means of balance weights. It closes up into an small-size case, it does not clutter the room when open and closing is fast and safe in emergency situations. The opening door is equipped with two symmetric handles in soft anti-shock material.

The totem integrates the power supply units both for medical and patient use. The latter are directed at 45° degrees from the wall and conveniently placed at different heights. The unit can be modified in accordance with the national standards.

The totem is equipped with an upper housing for the adjustable lights and is set up for the installation of chromo-therapy light sources. Also, the upper housing meets the space requirements for which the structure has been designed, .the totem allows indeed to activate a chromo-therapeutic programme for a patient without illuminating the ceiling of the nearby patient. The central lighting plan allows to increase the diffuse light coming from the bed heads. The chromo-therapy has been introduced on the basis of the input received by the patients: patients spend most of the time with their gaze towards the ceiling; patients need to perceive the passing of time. The chromo-therapeutic treatment can be planned on the basis of the light and colours circles that mark the hours of the day. The totem has a lower housing with integrated directional night lights. The night lights - directed towards the bedside floor area - can be both automatically and manually switched on by the patient/operator by means of the totem's special buttons.

The central totem, according to the initial tests, allows for a easy configuration of the two patients' personal and social space. Thanks to the implemented solutions, it makes the objects for treatment and daily entertainment accessible for the patients, thus creating a patient-friendly environment..

Keywords: Industrial Design, Ergonomics, Healthcare, Furniture, Hospital Room, Accessibility

INTRODUCTION

In the last fifty years medicine has made giant strides, both in terms of diagnosis and therapy. Health structures in western countries boast state-of-the-art technologies that, together with a fairly good autonomy of the patient, optimize the treatment conditions for several pathologies. Yet, such technical-scientific progress does not entail an analogous improvement in the satisfaction of patients and their relatives, because the evolution of medical sciences and the increased scientific knowledge has not brought about an analogous improvement from the ethical viewpoint and in the attention granted to the patients' emotional needs.

There is much talk about humanization of Hospitals nowadays, but we have to state that the latter can only occur if an attitude takes foot, that focuses on the ill person, bearing in mind his/her ideas, experience, fears and rights. All this requires overcoming the bio-medical model in favour of a more complex bio-psycho-social model, which acknowledges that the interaction of the psycho-social factors with the bio-chemical factors lies at the basis of the wellbeing of the individual. In other words, the focus of the medical intervention is no longer on the "disease to be treated" but on the "person to be treated and taken care of". Health is no longer regarded as lack of pathology, rather an objective that has to be positively achieved.

At this stage, the role played by Hospital Psychology – a recent, yet rapidly growing discipline – becomes decisive in improving the quality of life of hospital patients and of their relatives.

According to Hospital Psychology any hospital stay is bound to imply a degree of discomfort, due to the separation from the family, the need to adapt to the new rhythms, giving up privacy, the inevitable dependency on others and the resulting loss of personal autonomy. Such discomfort is accompanied by the "emotional distress" deriving from the specific pathology. The presence of an organic pathology always represents an interruption of the vital circle with consequences in the psychological sphere. Each organic condition implies a degree of distress, an additional distress – often even worse – is due to the strategies through which the patient, his/her relatives and hospital staff face the illness. The task of Hospital Psychology is to help reduce such additional distress as much as possible.

The physical environment, where the hospital staff operates and the patient lives in a time of suffering, is a crucial element. Indeed, the architectural spaces, the environment and the furniture can favour or hamper the humanization process.

Aim of the research has been to single out a different configuration of the patient room and to develop a new furnishing system that is capable of reducing both physical and mental fatigue for the patients.

The aspects linked with the physical fatigue in a patient room appear to be vital, because the users'/patients' health condition and strength are poorer than usual. Also from the mental viewpoint patients find themselves in a delicate position, because they feel they depend on others and on strangers.

The research focuses on the functional area around each bed in a hospital room with 2 or 4 beds.

The present conformation of a room does not take into consideration the possibility for the patient to access the bed straight from a wheelchair, because the necessary movements are not possible. Users (hospital staff or the patients' relatives) have to perform such movements in the area at the foot of the bed.

By examining a patient room it emerges that space is ill-organized, because the area around the bed are is without distinction by patients, hospital staff and relatives.

Our solution foresees the reorganization of the space by means of a multifunctional furnishing module that can concentrate the operations exclusively between the two beds, allowing the remaining area to be used for reception and socialization.

TARGETS OF THE PROJECT AND EXPECTED BENEFITS

The carried out studies found that new strategies are required to reorganize patients' rooms. As a consequence, it has become necessary to identify a room configuration with a new furbishing system that focuses on the functional areas around the beds in patients' rooms.

The result is the design of a system that concentrates all the activities between the two beds and allows the remaining areas to be used for reception and socialization.

Table 1: Needs to be met according to user's typology

Needs	Users
Comfort , privacy, socialization, colours, natural/artificial light, accessibility, total control of the bedside table and accessible electric plugs	Patients
Less furniture, more room, freedom, easy to clean and to maintain	Hospitals
More dedicated electric plugs, easy to open and close division screen in emergency situations	Hospital staff
Innovative and state-of-the-art design	Clients

HOSPITAL CARE

Hospital as "care place" toady have to guarantee high technological and welfare contents, but without overlook the "human side". Principles that should guide the hospital design are related to the life quality offered to the patients, health workers and visitors. The whole Elements that guarantee environmental, physical and psychic comfort (until the ergonomics furniture), as well as privacy, have to be really considered, in order to live and work in the best possible conditions and to increase the wellness and safety sensation of all the users which populate this "health case". Patient, in this "care place", is the more weak figure not only for its illness, but also for the emotional and psychological conditions due to the "seclusion" in interior domestic at all. The project for the stay in bed places are to form the daily spaces for people who live and spend their time with doctors, friends, relative in a single multifunctional space.

All this requires a wider and global planning, that the multidisciplinary approach of ergonomics can generally offer.

ROOM CONFIGURATION AND ACCESSIBILITY

Two critical aspects have emerged by the analysis carried out into the hospital room: the first one pertinent to the accessibility in relation to the room configuration; the second one pertinent to the use of furniture by patients and health workers.

Dimension shown in Table 1 have been defined through calculation of the necessary spaces for medical equipment and other furniture system, as well useful space for health activities and for circulation on both sides of the bed. Central space between the two beds (the double of dimension E*= 2.16 m.) has been fixed by the bed accessibility and by the right to use the bedside tables, medical equipment, sockets and séparé.

The necessary space to guarantee the accessibility and the patient circulation on wheelchair is defined by the manoeuvring area when séparé is half-open.

Table 2: Parameters and room dimensions

Parameters	mm
E Health activity and circulation, external side	720
E* Health activity and circulation, internal side	1062
F Bed	980
G Functional Width	2700
J Suggested Room Length	4400
J* Minimal Room Length	3300

Considering the actual configuration of the hospital rooms, the functional area of each bed, in a room equipped with two bed, is measured through a minimal width of about 3.3 m, under which it would not be wise to fall (J*); dimension G will be doubled for a total of 5.4 m. In the case of a single room, suggested dimensions are 2.7 m x 4.4 m.

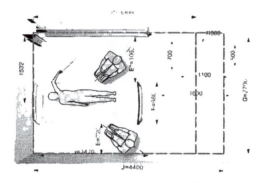

Figure 1. Parameters and room dimensions

Three "proximity spheres" are clearly recognizable in the bedside table.
Three sides accessibility is possible thanks to the double movement of the extensible arm: bedside table turns from b1 to b2 and vice versa, and extends from 0 to 200 mm. as to the column (from a0 to c0 and vice versa). This two axis movement allows to free lateral space required by the health worker.

Private sphere	Functional sphere	Communicative sphere

Figure 2. Proximity Spheres of Bedside Table

In fig.3 has been synthesized an extreme situation. The depicted patient belongs to the short population: woman 5%, height of 1.50 m. Bed is an horizontal position and the patient grabs the handle to turn the bedside table without efforts (verification performed with the SW Human Builder-CATIA).

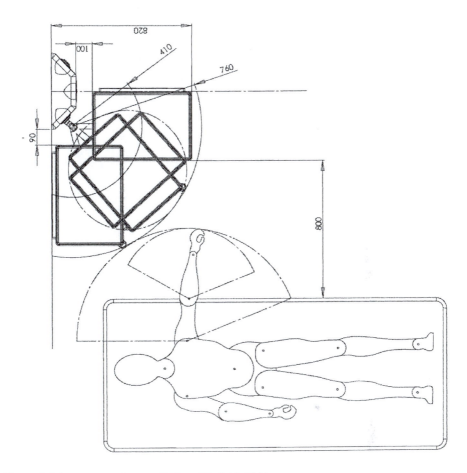

Figure 3. Movements and accessibility of Bedside Table

"Healthy design" has been the criterion adopted for the patient centred design, making also attention to the ergonomic aspects connected to the health. "Healthy Design" means to design the interaction with the hospitable environment in terms of wellness, comfort and above all perception. Our methodology collects input by the health workers experience and by their solutions to facilitate everyday activities and to reduce patients hardships. In this context, and as consequence of the society evolution, we point out that phenomena as nomadism, flexible jobs, population aging and small and less participating families, generate new needs. We speak, for example, about communication and necessity to be always connected to the world sharing information through different channels. Stay in hospital has changed, today day hospital therapies and less invasive interventions contribute to period in hospital reduction. Present patients need of electrical energy for portable computers and mobile phones, need of privacy in the communication and during relaxation, need to be more and more autonomous in order to make up for the lack of familiar assistance. The central totem made the ambient modular, flexible to the needs in the short and long terms of period in bed. According to the reachable and perception issues, three "proximity spheres" around the patient have been defined:

- the "private sphere", perceptively immediate, consisting of: chromoterapy buttons, water bottle shelf, handle for turning the bedside table and to approach the private side of the bedside table;

- the "functional sphere" consisting of: sockets, horizontal shelf of bedside table, drawers;

- the "communicative sphere", consisting of: chromoterapy beams, séparé with different transparency zones, external side of bedside table equipped with magazine rack and photograph holder.

One of the more interesting contribution has been the use of anti-trauma material which answers positively to the resistance and hygiene requirements and, at the same time, invites to be handled and presents itself with an organic geometry, soft as human skin.

PRODUCTS DESCRIPTION

FUNCTIONAL COLUMN

The functional column is a vertical element in two mirror-like extruded aluminum profiles, with an angle which looks toward the patient. The column is hanging from the floor for the cleaning activities and it's fixed to a wall through steel plates. The column is covered on the top by a plastic cap.

BEDSIDE TABLE

Bedside table has:

- a privacy side, made by open compartment and by an handle for the personal towel;

- a functional side: two slider boxes and a customizable extra tray equipped with wheels in order to release part of load supported by the mechanical arm;

- a communicative side: an elastic net which can be used as magazine rack or photograph older.

COLUMN SOCKETS

The sockets have been thought as box structures, the upper for the health workers and the lower for the in-patient. Both the boxes can be regulated in high, thanks to the slide guide obtained in the extrusion process of the column. Mobile Phones,

MP3 Readers, Medical devices could be used both suspended and placed on the bedside table.

CHROMOTERAPY DEVICE

The device is designed in order to light the ceiling area related to the single patient, avoiding, in this way, possible disturbs to the neighbouring in-patient. The in-patient looks to the ceiling for many hours during the day and the chromoterapy could stimulate the total relax

PARTITION ELEMENT

The partition element is a rolling wall made by different textiles. Thanks to the mechanic arms and wheels, the partition could divide the in-patient areas until the end of the beds. Three textiles areas, with different filling and colors, have been thought in order to answer to the privacy needs of the in-patients without natural light reduction in case of lateral windows. The front panel is equipped with an anti-trauma handle.

REFERENCES

Hall E.T., ed. (1996). La dimensione nascosta: vicino e lontano. Milano. Bompiani. - XII, 276 p.

Ulrich K.T., Eppinger S.D, ed. (2001) Progettazione e sviluppo di prodotto. Milano. McGraw-Hill.

Panero J., Zelnik M., ed. (1983) Spazi a Misura d'uomo. Milano. BE-MA Ed, 315 p.

Tosi F., ed. (2001) Progettazione ergonomica. Milano. Il sole 24 ore S.p.A., 382 p

J. TORNQUIST, Colore-Iuce. Applicazioni basic design, Hoepli, Milano 1983.

W. BERNASCONI, Cromoterapia. Per un rapporto equilibrato e terapeutico con i colori, Ottavino, Verona 1987

G. GUERRA, Psicosociologia dell'ospedale. Analisi organizzativa e processi di cambiamento, Caracci, Roma 2004.

Chapter 44

Toward the Standardization of Health Care Record Management Systems

Joyram Chakraborty, Linda Hansen, Aktta Patel, Anthony F. Norcio

Department of Computer Science
State University of New York, The College at Brockport
Brockport, NY 14420, USA

ABSTRACT

This paper reports preliminary study and its' findings concerning the Standardization of Health Care Records Management. A survey of the leading medical facilities in Baltimore, MD was carried out to collect the preliminary patient data forms. Using a meta-analysis of these forms, a hybrid form is proposed as a new instrument for data collection. This new data collection form will be pilot tested for validity with medical professions. The authors propose that this instrument demonstrates health care cost reduction potential through the standardization of health care records.

Keywords: Health care costs, Records management, Standardization

INTRODUCTION

Increasing costs of health care and the resultant effects on the general public in the U.S. have been well documented by researchers across various disciplines. A significant number of findings from these research efforts have called on the use of Information Technology as a possible solution to this problem (Ball et al, 2005; 2007, 2008). However, the research efforts have proposed limited solutions to address the ever rising costs of health care. It is the purpose of this study to propose a method of health care record standardization that can address this problem.

BACKGROUND

Research has indicated that the standardization of health care records could result in lower costs (Eichelberg et al, 2005; Bossen, 2006). However, significant variations in the current methods of operation at health care facilities have limited standardization efforts. According to literature findings, the most frequent challenges in standardization result from issues with interoperability, content structure, access services, multimedia support and security (Grimson et al, 2000; Fulcher, 2003, Eichelberg et al, 2005; Bossen, 2006, Hristidis et al, 2006).

Most health care providers have unique sets of records management practices that allow them to collect and manage patient records in an effective manner. These practices are usually developed with little standardization considerations. As a result, the record management system is developed specifically for the health care provider. As health care providers turn to information technology solutions to streamline their operations, they face significant costs of developing, implementing, training and maintaining proprietary, technology-based solutions (Grimson et al, 2000; Fulcher, 2003, Eichelberg et al, 2005; Bossen, 2006, Hristidis et al, 2006).

METHODOLOGY

A survey to collect empty patient check-in forms was carried out at 12 Pediatrician health care facilities in the Baltimore, Maryland region over the course of a month. These facilities were selected randomly to avoid bias. The headings from each provider's patient check-in form were entered into an excel spreadsheet. A second comprehensive spreadsheet was designed to include all the headings from all the forms combined together. A descriptive analysis of the comprehensive spreadsheet was carried out for significant findings.

RESULTS

The Red lines indicate that a facility requires information under that specific heading while the Black lines indicate that the facility does not collect information under the specific heading. The results indicate that there is a large degree of variation in the types of information collected at each health care facility. Figure 1 illustrates the demographic information collected. Similarly, figure 2 shows the variation in the emergency contact information collected while figure 3 displays the insurance information that is collected by the health care facilities.

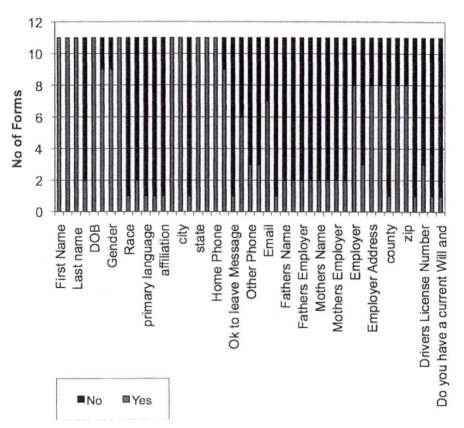

Figure 1. Demographics section of patient form

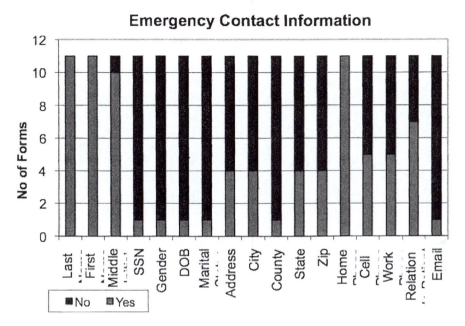

Figure 2. Emergency contact information section of patient form

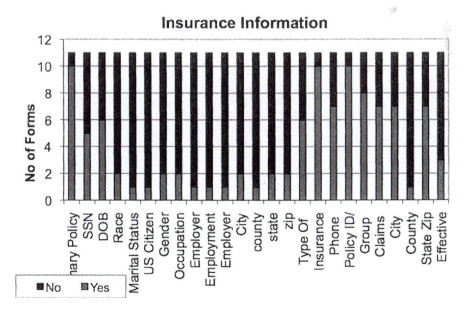

Figure 3. Insurance section of patient form

CONCLUSIONS AND FUTURE WORKS

The variation of information collected points to a lack of standardization of content structure. This is one of the challenges in the standardization of health care records Eichelberg et al, (2005). Our preliminary findings offer an opportunity to extend the literature and bridge the gap in the standardization of health care record content structure. The next logical step would be to carry out a more comprehensive meta-analysis of patient check-in forms from a larger sample. The result of this analysis could then lead to the creation of a comprehensive check-in form. This proposed form should then be tested using health care providers to test for ease of use and validity. If the proposed form gains acceptance, it could then be used as the basis for the standardization of content structure. The standardized check-in form could then be used to create off-the-shelf information technology solutions for health care providers. The new solution would reduce the information technology cost burden to health care providers.

REFERENCES

Ball, M. J., J. S. Silva, S. Bierstock, J. V. Douglas, A. F. Norcio, J. Chakraborty, and J. Srini, "Failure to Provide Clinicians Useful IT Systems: Opportunities to Leapfrog Current Technologies", Methods of Information in Medicine (Journal of the International Medical Informatics Association), vol. 47, no. 1, pp. 4-7, 2008.

Ball MJ, Douglas JV. Human Factors: Changing Systems, Changing Behaviors: Toward Healthe-People. New York: Springer2005. pp 60-70.

Ball MR, Bierstock SR. Clinician Use of Enabling Technology: The Missing Link. Journal of Healthcare Information Management 2007; 21 (3).

Blois SM. The Physician's Personal Workstation. MD Computing 1985; 2 (5): 22-26.

Bossen, C. 2006. Representations at work: a national standard for electronic health records. In *Proceedings of the 2006 20th Anniversary Conference on Computer Supported Cooperative Work* (Banff, Alberta, Canada, November 04 - 08, 2006).

Campbell EM, Sitting DF, Ash JS, Guappone KP, Dykstra RH. Types of unintended consequences relation to computerized provider order entry. JAMIA 2006; 13 (5): 547-556.

Eichelberg, M., Aden, T., Riesmeier, J., Dogac, A., and Laleci, G. B. 2005. A survey and analysis of Electronic Healthcare Record standards. *ACM Comput. Surv.* 37, 4 (Dec. 2005), 277-315.

Fulcher, J. 2003. The use of smart devices in eHealth. In *Proceedings of the 1st international Symposium on information and Communication Technologies* (Dublin, Ireland, September 24 - 26, 2003). ACM International Conference Proceeding Series, vol. 49. Trinity College Dublin, 27-32.

Grimson, J., Grimson, W., and Hasselbring, W. 2000. The SI challenge in health care. *Commun. ACM* 43, 6 (Jun. 2000), 48-55.

Horsky J, Gutnik L, Patel VL. 2006. Technology for emergency care: cognitive and workflow considerations. AMIA Annu Symp Proc 2006: 344-348.

Hristidis, V., Clarke, P. J., Prabakar, N., Deng, Y., White, J. A., and Burke, R. P. 2006. A flexible approach for electronic medical records exchange. In *Proceedings of the international Workshop on Healthcare information and Knowledge Management* (Arlington, Virginia, USA, November 11 - 11, 2006). HIKM '06. ACM, New York, NY, 33-40.

Kilman, D. G. and Forslund, D. W. 1997. An international collaboratory based on virtual patient records. *Commun. ACM* 40, 8 (Aug. 1997), 110-117.

Lehmann CU, Altuwaijri MM, Li YC, Ball MJ, Haux R. Translational Research in Medical Informatics or from Theory to Practice – A Call for an Applied Informatics Journal. Methods Inf Med 2008; 47: 1-3.

Mahotra S, Jordan D, Shortliffe E, Patel VL. Epub 2006 June 9. Workflow modeling in critical care: piecing together your own puzzle. J Biomed Inform 2007; 40 (2): 81-92.

Patil RS, Silva JS, Swartout WR. An architecture for a health care provider's workstation. International Journal of Bio-Medical Computing. Special Issue on The Health Care Professional Workstation 1994; 34: 285-299.

Silva JS, Ball MJ. The professional workstation as enabler: conference recommendations. International Journal of Bio-Medical Computing. Special Issue on The Health Care Professional Workstation 1994; 34: 3-10.

Chapter 45

Interactive System to Assist Rehabilitation of Children: Development of a System Employing New Rehabilitation Movement

Shuto Murai, Kenta Umehara, Hiroto Hariyama, Michiko Ohkura

Shibaura Institute of Technology,
3-7-5 Toyosu Koto-ku Tokyo
135-8548 Japan

ABSTRACT

For such symptoms as ambulation difficulty and paralysis caused by congenital diseases, the recovery is remarkable with rehabilitation from early childhood. However, long-term continuation of operative rehabilitation is difficult for children, especially judging whether the rehabilitation is appropriate without the support of a physical therapist in the home. We developed a new interactive system to encourage a physically disabled child to continue appropriate rehabilitation at home. This article reports the improvements to solve our previous system's problems.

Keywords: Rehabilitation, children, interactive system

INTRODUCTION

Ambulation difficulty and paralysis are examples of symptoms caused by such congenital diseases as Down Syndrome and brain paralysis." For such symptoms, symptomatic recovery is considered remarkable with rehabilitation from early childhood. However, rehabilitation is operative and monotonous. If the patient is a child, the continuation of long-term rehabilitation is even more difficult. In cases of domestic rehabilitation without the support of a physical therapist, judging whether the rehabilitation is appropriate is often impossible. Thus, the purpose of this study is the construction of an interactive system for children with physical disorders that appropriately assists continuous rehabilitation at home. Since the symptoms caused by congenital diseases differ among patients, we constructed an interactive system for one particular patient. In the future, we will expand the system to include other patients. Such trials are very useful to assist the rehabilitation of children with congenital diseases. However, few researches exist, probably because bringing them to the commercially-based development is very difficult. Our previous report described our first trial in which we built an interactive system using step motion and confirmed that the patient continued appropriate rehabilitation [1]. But problems were revealed, such as pain to maintain the step motion. This report describes our new system with which we addressed the previous system's problems.

METHOD

TARGET PATIENT

The target patient is identical as the first trial: an eleven-years-old boy suffering from Spina Bifida Aperta., which causes muscle force degradation and sensory disturbance in the periphery of the lower limbs as an after-effect of myelomeningocele.

SYSTEM EXAMINATION

Because the patient has difficulty putting pressure on his toes, he walks with excessive pressure on his heels. For his rehabilitation, a step motion was considered effective in which he continues to put pressure on his toes. However, since the first system's step motion was painful, we considered rehabilitation that employs a new movement where the target patient is sitting on the floor and presses the wall with each toe. We built a new system for this movement (Fig.1).

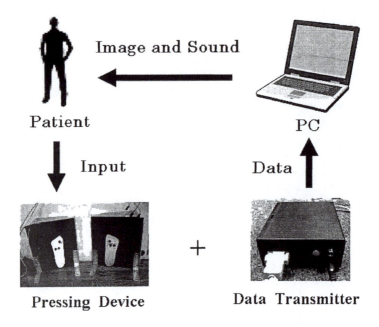

Fig 1 Diagram of system

PRESSING DEVICE

We made an input device for the new movement called the "pressing device" (Fig. 2). The first system's input device employed six switches, but the new device employs pressure sensors to control the system by the amplitude of the foot pressure.

Sensor of foot pressure

Fig 2 Pressing device

CONTENT

We made a new game in which the subject battled "zombies" and "dragons" by maintaining new movements for a certain period of time. During the game, the subject can see the conditions of the sensors on a game screen (Fig. 3). We prepared three degrees of difficulty to maintain the subject's interest. By increasing the degree of difficulty, the score for each victory increased and more power was required to knock down the monster. We also added an avatar function to change the method of attack. In addition, since we considered the patient's physical condition when playing the game, his greatest foot pressure was measured when a game started, and the sensor thresholds were set to detect sufficient amplitude of foot pressure. In addition, a sensor, which only measures the foot pressure of the base of the middle toe, lowers the threshold setting, because the foot shape is unique and it is difficult to emphasize its outside. However, we slowly set the threshold by repeating the number of times. The dates, the degrees of difficulties, the total scores, and other data are acquired as log data. Fig. 4 shows an example of the screenshot and Table 1 shows the threshold setting of the sensors.

386

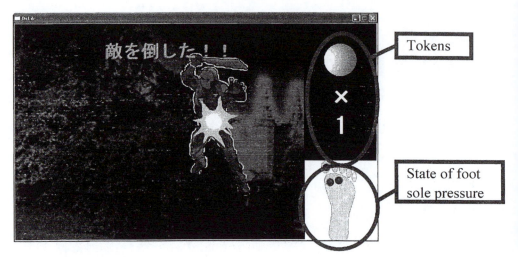

Fig 3 Screenshot (1)

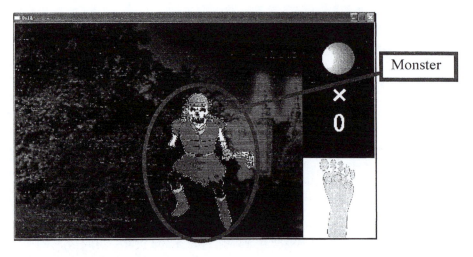

Fig 4 Screenshot (2)

Table 1 Threshold setting of sensors

	Ball of big toe	Base of big toe	Base of middle toe
Easy	40%	40%	20~40%
Normal	50%	50%	30~50%
Difficult	60%	60%	40~60%

ASSESSMENT EXPERIMENT

EXAMINATION OF ASSESSMENT INDEXES

Game scores and foot pressure data were used to evaluate the rehabilitative effect. Two kinds of foot pressure were measured at the following times:

· before and after using system
· while using system

In the former case, Footscan, a foot pressure measurement device, was used because it can visually indicate the center of gravity from the distribution of foot pressure. Fig. 5 shows an example of the data obtained by Footscan. In the latter case, the data were acquired during the game. This measurement shows the daily change of the foot pressure. In addition, data analysis of the game's records and simple questionnaire answers were used to evaluate the system's continuity.

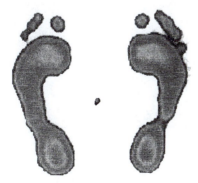

Fig 5 Obtained Footscan data

EXPERIMENTAL PROCEDURE

1. The foot pressure was measured just before using the system.

2. The game system was used daily for about one month.

3. The foot pressure was measured immediately after using the system.

Figure 6 shows the flow of the assessment experiment.

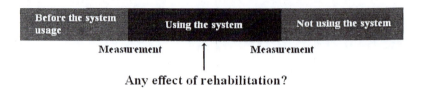

Fig 6 Flow of the assessment experiment

REFINEMENT DURING EXPERIMENT

While the subject used our novel rehabilitation method, we addressed its refinement and made the following three improvements:

1. We diagonally placed the game screen in front of the target patient.

First, we put a desk before him on which we put the game screen. However, he could not confirm his foot position or perform the game as expected. Therefore we moved the desk diagonally in front of him so that he could identify his foot position with the game screen.

2. Establishment of safeguard mat

When the rehabilitation was carried out on places with a little friction such as a wooden floor, the subject often slid. Maintaining correct posture was difficult, and practicing became impossible due to this posture problem. His posture was stabilized with a mat.

3. Assistance during system use

The power missing from the device during practice flew behind the subject's body, and we realized that he cannot stream down power. Therefore he could stream down the power well because another person supported his back.

EXPERIMENTAL RESULTS AND DISCUSSION

Figure 7 shows part of the foot pressure data of the target patient before using the system. Compared with Fig. 5, little pressure was found on his toes. Table 2 shows the number of knocked down monsters per game. For both the left and right legs, the results after using the mat were good. In addition, his practice posture improved. But the frequency at which he failed to continue pressing the sensor did not change: for example, from 38 at the beginning to 37 at the end.

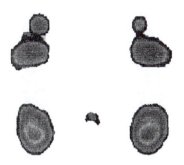

Fig 7 Foot pressure of subject before using system at Footscan

Table 2 Number of knocked down monsters per game

	Left leg	Right leg
Before using mat	2.6	4.7
After using mat	3.2	5.7

CONCLUSIONS

Our previous report described our first trial in which we built an interactive system using a step motion and confirmed that the patient properly continued rehabilitation. But problems existed, including pain that was caused while maintaining the step motion. In this article, we reported the construction of a new system that solved the previous system's problems. In our new system, the target patient strengthened his toes in a sitting posture. Evaluation results confirmed rehabilitation continuity. Because rehabilitation in a sitting posture is new, problems were found and solved immediately during the rehabilitation.

We will continue to improve the game system in the future based on the obtained knowledge.

ACKNOWLEDGMENT

We would like to thank Satsuki Amimoto and Masazumi Mizuma of Showa University Fujigaoka Rehabilitation Hospital.

REFERENCE

[1] Shuto M et al. 2009, "Interactive system to assist rehabilitation of children" HCI International, LNCS 5614, p. 588-593

Chapter 46

The Characteristics of Center of Pressure Progression for Elderly Adults

Min-Chi Chiu[1], Hsin-Chieh Wu[2], Min-Huan Wu[2]

[1] School of Occupational Therapy
Chung Shan Medical University
110, Sec. 1, Chien-Kuo N. Rd.
Taichung, 402, Taiwan

[2]Department of Industrial Engineering and Management
Chaoyang University of Technology
168 Jifong E. Rd.
Wufong Township Taichung County, 413, Taiwan

ABSTRACT

This study aims to elucidate the characteristics of center of pressure progression (COPP) for elderly adults during barefoot walking. The results indicated that the self-selected walking speed was 2.8 (SD=0.5) km/hr with the cadence of 83.1 (SD=7.5) steps/min. There were no significant difference between male and female in the walking speed and cadence (p>.05). For the percentage of time (time %) of COPP in the initial contact phase (ICP), forefoot contact phase (FFCP), foot flat phase (FFP) and forefoot push off phase (FFPOP) were about 5.6%, 4.9%, 57.4% and 32.1%, respectively. The progression angle (PA) was 4.9 degree (SD=2.6) with the inward curve and the average velocity of COP was 26.0 cm/s (SD=3.9). Moreover, gender effect only had significant difference in the time % of COPP (p<.05). These characteristics of gait patterns for elderly adults can provide some useful information for clinical rehabilitation in gait training, footwear and relative assistive devices design.

Keywords: Gait, center of pressure, elderly adults, gender, progression angle, velocity of center of pressure

INTRODUCTION

Evaluating and quantifying the normal gait pattern could be not only useful for the assessment and treatment of clinical rehabilitation, but also beneficial for footwear and relative orthoses design. The center of foot pressure (COP) is the momentary force on the planter surface which is a component of the resultant vertical ground-reaction force reacting with plantar surface of foot. The COP progression, or gait line, is a trajectory which formed by a series of the coordinate of center of pressure passes from hid-foot to forefoot. Even though, previous studies had provided valuable information to describe the characteristics of COP for normal subjects; however, how the age and gender effects on the pathway of COP are still unclear. Thus, the purpose of this study was to elucidate the characteristics of COP progression for healthy elderly adults by the measurements of time-distance parameter, progression angle and the velocity of COP. Moreover, the age effect on the COP progression will be discussed by comparing with healthy adults that we have done in our previous work.

Han et al (1999) used an in-sole pressure system to track the path of COP. It indicated that COP displacement was about 83% of foot contact length and 18% of forefoot contact width for middle-age (50~60 years old) adults. The velocity of COP was about 22-27 cm/s. They proposed that COP concept presents the reliable, constant, accurate values to clinicians and valuables to standardize, quantify the parameters of COP. Cornwall and McPoil (2000) investigated the velocity of COP for healthy adults during barefoot walking. It classified plantar regions into four zones (rearfoot, midfoot, forefoot and toes) and indicated that the velocity of COP has triple-peak pattern, which occurs at the percentage of time of 20%, 35% and 92% of stance phase, separately. Moreover, the averaged velocity of COP for healthy adults was 38 cm/s. They suggested that the velocity of COP may be the useful information for gait patterns. Moreover, Cornwall and McPoil (2003) determined the reliability ad validity of COP quantification and indicated that the intra-class correlation coefficient values ranged from 0.374 to 0.889 for the lateral-medial area index and from 0.215 to 0.905 for the lateral-medial force index. They further stressed the COP patterns was a successfully methods to describe the progression of plantar loading during walking. De Cock et al (2008) interpreted the COP trajectory during barefoot running for young adults. The results indicated that the displacement and velocity of the COP presented the information over the arch structure, a more lateral COP pathway were found for the low arch feet.

Furthermore, gender and asymmetry (right and left foot) factors didn't have significantly influence on the COP course.

Chiu et al (2009) reported the effects of walking speed and gender on the center of pressure progression for healthy adults. The study reported that the percentage of time (time %) of the stance phase in initial contact phase (ICP), forefoot contact phase (FFCP), foot flat phase (FFP) and forefoot push-off phase (FFPOP) were about 7.0%, 4.8%, 48.8% and 39.4%, respectively. For young adults, the self-selected walking speed was 3.7 km per hours. Moreover, the progression angle (PA) is 4.1degree (SD=1.6) with an inward curve and the average velocity of COP is 31.6 cm/s (SD=5.3). Walking speed significantly influence the time percentage of the COP progression and the velocity of COP. Furthermore, gender differ may affect the progression angle of COP.

Although previous studies had delineated the COP progression and discussed some factors that may impact on the COP trajectory, however, how the age and gender affect the pathway of COP are still unclear. Thus, the objectives of this study are: (1) to represent the characteristics of COP progression for elderly adults; and (2) to realize the aging effect on the progression of COP by comparing with healthy adults that we have done in our previous work.

METHODS

SUBJECTS

Thirty healthy elderly adults (15 women and 15 men) participated in this study. The average age was 70.8 (SD=4.1) years old with the average body mass index (BMI) of 25.2 (SD=3.7). None of them had a history of musculoskeletal disorders, or orthopedic injuries of lower extremity before the pass one year. All subjects were selected by the normal arch index (AI) which is the normal arch curve range between 0.21~0.26. The subject's basic information, foot anthropometric data including leg length, foot length, foot width, and relevant foot dimensions was displayed as the table 1. Although, there are significant differences in averaged age, body height, BMI and relevant foot anthropometric data between women and men (p<.05), however, there are no significant differences in self-selected walking speed and cadence between women and men (p >.05). The average self-selected walking speed is 2.8 km/h with the cadences of 83.1 steps/min.

EXPERIMENTAL DESIGN

A nested-factorial experimental design was employed. All subjects walk with a self-selected, comfortable speed. The measurements include: (1) the x-, y- coordinate of center of pressure (COP), (2) the progression angle (PA) and, (3) the velocity of the COP during stance phase of the gait cycle.

APPARATUS

Foot pressure measurement system (Footscan® system)

A dynamic pressure measurements system (footscan® system, RSscan INTERNATIONAL) was applied to record the coordinate of the COP, the progression angle, and the velocity of the COP. As fig. 1(a), the system comprises a 0.5m plate with 4 sensors/cm^2 and a 3D-Box interface. All data were recorded with a frequency of 500 Hz and processed by the software of Scientific footscan® (RSscan INTERNATIONAL).

Table 1 The basic information for all subjects.

Variables	All subjects (n=30)	Men (n=15)	Women (n=15)	p-value
Age(years old)	70.8(4.1)	72.3(4.0)	69.3(3.7)	.04*
Body height(cm)	157.9(8.2)	163.4(6.0)	152.3(5.8)	.00*
Body weight(Kg)	62.7(9.0)	63.7(7.7)	61.7(9.9)	.55
BMI(Body mass index)	25.2(3.7)	23.8(2.4)	26.6(4.1)	.03*
Foot anthropometric data (cm)				
Foot length	25.0(1.4)	25.8(1.3)	24.0(1.0)	.00*
Foot width	9.7 (0.6)	9.9 (0.5)	9.4 (0.5)	.01*
Leg length	80.0(3.8)	82.3(3.0)	77.8(2.9)	.00*
Heel-ankle circumference	31.8(2.1)	32.9(1.9)	30.7(1.6)	.00*
Instep circumference	23.0(1.6)	23.8(1.3)	22.2(1.5)	.00*
Self-selected walking speed (km/hr)	2.8(0.5)	2.9(0.5)	2.8(0.5)	.35
Cadences(steps/min)	83.1(7.6)	85.6(8.1)	80.5(6.2)	.07

*Significant level p< .05

Motion capture system

In order to monitor the consistent walking speed, a six-camera (Charge Coupled Device/CCD) motion capture system (VICON 460 Motion System, Oxford Metrics Ltd., UK) , as fig. 1 (b) ,was applied to record the displacement of the center of mass of the subject which was defined as self-selected walking speed. The marker protocol of Helen-Hayes was applied and three reflecting markers were placed at right anterior superior iliac spine, left anterior superior iliac spine and sacrum, separately. The sampling rate was 120 Hz with low-pass filtering at 6 Hz.

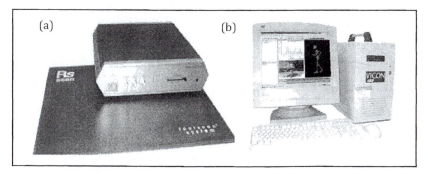

FIGURE 1 (a) Foot pressure measurement system, and (b) motion capture system

DATA ANALYSIS

Data reliability of the x-, y-coordinate of the COP was tested by Intra-class Correlation Coefficients (ICCs). Analysis of variance (ANOVA) was used to analyze the gender effect on the progression of COP. Moreover, statistical analyses were performed by using the statistical analysis software SPSS (v. 14.0).

RESULTS

RELIABILITY OF MEASUREMENTS

For data consistency, Intra-class Correlation Coefficients (ICCs) was used to calculate the absolute displacement of COP. For all subjects, the inter-trial reliability of the COP variables, for x-coordinate, the ICCx was 0.70, for y-coordinate; the ICCy was 0.95 (as table 2). It indicated a moderate to good reliability for the COP variables during the barefoot walking.

Table 2 The Intra-class Correlation Coefficients (ICCs) for the x- and y-coordinate of COP.

Variable	ICC_x	ICC_y
All subjects	0.70	0.95
Men	0.80	0.96
Women	0.67	0.95

THE CHARACTERISTICS OF COP PROGRESSION FOR ELDERLY ADULTS

The displacement of COP was recorded by the x- and y-coordinate from hid-foot to forefoot. Although all subjects have varied foot size, however the foot size was normalized by the "standardized foot" with the average foot length (25.0 ± 1.4 cm) and average foot width (9.7 ± 0.6 cm). The original absolute x-and y-coordinate were calculated into the relative x-and y-coordinate and were fitted on the trajectory of COP. As figure 2, the y-axis is the longitudinal foot axis which is the line from mid-heel (the first point of COP) to second metatarsal joint and the x-axis is perpendicular to the longitudinal foot axis. For older adults, total COP path is almost 93% of normalized foot length and 25% of forefoot width.

According to the software of Scientific footscan® (RSscan INTERNATIONAL), the period of the stance phase was divided into four sub-classified phases. The first sub-phase is initial contact phase (ICP) which was defined as the period from first foot contact until initial metatarsal contact. Secondary sub-phase is forefoot contact phase (FFCP) which was the period immediately following until the initial forefoot flat contact. Third sub-phase is foot flat phase (FFP) which was from immediately after the initial forefoot contact until the heel off; and the fourth sub-phase is forefoot push off phase (FFPOP) that was the period from heel off to last foot contact. For elderly adults, the percentage of time (time %) of COP progression on the phases of ICP, FFCP, FFP and FFPOP were about 5.6%, 4.9%, 57.4% and 32.1%, respectively.

Moreover, the progression angle (PA) was defined as the movement direction of COP. The right side of x-axis was signed as positive (+) angle (outward direction to the fifth toe). For y-axis, only one direction of COP progression is upward and was marked as positive (+) angle. For all subjects, the averaged progression angle was - 4.9 (SD=2.6) degree (with an inward curve) and the averaged velocity of COP was 26.0 (SD=3.9) cm/s.

THE EFFECTS OF GENDER

Table 3 displays the gender effect on the time percentage (time %) of COP progression, the progression angle (PA) and the velocity of COP. Gender effect only had significantly influence on the time % of COP progression during FFCP ($p < .05$). Men spent more time percentage (time %) on the phase of fore-foot contact than women. In addition, there were no significant difference between the men and women in the progression angle (PA) and the velocity of COP ($p > .05$).

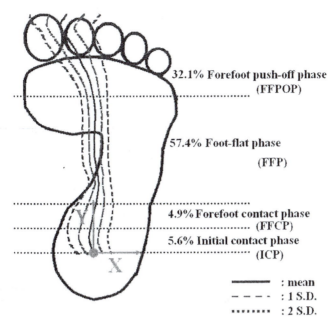

FIGURE 2 The center of pressure progression for elderly adults.

Table 3 Gender effect on the time % of COP progression, progression angle and the velocity of COP.

Measurements		variables		
		Men	Women	*p-value*
Time % of COP progression	ICP	5.8(1.8)	5.4 (1.7)	0.521
	FFCP	6.0(2.9)	3.8 (1.7)	0.016*
	FFP	56.7(5.2)	58.0(6.9)	0.565
	FFPOP	31.4(4.6)	32.8(6.5)	0.518
Progression angle (PA)	ICP	-12.1(7.3)	-8.1(6.5)	0.122
	FFCP	8.2(2.5)	8.0(2.1)	0.812
	FFP	-2.7(1.2)	-1.7(2.6)	0.218
	FFPOP	-14.6(4.8)	-10.6(7.1)	0.081
	COPP	-5.7(2.1)	-4.0 (2.8)	0.075
Velocity of COP (cm/s)	ICP	45.3(10.9)	46.0(9.0)	0.847
	FFCP	101.9(55.9)	113.0(38.0)	0.532
	FFP	18.4(3.6)	16.5(4.2)	0.179
	FFPOP	30.3(7.3)	29.8(10.2)	0.896
	COPP	27.4(3.7)	24.6(3.8)	0.053

*Significant level p< .05.

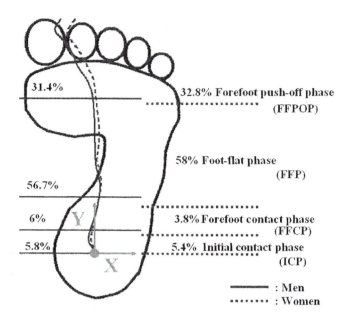

FIGURE 3 The COP progression for men and women.

DISCUSSION

For elderly adults, men have longer leg length, foot length and broader foot width than women; however, there were no remarkable difference in self-selected walking speed and cadences. The self-selected, comfortable walking speed was 2.8 km/h with the cadence of 83.1 (SD=7.5) steps/min. Comparing with young adults, elderly adults has slower self-selected walking speed and less average cadences than young adults (Chiu et al, 2009).

In this study, the course of COP progression under plantar region corresponded to 93% of normalized foot length and 25% of normalized foot width. The total displacement of COP was more extensive than Han's (1999) findings (83% of foot contact length and 18% of forefoot contact width); it may be due to the different measurement system. In addition, the average velocity of COP was similar with Han's (1999) results, the COP progression was 26.0 (SD=3.9) cm/s.

The percentage of time (time %) for COP processing from hid-foot to forefoot through the ICP, FFCP, FFP and FFPOP were about 5.6%, 4.9%, 57.4% and 32.1%, respectively. The progressing direction of COP from the middle heel to the second metatarsal joint combines several inward and outward curve and the total progression angle was -4.9 (SD=2.6) degree (with an inward curve). Moreover, gender effect only significantly affects the percentage of time (time %) for COP processing during FFCP. Men spent more percentage of time (time %) on the phase of fore-foot contact (FFCP) than women.

CONCLUSION

This study presents the characteristics of center of pressure progression (COPP) for elderly adults. In generally, older adults display slower walking speed, less cadences, and the slower velocity of COP than young adults. Moreover, gender effect had remarkably influence the time % of COP during FFCP. These features of COPP for elderly adults provide some useful information and benefits to clinical rehabilitation in evaluation, gait training, and footwear design.

ACKNOWLEDGEMENTS

This study was funded by National Science Council of Taiwan (grant no. NSC 97-2221-E-040-005-MY2) and Chung Shan Medical University (grant no. CSMU 97-OM-A-107). The authors specially acknowledge Global Advance Technology (Taiwan) for the instrument and technical support.

REFERENCES

Chiu, M.C., Wu, H.C., and Chang, L.Y. (2009), "The model of center of pressure progression for adults." The 17th Congress of the International Ergonomics Association (IEA). Beijing, China.

Cornwall, M.W., and McPoil, T.G. (2000), "Velocity of the center of pressure during walking." *Journal of the American Podiatric Medical Association*, 90(7), 334-338.

Cornwall, M.W., and McPoil, T.G. (2003) Reliability and validity of center-of-pressure quantification. *Journal of the American Podiatric Medical Association*, 93(2), 142-149.

De Cock, A., De Clercq, D., Willems, T., and Witvrouw, E. (2005), "Temporal characteristics of foot roll-over during barefoot jogging: reference data for young adults." *Gait & Posture*, 21, 423-439.

Han, T.R., Paik, N.J., and Im, M.S. (1999), "Quantification of the path of center of pressure (COP) using an F-scan in-sole transducer." *Gait & Posture*, 10, 248-254.

CHAPTER 47

An Application of the Intermittent Illumination Model for Measuring Individual's Corrective Reaction Time

Jui-Feng Lin[1], Colin G. Drury[2]

[1]Department of Industrial Engineering & Management
Yuan Ze University
Chung-Li, Taiwan 32003

[2]Department of Industrial & Systems Engineering
State University of New York at Buffalo
Buffalo, New York 14260, USA

ABSTRACT

While modeling hand-control movements, a corrective reaction time indicates how fast our brain can generate a movement order based on received visual information. The corrective reaction time was found to range from 190 to 290 milliseconds in the literature, but with no data on individual difference. This pilot study applies Drury's (1994) the intermittent illumination model and modifies his experimental designs to measure individual corrective reaction time. Four participants performed computer-based circular tracking movements by using a tablet as an input device. While conducting movements, the screen cursor blinked to generate predetermined visual delays. Measured movement speeds with the corresponding delays were utilized to calculate the corrective reaction times. The result of corrective reaction time ranges from 193 to 919 milliseconds longer than the reasonable range. Suggestions are given to deal with the potential issues.

Keywords: Corrective Reaction Time, Psychological Refractory Period, Hand-Control Movements, Tracking Movements, Intermittent Illumination Model

INTRODUCTION

A corrective reaction time indicates a time during which our brain receives visual feedback, programs a movement order, and sends the order to the controlled limbs. While performing a hand-control movement, such as pointing a finger to click a light switch on a wall, human behaves like a correction servo (Craik, 1947, 1948). This servo continuously performs ballistic movements (Lin, Drury, Karwan, & Paquet, 2009) to correct movement misalignment between the controlled object and the anticipated movement path or aimed target. To make the corrections, movement orders are made mainly based on visual feedback that provides the dynamic misalignment information. Although our eyes continuously capture the visual stimuli, the visual feedback on making corrections is intermittent, instead of continuous. This intermittent feature is due for the psychological refractory period (Welford, 1952) during which our brain is so busy for generating a new movement order that it need to temporally ignore the visual stimuli. Hence, the length of the corrective reaction time affects how rapidly and accurately we can perform hand-control movements.

HISTORY OF MEASURING CORRECTIVE REACTION TIME

The relevant findings on corrective reaction time come mainly from the studies of movement accuracy. The pioneer work on the accuracy of movements by Woodworth (1899) showed that movements made at a rate of 140 times/minute or greater were equally accurate with or without visual feedback. This led him to conclude that the time required to process visual feedback for movement control is about 450 milliseconds. This finding was further supported by Vince (1948) who used similar reciprocal movements that was tested in Woodworth's experiments. However, the experimental tasks conducted by Woodworth (1899) and Vince (1948) were reciprocal movements in which the measured movement time might include the time spent on reversing the movement direction after the targets were reached. Hence, Keele & Posner (1968) argue that the 450 milliseconds as the corrective reaction time was overestimated. To deal with the issue, instead of reciprocal movements, Keele & Posner (1968) asked their participants to perform discrete movements at different rates, comprising 190, 260, 350 and 450 milliseconds. Light-on and light-off conditions were manipulated to compare the effect of visual feedback on movement accuracy. Their results showed that visual feedback was helpful for all movement durations beyond190 milliseconds. This led Keele and Posner to conclude that the time required for the visual feedback loop to operate was somewhere between 190 and 260 milliseconds. Later, Beggs &

Howarth (1970) were also interested in examining time delays in processing visual feedback while performing sagittal-direction aiming movements. In contrast to the measurement methods used by previous investigators, Beggs & Howarth (1970) used an experimental paradigm in which the initial part of the movement trajectory was illuminated and the room lights were extinguished as the hand approached the target. Their idea to achieve the corrective reaction time was that aiming accuracy would diminish if vision is removed when the hand is less than one corrective reaction time from the target. Close to Keele & Posner's (1968) findings, a mean corrective reaction time of 290 milliseconds was reported by Beggs & Howarth (1970). Their finding of 290 milliseconds as the corrective reaction time was further applied by Drury, Montazer, & Karwan (1987) to build optimization models for self-paced tracking movements with good results.

More recently, 238 milliseconds as the corrective reaction time was predicted by Drury's (1994) intermittent illumination model. Based on the concept that paced tracking performance is disrupted by intermittent illumination of the course and the controlled element (Katz & Spragg, 1955), Drury (1994) integrated the models by Howarth, Beggs, & Bowden (1971) and Drury (1971) and then theoretically developed a model (Equation 1) that is able to obtain the duration of corrective reaction time.

Equation 1
$$\frac{1}{c} = K \times \sigma_\theta \times \left(t_r + \frac{d^2}{2(l+d)} \right)$$

where, c is the controllability (Drury, 1971), K is a constant (Howarth et al., 1971), σ_θ is the angular accuracy (Howarth et al., 1971), l is the light period, d is the dark period, and t_r is the corrective reaction time. The model predicts the linear relationship between the inverse of the controllability and the visual feedback cycle time, $t_r + d^2/[2(l+d)]$, which represents the sum of the corrective reaction time and the expected delay manipulated by the intermittent illumination of l and d. This model was tested with the data of three-intermittent illumination experiments conducted by Tsao & Drury (1975) on self-paced circular tracking movements. The results showed that the model explained over 90% of the variance in the slopes of the speed/width regression for a variety of dark and light intervals. Moreover, the model gave an estimate for the corrective reaction time as 238 milliseconds.

Although the role of corrective reaction time is important, unfortunately we know little about it. The corrective reaction time is not only a psychological element of our motion mechanism, but also plays an essential role in modeling hand-control movements. The self-paced tracking movement models by (Drury et al., 1987; Lin et al., 2009; Montazer, Drury, & Karwan, 1988) and self-paced aiming movement models by (Crossman & Goodeve, 1963/1983; Keele, 1968; Lin et al., 2009) have demonstrated that the length of corrective reaction time directly affects the movement speed and movement time if a given movement accuracy needs to be maintained. However, so far, the corrective reaction time has been assumed to be a general property – with no data on individual differences. The literature only tells that the reasonable value of the mean corrective reaction time ranges from 190 to 290 milliseconds. To understand the corrective reaction time better and study individual differences, the objective of this research is to apply Drury's (1994) intermittent illumination model and modified his experimental designs to directly measure individuals' corrective reaction time.

METHOD

PARTICIPANT AND APPARATUS

Two male and two female graduate students, aged from 25-30 years, were recruited to participate in this pilot study. They were all right-handed with normal or corrected-to-normal vision.

A personal computer (PC) with a 17" (432 mm) LCD monitor of resolution 1280 × 1024 pixels resolution was used. The PC ran Visual Basic (VB) using a self-designed experimental program that displayed experimental task and measured task performance. An Intous 3 305 mm × 488 mm drawing tablet with a tablet stylus was utilized as the input device. The movement distance ratio between the tablet and the computer screen was set as 1:1, equalizing visual & physical movement distances on the screen and the tablet.

EXPERIMENTAL SETTING AND PROCEDURES

While conducting the experiment, the participants sat alongside a dual surface adjustable table on which the monitor and the tablet were placed on the rear and the front surfaces, respectively. Both the monitor and the tablet were adjusted to heights where the individual participants felt comfortable. While performing movements, the participants wore a nylon half-finger glove and kept resting their hands on the tablet surface to keep the friction between moving hand and the tablet surface small and constant. A cardboard screen was placed between their eyes and the tablet to hide the visual feedback from their moving hands so that they only visual feedback was from the monitor screen.

To apply the intermittent illumination model, the experiment in this study was designed similarly to Drury (1994). However, instead of drawing circles on white paper, our participants moved the screen cursor to draw circles within circular courses shown on the screen. To conduct the tasks, they physically draw circles with the stylus on the tablet. The courses were defined by two concentric circles with a mean circle radius, 200 pixels (see Figure 1). A movement started by pressing down on the stylus cursor on the start point placed at the top location of the courses. Instead of controlling the intermittent illumination using the slide projector utilized in Drury (1994), the visual information of the cursor was intermittently displayed. Once the cursor was moved away from the start point, the start point disappeared and the cursor started to blink according to predetermined appearing/disappearing periods (see Table 1). However, the circular courses did not blink, eliminating any issues of dizziness and eyes fatigue.

For each circular course, the participants needed to draw one and three quarter continuous circles in which the movement time was measured from half a circle to one and a half circles, ensuring measured movements with consistent speeds. They were asked to draw as quickly as possible, but without moving outside the circular courses. If the cursor was moved outside the courses, that movement was considered as a failure trial. The participant had to repeat that course until it was successfully completed. Each participant had half an hour to practice before the formal measurements.

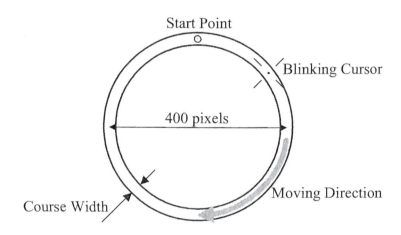

FIGURE 1 Demonstration of the experimental task.

EXPERIMENTAL VARIABLES

The independent variables were: course width and expected delay. The five values of course width were 24, 30, 36, 42, & 48 pixels (1 pixel $\cong$ 0.266 mm). The five values of expected delay with their combinations of light and dark durations determined by Equation 1 are listed in Table 1. The experimental combinations in this experiment were replicated two times, resulting in a total of 50 trials. All the trials were randomly conducted by each participant, taking about an hour to complete.

Table 1 Combinations of dark and light durations of the five values of expected delay

Expected Delay (millisecond)	Dark Duration (millisecond)	Light Duration (millisecond)
0.15	500	333
0.30	850	354
0.45	1150	319
0.60	1450	302

0.75	1750	292

The only dependent variable was speed measured for each trial. Due to the programming limitations, the time accuracy was about 16 milliseconds in task times averaging about six seconds.

RESULTS

ANALYSIS OF VARIANCE

Analysis of variance was performed on the speeds, using a mixed model with Width and Expected Delay Value as fixed effects and Participant as random, analyzing all the two-way and three-way interaction effects. There were significant main effects of Participant ($F_{3,100} = 5.85, p < 0.01$), Width ($F_{4,100} = 7.25, p < 0.01$), and Expected Delay Value ($F_{4,100} = 6.86, p < 0.01$). These main effects show that (1) the participants performed the movements at different speeds, (2) the increase of Width resulted in increased speed, and (3) speed decreased as Expected Delay Value increased. The two-way interaction effects were found with Participant × Width ($F_{12,100} = 11.03$, $p < 0.001$) and Participant × Expected Delay Value ($F_{12,100} = 5.07, p < 0.001$). The rates of increasing speed with increased width and decreased expected delay were different for the participants. Furthermore, the three-way interaction effect of Participant × Width × Expected Delay Value was also significant ($F_{48,100} = 1.54, p < 0.05$), indicating that the increase of Width had different sizes of effect on the two-way interaction effect of Participant × Expected Delay Value.

MODEL FITTING APPLYING DRURY'S (1994) METHOD

Since the main effects of Width and Expected Delay Value were found significant, the application of Drury's (1994) intermittent illumination model was able to be tested. Speed was first regressed on to Width for individual values of the expected delay to give the slopes, $1/c$ and intercepts shown in Table 2.

Table 2 Regression of speed on to width and calculated corrective reaction time

Expected Delay (s)	Intercept ($pixel \times s^{-1}$)	Slope (s^{-1})	$1/c$ (s)	r^2
0.15	-60.18	6.655	0.2474	0.985
0.30	-87.63	6.145	0.2742	0.917
0.45	-99.02	6.255	0.3317	0.984
0.60	-46.80	4.077	0.3396	0.868
0.75	-68.66	4.438	0.3877	0.937

The high linearity of data supported Drury's (1971) model, indicating a linear increase in speed with width. Then, as the method utilized in Drury (1994), the reciprocal of the slope (i.e., controllability) in Table 2 was regressed on to the intermittent illumination factor (i.e., the expected delay) to give

Equation 2
$$\frac{1}{c} = 0.2308 \times \left(0.919 + \frac{d^2}{2(l+d)}\right)$$

The model fitting of the data accounts for 95.4 % of the variance. The corrective reaction time obtained from Equation 2 was 919 milliseconds for all the participants on average. However, the obtained correction reaction time was much longer than the reasonable duration found in the literature (i.e., 190-290 milliseconds).

MODEL FITTING APPLYING A DIFFERENT METHOD

To calculate each individual participant's corrective reaction time, a modified method was utilized. Instead of calculating controllability ($1/c$) by regressing speed on width for individual values of the expected delay, the speed values of all experimental trials were divided by corresponding widths to obtain slopes and c values. This modified method provides more data sets, increasing available degrees of freedom. The obtained $1/c$ values specified to all participants and individual participants were regressed on to expected delay to give the intercept, slope, $1/c$ and corrective reactions time listed in Table 3. The regressions for individual participants are shown in Figure 2 below. As shown in Table 3, the model accounts for at least 81.1 % of the variance and the calculated corrective reaction time ranges from 193 to 792 milliseconds. Although, the model accounted for the data very well, only one of the calculated values of corrective reaction time is within the reasonable duration.

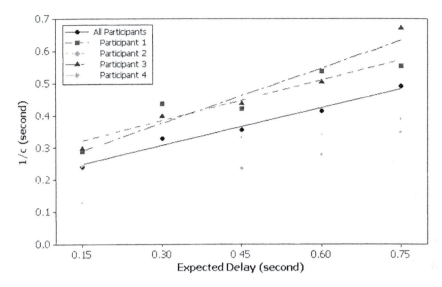

FIGURE 2 Relationships between controllability and expected delay.

Table 3 Regression of controllability on to expected delay and calculated corrective
reaction time

Participant	Intercept (s)	Slope (unit)	Corrective reaction time (s)	r^2
All	0.1898	0.3914	0.485	0.973
1	0.2391	0.4847	0.493	0.860
2	0.0741	0.3830	0.193	0.924
3	0.1942	0.6213	0.313	0.811
4	0.2049	0.2587	0.792	0.911

DISCUSSION AND FUTURE RESEARCH

No matter whether Drury's (1994) original method or the modified method was used, the calculated values of corrective reaction time were larger than the reasonable duration from 190 to 290 milliseconds. Specifically, when Drury's (1994) model was applied, the mean corrective reaction time of the overall participants was found to be 919 milliseconds; when the modified method was applied, the values were found to be 493, 193, 313, 792 milliseconds for participants 1, 2, 3, and 4, respectively and a mean value of 485 milliseconds.

The potential explanations of the unreasonable corrective reaction time include (1) inadequate measurement duration, (2) inappropriate manipulations of the expected delay and (3) indirect movement control. In the experiment, movement speeds were only recorded from half a circle to one and half circles. The participants might be still adjusting their movement speeds after passing the cursor to half a circle, especially for long-expected-delay trials. Also, the strategy that the participants stopped movements and then waited for the cursor to reappear was found for long-expected-delay trials. The restart of movements might add to the reaction time after the reappearance of the cursor, resulting in lower movement speeds. Furthermore, according to our experimental design, visual feedback was obtained from the computer screen, but not from the controlled limb. The indirect visual feedback may increase movement variability once the movements were stopped for the long-expected-delay trials, compared to the short-expected-delay trails in which kinesthetic feedback was available while continuously moving.

Future research is suggested with new experimental designs. First of all, a longer duration of measurement and longer adjusted distance are recommended. The duration of measurement could be increased to two circles and the measurement could start once the cursor completed one circle. Secondly, the values of expected delay should decrease to avoid strategy of stopping movements while tracking the circular paths. Finally, instead of using the drawing tablet, a touch-screen monitor is suggested to eliminate the movement variability issue.

CONCLUSION

This pilot study tested the application of Drury's (1994) intermittent illumination model with a modified computer-based experiment to measure individual corrective reaction times. The calculated corrective reaction times from four participants' data were longer than the reasonable range reported in the literature (193–919 milliseconds compared to 190–290 milliseconds). Three potential reasons, (1) inadequate measurement duration, (2) inappropriate manipulations of the expected delay and (3) indirect movement control, were proposed to explain the longer corrective reaction times. Corresponding solutions were also suggested for future research.

REFERENCES

Craik, K.J.W. (1947), "Theory of the human operator in control systems I : The operator as an engineering system." *British Journal of Psychology*, 38, 56-61; 142-148.

Craik, K.J.W. (1948), "Theory of the human operator in control systems II : Man as an element in a control system." *British Journal of Psychology*, 38, 142-148.

Crossman, E.R.F.W., & Goodeve, P.J. (1963/1983), "Feedback control of hand-movement and Fitts' law." *Quarterly Journal of Experimental Psychology*, 35A, 251-278.

Drury, C.G. (1971), "Movements with lateral constraint." *Ergonomics*, 14(2), 293-305.

Drury, C.G. (1994), "A model for movements under intermittent illumination." *Ergonomics*, 37(7), 1245-1251.

Drury, C.G., Montazer, M.A., & Karwan, M.H. (1987), "Self-paced path control as an optimization task." *Transactions on Systems, Man, and Cybernetics*, 17(3), 455-463.

Howarth, C.I., Beggs, W.D.A., & Bowden, J.M. (1971), "The relationship between speed and accuracy of movement aimed at a target." *Acta Psychologica*, 35, 207-218.

Katz, M.S., & Spragg, S.D.S. (1955), "Tracking performance as a function of frequency of course illumination." *Journal of Psychology*, 40, 181-191.

Keele, S.W. (1968), "Movement control in skilled motor performance." *Psychological Bulletin*, 70(6), 387-403.

Lin, J.-F., Drury, C., Karwan, M., & Paquet, V. (2009, August 9-14), In "A general model that accounts for Fitts' law and Drury's model." Paper presented at the Proceedings of the 17th Congress of the International Ergonomics Association, Beijing, China.

Montazer, M.A., Drury, C.G., & Karwan, M.H. (1988), "An optimization model for self-paced tracking on circular courses." *Transactions on Systems, Man,*

CHAPTER **48**

Systematic and Structured Development of Human-Included VE to Support Virtual Ergonomic Evaluation

Wen-Yang Feng[1], Chin-Jung Chao[1], Feng-Yi Tseng [1], Tien-Lung Sun[2]

[1]Chung-Shan Institute of Science and Technology
Long-Tan, Tao-Yuan 325, Taiwan

[2]Department of Industrial Engineering and Management
Yuan-Ze University, Tao-Yuan 320, Taiwan

ABSTRACT

Although virtual reality (VR) technology has been applied by many researchers for product design ergonomics evaluation, previous researches only consider a simple virtual environment (VE) in which the digital human model (DHM) could only touch the virtual objects but could not grab and manipulate them. Such simple VE is not sufficient for sophisticated ergonomics evaluation tasks. To build an interaction-rich, DHM-included VE, behavior scripts are used to program the desired interaction models. One problem in constructing such interaction-rich VE is that the interaction model is usually subject to change. When conducting ergonomics evaluation, the domain experts often want to change the interaction model. As the domain experts are usually not familiar with 3D programming, it is important to have an interaction model that is easy to be interpreted and modified by domain

*corresponding author, tsun@saturn.yzu.edu.tw

experts. This paper discusses a method to facilitate systematic and structured development of interaction-rich, human-included VE. The interaction model is divided into two parts. The first part contains the low-level, domain-independent behavior scripts that implement object grabbing and manipulation. The second part contains a Petri-net (PN) for domain experts to define the interaction scenarios. The PN-based graphic representation makes it easier for domain experts to understand and modify the interaction scenarios.

Keywords: Virtual environment, digital human model, interaction, design ergonomics evaluation

INTRODUCTION

Virtual ergonomic evaluation is useful at early design stage where the physical design mock up is not available but human-centered design must be considered (Jayaram et al. 2006, Kuo and Wang 2007, Duffy 2007). The virtual approach first constructs a digital human model (DHM) according to anthropometric data and then the DHM is put to a virtual environment (VE) that contains the virtual prototype of the product design. While the DHM works in the VE, its posture data are recorded and sent to ergonomics evaluation programs like SSP, NIOSH, and RULA. To make the DHM move naturally and accurately, the DHM is often derived by a human player using the motion capture devices.

Although previous researches have shown the feasibilities and benefits of using VR for design ergonomic evaluation, they only consider a rather simple VE, in which the DHM can only touch the virtual objects but can not grab and manipulate them. Such simple VE is suitable for evaluating interface design but not enough for evaluating ergonomics issues related to product operation or maintenance. For example, Figure 1 shows a DHM-included VE constructed to evaluate ergonomic issues related to the design of a control console. If the digital human can only touch the console, then the posture data that can be captured are restricted, as shown in Figure 1(a). To evaluate ergonomics issues related to maintenance tasks as shown in Figure 1(b), the DHM must be able to grab the screen, lift it, grab the cables and detach them.

(a)   (b)

FIGURE 1. DHM interacts with virtual objects (a) touch, (b) grab and manipulation.

Although many VR authoring tools supply behavior scripts to define and control the interactions between the DHM and the virtual objects, the interaction model constructed using the low-level behavior scripts are difficult to interpret and modify. During ergonomic evaluation, the domain experts often want to modify the interaction scenarios to try different design options. As the domain experts usually are not familiar with 3D programming, it is important to have an interaction model easy for domain experts to interpret and modify.

This paper discusses a method to facilitate systematic and structured development of interaction-rich, human-included VE. The interaction model is divided into two parts. The first part contains the low-level, domain-independent behavior scripts that implement object grabbing and manipulation. The second part contains a Petri-net (PN) for domain experts to define the interaction scenarios. The PN-based graphic representation makes it easier for domain experts to understand and modify the interaction scenarios.

LITERATURE REVIEW

Previous researches in virtual ergonomic evaluation could be classified into different categories according to the level of complexities to animate the DHM and to program its interactions. At the simplest level, the DHM in the VE is static, i.e., it is neither animated nor interacting with the virtual objects. Such DHM-included VE is suitable for visual examination or macro analysis where no posture data is required. For example, Duffy et al. (2003) develop a DHM-included VE for domain experts to examine a workspace layout design from different viewpoints to evaluate safety issues and identify possible dangerous areas. Zülch and Grieger(2005) develop a human-populated digital factory to let domain experts evaluate occupational health and safety issues from macro-ergonomic point of view.

At the next level, the DHM is discretely animated and the 'snapshot' of the posture data is captured for ergonomic evaluation. For example, (Chu and Kuo

2005) developed a web-based car interior design evaluation where customers at remote site could adjust the posture of the DHM and the system will calculate the forces/stresses for various body joints using the joint angle data of the DHM. Similarly, Hou et al. (2007) employ the discrete approach for workspace safety evaluation like view, strength index, low back spinal force, and so on. Di Gironimo and Patalano (2008) employ the discrete approach to study the correct positioning of new devices installed in the cab.

At the next level of complexity, the DHM is continuously animated using motion capturing devices. Such continuously animated DHM allows complicated and accurate posture data to be captured naturally and efficiently. Jayaram et al. (2006) discuss two approaches to implement this approach. The first one employs a commercial ergonomic evaluation tool JACK to model the DHM and to perform ergonomics evaluation. The JACK tool is integrated with a VE constructed using VR authoring tool. This approach saves time and efforts to construct the DHM and the ergonomics evaluations functions. But the commercial ergonomics tools are usually large and complicated. They are hence difficult to be customized. The second approach builds the DHM and the VE using the general-purpose VR authoring tool. This approach takes efforts to build the DHM and the ergonomics evaluation functions. But it is easy to build small-size, customized virtual ergonomic evaluation environments. In this paper the second approach is adopted to develop DHM-included VE for ergonomic evaluation. Kuo and Wang (2007) develop a continuous evaluation approach and apply it to evaluate the design of the controlling interface for a missile launching vehicle. The digital human model is real-time manipulated by user through magnetic and optical motion capturing system. Duffy (2007) employs the continuous approach to evaluate ATM machine interface design for disabled people.

INTERACTION-RICH, HUMAN-INCLUDED VE FOR VIRTUAL ERGONOMICS EVALUATION

This paper extends the continuous type of DHM-included VE (Jayaram et al. 2006, Kuo and Wang 2007, Duffy 2007) to develop an interaction-rich, human-included VE for ergonomic evaluation. The DHM-included VE considered in this paper differs from previous ones in two aspects: 1) it is implemented using the general-purpose 3D tools rather than special-purpose DHM tools, and 2) it is interaction-rich in that the DHM could grab the virtual objects and manipulate them.

The special-purpose DHM tools, e.g., Jack, Ramsis, Delmia, CATIA, etc., have been employed by previous researches to implement the DHM-included VE for ergonomic evaluation. These tools have built-in modules for automatic DHM construction from anthropometric data and quantitative ergonomics evaluation. They save the time and efforts to build the DHM-included VE. The drawbacks, however, are that they are inflexible for interaction programming. The general-purpose 3D VR tools, on the other hand, do not support automatic DHM construction and ergonomic evaluations. Extra efforts are required to build the DHM-included VE. Despite of this inconvenience, the general tools are used in this work to implement the interaction-rich, DHM-included VE due to their support for interaction programming.

The 3D modelling and animation tool 3ds Max (Autodesk 2008) is employed in this work to build the DHM and the virtual environment (Figure 2). The skeleton of the DHM, which drives the movements of the human model, is built using the Biped skeleton system in 3ds Max. The textured mesh of the human body is also built in 3ds Max and integrated with the skeleton using the Phyisque modifier. The DHM and the 3D scene constructed in 3ds Max are then imported to a VR authoring tool called Virtools (Dassault Systems, 2008), in which the DHM is connected to a motion capturing system called MOVEN (http://www.moven.com) (Figure 3). The motion capturing devices allow the human player to drive the DHM in real time to perform actions in the VE.

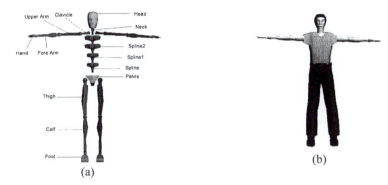

(a) (b)

FIGURE 2. DHM construction in 3ds Max (a) Biped skeleton, (b) textured mesh.

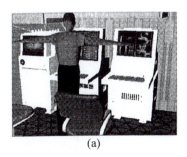

(a) (b)

FIGURE 3. (a) DHM-included VE in Virtools, (b) Moven motion capturing system.

Although the interactions between the DHM and the virtual objects could be specified using the behavior scripts provided in Virtools, interaction model constructed using behavior scripts are difficult to interpret and modify, especially for domain experts who are not familiar with 3D programming. To construct an interaction model that could be understood by the domain experts, we divide the interaction model into two parts. The first part contains the low-level behavior scripts that implement object grabbing and manipulation. These scripts are domain independent as they are not seen nor modified by the domain experts. The second part contains a PN-based graphic representation to control the domain-dependent interaction control logics. They are domain-dependent as they are seen and modified by domain experts. Details of this two-level interaction model are described in next sections.

LOW-LEVEL, DOMAIN-INDEPENDENT BEHAVIOR FUNCTIONS

The first part of the interaction model contains the low-level behavior scripts that implement interaction detection and animation for object grab and manipulation. Interaction detection determines whether an object is grabbed by DHM. It contains two main functions: collision detection and hand gesture recognition. Collision detection detects whether the DHM's hands collide with an object, and hand gesture recognition determines whether the touched object is grabbed or not. Collision detection is a common function available in most VR authoring tools. Hand gesture recognition, on the other hand, is usually not available in VR authoring tools and must be explicitly coded by the user.

The animation functions simulate the movement of the grabbed object. When the grabbed object moves without constrained, it moves together with the hand. If object deformation is not considered, then the movement could be simulated by setting the grabbed object as the child object of the virtual hand. Rigid-body transformations like translation and rotation applied to the virtual hand will also be applied to the grabbed object.

When the grabbed object moves with constraints, e.g., it is connected by a hinge joint, its movement is simulated by transition and rotation along a pre-defined local coordinate system. For example, in Figure 4 the monitor is constrained to rotate along the local x-axis, whereas the drawer is constrained to translate along its local y-axis.

FIGURE 4. Simulation of constrained movement of grabbed object.

HIGH-LEVEL, DOMAIN-DEPENDENT BEHAVIOR CONTROL

The high-level interaction control logics are represented by several types of PN-based graphic representations. The first type of PN defines the precedence constraints for object manipulation, e.g., the DHM has to open the door before it could pull out the power cables. The structure of this PN is shown in Figure 5, which is based on the workflow control PN proposed in (van der Aalst et al. 1994). The places P_i and P_o are the input and output places, the place P_{ni} represent the object manipulation, and the places at the right side, e.g., P_o-Request_resouce and P_i-Start_operation, etc., are the I/O places communicating with the resource management PN which will be discussed later. The transition T_i represents one-hand object manipulation, whose execution details are defined by another PN shown in Figure 6. The I/O places of the PN in Figure 6 are $P_{n,i-1}$ and $P_{n,i+1}$, i.e., the places before and after the transition T_i. The five transitions, c_1 to c_5, perform resource management and the actual object manipulations. For example, c_1 requests the required resources for this manipulation, which could be a tool or another object. The four places p_1 to p_4 serve as buffers between the transitions.

In Figure 5, the transition SP_i represents two-hand object manipulation, whose execution details are defined by another PN shown in Figure 7. The PN in Figure 7 contains two transitions marked as T to represent the operations performed by the left and right hand, respectively. The two transitions marked as C are control transitions. The PN could be further expanded if the two transitions marked as T are expanded using the one-hand operation PN shown in Figure 6. Figure 8 shows an expanded version of this PN.

418

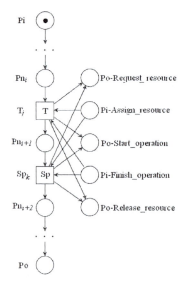

FIGURE 5. PN to control object manipulation sequence (van der Aalst et al. 1994).

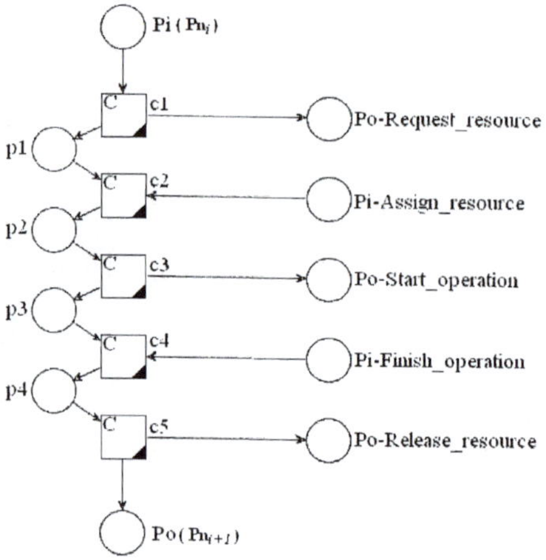

FIGURE 6. PN to control one-hand operation (van der
Aalst et al. 1994).

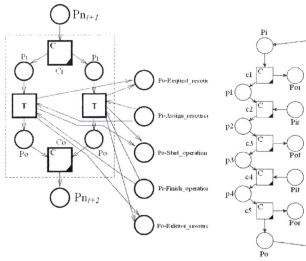

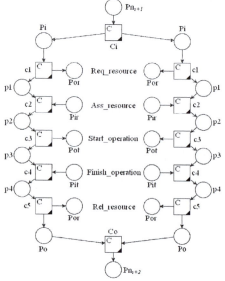

FIGURE 7. PN to control two-hand object
manipulation.

FIGURE 8. Expanded PN to control two-hand object
manipulation.

To control the hand coordination in object manipulation, another PN is used. This PN is constructed based on the coordination mechanism proposed in (Roposo et al. 2000). Figure 9 shows one example of such PN, which is integrated with the two-hand operation PN shown in Figure 8 to define an object manipulation that must be simultaneously performed by two hands. This coordination PN has a pool place P_n that stores two tokens, one for each hand. The arc that connects P_1 with T_1 has a weight of 2, which means that the transition t1 can not be fired unless P_1 gets two tokens. The place P_1 can get two tokens only when one token is required from req_resource1 (the left hand) and the other one is required from req_resource2 (the right hand). When t1 is fired, the left hand operation start_task1 and the right hand operation start_task2 will be simultaneously performed. Other types of coordination mechanism could be found in (Roposo et al. 2000).

Figure 10 shows two interaction models defined using the coordination PN shown in Figure 9. The monitor Figure 10(a) is programmed so that it must be opened with both hands. If the monitor is programmed so that it could be opened by one hand, then another control PN will be used. Figure 10(b) shows another interaction scenario in which the DHM has to use one hand to grab the upper part of the power cable and another hand to pull out the lower part to separate the power cable.

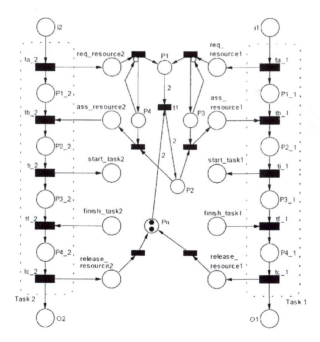

FIGURE 9. Coordination mechanism to define an object manipulation that must be simultaneously performed by two hands (Roposo et al. 2000).

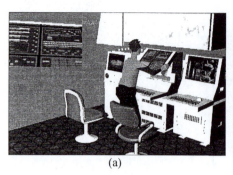
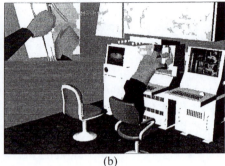

(a)   (b)

FIGURE 10. Hand coordination in object manipulation.

CONCLUSION

This paper discusses a method to facilitate systematic and structured development of interaction-rich, human-included VE. The interaction model is divided into two parts. The first part contains the low-level behavior scripts that implement object grab and manipulation. The second part is a PN-based graph to represent the high-level control logics. The first-level PN controls the object manipulation sequences, whereas the second-level PN controls the hand coordination. The PN-based graphic representation allows domain experts to develop interaction scenarios for virtual objects easily and rapidly.

The DHM-included VE is implemented using the general-purpose 3D tools, which facilitate the construction of small but customized virtual ergonomic evaluation environment. We use the proposed method in a project conducted for local industry in which an interaction-rich VE is to be built to evaluate ergonomics issues related to the design of control consoles installed in the ship control room. At the writing of this paper, we are at the stage of integrating the PN model with the VR model. Once the PN-based interaction specification interface is added on top of the DHM-included VE, we plan to conduct experiments to see how the proposed method could help domain experts quickly and easily develop interaction scenarios for ergonomics evaluation tasks.

REFERENCES

van der Aalst, W.M.P., van Hee, K.M., and Houben, G.J. (1994) "Modelling and Analysing Workflow Using a Petri-net based Approach", proceedings of the Second Workshop on Computer-Supported Cooperative Work, Petri Nets and Related Formalisms, Zaragoza, Spain.
Autodesk Inc. (2005), *3ds Max 8 Reference*. California.

Chu, C. H., and Kuo, C. F. (2005), "An Online Ergonomic Evaluator for 3d Product Design,." *Computers in Industry*, 56, 479–492.

Di Gironimo, G., and Patalano, S. (2008), "Re-design of a railway locomotive in virtual environment for ergonomic requirements", *International Journal of Interact Design and Manufacturing*, 2, 47–57.

Duffy, V. G., Wu, F .F. and Ng, P. P. W. (2003), "Development of an Internet Virtual Layout System for Improving Workplace Safety." *Computer in Industry*, 50, 207–230.

Duffy, V. G. (2007), "Modified Virtual Build Methodology for Computer-Aided Ergonomics and Safety." *Human Factors and Ergonomics in Manufacturing*, 17(5), 413–422.

Hou, H., Sun, S. and Pan, Y. (2007), "Research on virtual human in ergonomic simulation" *Computers & Industrial Engineering*, 53, 350–356.

Jayaram, U., S. Jayaram, I. Shaikh, Y. J. Kim, and Palmer, C., (2006) "Introducing quantitative analysis method into virtual environments for real-time and continuous ergonomic evaluations", *Computers in Industry*, 57, 283–296.

Kuo, C. F., and Wang, M. J. (2007) "Interactive Virtual Evaluation for Interface Design." *Human Factors and Ergonomics in Manufacturing*, 17(5), 485–495.

Roposo, A. B., da Cruz A.J.A., Adriano, C.M., and Magalhaes, L.P. (2000) Coordination Components for Collaborative Virtual Environments, COMPUTERS & GRAPHICS, Volume 25.

Zülch, G., and Grieger, T. (2005) "Modelling of Occupational Health and Safety Aspects in the Digital Factory." *Computers in Industry*, 56, 384–392.

CHAPTER 49

Multi-Scale Entropy Analysis for Postural Sway Signals with Attention Influence for Elderly and Young Subjects

Wen-Hung Yang, Bernard C. Jiang

Department of Industrial Engineering and Management
Yuan Ze University
135 Yuan Tung Road, Taoyuan, Taiwan

ABSTRACT

The role of attention in postural control is important in a balanced task. The objective of this study is to compare the complexity of postural stability between healthy young and elderly faller subjects, with and without the influence of an additional attention task (dual-task). The "complexity" was used to reflect adaptability to change and to determine postural balance stability from the center of pressure (COP) signals. The COP data were collected from 13 elderly fallers and 15 healthy young subjects under standing still with or without additional attention task. The COP signals were pre-processed by empirical mode decomposition (EMD), and high frequency signals were calculated as "complexity" through multi-scale entropy (MSE) analysis. With the view of complexity, our results showed that attention certainly has an influence on postural control for elderly fallers. In that group, the complexities in dual-task COP signals were lower than those when standing still, and they exhibited less postural stability under an attention influence. Those results indicated a poor adaptability to attention influences for elderly fallers.

Keywords: Multi-scale entropy (MSE); Complexity; Center of pressure (COP); Postural stability; Attention influence

INTRODUCTION

Postural control is an issue for slips and falls during gait or static attention perturbations. Recent studies have reported that attention may have an influence on postural control (Rubenstein, 2006; Jamet et al., 2008; Yang, 2008). Environmental changes tend to have a greater destabilizing effect on older adults (Deviterne et al., 2005). The above studies have indicated that attention plays an important role on postural (balance) control stability for elderly people when standing still. In static tests, a force plate may be used to measure the displacement of the center of pressure (COP), which represents in both anteroposterior (AP) and mediolateral (ML) directions. Figure 1(b)–(d) illustrates the displacement of the COP in the AP and ML directions.

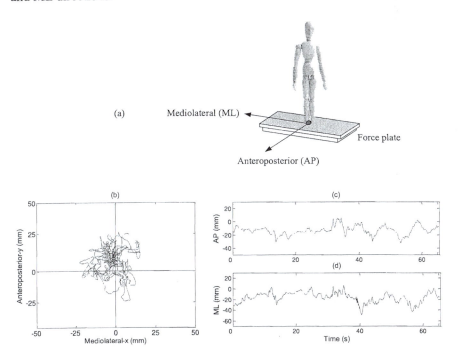

Figure 1. The center of pressure (COP) signal measurement and trajectory. (a) is a diagram of a force plate test system; (b) is displacement of COP within the force plate; and (c) and (d) are COP displacement in the AP and ML directions over time, respectively.

The COP signals have been demonstrated to be both nonlinear and non-stationary and empirical mode decomposition (EMD) is a suitable approach for pre-processing COP data (Huang et al., 1998). Recently, a multi-scale entropy (MSE) approach was developed to measure the complexity of physical and physiological time series with limited data (Costa et al., 2005). Based on the MSE approach, a complexity concept was adopted in this study to evaluate postural stability and variation in COP signals. Here, higher complexity means better adaptability to an external environment, while lower complexity indicates poorer adaptability (Peng et al., 2009; Costa et al., 2007; Lin, 2009; Su, 2009). There is an assumption that a healthy and adaptive system (i.e., healthy young subjects) can handle attention stress or challenge better than a less adaptive system (i.e., elderly fallers).

In our work, a static (standing still) experiment was designed to collect COP signals. The aim of this study is to quantify and compare the variability in complexity of postural stability in healthy young and elderly faller subjects under the conditions of standing still and with and without an attention influence. Through the experiment and MSE analysis, the influence of an attention task is investigated for elderly subjects. In particular, in order to understand the dynamic properties of postural control, the variability in COP position time series was quantified, along the horizontal plane, in anteroposterior (AP) and mediolateral (ML) directions.

EXPERIMENT PROTOCOL AND COMPUTATIONAL METHOD

In order to evaluate the complexity of postural dynamics, this study has conducted a dual-task experiment that involved the addition of attention tasks to collect COP signals from healthy young and elderly faller subjects. Figure 2 shows the four stages of COP signal analysis used here. In the second stage, an EMD technique was adopted to decompose the dynamic fluctuations from the COP signals. In addition, high frequency signals were reconstructed to reveal differences between the two groups. Following the EMD procedure, the reconstructed signals were analyzed by MSE. Finally, complexity based on the MSE approach was evaluated and compared between the two groups and between the single- and dual-task conditions.

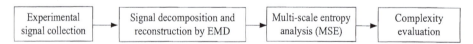

Figure 2. A functional module flowchart for COP signal analysis.

SUBJECTS

The COP data obtained from standing still subjects for 65 seconds were collected from two groups: a group of 13 elderly fallers (individuals with more than three falls in one year) in a senior home; and a group of 15 healthy young people. The average ages of the elderly and young groups were 80.2±4.3 y and 24.5±2.6 y, respectively.

EXPERIMENT PROTOCOL

Based on the research objective, the process of the entire experiment was divided into two stages:

Stage 1: Subjects stood barefoot on a force plate platform with their eyes open and with their arms hanging loosely by their sides, for 65 sec.

Stage 2: Subjects stood still on a force plate with their eyes open for 65 seconds; however, while standing, interference in the form of an attention influence was added (the dual-task in this study). The dual-tasks tested included mental arithmetic calculation, visuospatial tasks, visual reaction time tasks, and word recall.

EMPIRICAL MODE DECOMPOSITION (EMD)

EMD decomposes the signal into a set of intrinsic mode functions (IMF) representing oscillatory modes embedded in the signal. Each IMF component is extracted by a series of sifting procedures, and all IMFs are iteratively extracted. In this study, the COP signal $x(t)$ is represented as the sum of the IMF components and the final residual, as equation (1)

$$x(t) = \sum_{j=1}^{n} IMF_j + r_n \tag{1}$$

where IMF_j is the jth IMF, n is the number of extracted IMFs, and r_n is interpreted as a trend or as a residual having only one extremum in the signal $x(t)$. In our study, COP signals are decomposed into six IMF components ($n = 6$) in this step. Costa et al. (2007) reported that differences in dynamic fluctuation between young and elderly subjects are usually demonstrated in short duration, high frequency COP time series signals. Thus, we summed high frequency components from IMF1 to IMF3 for subsequent analysis.

Figure 3 presents reconstructed high frequency COP signals for an elderly faller and a healthy young subject under the two experimental conditions. Differences in the COP movement ranges are prominent between the elderly faller and healthy young subject for both single and dual-task situations (compare Figure 3(a) with 3(b) and Figure 3(c) with 3(d)). More importantly, from Figure 3(b) and 3(d), it is clear that the significant variation of the dual-task is exposed by the reconstructed COP signal between the elderly faller and healthy young subject. It is evident that the extraction of high frequency COP signals by EMD can reveal useful characteristic fluctuations during a short time.

426

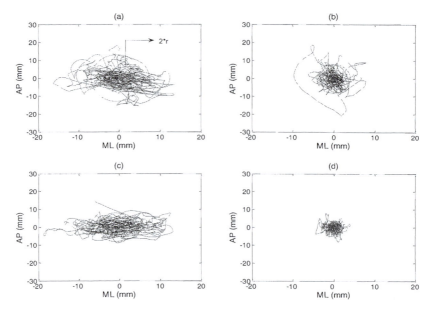

Figure 3. The reconstructed COP signal for an elderly faller and a healthy young subject while standing still (single task) and standing with a dual-task. (a) is an elderly faller for standing still with dual task; (b) is an elderly faller for standing still; (c) is a healthy young subject for standing still with dual task, and (d) is a healthy young subject for standing still. Each reconstructed COP signal was obtained by summing the three highest frequency IMF components in the EMD procedure ($IMF_1+IMF_2+IMF_3$). The red solid point is the central point of COP trajectory. The red dash-line circle represents a subject's rough sway range based on the radius of $2r$, where r is the standard deviation of the distance from each displacement point to central point of the COP data.

MULTI-sCALE ENTROPY (MSE) APPROACH

Multi-scale entropy (MSE) method is developed to calculate entropy and complexity over multiple time scales for physiological signals (Costa et al., 2005). Here, we briefly described the MSE approach. Given a signal data time series $\{x_1, x_2,...,x_n\}$, the MSE method first constructs a consecutive coarse-grained time series by averaging a successively increasing number of data points in non-overlapping windows. Figure 4 presents an illustration of a coarse-graining procedure for scale 2 and scale 3.

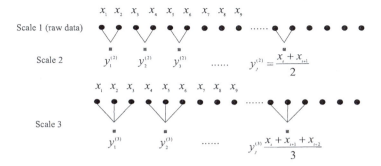

Figure 4. Schematic illustration of the coarse-graining procedure used in multi-scale entropy (Costa, 2005).

In Figure 4, each element of the coarse-grained time series, $y_j^{(\tau)}$, is calculated according to equation (2).

$$y_j^{(\tau)} = \frac{1}{\tau} \sum_{i=(j-1)\tau+1}^{j\tau} x_i \quad , 1 \le j \le \frac{N}{\tau} \tag{1}$$

where τ represents the scale factor, and the length of each coarse-grained time series is N/τ. For scale 1, the time series $\{y^{(1)}\} = \{x_1, x_2, x_3, \cdots x_N\}$ is simply the original time series. Since MSE was developed on the basis of sample entropy, match points (m) and match tolerance (r) need to be considered. For all cases shown in our paper, sample entropy is presented with $m = 2$ and $r = 0.15$. In addition, each of the coarse-grained COP signals was plotted as a function of the scale factor.

RESULTS AND DISCUSSION

The reconstructed COP signals (IMF1+ IMF2+ IMF3) were analyzed by MSE approach, and the variations of degree of complexity were evaluated. The complexity index (CI) was compared to determine the degree of complexity of signals when the MSE curves crossed. The CI is the area under the MSE curve as obtained by $\sum_{i=1}^{sn} SampEn(i)$, where sn is the presented scale number. A greater CI value indicates higher complexity of the COP signal. The complexity variability of COP signal was quantified in both AP and ML directions.

Figure 5 illustrates MSE curves for reconstructed COP signals from elderly fallers and healthy young subjects. The curves monotonously increase with scale factors. For scale one, corresponding to traditional sample entropy, the COP signals produce the lowest value of entropy for elderly fallers and healthy young subjects. For larger scales, the difference in entropy between standing still and dual-task situations becomes larger. The above results imply that characteristic oscillations may be revealed in the COP signal at different scales. In health young subjects, higher CI values were observed in the AP direction while in a dual-task situation (Figure 5 (a)). However, for elderly fallers (Figure 5 (b)), the CI values in both ML and AP directions were higher when standing still than when under dual-task conditions.

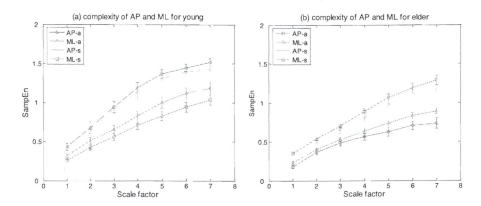

Figure 5. The MSE analysis of reconstructed COP signals for healthy young (a) and elderly fallers (b) ($m = 2$, $r = 0.15$). The symbols illustrate mean values of sample entropy (SampEn) for subjects, and error bars illustrate standard errors. The symbols present sample entropy (SampEn) value for a subject. Notations are presented as follows. AP-a: in AP direction for standing still with dual task; AP-s: in AP direction for standing still; ML-a: in ML direction for standing still with dual task; ML-s: in ML direction for standing still.

Table 1 presents means and standard deviations of CI values for 13 elderly fallers and 15 healthy young subjects under standing still and dual-task conditions. The difference in CI values in the ML and AP directions was significant for elderly fallers (5.98 ± 1.36 vs. 4.21 ± 0.79, respectively, when standing still, and 5.45 ± 1.20 vs. 3.45 ± 1.50, respectively, when dual-tasked).For elderly fallers, CI values when dual-task were lower than when standing still, indicating that elderly fallers are less adaptive system to environment change. In other words, the elderly fallers exhibit less postural balance stability with attention influence. For elderly fallers under dual-task conditions, the AP direction shows lower CI values (3.45 ± 1.50); under standing still, the ML direction shows higher CI values (5.98 ± 1.36). These results point out that under the influence of additional attention stimuli, elderly fallers show the poorest control of balance in the AP direction. While standing still, better stability is evident for postural balance in the ML direction.

However, there were no significant differences in the results of such comparisons in the healthy young subject group. Young people showed no significant differences in the CI values in the AP direction when dual-task or when standing still (7.57 ± 1.19 vs. 7.24 ± 1.71, respectively). In other words, for young people, postural balance stability in the AP direction was not affected by external attention. Furthermore, the CI values for healthy young subjects were mostly higher than those for the elderly, whether in standing still or dual-task situations. The exception was in the ML direction when standing still (4.76 ± 1.28 in healthy young vs. 5.98 ± 1.36 in elderly fallers). This implies that elderly fallers show better postural stability in the ML direction than healthy young subjects when standing still.

Table 1. Comparison of MSE analysis results for reconstructed COP signals in elderly fallers and healthy young subjects. The p values were obtained using a one-tailed, paired t-test with unknown variance. CI represents the complexity index; Sway direction abbreviations: ML, mediolateral; AP, anteroposterior.

CI	Elderly fallers		Healthy young	
mean ± SD	ML	AP	ML	AP
Dual task	4.21 ± 0.79	3.45 ± 1.50	5.62 ± 1.62	7.57 ± 1.19
Standing still	5.98 ± 1.36	5.45 ± 1.20	4.76 ± 1.28	7.24 ± 1.71
p value				
Dual task vs. Standing still	0.005	0.008	0.032	0.306

CONCLUSIONS

To understand the influence and variation of postural stability as affected by attention between elderly and young, an experiment was designed to collect COP signals for 13 elderly fallers and 15 healthy young subjects under static standing still and dual-task (i.e., an attention influence change) conditions. Complexity was adopted to evaluate the postural balance stability in our research. According to the MSE concept, the CI was used to quantify the complex variability of COP signals in the subjects' AP and ML directions. The experimental results reveal that COP signals measured from elderly fallers were less complex than those from healthy young subjects. Attention influence did affect postural stability, as indicated by the differences between elderly fallers and healthy young subjects during dual-tasking.

For elderly fallers, dual-task CI values were lower than those in a single-task situation. This indicates that elderly fallers have less adaptive systems when the environment changed or while handling stress. On the contrary, for healthy young subjects, complexity with dual-task was higher than in the static standing still task. This suggests the presence of attention adaptability for healthy young subjects and indicates that external attention stimuli may not significantly affect their postural stability. When dual-tasked, the ability to adjust physical balance in the AP direction of the elderly was significantly inferior to that of the young. The results suggest that the human aging process may lead to greater variation in balance ability in the AP direction than in the ML direction.

ACKNOWLEDGEMENTS

We thank Dr. C.K. Peng from Beth Israel Deaconess Medical Center of Harvard Medical School for his valuable suggestions, and gratefully acknowledge the support from the National Science Council of Taiwan (NSC 96-2628-E-155-008-MY3).

REFERENCES

Costa, M., Peng, C.-K., and Goldberger, A. L. (2005), "Multiscale entropy analysis of biological signals." Physical Review E, 71, 021906-1 – 18.

Costa, M., Priplata, A. A., Lipsitz, L. A., Wu, Z., Huang, N. E., Goldberger, A. L., and Peng, C.-K. "Noise and poise: Enhancement of postural complexity in the elderly with a stochastic-resonance–based therapy." A Letter of Journal Exploring the Frontiers of Physics, 68008, 1-5.

Deviterne, D. Gauchard, G. C. Jamet, M. Vançon, G., and Perrin, P. P. (2005), "Added cognitive load through rotary auditory stimulation can improve the quality of postural control in the elderly." Brain Research Bulletin. 64, 487–492

Huang, N. E., Shen, Z. Long, S. R. Wu, M. C. Shih, E. Zheng, H. Q. Tung, C. C., and Liu, H. H. (1998), "The empirical mode decomposition method and the Hilbert spectrum for non-stationary time series analysis." Proceedings A of Royal Society, 454, 903-995.

Jamet, M., Deviterne, D. Gauchard, G. C. Vançon, G., and Perrin, P. P. (2004), "Higher visual dependency increases balance control perturbation during cognitive task fulfillment in elderly people." Neuroscience letters. 359, 61–64.

Lin, Y. T. (2009), "Postural balance ability analysis for inpatients after laryngeal microsurgery in anesthesia using multiscale entropy." Master's Thesis, Department of Industrial Engineering and Management, Yuan Ze University, Chung-Li, Taoyuan, Taiwan.

Peng, C.-K. Costa, M., and Goldberger, A. L. (2009), "Adaptive data analysis of complex fluctuations in physiologic time series." Advances in Adaptive Data Analysis, 1(1), 61-70.

Rubenstein, L. Z. (2006), "Falls in older people: epidemiology, risk factors and strategies for prevention." Age and Ageing, 35-S2, ii37–ii41.

Shumway-Cook, A., and Woollacott, M. (2000), "Attentional demands and postural control: the effect of sensory context." Journal of Gerontology: Medical

Science, 55A, M10.

Su, H. Y. (2009), "Postural balance ability analysis for outpatients after Port-A surgery in anesthesia using multiscale entropy." Master's Thesis, Department of Industrial Engineering and Management, Yuan Ze University, Chung-Li, Taoyuan, Taiwan.

Yang, T. H. (2008), "The stability for different effect and aging on multiscale entropy." Master's Thesis, Department of Industrial Engineering and Management, Yuan Ze University, Chung-Li, Taoyuan, Taiwan.

Chapter 50

Human Reliability of Prescription Preparing Processes

Chih-Wei Lu[1,2], Wei-Jen Chang[1], Jen-Pin Cheng[1],
Sung-Yen Tsai[1], Chi-Ling Huang[1]

[1]Kaohsiung Medical University
[2] Chung Yuan Christian University

ABSTRACT

Now, the prescription safety always has been a key part of patient safety and the human reliability is a key factor of pharmacist performance during prescription in the medicine station. Prescription errors would make patient hurt or death. After 2002, some serious patient safety events happened in Taiwan. The reliability of prescription has been a very trendy topic of conversation. Under the pressure of cost-down from the healthcare insurance company, the number of pharmacists has been minimized in hospitals and the duty of job has been over loaded. The objects of this proposal are using ergonomics approaches to investigate the human reliability at medication workstations. Participates were 51 pharmacists. The used methods were: in-plant survey, ergonomic job analysis, workstation measurement, anthropometry and questionnaires. The incident rate and risk factors of human error of medication processes has been put together to generate advance reengineering device of work environment. Based on the result of self-report questionnaire, 100% people knew that miss would happened during the prescription prepare processes. There were 93.3% pharmacists thought they had made miss after they had done their jobs, 63% pharmacists made miss during the last 3 months, and 97.8% pharmacists had heard officemates talked about miss in the office room. About what type of jobs were problem maker, 55.1% thought prescription prepare process had highest miss rate, the next high (20.5%) was pharmacy machine processes, and 11.5 % was bagging process. Based on the Ramussen principle of motion levels of prescription, and the highest probability of miss have been ranked by the 50% pharmacists of this study. The sequences of miss rates of scopes were regulation (49.0%), technology (31.0%), and knowledge (19.6%). Based on Swain principle to classify the types of miss and the highest probability of miss have been ranked by the 50% pharmacists of this study The miss rates of scopes were task (49.0%), loss (25.5%), miss of opportunity (19.6%), and order miss(5.88%) in sequences. Based on error report system of pharmacy, the human reliability of prescription by season from 2004 to 2008, the heighted error rate happened in Oct. 2005 and rate was 0.19% (reliability=99.89%). And, the lowest was 0.057% (reliability=99.943%) in Feb. 2007. During June 2004 ~ May 2008, the trend of error rate was decreasing from 0.002 (0.2%) to 0.0005 (0.05%). And, the reliability of preparing prescription was 99.8% to 99.95% by month. Based on the results of this study and subjective self-report, motion in regulation scope and task miss was the major human errors. Pharmacists were very familiar the procedure of their jobs. However, sometimes, they would make mistakes for some reasons in the well-known tasks. It might be carless, miss one or two steps or made un-appropriate decisions.

Keywords: Ergonomics, human reliability, prescription, patient safety

INTRODUCTION

Background

Prescription preparing and the job of pharmacist both are key jobs in healthcare industries. People believe they will be cared very safe when they go to the hospitals. However, events have happened in hospitals and made people worry and concern in current years (Abood, 1996). Since 2002, there are many serious of accidents about the medical errors happened in Taiwan. One nurse made a wrong injection in the North Town Woman and Children's Hospital at 2000\11\29; 2). In the Chong Ai Hospital, A pharmacist gave a wrong medication at 2000\12\10. The emergency doctor gave a wrong medication in the Cheng Hsin Hospital at 2007\11\18 (Chen, 2003).

According to Taiwan Health Reform Foundation (THRF), there were 476 medical disputes, 56 cases were prescribing errors and about 10% of all errors (Jen, 2003). Types of errors included giving wrong medications; worsen after using the medication, giving a wrong dosage. In 1999, The USA Healthcare Society pointed that there 44,000 to 98,000 person's death was due to medical negligence (Allan & Barker, 1990; Dean, 2002). Health care industry has started to face up and pay more attention to patient safety and prescription safety. All of medication errors were contributed to "human errors". Based on the database of industrial accidents, more than 70% accidents were related to human errors. In all over the world, the safety engineer and enterprises work hard to control human errors (Ind. Safety & Health 2002, 2003).

434

Objective

The objectives of this study were collecting the basic data of pharmacists, self aware of error and using error model to find potential risk factors of human error to increase the reliability of prescription and promote patient safety.

METHODS

The methods of this study were questionnaires, human error analysis and work environment survey. The questionnaire was designed to collect the information of subjects. These were demography, job conduction, major error evaluation and physiology and psychology conduction. Human error analysis methods included job analysis methods and human error analysis method. The Hierarchical Task Analysis (HTA) was used to analysis the tasks in the job of prescription. And, the systematic human error reduction & prediction approach (SHERPA) was used to analysis the human errors in the pharmacy. To collected data of work environment in pharmacy, ruler and watch or Digital video had been used. Moreover, the error events summarized from the pharmacy error report system had been collected and analyzed.

Figure 1 shows flow-chart of activity and human error analysis. First step, the standard operation procedures (SOP) was reviewed to generate the tables of task. Then, HTA and SHERPA were constructed. Finally, the error junctions had been classified and identified.

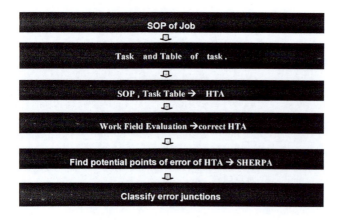

Figure 1. flow-chart of activity and human error analysis

Type of human error

The Swain and Rasmussen methods had been taken & discussed to classify the types of errors (Allan & Barker1990, Allan & Barker et al 1995).

The Human Error Classification of Swain

Based on the Swain's human error classification, four types of human activity error are omission error, commission error, sequence error and times errors. The omission error was generated by careless mistake or omitting; the commission happened by un-appropriate actions; the sequence error was generated by wrong operating steps or process; the timing error was generated by actions in un-appropriate timing.

The Human Error Classification of Rasmussen

The Rasmussen's human error classification was based on behavior types of human and separated three types: The skill-based was people make error in a skilled condition; the ruled-based was people understand the condition but take wrong action scheme or program, then generate un-appropriate action; the knowledge-based was people do not recognize condition to make wrong decision, then generate un-appropriate action.

The Error Report System of Pharmacy

The hospital already has a report system of error collecting in the pharmacy department for patient safety. The study group has got the data and analyzed to find risk factors of human error about prescription. The human reliability data about prescription preparing has been compared with the information of job analysis and HTA & SHERPA.

Data Analysis

All the data had been collected and keying , Then analyzed by the software of Microsoft office Excel and SPSS 12.0.

RESULTS

There were 46 pharmacists participated in this study at the medical center in the South of Taiwan. All of them wrote questionnaire, the return rate was 100%. Most pharmacist had been graduated from university (high school 4.4%, university 53.3%, graduated degree 42.2%), The basic information of pharmacists shows on Table 1. The average age (SD) was 35.2(7) years old; body weight (SD) was 53.4 (8.9) Kg; body height (SD) was 160.3 (8.4) cm. Ten persons (21.7%) were male and 36 (78.2%) were female.

Table 2 lists the self-report human error rate from pharmacists. All the people in the group had knew their job may slip up; 97.8% of them had heard colleague had mentioned mistake; 91.3% had even slip up during their job and 63% had ever slip up within three months.

Table 1. Basic information of pharmacists

Item		mean	SD
Age (Yr)		35.2	7
Body weight (KG)		53.4	8.9
Body height (cm)		160.3	8.4
Education	High school	4.4%	
	University	53.3%	
	graduated degree	42.2%	

Table 2 Self-report Human Error (N=46)

Type		N	(%)
I heard my colleague mentioned mistake			
	No	1	2.2
	Yes	45	97.8
I've ever slip up within three months.			
	No	17	37
	Yes	29	63
I've ever slip up during my job.			
	No	4	8.6
	Yes	42	91.3
I know the job may slip up.			
	No	0	0
	Yes	46	100

Based on HTA and SHERPA methods to analyze prescription preparing human reliability, the table of error type had been summarized and been passed to pharmacies to judge the priority of error types. The Swain's error classification method was based on the activity. The activity error types are: omission errors, commission errors, sequence errors, times errors. Figure 2 shows human error types of Swain. Among the 46 pharmacist 25 voted the omission error was the most common error in their job. The commission error was the highest error rate (25/46) and the omission error was the second high (12/46); the third was timing error (9/46) and the fourth was sequence error (2/46). The Ramussen's error classification method was based on the human behavior. The behavior levels were: skill-based, ruled-based and knowledge-based. Figure 3 shows human error types of Ramussen. The rule- based error was the highest error rate (25/46) and the skill-based error was the second high (15/46); the third was knowledge-base error (9/46).

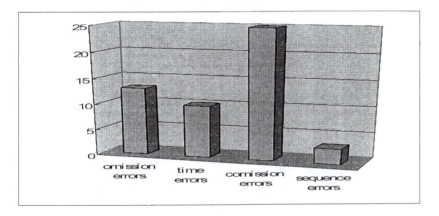

Figure 2 Human error types of Swain

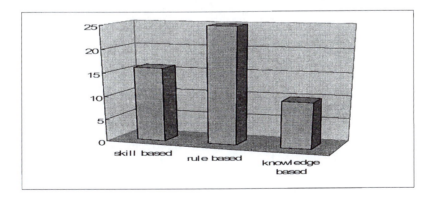

Figure 3. Human error types of Ramussen

Based on error report system of pharmacy, the human reliability of prescription preparing has been generated and shown in on Figure 4 lists the data of error distribution by season from 2004 to 2008. The heighted error rate

438

happened in Oct. 2005 and rate was 0.19%. And, the lowest was 0.057% in Feb. 2007. During June 2004 ~ May 2008, the trend of error rate was decreasing from 0.002 (0.2%) to 0.0005 (0.05%). And, the reliability of preparing prescription was 99.8% to 99.95% by month.

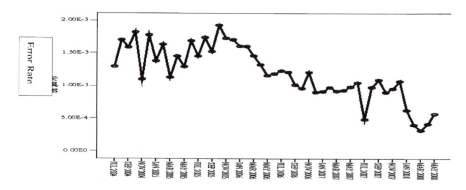

Figure 4 Error distributions by month (2004-2008)

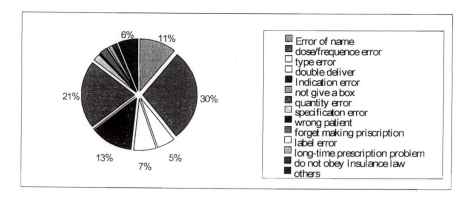

Figure 5. the types of error during prescription preparing (2005.4~2008.5)

(N=14,075,165)

Figure 5 lists the he types of error during prescription preparing from April 2005 to May 2008 (N=14,075,165). The error rate of dose/ frequency was 30% and was the height of all types of error. The second height error rate was quantity (25%). The third height error rate was indication (13%). The fourth height error rate was name of drug (11%). The error rates of double deliver, others, typing were 7%, 6% and 5% in sequence.

DISCUSSION

Based on the result of questionnaire, this was a yang work group (mean of age= 35.2 years). The type of commission error was the major type of error and the type of omission error was few. Because the pharmacist is a high education required professional job, most pharmacist had been graduated from university (high school 4.4%, university 53.3%, graduated degree 42.2%), the error had been accrued by short of knowledge was very few. The classification of Swain was based on the activity of human and the most common types of errors were commission, omission, timing and sequence errors consecutively. The Ramussen's human error classification was based on the behavior of human. And the most common types of errors were rule-based, skill- based and knowledge-based errors sequentially. Most pharmacists' considerate the commission errors and ruled-based error were the most common types of error in the prescribing process. Maybe most pharmacists have been very familiar with their jobs. However, they might rarely careless or did un-appropriate activities then made an error in their already familiar job.

Chen had done a study about prescription slips in healthcare enterprise. He mentioned that there were 1,742 prescription slips events in 2003 and the error rate was 0.24% (Chen, 2003). Regarding types of slip, the most common error types were writing error prescription (42%), un-follow forming of insurance (33%), typing error (24%) and prescribing error (1%) (Chen, 2003). Although the philosophy of error type characterization of Chen was different from this study, the error rates of dose/ frequency (30%) and name of drug ((11%) had added up to 41% which was similar to the rate of writing error prescription (42%). The error rate of Chen's study was like this study in 2004.

CONCLUSION & RECOMMENDATION

Most pharmacists recommended commission and rule were major type of errors. Pharmacists had high education level and were very familiar their jobs, but errors happened during unusual tasks or careless conditions. The cause of human error might be work stress, over loading and lost attention.

To increase human reliability of prescribing, good job rotation, task scheduling of pharmacy department in hospital, and training to reinforce basic concepts of patient safety & prescription procedure were recommended.

ACKNOWLEDGMENT

This study has been funding by the National Science Council, Taiwan (NSC 95-2221-E037-005). We specially thank the pharmacists at the Kaohsiung Medical University Hospital for lending us the experimental apparatus, and giving us many valuable suggestions.

REFERENCES

Chen, M.H. (2003); "Prescription slips in healthcare enterprise", TZU CHI Medical Journal, 15:4 247-58.

Jen, W.P. (2003), "Introduction of basic cause analysis", Communication of Taiwan Joint Commission on Hospital Accreditation, Volume 4, No. 03.

Abood R.R.(1996). " Errors in pharmacy practice. " U.S Pharmacist 21:122_132.

Allan EL, Barker KN. (1990), "Fundamental of medication error research", Am J Hosp Pharm 47(3):555-71.

Allan EL, Barker KN, et al(1995). "Dispensing errors and counseling in community practice " American Pharmacy NS 35(12):25-33.

Chapter 51

Spine Load in the Context of Automotive Seating

R. Zenk[1], M. Franz[2], H. Bubb[3]

[1]Brose Fahrzeugteile GmbH & Co.
Kommanditgesellschaft, Ketschendorfer Straße 38-
50, D-96450 Coburg, Germany

[2]BMW AG Forschungs und Innovationszentrum,
Knorrstrasse 147, D-80788 München, Germany

[3]Technische Universität München, Lehrstuhl für
Ergonomie Boltzmannstrasse 15, D-85747 Garching
bei München, Germany

ABSTRACT

In our present society for many people the automobile is an essential companion in everyday life. Be it commuting to and from work or during our leisure – every week most of us spend many hours sitting in their car. In this context the seat is the main interface between the human being and the automobile itself. Functioning well, this close relationship can foster the well-being of the passenger and raise his spirit; being flawed it can otherwise cause severe pain in the back after a longer journey. Thus, for car manufacturers, the aspects of seat comfort are becoming more and more prominent in distinguishing themselves from their competitors. Despite its importance the development of comfort parameters in automotive seating is still being consigned to the subjective judgements of a poor number of seating experts or randomly selected test subjects. Easily influenceable by parameters like affection, physical conditions or environmental factors and a lack of standardization these subjective evaluations have a big intra individual and inter individual range.

A first approach to improve the objective judgement of an automotive seat was made by the concept of "optimal" load distribution, based on the identification of a close relationship between the pressure on the seat and the discomfort felt by the person sitting. In the next step a further objectifying of seat comfort will be made with an in vivo measuring of the pressure in the spinal disc, which is a reliable indicator for the load in the spine. For this research program the pressure sensor is implanted with a canula in the middle of the disc intervertebralis of a voluntary subject. This way the local pressure on the disc is defined in an automobile, concentrating on the very seat settings which cause pain in the upper back. From these results we will be able to improve our knowledge on how to avoid uncomfortable seating in automobiles and start making well-founded recommendations on seat construction. So an intelligent seat can be created, which can adjust the seatcomfort to each individual automaticly. Especially, for long term driving those systems are very interesting and can produce comfort on high level.

Keywords: Comfort; Spine Load Measurement; Intelligent Car Seat; Automatic Seat Comfort Adjustment

INTRODUCTION

In common parlance comfort may refer to both comfort and discomfort. But speaking precisely the item 'comfort' is associated with feelings of relaxation, well-being, satisfaction, aesthetics and luxury (Bubb, 2003). While 'comfort' is connected with aspects of 'favor', the item 'discomfort' characterizes the aspects of 'suffering'. Discomfort is associated with biomechanical factors that produce feelings of pain, numbness and stiffness. These feelings increase with time on task and fatigue (Zhang et al., 1996).

In 2003 de Looze reviewed the literature to determine the relationships between objective measures and subjective ratings of comfort and discomfort: 21 studies were found in which simultaneous measures of an objective parameter and a subjective rating of comfort or discomfort were obtained. Pressure distribution appears to be the objective measure with the clearest association with the subjective ratings. Other variables were less clear and usually not statistically significant.

So an approach for objectifying the comfort of car seats was made by the concept of "optimal" load distribution (Mergl, 2006; Zenk, 2006), based on the identification of a close relationship between the pressure distribution on the seat and the discomfort felt by the person sitting on it.

Three parameters turned out to be essential: percentage of load, maximum pressure and gradient per body region. In this experiment always the same seat model was used. There were no differences in seat construction and design application. So there were no significant changes of the maximum pressure and gradient. Therefore in this study only the percentage of load of the seat pan will be treated. In figure 1 the values are shown on the Body Map (Hartung, 2006).

For the short term this load distribution of the ideal model produces the least discomfort according to an interview method. The forecast of the discomfort after long term driving is also approvable, as we learned from a three hour journey in a BMW 7-Series with several subjects (Zenk et al., 2006). The result of this analysis showed, that ideal distribution is accepted as the significantly best seat position – also for long term.

Although being validated by many various subjects for short and long term, all these results were based on interview (subjective values) and objectification on real measurement values was missing. In this study an important step for objectifying the seat comfort is made with an in vivo measuring of the pressure in the spinal disc, which is a reliable indicator for the load in the spine (Nachemson, 1966). The essential question of this paper is to find out if the ideal pressure distribution is also the best seat setting according to the intervertebral disc pressure.

In the automotive industry, drivers are interested not only in comfort but also in keeping their body in a healthy shape (Vink, 2005). There is an increasing demand for information from the end user. Therefore an in vivo measuring of the pressure in the spinal disc with a real subject in an automotive seat was done with different seat settings. The experiment was part of a larger medical study analysing different products and their impact on spinal load. This invasive medical research was the subject's own and voluntary accord. A specially initialized surgery on a human being for the sole validation of the spinal load would not have been appropriate. So it was a one-time opportunity to deduce scientific knowledge for an ideal seat design.

EXPERIMENTAL DESIGN

The operation was conducted in a clinic located in Munich renowned for disc operation in the spine.
The subject for this examination is a professional sportsman (42 years old, 176 cm, 83 kg) – which is equivalent to the 50th percentile man. This study was voluntary and additionally the whole test was approved by the ethics commission in Berlin.

Implantation of the sensors

The intervertebral discs consist of two elements: a ring of collagen fiber, the anulus fibrosus and a colloidal gel in the nucleus pulposus (cp. figure 2). The feeding of the intervertebral disc is done by the diffusion of the synovial fluid during the compression and decompression of the whole spine (Tittel, 1994).

For this research study two pressure sensors were implanted minimally invasive with a canula into the middle of the intervertebral disc (nucleus pulposus). The surgery was running the same way as an intervertebral disc operation using the endoscope technology (Wilke, 2004).

The sensors were implanted in the most delicate area of the spine; in the disc L4-L5 and L5-S1. In figure 3 the pressure sensors are shown in an X-ray picture. In the disc L4-L5 it is possible to identify the canula during implanting the sensor.

Both implanted sensors (type: K4121-00-1629-D) depicted in figure 4 are from the Mammendorfer Institut for physics and medicine and have a length of about 640 mm and a diameter of 1.45 mm. The sensitivity of the sensor is 0.15 mV/VmmHg. The accuracy of the whole measurement system (sensor, measuring amplifier, PC) is +/-0.1 bar and the working range is 0-50 bar. Both sensors were connected by two flexible cables with the measurement system – running out of the back of the subject.

The sensors were implanted in the morning and the whole study took one day. After the measuring both sensors were removed by another surgery.

Experimental Setup

The essential test equipment for this study is shown in the figure 5. The test car was a BMW 7-Series, equipped with a multifunctional leather seat. It is possible to adjust the seat in 16 degrees of freedom allowing to easily achieving the ideal load distribution. The seat was specially prepared for this experiment enabling the exact measurement of the seat position and angles. By using the software Ediabas it was ensured, that the seat settings could be easily reproduced. Every part of the seat adjustment and each seat motor rotation were digitally recorded.

The second measurement tool used for this experiment is a pressure mat system (Tekscan). The 1024 single load sensitive sensors are evenly distributed on a measuring area of 470 mm x 470 mm. The pressure mat system is only placed on the seat pan and not on the backrest, because there is a minimal electric current running in the mat itself. To avoid any interference between the two test methods, the direct contact between the pressure mat and the cable of the intradiscal sensors – coming out of the back of the subject – was prevented. In a first test cycle the implanted pressure sensors were used without pressure mat system for one seat setting. In the next step the implanted pressure sensors were used in combination with the pressure mat system and the measured values of the implanted sensors were exactly the same. So interferences between both measurement tools are excluded.

From each seat setting a picture of the subject and of the pressure distribution was taken, which is shown in figure 6.

According to the ideal pressure distribution (Mergl, 2006; Zenk, 2006) an optimal seat setting was configured (cp. figure 1). Here the real load of the seat pan of the subject in the test car is 54.0 % on the buttocks and 6.4 % on the front thighs. These corresponding pictures of seat setting and pressure distribution are highlighted in the middle of figure 6. In contrast to this "IDEAL" seat setting two other settings were produced.

On the one hand there is the setting "MAX" – here a maximal support of the frontal thighs (10.6 %) and consequently a smaller load on the buttocks (47.6 %) is given. Therefore the seat-cushion was tilted upwards and the seat-cushion depth was extended. On the other hand there is a setting "MIN" – in this case a minimal support of the frontal thighs (0.4 %) and consequently more load on the buttocks (66.5 %) is given. Therefore the seat-cushion was tilted downwards until to the stopper and the seat-cushion depth was minimized. For all the tests the backrest position was kept constant in relation to the seat pan during the test procedure to avoid influence of the backrest angle on the load in the spine. Thus the dependent variable of this trial was the load distribution on the seat pan.

The subject was also asked to give a small oral review for the discomfort feeling of each seat setting (MAX, MIN and IDEAL), without being informed which seat setting is adjusted.

RESULTS

For each seat setting the load in the intervertebral disc was recorded for 15 sec with the software ARGUS junior. The mean and maximum values for the three different seat settings are shown in figure 7.

In the first picture (cp. figure 7) there is a maximum support of the frontal thighs (indicated with the arrow – upwards). In this case the mean value is

0.95 bar and the standard deviation during 15 sec record time is 0.08. The lowest pressure values are achieved in the highlighted "IDEAL" seat position. The mean value during this measurement is 0.5 bar and the standard deviation is 0.04. In a final step a seat adjustment is created with no support of the frontal thighs by the seat (indicated with the arrow – downwards). Here the highest pressure values are achieved because of the upholding of the frontal thighs by the subject himself. The mean value for this seat position is 1.5 bar and the standard deviation is 0.05.

For each seat setting ("MAX", "IDEAL" and "MIN") the signals of the implanted sensor were recorded with a frequency of 50 Hz, so there is a result of 750 measurement values after a analysis time of 15 sec. After analyze via t-test of the measurements in detail, it can be considered, that the lowest pressure values are significant achieved in the "IDEAL" seat position. There exists also a significant difference between the mean values of the "MAX" and "MIN" seat setting.

The test person gives the feedback, that there is no discomfort for the seat setting "IDEAL" and the subject could imagine driving in this seat adjustment for a long time. According to the interview, the subject felt a small discomfort for the seat settings "MAX" and also for seat setting "MIN" after a few seconds. A more exact classification between the two different seat settings ("MAX" / "MIN") was impossible for the test person.

DISCUSSION

For the discussion it is essential to take the anatomical influences of the intradiscal spine load into consideration. Pressure variations in the intervertebral discs in the lumbar spine can be traced back to different influences. The general reason a muscle effort for all postures is required and so the pressure disc alteration is increased. Another possibility for a pressure distribution by spine movement could be the anatomical structure around the vertebral body.

Kendall (2001) and White, Panjabi (1978) described that one reason for pressure distribution in the disc is the muscle iliopsoas – especially the muscle psoas major (cp. figure 8).

The insertion place from the muscle is the lateral side of the vertebral body and the lateral side of the discs L1-L5. The muscle has its seeds at trochanter minor thigh bone. It is a fact that the form of lumbar spine is changed by the contraction of the muscle iliopsoas (Kendall, 2001). Every flexion of the spine makes a pressure alteration in the disc.

Another important issue for pressure variation in the discs is given by the ligaments (Kapandji, 1985). They have many different functions, some of which may seem to be in opposition to the others. All vertebral bodies are connected by ligaments. First they must allow adequate physiologic motion and fixed postural attitudes between vertebrae. Secondly the ligaments must protect the spinal cord by restricting the motions. This is also a reason for pressure alteration in the disc; the ligaments are much like rubber bands.

White and Panjabi (1978) described that the ligamentum flavum have pre-tension when the spine is in neutral position. The "resting" tension in the ligaments – especially the ligamentum flavum – produces "resting" compression of the disc and so stability is given to the backbone. If the spine – in particular the lumbar spine – is leaving the neutral position for example by moving the pelvis, the pressure in the disc is growing up.

All vertebral bodies are jointed by facet. White and Panjabi (1978) showed that the area L4-L5 and L5-S1 bear the highest loads and tend to undergo the most motion. Helander (2003) described pressure distribution in the discs by spine movement and Wilke (2004) measured pressure distribution in the intervertebral discs in different body postures. So it is assumed that the pressure in the discs varies in different sitting positions in a car seat. The reason for an increase in pressure is a combination of bending and stabilisation of the spine by muscle contraction and ligaments. If the seat position is not optimal more muscular activity is needed for stabilisation resulting in a distinctly larger pressure in the disc. In this case the pressure increased from 0.5 bar ("IDEAL" – seat position) to 1.5 bar ("MIN" – seat position). In the first position the muscle iliopsoas is relaxed, during in the second posture it has to do static muscle work to keep the thighs in a comfortable position. In a preliminary study the muscle tension of the musculus erector spinae was measured with an EMG in the aforementioned three different seat positions. Here the muscle tensions of the differing seat settings from the 'ideal' seat position were significant higher. In a well supported seat position with an "optimal" load distribution as described above, the muscle activity is low and therefore the discs bear fewer loads.

Although the surgery and experiment was only conducted for one person the results are significant, because they correspond very well with the studies of Wilke in 2004. The studies were performed at the same clinic and the subject took place on an office chair in different seat positions and on a pezzi ball. The range of the pressure in different postures in the intervertebral disc L4-L5 varied from 1.0 bar to 9.0 bar during sitting. Of course it is difficult to make a sweeping conclusion from that data (50th percentile man) to a 5th percentile female or a 95th percentile male. But in this experiment there is a significant link between the subjective and objective determination of seating comfort detected.

Reaching only 0.5 bar the "IDEAL" load distribution in this study produces the smallest load on the disc. This seat setting is based on the ideal pressure distribution: The load of the seat pan should be 50-65 % on the buttocks and 6 % on the front thighs. These values were achieved in a former study (Mergl, 2006; Zenk, 2006) with different subjects of different body proportion and body height.

The highest pressure values in this experiment are achieved with the seat setting "MIN"; the mean value is 1.5 bar and the maximum value 1.6 bar, because of the upholding of the frontal thighs – here there is no support from the seat.

CONCLUSION

Medical evidence was provided that pressure in the intervertebral discs in the lumbar spine can be varied by changing the pressure gradient of the seat pan.

The data on this one subject might suggest that there is a relation between discomfort and load distribution on the seat and posture. The results of the in vivo measuring of the intervertebral disc pressure of one person correlate with the pressure distribution: Ideal pressure distribution means lowest load on the discs.

For the first time the results of the ideal pressure distribution (Mergl, 2006; Zenk, 2006) – resulting from subjective interviews – were objectively validated. So the comfort in an automotive seat becomes measurable and assessable.

The resulting consequences for seating in automotive: A maximum of comfort (also for long term) is provided in the ideal seat position by an optimal load distribution. It is possible to adjust comfort, which is furthermore validated for long term.

REFERENCES

Bubb, H.; 2003. Quality of Work and Products in Enterprise of the Future, Product Ergonomics, pp.3-6, ergonomia Verlag oHG, Stuttgart

De Looze, M.P.; Kuijt-Evers, L.F.M.; van Dieen, J.; 2003. Sitting comfort and discomfort and the relationships with objective measures, Ergonomics, Vol. 46, No. 10, pp. 985-997

Hartung, J.; 2006. Objektivierung des statischen Sitzkomforts auf Fahrzeugsitzen durch die Kontaktkräfte zwischen Mensch und Sitz,

Dissertation am Lehrstuhl für Ergonomie, Technische Universität München

Helander, M.G.; 2003. Forget about ergonomics in chair design? Focus on the aesthetics and comfort!, Ergonomics, Vol. 46, Nos. 13/14,   pp. 1306-1319

Kapandji, I.A.; 1985. Funktionelle Anatomie der Gelenke, Band 3 – Rumpf und Wirbelsäule, Ferdinand Enke Verlag, Stuttgart

Kendall, F.P.; McCreary, E.K.; Provance, P.G.; 2001. Muskeln, Funktionen und Tests, 4. Auflage, Urban & Fischer Verlag, München Jena

Mergl, C.; 2006. Entwicklung eines Verfahrens zur Objektivierung des Sitzkomforts auf Automobilsitzen, Herbert Utz Verlag, München

Nachemson, A.; 1970. The load on lumbar disks in different positions of the body, Clin. Orthop. 45: 107-22

Tittel, K.; 1994. Beschreibende und funktionelle Anatomie des Menschen, 12. Auflage, Urban & Fischer, Jena Stuttgart

White, A.A.; Panjabi, M.; 1978. Clinical biomechanics of the spine, J.P. Lippincott Company, Philadelphia Toronto

Vink, P., (ed) 2005. Comfort and design: principles and good practice. Boca Raton: CRC Press.

Wilke, H.J.; 2004. Möglichkeiten zur Bestimmung der Wirbelsäulenbelastung und Konsequenzen für die Empfehlungen für das Sitzen, ergo mechanics, interdisziplinärer Kongress Wirbelsäulenforschung, Shaker Verlag, Aachen

Zenk, R.; Mergl, C.; Hartung, J.; Sabbah, O.; Bubb, H.; 2006. Objectifying the Comfort of Car Seats, SAE Conference 2006, SAE no 2006-01-1299

Zhang, L.; Helander, M.G.; Drury C.G.; 1996. Identifying Factors of Comfort and Discomfort in Sitting, Human Factors, Vol. 38, No. 3, pp. 377-389

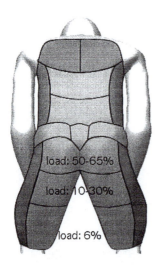

Figure 1: Ideal load distribution plotted in the Body Map (Hartung, 2006)

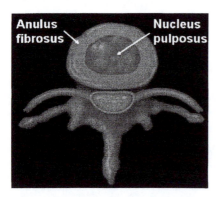

Figure 2: Transversal view of the vertebral body

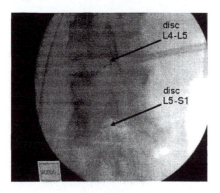

Figure 3: Dorsal view of lower lumber spine (X-ray picture)

Figure 4: Schematic illustration of the sensor

Figure 5: Equipped test car

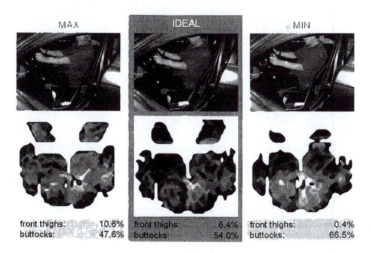

MAX	IDEAL	MIN
front thighs: 10.6%	front thighs: 6.4%	front thighs: 0.4%
buttocks: 47.6%	buttocks: 54.0%	buttocks: 66.5%

Figure 6: Different seat settings with corresponding load distribution

452

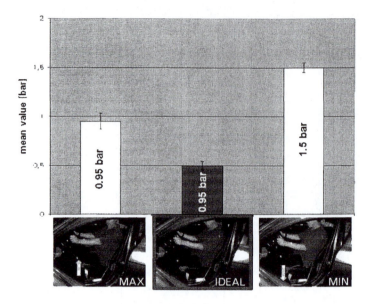

Figure 7: Pressure in the disc according to the different seat settings

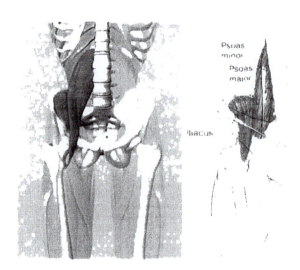

Figure 8: Musculus iliopsoas (Psoas minor, Psoas major, Illiacus) – Kendall (2001)

CHAPTER 52

Relevant Analysis on Rollover and Physical Properties of Mattresses

Manami Nozaki[1], Kageyu Noro[2], Takayuki Sasaki[3]

[1]Toho University
Tokyo, Japan

[2]Wasea University / ErgoSeating Co., Ltd.
Tokyo, Japan

[3]Asahi Glass Co., Ltd.
Kamisu, Ibaraki, Japan

ABSTRACT

A rollover action, which is defined as a body movement accompanied by entire body turns taking place during sleep, is an important body movement viewed from prevention of disuse syndrome caused by continued bed rest. Some bedding materials may hinder patients' rollover actions. This research is aimed at clarifying relationships between rollovers and characteristics of mattresses to obtain suggestive ideas for provision to patients of bedding conducive to easier rollovers. Three types of urethane mattresses with mutually different physical properties were used as samples. EMGs were taken from rectus abdominis during rollovers of a specified pattern. Based on analysis of the EMG records, peak EMG signal values and time lengths of rollover to lateral position were identified, and used for comparing sample mattresses. It was learned as a result that a mattress with its physical properties of density, hardness and impact resilience holding the middle course among three samples was the best in terms of ease of rollover. A subjective survey also was performed on ease of rollover and sleeping comfort among others with subjects sleeping in the sample mattresses for 5 nights each in their own home. Its result showed a mattress with physical properties holding the middle course gained higher marks in terms of ease of rollover compared with another mattress that was felt as soft and sinking down as the first one. It was suggested as a conclusion that ease of rollover is provided by a mattress with the physical properties of density, hardness, and impact resilience holding the middle course, while also it has some connection with subjective feelings of resilience and possibly restoring ability as well.

Keywords: Rollover, Mattress, Physical property, Tactile impression

INTRODUCTION

There are two types of body movements while in bed: one is a movement taking place mainly at the boundary between rapid eye movement (REM) sleep and Non-REM sleep, a movement happening in most cases without your realizing it; the other a movement most likely seen before falling asleep or after wake-up, a movement typically characterized by leg movements. These body movements are called rollovers when they are accompanied by entire body turns, most of which are conducted consciously. Since continued bed rest in treatment of diseases causes a decline in physical and mental functions, leading to a higher risk of disuse syndrome, rollovers in bed are vitally important for maintenance of health. The reason for patients' inability of rollover stems from motor deterioration of their whole body along with some features of bedding that work to hinder their rollovers.

Provision to patients of bedding that facilitates rollovers will promote their rollover actions to ensure their improved physical functions and self-reliant way of life. There are also reports that the ease of rollovers contributes to improved quality of sleep and sleep comfort. With good sleep prone to be disturbed in today's society, it will be of great importance for not only those patients but also ordinary people to be provided with bedding conducive to easier rollovers. As Buckle and Fernandes (1998) point out, however, little or no literature surprisingly exists in the field of ergonomic study of mattresses exemplified by that of human-mattress interface.

Against the backdrop mentioned above, this paper is aimed at clarifying relationships between ease of rollovers and characteristics of mattresses to obtain suggestive ideas for provision to patients of bedding conducive to easier rollovers.

METHODS

Sleeping in bed is quite a personal action continuing for many hours conducted in an unobservable environment. With the aim of pursuing relationships between rollovers taking place during sleep and properties of mattresses, we developed the following two types of experiments: Experiment-1 was designed to have subjects make rollovers in a specified manner under a controlled experimental condition, with their muscular stress measured; and Experiment-2 aimed at subjective evaluations performed in a poorly-controlled environment where subjects were requested to give their subjective ratings on sample mattresses in terms of sleep comfort and other items after using them for a relatively long period of time at their own home. A participatory ergonomics approach was taken for administration of the research project as a whole (Noro et al, 2009).

Experiment 1

This experiment was designed to have subjects rollover on sample mattresses in a specified way under a controlled experimental environment in a bid to make a measurement of muscle stress in that condition.

Sample mattresses. Three types of mattresses were selected, which have mutually different physical properties of density, hardness, impact resilience and hysteresis loss, in accordance with the aforementioned research objective. Table 1 summarizes those physical properties.

Table 1. Physical properties of samples

item	unit	Sample A	Sample B	Sample C
Density	kg/m³	72.1	87.1	27.4
Hardness(ILD)	N/314cm²	80	55	160
Impact resilience	%	9	2	38
Hysteresis loss	%	27.4	78.9	43.7

Size Sample A 940×1900×100mm. B 970×1950×70mm, C 970×1980×110mm

Sample A holds the middle course in terms of three properties concerned of density, hardness, and impact resilience, while marking the lowest value for hysteresis loss. Sample B, a visco-elastic product, has the highest density, though being lowest among three in terms of both hardness and impact resilience, still posing the largest hysteresis loss. As for Sample C, density is lowest, while both hardness and impact resilience are highest.

Subjects. Following directions from the ethics committee in our organization for a human-body-related experiment, only two people were allowed as healthy subjects expressing their consent to be engaged in the envisaged experiment. The main attributes of the subjects are given in Table 2.

Table 2. Physical attributes of subjects

Subjects	Gender	Height(cm)	Weight(kg)	BMI (kg/m2)
A	Female	159	59	23.3
B	Male	173	66	22

Procedure. EMGs were measured from relevant muscles during rounds of rollovers conducted in a specified way.

Richter (1989) classified rollover actions of healthy adults into 11 patterns depending on combined ways of movements of upper limbs, lower limbs, and body trunk. Nozaki (2004) roughly divided rollovers into 4 patterns based on the timing of pelvis rotation: i.e., upper limb type, lower limb type, raised knee type, and compound type, showing the existence of kinematically different characteristics between these 4 patterns.

The type of rollover employed in our experiment was the upper limb type defined by Nozaki. This is because the upper limb type is characterized by an upper limb taking a lead off the hip, during which rectus abdominises function as major muscles facilitating measurement of EMGs reflecting muscle stress needed for rollovers. The instrument used was an electromyograph, PowerLab / 8SP, Model ML785, by ADInstruments, with a measuring range of 0-2 mV equipped with high- and low-pass filters of 500 Hz and 10 Hz, respectively. Right and left rectus abdominises were chosen for the EMG measurement. Analysis on the recorded waveforms gave us numerical values of peak EMG signals and time lengths for rollover to lateral position. Ten rounds of rollovers were performed with their average values serving as the base data for our analysis.

Experiment 2

Experiment 2 was performed with the aims of obtaining data for subjective evaluation on ease of rollover. The experiment was subjective surveys on tactile impressions of mattresses after being used for a relatively long period of time in a certain uncontrolled living environment.

Samples. Three sample mattresses and a pillow used were the same as those for Experiment 1.

Subjects. Six people (three males, three females) participated as subjects, whose physical attributes are listed in Table 3.

Table 3. Physical attributes of subjects

Subjects	Gender	Height(cm)	Weight(kg)	BMI (kg/m2)
A	Male	172	72	24.3
B	Female	160	53	20.7
C	Female	159	59	23.3
D	Female	162	52	20
E	Male	173	66	22
F	Male	182	85	25.7

Procedure. We had the subjects use 3 sample mattresses to sleep in their own room for 5 nights in a row for each sample, with their subjective evaluations made just after wake up every morning during the survey. A specially designed questionnaire form was used comprising such items as: room and bedding condition, easiness into falling asleep, sleeping hours, whether to have awaken up midway through sleep, with or without memory of rollovers during sleep, tactile impressions of mattresses. Questions about tactile impressions included paired items of: hard — soft, restoring immediately — bearing traces, resilient — non-resilient, sinking down — not sinking down, wrapped around — not wrapped around. Answers were marked on a 7-point interval scale.

RESULTS
Results of Experiment 1

Figure 1 gives rollover time lengths and peak EMG signals.
1) Rollover time lengths: Regarding time length needed for one round of rollover, Sample B took significantly longer time than Samples A and C (p < 0.01), meaning Sample B needs more time to rollover.

2) Peak EMG signals: Peak values indicated during rollover for Sample A was significantly smaller than for Sample B (p < 0.01), which means smaller muscle strength is needed for rollover on Sample A.

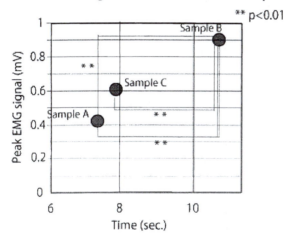

Figure 1. Peak EMG signals and rollover time to lateral position by mattress type

Results of Experiment 2

Tactile impressions. Tactile impressions received from sample mattresses are demonstrated in Figure 2. A tendency is seen from an overview of the figure that Sample C gets higher scores for items related to hardness while Sample B gaining higher scores for items related to softness. Sample A, on the other hand, is seen receiving relatively high scores for all the items in question.

To be specific, concerning hardness, Samples C gave significantly harder impressions than Samples A and B (p < 0.01). Samples A and C presented significantly stronger impressions in terms of ability of restoring immediately (p < 0.05). Regarding resilience, Sample A gained significantly higher marks than Sample B (p < 0.05). Sample B was given significantly higher marks in the feelings of 'sinking down' than Sample C (p < 0.05), while Sample C was given significantly lower marks in the feelings of 'wrapped around' than Sample A (p < 0.05) as well as than Sample B (p < 0.01).

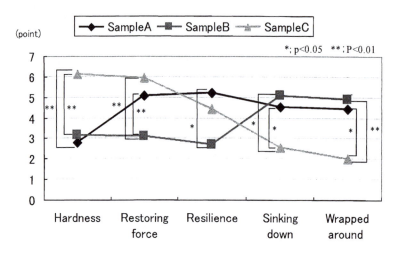

Figure 2. Tactile impressions of mattresses

Ease of rollover. Table 4 gives the number of days by subject when ease of rollover was felt during the 5-day survey period. With respect to ease of rollover, Sample A and Sample B received answers from two subjects saying they felt ease of rollover every night during the 5-day survey, while also ease of rollover being felt for two to four nights by one subject. None of subjects felt ease of rollover from Sample B.

Table 4. Number of days when ease of rollover was felt

Subjects	Sample		
	A	B	C
A	2	0	4
B	0	0	0
C	5	0	5
D	0	0	0
E	5	0	5
F	0	0	0

DISCUSSION

Figure 3 demonstrates by sample the feelings of hardness versus feelings of 'sinking down' obtained from the subjective survey. Samples A and B pose a similar tendency of 'soft' and 'easily sinking down', while Sample C showing features of hardness and 'not sinking down'. Sample A proved to assure a rollover with the least muscle stress and got high scores in terms of ease of rollover in the subjective survey. However, none of subjects felt ease of rollover from Sample B, which received tactile feelings of the same degree of softness and senses of 'sinking down' as Sample A.

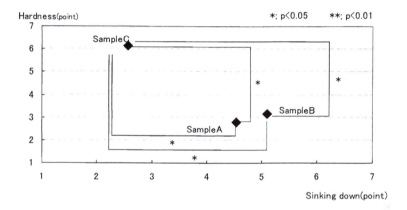

Figure 3. Feelings of hardness vs sinking down

This implies that, in addition to hardness and 'sinking down', there may be some other features closely related to ease of rollover on a mattress. Actually, Sample A got significantly higher marks in resilience and restoring ability than Sample B. Figure 4 indicates by sample the scores from the subjective survey in terms of resilience and restoring ability. Given this, it was presumed that ease of rollover would be associated with feelings of resilience and possibly restoring ability as well.

460

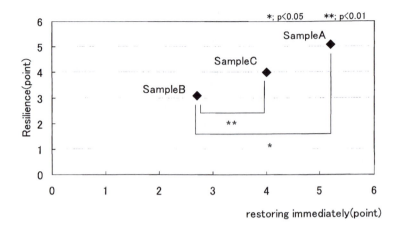

Figure 4. Feelings of resilience vs restoring ability

Implications toward future research

As sleeping is part of our lives, evaluation of mattresses should be conducted in our daily lives. We would like hereafter not only to continue kinematic studies on rollover actions under controlled conditions, but also to make a further study on a mattress in combination with the floor to put it on, covering cloth, and a pillow in the actual sleeping environment.

CONCLUSION

It has been implied that Sample A is useful as a mattress providing ease of rollover with its physical properties holding the middle course among three sample mattresses. We also have gained suggestions that mattress' properties of resilience and restoring ability could serve as effective indices to evaluate a mattress as a bedding material.

REFERENCES

Buckle, P. and Fernandes, A. (1998). Mattress evaluation—assessment of contact pressure, comfort and discomfort, *Applied Ergonomics,* 29(1), 35-39.

Noro, K., Sasaki, T. and Kaku, D. (2009). Mattress Development through a Participatory Ergonomics Approach, *HFES 2009 San Antonio.*

Nozaki, M., Fujimaki, G., Ebara, Y., and Noro, K. (2004). The Kinematic Characteristics of Rolling Motion at Awakening, *JAPAN Ergonomics Society 2004, Saitama.* (in Japanese).

Richter, R.R., VanSant, A.F., and Newton, R.A. (1989). Description of Adult Movements and Hypothesis of Developmental Sequences, Physical Therapy, 69(1), 66.

Chapter 53

Information Management and Decision Processes in Emergency Departments

Gabriella Pravettoni[1], Claudio Lucchiari[1],
Gianluca Vago[2], Robert L. Wears[3]

[1]Social and Political Studies Department
University of Milan
Milano, Italia

[2]Clinical Sciences Department
University of Milan
Milano, Italia

[3]Department of Emergency Medicine
University of Florida
Jacksonville, FL 32209, USA

ABSTRACT

Information management, situation awareness and shared situation awareness are considered key points to success in complex environments. The present study addresses this issue by reporting and discussing data collected in an empirical study conducted in two emergency departments.

Our data show that to keep situation awareness, physicians and healthcare personnel need to integrate pieces of information may receive from different sources, at different times, and through different modalities. It seems that tools for supporting situation awareness, and decision-making are needed to enable the emergency staff to complete their work efficiently and to improve patient safety . However, computer mediated communication must be integrated with face to face communication, and ergonomics issues have to be addressed in organizing the clinical workspace. In fact, clinical workers' mental models and situation awareness arise in physical spaces, which features may support or limit cognitive and communication processes.

Keywords: Medical decision-making, shared situation awareness, information management, emergency medicine

INTRODUCTION

Emergency departments are complex environments in which decisions are not taken in isolation but within a context that is dynamically, sometimes abruptly, changing. In this setting, information management (identification, collection and integration of data) is instrumental to success.

The predominant individualistic and mentalistic perspectives take the individuals mental capacity as the starting point. Failure to retrieve Information from long-term memory, or the fact that people misunderstand the meaning of representations and the limited capacity of working memory are all taken to be the innermost explanation of human error. In this framework one of the most important explaining mechanism used is Situation Awareness (SA).

This study's aim was to describe in detail the dynamic nature of decision-making in emergency medicine and the crucial role of situation awareness. Developing and maintaining situation awareness and shared situation awareness (SSA) must be seen as a key factor in those contexts where it is necessary to make crucial decisions in a very short space of time. Decision-making, in this case, takes place in dynamic situations where time and resource variables have to be taken into consideration so as to evaluate the success of the decision that has been made. It is not, in fact, sufficient to use rational choice processes because of the speed with which actions must be taken, and the dynamically changing nature of the problem. It is, therefore, necessary for there to be a constant mental updating of the mental representations of what is taking place. This constitutes a pre-requisite for the maintaining of SA. In fact, Endsley (1995) defines SA as the perception of the elements present in a certain context as a mass of data whose significance must be understood in terms of its temporal evolution so as to be able to make dynamically organized decisions.

Critical decisions in the spheres of the military, of aviation and of emergency departments are clear examples of dynamic decision-making in which SA is vital for the successful functioning of the organization in question. In fact, an understanding of the decision-making and organizational dynamics which allow the maintaining of SA is, in these contexts, fundamental to the planning and the setting up of systems, from both organizational and technological points of view, particularly systems able to support the cognitive activities that enable the decision-maker to function efficiently and effectively. Studies of this kind have dealt predominantly with military situations (Jentsch et al., 1999; Wickens, 2000) but less commonly with medical situations (e.g. Blandford & Wong, 2004).

From her first studies, Endsley has proposed a three-level SA model: perception, comprehension and projection. Perception refers to the process of data detection, which is obviously a pre-condition for any awareness (by definition, we cannot be aware of that which we do not perceive). This first level involves not only a perception of the sensorial aspects of the elements but also a perception of their dynamic aspects, their attributes and their status within a complex environment. The second level is that of comprehension, which implies the interpretation and integration of the information perceived (from which, therefore, we develop an initial form of awareness) so as to be able to explain what is happening in a certain situation, characterized by a space/time perspective. Finally, level 3 is the process of projection into the immediate future of that which has been integrated at level 2, which enables a sort of forecast of how the situation is evolving. This third point is important because it underlines how SA is a dynamic, continuously changing process, how the dynamic decision-making process cannot ignore the fact that the situation is continuously changing and that decision X will have consequences on the evolution of the situation, consequences that must be taken into consideration before making any decision. The entire decisional flow is influenced by the temporal succession of decisions, which cannot be considered as single decisions but as a chain of decisions able to influence each other (Brehmer, 1987). Not taking this aspect into consideration thus means losing much of the SA necessary to be able to make effective decisions in dynamic contexts.

Jentsch and colleagues (1999) have shown how a loss of SA can have dangerous consequences in critical contexts like aviation. This is even more evident in a military context, where the decisions of a pilot during air combat, for example, are based on a representation of the situation which the pilot constantly updates so as to assess the development of the combat. However, the role of SA is also very evident in other contexts, such as competitive sports, where it is necessary to make very rapid decisions (though this could also be applied to other less time-critical situations), or emergency medicine. In a hospital accident and emergency department, activities normally follow a certain degree of routine, which makes it possible to generally keep the situation under control. There are, in fact, protocols and hierarchies that regulate the flow of activities and decisions. For example, the seriousness of a patient's condition is evaluated by a nurse whose job is to assess, in accordance with agreed criteria, the extent of the injury or illness and thus establish the urgency of the need for a doctor. This constitutes the first phase of decision-making, carried out in *triage*, which will in some way influence the entire decisional flow within the accident and emergency department. Nevertheless, this routine is often interrupted by urgent cases (the so-called 'Code Red' cases) that require a total reorganization of work. Medical staffs have to dedicate themselves to the person whose life is in danger and decisions (concerning exams, consultations, emergency operations and so on) have to be taken quickly. The situation changes from routine to emergency; it is no longer simply a question of keeping the situation under control. In these cases, maintaining SA is critical. The doctor receives information from a variety of sources. The information has to be integrated coherently and interpreted in order to understand the evolution of the patient's health and the action that will need to be taken. However, it is our opinion that the study of factors which allow the maintaining of SA in a hospital accident and emergency department is not only important in understanding the dynamics in cases of emergency. In fact, there are many situations in which the doctor is called upon to make critical choices, even in the absence of clear signs of urgency or of life being danger. Many patients go to accident and emergency departments in what could be defined as 'ambiguous' conditions. Symptoms and signs may be interpreted differently and may require particular medical examinations or the consultation of a specialist. These sources may then provide new information that the doctor has to integrate with previously acquired information. Then it is necessary to decide whether to admit the patient to hospital or to discharge them with a course of treatment to follow, or whether simply to reassure them and arrange follow-up or future exams. All of this process is included in the organization of the accident and emergency department and of the whole hospital, and there are times, technological and organizational constraints that influence the decision of the doctor. The doctor must, then, constantly monitor the situation of each and every patient, and must have a cognitive representation on which to base his or her decisions. Consequently, the doctor must be able to maintain sufficient SA for each patient. If one thinks that, even in routine situations, a doctor has to follow a large number of patients (and therefore receive a huge amount of variably accurate information coming from a series of different sources); it is easy to see how maintaining SA is a crucial and far from simple task. Such a situation can be considered as a critical case of information and staff management (Klein, 1998). How is it possible to support the maintaining of SA? Various studies have attempted to answer this question by means of technological experimentation. In the field of military aviation, for example, different kinds of technology have been tested to try and improve the pilot's visualization of the environment surrounding the aircraft (Williams, 2000). Similarly, in medicine, in particular in the field of anesthesiology (Zhang et al., 2002), there has been experimentation with the use of advanced monitors and representation forms. The use of technology can certainly contribute in the management of SA, but cannot guarantee it. In fact, the cognitive and communicative aspects of the decision-making process are equally important. It is by optimizing the interaction of these aspects and formats of information (including technological aspects) that it is possible to facilitate success in maintaining SA and a positive outcome of the decision-making process. Furthermore, in most situations involving dynamic decision making and control of dynamic systems the task is conducted by a team of people. Thus, predominant models of Situation Awareness are inadequate for the study of systems operated by teams. Situation Awareness has to be approached by a different perspective to use in team contexts. In fact, much of the cognitive content and co-ordinate negotiations are not bound to a single person but rather distributed between individuals. The distributed cognition approach (Hutchins, 1990, 1995, 1996) implies that we shift our focus from the individual actor to how information is represented and how the representations are transformed and propagated through the system. People as well as the artifacts which they use are regarded as constituting a cognitive system where cognition is the product constructed as a consequence of the coordinated work practices between the units of the system. In particular, SSA may be defined as the active construction of a model of a situation partly shared and partly distributed between two or more agents, from which one can anticipate important future states in the near future (Artman & Waern, 1999).

METHODS

We performed systematic non-participant observations during several work sessions in the emergency departments of two Italian hospitals. We also collected data through semi-structured interviews of physicians and nurses after the observations. Observation were carried out during turns of 4 hours each. All observations and interviews were audio recorded and hand-written notes were taken to describe non-verbal activities. All audio data was transcribed for analysis. Data was systematically analysed qualitatively. To understand situation awareness, all transcript extracts that related (positively or negatively) to SA were extracted and further analysed also using interview content.

RESULTS

On the basis of the analysis, we can say that the shared SA of doctors working in the four emergency departments was partial and, therefore, not optimal. Our observations showed that information management in emergency departments is a complex task for healthcare personnel. The information flow is nominally organized by the physician in charge, who has control of the overall situation and thus must develope a situation awareness. However, many factors interfere with this process, contributing to a partial or misleading representation of the situation. In particular, we have analyzed the role of technological media (electronic data base and computer mediated communication) in giving rise to and sustaining situation awareness. Shared experience seem to be the more important factor (able to influence the whole situation).The greater the team experience, the greater seems the ability to maintain an adequate level of shared SA. The more experienced teams seem not only to be able to give a more accurate description of the series of decisions they make, defining the factors which guided them, but also seem more aware of the need to maintain a global representation of the situation, precisely that which we refer to as SA

Our data showed that in one ED, electronic media were systematically used, and situation awareness was easily and rapidly constructed by the physician in charge. Conversely, in the second ED computer assisted data managment was only marginal, with reliance on human memory instead of recording devices. The use of poorly structured data sources, for example, may give rise to information biases, leading to decisions based on partial knowledge about the patient's situation and thus contribute to a sub-optimal decision, process, or outcome. Furthermore, physicians showed a low level of awareness of the decisional flow, since the explanations collected seem to rationalize more than describe the actual decision process. However, the systematic use of electronic media interferred with intra-staff communcation, limiting information sharing. Furthermore, the particular organization of the workspace was sub-optimal, since physicians could not use the electronic database without losing sight of patients and the clinical environment, thus limiting shared situation awareness.

CONCLUSIONS

Taken as a whole, this body of data seems to indicate that information management is a key point to address in the analysis of the emergency department setting. It seems that tools for supporting situation awareness, and decision-making are needed to enable the emergency staff to complete their work efficiently and to improve patient safety. However, computer mediated communication must be integrated with face to face communication, and ergonomics issues have to be addressed in organizing the clinical workspace. In fact, clinical workers' mental models and situation awareness arise in physical spaces, which features may support or limit cognitive and communication processes. Our data may be partly related to the Italian hospitals involved in the study; however, it is our opinion that our conclusions are more generally applicable, if one considers previous research in other environments (e.g. Henderson & Mason, 1999). Future researches will have to address these issues with more quantitative measures so to find and test decision aids and supports to be applied.

REFERENCES

Artman H. & Waern Y. (1999). Distributed Cognition in an Emergency Co-ordination Center. *Cognition, Technology & Work*, 1, 4, 237-246.

Blandford, A. & Wong, W.B.L. (2004). Situation awareness in emergency medical dispatch. *International Journal of Human-Computer Studies*, 61, 421–452.

Brehmer, B. (1987). Development of mental models for decision in technological systems. In J. Rasmussen, K. Duncan & J. Leplat (Eds.), *New Technology and Human Error* (pp. 111 - 120). Chichester, UK: John Wiley & Sons.

Endsley, M.R. (1995). Toward a theory of situation awareness in dynamic systems. *Human Factors*, 37, 32–64.

Endsley, M.R. & Smolensky, M.W. (1998). Situation awareness in air traffic control: the picture. In M.W. Smolensky & E.S. Stein, (Eds.), *Human Factors in Air Traffic Control* (pp. 115–150). San Diego: Academic Press.

Henderson, S.G. & Mason, A.J. (1999). Estimating ambulance requirements in Auckland, New Zealand. In P.A. Farrington, H.B. Nembhard, G.W. Evans and D.T. Sturrock (Eds.), *Proceeding of the 31th Winter Simulation*

Conference on Simulation: a bridge to the future (pp. 1670-1674). New York: ACM press.

Jentsch, F., Barnett, J., Bowers, C.A. & Salas, E. (1999). Who is flying this plane anyway? What mishaps tell us about crew member role assignment and air crew situation awareness? *Human Factors*, 41, 1–14.

Klein, G.A. (1998). *Sources of Power: How People Make Decisions*. Cambridge: The MIT Press.

Wickens, C.D. (2000). The trade-off of design for routine and unexpected performance: implications of situation awareness. In M.R Ensley, and D.J. Garland (Eds.), *Situation Awareness Analysis and Measurement*. Mahwah, NJ: Lawrence Erlbaum Associates, Inc. Publishers.

Williams, K.W. (2000). Impact of aviation highway-in-the-sky displays on pilot situation awareness. Washington, DC: U.S. Department of Transportation.

Zhang, Y., Drews, F.A., Westenskow, D.R., Foresti, S. & Agutter (2002). Effects of Integrated Graphical Displays on Situation Awareness in Anaesthesiology. *Cognition, Technology and Work*, 4, 91–100.

CHAPTER 54

A Human Factors Approach to Evaluating Intravenous Morphine Administration in a Pediatric Surgical Unit

Avi Parush[1], Catherine Campbell[1], Jacqueline Ellis[2], Regis Vaillancourt[2], Jean Lockett[2], Daniel Lebreux[2], Elaine Wong[2], Elena Pascuet[2]

[1]Carleton University, Ottawa, ON, Canada

[2]Children's Hospital of Eastern Ontario (CHEO)

ABSTRACT

Medication errors in pediatric patients are well recognized and there is evidence that potentially harmful medication errors may be three times more common in the pediatric population than in adults. However, it is not clear where errors are most likely to occur in the sequence of prescription, dispensing, administration, and monitoring, and what factors contribute to their occurrence. The goal of this project was to understand the context and tasks associated with the administration of IV bolus morphine to pediatric patients. A human factors approach was adopted whereby the relationships between individuals, technology, artifacts, and the physical environment were studied. A total of 51 observations of morphine administrations were conducted in a post-surgical unit, all events and communications were captured, and then analyzed to identify a total of 75 influencing factors. Four major categories were identified: 1. Environmental, including interruptions, unexpected events, noise, work patterns, etc; 2. Tools, devices, and resources, including issues with information display, availability of critical information, equipment design, etc.; 3. Operating characteristics, including fatigue, experience and risk-taking behaviors; and 4. Organizational and social factors, including communication, clarity of responsibilities, distribution of workload, etc. The findings suggest that the error-likely points in the process are more frequently related to environmental and/or physical factors of the post-surgical care context.

Keywords: human factors, medication process, medication errors, influencing factors

INTRODUCTION

Children pose a unique set of risks when trying to ensure the safe preparation and administration of medicines. In fact, the rate of medication errors resulting in harm or death is significantly greater in pediatric patients (31%) than it is in adults (13%) (Stucky et al., 2003). Due to the wide variation in body mass among children, drug doses are typically calculated individually based on the patient's age, weight, body surface area, clinical condition, and specific pharmacokinetic parameters (Conroy et al., 2007). In addition, dosage formulations are often extemporaneously compounded to meet the need for small doses in pediatric patients (Ghaleb et al., 2006). These factors contribute to the potential for medication errors at each stage of the medicines management process (Conroy et al., 2007; Engum & Breckler, 2008; Ghaleb et al., 2006).

Morphine is a 'high alert' drug that is associated with drug errors, including tenfold errors that have the potential, if undetected, to cause harm or even death to the patient. Morphine is the preferred drug for treating moderate to severe postoperative pain in children and is one of most common drugs implicated in drug errors (Cousins et al., 2005; Dibbi et al., 2006; Wong et al., 2004). An observational study of randomly sampled morphine infusions over a 7 month period in the Critical Care Unit of a large hospital, found that 65% of the 232 infusions were more than a 2 fold error (Marshman et al., 2006). In pediatric literature, the prevalence of morphine adverse events ranges from 1% to 43% (Esmail et al., 1999). Finally, a recent study found that morphine was linked to the most incidents, with 8.8% of the 305 reported incidents with an outcome of harm (ISMP-Canada, 2009).

The intravenous (IV) administration of morphine is a complex process that has many components and requires preparation by both the pharmacist and the nurse prior to administration to the patient (Cousins et al., 2005). Errors occur across the entire spectrum of prescribing, dispensing, and administering of morphine (Dibbi et al., 2006; Wong et al. 2004; Parshuram et al., 2008). Although protocols are in place to prevent errors, there has been no consistent decrease in the number of medication errors. The arising question is: what are these influencing factors that play a role in the occurrence of medication errors and adverse events? There is a critical need to uncover factors that influence the process and utilize this knowledge to mitigate and avoid 'high alert' medication errors.

STUDY METHOD

PARTICIPANTS

Observations of the IV bolus morphine and HYDROmorphone preparation and administration process involved direct observation of 67% of the post-surgical unit's staff. The range of experience of participants varied from a few months, including nurses having recently completed their final practicum, to 28 years working on a post-surgical unit. Not all staff observed during the first phase of the study participated in the interviews and some staff that participated in the interviews had not been observed preparing or administering IV bolus morphine medication. However, all staff that participated in the interviews had previously been involved in IV bolus morphine or HYDROmorphone preparation or administration commensurate with their position and years of experience on the unit. According to the staff, the frequency with which they perform the process is irregular and directly dependant on the type of patients assigned to them. Any one nurse may not be assigned patients with IV bolus morphine for weeks at a time and then have multiple patients receiving IV bolus morphine a number of times over one shift.

OBSERVATIONS

There was a single observer with a background in human factors and safety analysis. A total of 51 observations of IV bolus morphine/HYDROmorphone (45:6) were conducted in a pediatric post-surgery unit. The length of each observation was generally just over 20 minutes, including the time between administration and follow-up, where the nurse could be performing other tasks. The majority of the observations took place during day shifts (7:30am – 3:30pm) and evening shifts (3:30pm – 11:30pm) with only one during the night shift (11:30pm – 7:30am).

A typical observation was initiated when a nurse indicated to the observer that they intended to give IV bolus morphine, or the observer overheard one nurse asking another nurse if they would cosign a morphine preparation in the med room. The observer followed the nurse into the medication room, silently observing drug preparation and documenting such variables as location, start time, noise levels, lighting, interruption to the nurse preparing the morphine, presence or lack of appropriate supplies, etc. The observer then followed the nurse as they collected supplies, still taking notes, and entered the patient's room where the drug is actually administered and the patient is monitored for adverse effects. Variables noted while in the patient's room included entry and exit time, communications, noise levels, allergy check, ID band check, proper sterile technique, drug injected over 3 to 5 minutes, interaction with equipment in the room and at the bedside, etc. The observer typically and re-entered only when the nurse returned to monitor vital signs, at which point similar variables were noted including time.

FINDINGS

PRELIMINARY ANALYSIS AND DATA SATURATION

An initial set of 54 unique influencing factors, grouped in four categories, was compiled from literature including process industry guidelines, management systems guidelines, and contributing factors identified in previous medication error studies. These unique factors were used to guide the ongoing analysis of the observations in order to assess data saturation (Sandalowski, 1995; Wolfe, 2003). The key criterion for data saturation was the identification of new influencing factors, either from the prepared list of factors or new factors that were discovered in the observations.

Coding was done by tagging each observed event with a unique influencing factor taken from the prepared list. For example, increased noise of a crying baby from one of the rooms was tagged as Distraction, or when a nurse dropped the Medication Record (MAR) in the slot outside the medication room was tagged as Availability of Critical Information. The number of influencing factors identified in each sequential observation was counted and plotted as a function of the observation sequence. These are plotted in Figure 1.

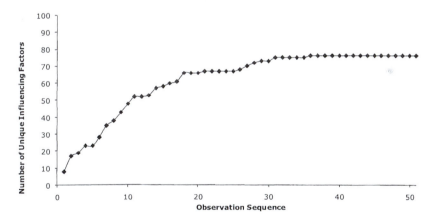

Figure 1. The cumulative number of identified unique influencing factors as a function the observation sequence.

It can be seen in Figure 1 that the number of identified influencing factors increased rapidly during the first 18 observations. The rate of identification of unique factors slowed down after the 18th observation, and no new types of influencing factors were identified after the 36th observed IV bolus morphine/HYDROmorphone administration events.

CATEGORIZATION OF INFLUENCING FACTORS

The objective of the observations was to identify influencing factors that may play a role during the IV bolus morphine administration process, and in turn, could be part of adverse events. The objective was to look for any type of an influencing factor, be it individual, environmental, task-related, or organizational/systemic factors. As was explained above, each observed event related to the medication administration process was coded as one or more of the unique influencing factors. This coding process resulted in 75 unique influencing factors. All 75 factors were grouped into four main categories and 20 sub-categories. The four main categories were: 1. Environmental factors; 2. Equipment, devices & resources; 3. Operating characteristics; and 4. Organization and social factors. All main categories and sub-categories of the identified influencing factors, along with an explanation of each, are presented in Table 1.

Table 1. The main categories and sub-categories of the influencing factors, along with a brief explanation of each factor

Environmental factors	
Interruptions	An event that takes the nurses attention away from the primary task (prep/administration of narcotic) for a period of time, requiring the nurse to stop what she is doing to conduct or attend to something else.
Distractions	An event that could potentially take the nurses attention away from the primary task but does not require stopping the primary activity.
Unexpected event	A situation where the nurse delivering morphine is required to act quickly with little or no prior knowledge of the patient.
Noise	Ambient noise above relative silence, including talking, crying/screaming, beeping, alarms, etc. Ambient noise is documented using a subjective scale (see below). Anything above moderate is counted as an occurrence.
Lighting	Ambient or task lighting in the work environment that come from artificial or natural lighting sources. Documented using a subjective scale of lighting conditions. Considered a potential influencing factor if lighting conditions are less than enough to read by.
Workplace hazards	Events related to conventional Occupational Health and Safety (OHS) hazards: slip/trip infection control, etc
Equipment & tools	
Display design (information Retrieval)	Usability issues related to the presentation of information. Includes events related to the availability/accessibility of information, labeling of fields and forms, legibility of hand-written information. In all cases information is being given to the user

Data entry design (information Entry)	Usability issues related to the design and arrangement of individual forms or interfaces for data entry. Includes events related to the usability of the MAR, Charts, Narcotic Signature Sheet, and others. In all cases information is being recorded by the user
Clarity of instruction	Events specifically related to job aids and procedures designed to communicate instructional information to the users.
Availability of critical information	Issues related to the availability/recognition of critical information required for decision-making that has the potential to impact patient safety. This category refers to events where critical information is (or could be) not immediately recognized, or misinterpreted.
Equipment design	Usability issues related to the design of equipment. Includes issues related to the presentation of information on displays or controls. Includes intuitiveness of use and material properties.
Equipment availability	Issues or potential issues related to the placement, availability of shared equipment
Equipment location/access	Issues or potential issues related to the placement or accessibility of equipment required for the task. Also includes the placement of other equipment that may interfere with the work by virtue of its location.
Operating characteristics	
Experience	Events or anecdotal evidence that a lack of experience, or lack of familiarity with a process or environment, impacted or could have impacted the time or quality of the task conduct.
Risk-taking behavior	Events or anecdotal evidence that staff are consciously not adhering to procedure or not heeding warnings provided by events, equipment, or received information.
Fatigue	Events or anecdotal evidence that an individual is tired; has not slept enough. Cum hours work / shift, or hours slept in past 24h.
Organization & social factors	
Communication	Events or anecdotal evidence that there was a miss communication of information between nurses, doctors, residents, specialists, admin, etc. or that one party was missing information held by another party. Related to situation awareness.
Clarity of responsibilities	Unclear assignment or understanding of responsibilities. Indicated by events where confusion, delay or duplication of effort did or could have resulted. In this case 'responsibilities' may include assigned duties, job requirements or patient care tasks.
Group/individual planning & orientation	Events related to the workflow of an individual or group. Includes unclear or disorganized distribution of tasks between staff, disorganized planning or conduct of a task or set of tasks by one staff and multi-tasking.

474

Distribution of workload	The formal distribution of tasks and level of effort between staff. Indicators include the number of patients assigned to staff, the number and type of staff cover assignments, performance of a teaching function (collected in demographic data) as well as anecdotal evidence of overwork or significantly delayed/missed breaks.

FREQUENCY OF OBSERVED INFLUENCING FACTORS

In order to assess the potential weight, and thus potential impact, of each of the identified influencing factors, two analyses were performed. The first analysis ranked the influencing factors as a function of the frequency with which they were observed. The frequency of each of the factors is presented in Figure 2.

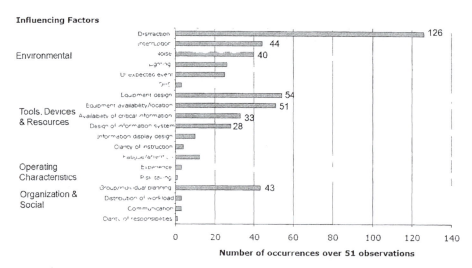

Figure 2. The frequency of observing each type of the influencing factors for the 51 observations.

Relative to the total of 492 observed events that were categorized as influencing factors more than 50% were observed less than 15 times. Following are some of the factors that were observed more than 15 times throughout all the observations: interruptions by another staff, not enough work space in the medication room, too much traffic in the medication room, distractions (by noises, other staff or family members talking), poor information presentation in forms, no designated work space in patient room, and clutter around the patient bed.

The objective of the second analysis was to verify the observed frequency of the influencing factors. Four nurses from the unit rated their perceived frequency of each of the 75 factors on a scale from 1 to 5, with 1 being not frequent and 5 being very frequent. Inter-rater reliability of these ratings was high (Intra-class Correlation coefficient = .814, p<.001). The ratings of perceived frequency confirmed some of the frequently observed factors. More specifically, the following factors that were observed frequently were also rated as occurring frequently by the nurses: interruptions by other stuff, interruptions by family members, not enough work space in the medication room, too much traffic in the medication room, and distractions by irrelevant conversations. It is interesting to note that factors such as poor lighting and poor data entry design were not rated by the nurses as frequent influencing factors. This could be attributed to the fact that the observer was a human factors specialist and more sensitive to such issues.

DISCUSSION

The unfortunate pervasiveness of medication errors with children is well reported and documented (e.g., Alexander et al., 2009; Lesar, 2002; Miller et al., 2007). Whereas many studies focus on capturing and documenting the error rates and error types, few look for the potential influential factors associated with pediatric medication errors in hospitals (James et al. 2009). In the study reported here, a typical human factors approach was adopted to study and identify various influencing factors associated with the process of administering morphine in a pediatric post-surgery unit.

The process was initially guided by a generic list of known influencing factors in order to identify events taking place during the medication process. Through observations of various events in the process of morphine administration and categorization of these events in terms of influencing factors, four main categories were defined: 1. Environmental, including interruptions, unexpected events, noise, work patterns, etc; 2. Tools, devices, and resources, including issues with information display, availability of critical information, equipment design, etc.; 3. Operating characteristics, including experience and risk-taking behaviors; and 4. Organizational and social factors, including communication, clarity of responsibilities, distribution of workload, etc. This categorization is partly in line with a number of studies that identified some contributing factors to medication error (James et al. 2009). These include look-alike, sound-alike drugs, low staffing, computer software, high workload, interruptions, distractions, inadequate lighting, and others.

Taken together the frequently observed factors and the subjectively rated frequent factors, it can be suggested that environmental and contextual factors occur more frequently than factors in the other categories. This finding by itself is insufficient to conclude that such factors are the risk factors associated with potential adverse events. However, the frequent occurrence of such factors can suggest that they are more likely to play an influential role in medication errors. The identified factors emphasize that when examining causes of medication errors and ways to mitigate them, one should look into the systemic factors rather than only individual factors and individual human errors. In other words, frequent influencing factors such as interruptions and distractions combined with physical layout limitations suggest that the overall physical and organizational/human environment should be the objective of any preventative and mitigation strategies. For example, it may be hard to completely eliminate distractions and interruptions. Yet, reorganizing the physical environment (e.g., the medication room) or using other means of medication dispensing or improving information access and display can significantly reduce the detrimental impact of distractions and interruptions.

The involvement of nursing staff in the assessment of relative frequency not only supports the observational findings, but by virtue of reviewing the list of observed factors, also supports the awareness of the staff with respect to potential risk factors. This in itself may contribute to the reduction of practice-related factors, such as interruptions by other staff during dose preparation.

Finally, the human factors method adopted here is such that can be replicated and validated in various ways. The method, composed primarily of using known human factors to guide a field observation along with obtaining subjective ratings of subject matter experts, resulted in a specific, context-oriented list of influential factors. The health-care institution can use these findings to adopt and implement a prevention strategy that will be evidence-based.

REFERENCES

Alexander, D.C., Bundy, D.G., Shore, A.D.., Morlock, L., Hicks, R.W., and Miller, M.R. (2009). Cardiovascular Medication Errors in Children, *Pediatrics*, July 1, 2009; 124(1): 324 - 332.

Conroy, S., Sweis, D., Planner, C., Yeung, V., Collier, J., Haines, L. and Wong, I.C. (2007). Interventions to reduce dosing errors in children: a systematic review of the literature. *Drug Safety*, 30, 1111-25.

Cousins, D.H., Sabatier, B., Begue, D., Schmitt, C. and Hoppe-Tichy, T. (2005). Medication errors in intravenous drug preparation and administration: a multicentre audit in the UK, Germany and France. *Qual Saf Health Care*, 14, 190-5.

Dibbi, H.M., Al-Abrashy, H.F., Hussain, W.A., Fatani, M.I. and Karima, T.M. (2006). Causes and outcome of medication errors in hospitalized patients. *Saudi Med J*, 27, 1489-92.

Engum, S.A. and Breckler, F.D. (2008). An evaluation of medication errors-the pediatric surgical service experience. *J Pediatr Surg*, 43, 348-52.

Esmail Z, Montgomery C, Courtrn C, Hamilton D, Kestle J. (1999) Efficacy and complications of morphine infusions in postoperative paediatric patients. *Paediatric Anaesth*, 9 (4), 321-27.

Ghaleb, M.A., Barber, N., Franklin, B.D., Yeung, V.W., Khaki, Z.F. and Wong, I.C. (2006). Systematic review of medication errors in pediatric patients. *Annals of Pharmacotherapy*, 40, 1766-76.

Institute for Safe Medication Practice. 2009. ISMP Canada Safety Bulletin – August 31st, 9(6).

James, K. L., Barlow, D., McArtney, R., Hiom, S., Roberts, D., & Whittlesea, C.

(2009). Incidence, type and causes of dispensing errors: a review of the literature. *International Journal of Pharmacy Practice*, 17(1), 9-10-30

Lesar, T.S. (2002). Tenfold medication dose prescribing errors. *The Annals of Pharmacotherapy,* Vol 36, No. 12, pp. 1833-1839.

Marshman, JA, UDK, Lam, RW K, and Hyland, S. (2006) Medication Error Events in Ontario Acute Care Hospitals. *CJHP.* 59(5), 243-250.

Miller, M.R., Robinson, K.A., Lubomski, L.H., Rinke, M.L., and Pronovost, P.J. (2007). Medication errors in paediatric care: a systematic review of epidemiology and an evaluation of evidence supporting reduction strategy recommendations. *Qual. Saf. Health Care*, April 1, 2007; 16(2): 116 - 126.

Parshuram, C.S., To, T., Seto, W., Trope, A., Koren, G. and Laupacis, A. (2008). Systematic evaluation of errors occurring during the preparation of intravenous medication. *Canadian Medical Association Journal*, 178, 42-8.

Sandalowski, M. (1995). Sample Size in Qualitative Research. *Research in Nursing and Health,* 18, 179-183.

Stucky E.R, American Academy of Pediatrics Committee on Drugs, American Academy of Pediatrics Committee on Hospital Care (2003). Prevention of medication errors in the pediatric inpatient setting. *Pediatrics*, 112(2), 431-436.

Wong, I.C., Ghaleb, M.A., Franklin, B.D. and Barber, N. (2004). Incidence and nature of dosing errors in paediatric medications: a systematic review. *Drug Safety,* 27, 661-70.

Wolfe, M.S., et.al. (2003). Human factors in healthcare: combining quantitative and qualitative methods. In the *Proceedings of the Annual Conference of the Human Factors and Ergonomics Society*, 2003, pp. 1411-1414.

Chapter 55

Linguistic Etiquette in Social Robot Interaction with Humans in Medicine Delivery

Biwen Zhu, David B. Kaber

Edward P. Fitts Department of Industrial
and Systems Engineering
North Carolina State University
Raleigh, NC 27695-7906, USA

ABSTRACT

Social service robots have been used in hospitals as a technological solution for routine patient tasks, such as medicine delivery. In such application, robot speech capability has been found to be a critical interface feature in the effectiveness of patient robot interaction (PRI). The objective of this study was to review a human linguistic etiquette model in the literature (by Brown and Levinson) and investigate its applicability to PRI. The mediating effect of robot physical humanoid features on user perceptions of linguistic etiquette strategies (derived from the model) was also assessed. Results revealed the etiquette model could be partially extended to PRI. Subjects interpreted and understood robot negative utterances consistent with the original model; however, consistency was not found for positive etiquette strategies. With respect to user overall perception of robot etiquette (PE), negative and positive strategies resulted in the highest and lowest PE scores, respectively. However, the effects of strategy on PE were not mediated by simple humanoid feature changes in robot appearance. Overall, this study provides a basis for determining appropriate robot etiquette strategies to enhance/improve user experiences in collaborative task scenarios with robots in a healthcare context.

Keywords: Etiquette, Human-robot interaction, Service robot, Nursing robot, Medicine delivery

INTRODUCTION

Robots have been developed to assist nurses in medicine delivery tasks in hospitals (Krishnamurthy & Evans, 1992; Zhang, Zhu, Lee & Kaber, 2008). Typical applications include transport of medicines from pharmacists to a nursing station with no direct interaction with patients. However, improvements in intelligent control systems and precision sensors are expected to support direct interaction with patients in the future. For such applications, robot speech should meet patient expectations of nurse behavior in delivery tasks (e.g., being respectful and polite) in order to promote perceptions of healthcare service quality.

In the present study, linguistic etiquette was investigated in the context of patient robot interaction in a medicine delivery task. The motivation for studying linguistic etiquette was: (1) language or speech has been identified as one of the primary features for causing user perception of humanness in interacting with machines and may trigger etiquette responses (Nass, 2004); and (2) there exist linguistic models of human-human etiquette in the social-linguistic area (e.g., Brown & Levinson, 1987), which may serve as basis for study of etiquette in PRI.

A MODEL OF LINGUISTIC ETIQUETTE

Brown and Levinson formulated a "face-saving" theory of how individuals from different cultures produce linguistic etiquette (Brown & Levinson, 1987). They assumed that every individual has two types of "faces", including positive and negative. Negative face is a basic claim to territory, personal preservation, and right to non-distraction (e.g., a human's desire for freedom of action and freedom from imposition); whereas, positive face concerns a consistent self-image, value or personality (e.g., a human's desire for his value to be appreciated and approved in a social setting). The model also assumed that a "rational" person has the ability to weigh different means to an end, and choose the one that most satisfies the desired goals and saves face.

Given the assumptions of the universality of face and rationality, it is possible that virtually all interactions between social agents involve some degree of face-threatening acts (FTAs). To mitigate the adverse effect of a FTA, five etiquette strategies were proposed as the main components of the model including: (1) Do not do a FTA, if the risk of face-loss is great; (2) use indirect requests by means of innuendo and hints (off-record strategy); (3) use strategies aimed at the addressee's negative face, usually by means of offering apologies and deference (negative strategy); (4) use strategies aimed at supporting the addressee's positive face, usually by means of exaggerating interest or providing approval (positive strategy); and (5) do the FTA with no redressive actions (bald strategy).

This qualitative model of linguistic etiquette in human-human interaction (HHI) has been applied in the human computer interaction (HCI) domain. For example, Wang, Johnson et al. (2005) investigated the direct (bald) and indirect (polite) strategies in a pedagogical system and found that polite tactics can affect student motivational state and help them learn difficult concepts in interacting with a computer tutor. Most relevant to the present research, Miller et al. (2004) conducted a field study with persons over 65 yrs. of age, in which medical reminder systems following various etiquette strategies (positive, negative, bald, off record) were tested. Results showed that Brown and Levinson's model provided reasonable predictions of user perceived politeness except for the "off-record" strategy. In fact, the "off-record" strategy was regarded as the most inappropriate, since indirect requests were fuzzy and hard for machines to reproduce. They also found that a negative strategy was the most appropriate for medicine reminding purposes. Related to human interpretations of specific etiquette strategies, Wilkie, Jack et al. (2005) reported that a negative strategy was regarded as more formal and apologetic; whereas, a positive face redressive strategy was considered more manipulative, patronizing and intrusive. These results are all consistent with the definitions from the original model by Brown and Levinson, thus providing some validation for extending the model from HHI to HCI.

However, much less research has been done to study "etiquette effects" in human interaction with social service robots. Some studies have investigated how service robots should respect human social spaces and shared workplace preferences (Walters, et al., 2005; Walters, Dautenhahn, Woods, & Koay, 2007), but not how robots should properly use language to meet user etiquette expectations in a task. Although HCI etiquette research might offer some insights for designing optimum etiquette for HRI, significant differences exist between the two domains (Breazeal, 2004). For example, there may be no observable interface that mediates interaction between a human and robot (e.g., a display screen or input device) and a robot may intentionally initiate interaction with a human. More importantly, when interacting with humans, robots also bring a set of particular affordances (Norman, 2002) due to their physical embodiment (such as grippers for manipulation). These affordances may be richer with respect to conveying system functionality versus affordances of software agents or computers. Such additional cues to a robot's purpose may further trigger human use of "social rules" commonly applied in HHI (Nass & Moon, 2000). For example, prior studies have found that machines (or computers) can be developed to exhibit politeness (Nass, Moon, & Carney, 1999) and reciprocity (Fogg & Nass, 1997). Such phenomena are expected to be more prominent in HRI due to additional affordances provided through robot physical appearances.

ROBOT ANTHROPOMORPHISM AND ETIQUETTE

Anthropomorphic features in robot interface design may provide affordances and, consequently, affect human psychological interaction with robots (e.g., attention to robot requests and level of engagement in tasks (Te Boekhorst, Walters, Koay, Dautenhahn, & Nehaniv, 2005)). Humanlike appearances or behaviors of robots are expected to affect people's expectations of social etiquette in interacting with technology (computer or robot). Prior research has suggested that humanlike communication is more desirable than humanlike behavior for supporting human interaction with social robots (Dautenhahn, et al., 2005), and that the degree of humanness required depends on user expectations in the specific context (Goetz, Kiesler, & Powers, 2003). In the context of PRI in a medicine delivery task, simple changes in service robot appearance may mediate the effectiveness of communicative etiquette strategies; thus, indirectly affecting user overall perception of robot etiquette.

RESEARCH GOAL AND HYPOTHESES

The aim of this study was to apply the linguistic etiquette strategies in Brown and Levinson's model to social robot interface design and to assess its applicability in the PRI domain in simulated medicine delivery tasks. Negative and positive strategies were modeled along with a mixed-etiquette strategy representing a higher degree of language etiquette, as well as a bald strategy representing a lower degree of language etiquette. (The degree or level of etiquette is quantified later.) The "*off-record*" strategy was excluded from this study because of its vague nature, making it difficult or even impossible for machines to produce (Miller, et al., 2004). The strategy of "*simply do not do the FTA*" was also excluded since it would result in an incomplete robot task, which did not support the planned experiment. The mediating affect of robot physical humanoid features on the effectiveness of these strategies was also assessed. In general, it was expected that: (1) a negative etiquette strategy would evoke more feelings of robot respect for a user's freedom to make decisions versus positive and bald strategies; (2) a positive etiquette strategy would evoke more feelings that a user's values or wants were appreciated versus negative and bald strategies; (3) user overall perceived etiquette would be higher under a high-etiquette condition versus a low-etiquette condition; and (4) the presence of humanoid features in robots would facilitate perceived etiquette of linguistic strategies.

METHODOLOGY

To address the study aims, a "wizard of oz" experiment was designed in which two assistive service robots with different levels of human physical likeness were remote controlled by a researcher (see Figure 1, left panel) to perform medicine delivery tasks for 32 student subjects (15 male and 17 female; $M = 24.3$, $SD = 5.04$). The two prototype robots presented the various linguistic strategies to subjects. The primary task for subjects was to watch a Sudoku instruction video on a laptop and, at the same time, complete a puzzle on paper to the best of their ability within 15 minutes. The secondary task for subjects was to attend to and receive medicine for a robot. The affect of the etiquette strategy was expected to be more pronounced in situations when one interactant (the robot) interrupted another (the human). In other words, the robot delivery task was regarded to be an interruption to subject primary task performance and the interruption was expected to amplify the affect of robot behavior (e.g., use of language) on perceived etiquette. Qualified participants had minimal or no experience in playing a Sudoku puzzle game. Since this was a study of initial etiquette expectations (and not post-familiarity expectations), we also ensured that none of the subjects had seen the particular mobile robot platform prior to the experiment.

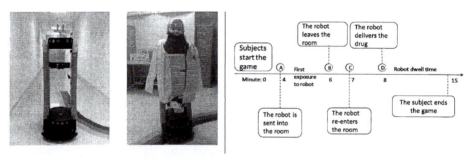

Figure 1. Left panel presents robot prototypes (left side represents low physical humanness; right side represents high physical humanness). Right panel represents robot task flow.

The right panel of Figure 1 also presents a flow chart of the tasks that occurred in the experiment. Subjects were initially exposed to a robot between 4 to 6 minutes into a trial. During this time, the robot simply wandered about the room without direct communication with a subject. This event was similar to a real hospital situation in which a patient might see a robot in a common space before interacting with it in their room for healthcare. The initial exposure event was also used to increase participant curiosity and state of arousal in advance of the medicine delivery. The robot made its first delivery attempt 8 minutes into the trial, with a maximum of three attempts until it left the room. During the robot dwell time, if a subject did not respond, this was considered a failure of robot performance. A trial ended when the subject finished his/her primary task or after 15 minutes elapsed.

INDEPENDENT VARIABLES AND RESPONSE MEASURES

Robot language etiquette strategy was the primary independent variable manipulated in this study. Four types of etiquette utterances were scripted based on Brown and Levinson's model (see Table 1), representing a bald, positive (P), negative (N) or mixed strategy. The mixed strategy was a combination of positive and negative strategies (P+N), representing a high level of etiquette with a maximum count of polite sentences. Using a method of etiquette feature counts, we categorized the bald strategy as representing a low level of etiquette, and the positive or negative strategies as representing moderate etiquette levels. The level of robot physical humanness was also controlled as an independent variable.

Table 1. Bald, positive, negative and mixed etiquette utterances

Utterance scripts
You have missed a dose of your medication. Accept and take your medicine now. **(bald)**
Hello, I know you must be busy, but your health is important. I have come to deliver medication to you. This should only take a second. Please accept and take your medicine now. **(positive)**
Excuse me, I am sorry to interrupt, but my nurse supervisor has indicated that you have not taken your medication scheduled for today. Here is the medication you need. I just want to ask you to confirm receipt of the medicine for me. If you have time now, could you please accept and take the medicine? **(negative)**
Excuse me; I know you must be busy. I am sorry to interrupt, but your health is important. My nurse supervisor has indicated that you have not taken your medication scheduled for today. I have come to deliver the medication to you. This should only take a second. Here is the medication. I want to ask you if you would confirm receipt of the medicine for me. If you have time now, could you please accept and take the medicine? **(mixed)**

The primary response measure was subject perceived etiquette (PE) score. Participants were asked to evaluate the robot speech in terms of its relative disruptiveness, politeness (both positive and negative aspects), length of the message, usefulness/relevance, ease of understanding, and trustworthiness. These factors were derived and revised from measures used in previous HCI studies (Wilkie et al., 2005). Subjects made pair-wise comparisons of etiquette components to establish rank weights. These weights were applied to the component ratings to determine the PE score. As part of the PE scores, subject negative and positive ratings of robot etiquette were collected. After each trial, a subject was also asked to evaluate the degree to which the speech evoked more feelings that the robot respected their freedom to make decisions (negative rating), as well as the degree that the utterance evoked more feelings that their value was appreciated by the robot (positive rating). These two responses were used to test subject interpretations of the etiquette strategies.

EXPERIMENTAL DESIGN AND PROCEDURES

Since this study was a test of robot etiquette in an initial exposure scenario, a two factor between-subject design was adopted (i.e., each subject only saw one experiment condition). From the sample of 32 subjects, groups of four were randomly assigned to one of the eight test conditions (4 etiquette strategies * 2 levels of robot humanness). Before the formal test, each participant also experienced a "control condition", in which they first listened to a medicine reminder message presented with a laptop computer and then completed the PE questionnaire. Participants did not complete a Sudoku puzzle or interact with robots in the control condition. The condition was used to test whether the experiment scenario influenced participant perceptions of etiquette in the reminding aspect of the medicine delivery task (i.e., computer vs. robot).

Subjects were told that they would be awarded a gift certificate if their Sudoku performance was the best among all subjects. We also informed subjects in advance that a robot would come into the room and attempt to provide them with a patient service. However, the subjects did not know in advance the specific task (medicine delivery) the robot would perform. Instead, they were told the robot would give them some hints on solving the Sudoku puzzle (e.g., the robot might say "Put the number 7 in Row 1, Column 6") only after the robot finished its main task. Subjects were told they could choose to respond to the robot or not. If a subject responded to the robot, (s)he received the hint but lost some time in the primary task. Therefore, subjects needed to balance between time and the hint and make the proper decision during a trial. Such a dependence of the primary task on the secondary task was created to avoid the situation of subjects totally ignoring the robot and simply concentrating on the puzzle.

RESULTS

Data from one subject was excluded for statistical analysis as the subject failed to follow instructions and rated all items on the PE questionnaire with the maximum value (=7), both in the control condition and the robot test condition. The mean PE score in the laptop condition was 5.29 (SD=0.99), which was higher than for the robot conditions in general (M=4.93, SD=0.97). This indicated that the robot condition did have an adverse effect on user perception of robot etiquette. The effect is attributable to subject performance of the Sudoku task in the test trial and the disruptive nature of the robot reminding utterances. This finding motivated the application of etiquette strategies in the robot delivery task.

Friedman's two-way nonparametric analysis was used to test the effect of robot appearance and etiquette strategy on subject PE. Results indicated that there were significant differences in PE score among the four etiquette strategy types (Chi-square = 93.41, p < 0.0001). However, there was no evidence that PE scores increased as the number of etiquette features in an utterance increased.

Post-hoc analysis using pairwise comparisons showed that PE scores for the negative condition were significantly higher than those for the other three conditions. Utterances in the bald and P+N conditions resulted in the second highest PE scores, which indicated that humans might prefer concise and clear requests during the interaction with social robots (even though it is not polite). Surprisingly, the PE scores for the positive condition were the lowest among four groups according to the post-hoc results. It is possible that the language in this condition was not sensitive enough to subject needs and hence, did not succeed in addressing, supporting and enhancing the positive face when applied to robots. (This explanation directly relates to the applicability of the etiquette model to the PRI domain and will be discussed later.) Another explanation is that the subjects simply did not like the way the robot addressed their positive needs. More specifically, applying negative strategies (in terms of apology) may be more expected from service delivery robots, as compared to applying positive strategies by way of exaggeration and claiming common ground. For example, one subject reported in a post-experiment survey that he thought the robot was not capable of "sensing" the situation and its "interests" in the user's primary job were "fake".

With respect to the effect of robot physical appearance on PE, no significant differences were found (Chi-square = 0.009, p = 0.92). This may be due to the changes in robot appearance not being dramatic enough to trigger subject psychological state changes. This is also likely because the subjects were concurrently involved in a cognitively complex dual-task scenario.

USER SENSITIVITY TO ETIQUETTE STRATEGIES

Friedman's two-way nonparametric analysis was further used to evaluate subject sensitivity to both positive and negative etiquette strategies. This analysis addressed whether negative etiquette utterances evoked more feelings that the robot respected a users' freedom to make decisions (higher negative ratings), as well as whether positive etiquette utterances evoked more feelings that the user's value was appreciated by the delivery robot (higher positive ratings). We found significant differences among negative ratings across the four strategy types (Chi-square = 112.07, p < 0.0001).

As expected, post-hoc results showed that the negative ratings for the negative condition and P+N condition were significantly higher than those in the other two conditions. This indicates that the design of the negative utterance in this study was supportive of Brown and Levinson's model. However, only marginally significant differences in positive etiquette ratings were found among the four strategy types (Chi-square = 7.054, p = 0.07). This result suggested that subjects did not interpret the positive strategy consistently with the original model (low sensitivity), as compared to a negative etiquette strategy (high sensitivity). Such phenomenon is likely a cause of the low PE scores in the positive etiquette trials and should be considered with caution from a design perspective.

DISSCUSSION AND CONCLUSION

Results suggested that Brown and Levinson's etiquette model may be partially extensible to the HRI domain. In specific: (1) positive linguistic strategies should be avoided due to human sensitivity issues; (2) negative strategies should be encouraged (in line with (Miller, et al., 2004)) as a result of high PE; and (3) bald and P+N strategies should be applied with caution. In fact, differences in individual speech preferences may be significant for bald and P+N strategies, rendering them less effective than negative strategies. In addition to this, strategies with higher (or lower) etiquette levels, do not necessarily result in higher (or lower) PE scores, which indicates that subject perception of robot etiquette may be more dependent on types of etiquette versus levels.

With respect to robot anthropomorphism, no effects of robot physical humanoid features on PE scores were found. This may be due to our manipulations of robot appearance being static and basic; that is, the face, clothing, etc. did not trigger specific human etiquette expectations for the robot. More dynamic humanoid features, such as facial expressions, robot head movement and eye contact need to be further investigated for mediating effects on PE. Additional research should also examine an off-record etiquette strategy to determine the influence of robot "body language" on human perceptions in medicine delivery tasks.

In general, this study provided a preliminary basis for studying linguistic etiquette strategies in the context of HRI. More work needs to be done to investigate how each etiquette strategy might influence user task performance as well as robot task performance. In medicine delivery tasks, how these strategies may affect user compliance or quick responses to robot requests is another interesting topic for study. Other future research questions include: (1) how to quantitatively measure effectiveness of a particular etiquette strategy in HRI; (2) how to better implement combinations of mixed-etiquette strategy types for effective HRI; and (3) what are the long-term effects of etiquette for HRI. This list is by no means comprehensive, but provides some ideas for this new area of HRI.

ACKNOWLEDGEMENT

Biwen Zhu's work on this project was supported in part by The Ergonomics Center of North Carolina (ECNC). We thank Nelson Couch, Haig Khachatoorian and Jeff Thompson for their valuable inputs and suggestions on the research.

REFERENCES

Breazeal, C. (2004). Social interactions in HRI: the robot view. In *Proceeding of IEEE Transactions on Systems, Man, and Cybernetics*, Part C: Applications and Reviews, 181-186.

Brown, P., & Levinson, S. C. (1987). *Politeness: Some Universals in Language Usage*. Cambridge, UK: Cambridge University Press.

Dautenhahn, K., Woods, S., Kaouri, C., Walters, M. L., Koay, K. L., & Werry, I. (2005). What is a Robot Companion-Friend, Assistant or Butler? In *Proceedings of IEEE IROS*, Edmonton, Canada, 1488 - 1493.

Fogg, B. J., & Nass, C. (1997). Do users reciprocate to computers? *Proceedings of the CHI Conference* (Atlanta, GA). New York: Association of Computing Machinery.

Goetz, J., Kiesler, S., & Powers, A. (2003). Matching robot appearance and behavior to tasks to improve human-robot cooperation, In *Proceedings of the 12th IEEE International Workshop on Robot and Human Interactive Communication* (ROMAN), (pp. 55-60).

Krishnamurthy, B., & Evans, J. (1992). HelpMate: a robotic courier for hospital use. *Proceedings of IEEE International Conference on Systems, Man and Cybernetics, Chicago, IL, USA*.

Miller, C. A., Wu, P., & Chapman, M. (2004). The Role of "Etiquette" in an Automated Medication Reminder. *Proceedings of the AAAI*, San Jose, California.

Nass, C. (2004). Etiquette equality: exhibitions and expectations of computer politeness. *Communications of the ACM, 47*(4), 35-37.

Nass, C., & Moon, Y. (2000). Machines and Mindlessness: Social Responses to Computers. *Mindfulness Theory and Social Issues, 56*(1), 81-103.

Nass, C., Moon, Y., & Carney, P. (1999). Are respondents polite to computers? Social desirability and direct responses to computers. *Journal of Applied Social Psychology, 29*(5), 1093-1110.

Norman, D. (2002). *The design of everyday things*. New York: Basic Books New York.

Te Boekhorst, R., Walters, M. L., Koay, K. L., Dautenhahn, K., & Nehaniv, C. L. (2005). A study of a single robot interacting with groups of children in a rotation game scenario. In *Proceedings of IEEE CIRA*, Finland.

Walters, M. L., Dautenhahn, K., te Boekhorst, R., Koay, K. L., Kaouri, C., Woods, S., et al. (2005). The Influence of Subjects? Personality Traits on Personal

Spatial Zones in a Human-Robot Interaction Experiment. In *proceedings of 14th IEEE Int. Workshop on Robot & Human Communication* (RO-MAN), Nashville, USA.

Walters, M. L., Dautenhahn, K., Woods, S. N., & Koay, K. L. (2007). Robotic etiquette: results from user studies involving a fetch and carry task. *Proceedings of ACM/IEEE International Conference on Human-Robot Interaction*, Arlington, Virginia, USA

Wang, N., Johnson, W. L., Rizzo, P., Shaw, E., & Mayer, R. E. (2005). Experimental evaluation of polite interaction tactics for pedagogical agents. *Proceedings of International Conference on Intelligent User Interfaces*, New York.

Wilkie, J., Jack, M. A., & Littlewood, P. J. (2005). System-initiated digressive proposals in automated human-computer telephone dialogues: the use of contrasting politeness strategies. *International Journal of Human-Computer Studies, 62*(1), 41-71.

Zhang, T., Zhu, B., Lee, L., & Kaber, D. (2008). Service robot anthropomorphism and interface design for emotion in human-robot interaction. *Proceedings of IEEE Conference on Automation Science and Engineering* (CASE), Washington DC, USA.

Chapter 56

Fuzzy Based Risk Assessment of a HIS Implementation: A Case Study from a Turkish Hospital

Gulcin Yucel[1], Selcuk Cebi[1], Ahmet F.Ozok[2]

[1]Istanbul Technical University
Turkey

[2]Istanbul Kultur University
Turkey

ABSTRACT

According to the Institute of Medicine Report *To Err is Human*, there are 98,000 medication errors per year and 44,000 Americans die as a result of medical errors. One possibility to have a safer health care system is to support the processes by Information Technology (IT). This article discusses risk assessment of a hospital information system (HIS) by using fuzzy based decision making methodology. A case study was conducted at a Teaching and Research Hospital in Istanbul, Turkey. For this purpose, Zeng et al. (2007)'s fuzzy risk assessment model is used which quantifies linguistic variables and which is based on fuzzy inference. In the other risk assessment methods, risk magnitude is defined as risk probability and risk severity, but there are many factors which affect the possible risk in the implementation of HIS. Therefore, in this model, these factors are defined and integrated into the decision making process of risk assessment. Another important issue in risk assessment is to have high quality data. Since the risk assessment is conducted before the implementation, there is no existing data. Fuzzy reasoning techniques provide an effective tool for handling uncertainties and subjectivities. In the case study, the risk for HIS implementation is measured based on experts' evaluations by using fuzzy analytic hierarchy process (AHP) and fuzzy reasoning. The risk magnitude of the new HIS implementation for the hospital is found as major with a belief of 100%.

Keywords: Hospital Information System, Risk Assessment, Influencing Factors, Decision Making, Fuzzy Reasoning Approach

INTRODUCTION

According to the Institute of Medicine Report *To Err is Human*, there are 98,000 medication errors per year and 44,000 Americans die as a result of medical errors. This number is even higher than the number of people who died because of AIDS, breast cancer, or motor vehicle accidents per year. One possibility to have a safer health care system is to support the processes by Information Technology (IT). In literature, serious reasons can be found to scrutinize IT applications in the domain of healthcare. There is also research which shows the unintended effects of IT. Risk assessment of the process under the new technology would reduce the causes of unintended results of new applications (Win et al., 2004). The most common risk assessment techniques in literature are *Event Tree Analysis* (ETA), *Fault Tree Analysis* (FTA) and *Failure Mode and Effects Analysis* (FMEA).

In ETA, risk is assessed based on the probability of an event's success and failure. However, in many systems it is very difficult to obtain the probability of events. It is also not applicable to assess the risk of a new system before it is implemented (Win et al., 2004). FTA involves analyzing the root causes of a top event; therefore it is not applicable for predictive risk analysis either (Win et al., 2004; Bonnabry et al., 2005). For critical processes especially related with patient safety, it is not acceptable to wait for an incident to decide the safety requirements. As well as high risk areas such as aviation, aerospace, nuclear power plants/generation and the food industry, in healthcare area more proactive risk analysis techniques should be applied. Since April 1st, 2001, it is an obligatory for accredited hospitals to conduct at least one of the proactive risk assessment techniques by Joint Commission on Accreditation of Health Organization (JCAHO) (Bonnabry et al., 2005). FMEA involves identifying the possible failure modes and effects before the failure can happen. Therefore, FMEA was proposed as a risk assessment method of Electronic Health Record (Win et al., 2004). Also, Bonnabry et al. (2005) proposed *Failure Modes, Effects and Critically Analysis* (FMECA) to assess the risk of a new process in pediatric parenteral nutrition solutions. FMEA and FMECA define risk assessment based on probability, severity or criticality of the failure. However, there are many possible risk factors such as human factors, technological factors and organizational factors. Therefore Zeng et al. (2007) proposed a new risk assessment method in which these factors are evaluated and integrated to the decision making process of risk assessment beside risk probability and risk severity. In the model, possible risk factors' impacts, risk likelihood and risk severity are evaluated by experts. Subjective risk assessment methods, in which data is obtained by experts, can provide combining multiple attributes for the prediction of a project success or failure. Subjective risk assessment models have been used to develop prediction of medication adherence, organizational success and health care costs (Molfenter and Gustafson, 2007). Another important issue in risk assessment is to have high quality data. Since the risk assessment is conducted before the implementation, there is no existing data. Fuzzy reasoning techniques can provide an effective tool to handle uncertainties and subjectivities.

In this study, before implementation of new HIS, it is aimed to construct risk assessment for avoiding potential failures and threads. Since it is a new system, there is no data about failure modes and their probabilities. When there is no sufficient data, analysis input can be obtained through experts' opinions e.g. by using linguistic variables. One advantage of using linguistic variables is that this kind of expression is more intuitive and easy for experts to give their opinions in an ambiguous situation where numerical estimations are hard to get (Lin and Wang, 1997). According to Zadeh (1965), fuzzy set approaches have always been one of the most appropriate tools when it is necessary to model the human knowledge. Therefore, we use a risk assessment model which consists of fuzzy analytic hierarchy process and fuzzy reasoning to measure the risk of HIS implementation.

HOSPITAL INFORMATION SYSTEM

A Hospital Information System (HIS) is an integrated information system designed to manage the administrative, financial and clinical aspects of a hospital. Under a HIS, administrative, financing and patient care processes such as appointment scheduling, patient acceptance, examination, surgery, treatment, hospital returnee are done electronically. It provides recording all patient services and given medications as well as billing and connecting to the Social Security Institution (SSI) (Ulgu, 2008). In order to be paid for the expenses of patient care, the services and medications are sent to SSI's system called MEDULA. After MEDULA approved the expenses, hospitals can be paid. In the last 10 years, many government hospitals have implemented HIS. The first motivation behind implementations of HIS was to prevent losses due to misuse, overuse and illegal use, but now HIS implementation is also believed to provide better patient care and better hospital management (Ministry of Health of Turkey).

RISK ASESSMENT MODEL

We use a risk assessment model proposed by Zeng et al. (2007) in order to measure the risk for the implementation of HIS. The main structure of the method consists of fuzzy reasoning and fuzzy analytic hierarchy process (AHP). The proposed model comprises five steps: (1) evaluation of influencing factors (IFs), (2) evaluation of risk likelihood (RL), (3) evaluation of risk severity (RS), (4) fuzzy inferencing, and (5) calculation of risk magnitude. The structure of the proposed methodology is given in Figure 1.

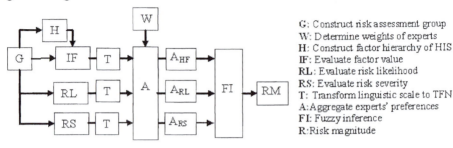

G: Construct risk assessment group
W: Determine weights of experts
H: Construct factor hierarchy of HIS
IF: Evaluate factor value
RL: Evaluate risk likelihood
RS: Evaluate risk severity
T: Transform linguistic scale to TFN
A: Aggregate experts' preferences
FI: Fuzzy inference
R: Risk magnitude

FIGURE 1 Structure of the algorithm

First, risk assessment group, influencing factors, weights of experts (w_{ei}) and fuzzy membership functions that are used in calculations are determined. The priorities of IFs are calculated by using fuzzy AHP. Experts are asked to decide the relative importance of each factor. The scale for pairwise comparisons is given in Figure 2. This scale is defined as follows: *equally important* (Eq), *weakly more important* (Wk), *strongly more important* (St), *very strongly more important* (Vs), and *absolutely more important* (Ab). By applying AHP, the weight of each factor is obtained.

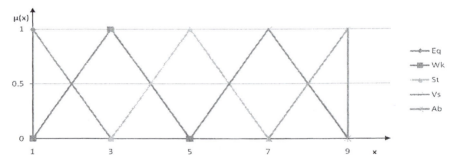

FIGURE 2 Linguistic scales for pairwise comparison(Hsieh et al.,2004)

IFs' impacts on HIS implementation success $\left(\tilde{A}_i^k\right)$ are also evaluated by experts using intangible linguistic terms given in Figure 3 (very poor (VP), poor (P), fair (F), good (G), very good (VG)). Evaluation of RL and RS is done by using tangible linguistic scale where linguistic terms are defined as *very low* (VL), *low* (L), *medium* (M), *high* (H), *and very high* (VH). Triangular fuzzy numbers of the tangible and intangible linguistic scales are given in Figure 3 (Olcer and Odabasi, 2005). To evaluate the calculated Risk Magnitude (RM), a linguistic scale as defined in Figure 4 is required (Zeng et al., 2007).

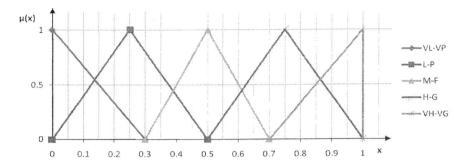

FIGURE 3 Triangular fuzzy numbers for evaluation of RL, RS and IFs

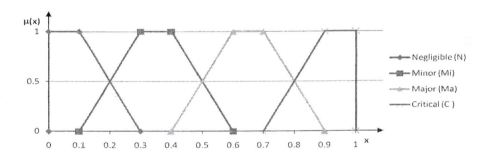

FIGURE 4 Triangular fuzzy numbers for RM

Each member of the risk assessment group gives his/her preferences. Therefore, expert weights are used to aggregate the assessments of the group. In literature, the most used aggregation operator is *weighted mean* (Eq.1).

$$\tilde{A}_i = \tilde{A}_i^1 \otimes w_{e1} + \tilde{A}_i^2 \otimes w_{e2} + \ldots + \tilde{A}_i^m \otimes w_{em}$$
(1)

where w_{ei} is the weight of ith expert and m is the number of experts. The weights of the factors in the pairwise comparison matrix are calculated by using *arithmetic averaging method* (Zeng et al., 2007).

$$w_i = \frac{1}{n} \sum_{j=1}^{n} \frac{a_{ij}}{\sum_{k=1}^{n} a_{kj}} \quad i, j = 1, 2, \ldots, n$$
(2)

where w_i is the level weight of FIs hierarchy. l is the number of level of HIS w_{il} is the level weight of lth upper section. So, the final weight is calculated for each bottom level factor by

$$w'_i = w_i \prod_{i=1}^{t} w_{il} \tag{3}$$

The total score of IFs is calculated by

$$PM\widetilde{S} = \sum_{i=1}^{n} \widetilde{A}_i w'_i \quad i=1,2,3,....n \tag{4}$$

and the next step, RL and RS are evaluated by the risk assessment group members and Eq. 6 and Eq. 7 are used to aggregate their evaluations.

$$R\widetilde{L} = R\widetilde{L}^1 \otimes w_{e1} + R\widetilde{L}^2 \otimes w_{e2} + ... + R\widetilde{L}^m \otimes w_{em} \tag{5}$$

$$R\widetilde{S} = R\widetilde{S}^1 \otimes w_{e1} + R\widetilde{S}^2 \otimes w_{e2} + ... + R\widetilde{S}^m \otimes w_{em} \tag{6}$$

where $R\widetilde{L}$ and $R\widetilde{S}$ is the fuzzy aggregated results of the group preferences of $R\widetilde{L}$ and $R\widetilde{S}$ and $\otimes$ and $\oplus$ are the fuzzy multiplication operators and the fuzzy addition operator, respectively. In the fuzzy inference step, the aggregated values of $PM\widetilde{S}$, $R\widetilde{L}$ and $R\widetilde{S}$ are entered to the fuzzy inference system to decide which rules are relevant for the current situation. Then, the fuzzy output of $R\widetilde{M}$ is calculated (Eq. 7-8). To calculate the crisp RM value, Eq. 9 is used. The output modification step sometime is necessary for securing a reliable decision (Zeng et al., 2007).

$$\mu_{RM_i}(y) = \bigcup_{i=N,Mi,Ma,Ci} \left\{ \mu_i \left(\mu_{PMS}(x) \cap \mu_{RS}(x) \cap \mu_{RL}(x) \right) \right\} \tag{7}$$

$$R\widetilde{M} = \left(\mu_{RM_N}(y), \mu_{RM_{Mi}}(y), \mu_{RM_{Ma}}(y), \mu_{RM_{Ci}}(y) \right) \tag{8}$$

$$RM = \frac{\sum_{i=1}^{q} Y_i \mu_{RM}(y_i)}{\sum_{i=1}^{q} \mu_{RM}(y_i)} \tag{9}$$

A CASE STUDY

A case study of risk assessment of a HIS implementation is presented. A Turkish training and research hospital wants to implement new HIS because of problems with old one such as no integration between HIS and other ITs in hospital, no decision support for MEDULA payment rules. In MEDULA system, all medication type, doses, necessary tests, etc. for each diagnostic are defined. If there is mismatch between treatment and diagnostic, the hospital has to explain the reason. If it is not accepted by MEDULA, then hospitals won't be paid for these treatment and inspection.

ALLOCATE EXPERTS' WEIGHTS

The risk assessment group consists of the head of the IT department of the hospital, head of department of Pulmonology and associate head of the hospital since these people had experience and knowledge about the hospital processes and health IT applications. Weights of experts are assigned based on expert's individual background and experience as 0.35, 0.2 and 0.45 respectively.

CONSTRUCT INFLUENCING FACTORS HIERARCHY

Many factors have an impact on IT implementation success. Several models can be used to construct influential factors regarding as a risk such as Brainstorming, Checklist, What-if, Failure Mode and Effect analysis. In this study, influential factors are constructed based on previous researches. There are many models which provide system factors than can be useful to predict IT implementation success. Karsh and Holden (2007) identified the factors which have shown to impact on successful IT implementation based on literature about technology acceptance, technology satisfaction and new technology implementation. The influential factors on successful implementation of HIS given in Table 1 are decided by risk assessment group based on whole system organization predictors of IT implementation success (Karsh and Holden, 2007).

Table 1 IFs Hierarchy (based on Karsh and Holden, 2007)

Technology	Individual/Person	Organizational
-Ease of use	-Age	-Training
-Usefulness	-Gender	-Technical Support
-Compatibility	-Personality	-Organizational Support
-Impact	-Education	-Organizational Commitment
-Enjoyment	-Skills	-User involvement
	-Learning	
	-Prior experience	

OBTAIN EXPERTS' OPINIONS

Experts are asked to decide the relative importance of each factor by using scale given in Figure 2. The weights of IFs are calculated based on pairwise compassions by using Eqs (1-3). IFs' impacts on HIS implementation success are also evaluated by experts using intangible linguistic terms given in Figure 3. The weights and the impacts of IFs are presented in Table 2. By using Eq.(4), FI value is obtained as (0.588, 0.853, 0.939).

Table 2. Weights of factors

Factors	Sub-Factors	Impacts	Local	Global
Technology			0.065	
	Ease of use	(0.46,0.7,0.94)	0.302	0.020
	Usefulness	(0.5,0.75,1)	0.128	0.008
	Compatibility	(0.66,0.95,1)	0.245	0.016
	Impact	(0.53,0.7875,0.94)	0.232	0.015
	Enjoyment	(0.4,0.6,0.86)	0.093	0.006
Individual/Person			0.306	
	Age	(0.375,0.55,0.695)	0.185	0.057
	Gender	(0,0,0.3)	0.116	0.035
	Personality	(0.62,0.9,0.94)	0.140	0.043
	Education	(0.1,0.2375,0.51)	0.053	0.016
	Skills	(0.66,0.95,1)	0.228	0.070
	Learning	(0.48,0.725,0.865)	0.125	0.038
	Prior experience	(0.5,0.75,1)	0.154	0.047
Organizational			0.628	
	Training	(0.66,0.95,1)	0.257	0.161
	Technical Support	(0.7,1,1)	0.144	0.091
	Organizational Support	(0.7,1,1)	0.232	0.146
	Organizational Commitment	(0.7,1,1)	0.232	0.146
	User involvement	(0.5,0.75,1)	0.134	0.084

MEASURE RL AND RS

The measurement of RL and RS are done in the following steps: (1) RL and RS are decided by multiple experts by using intangible linguistic scales (given in Figure 3), (2) their evaluations are converted to triangular fuzzy numbers (TFNs), (3) TFNs are aggregated by using Eqs (5,6). The evaluations of experts, aggregated risk likelihood and aggregated risk severity are given in Table 3.

Table 3. Evaluations and TFNs of RL and RS

	Linguistic Evaluations		TFNs	
Weights	RL	RS	RL	RS
E1 (0.35)	VL	VH	(0,0,0.3)	(0,7,1,1)
E2 (0.2)	L	H	(0,0.25,0.5)	(0.5,0.75,1)
E3 (0.45)	M	VH	(0.3,0.5,0.7)	(0,7,1,1)
Aggregated values			(0.14,0.28,0.52)	(0.66,0.95,1.0)

RISK ASSESSMENT BY FUZZY INFERENCE MECHANISM

The aggregated values of FI, RL and RS are required to convert into fuzzy sets to be used in the fuzzy inference. Fuzzy sets of FI*, RL* and RS* obtained by using linguistic scales given in Figure 3 are as follows:

FI*= {(fair, 0.25), (Good, 0.8), (Very Good, 0.63)},
RL*= {(Very Low, 0.47), (Low, 0.94), (Medium, 0.5), (High, 0.05)},
RS*= {(Medium, 0.08), (High, 0.63), (Very High, 0.86)}.

Then, all of the control outputs are calculated as given in Table 4. By using Eqs. (7) and (8), the RM* is obtained as RM*={(μN(RM), 0.47), (μMi(RM),0.8), (μMa(RM), 0.8), (μC(RM),0.5)}.

Table 4. The control outputs of fired rules

Factor Index (FI)		Risk Severity (RS)		Risk Likelihood		(RL)			
				L	0.08	M	0.63	H	0.86
VG	0.63	H	0.05	Ma	0.05	C	0.05	C	0.05
		M	0.5	Ma	0.08	C	0.50	C	0.50
		L	0.94	N	0.08	Ma	0.63	Ma	0.63
		VL	0.47	N	0.08	N	0.47	Mi	0.47
G	0.8	H	0.05	Mi	0.05	C	0.05	C	0.05
		M	0.5	Mi	0.08	Ma	0.50	Ma	0.50
		L	0.94	N	0.08	Mi	0.63	Ma	0.80
		VL	0.47	N	0.08	Mi	0.47	Mi	0.47
F	0.25	H	0.05	Mi	0.05	Ma	0.05	C	0.05
		M	0.5	N	0.08	Mi	0.25	Ma	0.25
		L	0.94	N	0.08	Mi	0.25	Mi	0.25
		VL	0.47	N	0.08	N	0.25	Mi	0.25

The centre-average defuzzification operator given by Eq. (9) is used to convert the fuzzy output RM* into a numerical value. The overall risk magnitude is obtained as 0.65 under the defined scale system of RM. This value is equal to major risk with a belief of 100%.

DISCUSSION

This model takes in to account multiple factors simultaneously beside risk level and risk severity to predict the risk of HIS implementation failure. Risk magnitude of a new HIS implementation for the hospital is found as major with a belief of 100%. Therefore it is necessary to make more detailed analyses before the implementation. Also, most influencing factors on risk are found as training, organizational support and organizational commitment, by investigating these factors, risk magnitude can be decreased. The failure of the implementation's results will cause too much chaos with patient serving and too much revenue loss in this big hospital which serves daily 4000-5000 outpatients. Therefore, for a case of a system implementation failure, a backup plan should be defined and provided to all users.

Finally, the authors would like to thank Bo Hoege, Arslan Tasdemir, Sergun Guloglu and Levent Kart for their support through the research.

REFERENCES

Bonnabry, P., Cingria, L., Sadeghipour, F., et al. (2005), "Use of a systematic risk analysis method to improve safety in the production of pediatric parenteral nutrition solutions." *Qual Saf Health Care*, 14, 93-98.

Hsieh, T. Y., Lu, S. T., Tzeng, G. T. (2004), "Fuzzy MCDM approach for planning and design tenders selection in public office buildings." *International Journal of Project Management*, 22, 573–584.

Karsh, B.T. and Holden, R.J. (2007), *"New technology implementation in health care."*, in P. Carayon (Ed.).: Handbook of Human Factors and Ergonomics in Healthcare, Lawrence Erlbaum Associates: New Jersey. pp. 393-410.

Lin, C.T. and Wang, M.J.J. (1997). "Hybrid fault tree analysis using fuzzy sets." *Reliability Engineering and System Safety*, 58, 205-213.

Ministry of Health of Turkey, (2010), Hastaneler ve hastane bilgi sistemleri, Ministry of Health of Turkey Web Site : http://www.sagliknet.saglik.gov.tr/portal_pages/notlogin/saglikcilar/saglikcilar_hastaneler.htm

Molfenter, T. D. and Gustafson, D. H. (2007) *"Quality improvement in health care."*, in: Handbook of human systems integration, Carayon, Pascale (Ed.). pp. 771-787.

Olcer, A.I. and Odabasi, A.Y, (2005), "A new fuzzy multiple attributive group decision making methodology and its application to propulsion/maneuvering system selection problem", *European Journal of Operational Research*, 166, 93-114.

Ulgu, M.M. (2008),*Hastane bilgi sistemi tedarik süreci başlatan hastane idarecileri için öneriler,* Istanbul Heath Office Web Site : http://www.istanbulsaglik.gov.tr/w/mev/mev_gen/gen_bilgi_islem/HBSACI_2008_EK.pdf

Wing, K.T., Phung, H., Young, L. et al. (2004), "Electronic health record system risk assessment a case study from the MINET", *Health Information Management*, 33(2), 43- 48.

Zeng, J., An, M., Smith, N.J. (2007), "Application of a fuzzy based decision making methodology to construction project risk assessment." *International Journal of Project Management,* 25, 589 – 600.

Zadeh, L.A. (1965), "Fuzzy Sets.", *Information and Control*, 8, 338–353.

Chapter 57

A New Model-Based Approach for the User Interface Design of Medical Systems and Devices

Armin Janß, Wolfgang Lauer, Fabrice Chuembou, Klaus Radermacher

Chair of Medical Engineering
Helmholtz-Institute for Biomedical Engineering
RWTH Aachen University, Germany

ABSTRACT

Rapidly evolving technological progress and automation in the field of medical devices and systems (e.g. in the field of Ambient Assisted Living or in clinical context) not only lead to an enhancement in efficiency and effectiveness concerning therapeutic results but also to a change of the Human-Machine-Interaction characteristics. To ensure safe and reliable user interfaces not only ergonomic but also error-tolerant interface design has to be taken into consideration by the design engineer. In this context, the Chair of Medical Engineering (mediTEC) at the RWTH Aachen University has developed a novel modeling and analysis tool (mAIXuse) in order to create standardised representation and assessment of the user interaction process, even with complex graphical interfaces of medical devices and systems. Adapted from two model-based methodologies the tool uses formal, normative models to predict and assess user-, interaction- and system-behaviour. The software-based tool can be used from the very early developmental phases (definition and specification) up to the validation of existing user interfaces.

Keywords: User Interface Design, Human Error, Risk Analysis, Usability Evaluation, Model-Based Assessment, Cognitive Task Analysis

INTRODUCTION

The overall risk for manufacturers of risk-sensitive products (e.g. in aeronautics, nuclear engineering, medical technology and pharmaceuticals) has increased in recent years. This increases not only the cumulative potential damage, but also the potential consequences. Especially in the medical field of diagnostic and therapeutic systems undetected remaining failures or residual risks, occurring in the development process, often cannot be tolerated. Bringing defective and erroneous products to the market means a high potential of severe consequences not only for the patients but also for the manufacturers. Apart from ethical considerations, related costs can endanger the livelihood of enterprises. The medical branch is, similar to almost all manufacturing industries, under a high cost and time pressure, which is characterized by globalization and dynamic markets, innovation and shorter product life cycles.

The risk management process can be seen as a "systematic application of management principles, procedures and practices to identify, analyze, treat and monitor understanding of risks, allowing organizations to minimize losses and maximize opportunities" (Hoffmann, 1985). However, these goals are often missed in practice (Hansis, 2002). There are about 40.000 malpractice complaints and the evidence of more than 12.000 malpractice events per year in Germany (Kindler, 2003). More than 2.000 cases of this so-called "medical malpractice" can be traced to medical and surgical causes, leading in the end to the death of the patient (Hansis, 2001). The expert's opinion on the development of health services since the year 2007 is that in Germany approximately 17.000 deaths each year are due to "preventable adverse events". In addition to false decisions and medication errors, use errors are a major cause of deaths in the medical context (Merten, 2007). The Australian Health Care Study e.g. stated that in 16.6% cases of patient treatment adverse events have been recorded, 51% have been classified as avoidable (Wilson et al., 1995).

Moreover, Harvard Medical Practice Studies confirmed the important role of human factors in safety aspects of human-machine-interaction in clinical environment. Following their statement, 39.7% of avoidable mistakes in operating rooms arise from "carelessness" (Leape, 1994). Avoidable mistakes, especially in combination with the use of technical equipment in the medical context, are, with a disproportionately high rate, due to human-induced errors (Leape, 1994). Results from the Harvard Medical Practice Study were confirmed by a number of other epidemiological studies. In the retrospective study by Gawande et al. regarding 15000 surgically treated patients in 3% of the cases user-based injuries have been found, of which 54% have been preventable errors (Gawande et al., 1999). According to incident reports in orthopedic surgery avoidable mistakes with technical equipment are in 72% due to human error (Rau et al., 1996).

Relating to these studies, human-oriented interface design plays an important role in the introduction of new technology and special devices into medical applications. To ensure safe and reliable user interfaces not only ergonomic but also error-tolerant interface design has to be taken into consideration for the acceptance as well as for the routine application of medical devices (Radermacher, 2004).

From the cognitive psychological point of view, inadequate coordination of the design of the work equipment and environment with the task and situation, the erroneous interpretation of information in user interfaces and situational factors as well as external environmental factors and extrinsic stress are the main causes of "human error" in medicine (Bogner, 1994). Statistics on malpractice incidents and the handling of medical devices seem to prove this fact (Kohn et al., 2000).

STATE OF THE ART AND BOTTLENECKS

The effective implementation of risk management and usability engineering requires not only a systematic approach concerning management principles but also appropriate methods and tools for the different stages of implementation.

Based on the Failure Mode and Effects Analysis (FMEA), a system or a process can be systematically investigated in the early development stages in order to initiate prevention measures (Pfeifer, 2001). The disadvantages of the methodology, mentioned by many practitioners, are high effort and the complexity of the analysis, which increases significantly with a rise in the range of functionality of the analyzed system (Pfeifer, 2001). With regard to the triple "error, consequence, cause", which defines a sequential analysis, the FMEA doesn't allow to model complex causal failure chains within the investigation.

Besides the FMEA, the fault tree analysis (FTA) is one of the most commonly used methods in risk and quality management. The core of the method is the development of the fault tree, based on a previously performed system analysis. In this inductive approach, initially on the basis of potential error, all possible consequences and failure combinations are analyzed. Finally the detected errors or malfunctions are rated with regard to the probability of occurrence (Pfeifer, 2001). As with the FMEA, complex relationships between failures, which seem to be time and space independent in their appearance at first sight, cannot be identified. The calculation of component-based error probabilities can be done in product-based experiments, but the difficulty to evaluate human-induced errors in a realistic dimension to capture and quantify them in a sufficient way, however, remains unanswered.

Experimental usability testing and heuristic analysis offer many advantages for manufacturers, but are still connected to method-related disadvantages. The experimental usability testing provides, in contrast to formal analytic procedures, a potentially high validation level as, in addition to tasks and system-specific characteristics, also (partly implicit) complex environment and user-specific factors (such as stress, fatigue, awareness, temperature, climate) can be included within the laboratory and field tests by the analysis of e.g. video analysis, physiological data, observations and interviews. However, the existence of an interactive mock-up (also necessary for heuristic analysis) and the presence of a representative user group are necessary for the conduction of these tests, which usually causes high time-related and infrastructural costs. Results from usability tests are often acquired too late for a proper integration in the ongoing development process without inducing high costs. The design engineer has to examine and determine the usability of an envisioned system already in the very early developmental phases, often without being able to interact with a first mock-up.

In various stages of the developmental process a multiplicity of appropriate methods for the support of usability evaluation can be applied. Model-based approaches can be utilized in the definition and specification phase. Approaches for modeling user, interaction and system behavior are e.g. cognitive modeling and task analysis. Due to the lack of efficient and easy-to-use modeling tools, cognitive architectures are currently mainly used for fundamental research in the area of cognitive psychology.

Pertaining to the (cognitive) task analysis, there are several methodologies (e.g. hierarchical task analysis) which can be applied depending on the specific questions and cases. The application of task analysis for supporting user interface design of envisioned and redesigned interactive systems can be a fundamental contribution to enhance human-centered development (Diaper, 2004). Regarding model-based usability examination approaches, there is often the lack of an application-oriented software tool, although the theoretical framework behind shows well-developed syntax models and structures.

The majority of commercially available software tools concerning (cognitive) task analysis enables interactive data storing and provides different task modeling options. Unfortunately none of these tools offer a subsequent failure analysis, in order to analyze potential risks in the use process and to derive accurate information concerning proper design of the user interface.

Furthermore, with existing software-based tools/methodologies the detailed modeling of the complete use process with a complex human-computer-interface (e.g. with reference to the task taxonomy and time dependencies) is not accomplishable.

APPROACHES

Based on an initial concept, developed in the framework of the BMWi funded AiF/FQS project INNORISK, a model-based usability and risk analysis method has been created at the Chair of Medical Engineering at the RWTH Aachen University. The mAIXuse method allows manufacturers of risk-sensitive equipment/systems to conduct a formal-analytical usability evaluation and a use-oriented risk analysis in very early developmental stages. The mAIXuse tool can be applied prospectively in the development process as well as in the framework of the analysis respectively redesign of existing human-machine-interfaces and their validation.

On the basis of an initial risk analysis (e.g. according to ISO 14971 for medical devices) user interaction sequences and their potential impact on the overall process are initially evaluated. The proposed formal-analytic mAIXuse method is based on a twofold strategy (Figure 1).

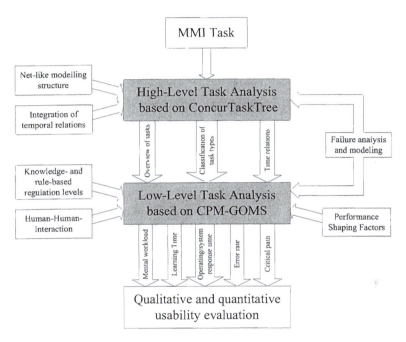

Figure 1. Method for model-based usability assessment and prospective risk analysis

The first part of mAIXuse is a methodological extension of the ConcurTaskTree approach (Paternò et al, 1997), which provides a graphical syntax for the specification of task models, taking into account temporal relations. Within this initial high-level task analysis, the usability investigator gains a systematic overview of the operations of the user, the system and the interactions which are required to achieve a specific task objective within the task fulfillment. As part of the graphical notation, it is possible to describe different task categories and types as well as various attributes and objects. Within the mAIXuse modeling, there are five task categories ("Abstraction", "Perception", "Cognition" "System", "Action" and "Human-Human-Interaction"). The category "Abstraction" provides a higher-ranking task type (mathematically a parenthesis).

Unlike traditional task analysis (Diaper, 2004), within the mAIXuse modeling not only top-down dependencies are presented, but with the help of temporal relations (e.g. sequence, concurrency, disabling, sharing, choice, etc.) also coequal and other depending tasks are linked with temporal interconnections.

Additionally, a human error analysis, which is based on various failure taxonomies, accompanies the task analysis process. Human erroneous actions are classified by their outer appearance, in order to discover potential problems in the interaction task (Swain, 1983). On the one hand the underlying causes are identified by a failure classification, according to GEMS (Generic Error Modeling System) (Reason, 1987) in order to analyse handicaps in the cognitive regulation levels (skill-, rule- and knowledge-based behaviour). On the other hand, failure causes within the perception phase (acoustic, haptic, visual) can be detected by various guidelines and checklists which have been derived from several technical standards.

In the second part of the mAIXuse analyzing process, a cognitive task model based on the initial high-level task analysis is created. Here, parallel tasks concerning cognitive, perceptual and motor low-level operators are decomposed. Subdividing the previously detected high-level decompositions into low-level operators eases the acquisition of cognitive processing steps involved. However, the success of this step is depending on the accuracy of the previous high-level task modeling.

Following this approach, a methodological extension of CPM-GOMS (Cognitive Perceptual Motor – Goals Operators Methods Selection Rules) (John and Gray, 1995), a cognitive task analysis, based on the Model Human Processor (MHP) (Card, Moran and Newell, 1983), is actually being developed.

The MHP architecture is developed on the basis of a computer-oriented architecture and integrates various categories of repository (visual and auditory cache, working and long-term storage). Furthermore, the average processor cycle times are specified by several rules (e.g. Hick's law, power law of practice, Fitts' law etc.). Here, concurrent-working processors are able to model low-level operators (perceptual, cognitive and motor-driven) in a parallel mode, in order to generate an exact and detailed overview of the information processing tasks of the human operators. Especially the parallelism between the processors/operators can be helpful in order to model and analyse multimodal human-machine-interactions, and human-human-interaction.

Dependencies between the users' perceptual, cognitive and motor-driven activities are mapped in a schedule chart (PERT - Program Evaluation Review Technique). Here, the critical path represents the minimum required execution time. John and Gray have provided templates for cognitive, motor-driven and perceptual activities alongside their dependencies under various conditions (John and Gray, 1995). In project Ernestine CPM-GOMS has been validated to real-world task performance by assessing Human-Machine-Interfaces of telephone companies' workstations (Gray, 1992).

Additionally, the second part of the methodology developed includes, in contrast to original CPM-GOMS, the distinction of skill-, rule- and knowledge-based behaviour according to Rasmussen (Rasmussen, 1983) in the PERT-Chart. Moreover, external performance shaping factors are currently being integrated, in order to provide supplementary qualitative evaluation criteria. In addition, a process bar is integrated into the PERT Chart in order to transcribe human, system and interaction activities (obtained in the high level analysis) to facilitate the decomposition of high-level tasks and to create additional help for the user.

In combination with existing quantitative statements on learning and operating time, as well as on failure rates and efficiency, the inclusions mean a powerful extension of CPM-GOMS. In the course of these analyses potential contradictions and conflicts in concurrent cognitive resource utilisation can be detected and risks and hazards in human-machine-interaction can be identified and assessed. In addition, a failure analysis on the basis of above-mentioned human failure taxonomies will also be also possible after the cognitive task analysis.

EVALUATION

The mAIXuse method and the corresponding software tool have been evaluated with medical industrial partners. Furthermore, the developed prospective usability evaluation tool has been investigated in comparison to conventional approaches. Here, the results of the mAIXuse application in a two days workshop with different test groups have been compared to the results of a classical Process-FMEA (Failure Mode and Effect Analysis). Each group was limited concerning the application time to 2 ½ hours. Two consecutive GUI dialogues concerning computer-assisted planning of the implant-based resurfacing of an arthrotic femoral head have been investigated. The corresponding planning and navigation system is currently being developed at the chair of medical engineering. The "Positioning of the safety zone" for the femoral neck and the "Positioning of the implant" have been modeled and analyzed concerning potential human failures. Finally the subjects had to answer a questionnaire in order to evaluate both methods relatively and absolutely concerning user-satisfaction, understandability and learnability.

Results show the advantages of the mAIXuse method in contrast to the conventional Process-FMEA. The new modeling and evaluation tool shows better results regarding all subjective and objective criteria. Even if the subjects had worked in the first session with the mAIXuse tool, they were unable to find an equal number of risks respectively critical risks with the process FMEA. Further analysis of the questionnaires and the discussion with the subjects revealed, that the predefined lowest level of the modeling technique and the use of temporal relations in the mAIXuse approach are responsible for its higher effectiveness.

Objective results are corroborated by the results of the questionnaire. Even the cost benefit evaluation has been rated as "good", because almost two times more potential system-inherent failures have been detected with the mAIXuse method than with the Process-FMEA. Additionally, the application of the software-assisted tool doesn't show a reduction of the communication or the collaboration of the test subjects. This before-estimated negative effect could not been observed during the evaluation. On the contrary, the software tool supports the efficient and effective analysis and the comprehensive overview of the modeling. Objective and subjective test data are shown in Table 1.

Table 1: Evaluation results of the comparison between the mAIXuse and the Process-FMEA application

Objective Criteria	Group A		Group B	
	FMEA	mAIXuse	mAIXuse	FMEA
Detected risks	14	27	29	16
Critical risks	5	9	12	7
Application time	2 ½ hours	2 ½ hours	2 ½ hours	2 ½ hours

Subjective Evaluation
(1 = very good, 2 = good, 3 = satisfactory, 4 = adequate, 5 = poor)

User satisfaction	3-4	2	1-2	4
Understandability	2-3	1-2	2	3
Intuitiveness	3-4	2	1-2	4
Learnability	3	2	1-2	3

DISCUSSION

The new method for prospective usability-evaluation and the corresponding software tool have proven their applicability and practicability in various investigations. After a short introduction in the methodology all test subjects have been able to successfully apply the mAIXuse method and the corresponding software tool on their own. A comparison of the modeling results of the interaction process concerning the same user interface with different test groups shows nearly identical outcome. This leads to the hypothesis that, because of the predefined lowest level of modeling and the integration of temporal relations, an almost unique modeling structure has been developed. The way of coding and presenting is intended to support designers and engineers with the internal communication and to ease the understanding of the human-machine-interaction process. Furthermore the mAIXuse tool enables medical device developers and manufacturers to conduct a model-based usability-evaluation and a usage-oriented risk analysis on their own. The documentation required for the risk management and the usability-engineering process, is automatically generated as FMEA form-sheets.

When using the mAIXuse technique for the modeling of a prototypical interface an interesting and productive side-effect could be observed. In order to model the current state of the use process with an existing system/interface with the mAIXuse tool, the test subjects showed explorative test behaviour. The combination and integration of explorative user tests in the framework of the mAIXuse modeling process show a promising approach for an enhanced usability evaluation method, extending the conventional Cognitive Walkthrough approach.

The objective of this human-centred modeling approach is to optimize the usability of complex medical devices. Nevertheless, interface designer and technical designer shall be supported methodologically with the help of a software-based analysis tool in order to systematically identify potential human-induced hazards with human-machine-interfaces. Especially small and medium enterprises, mainly developing medical devices, shall be supported within the application of the usability-engineering process and therefore within the accreditation process.

ACKNOWLEDGMENT

Parts of the work presented have been developed in the framework of the INNORISK project. The AiF/FQS project has been funded by the Federal Ministry of Economics and Technology (BMWi).

REFERENCES

Bogner, M.S. (1994): Human Error in Medicine. Lawrence Erlbaum Associates.

Card, S.K., Moran, T.P. & Newell, A. (1983): The psychology of Human-Computer-Interaction. Hillsdale, New Jersey: Lawrence Erlbaum Associates.

Diaper, D. & Stanton, N. (2004): The Handbook of Task Analysis for Human-Computer Interaction. London: Lawrence Erlbaum Associates.

Gray, W. D., John, B. E., Atwood, M. E. (1992): The precis of Project Ernestine or an overview of a validation of GOMS. In proceedings of CHI, (Monterey, California, May 3- May 7, 1992) ACM, New York. pp. 307-312

Hansis, M. L., Hart, D., Hansis, D. E & Becker-Schwarze, K. (2001): Gesundheitsberichterstattung des Bundes. Heft 04/01: Medizinische Behandlungsfehler. Robert-Koch-Institut

Hansis, M. L. (2002): Medizinische Behandlungsfehler in Deutschland. Schleswig-Holsteinisches Ärzteblatt. 06 / 2002 S. 51 – 59

508

Hoffmann, K. (1985): Risk Management – Neue Wege der betrieblichen Risikopolitik. Verlag Versicherungswirtschaft, Karlsruhe

John, B.E. & Gray, W. D.: CPM-GOMS (1995): An Analysis Method for Tasks with Parallel Activities. In: Conference companion on Human factors in computing systems, pp. 393-394, NY: ACM Press.

Kindler, M. (2003): Risikomanagement von medizintechnischen Geräten bei Entwicklung, Zulassung und Anwendung In: Weiterentwicklung der Arbeitsprozesse in der Krankenhaustechnik. Strategien, Administration, Kooperation. Kongress TK.

Kohn, L.T., Corrigan, J. & Donaldson, M.S. (2000): To Err Is Human: Building a Safer Health System. Washington, DC: National Academy Press.

Leape, L.L. (1994): „The Preventability of Medical Injury". In: M.S. Bogner (Hrsg) Human Error in Medicine. Hillsdale NJ, Erlbaum Publ.

Merten, M. (2007): Risikomanagement: Den Ursachen auf der Spur. Deutsches Ärzteblatt. Deutscher Ärzte-Verlag GmbH.

Paternò, F., Mancini, C. & Meniconi, S. (1997): ConcurTaskTrees: A Diagrammatic Notation for Specifying Task Models. In Proc. of IFIP Int. Conf. on Human-Computer Interaction Interact '97 Sydney, Chapman & Hall, London, pp. 362–369.

Pfeifer, T. (2001): Qualitätsmanagement: Strategien, Methoden, Techniken. 3. Aufl., Carl Hanser Verlag, München.

Radermacher, K.,Zimolong, A.,Stockheim, M.,Rau G. (2004): Analysing reliability of surgical planning and navigation systems. In: H.U. Lemke, M.W. Vannier et al. (eds.), International Congress Series 1268, CARS'04, pp 824-829

Rasmussen, J. (1994): Skills, Rules, Knowledge: Signals, Signs, Symbols and other Distinctions in Human Performance Models. IEEE Trans. Systems, Man and Cybernetics. Vol. SMC-3, pp. 257-267.

Rau, G., Radermacher, K., Thull, B. & v. Pichler, C. (1996): Aspects of Ergonomic System Design Applied to Medical Worksystems. In: Computer-integrated surgery: technology and clinical applications. (eds. Taylor, R.H.). MIT Press, pp: 203-221.

Reason, J.T. (1987). Generic error-modeling system (GEMS): A cognitive framework for locating human error forms. In: Rasmussen, J., Duncan, K. and Leplat, J., Editors, 1987. New technology and human error, Wiley, London.

Swain, A. D.; Guttman, H. E. (1983): Handbook of Human Reliability Analysis with Emphasis on Nuclear Power Plant Applications NUREG/CR-1278, Sandia Laboratories, Albuquerque, NM 97185.

Wilson, R.M., Runciman, W.B., Gibberd, R.W., Harrison, B.T., Newby, L. & Hamilton, J.D. (1995): The Quality in Australian Health Care Study. Med J Aust. Nov 6; 163(9):pp. 458-471.

CHAPTER **58**

Integrated Analysis of Communication in Hierarchical Task Models focused on Medical Critical Systems

Tomasz Mistrzyk

OFFIS - Institute for Information Technology
Escherweg 2, 26121 Oldenburg
GERMANY

ABSTRACT

Health care belongs to socio-technical systems which are complex and technological prone to accidents. In such systems communication is a major factor for accidents. This paper shows a concept for the documentation and analysis of communication in hierarchical task models. It provides an overview of an approach, which documents communication in an extended hierarchical task model. The proposed method offers a systematic heuristical approach to find weak points with the communication of a socio-technical system especially within medical environment. As a modeling environment, taking into account the special needs for such systems, AMBOSS is used. It allows the specification of rules, objects, topology of tasks, risk assessment factors, barriers and most notably the information flow between tasks.

Keywords: Communication, Hierarchical Task Models, Socio-technical and Safety Critical Systems, Medical Environment

INTRODUCTION

While examining the development of socio-technical systems it was conspicuous that their complexity denotes a multi-plane continuous growth. This results in high demands on the analysis as well as the correct composition of such systems. Communication is one of the key positions because it is responsible for the adequate coordination of the tasks within a socio-technical system. Weaknesses within communication between actors in socio-technical safety-critical systems are considered to be the main cause for critical events or accidents (Bellamy, 2007). There are several examples which present the importance of correct communication in a safety-critical system. Most prominent, an example in the medical area describes cases in which the wrong patient undergoes an invasive procedure. A mix-up of patients who are to be operated was an effect of latent system conditions such as wrong communication (Chassin, 2003).

Health care belong to industry which is complex, reliant on high technology and team based prone to accidents. Kohn et al. argue in the book *To Err is Human* that there is an important difference between health care and other safety critical industries. In healthcare, an accident happens usually to a third party and not directly to a worker in a company. The health professionals are rarely affected by damage. However, most of the time harm occurs to only one patient at a time, not to a group of individuals, which promotes the hiding of an accident or incident (Kohn et al, 2005). Wrong communication is one of the crucial parameters which lead to latent failure. That is why discovering and fixing such failures has a remarkable effect on building safer systems.

Aim of the article is to give an overview of an approach which documents communication in an extended hierarchical task model and therefore offers a systematic analysis which could be used in a safety medical environment. An appropriate specification of communication in combination with the hierarchical structure of a task model plays an important role at this place.

The first section presents a short introduction into task modeling and the characterization of the systems that are involved. Then the task modeling environment will be described. Additionally, the communication model which is used for specification and analysis is presented. Further the integrated method for specification and analysis of communication in a task model will be presented.

TASK MODELING

Task modeling transfers the relevant knowledge of the user in structured task knowledge during the task analysis. This knowledge can be specified in hierarchical task models. Task modeling approaches (such as Baron et al., 2006, Giese et al.,

2008., Lu et al., 2002, Stuart, 2004) typically concentrate on the hierarchical decomposition of tasks into subtasks and their relative ordering with temporal relations. Task models are used in the early phases of a system design or as documentation method for the result of task analyses. In both cases they are semiformal in nature in the sense that they formally structure (hierarchy, temporal relations) informal elements (task descriptions). They have been used to some extent in system design (Paris et.al., 2003) or user interface design (Puerta, 1996), (Szekely et al., 1992), but increasingly are also used for the design and analysis of workplace situations (Stuart, 2004).

An important concept in most task modeling approaches concerns the actors performing the tasks. In CTTE (Mori et al., 2002), for example, explicitly identified roles co-operate in the concurrent performance of a highly structured task. An appropriate role or user model is included in all approaches mentioned above. Using these models, it is possible to describe the achievement of goals in complex so-called socio-technical systems.

On the one hand these systems include technical components, such as computers, transport and telecommunication devices as well as production machines. On the other hand human actors, who are performing manual tasks and interactive tasks, use machines to support their work. The socio-technical theory is based on the general system theory (Bertalanffy, 1968). A system in this theory is a composition of several autonomous parts which are interdependent and work together to reach the same goal of the whole system. Examples are the case of a nurse using an infusion pump or a crew in an aircraft performing a landing procedure.

The task model of such a system can be used to specify the order of tasks and the rationale behind the planning and structuring of the job to be performed. This can be useful in order to analyze the technical system for optimization purposes, or to start the design of a new not yet existing process. For instance designing procedures used for a user interface of a new MRI (Magnetic Resonance Imaging) System. Another field could be to find hazards in already settled procedures (Mioch, et al., 2010).

Specifying task models of socio-technical systems are especially relevant when safety-critical processes and systems are concerned. Examples of these systems can be found in health care (e.g. hospital wardens), industrial production (e.g. power plants), or traffic systems (e.g. airplanes, vessels). If a sufficiently rich specification technique is available, the task model can be helpful in detecting potential problems created by, for example, inadequate task order, inadequate mapping task to actors, lack of time or correct information especially for critical tasks.

TASK MODELING ENVIRONMENT AMBOSS

The proposed integrated methods to specify and analyze communication are considered task models that have been represented using the AMBOSS notation. Amboss (Giese et al., 2008) is a free modeling environment following the traditional approach of hierarchical task structure. Amboss (Giese et al., 2008) is a tool developed at the University of Paderborn supporting hierarchical task modeling

512

especially for safety critical systems. It can be downloaded from the university's homepage (AMBOSS, 2009). The software provides the typical tree-based editing functions, such as editing a child node or a sibling node; apart from that, however, AMBOSS allows direct editing and manipulation of nodes and connections for simple structural manipulation tasks. AMBOSS allows the description of tasks at different levels of abstraction in a hierarchical manner, represented graphically in a tree-like format. In Figure 1 the meta model of the task model is shown.

AMBOSS provides a set of temporal relations between the tasks such as; sequential: The subtasks are performed in a fixed sequence, serial: the subtasks are executed in an unsystematic sequence, parallel: in this relation the subtasks can start and end at random relation to each other, simultaneous: the subtasks start in an arbitrary sequence with the constraints that there must be a moment when all tasks are running simultaneously before any task can end, alternative: just one randomly selected subtask can be executed. There are almost the same temporal relations that can be found in TOMBOLA (Uhr, 2003) or in CTTE (Mori et al., 2002). A task node without any subtasks is automatically noted as an atomic task.

The modeling environment has additional distinct views of a task model, which can be used for inspecting particular attributes of the tasks. For example, if an analyst likes to observe what kind of objects are manipulated in a system by a particular task, he can switch to the object view, take a look over the model, and analyze the dependencies between tasks and objects. It is also possible to review what kind of object is associated to a particular task, what kind of access rights (read or write) the task has and in which room the objects are located.

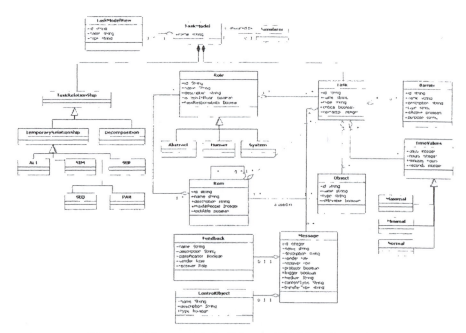

Figure 1: Meta model of the AMBOSS task model

AMBOSS also enables the handling of topological aspects and relationships. To capture dependencies linked to the topology of a system, AMBOSS is able to build a relation between a room and a task which is performed in this room. The user is able to specify, for example, that the performance of critical tasks is only allowed in a particular room. An expert can specify the relationship either from the point of view of a task or a room. Additionally, every object of the task model can be assigned to a room in a similar way as the task itself. In addition there is an option to specify if a particular actor has a permission to enter a room or not.

Analogous to other environments (Biere et al., 1999), (Mori et al., 2002) AMBOSS contains the concept of roles and actors. Basically there are three types of actors used: *system, human* and *abstract*. The user can specify additional actor subtypes, and instantiate each actor, to make it more concrete and to fit the role to a particular socio-technical system. If an actor is linked to a task, his role can also be specified as being the person who is responsible for a particular task. This concept helps to organize areas of responsibility for each actor involved in a task model.

The main purpose during the development of AMBOSS was to provide a modeling environment that provides support for the development and analysis of task models in safety critical domains. For modeling tasks in such an environment, the model needs to be enhanced with more adequate parameters. For safety purpose, AMBOSS contains the concept of barriers. This concept allows specifying parameters helping to protect human life and/or computer infrastructure. In a production process, barriers could be used to protect workers from hot temperatures or lead apron which offers personnel radiation protection. Barriers are in general a special kind of dynamic object which can be activated or deactivated by tasks.

Similar to other modeling approaches, (Baron et al., 2006), (Mori et al., 2002), (Uhr, 2003) AMBOSS is able to simulate a task model. The simulator presents the user exactly what happens in a task environment at a particular moment. A finished task model can be simulated by taking into account the task hierarchy, temporal relations which provide the task execution order, and the communication flow which shows messages along with their parameters. Additionally, the user is able to observe the activation and deactivation of barriers during the simulation. With the simulator the model can be proved with regard to the interdependency between tasks and their parameters. If discrepancies are detected, they can be corrected immediately. In that way, the user is able to validate his semi-formal task model (Van Welie et al., 1998). Using parameters like objects, barriers, risk factors, roles, timing and communication, it is possible to describe a task model in more detail and to create a good impression of the tasks with different views of the same model.

MODELING COMMUNICATION

As mentioned before, communication plays a key role for the correct cooperation between human beings and interaction with technical systems. Communication can become itself a critical factor in the analysis of safety-critical systems (Leveson, 2004). To propose an adequate integration of communication into the model, a closer look into the relevant parameters of communication in a safety environment

514

is to follow. As a fundamental theoretical model, the information-communication model of Shannon and Weaver (Shannon, 1949) was chosen. The Shannon and Weaver communication model originates in the information theory and proposes that every communication process has to include six elements: a source, a sender, a message, a channel, a decoder and a receiver. In this model, communication is seen as the exchange of information (signals) between transmitter and receiver. This excludes a priori a large part of aspects otherwise coming under the term of communication, such as psychological, social, or ethical issues. Basing the following method on information theory restricts communication in that model to the flow of information. In this context, every task and its associated role in a model can be sender and recipient of information. Hence, the user is able to describe a communication flow while taking into account how the actors communicate with each other, who is communicating with whom, and what kind of role a piece of particular information plays regarding a specific task.

One of the relevant parameter of communication is the information itself and the medium which is used for the transfer of a message. The type of medium can be characterized as electronic, physical or a mix of both. Additionally, the modeler can specify the type of transfer as synchronous or asynchronous and the type content which can be numerical, textual, and graphical or gesture.

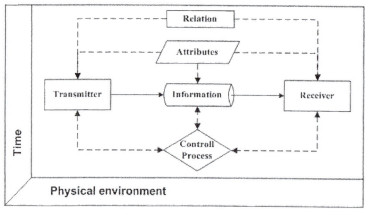

Figure 2: Communication model

Any communication process includes a control structure. Such structure is implemented to ensure the timeliness, completeness, and correctness of the information. According to that it is possible for each communication process to model a feedback, which is a sub-process in the communication between receiver and sender. A feedback is a helpful parameter which provides information for the sender if his message was received and understood. Expected but not received, feedback, for instance could leave an actor uncertain about whether a necessary condition is fulfilled to start another action or repetition of the same action is required. In the field of health care is in the focus especially the communication between doctors and nurses (Burd et al., 2002) as well as health personal and patient (Brattstörm, 2005).

Just like every other process, communication takes place within some sort of time frame. This time frame percolates the entire communication process and interferes with its objects. For instance, attributes of actors can change during an executing time or information might lose or gain importance. In safety critical systems communication often underlay strict time constraints. In addition to the feedback, the extended communication model allows to express also the criticality of a message and mark it as critical or not. If a message is modeled as critical, the modeler is able to link such a message with a control object.

A control object is another concept of helping to ensure safe communication. This could be for example a physical object like an envelope, but also a procedure or a convention. Dependent to the quality of the control object such object can be specified as safe or not. The last concept to ensure the communication in this model is to specify if a message is controlled by a protocol. This could be for example a procedure which has to be followed by a fellow preparing patient for a surgery. Taken as a whole, the advanced communication model of AMBOSS allows specifying many details of the communication. This can be seen as a new dimension beside the traditional dimensions like task, actor and object of a task model. Because of this advantage AMBOSS is used as a base for the analysis of communication in a socio-technical system.

APPROACH

The main objective of the provided method is to integrate communication analysis into a hierarchical task model. Using this heuristic method the analyst can systematically search for possible patterns of weak constellations in the communication flow and latent communication errors. As was already mentioned such errors have got the potential for provoking conditions within the local workplace and with active failures and local triggers create an accident opportunity (Reason, 2000). After an erroneous communication has been found the method can also be used to find a solution for the correction of incorrect communication processes. The method can be started after a task model has been created. It contains the following four steps. In the **first step**, all communication sequences are classified into one of three groups of criticality.

Table 1 Criticality Groups of Communication

Group criticality	Transmitter status	Receiver Status
A	critical	critical
B	non critical	critical
	critical	non critical
C	non critical	non critical

This can be done based on the critical status of the sender and the receiver which provides a basis for the priority of the fact-finding, or based on critical status which the communication process already got, an example of such classification is shown in **table 1**. In the **second step** an expert looks closer on each message's sequences, starting with the highest priority class A. This exploration contains a binary evaluation of each sequence from the point of view of both communication partners. This part of the analysis already shows the most important weaknesses of the communication flow.

Table 2 Combination of the characteristic using the parameter protocol

Using Communication Protocol		
possible consequences	Transmitter	Receiver
No consequences	always	always
Possible latent failure	always	sometimes
	sometimes	always
Latent Failure	sometimes	sometimes
Deviation	ignore	...
	...	ignore

An example of the binary evaluation with the parameter protocol is shown **table 2**. After the communication weaknesses have been identified in the system the expert starts with the qualitative judgment. The objective it to judge how good/bad the current using type of e.g. *communication medium* fits to the communication process. The expert can choose between *very good, good, sufficient, not suitable* and *not available*. Based on this judgment he/she can give a more informative statement about each parameter within the different sequences of the communication flow. A good example of a medium which is still used even on an intensive care station, are results of blood tests which are handwritten on a paper form. Even if the medium would be rated as sufficient the protocol handwritten should be rated as not useable.

In the **third step**, the aggregation of the identified weak spots in the system takes place. The analyst is able to recognize which parameter is most affected by the communication process and how often it does. Furthermore it can be seen when a particular parameter is used incorrectly and weakens the system. Our paper form example from step two started probably in the laboratory of the hospital or on the intensive care unit when a nurse sent the form together with the patient's blood and the request to the laboratory. In addition the most endangered communication process can be detected. Finally, the analyst can find out at which position of the communication flow the system shows deviations from the expected procedure.

The last **step (four)** is used to define suggestions that are classified along the single communication parameters for a proposed improvement.

By applying this analysis method to real-world case studies, it has been found out that it is especially valuable for the detection of latent communication errors in a socio-technical system. These errors occur somewhere in the system but not next to the person that suffers from the consequences when actually performing a task. The author observed such a failure during a visit on an intensive care unit. An example was the erroneous adjustment of a perfusor which happened due to a blurred communication protocol between the responsible actors. The definition of the protocol has been identified as a latent error during the communication analysis which had been done together with the personal on the intensive care unit.

CONCLUSION

This paper concentrates on the issue of modeling analysis of communication in hierarchical task models. That communication plays a vital role in socio-technical systems is revealed within the discussion above. Other task modeling approaches allow only a limited specification of information flow (Mori et al., 2002). In the modeling environment within AMBOSS, communication is treated as structural information, creating additional relations by introducing message objects between tasks. The proposed heuristic method used this advantage and provided a systematic technique to document, analyze and correct communication processes in socio-technical safety-critical systems.

ACKNOWLEDGMENTS
I particularly wish to thank Prof. Gerd Szwillus who provided helpful comments during the finding phase and finally provided the right amount of inspiration needed on my way to my dissertation.

REFERENCES

AMBOSS, (2009). http://mci.cs.uni-paderborn.de/pg/amboss/index.php

Baron, M., Lucquiaud, V., Autard, D. & Scapin, D.L (2006).: K-MADe: un environement pourle noyau du modèle de description de l'activité. Proceedings of the 18th Frenchspeaking conference on Human-computer interaction, Montreal, Canada.

Bellamy, L.J. Geyer, T.A.W. (2007) Development of a working model of how human factors, safety management systems and wider organisational issues fit together, Health and Safety Executive Report, RR543, http://www.hse.gov.uk/research/rrpdf/rr543.pdf

Bertanalanffy, L. (1968). General System Theory. (Ed.) New York, George Braziller

Biere, M., Bomsdorf, B. and Szwillus, G. (1999) Specification and simulation of task models with VTMB. In CHI '99 Extended Abstracts on Human Factors in Computing Systems, Pittsburgh, Pennsylvania, ACM Press, New York, NY.

Brattstörm, M. (2005) Communication Problems in Health Care, In Clinical Rheumatology, Vol. 6, pp. 158-161, Springer London

Burd, A., Cheung, K.W., Ho, W.S., Ying, S.Y and Cheng P.H. (2002) Before the paradigm shift: concepts and communication between doctors and nurses in a burns team, In Journal of the International Society for Burn Injuries, Vol. 28,

518

Isssue 7, pp. 691-695

Chassin M.R. Becher, E.C. (2003) The wrong patient. Ann Intern Med., Nr.136, pp.826-33

Giese, M., Mistrzyk, T., Pfau, A., Szwillus, G. and von Detten M. (2008) AMBOSS: Task Modelling Approach for Safety-Critical Systems, Inn Proceedings of the 7th Int. Workshop on Task Models and Diagrams TAMODIA' 2008, Springer-Verlag

Kohn, L.T., Corrigan, J.M and Donaldson, M.S. (2005) To Err Is Human Building a safer Health System, Institute of Medicine, Committee on Quality of Healthcare in Americe, National Academy Press, Washington, D.C.

Leveson, N. (2004) A New Accident Model for Engineering Safer Systems, Safety Science, Elsevier Science Ltd

Lu, S., Paris, C., Vander Linden, K. (2002) Tamot: Towards a Flexible Task Modeling Tool. In The Proceedings of Human Factors 2002, Australia.

Mioch, T., Mistrzyk, T. and Rister, F. (2010) Integrated Procedure Designing and Validation by Cognitive Task Model Simulation, In proceeding of the 19th Annual Conference of Behavior Representation In Modeling Simulation (BRIMS), USA, SC, Charleston, pp.

Mori, G., Paternò, F. and Santoro, C.: (2002) CTTE: support for developing and analyzing task models for interactive system design. In IEEE Trans. Softw. Eng. Vol. 28, pp.797-813.

Paris, C., Lu, S. and Vander Linden, K. (2003) Environments for the Construction and Use of Task Models. In: The Handbook of Task Analysis for Human-Computer Interaction. Edited by Dan Diaper and Neville Stanton. Lawrence Erlbaum Associates. pp 467-482.

Puerta, A. R. (1996) The Mecano Project: Enabling User-Task Automation During Interface Development. AAAI96: Spring Symposium on Acquisition. Learning and Demonstration: Automating Tasks For Users. pp. 117-121

Reason, J. (2000) Human Error: Models and Management, In British Medical Journal, Vol. 320, pp. 768-770.

Shannon, C. (1949) Weaver, M.: The Mathematical Theory of Communication, University of Illinois

Stuart, J. & Penn, R. (2004) TaskArchitect: taking the work out of task analysis. In Proceedings of the 3rd Annual Conference on Task Models and Diagrams TAMODIA '04, vol. 86. ACM Press, New York, NY, pp.145-154

Szekely, P., Luo, P. and Neches, R. (1992). Facilitating the Exploration of Interface Design Alternatives: The HUMANOID Model of Interface Design. In Conference Proceedings of CHI '92, pp. 507-515

Uhr, H. (2003). TOMBOLA: Simulation and User-Specific Presentation of Executable Task Models. HCI International 2003

Van Welie, M., Van der Veer, G.C., and Eiens, A. (1998). Euterpe - Tool support for analyzing cooperative environments. In: proceedings of the Ninth European Conference on Cognitive Ergonomics, Limerick, Ireland

CHAPTER 59

Effects of BCMA on Clinicians' Communication, Coordination and Cooperation

Gulcin Yucel[1,3], Bo Hoege[2], Vincent G. Duffy[3]

[1]Istanbul Technical University
Turkey

[2]Technische Universität Berlin,
Germany

[3]Purdue University, IN
USA

ABSTRACT

In many hospitals various kinds of information technologies are implemented to improve patient safety. In this article we focus on usage of bar code medication administration (BCMA) as one application of information technologies. The study was conducted 2009 in two US-American hospitals; one was planning to implement BCMA and one was already using BCMA for 10 years. The study aimed to identify patterns of communication, coordination and cooperation between nurses and pharmacists, to compare work processes of medication administration and to identify weak points. As a result of the study it was observed that BMCA software reduced flexibility end error potential of medication administration and by this supported patient safety. But less flexibility constrained spontaneous treatment and increased administrative interactions with the system. Since the data analysis has not been finished yet, only parts are introduced and more results are expected in future.

Keywords: Healthcare IT, BCMA, Work Analysis, C3-Modeling

INTRODUCTION

In many hospitals various kinds of information technologies are implemented to improve patient safety. The study which we are going to introduce in this article is focusing on usage of bar code medication administration (BCMA) as one application of information technologies. Although the system of BCMA provides the "Five Rights" of medication administration: right patient, right medication, right dose, right time, and right route (Payton et al., 2007), in literature serious reasons can be found to scrutinize IT applications in the domain of healthcare. Side effects of IT usage could not only disturb the general administration in hospitals but also the health of the hospitals' customers – the patients.

In literature, different studies can be found which point out that the use of BCMA can have negative effects on patients' health. We are going to introduce some of these studies to give reasons for our motivation, and then we are going to introduce our approach on analyzing effects of IT on clinicians' communication, coordination and cooperation.

On the one hand, different studies show that dispensing errors and potential ADEs decrease after implementing bar code technology (Poon et al., 2006). The implementation of bar code technology seems to reduce the amount of medical errors (Oren et al., 2003, Rough et al., 2003, Sakowski et al., 2005 in Carayon et al., 2007, Coyle and Heinen, 2002; Johnson et al. 2002; Puckett, 1995 in Koppel et al., 2008). Hence research gives scientific reasons for the healthcare sector to implement BCMA systems for prevention of medication errors (IOM, 2007; Bates, 2000 in Koppel et al. 2008). On the other hand there is research which reveals many unintended consequences of BCMA (Bates et al., 2001; Patterson et al., 2002; McDonald, 2006; Mills et al., 2006). We found that many studies were conducted to identify the side effects of health information technology (HIT) but limited researches have been undertaken to find the reasons for these side effects (Pirnejad et al., 2008; Koppel et al., 2008).

At HCII2009 we introduced an approach (Yucel et al., 2009) to identify potential reasons of the side effects of BCMA by analyzing communication-coordination-cooperation (C3) among clinicians. In this article we want to show first results of the study we conducted in two hospitals. We followed the request by Pirnejad et al. (2008) who suggest analyzing the problems in current intra-organizational communication, current IT usage in healthcare communication and its potential benefits and pitfalls in order to find out the reasons of the side effects.

CONCEPTUAL MODEL

To prepare our analysis on effects of IT on clinician's C3 we designed a conceptual model which demonstrates our hypotheses (see Figure 1). The model contains not only BCMA effects on C3 mechanisms but also the relationship of the C3 mechanisms with side effects and success. In our model, a BCMA implementation could result in two ways: success or side effects. Also, it affects both patients and clinicians. Therefore, in the conceptual model both clinician side and patient side are taken into account.

Success and side effects for clinician and patients can be defined separately. In the main hypothesis, it is aimed to find out if BCMA affects C3 mechanisms between nurse, pharmacists and physicians. According to Varpio (2008) the links between inter-professional communication and medical errors are slow. Finding out the links between them will contribute to understand how and why medical errors occur. Therefore in the second hypothesis, it is aimed to find out the relationship between C3 mechanisms and violations and workarounds. In hypothesis 3 it is intended to find out the relationship between C3 mechanisms and harm and hazardous to patient. Hypothesis 4 and 5 are constructed to find out the relationship between C3 mechanisms and the success of BCMA implementation.

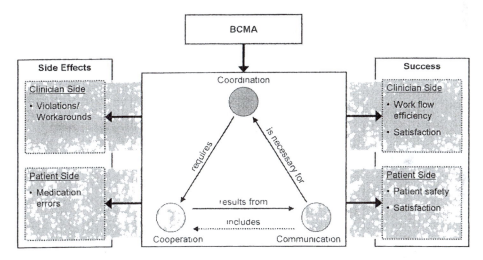

FIGURE 1 Conceptual model based on C3 model by Müller (2005)

To measure the effects of BCMA on C3 mechanisms and the relationship between C3 mechanisms and the outcome we used different methods. For analyzing the effects of BCMA on C3, observations combined with interviews were conducted to obtain an understanding of the medication administration process and to design a so called C3-model by the UML based C3 modeling technique (Foltz et al., 2000).

For measuring the errors and violations a survey was developed and sent to the Non-BCMA hospital. The survey was divided into sections about background (e.g. demographic data, work features), communication (e.g. channels, patterns) and a violations and medication error section where participants gave answers about frequencies and contributors.

The study was conducted in two US-American hospitals in 2009; one was planning to implement BCMA (Non-BCMA-hospital) and one was already using

BCMA (BCMA-hospital) for 10 years. The exploratory study aimed to identify patterns of communication, coordination and cooperation among nurses, pharmacists and physicians, to identify weak points, and to compare work processes of medication administration. The interviews were conducted in both hospitals, the C3 diagrams are only drawn for the hospital which implemented BCMA during the study.

METHOD

For obtaining results regarding clinicians' C3 mechanisms, observations, interviews and surveys were conducted. Each hospital was visited once for a general introduction into the task of medication administration. Specific hospital units of nursing and pharmacy were introduced as well as specific HIT (e.g. Pyxis MedStation System).

For the preparation of the interviews, four observations in specific nursing units were conducted, each of one hour duration. Focus of the observations was the medication administration. As a result, C3 models were drawn to visualize the processes for paper based and barcode based medication administration.

INTERVIEWS

The interviews for the staff of the Non-BMCA-hospital were only about present events or procedures and contained questions regarding IT usage, problems in medication administration process, and features of communication between nurses, pharmacists and physicians. Before the interviews were applied, cognitive interviews were conducted with one nurse and one pharmacist in order to find out the understandability as well as the difficulty of the questions (Willis, 2004). Based on the responses, the questions were revised.

Interviews were done with an ad hoc sample of 6 nurses and 10 pharmacists in the Non-BCMA-hospital. In the BCMA-hospital, 10 nurses and 4 pharmacists were interviewed. The participants were chosen from different nursing units including surgery, ICU, AICU, high risk pregnancy, clinical education, psychiatry. The numbers of experienced and inexperienced nurses (i.e. from 30 years to 11 months) were approximately balanced.

The interview results were also used to design a survey. The survey was developed based on previous research and based on the answers of the interviewees. Since the survey data is not yet completely analyzed it will be introduced in a future article.

C3 MODELING

The C3-notation is a semi-formal notation form and is based on the Unified Modeling Language (UML). This notation form allows the modeler to describe and visualize weakly structured work activities for generating a mutual understanding of the process itself. It is used to describe a present process as well as to design theoretical processes. Furthermore, if best practice was obtained, the C3-notation could be used for keeping this knowledge.

Highly collaborative C3 mechanisms can be more understandable in a scheme where all involved interaction partners are put into parallel columns. The actions of the interacting partners are noted in symbolic shapes. Furthermore, the notation form allows visualizing received information, used tools, decisions and connections.

For this study, the technique was chosen since we expected to identify weak points like lacks of information, communication errors and to understand how the interaction between all involved C3 partners happens.

RESULTS

INTERVIEWS

The interviews focused on the experiences of the nurses regarding the medication administration process. General problems, reasons for contacting interaction partners such as other clinicians, and advantages and disadvantages of the BCMA system were of major importance.

Problems of the paper based system

From the paper based system mostly procedural mistakes were reported. When a patient is transferred to another nursing unit not always the patient's medication is transferred as well. Also the nursing unit sometimes does not know which medications were given to the patient by the previous unit since the information was not transferred.

In general, pharmacists can not check if and which medications were given to patients. This is a system immanent problem since pharmacists do not have access to this information. Also system immanent is that, as reported, pharmacists are often interrupted by calling nurses while putting orders into the pharmacy system. As typical is mentioned that doses are not charged after administration which can lead to double medication administration.

Contact reasons

The main difference in the way of how nurses contact pharmacists in the BCMA based system is that the system enables sending requests for missing doses. About both systems it is reported that most of the reasons are equal like calling for wrong or missing medication, questions about medications, informing about new admissions or updating physicians' medication orders.

In the BCMA based system problems with the scanning of barcodes are generally reported as a reason to contact pharmacists. The pharmacists in the paper based system report that nurses are also contacted for giving information about drug levels, for caring patients, for learning patient information (e.g. weight, age) and if a medication was given or not. By contrast pharmacists in the BCMA system report not to have any personal contact with nurses.

Advantages of BCMA

Nurses reported that BCMA has several advantages regarding affects on patient safety. The system prevents medication errors and allows tracking medications and patients. It gives a schedule to follow and avoids giving doses twice or more times. It is beneficial that early or late administrations have to be reasoned. BCMA clarifies not only doses and medications but also gives accurate information about medication and time.

Affects on work flow were also reported as mostly positive. The automatic charging gives accuracy to the records. Although the charging time was reported to be shorter compared to the paper based system, the process over all is mentioned not to save more time.

Affects on the communication between nurses and pharmacists were described as more straight-line. Pharmacists share the same software and use the same interface. Wrong or missing medications can be requested by the BCMA system's interface. The floor pharmacist who is probably busy with inpatient care has not to be contacted for these reasons anymore. In summary, the BCMA system was mentioned to clarify the medication administration and to help timing the administration.

BCMA was mentioned to have positive effects on nurse to nurse communication by visualizing the schedules of medication administration. Another benefit which was reported is that the system allows running reports for 24 hours about medication administration. This function allows seeing an overview what medications were given and which not.

Complaints and disadvantages of BCMA

On the side of negative answers, wireless barcode scanners were mentioned to be more helpful than the used wired ones. Scanning errors also occur. Regarding the administration of single-doses, it was reported that it is not obvious if the dose was administered. Only the medical history check allows proving this. Single-doses are not marked to be given, refused or not given. When a physician changes medication orders, it can take up to three hours to put the changes into the BCMA system by Pharmacy Department which constrains immediate reactions. Also the system differentiates between combinations of doses (e.g. $3 \times 25mg \neq 1 \times 50mg + 1 \times 25mg$).

BCMA can also cause new violations. It was reported that extra armbands can be printed and scanned in the nursing station instead of directly scanning patient's barcode. Scanning of armbands can also be by-passed because of scanning failures. However it is also reported that after the system update, it will not be possible to print an extra armband. Nurses can only write patient's SSN if the patient is not there; they enter SSN and then write why medications are hold. Finally, it was mentioned that the BMCA system can validate or refuse medication but it does not provide rules which could be necessary to check (e.g. if blood pressure and heart rate are low, do not give this medication).

526

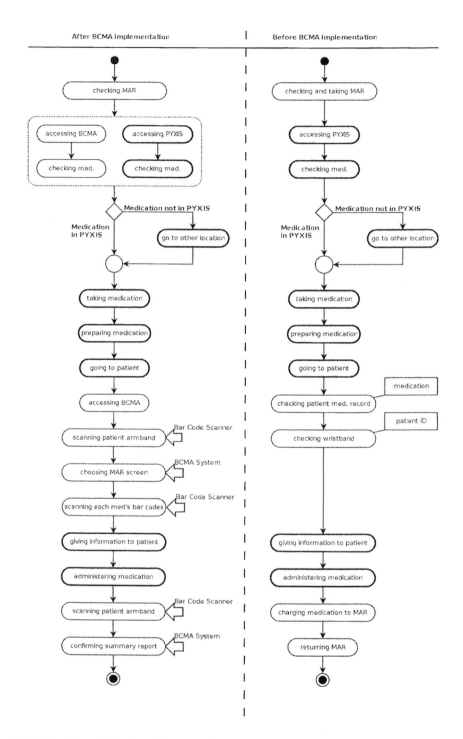

FIGURE 2 C3 models of medication administration processes after (left) and before

(right) BCMA implementation. Round boxes represent actions; bold boxes are actions which occur in both processes. Actors in both models are nurses only. For lack of space, possible interactions with pharmacists or physicians are not shown.

C3 MODELS

In Figure 2, C3 models of the same hospital are shown. The left model contains the new implemented BCMA system; the right model was built for the former paper based system.

Usually, there is no interaction with other clinicians during medication administration but in both processes there are many possibilities for interruptions after and during each step (e.g. other nurse/patient requests help, etc.). For all exceptional situations such as missing dose, wrong medication and new order all the interactions between physicians, pharmacists and nurses were drawn into C3 models and will be introduced in a future article.

Overall, both processes are very similar. Half of the BCMA actions are equal to the actions which were observed in the prior paper based system. The differences are found in the preparation phase of medication administration: parallel to the Pyxis MedStation system the BCMA system has to be used. Although this action should not be interrupted, interruptions were often observed.

The administration itself differs in the steps of identifying the patient, proving and charging the medication. After the action *going to patient* six new actions replace four old ones. So to administer the medication in the BCMA based system, more steps are necessary and lead to new difficulties. The interaction with the barcode system in the patient room was mentioned to be unfamiliar and uncomfortable if only a touchpad was available as an input device. Due to limited space to handle and scan the medication, often the patients' beds or their personal tables are used. These tables have to be cleaned first which is reported as time consuming. In general, if a medication package contains more than the prescribed dosage (e.g. liquids), the rest of the medication has to be returned to the Pyxis MedStation accompanied by a witness. In this case, it was reported that sometimes the rest of medication together with the barcode was returned before scanning the barcode.

DISCUSSION

This article introduced why it is necessary to concentrate research on the changing area of HIT systems such as BCMA. A conceptual model was described which aims to measure influences of HIT systems on C3 mechanisms and to evaluate its outcomes. In parts, this model was used in two field observations: one hospital with a paper based medication administration, one with a barcode based system. The fact, that the hospital which used the paper based system changed to a BCMA system during the study allowed to investigate influences on the clinicians work tasks caused by this change. At the current state of our analysis, we can conclude that (1) BMCA software reduced flexibility end error potential of medication administration and by this supported patient safety. (2) Less flexibility constrained spontaneous treatment and increased administrative interactions with the system. (3) BCMA reduced the interruptive calls by nurses and improved the communication between nurses and pharmacists by providing a shared interface. (4) Handling and usage problems were observed (which possibly disappear after a learning phase).

We are positive that the ongoing analysis of the data will provide an opportunity for recommendations regarding the elimination of unintended side effects of BCMA and the identification of supportive effects of BCMA on C3 mechanisms and work flows.

ACKNOWLEDGEMENTS

The authors would like to thank Prof. Steve Abel, Prof. Carol Birk, Dr. Kyle Hultgren, the American Society of Health-System Pharmacists (ASHP) and Regenstrief Center for Healthcare Engineering at Purdue University for their support through the project. We thank all interviewees and the personnel of both hospitals for their friendly and patient cooperation.

REFERENCES

Bates, D.W. (2000), "Using information technology to reduce rates of medication errors in hospitals." *BMJ*, 320, 780-791.

Bates, D.W., Leape L.L., Shabot, M.M. (2001), "Reducing the Frequency of Errors in Medicine Using Information Technology." *J Am Med Inform Assoc.*, 8 (4) 299-308.

Carayon, B., Wetterneck, T., Hundt, A.S. et al. (2007), "Evaluation of nurse interaction with bar code medication administration technology in the work environment" *Journal of Patient Safety*, 3(1), 34 – 42.

Coyle, G.A. and Heinen, M. (2002), "Scan your way to a comprehensive electronic medical record. Augment medication administration accuracy and increase documentation efficiency with bar coding technology." *Nurs Manag.*, 33, 56-59.

Douglas, J. and Larrabee, S. (2003), "Bringing barcoding to the bed side-implement information technology to track and reduce medication error." *Nurse Management*, 34, 36-40.

Foltz, C., Killich, S., Wolf, M., (2000), K3 User Guide, IAW Aachen, Internet: http://www.iaw.rwth-aachen.de/download/produkte/k3_userguide_2000-11-21.pdf, last access: 1.3.2010.

Institute of Medicine (2007), *Preventing Medication Errors*. The National Academies Press, Washington.

Johnson, C.L., Carlson, R.A., Tucker, C.L. and Willette, C. (2002), "Using BCMA software to improve patient safety in Veterans Administration Medical

Centers." *J Healthc Inform Manag,* 16, 46 –51.

Koppel, R., Wetterneck, T., Telles, J.L. Karsh, B.T. (2008), "Workarounds to Barcode Medication Administration Systems: Their Occurrences, Causes, and Threats to Patient Safety." *J Am Med Inform Assoc.*, 15,408 - 423.

McDonald, C. J. (2006), "Computerization Can Create Safety Hazards: A Bar-Coding Near Miss." *Annals of Internal Medicine.*, 144, 510 - 516.

Mills, P.D., Neily, J., Mims, E., et al. (2006), 'Improving the Bar-Coded Medication Administration System at the Department of Veterans Affairs." *Am. J. Health Sys. Pharm.*, 63, 1442-1447.

Müller, E., (2005), "Kooperation und Koordination. (unpublished lecture notes)". TU Chemnitz - IBF - Professur Farbikplanung und Fabrikbetrieb. Internet: http://chemie.tu-chemnitz.de/mb/FabrPlan/ Kooperation.pdf, last access: 20.01.2008

Oren, E., Shaffer, E.R., Guglielmo, B.J. (2003), "Impact of emerging technologies on medication errors and adverse drug events." *Am. J. Health Sys. Pharm.*, 60(14), 1447 - 1458.

Patterson, E.S., Cook, R.I., Render, M.L. (2002), "Improving Patient Safety by Identifying Side Effects from Introducing Bar Coding in Medication Administration." *J Am. Med. Inform. Assoc.*, 9(5), 540 – 553.

Payton, J., et al. (2007), "Bar code medication administration system improves patient safety." *Journal of Arkansas Medical Society*, 104(4), 84-85.

Pirnejad H., Niazkhani, Z., Berg, M., Bal, R. (2008), "Intra-organizational communication in healthcare - Considerations for standardization and ICT application." *Methods Inf in Med.*, 47(4), 336 – 345.

Poon, E.G., Cina, J.L., Churchill, W., et al.(2006), "Medication dispensing errors and potential adverse drug events before and after bar code technology in the pharmacy." *Annals of Internal Medicine.* 145(6), 426-438.

Puckett, F. (1995), "Medication-management component of a point-of-care of system." *Am J. Health Sys Pharm.*, 52, 1305-1309.

Rough, S., Ludwig, B. and Wilson, E. (2003), "Improving the medication administration process: the impact of point of care bar code medication scanning technology." Proceedings of the ASHP Conference, New Orleans. LA.

Sakowski, J., Leonard, T., Colbrun S., et al. (2005), "Using a bar-coded medication administration system to prevent medication errors in a community hospital network." *Am J Health Syst Pharm.*, 62, 2619 - 2625.

Varpio, L., Hall, P., Lingard, L. et al., (2008), "Interprofessional communication and medical error: A reframing of research questions and approaches." *Academic Medcine*, 83(10), S76 - S81.

Willis, G. (2004), *Cognitive Interviewing A tool for improving questionnaire design,* Sage publications, CA

Yucel, G., Hoege, B., Duffy, V.G., Roetting, M. (2009), "Analyzing the Effects of a BCMA in Inter-Provider Communication, Coordination and Cooperation", proceedings of the thirteenth International Conference on Human-Computer Interaction, San Diego, CA.

Chapter 60

Designing Medical Device Human Performance Validation Protocols for FDA Clearance: A Case Study

Robert A. North

Human Centered Strategies
Colorado Springs, CO

ABSTRACT

Designing successful user performance validation study protocols for medical device clearance at the US Food and Drug Administration (FDA) entails a multi-faceted approach that combines methods of observation, subjective measurement, and contextual inquiry techniques. In this paper, the essential elements of a user performance validation will be presented via a case study example involving user interaction with an automatic electronic defibrillator (AED). The major product of the validation study is an exploration and identification of safety risks during user-device interaction by means of testing performance of a sample of users from each user group that will use the device. The case presented provides an example of study design factors including participant selection, study task selection, facilities and equipment, objective and subjective data collection methods, analysis of data, mitigations to reduce risk, and retesting of those mitigations for their effectiveness.

Keywords: Healthcare Systems, Medical Device Human Factors, use error, use-related hazards

INTRODUCTION

Medical device manufacturers seeking to market devices in the US are subject to compliance with the US Code of Federal Regulations Quality Systems Regulation (QSR). Within this set of regulations is the Design Controls Process, which lays the foundation for the device development process from design concept through market release and post-market surveillance. Throughout the course of the Design Controls process, the manufacturer must ensure that the device "meets the needs of the user" and, through appropriate testing, validates the design of the device for "safe and effective use in real or simulated use environments." The Design Controls Process was initiated in the late 1990's primarily due to the number of medical device recalls and adverse events that could be attributed to design flaws.

The FDA estimates that as many as forty percent of these design-related recalls and adverse events have their roots in user-device interaction. As medical devices and systems have grown more complex placing more demands on the end user's cognitive abilities, use-related hazards leading to "use error" have risen dramatically. For this reason, in 2000, the FDA's Center for Devices and Radiological Health (CDRH) published its guidance entitled *Medical Device Use Safety: Incorporating Human Factors Engineering in Risk Management* relating human factors, ergonomics, and usability methods directly to the process of risk management in device design. Figure 1 below illustrates the process discussed in this FDA guidance.

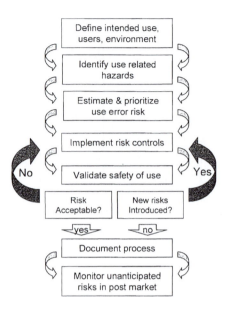

Figure 1. Use Hazard Risk Management Process

HUMAN FACTORS IN USE-HAZARD RISK CONTROL

The role of human factors in the device use-hazard risk control process is the systematic identification and elimination of use errors during user-device interaction that lead to patient or user harm. Eliminating use error risk begins in the earliest stages of design by applying techniques such as task analysis, contextual inquiry, heuristic evaluations and inspections, and formative usability evaluations. Examining the history of use errors for similar devices is also essential in identifying use hazards and eliminating these problems through design. Throughout this process, the standard practices of the human factors, usability engineering, and ergonomic research disciplines are essential for designing unbiased evaluations, recording and analyzing data, and developing subjective and objective performance measures.

VALIDATION TESTING: ESSENTIAL EVIDENCE OF USE HAZARD RISK CONTROL

Validation testing of user performance in real or simulated use settings is the single-most important element of the FDA device clearance submission process regarding human factors. The goal of validation testing for medical device user interfaces is the identifcation any remaining features of the design that may present opportunities for use error. This presents a different philosophy for the test designer centered on "forensic" inquiry as opposed to proving statistical reliability of user success. Advice from FDA Human Factors reviewers points to five critical points to remember regarding user performance validation:

1. **Prepare for validation success.** Systematic user interface testing and analysis during early design stages should reduce and eliminate use error risks to acceptable levels, so that validation reveals few if any problems.

2. **Focus on use error identification, not measuring task success/failures.** Tests with 15 or more users should enable identification of remaining use errors in user-device interactions. Determine and clarify use error root causes in post-test interviews. Explain why the current design cannot be reasonably modified to reduce a use hazard or how warning methods will communicate the risks.

3. **Include the <u>high-risk</u> tasks.** Testing should include the device interaction sequences that represent high risk to patient or user safety as identified in analysis and testing in early phases. (Context of high-risk tasks should be maintained, however.)

4. **Test appropriate user populations.** Enlist participants from the user population that reflect the expected education, experience, training, and skill sets. Participants should not be current employees of the manufacturer.

5. **Describe and test mitigations**. If use errors were revealed in validation testing, and these errors were serious enough to implement design changes, labeling changes, or other modifications, then further testing focusing specifically on those proposed use hazard mitigations is warranted as evidence of their effectiveness.

VALIDATION CASE STUDY: AMBULATORY INFUSION PUMP

Automatic Electronic Defibrillators (AEDs) are devices used to rescue victims of sudden cardiac arrest. Their availability has been extended to public places including golf courses, airports, and sports stadiums allowing access to a wide spectrum of users, especially emergency response teams with training in cardio-pulmonary resuscitation (CPR). Recently, a device manufacturer made several significant changes to their existing AED, including a voice prompting system and redesign of the electrode pads used to shock the patient. In light of these changes to the user interface, the manufacturer conducted several formative tests of the redesign of the device, followed by the planning and execution of a validation study of user performance in support of evidence of safe and effective use as required by FDA/CDRH. An outline of the overall protocol developed for this study is presented below.

STUDY TASK SELECTION

The selection of tasks for user performance followed the sequence of required tasks leading to successful administration of a shock to the patient. In a linear sequence of actions, these tasks include:

- *Task 1: Initialize Device:* Set the AED next to the victim, open the lid, and verify that the AED is working properly

- *Task 2: Remove victim clothing:* Remove the victim's clothing including undergarments to ensure that the chest is bare for placement of pads

- *Task 3: Place Electrode Pads on Chest:* Remove the electrode pads from their pouch, separate each pad from liner and place with elastic side down on victim's chest in position shown on pad labels.

- *Task 4: Administer Shock to Patient:* Wait for voice prompt to indicate that AED is analyzing heart rhythm, move away from contact with patient until directed to press button to initiate shock, press button, avoid patient contact until notified that shock has been delivered.

PARTICIPANT SELECTION

Fifteen (15) participants were selected from the two current user populations of this AED which represent 99% of the current user base. The first user group was comprised of individuals known as "first responders", or individuals with training on AED usage and life-saving techniques, but who are not a part of a professional paramedical team. The second group of 15 participants represented individuals from life-saving units such as police, fire, and paramedic teams. The choice of fifteen as the sample size for validation studies represents an FDA-recommended minimum number acceptable for the purpose of identifying usability problems with use error implications. To control for familiarity with the device, the participant population inclusion criteria included training on the operation of an AED between 12-36 months ago. This criteria provides for a participant pool that has limited recent familiarity with AEDs, and is more representational of the skill-retention properties of the users most likely to be required to use the AED being tested.

TRAINING SIMULATION

Validation studies should provide user training representative of how users become familiar with device operation in the real world. With hospital devices such as infusion pumps, the pre-validation test training should emulate the anticipated in-service training that would be provided by manufacturers followed by at least a 24-hour time period prior to validation performance in the study.

For the AED validation study, the manufacturer chose to employ the most representative aid in guiding the user—the Quick Reference Guide, supplied with the device and found inside the cover of the AED upon opening the latch.

TESTING ENVIRONMENT AND SCENARIO

The testing environment consisted of a testing room simulating a busy airport terminal with representative ambient noise and lighting. Because testing of the use of the AED does not include the time or ease of obtaining the AED, the test moderator provided the AED to the participant and directed the participant to begin the rescue procedure on the simulated victim, represented by a manikin lying on the floor in the testing room.

DATA RECORDING

A digital video recording was made for each participant's performance. The video served two purposes: (1) post-test performance scoring of the session including timing and task completion criteria, and (2) post-test retrospective analysis of performance via playback and pause to focus on particular user behaviors during the test session. A digital picture was taken of the user's placement of the electrode pads on the simulated victim's chest for subsequent scoring as to effective placement regarding shock effectiveness.

POST-TEST QUESTIONNAIRES

Perhaps the most important diagnostic tool for the validation study investigator is the administration, interpretation, and follow-up regarding responses to post-test questionnaires regarding aspects of user performance. This is due to the fact certain use errors may not manifest themselves in performance, but still present difficulties to the user. These non-observable task performance difficulties can be assessed via subjective rating scales that (1) assess the overall perceived difficulty of each task, and (2) probe for specific aspects of the user interface that are responsible for the difficulty.

For example, in the AED study, participants may have exhibited correct pad placement on the victim's chest, but have scored the pad placement task as "difficult" on a rating scale for that task. When asked for specifics about what contributed to that difficulty, it was reported that users expected that they expected a specified order of pad placement on the victim, i.e. "place Pad 1 above the right nipple of the victim, and "Pad 2 below the left nipple" instead of leaving this placement to the discretion of the user.

ANALYSIS OF USER PERFORMANCE

The analysis of user performance in a validation study requires a combination of the examination of objective measures, subjective responses, and directed inquiry focused on understanding root causes of either observed or reported difficulties in performance.

In the table below, typical outcomes regarding observed performance, subjective rating of difficulties, and potential responses are linked to investigative responses. The specific example for the AED validation study pertains to the electrode pad placement task (Task 3).

Table 1. Post-test investigative responses based on observed participant behaviors.

Observed Behavior	Investigative Responses (examples)	
	If Task Rated Difficult	If Task Rated Easy
Correct performance: pads placed on chest within acceptable position for adequate shock administration	Probe for source of difficulty (understanding of instructions, physical effort required, initial confusion with meaning of graphics, etc.)	Probe for source of ease of task execution: i.e. prior knowledge, voice instructions, graphics on pads, etc.
Marginal performance: pads placed slightly out of correct alignment, hesitation in final placement, correcting of wrong placement	Probing for perceptual, cognitive, or physical task aspects contributing to the difficult rating, review of video to aid participant in recalling difficulty source	Review of video performance segment to prompt participant, probing for perceptual, cognitive, or physical difficulties in task execution
Incorrect performance: pads placed incorrectly leading to ineffective shock administration	Probing for perceptual, cognitive, or physical task aspects contributing to the difficult rating, review of video to aid participant in recalling difficulty source, inquiry as to why participant believed they were performing the task correctly	Review of video related to specific error(s), inquiry as to why participant believed they were performing task correctly (interpretation of instruction, prior experience, etc.)

RESULTS IMPLICATIONS

The major outcome of the validation study is the identification of consistent patterns of observable use errors, observed near-misses including hesitation during performance, errors that were subsequently corrected, and consistently reported difficulties with certain tasks, especially when those tasks are critical in the execution of a particular high-risk step. The implication of findings should be based on the risk severity of the various outcomes. For instance, in user performance with an AED, where every second is critical to saving the life of the victim, a delay of more than ten seconds before placing pads on the victim might be deemed a "use error", even though no observed "overt error" was committed. The implication of this outcome may warrant redesign of the graphic symbology on the pads to indicate clear pad placement. Redesigns addressing high-risk use error should be revalidated in subsequent user performance studies.

REFERENCES

Medical Device Use Safety: Incorporating Human Factors Engineering in Risk Management, Food and Drug Administration, Center for Devices and Radiological Health, July, 2000.

CHAPTER 61

A Qualitative Assessment of Medical Device Design by Healthy Adolescents

Alexandra R. Lang[1,3], Jennifer L. Martin[1,3],
Sarah Sharples[1,2], John A. Crowe[1,3]

[1] Multidisciplinary Assessment of Technology Centre for Healthcare

[2] Human Factors Research Group
University of Nottingham, UK

[3]Electrical Systems & Applied Optics Research Division,
University of Nottingham, UK

ABSTRACT

Medical devices are primarily designed for adult use, resulting in adolescents with chronic conditions using products which have been designed with little or no consideration for their specific needs. This can lead to ineffective use and low compliance. This paper investigates such issues via the analyses of qualitative data obtained through a workshop involving healthy adolescent proxy users. The results show that adolescents are interested in the aesthetics, usability and acceptance of devices. It is evident from this study that if given the opportunity and appropriate methods, adolescents can be useful and enthusiastic participants in research and design.

Keywords: Adolescent, Teenagers, Medical device, User Needs, Proxy.

INTRODUCTION: ADOLESCENTS IN RESEARCH

Although children and young people are consumers of healthcare services, there has been "little work involving them to address their needs, prioritize their concerns and provide the kind of quality health care they want" (Moore & Kindness, 1998). With regard to medical device design and manufacture, whilst there is usually consultation with adult and child user groups, teenage users are generally 'shoehorned' into using devices made for one or other of these groups. As a result, it is likely that the needs of adolescent users of medical equipment are not currently being met, which may be leading to ineffective use and a lack of compliance with treatment regimes (Cameron 1993; Fielding & Duff 1999; Suris, Michaud *et al.* 2004).

Ingersoll (1989) and Kyngäs (2000) discuss how up "approximately 50% of adolescents with long term conditions do not comply with care recommendations". This dramatically high figure indicates how challenging it is to work with adolescents with regard to promoting good adherence to treatment regimens. By encouraging adolescent involvement in research and utilising participatory methods it is anticipated that a better understanding of teenage needs can be obtained.

In all healthcare research, the use of proxies is well documented and accepted (Nelson, Longstreth *et. al.* 1994; von Essen 2004; Varni *et. al.* 2007) as a valuable way of obtaining information from hard to reach populations. In this study, it was decided to consult healthy teenagers as proxies rather than parents, teachers or other adults with roles of responsibility that is often the case. Although they will be naïve regarding the realities of living with a chronic condition they will be able to empathize better with the everyday stresses experienced than an adult proxy

This paper describes and analyses the qualitative data which was obtained through a workshop method involving healthy adolescents (Lang *et. al*, 2009) and discusses the potential of using the information gained to inform the medical device design process so that it is more inclusive of adolescent users.

METHODS

Four workshops were conducted to elicit information from 71 healthy, adolescents who were naïve regarding their perspectives about the design of existing medical devices. To perform this within a familiar environment, workshop sessions were designed around the curriculum and carried out in schools. By utilizing school curricula and involving educational experts the study was carefully tailored to the specific needs of young people. The workshop involved three activities: an individual task, during which students provided personal responses to posters of medical devices; a team exercise where the students had to respond to good points and bad points for a device; and a questionnaire. The workshop was conducted four times allowing its refinement, resulting in a method that engaged a young audience and elicited valuable qualitative data on medical device requirements.

RECRUITMENT AND SAMPLING

Two secondary schools, located in the East Midlands region of the UK, were contacted through science and technology staff who were given the outline and aims, along with potential benefits to the school, students and researchers. Four classes with 71 students in total aged between 14 and 18 yrs old were selected. The participant population was split with approximately two thirds of the participants falling within the 15-16 year old age group and one third being female. The wellbeing of the participants was of paramount importance during the study and therefore ethical approval was obtained from the University of Nottingham, School of Mechanical, Materials and Manufacturing Engineering Ethical Review Committee.

As the workshop would be carried out within the school environment and timetable it was considered acceptable by the ethics committee for the teacher in charge of each of the classes to provide 'in loco parentis' consent for the involvement of the children. This enabled parental consent to be waived; to the benefit of the study given the feedback from school staff concerning the poor response rate for consent form completion, mainly due to their loss or forgetfulness by the students.

WORKSHOP METHOD

The workshops were designed in collaboration with teachers and training professionals to ensure that the study was appropriate for the teenage participants. It was stressed during the workshop that previous knowledge of devices was not required for successful participation in the workshop, and that responses should be in the context of a persona rather than personal experience. It was also emphasized that there were no correct or incorrect answers, with all opinions being equally valid. A range of workshop activities encouraged the students to write down their own personal thoughts and also report the points and discussions raised during the team exercises.

ANALYSIS AND RESULTS

Overall, there was a positive reaction to the workshop; evidenced by the enthusiasm shown by the participants and the volume of information gathered. This supports the theory that teenagers can be included in research and will participate willingly if they believe that their views are being valued and respected. Within the workshop eight devices used for the treatment and management of endocrine conditions and five devices from the field of respiratory medicine were scrutinized (see Table.1.0).

The coding of the data was based on the written content of the workshop activities as transcripts of the verbal discussion were not feasible in the classroom environment. The data set was analyzed in QSR NVivo using a qualitative grounded theory approach as defined by Strauss and Corbin (1998). This method allows the coding task to be driven by the data (Gibbs 2002) from which themes emerge that can then be coded themselves. The following sections describe the information obtained and the themes and ideas that emerged from the analysis of the data.

Table 1.0 Workshop Device Selection

Endocrine Devices	Respiratory Devices
One Touch® Glucose Meter	
Medtronics® Insulin Pump	
Nutropin® Growth Hormone Pen	Positive Expiratory Pressure (PEP) Mask
Humatropin® Growth Hormone Pen	Acapella® Vibratory PEP System
Freestyle® Glucose Meter	Coach2® Incentive Spirometer Kids
Easypod® Growth Hormone Delivery	I-neb® Adaptive Aerosol Delivery System
Animas® Insulin Pump	'Puffer' Asthma Inhaler
Accu-Chek® Glucose Meter	

PRELIMINARY ANALYSIS

Preliminary analysis considered the positive and negative feedback about the devices, in order to gain an indication about the participants' general views. This was achieved through simply coding responses as positive or negative and then extracting themes from these initial groupings. The results showed that the participants generated more positive descriptive statements than negative for the set of endocrine devices (positive 125/negative 98), whilst the respiratory devices produced more negative statements (positive 55/negative 120). This was supported by the fact that the participants suggested nearly twice as many improvements for the respiratory as for the endocrinology devices (81 versus 45). This indicates that the current devices used for the treatment and management of endocrine conditions are more ably meeting adolescent needs than the respiratory devices.

With regard to the endocrine device data, the positive comments were split between the aesthetic aspects of the devices (64) and practical use aspects (58). The aesthetic comments largely cited positive aspects concerning the colours, good general design and produced descriptive expressions such as *"modern"* or *"cool"*, these contributed to 18 references. Interestingly, the negative comments (60) were more directed towards the usability and practicality of the devices, with only 20 negative comments relating to aesthetics. The main exception to this was the teenagers' use of *"it's boring"* or *"dull"* to describe the devices; which occurred 17 times. These comments to the assessment of the devices might imply that a more interactive device, which provides better engagement, might be desired by the teenage participants. This idea was supported by the follow up activity when the students were instructed to work in groups and present potential improvements to the devices. Although there were a range of suggested aesthetic changes, the majority of suggestions concerned changes to the device in terms of the usability or functionality. Example statements such as *"button shape is poor considering how technological it is"* and *"it has a number of different components which might make it quite difficult and long to use"* demonstrate the importance that the adolescents put on the practicality of using a device and not just its appearance.

For the set of respiratory devices, overall there were much fewer positive coded statements than negative (positive 55/ negative 120). Within the collection of negative observations, 64 referred to use or function of the device, whilst only 28 comments related to aesthetic issues. Once again; this showed that although this age group are concerned with visual aspects of device design, their needs encompass much more than this and that they are concerned with the use, functionality and practicality of devices in respects to their lifestyle. Again there was a high incidence of statements including words such as *"boring and dull"* (28 refs) whilst within the positive category there was much less use of words such as *"cool"* and *"funky"* (5 refs). This could indicate that there was less satisfaction with the aesthetic aspects of the respiratory devices, as the teenagers made less positive comments.

Interestingly, for both the endocrine and respiratory devices, the participants portrayed more emotive responses to the devices from a negative perspective. Words such as *"scary"*, *"depressing"* and *"intimidating"* along with illustrated unhappy faces next to comments provided 27 coded references in this preliminary assessment. In contrast there were only a few definitive emotional statements that were positive.

IN DEPTH QUALITY ANALYSIS

Following on from the initial exploration of the data, the information was examined and coded more thoroughly to identify main themes and sub themes, depicting the breakdown of concepts into issues, attitudes and outcomes (improvements).

Acceptance was a key theme that emerged throughout the workshops in a variety of ways. The issue of use both in public and amongst friends was a recurring issue (16 respiratory, 12 endocrinology) and overlapped strongly with the category of **Everyday Life**. The concept of acceptance was raised more in relation to the endocrine (86 refs) than the respiratory devices (54) and comments such as *"it's good because it looks like an MP3 player"* and *"looks like a pen more than a scary injection"* highlighted that the devices designed to look more like other items were more readily accepted by this age group. This idea was reinforced by the fact that there were 20 instances where the devices were specifically compared to popular items of technology. Despite this, there were still misgivings about the public use of the devices, indicated by the fact that **Discreteness** emerged as a sub theme in its own right with nine instances relating to the **Public Use** subtheme. The sub theme of **Age** (37 endocrinology and respiratory 22) was quite dominant in terms of the number of coded references and illustrated that the adolescents felt that many devices were more appropriate for adult or child users. Devices which stood out with respect to this were the Spirometer; *"it would be embarrassing for teens!"* and *"good for smaller children but it's too childlike"* whilst at the other extreme an example of device which they felt was too old for them was the Accu-Check®; *"better for older generation"* and *"it's more for old men"*.

Aesthetics played a large role in the students' assessment of the devices with more criticism than compliments received. There was also significant correspondence between this theme and the coded references in the **Suggested Improvements** categories, with customisation of the design being a popular proposal from the students. Most of the comments on aesthetics were unenthusiastic about the current designs and colours of the devices (63 endocrine references and 40 respiratory references). Colours such as grey and muted blues were not appreciated; *"the dull colour makes it seem old"* and some students likened them to 'hospital' colours, stating that *"you wouldn't want to be reminded of hospital whilst using it"*. Brighter colours received more favourable comments although the orange of the Humatropin® Pen was deemed *"too bright"* and therefore attracted too much attention. Interestingly black split the participants, with some liking the fact that dark colours made a device more discreet whilst others thought it too *"daunting and depressing"*.

The **Design and Images** subtheme which generated 47 coded notes from the endocrinology group and 26 from the respiratory devices was very broad in its assessment of the visual design aspects, with students commenting positively on aspects such as glittery or metallic finishes to a range of sticker and sheath designs. All of these options which help a user to 'customise' their device were encouraged by the student participants and they went further to suggest that people should have *"interchangeable covers, like phones...so you can change them with your mood"* and that *"it would be nice if you could send in your own picture and have your pet printed on it, then it would be nicer"*. Another main message that this subtheme highlighted was the link to **Acceptance** and that the current options for users to personalise devices were aimed at younger users and were not appropriate for teenagers.

Everyday life is a main theme that has many connections to others. In total from both medical categories, this field was created from 104 coded comments. It is evident that the overlap with the **Public Use** subtheme and subsequently **Acceptance** is very prominent, with examples such as *"it looks simple to use and it's not so huge so could easily slip into your pocket without being noticed"* and *"wouldn't want to use it in front of my friends, it's bulky"* demonstrating that this age group are able to appraise the equipment in relation to their teenage lifestyle. This section was particularly important for the respiratory devices as it generated a cluster of 15 notes, all of which were negative in relation to **Everyday Life**. Other subthemes to emerge from this category were **Nuisance and Hassle** and **Pockets and Bags**, most of which related to the practicalities of using these devices on an everyday basis and ensuring that the design enabled their use to be incorporated without unnecessary inconvenience. Although it was acknowledged earlier in the paper that the proxies used in this study; healthy adolescent students, would be naïve of the everyday realities of living with a chronic condition, this theme of **Everyday Life** offers real insight into teenage life perspectives and priorities. Whether these would be markedly different from a teenager living with a chronic condition needs to be explored, however as teenagers strive to be accepted socially by their counterparts then this information can only be constructive in its application to device design.

It was evident early on during coding that **Size and Shape** would be a key issue for the adolescents, with 68 notes coded from the respiratory group and 52 from the endocrine set; particularly relating to the themes within **Everyday Life**. This topic manifested itself into two distinct subdivisions, some being described as too big and some being described as too small. Interestingly, the endocrine devices were split, with a couple; namely the Freestyle® and One Touch®, being described as small with concerns of *"it's small enough to get lost"* whilst others thought it a benefit *"it's nice and compact"*. In contrast the respiratory devices were generally thought to be too cumbersome, with the majority out of 45 references remarking on the fact that *"it's too chunky, you wouldn't be able to carry it around with you"*. **Shape** was specifically an issue with regard to the respiratory devices, with the Acapella® being praised for its *"attractive shape"* and the fact it felt *"comfortable and strong to hold"*, whilst others, specifically the Spirometer and PEP Mask were felt to be poor with respect to their shape. This aspect was then related directly to the emergent subtheme of **Portability**, as was the size of the devices, with students bringing into discussion the importance of a case or carrier for the device, which would not only help hygiene but also provide a canvas for personalization of the device which would also then ultimately help the issue of **Public Use** and **Acceptance**.

Usability emerged as a key topic and, as anticipated, was the umbrella term to cover a range of subthemes which detail the comments that address a range of devices features. In total 158 notes were coded from the respiratory device selection and 154 from the endocrine group. This was divided down between the following subthemes with coded references ranging fairly equally between them. **Complexity** was an issue raised for both medical sets of devices, the majority of the 26 notes for the respiratory devices stating terms such as *"confusing"* and *"complicated"* to describe the equipment, with some of the more detailed comments overlapping with the subtheme of **Intuitiveness**; *"it looks complex and not easy to work out"*. This general perspective was mirrored within the review of the endocrine devices. However, within this subtheme many of these comments were not explicit in determining what was 'complex' about the device and so were not as informative as other subthemes. It may be that the teenagers found it more difficult to articulate what they found 'confusing' about the device as opposed to making observation about physical features. This is where familiarity and experience with the device by real users would provide invaluable insight for developers, something which is much more difficult to obtain via proxy users.

Despite this, the healthy adolescents did provide useful feedback regarding other aspects of the devices. A similar situation was experienced with the subtheme **Easy to Use**. Participants often used this phrase to describe the devices without providing justification for this statement. Several related the ease of use to the design of the **Buttons**, how they were easy to manipulate and use to navigate the menus in the device interface; however these were exclusively within the critiques of the endocrine devices (7 refs). There was also a general conflict at this point in the data with some individuals feeling that the devices were easy to use whilst others thought them to be confusing and complex. This aspect of the device assessment would have to be carried out to elicit additional detail and clarify these issues.

  Intuitiveness of device use was interesting in that some devices, such as the Spirometer prompted statements such as *"it looks easy to use and you know what it's for"*, whilst the Acapella® prompted several versions of *"what is it?"* and *"the use is less clear because there is no display"*. This was quite significant as the **Information** subtheme provided evidence suggesting that the adolescents liked the device to provide them with information about what they were doing and how well it was being achieved. This was particularly important to the set of respiratory devices, where 18 coded references were established. It was also important in the fact that many references were doubly coded with the subtheme of **Interaction and Feedback**. Accounts from the teenagers such as *"why isn't there a gauge on it like the others?"*, *"have a screen rather than a dial with more information"* and *"...why don't you have a better thing to tell you how you are doing"* point to the fact that they felt that feedback and information from some of the devices was insufficient. This was particularly evident with the respiratory devices, whereas those used to manage and treat endocrine conditions were generally more modern and interactive and this was reflected in the data; *"the screen is good with numbers and stuff...that's a positive"*.

  Interaction and Feedback gradually established itself as an important theme in its own right and was interrelated to **Usability** and the category assigned **Improvement Suggestions**, from which there were approximately 24 suggestions specific to the interaction of the device. Within this grouping screen clarity was a big issue for the endocrine devices with 15 refs, the majority of which were positive. The majority related to good clarity and brightness with comments such as *"good, easy, readable screen"*. The main issue with the screen was the size and this was device dependent. Some were thought to be appropriate in relation to the device e.g. the Animas® Insulin Pump *"good screen size"*, *"easy to read"*, whilst the Accu-Check® and the Easypod® screens were described as *"small and not easy to see or recognise"* and *"the screen is too small for the size of it ☹"*. This could be related to the comparisons that the participants make between these technologies and familiar devices such as mobile phones and cameras, on which the screens are getting increasingly large in relation to the size of the gadget

  Other subthemes identified during the coding of the **Usability** references were **Comfort, Multiple Components, Practicality and Storage.** The first of these, **Comfort** was only identified in association with the data from the respiratory devices. This may be due to the fact that the duration of use at any one time extends to up to half an hour and therefore the issue of comfort is perhaps more crucial for these than the short space of time within which the endocrine ones are used. This produced 13 references and although fit to hand was important the main concerns raised by participants was the comfort of the mouthpieces on these device. The ability to change the orientation of the mouthpiece during use was suggested, as were improvements for the cushioning and thickness of the rubber that constitutes most of the mouthpieces. **Multiple Components** was exclusively coded from the range of endocrine devices. The variety of components was generally deemed to be negative, providing more hassle during use and *"it could be fiddly if you are out"*. In relation to this, the Accu-Check® was singled out as being *"less complicated and fiddly because its mostly all in one"*, indicating that this type of device (along with the coded data in the **Nuisance and Hassle** subtheme) might better meet teenagers needs than the devices with many separate parts. In conjunction with **Storage of Device** the students felt that cases and bags were good for protection of the device as well as practical reasons of portability and hygiene. The only exception to this was the large, rigid case storing the Easypod® which was described as *"a giant inconvenient box"* and *"ridiculous for the size and weight of the device"*, demonstrating that the levels of acceptance for this feature may depend upon the size of the device and its components.

Improvement Suggestions was a theme which overlapped with many of the other categories and has been touched upon in the previous sections. **Customisation** and **Interaction and Feedback** are core topics within this theme and present definitive areas for further investigation and present, along with the **Aesthetic** and **Usability** fields, a clear starting point for identifying adolescent user needs and priorities with regard to medical devices.

As a result of this analysis and the information obtained through discussions with clinical staff the Acapella® physiotherapy device will be taken forward as a case study to investigate how user needs assessment can assist the redevelopment of a device.

CONCLUSIONS

Involving minors of any age in research has many barriers, but as this study demonstrates there are ways of successfully achieving this. From the workshops it is evident that considerable attention was paid by the participants to the aesthetics of the devices, providing evidence to suggest that this user group are heavily concerned with the 'look and style' of a device. This is particularly important when considering social acceptability and identity, both of which are very important to young people and are likely to affect compliance. Another important theme that emerged; particularly from the respiratory devices data set, was the apparent lack of interaction and feedback from this range of devices. It was evident that the adolescent participant group expected more feedback during device use and felt that the devices were deficient in regard to this.

This study indicates that the design of current medical devices does not always meet adolescent specific needs. Additional focus and development is required to ensure that the aesthetics and usability of devices meet the specific requirements of teenagers. If these issues can be addressed, and device design can be more inclusive of the needs and priorities of adolescent users, then this should lead to improved use of these devices as well as improved health outcomes for the young people who have to use these devices.

It is evident from the data collection workshops that there is scope for further investigation into the involvement of adolescents in medical device development. Further research is planned that will involve real adolescent users to comprehensively assess and redevelop their devices in accordance with their user needs.

REFERENCES

Cameron, K. T. (1993). Treatment compliance in childhood and adolescence. *Cancer* 71(S10): 3441-3449.

Essen, L. v. (2004). Proxy ratings of patient quality of life Factors related to patient proxy agreement. *Acta Oncologica* 43(3): 229-234

Fielding, D. & Duff, A. (1999). Compliance with treatment protocols: interventions for children with chronic illness. *Arch Dis Child* 80(2): 196-200.

Gibb, G. R. (2002). Qualitative data analysis: explorations with NVivo. Maidenhead, UK, Open University Press.

Ingersoll, G. M. (1989). *Adolescents.* Englewood Cliffs, NJ. Prentice Hall.

Kyngäs, H. (2000). Compliance of adolescents with chronic disease. *Journal of Clinical Nursing* 9(4): 549-556

Lang, A. R., Martin, J. L., Sharples. S. & Crowe, J. A. (2009). Enabling adolescents to participate in the design and improvement of medical devices. International Ergonomics Association, 17th World Congress on Ergonomics. Beijing, China. 9th-14th August 2009.

Moore, H. L. & Kindness, L. (1998). *Establishing a research agenda for the health and wellbeing of children and young people in the context of health promotion. Promoting the health of children and young people. Setting a research agenda.* London, Health Education Authority (HEA).

Nelson, L. M., Longstreth, W. T. J. et al. (1994). Completeness and Accuracy of Interview Data from Proxy Respondents: Demographic, Medical, and Lifestyle Factors. *Epidemiology* 5(2): 204-217.

Strauss, A. L. and J. Corbin (1998). Basics of Qualitative Research Techniques and Procedures for Developing Grounded Theory London, Sage Publications.

Suris, J. C., Michaud, P. A. & Viner, R. (2004). The adolescent with a chronic condition. Part I: developmental issues. *Arch Dis Child* 89(10): 938-942.

Varni, J., Limbers, C. & Burwinkle, T. (2007). Parent proxy-report of their children's health-related quality of life: an analysis of 13,878 parents' reliability and validity across age subgroups using the PedsQLTM 4.0 Generic Core Scales. *Health and Quality of Life Outcomes* 5(1): 2

Chapter 62

Contextual Inquiry for Medical Device Development: A Case Study

Jennifer L. Martin, John A. Crowe

MATCH (Multidisciplinary Assessment
of Technology Centre for Healthcare)
Electrical Systems and Optics Research Division, Faculty of Engineering,
The University of Nottingham, Nottingham, NG7 2RD, UK

ABSTRACT

To provide high quality care for patients the healthcare industry is dependent upon the provision of well designed medical devices. To achieve this, ergonomists recommend taking a user-centered and systems approach to design. In this study Contextual Inquiry was used to study 12 clinical participants whilst using a prototype new medical imaging device during patient consultations. Contextual Inquiry was found to be effective for evaluating a prototype device within a clinical environment. A number of contextual issues were identified which were likely to affect the use and uptake of the new device. These included the physical and organizational structure of the environment, as well as the characteristics and capabilities of clinical users and patients.

Keywords: Medical device, contextual inquiry, usability, healthcare, observation, interview.

INTRODUCTION

To provide effective and safe care for patients the healthcare industry is dependent upon the provision of high quality medical devices that meet the requirements of the people that will use them, as well as those who will be treated by them. For any one medical device there may be multiple types of users including a variety of healthcare staff, as well as patients and the people that care for them. It is essential that the needs, skills and capabilities of these people are taken into account during development. Adopting a user-centered design approach has long been shown to contribute to better and safer medical device design, particularly in fields such as drug delivery and anesthesiology (Webb et al., 1993).

Development and evaluation should also take place against an understanding of the healthcare system in which the device will be used. Medical devices and their users do not exist in a 'stand-alone' relationship to other aspects of patient care, and the introduction of a new device can have significant, and sometimes unanticipated, effects on patient care (Sawyer et al., 1996). Therefore, in addition to a user-centered design approach, a holistic or systems approach should also be taken.

However, medical devices are often developed in isolation from the treatment pathway and this has been identified as an explanation for why many devices fail to truly meet the needs of the intended users (Morris, 1999). Conducting user research and device evaluations in the clinical environment is not a straightforward undertaking. Developers must comply with strict research governance procedures as well as being sensitive to the personal and sensitive nature of the clinical issues being investigated.

As a result, when conducting user research developers often resort to discussion based methods which can be applied outside of the clinical environment. This is problematic as contextual information will not be collected and data will be restricted to what the users are consciously aware of and able to recall and articulate at the time. There is a need therefore, for practical, scientifically rigorous, observational methods which can be applied within a clinical environment.

Contextual Inquiry (CI) is a human factors research method that combines the questioning and observation of users within the environment of use. It was developed by researchers in the field of Information Technology (IT) and has been widely and successfully applied during the evaluation of computer systems (Holtzblatt and Beyer, 1993). Contextual Inquiry has its roots in ethnography and aims to collect contextual information by observing people in their work or living environment. The developer observes the user as they perform tasks and asks questions about what is happening and why, as well as how tasks could be improved. It has been identified as a useful way of identifying "taken-for-granted information that the user may not recognize as relevant or significant" (Martin et al., 2006)

There are a number of barriers to conducting observational studies in a clinical environment. Researchers must consider: patient confidentiality; infection control; as well as any risks to patient safety that may arise from having an external observer present. In addition, researchers must be mindful that patient consultations will often cover sensitive issues. Factors such as these may explain why, although CI has been successfully used to evaluate healthcare IT systems, (Coble et al., 1997) it has not yet been applied to the development of medical devices.

AIMS

The aim of this study was to investigate the effectiveness of Contextual Inquiry for medical device development, specifically, to establish the barriers to performing this method in a clinical environment and to adapt it accordingly.

METHODS

An adapted observational interview method was used to study healthcare staff from a UK hospital whilst they used a working prototype medical device. Before commencing the study ethical permission was obtained from the UK NHS.

THE DEVICE

The device described in this study uses laser Doppler technology (Briers, 2001) to produce images of patients' blood vessels. The aim is to produce a device that will enable a range of different types of healthcare staff to perform common procedures more successfully and efficiently, and with less pain and discomfort to patients. For reasons of commercial confidentiality it is not possible to describe the device in detail. This study describes the evaluation of an early working prototype of this device; the aims of the evaluation were to:

- Evaluate the usability of the device in the clinical environment
- Evaluate the perceived utility of the images produced by the device
- Identify design problems and barriers to adoption
- Collect the opinions of clinical users on the design of the device

PARTICIPANTS

A total of 12 participants took part in the study (9 female and 3 male). A purposive sample was used, to include a range of healthcare staff who were required, as part of their clinical role, to locate and access the blood vessels of patients. All participants were healthcare staff working in the renal dialysis department of a UK National Health Service (NHS) hospital. 10 of the participants were Registered Nurses with clinical experience ranging from 2 to 23 years. 2 of the participants were renal support nurses (unqualified nursing assistants) with experience ranging from 1 to 4 years.

ADAPTATION OF CONTEXTUAL INQUIRY

This study took place in a clinical environment. Clinical participants were observed and questioned whilst they used the prototype medical device during patient consultations. This meant that it was necessary to adapt the Contextual Inquiry method to suit this environment.

The rationale behind a contextual inquiry is that the participant and the researcher are partners in the process of discovery and, therefore, to successfully perform this method it is essential that an open and detailed dialogue develops between them. However, in this study it was possible that either the participant or the researcher may have wanted to raise subjects that would be difficult to discuss in front of the patient, for example issues that may have worried or embarrassed the patient. For this reason it was decided that if either the participant or the researcher felt that information should not be discussed during the consultation they would make a written note of these issues and they would be discussed once the patient had left. Due to their greater experience of patient consultations the researcher took the lead from the clinical participants in terms of which topics were appropriate for discussion in front of patients.

It was made clear to the clinical participant that if they felt that the presence of the researcher was not appropriate for any reason then they were free to ask them to leave. In addition, whilst informed consent was being taken, it was explained to the patients that they were free to ask the researcher to leave at any time without giving a reason.

PROCEDURE

Twelve observational interviews – contextual inquiries - were conducted in the renal clinic of one UK National Health Service (NHS) hospital. The audio of the study was recorded using a digital recorder. Each contextual inquiry took between 1 and 2 hours, during this time each of the clinical participants treated between 2 and 5 patients.

Before beginning each contextual inquiry the purpose of the study was described in detail to the clinical participants. It was made clear to them that, as they were the experts in this clinical area, we wanted to learn as much as possible from them. The researcher stressed that it was not an evaluation of their performance but the performance of the device. It was explained that the aim of the study was to identify problems with the current design and that we wanted to produce a device that would fit in with the ways that they worked, not a device that would require them to change how they worked. They were encouraged to be critical of the device. The device was demonstrated to the clinical participants and they were encouraged to fully familiarize themselves with it and to ask any questions about the study or the format of the study.

The clinical participant then conducted their patient consultations. This involved them performing a physical examination of the patient in accordance with normal clinical practice. The participant then used the prototype device to take an image of the patients' blood vessels and used this information combined with the information from the physical examination to treat the patients.

During this process the researcher worked to develop an open and comprehensive dialogue with the clinical participants. The researcher asked questions when it was not clear what the participant was doing or why they were doing it. In addition, they prompted the clinician to describe their interactions with the patients and the device, as well as any difficulties that they experienced. Prompts included:

- "It looks like you're finding it difficult to get the device in the right place to take the image. Where exactly do you want to place it, does it need to be at a different angle?"
- "Could you explain why you stopped examining the patient's hand and started examining their arm instead?"
- "Why have you decided to use the device? What help are you hoping it will give you?" and afterwards, "Did the device provide useful information in this case?"
- "To what extent are you basing your treatment on the information you got from the physical examination of the patient and how much is based on the image from the device?"
- "You look like you're a bit frustrated there, what is it that you're trying to do?"

Once the consultation was finished and the patient had left, any issues which had not been raised during the consultation were discussed.

Immediately after the end of each contextual inquiry a de-briefing session was held. The researcher summarized what they understood to be the findings from the session and checked with the participant that these were correct. Participants were encouraged to raise any additional issues, specifically with regard to: the usability of the device; the anticipated benefits to patients and clinicians; and any factors that may prevent the safe and effective uptake of the device.

RESULTS

APPLICATION OF CONTEXTUAL INQUIRY

All of the individual Contextual Inquiries were completed successfully. A total of 39 patient consultations were observed with a further 2 patients declining the invitation to take part in the study. A number of the clinical participants reported that this was the first time that they had been invited to take part in a study of this type and when asked, all of the participants reported that they thought that it had been a useful exercise.

The participants varied in how comfortable they appeared at being observed whilst performing clinical tasks. The qualified nurses were more confident and offered significantly more unprompted opinions and information. In contrast, the two healthcare assistants were noticeably more reticent and in both cases the researcher had to reassure them of the particular importance of their participation. As the sessions progressed the dialogue with these participants became more collaborative, detailed and open.

The researcher noted a general reluctance by the majority of participants to be critical of the device. Only one participant (a staff nurse with 14 years experience) appeared comfortable with offering open and frequent opinions of the device. On a number of occasions the researcher noticed that the participants were apparently having problems operating or maneuvering the device and had to prompt them to acknowledge and describe these.

With regard to conducting the study during patient consultations, all of the patients appeared comfortable with the researcher being present during their consultation and with the discussions that took place. At no point during the sessions was the researcher asked to leave by either the clinical participant or the patient.

DEVICE EVALUATION

During the contextual inquiry sessions a number of usability and contextual issues with the device were identified. The physical environment had a significant effect on how the device was used during the patient consultations. The consultation rooms varied in size with the smaller rooms leading to difficulties with moving the device around the patient and into the correct place to take the required image. The presence of other necessary medical equipment also made it difficult to move the device around the room. Other issues with the physical environment included variable lighting and access to electrical sockets.

Valuable data on how the participants carried out their clinical tasks was collected such as: the points during the patient consultation when the participants required an image of the patients' blood vessels and from which part of the body. A lot of this information was tacit; it was often unclear from simply observing the participants why they decided to use the device at a particular time or why they were examining a certain area of the patient. It was at these points that the Contextual Inquiry method was particularly useful as it enabled the researcher to prompt the clinician to explain what they were doing and why, and to describe how the device was aiding the process (or not).

The characteristics of the patients being treated affected how the device was used during the study. Patients varied considerably in terms of their size and shape and a number also had mobility difficulties. This made it difficult to get these patients into the correct position for the device to be able to take an image of their blood vessels. This was identified as a design problem; the device required the patient to move into a position that could be uncomfortable and then to stay in that position for as long as five minutes whilst images were taken.

The study revealed that a priority of the clinical participants was to complete the patient consultations quickly and efficiently. During the consultations patients frequently expressed their desire to be treated quickly so that they could go home. This meant that the start-up time for the device, as well as the time it took to maneuver the device into position, and process the images would be factors that would affect the uptake of the device.

A number of smaller usability issues with the prototype device were identified. These included: the complexity of the set-up procedure; confusing information in the operating instructions; a lack of feedback provided to the user; and redundant functionality.

DISCUSSION

Contextual Inquiry was found to be effective for evaluating a prototype imaging device within a clinical environment. This study supports the theory that the characteristics of medical device development make it well suited to a method such as CI. Tacit knowledge played a significant role in the patient consultations, the clinical participants were observed making quick decisions about treatment based upon multiple sources of information. The dialogue revealed that this information was a mixture of a wide range of patient information combined with knowledge and skills developed during years of training and clinical practice. By prompting participants to talk about these processes as they happened, the researcher could begin to understand how they affected patient care and how the image provided by the medical device could support the clinician. The participants appeared surprised at the importance that the researcher placed on this data, indicating that this information would not have been collected from a discussion-based method such as an interview or focus group.

None of the participants had taken part in a study of this type before and the researcher had to take care to ensure that they felt comfortable with the style of the inquiry and to develop an open dialogue with them. In particular, the more junior participants required repeated reassurance of their importance to the aims of the project. These findings may be symptomatic of the fact that device developers frequently rely on the opinions of senior doctors to represent the opinions and experiences of clinical users, even when a wide range of healthcare staff will operate a device. This study shows that, if given the opportunity, a wider range of clinical staff can be effective and enthusiastic contributors to design and development.

The majority of the patients involved in this study were comfortable with having an external observer present during their consultations. This was likely to be due to the fact that they were all long-term patients, who attended the clinic a number of times each week for regular treatment. As a result, they were comfortable with the clinical participants and trusted them with their health and well-being. It is possible, however, that other patient groups would be less comfortable with having their consultations observed, for example patients with

556

newly diagnosed or acute conditions. The characteristics of the patient populations to be studied should be carefully considered before engaging in similar research.

With regard to how the clinical environment may affect the use of the device, valuable information on both the physical and organizational structure of the clinic was captured during the study. Of particular note was the degree to which these factors varied within one department in a single hospital. This indicates that in order to develop an understanding of the full range of potential environments for the new device further research should be conducted in a variety of hospitals and clinics.

CONCLUSION

It is well established that to produce a medical device that is both clinically effective and safe it is essential that the contexts in which the device will be used must be specified and considered. Clinical environments are frequently complex and it is difficult for developers to identify and understand the range of factors that will affect how medical devices may be used. This study has shown that, with some adaptation, Contextual Inquiry can be an effective way of understanding both clinical environments and clinical users.

ACKNOWLEDGEMENTS

The authors acknowledge support of this work through the MATCH Programme (EPSRC Grant GR/S29874/01), although the views expressed are entirely their own. The technology development aspects of this research have been funded by the Technology Strategy Board, UK. The authors would like to express their gratitude to the staff and patients at the Nottingham University Hospitals NHS Trust, UK.

REFERENCES

Briers, J.D., (2001) "Laser Doppler, speckle and related techniques for blood perfusion mapping and imaging." *Physiological Measurement*. 22(4) R35-66.
Coble, J.M., Karat, J. and Kahn, M.G. (1997), "Maintaining a focus on user requirements throughout the development of clinical workstation software reports." Proceedings of the 1997 Conference on Human Factors in Computing Systems, CHI, Mar 22-27 1997. ACM, New York, NY, USA, Atlanta, GA, USA, 170.
Holtzblatt, K., and Beyer, H. (1993) "Making customer-centered design work for teams." *Communications of the ACM*, 36(10), 93–99.
Martin, J.L., Murphy, E., Crowe, J.A. and Norris, B.J. (2006) "Capturing user requirements in medical device development: the role of ergonomics." *Physiological Measurement*, 27(8):R49–62.

Morris, C. (1999), "Patient pull: the changing influences on medical device design, Part I." *Medical Device Technology*, 10(4), 30–33.

Sawyer, D., Aziz, K.J., Backinger, C. L., et al. (1996), "Do it by Design: An Introduction to Human Factors in medical Devices". In: US Department of Health and Human Services, Public Health Service, Food and Drug Administration, Center for Devices and Radiological Health.

Webb, R.K., Russell, W.J., Klepper, I. and Runciman, W.B. (1993), "The Australian Incident Monitoring Study. Equipment failure: an analysis of 2000 incident reports." *Anaesthesia and Intensive Care*, 21(5), 673–677.

CHAPTER 63

An Intelligent Community Care System Using Network Sensors and Mobile Agent Technology

Su Chuan-Jun, Chen Bo-Jung

Department of Industrial Engineering and Management
Yuan Ze University
Taoyuan, 135 Far East Road, R.O.C.

ABSTRACT

Planning on living arrangements for senior citizens such as welfare, medical care, nursing homes, manpower development and career counseling should be started as early as possible to reduce their dependency on society. This also means that more emphasis will be placed on improving the quality of life and care for the elders. Most long-term care services are provided on an institutional basis by nursing homes due to the structural changes in society and family. Traditional family-based home care system is perishing. Once an elderly people are unable to care for themselves, the family has no choice but place them in a nursing home. The majority of the elderly, however, would much prefer "age in place" and live out their lives where they belong.

To provide more responsive and personalized care services, we establish an "Intelligent Community Care System (ICCS)" by using Network Sensors and Mobile Agent paradigm. The Network Sensors Environment basically involves the deployment of a full ZigBee network, so that every user with a tag in the environment would be accessible to multiple sensor nodes simultaneously in order for information to be relayed to the Server for processing. The Agent Environment encompasses the entire Agent Network, which should be able to search for appropriate services and information through the database and relevant service providers in order for caregivers to utilize when carrying out community care. At the same time, the system would also keep track of the community care performed by each caregiver. In other words, the system would allow caregivers to take the initiative by visiting those who require care and administering appropriate community care services.

Keywords: Community Care, Network Sensor, Mobile Agent

INTRODUCTION

With the rapid development of medicine, science and technology, many diseases can now be cured. Medical science strengthens the vitality of the elderly even in the case of incurable diseases. Therefore, aging population is an inevitable trend of human society. According to the "Survey of Health and Living Status of the Middle Aged and Elderly in Taiwan Survey Report" from the Bureau of Health Promotion, Department of Health, Taiwan, 88.7% of elders acquire at least one chronic disease as diagnosed by doctors and 51.3% of them acquire at least three or more chronic diseases. Thus, with the aging population or elders acquiring chronic diseases in this manner, the needs of these people who are incapable of normal daily living or needing care will increase. More people consequently attach great importance to elderly care and quality of life (Chung, W.C. 2007).

The care services based on different demands can be divided into home care, community care and institutional care. Community care was started from the criticism in England during nineteenth century, which was about the institutional care of the Poor Low announced by Queen Elizabeth in 1601. The institutional care was originally for improving quality of institutionalization, yet complaints arise due to dull life in institutionalization caused by the loss of privacy and options (Chen, C.Y. 2003). In fact, not only community care sites but long-term care centers can still be improved:

1. **Care service being provided passively:** Not all elders could participate in care services or medical care by themselves, and some elders could not always stay at home to wait for the caregiver. And community care service can't be provided for all elders. Hence, we hope the service should be provided actively. In other words, the caregivers could know an elder's location and visit the elders for providing care service actively and personally.

2. **The load of caregivers:** For caregivers, voluntary medical staffs especially, should take some materials for making the decisions of each elder, such as basic information, favorites, habits or record of chronic disease, etc. Therefore, we hope the caregivers don't need to take more materials, but they can know any related information clearly to provide care service completely by table PC or PDA.

3. **Records are overlooked, lost or broken:** For providing more complete and

personal care service, caregivers have to analyze the result of care service for the elders. But sometimes caregivers forget to record or the records are lost or broken. Hence, we hope the records should be Electronic Records, and caregivers can record easily and fast by table PC or PDA.

For realizing and improving the community care, we propose to develop a smart community care system based on the concept of ambient intelligence using RFID and mobile agent technologies. It's a way to enhance community care efficiency, provide community service actively and reduce the workload of caregivers. This paper aims to develop a "Intelligent Community Care System (ICCS)" based on the concept of Ambient Intelligence using Radio Frequency Identification (RFID) and Mobile Agent (MA) technologies. Caregivers may locate care-receivers easily in a community with RFID while MA furnishes context-aware, timely and accurate information for the care provision. The care historical data can be automatically recorded as well. ICCS is expected to be able to promote the pervasive community care services with convenience, context awareness and accuracy.

The paper is organized as follows. In section 2, we research the issues about AmI, context-aware, context-aware mobile agent in healthcare and discuss the implication of social. Section 3 presents system design and architecture of ICCS. Section 4 covers usage scenarios and the implementation. Finally, Section 5 provides the conclusions and suggestions to future studies.

LITERATURE REVIEW

AmI

The early developments in Ambient Intelligence took place at Philips. In 1998, the board of management of Philips commissioned a series of internal workshop to investigate different scenarios that would transform the high-volume consumer electronic industry from the current "fragmented with features" world into a world in 2020 where user-friendly devices support ubiquitous information, communication and entertainment. Ambient intelligence implies a seamless environment of computing, advanced networking technology and specific interfaces. It is aware of the specific characteristics of human presence and personalities, takes care of needs and is capable of responding intelligently to spoken or gestured indications of desire, and even can engage in intelligent dialogue with the user (Mendes, M., R. Suomi, & C. Passos, 2004). Basically, context-aware is central issues to ambient intelligence. The availability of context-aware and the use of context-aware in interactive applications offer new possibilities to tailor applications and systems to the current situation. However, context influences and often fundamentally changes interactive systems (G. Riva et al. 2005).

Context-aware

Context-aware is a concept created by Schilit and Theimer in 1994. It is the definition of "Context-aware based on the user's location and the surrounding persons, environment or devices to change the status and adjust the services provided by (Kikhia, B. 2008)." Kaasinen (Kaasinen, E. 2003) consider that how to satisfy the users can be divided into five sections to discuss: User attitudes, Contents, Interaction, Personalization, and Seamless service entities. One of the most important is Contents. In order to satisfy the users, the contents must provide with real-time and regional information; for that reason, it is necessary to be able to identify the location even environment the user in, and provide available personal services to users. As mentioned above, RTLS based on RFID could be the main input of context-aware for identification, tracking or locating. Context-aware could be consider that the location of the users should be sensed to decide the behavior of the service; moreover, it will provide users with related services, so we use mobile agent that can be able to furnishes context-aware, timely and accurate information for the care provision.

Context-aware Mobile Agent in Healthcare
Context aware is a concept that has been described for some time, but technologies (e.g. wireless technologies, mobile tools, sensors, wearable instruments, intelligent artifacts, handheld computers) are now available to support the development of applications. Such technologies could help health care professionals to manage their tasks while increasing the quality of patient care. (Bricon-Souf, N. & C.R. Newman 2007). Hence some researchers research this topic and use context-aware mobile agent in healthcare. In 2003, Munoz, M.A., et al. (Munoz, M.A. et al. 2003) applied context-aware mobile communication in hospitals. The authors wanted to empower mobile devices to recognize the context in which hospital workers perform their tasks and proposed an extension of instant messaging to add context awareness as part of the message such as circumstances that must be satisfied before the system delivers the message. Contextual elements used include location, delivery timing, role reliance, artifact (particularly the device) location and state.

Burstein, et al. in 2004 (Burstein, F. et al. 2005) applied agents to the highly dynamic and variable context of healthcare emergency decision-support domain and advocated the use of mobile agents to support the deployment of an ambulance service in real-time. Eungyeong Kim, et al, in 2008 (Eungyeong, K. et al. 2008), proposed a u-Healthcare system that can perceive emergency situations in chronic hypertension patients and initiate emergency action. The system considered patient mobility and integrates a medical recommendations knowledge base and communication protocols between system agents in an RFID and Cell phone architecture.

The Impact of the Social Dimension
When new technology or more comprehensive applications of technologies are developed, people usually hope that their life can be better or more convenience. On the other hand, some of the technologies often give certain effect or impact on society. Privacy, identity, security and trust are central key issues in ambient intelligence visions, and the success of ambient intelligence (AmI) will depend on how secure it can be made, how privacy and other rights of individuals can be

protected and how individuals can come to trust the intelligent world that surrounds them and through which they move (Friedewald, M. et al. 2007). From a principal point of view, privacy is generally considered to be an indispensable ingredient for democratic societies. Bohn et al. (Bohn, J.g. et al. 2004) consider that people perceive their privacy being invaded when borders are crossed.

In the same way, security concerns revolve around vulnerabilities and the protection of confidential data from unauthorized access and manipulation. As with all wireless communication systems, such as RFID systems are also subject to a number of problems, one of the most important being the illicit tracking of RFID tags. Other fundamental security issues, such as confidentiality, integrity, authentication, authorization, non-repudiation, and anonymity, can often not be overcome unless special security mechanisms are built into the system. For information security, Karen (Loch, K.D., Carr, H.H, & Warkentin, M.E. 1992) divided the threats into four categories: Disclosure, Modification, Destruction and Denial of use. No matter how to define information security, it usually responds to information confidentiality, information integrity and information availability. Provided that AmI will be introduced into life or any new technologies try to introduce into society to apply any applications, there are also necessary to make the evaluation of any critical impact on society, as well as the understanding of public acceptance.

RESEARCH METHODOLOGY

The ICCS system architecture

The purpose of this research is to establish a Intelligent Community Care System (ICCS) based on the concept of ambient intelligence using ZigBee based RFID and mobile agent, JADE as shown in Figure 1. We expect the system to be capable of locating the caregivers and care receivers in the community and identifying their needs. Therefore, the ICCS should be able to search for appropriate services and information through its own database and relevant external service providers in order for caregivers to utilize when carrying out community care. At the same time, the system would also keep track of the community care performed by each caregiver. In other words, the system would allow caregivers to take the initiative by visiting those who require care and administering appropriate community care services. Whenever care receivers have doubts that caregivers do not fully understand, caregivers could refer to the information displayed on their Tablet PC or PDA to adequately answer to the questions.

In the design of ICCS, a large or complex task is divided into several modules. Each module need to deal with a variety of tasks. ICCS can be divided into three major modules as shown in Figure 1: The RFID Environment basically involves the deployment of a full ZigBee network, so that every user with a tag in the environment would be accessible to multiple RFID nodes simultaneously in order for information to be relayed to the Server for processing. The Agent Environment encompasses the entire Agent Network, which is comprised of an Information

Agent (IA), Nursing Agent (NA), Scheduling Agent (SA), User Agent (UA) and External Service Provider (ESP). The ICCS Server and Database is responsible for the management of the RFID Environment and Agent Environment while serving as the bridge between the two environments. In other words, the ICCS Server would process and compute massive amounts of original positioning data from the RFID Environment. After computation, the results would be stored in the database for Mobile agents in the Agent Environment to use.

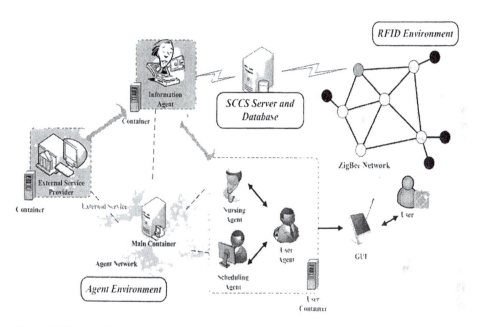

Figure 1. The architecture of ICCS.

RFID Environment in ICCS

In order to allow the RFID Environment of ICCS to acquire the relevant data for computation, the ZigBee Network with interconnected ZigBee RFID equipment refers to the Mesh Network constructed, and it must comprise Tags, Sensor Node, and Router or Gateway as shown in Figure 2. The ZigBee Network allows rapid access of original positioning data of all personnel wearing Tags within the environment, and then use a measurement of Received Signal Strength Indication (RSSI) via the power present in a received radio signal to calculate and make Real Time Location System (RTLS).

Within the environment, a Tag will be accessed by multiple Sensor Nodes simultaneously to obtain varying Received Signal Strengths (RSS). Through the ZigBee Network, the data would then be collected to the router or gateway, which in turn, sends the data to the ICCS Server via wired or wireless network as shown in Figure 2. The ICCS Server would process the positioning data by an RSSI positioning algorithm and allow the system to point the location area of every personnel wearing tags within a community so as to support other modules (i.e. demands of usage in the Agent Environment) in the system.

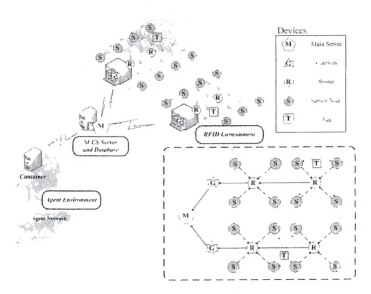

Figure 2. The Architecture of RFID Environment in ICCS.

Agent Network Environment in ICCS

An agent environment consists of hosts running the agent containers and hosts supporting the external service providers. Agent containers can be added and removed dynamically to an agent environment that allows for large-scale, even network-wide, installations. In this research, JADE (Java Agent DEvelopment Framework) was used to develop a multi-mobile agent-based system in ICCS. Four primary architectural features discriminate agents in ICCS as follows: *Information Agent* (IA), *User Agent* (UA), *Scheduling Agent* (SA), *Nursing Agent* (NA) and *External Service Provider* (ESP).

Information Agent (IA)

The IA is encapsulated in a data container and connected to the "ICCS Server and Database" as shown in Figure 1. The IA is a stationary agent and operates at a higher level of trust and mediates access resources from mobile agents to database. The IA can be 'static' because it doesn't have the ability to migrate; it usually resides on the data container to provide relevant services, service information, care-receiver data and persons' locations. Therefore, they are considered as secure and granted to access the resources from the "ICCS Server and Database". In the proposed ICCS, the IA plays an important role to dynamically interface with the "ICCS Server and Database". It is through the IA that mobile agent can have direct access to the ICCS server and database. And the IA would take the place of the web sites to bridge the NA and SA to the databases.

External Service Provider (ESP)

The ESP is a type of stationary agent that is used as a provider. It is encapsulated in each external data container and connected to their database as shown in Figure 1. The ESP also operates at a higher level of trust and mediates access resources to their own database. The ESP resides on the own container to provide NA with its specific information. Therefore, ESP will serve as a bridge between the ICCS and external data exchange so that ICCS can cooperate and stay connected to other information providers. When ESPs log onto the Main Container of ICCS, the NA would be able to search and migrate to the ESP data container that provides relevant information to request for such information. ESPs can be dedicated providers of meteorological information or drug information and so forth.

User Agent (UA)

The User Agent is the agent used by caregiver, which can help caregivers to perform their tasks. The UA serves as a bridge to interface with a host's computer and applications. It acts as a mediator for users and applications and invokes application services when receiving user requests. The UA collects a request from GUI, submits the request to SA or NA, and presents the final results from SA or NA to the user. Therefore, the UA is responsible for providing GUI to users. It manages all the elements used in the interface, such as frames, tables, and buttons. All of the related values of parameters, layouts of the forms, and styles of the elements are also controlled by the UA. UA also acts as an interpreter of presented data and specific message format - ACL.

Scheduling Agent (SA)

The SA is a mobile agent migrating between a client container and IA's data container. It is responsible for notifying caregivers of their schedule, including where are they going and doing. The caregiver's schedule encapsulates the shortest care service path, care-receiver's brief data and location, etc. of the schedule. The SA is invoked when a caregiver intends to know his/her personal schedule and to relevant information. Hence, the SA's process has two steps:

1. Based on the identity of caregiver who has just logged onto the system, the SA will migrate to the Container where IA is located to retrieve relevant information such as basic profile of communities in need of care, list of basic care receivers and reminders for caregivers on the items they need to bring on top of arranging the schedule of the caregiver on the particular day. The SA will then send this information to the UA, who will display the information on the Tablet PC/PDA used by the caregiver through a GUI.
2. When the ZigBee Nodes locate signals that indicate the caregiver has arrived at the community, the SA will once again migrate to the data container where IA is located to retrieve further information, including profile of care receivers and their current locations within the community. The SA will also notify the caregiver regarding the care receiver who is located closest to him so that the caregiver would be able to take initiative and approach the care receiver within

the shortest time possible to administer community care.

Nursing Agent (NA)

The NA is also a mobile agent migrating between client and each data container, and it acts as a commissioner to explore the possible information in each data container, including IA and ESPs. The NA is responsible for providing the caregiver with proper services and profiles, and is also responsible for updating and managing the care-receiver profiles. The care-receiver profiles are including the basic information, history records, and external information like diseases or medicines, etc. So the CA is invoked when a caregiver submits a request to form the ICCS.

In the ICCS, when a caregiver arrives at a care receiver's location, the ICCS will determine if both of them are in the same area. If such is the case, the NA will be invoked and migrated to the container where the IA is located to retrieve relevant information on the care receiver such as basic profile, previous medical records and the services he should receive based on the identity of the caregiver and care receiver. Later, according to the information from the IA, the NA will migrate to each relevant external services provider's container and search all the relevant information. For example, cold weather is known to hike blood pressure and could be potentially dangerous for patients suffering from hypertension. The NA would seek the ESPs of meteorological data to obtain the latest weather reports. Or, if the care receiver is known to take specific drugs to suppress, the NA would migrate to the ESPs of drug information to search for relevant information (i.e. ingredients, side effects or safety precautions) on the aforementioned drug.

The NA would then return to the caregiver's client container and relay the data to the UA, which would in turn display the information on the Tablet PC or PDA used by the caregiver through a GUI. As the caregiver performs his services, the NA would keep track of the event and migrate once again to the data container where the IA is located to relay the record to IA. The IA would store the record in the Database to conclude the care service. At this point, the SA would remind the caregiver as to the locations of care receivers who are still awaiting assistance.

CONCLUSIONS

The aging global population has seen community nursing gain in importance. To provide the elderly with more comprehensive and personalized nursing services, caregivers providing community care need to know about their care-receivers' condition and requirements more quickly. The burden on each caregiver must also be reduced so they can spend more time and effort on care-receiver. To provide better and more personalized community care services, each care process and outcome must be recorded for later analysis and discussion.

To this end, this study used RFID and mobile agent to realize the ambient intelligence and implement it in the "Smart Community Care System". For caregivers, the ICCS allows even new volunteers to provide each care-receiver with comprehensive and personalized care services based on the system provided services and information. It also helps new volunteers to familiarize themselves with their tasks more quickly. At the same time, it reduces the workload on community caregivers and the chance of recording errors so each caregiver can spend more time on the care-receiver.

For the care-receiver, they can live in a community they are comfortable with and caregivers will visit them at their current location. This means he or she will receive the care they need even if they should be out of their room.

The ICCS can also integrate all kinds of useful information and search results from the Internet based on the physical or psychological condition of the caregiver. Relevant information can then be provided to the caregiver during the care delivery process and help they understand the care-receiver's condition. Finally, each care delivery process and outcome can be quickly and promptly recorded. Everything from the physiological indices to drug usage can be tracked for analysis and discussion in order to provide the care-receiver with better care services.

REFERENCES

Bohn, J.g., Coroama, V., Langheinrich, M., Mattern, F., Rohs, M., et al., (2004). Living in a world of smart everyday objects--Social,economic, and ethical implications. Human and Ecological Risk Assessment. Volume 10 No.5, pp. 763-785.

Bricon-Souf, N. & C.R. Newman, (2007). Context awareness in health care: A review. International Journal of Medical Informatics, Volume 76 No.1, pp. 2-12.

Burstein, F., Zaslavsky, A., Arora, N., et al., (2005) Context-aware mobile agents for decision-making support in healthcare emergency applications, The 1st Workshop on Context Modeling and Decision Support, A.G. T. Bui, Editor. (pp. 1-16). Paris, France

Chen, C.Y., (2003). Long-Term Care for the Elderly. Formosan Journal of Medicine, Volume 7 No.3, pp. 404-413.

Chung, W.C., (2007). A Study on Institution Developing Community Care and Integrating Resource, in Graduate School of Social Informatics 2007, Yuan-Ze University.

Eungyeong, K., Hyogun, Y., Yupeng, Z., Lee, M., Jaewan, L., et al. (2008). A Hypertension Management System with Emergency Monitoring. Information Security and Assurance.

Friedewald, M., Vildjiounaite, E., Punie, Y., Wright, D., et al., (2007) Privacy, identity and security in ambient intelligence: A scenario analysis. Telematics and Informatics. Volume 24 No.1, pp. 15-29.

G. Riva, F. Vatalaro, F. Davide, M. Alcañiz, et al., (2005) Interactive Context-Aware System Interacting with Ambient Intelligence. In Schmidt, A. (Ed.), Ambient Intelligence (pp. 107–123). IOS Press.

Kaasinen, E., (2003). User needs for location-aware mobile services. Personal Ubiquitous Comput, Volume 7 No.1, pp. 70-79.

Kikhia, B., (2008). Acceptance of Ambient Intelligence (AmI) in Supporting Elderly people and people with Dementia, in Department of Business Administration and Social Sciences. Lulea University of Technology: Sweden.

Loch, K.D., Carr, H.H, & Warkentin, M.E., (1992) Threats to information systems: Today's reality, yesterday's understanding, MIS Quarterly, Volume 16 No.2, pp. 173–186.

Mendes, M., R. Suomi, & C. Passos, (2004). Digital Communities in a Networked Society. Baker & Taylor Books

Munoz, M.A., Rodriguez, M., Favela, J., Martinez-Garcia, A. I., Gonzalez, V. M. et al., (2003). Context-aware mobile communication in hospitals. Computer, Volume 36 No.9, pp. 38-46.

Rifartek Technologies. Retrieved from the Web: http://www.rifartek.com/Web

CHAPTER 64

Mobile Agent Based Ubiquitous Health Care (UHC) Monitoring Platform

Su Chuan-Jun, Chiang Chang-Yu

Department of Industrial Engineering and Management
Yuan Ze University
Taoyuan, 135 Far East Road, R.O.C.

ABSTRACT

Ubiquitous health care monitoring systems in the field of e-health enable immediate analysis of individual physiological data and personalized patient feedback in real time using alarms and reminders. The systems monitor vital signs such as ECG (electrocardiography), EMG (electromyography), oxygen saturation, respiration, activity, and temperature. Patients can be remotely and ubiquitously assessed, diagnosed, and treated. In the case of rapidly deteriorating medical conditions, the systems can automatically notify associated medical staff by making mobile phone call or sending an SMS alarm to provide the patient with first-level medical support. In this research we have developed a highly distributed information infrastructure for ubiquitous health care (UHC) by using Intelligent Agent paradigm, which is able to notify the responsible care-provider of abnormality automatically, offer distance medical advice, and perform continuous health monitoring for those who need it. To confront the issues of interoperability, scalability, and openness in heterogeneous e-health environments, a FIPA2000 standard compliant agent development platform - JADE (Java Agent Development Environment) was adopted for the implementation of the proposed intelligent multi-agent based UHC system.

Keywords: e-health, JADE (Java Agent Development Environment), Mobile Agent, Health Care Monitoring

INTRODUCTION

Mobile ubiquitous monitoring systems in the field of e-health enable immediate analysis of individual physiological data and personalized patient feedback in real time using alarms and reminders. The systems monitor vital signs such as ECG (electrocardiography), EMG (electromyography), oxygen saturation, respiration, activity, and temperature. Patients can be remotely and ubiquitously assessed, diagnosed, and treated. In the case of rapidly deteriorating medical conditions, the systems can automatically notify associated medical staff by making mobile phone call or sending an SMS alarm to provide the patient with first-level medical support.

Furthermore, scarce medical personnel are traditionally occupied with observing physiology parameters on the monitor screens sitting in the monitoring center on a 24-hours-per day, 7 days-per week basis. The task of monitoring is a tedious duty for medical staff to simultaneously audit and interpret the massive amounts of information regarding a patient during the process of monitoring, performing diagnostics, and verifying therapeutic intervention. Some of complications have arisen from this situation because each medical personnel may be in charge of many patients. The development of multi-agent based mobile monitoring systems is capable of releasing medical staff from routine monitoring tasks and focusing on his/her daily works. With such a mobile monitoring system, patients benefit from better accessibility while medical staff can be more efficient and accurate in following up patient histories with easily available patient data. Families can reduce the time lost in visits to the hospital, which in turn reduces the number of occupied beds that require monitoring, making room for more critical patients. Governments also stand to benefit from reduced hospitalization time for non-critical patients. Healthcare institutions gain through reduced patient treatment costs, better resource management and significant health-economic improvements.

With the growth of the population, now e-health care is facing a big challenge to provide better public or private health care especially in monitoring aspect. For this reason, a greater solution to enhancement of the quality of e-health care is in urgent need. During the last few years several initiatives were undertaken addressing different applications of e-health issues, ranging from doctor (e.g. remote access to medical data) to patient (e.g. remote monitoring of vital signals, medical record) up to Internet based medical data access. The developments of cost-effective, automated, distributed health-care monitoring systems are urgently needed.

Due to medical resources are insufficient, some savants use agents to solve e-health care problems especially in using mobile agent. Mobile agent (MA) has attracted much attention in this decade because of the unique characteristics of the technology. A mobile agent is a program that represents a user (or user take) and can autonomously migrate between the various nodes of a network to perform computations on her behalf.

The mobile agent environment in which mobile agents execute is a software system distributed over a network of heterogeneous computers. The rationale behind the idea of developing a Mobile Agent Based Ubiquitous Health Care (UHC) platform for wide-area vital sign monitoring is its various advantages over traditional client-server approach (Graat, 2003) (Kinshuk, Hong and Patel 2002).

Java Agent DEvelopment Framework (JADE) is a framework that facilitates the development of agent applications in compliance with the FIPA specifications for interoperable intelligent multi-agent systems. The JADE agent platform tries to keep the high performance of a distributed agent system implemented with the Java language. It is also a middleware for developing distributed applications through leveraging state-of-the-art distributed object technology embedded within the Java runtime environment. Therefore, the goal of JADE is to simplify the development while ensuring standard compliance through a comprehensive set of system services and agents. JADE uses an agent model that allows high runtime efficiency, software reuse, agent mobility, and the realization of different agent architectures.

LITERATURE REVIEW

It is our conviction that agents have the potential to assist in a wide range of activities in an e-health care environment. They can maintain the autonomy of the collaborating participants, integrate disparate operating environments, coordinate distributed data, and other organization involved in health care. As explained in (Shankararaman et al., 2000), Agent Technology allows:

- To proactively anticipate the information needs of a patient, and deliver it in a periodical basis.
- To support communication and coordination, either synchronous or asynchronous, among members of a medical team, to enable the share of distributed information and knowledge sources, and to provide distributed decision making support.
- To adapt medical services to patients' needs (personalization).

A multi-agent system consists of multiple autonomous agents with the following characteristics (Jennings et al., 1998):

- Each agent cannot solve a problem unaided.
- There is no global system control.
- Data is decentralized.
- Computation is asynchronous.

Most of these characters have already been investigated in the medical domain through the use of multi-agent system architectures. For example, the GUARDIAN system (Hays-Roth and Larsson 1996) considered patient monitoring in a Surgical Intensive Care Unit. Support is provided for collaboration among specialists, each an expert in a specific domain but fully committed to sharing information and knowledge among each other and the nurses that continuously monitor the patient in the physicians care. Another example of the MAS-based system for healthcare has also been described in (Huang, 1995), in which a multi-agent system was designed to support collaboration among general practitioners and specialists about patient healthcare In patient appointment scheduling where medical procedures have become more complex and their tests and treatments more interrelated, manual and traditional software solutions have been shown to be inadequate while a multi-agent solution gave significantly improved results (Decker and Li, 1998).

A European IST project aims to design and develop a configurable agent-based framework for virtual communities focused on supporting assistance to elderly people employing tele-supervision and tele-assistance (Shankararaman et al., 2000). A distributed decision support system based on the multi-agent paradigm can support cooperative medical decision-making (Lanzola et al., 1999). There have also been developments in general patient care management (Huang, 1995) and medical training (Farias and Arvanitis, 1997). The deluge of medical information available on the Internet has led to the development of information agent to collect and organize this information, such as the Multi-Agent Retrieval Vagabond on Information Network (MARVIN) (Baujard et al., 1998), developed by the Health On the Net Foundation and the Swiss Institute Of Bioinformatics. The Independent LifeStyle Assistant (I.L.S.A.) (Karen et al., 2002) is also a multi-agent system that aid elderly people to live longer in their homes, increasing the duration of the independence from round the clock while maintaining important social connectedness and reducing caregiver burden. Those multi-agent systems have to be devised to provide aid in carrying out activities of daily living, and health care maintenance.

RESEARCH METHODOLOGY

Requirement Analysis

Traditional health monitoring process consumes numerous time and traveling expenses for care-receivers to visit hospitals or health care sectors frequently. It is quite inconvenient for them, especially elders or long-term care-receivers, to be present at the hospitals on a long trip regularly. However, health monitoring usually requires frequent vital signs check-up to ensure their health condition. With the growth of this population, e-health monitoring is now facing a profound challenge to provide better public health care nationwide. There is an urgent need to develop a system that is capable of performing ubiquitous electronic health monitoring automatically and autonomously to users who are usually mobile and situate in a low bandwidth, high latency, asynchronous transaction, and unstable connection environment.

- The system requirements for the proposed UHC are therefore at least the following: Openness: Each instance of UHC installation situates at home with a technophobic client and a typically non-standard monitoring device. The UHC thus must be easily deployable configurable, and updated. UHC is required to facilitate the evolution of a specific installation by providing an open architecture in which new devices and knowledge based modules may be integrated.

- Modularity: Due to the complexity of health monitoring domain, distributed and encapsulated expertise will be critical to the viability of UHC. "Modularity"is thus another crucial requirement for UHC to achieve the extendability by partition the functions into smaller logical units that can be modified, enhanced, or added functionality.

To satisfy these requirements, a mobile multi-agent information platform UHC that is implemented by JADE and allows MAs to work on behalf of health care professionals, to collect distributed users' vital sign data, and to spontaneously inform abnormal situations to associated health care professionals in real time is proposed in this research.

UHC System Design

E-health care is a complex task which involves the sharing of expertise about medical knowledge, medical data, and services among care-receivers, specialists, as well as medical personnel. Moreover, an e-health care platform is composed by an open architecture, distributing physiological information and resources, multi-systems with heterogeneous components, storing bio-signals acquired from patients, and secure infrastructure. Consequently, a significant desire in the health care monitoring practice is stimulated to integrate several disparate and stand-alone subsystems and corresponding information repositories.

In this research, JADE has been adopted among various multi-agent platforms as the underlined architecture and implementation of UHC due to the following advantages (Kinshuk, 2002):

1. Distributed autonomous applications development - In order to achieve the objectives of UHC, agents that are autonomous, intelligent, and capable of communicating and collaborating need to be implemented. JADE simplifies such a development.
2. Negotiation and coordination - In UHC, JADE provides easy-to-use software libraries (i.e. patterns of interaction between agents) to solve negotiation and coordination among a set of agent, where the resources and the control logics are distributed in the environment.
3. Pro-activity - JADE agents have been designed to control their own thread of execution. These agents can be easily programmed to initiate the execution of actions without human intervention just on the basis of pre-defined goals and state changes. The property of proactivity is essential in designing physician agents of UHC, which requires controlling their own actions guided by regulations.
4. Multi-Party applications - Peer-to-peer architectures that JADE used are more efficient than client-server architectures for developing multi-party applications. Sometimes, the server might become the bottleneck and the point of failure in the entire system. The implementation of UHC based on JADE architecture that allows clients (medical staff or patients) to communicate each other without the intervention of a central server and subsequently reduces the network traffic.
5. Interoperability - JADE complies with the FIPA standard that enables end-to-

574

end interoperability between agents of different agent platforms.

6. Versatility - JADE provides a homogeneous set of APIs that are independent from the underlying network and Java version. It also provides the same APIs for J2EE, J2SE, and J2ME environments. This feature makes UHC a heterogeneous client (PC, PDA, mobile phone, etc.) environment..

7. Ease of use - JADE APIs and ready to use functionalities can shorten the system development cycle (some estimations have been give that indicates the reduction of development time cab be up to 30%).

The JADE-implemented UHC is hence expected to be capable of integrating disparate information sources and isolated heterogeneous components to perform autonomous health monitoring. The UHC is composed of six types of architectural components as depicted in Figure 1: (1) User Agent, (2) Resource Agent, (3) Physician Agent (4) Diagnostic Agent (5) Knowledge-based Data Server, and (6) External Services.

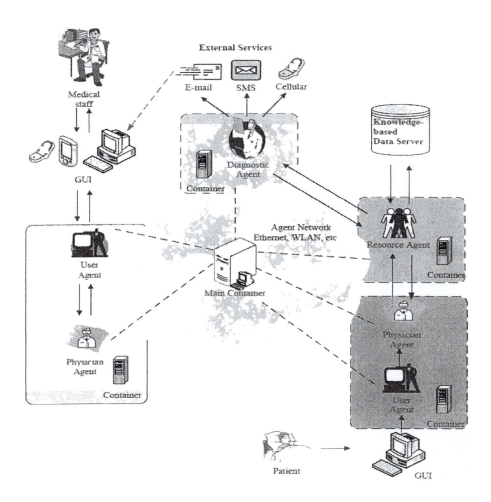

Figure 1. The UHC structure.

User Agent

The user agent is a type of stationary agent and it is a user's intelligent gateway to the platform. It is regarded as a bridge to interface with physician agents and users. The user agent acts as an interpreter of users and heterogeneous agent communications. When receiving the requests from users, it will to drive internal services in the platform. .Therefore, the user agent is a mechanism of access control resources and authenticates users before starting physician agent. The user agent is responsible for the final presentation of results by physician agent or by diagnostic agent to users.

Resource Agent

The resource agent operates at a higher level of trust and mediates access resources from physician agent to host computers and it's also another type of stationary agent.

It can be "static" because they don't have the ability to migrate between hosts. The resource agent usually resides on the host (container) to provide expert advice or services locally. Hence, it is deemed as a secure agent and granted to access the resources of data server. The resource agent plays an important role in dynamically accessing the resources in our recommend platform. Therefore, the physician agent can ask resource agent to directly access the resources of the data server, since physician agent are restricted to communicate only with agent container and also other agents in the platform. The resource agent would take the place of the interface to bridge the users to the databases.

Physician Agent

The physician agent is a mobile agent and it is used by the medical staff or patient. It can help the medical staff perform monitoring tasks and help patient perform the data transmit. The physician agent enables medical staff to monitor the conditions of patients in real time without being around the patients. Besides it also helps the diagnostic agent to examine the status of outpatients at remote locations. In this first phase, medical staff can use their devices (e.g. computer, PDA, cell phone) to trigger physician agents as delegates for them (see figure 1). Medical staff have tasks of monitoring patients to finish while they are fully occupied with other emergency situations, then they initiated the physician agent to help them. In order to activate a physician agent, an order from the medical staff via user agent must be dispatched to agent main container. As soon as the agent main container received such an order, a physician agent would be subsequently initiated. Before the physician agent migrates to another agent container, it has already authenticated, gotten agent identifiers (AID) and determined the privileges to be granted by the agent main container. The physician agent then will be sent to the Internet through Ethernet, wireless local area network (WLAN), etc. After connected to Internet, the agent will continue doing the assigned monitoring job depending on the characteristics of a given clinical case. The physician agent is devoted to data acquisition, after roaming in the networks to collect related physiological data of patients. Eventually, the physician agent carries the collected information with it and returns to the medical staff's devices.

Diagnostic Agent

The diagnostic agent is another type of stationary agent. In the proposed platform, the diagnostic agent can be considered a data-analyzed engine. It is capable of analyzing collection of short-term changes and health status data of patients. The main task of diagnostic agent is to (1) identify receive patterns from patients (2) indicate or predict a sudden change in patients' status. After detecting the collected data, the diagnostic will supply medical staff with the periodic information. If detecting abnormalities in collected data, the diagnostic agent will send to medical staff with emergency information via the external services. The next step, when it comes to some critical life-death situations, it will provide patients with needed automation assistance (e.g., call ambulance).

Knowledge-based Data Server

Efficient and effective patients monitoring is a complex endeavor that is highly dependent on disseminating time-critical information (Lee et al., 2000). The time-critical information involves not only the records of patients' conditions but also knowledge sharing of expertise and medical practices. The physiological data collected by physician agent are the results of observation or examination forming the basis of further diagnoses and therapies. Each patient is an individual case with individual patterns, complications, and disease. Therefore, the knowledge-based data server is considered in this section in order to promote the performance of the proposed platform. The knowledge-base data server is split into two main information repositories: (1) PatientStatus and (2) PatientProfiles. These two repositories are the storage units that are utilized to deposit the physiological information collected by physician agent. The "PatientStatus" repository involves every monitoring physiological data from patients (e.g. HR, SpO2, etc). The "PatientProfiles" repository contains up-to-date electronic records for patients: the limits of monitored parameter values and some basic information of patients. When user agent delivers patients' monitoring data to resource agent, the resource agent stores the data in the "PatientStatus" repository; at the same time, it copies these data and sends to diagnostic agent. The diagnostic agent will analyze patients' physiological information in real time.

External services

The external services will be initiated by diagnostic agent after analyzing the incoming physiological data of patients. The external services embrace the hardware and services in the platform including E-mail and short message services (SMS), etc. It is an extensible component varied on different scenario of applications. The result of the physiological data analysis is the prediction for diagnostic agent to consider while determining the provision of next action for medical staff. If the prediction of the patient's signals results in abnormal situation, the diagnostic agent will take proper actions such as issuing an alert message according to the analyzed outcomes. Moreover, if some emergency situation of patients occurs, the external service for automatically dialing ambulance is going to be launched.

CONCLUSIONS

Multi-agent systems have a set of properties that make them suitable for use in solving the many problems that are encountered in the healthcare domain. In previous studies, as described in Section 2.4, many agent technologies have been applied in the healthcare field. Nevertheless, not all existing multi-agent systems can provide good solutions for the healthcare monitoring domain.

In this research, we have developed a UHC system based on the JADE platform to address the specific requirements of a distributed healthcare environment. From the different aspects of usage requirements, UHC provides different functions to cater to each user. From the viewpoint of the medical staff, the workload is reduced by the substitution of routine patient-supervision tasks. Using a personal device, medical staff can, at any time and from any location, view the vital signs of patients, who are generally widely spread across many locations. Furthermore, in UHC, the diagnostic agent substitutes for a physician, performing the first step of diagnosis. While from the point of view the patients, patient care is improved by the immediate transits as well as by the communication between agents, which provide more reliable information, delivered in a more convenient manner.

From the viewpoint of the system, UHC provides a patient monitoring environment that is simple and efficient. Furthermore, it provides a ready exchange of information, and as permanent data is collected only once, less time is spent in searching for mislaid data, thereby increasing efficiency. The communication via asynchronous message passing between agents can solve the problems that would arise when using blackboard; this is because blackboard is not suitable for a scalable system (Graat, 2003). In UHC, each agent has its own mailbox through which it can receive messages sent by other agents; this is a better mode of communication than blackboard. The method of asynchronous message passing does not require centralized management information and also provides good privacy for patient personal information. Ultimately, documentation is diminished as a consequence of the elimination of hand-written records, clinical reports, medical errors, etc.

UHC provides a new opportunity to integrate and analyze the immense amount of data encountered in patient monitoring. It can not only respond to inquiries for medical information relevant to a person's medical history but also monitor the patient status and alert its owner about unhealthy trends as well as inconsistencies in the health records. It clearly presents an innovative technique to assist healthcare practitioners in collecting, filtering, and examining the relevant information for a patient, providing basic diagnosis and suggesting actions in an efficient manner. The development of UHC will be critical for the future of healthcare monitoring.

REFERENCES

Baujard O., Baujard V., Aurel S., Boyer C. and Appel, R. D. (1998). A Multi-agent Softbot to Retrieve Multilingual Medical Information on the Web. Medical Informatics. Volume 23 No. 3, pp. 187-191.

Decker, K S and Li, J. (1998). Coordinated Hospital Patient Scheduling. International Conference on Multi-Agent Systems - ICMAS_98, Paris, France.

Farias, A. and Arvanitis, T. N. (1997). Building Software Agents for Training Systems: A Case Study on Radiotherapy Treatment Planning. Knowledge-Based Systems. Volume 10. pp. 161-168.

Graat, G. (2003). Agent-based Information Retrieval Supported by Information Markets. M.S. Thesis, Faculty of General Sciences, Maastricht University, Netherlands.

Hays-Roth, B. and Larsson, J. E. (1996). A domain specific architecture for a class of intelligent agent monitoring systems. Journal of Experimental and Theoretical Artificial Intelligence. Volume 8. pp. 149-171.

Huang, J., Jennings, N. R, and J. Fox J. (1995). An Agent-based Approach to Health Care Management. Applied Artificial Intelligence: An International Journal, London, Volume 9 No.4, pp.401-420.

Huang, J., Jennings, N. R., and Fox, J. (1995). An Agent Architecture for Distributed Medical Care. Wooldridge, M. J. and Jennings, N.R. Editors., Intelligent Agents, Lecture Notes in Artificial Intelligence, Springer-Verlag, pp. 219-232.

Jennings, N. R., Sycara, K. and Wooldridge, M. (1998). A roadmap of agent research and development. Autonomous Agents and Multi-Agent Systems. Volume 1 No.1, pp. 275-306.

Karen, Z.H., John, P., and Christopher, W.G. (2002). An Open Agent Architecture for Assisting Elder Independence. The First International Joint Conference on Autonomous Agents and MultiAgent Systems (AAMAS). pp. 578-586.

Kinshuk, K., Hong, H. and Patel, A. (2002). Adaptivity through the Use of Mobile Agents in Web-based Student Modelling. International Journal on E-Learning. Volume 1 No.3, pp. 55-64.

Lanzola, G., Gatti, L., Falasconi, S. and Stefanelli M. (1999). A Framework for Building Co-operative Software Agents in Medical Applications. Artificial Intelligence in Medicine. Volume 16 No.3, pp. 223-249.

Lee, R., Conley, D., and Preikschat, A. (2000). eHealth 2000: Healthcare and the Internet in the New Millennium. Wit Capital, Retrieved from the WorldWideWeb: http://www.witcapital.com/.

Shankararaman, V., Ambrosiadou, V., Panchal, T. and Robinson, B. (2000). "Agents in health care", Workshop on Autonomous Agents in Health Care. pp. 1-11.

Shankararaman, V., Ambrosiadou, V. and Robinson, B. (2000). Agents in Medical Informatics. Proc. of IASTED International Conference on Applied Informatics, Austria.

CHAPTER 65

Human Factors in a Voluntary Medical Incident Reporting System

Yang Gong

Department of Health Management and Informatics
University of Missouri
Columbia, MO 65212, USA

ABSTRACT

Currently, reporting of medical errors or near misses is one of the leading initiatives proposed to enhance patient safety(Cohen, 2000). It is believed that medical error reporting systems could be a good resource to learn from errors if medical error data are collected in a properly structured format and are useful for the detection of patterns, discovery of underlying factors, and generation of solutions(R. J. Holden & Karsh, 2007). However, currently medical incident reporting systems show great inconsistency in terms of reporting terminology and data granularity (Y. Gong, Richardson, Zhijian, Alafaireet, & Yoo, 2008; Karsh, Escoto, Beasley, & Holden, 2006). We propose to investigate human factors which may greatly affect the completeness and accuracy of voluntary reports and not being adequately addressed in the current reporting system. The overall objective of this project is to understand, identify and eliminate the technical barriers to complete and accurate incident reporting through a human-centered consideration.

Keywords: Voluntary reporting, medical errors, human factors

INTRODUCTION

Patient safety has become a major public concern since the publication of the Institute of Medicine report, To Err is Human, in 1999. The creation of mandatory and/or voluntary reporting systems is suggested a helpful approach in reducing medical errors(Kohn, Corrigan, & Donaldson, 1999). In response to the report, the University of Missouri Healthcare (UMHC) created a voluntary Patient Safety Network Incident Reporting System (PSN) in 2002. The PSN is web-based data entry system developed based upon a reorganization and upgrade of seven paper-based reporting systems. It enables patients/families to report compliments, complaints, or suggestions via the web. Staff will be able to report variances, close-calls/near misses, and adverse events. Additionally staff may enter comments on behalf of patients or themselves(Unkown, 2006).

Each year, the PSN system collects approximately 3000 cases reported from five facilities across UMHC where it carries about 500 staffed beds and 19000 patient admissions annually. Such a system is aiming at a systematic, comprehensive approach to reporting and reviewing incidents and near misses for the prevention of future occurrence. The PSN database contains one table with 26 required fields. Each case is automatically assigned a unique event ID. Of the fields corresponding to user interface, some require reporters to type in and some require reporters to select from pre-defined menus (pull-down lists, radio groups, check boxes, etc). Examples of required fields displayed in pre-defined menus include harm score, event type, error description and brief description. Other information such as profession of reporter, patient age, report date, event date, patient unit ID, patient unit ID, reporter's department and so on are typed in by reporters. There is an event short summary comprised of two parts, a "harm score" and a short description predefined in a dropdown list. The harm score is a subjective rating of severity by the reporter from 0(no clinical changes) to 5(death). Each harm score corresponds to a different number of items provided to reporters as a reference for evaluating the severity of events. Based on the hospital policy, a preliminary case review of all cased assigned a harm score of 4 or 5 must be conducted within ten working days of the report. For all reports, reporters may choose to report them anonymously or retain their IDs. Besides the harm score, all reporters are required to describe the event using a brief message which should contain a little detail as to why the harm score is applicable. Then, hospital administrators, service medical directors or department review the reports and fill in the solution, review, and additional information fields. Therefore, all cases used in this project have been reviewed and responded.

Incompleteness, inconsistency and lack of expression of events are the three main reasons greatly impede the usefulness of reports for root cause analysis. (Y. Gong, et al., 2008). Earlier studies have identified some barriers and incentives to reporting at the individual, organizational and society levels(Barach & Small, 2000). Despite large amount of studies suggest instituting a "just culture" that encourages learning, non-punishment(Arroyo, 2005; Barach & Small, 2000; Bates et al., 1995; Cullen et al., 1995; Kaplan & Rabin Fastman, 2003; Kingston, Evans, Smith, & Berry, 2004; L. L. Leape, 1994; Lucian L. Leape, 2002; Marx, 2001; Wakefield, 2005), few studies have investigated the system difficulty and inefficiency regarding ease of use, ease of understanding and their relations with the level of details in reporting(Beasley, 2004; C. W. Johnson, 2003; Kaplan & Rabin Fastman, 2003; Runciman, Merry, & Smith, 2001). As Johnson argued that it is central to efforts to include interface specialist in the design process in order to design intuitive and usable reporting forms with clear instructions. Also, as Holder and Karsh stated that reporting system components associated with system ease of use and time efficiency should fit the busy and fatiguing work flow of health care practice(R. J. Holden, & Karsh, B. , 2005). To date the research addressing reporting system inefficiency has been limited and fragmented, with findings not always broadly disseminated. Although general rules of human-centered designs have been introduced in many other fields, currently there is a lack of design framework for medical voluntary incident reporting systems to effectively collect, catalog, and analyze the reports.

The purpose of this pilot project is to explore the degree of incompleteness and inconsistency of the reports as well as the associated human factors, based upon which we have proposed potential solutions through a human-centered design in health care. The human-centered design is expected to improve data quality and reporting rate in voluntary systems.

METHODS

We examined a total of 2919 PSN reports generated within a 12-month period during 2005-2006 at the University of Missouri Hospital and Clinics. The reports cover ambulatory, in-patient and nursing home cases. To systematically examine the quality of reports, we conducted a content analysis by systematically recoding the harm scores and adjusting the improper event categories which do not match event descriptions. We also tried to identify additional reportable factors such as delay duo to incidents or near misses. During the process, we further identified the users' characteristics and their tasks in the PSN system. All the methods we employed in this project helped us develop a set of design requirements for a human-centered voluntary reporting system.

IMPROVING CONSISTENCY BY RECODING

The purposes of recoding were to reduce the individual reporter's bias and improve the consistency among all types of events in the voluntary system. The process of recoding was started by systematically examining the harm factor through a delay and harm consideration instead of simply choosing 0 through 5. The harm and delay matrix produced 8 possible ratings, ranging from No Delay, No Harm to Delay, Death. Each case was evaluated further by two independent coders who categorized the effect of the event on the patient baseline information located in the event description, follow-up information, and solution fields in the original reports. Each of the categories has a numeric value stored in a database. We examined the *Kappa* and percent agreement using statistical software. Based upon our analysis, we proposed the human factors which potentially affect the quality of voluntary reports.

MEASURING LENGTH OF NARRATIVES

We also measured the length of each free text field. Based upon our initial observation, some reports carry a much longer description than others. The free text narratives in a report are comprised of reporter-generated and reviewer-generated sections. The Description field stores reporter-generated data, whereas, the Solution, Review and Information fields contain reviewer-generated data. All type of incidents used an identical reporting template, which became our motivation to know if the cases with higher harm score contain longer descriptions.

USER ANALYSIS

User analysis is to identify the characteristics of PSN users such as expertise and skills, knowledge bases, educational background, cognitive capacities and limitations, perceptual variations, age-related skills, and time available for learning. This is the first and essential step towards human-centered design(Yang Gong & Jiajie Zhang, 2005). We reviewed the PSN tutorial and profession field stored in the system during the analysis and then contacted the Office of Clinical Effectiveness of UMHC to discuss and confirm the results of user analysis.

TASK ANALYSIS

Task Analysis is to identify the procedures and actions to be carried out and the information to be processed to achieve task goals for the PSN reporting tasks. One important function of task analysis is to ensure that only the features that match users' capacities and are required by the task will be included in the system specifications. Sophisticated features that do not match the users' capacities or are not required by the task will only generate additional processing demands for the user and thus make the system harder to use(Yang Gong & Jiajie Zhang, 2005). This analytic approach will help identify how different users interact with the same medical incident reporting system. Through tasks analysis, we are able to identify the discrepancies between the current system and the ideal design.

We analyzed the reporting procedure of a typical reporter and presented it in a flow chart. We then analyzed the users' actions based upon four interface sections (Answer initial questions, Event common questions, Event details, Report summary) of the system using GOMS Model(Kieras, 1994) and proposed alternatives for reducing unnecessary human inputs and cognitive workload in future prototypes.

RESULTS

The majority of the reports (66.2%) were reported by registered nurses. However, 10.2% of the reporter's roles were anonymous. In addition, an index of error description and incident types was generated which indicates the leading categories are "other" (41.3%), and "miscellaneous" (32.8%), selected by the reporters. This implies that the predefined categories in the reporting system may not properly support the user's reporting needs and further investigation is needed to determine the exact categories required. The harm score changes are shown in table 1. We found that the total number of cases reduces when harm score is getting greater.

An inter-coder reliability coefficient test on harm scores resulted in Cohen's *Kappa* 0.83, percent agreement 0.82. The supplementary harm & delay matrix was used, however, we could not identify much direct information regarding the degree of delay in the database. Due to too many assumptions were needed for using the matrix, we suspect the harm & delay matrix may not be a good instrument in such a dataset.

Table 1 Harm Scores Changes over Recoding
Based upon Recoding Results of 2919 reports

Harm Score	Before		After		% change
Zero	1766	60.47%	1907	65.34%	↑4.87%
One	626	21.46%	483	16.56%	↓4.90%
Two	336	11.52%	274	9.36%	↓2.16%
Three	144	4.94%	206	7.06%	↑2.13%
Four	36	1.23%	27	0.93%	↓0.31%
Five	11	0.38%	22	0.75%	↑0.38%

We found over 30% unmatched (mismatched) incident types and error descriptions in the database. Through our recoding process, those "miscellaneous" or "other" cases usually could be coded by the categories offered in the system. Matching issues may have various reasons. According to human factors perspective, we proposed the following three main reasons:

1. No matches: A small portion of the reports does not match the incident types in the pre-defined menu. For example, rarely occurred cases (outdated medication applied to patient) happen less than one time in every three years. Under such a situation, "miscellaneous" or "other" is an accurate reflection.
2. Multiple matches: A number of incidents could fit into multiple categories. As the system only allows a single category to be reported, this may confuse reporters. For example, a patient complained about extreme pain without relief when pain medication had been given. When given larger

amounts of pain medication, the patient stopped breathing and medication had to be reversed with Narcan(Naloxone). This case had two choices of code "procedure/treatment", and "drug reaction". In the original report, it was coded "miscellaneous". Allowing reporters to choose a secondary error type appears to be reasonable in voluntary reporting systems.

3. Partially matched without immediate outcome or cause: reporters of such a voluntary system (majority of them are nurses) are typical multitasking people, they usually lack of enough time or effort choosing the optimal categories because of limited evaluation of outcomes or causes, instead "miscellaneous" or "other" may serve as a safe choice. We must realize that this kind of "safe" choice will soon become a barrier for aggregating cases at all levels.

The length of free text description shows a trend for both the reporter-generated data and the total data generated by both reporters and reviewers (figure 1). The cases with harm score between zero and four show an increase in terms of total words. The total words show a big drop when harm score at five (death). The means reporters often use a much shorter description in a case with harm score five than that in harm score four.

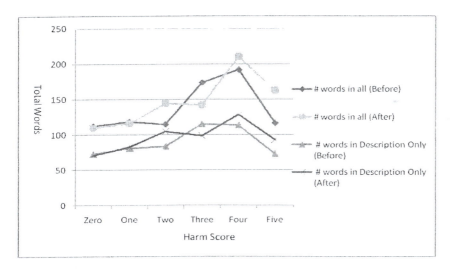

Figure 1 Average Free Text Length (Description, Solution, Review, and Information) based upon Recoding Results of 2919 reports

User analysis results show that about 2/3 reporters are nurses or staff with nursing background. They work closely with physicians in clinical departments.
Our descriptive analysis shows that the top reporting professionals (95.0%) are registered nurses (66.2%), anonymous (10.2%), pharmacists (4.1%), physicians (4.0%), respiratory therapists (3.8%), lab technicians (2.6%), other (2.2%), and manager (1.9%). Other reporting professionals, such as unit clerks, physical

586

therapists, contributed 5.0% of the total reports.

We identified three types of users in the PSN system. (1) Individual reporters who are nurses and departmental staff generate the majority of reports. They are multi-tasking people when they work; (2) Reviewers at the departmental level and at the health care system level, who review and follow up each report and append their comments/solutions/feedback to the original report. Reviewers may be nurse managers, physicians, administrators or quality assurance staff in the Office of Clinical Effectiveness at UMHC.

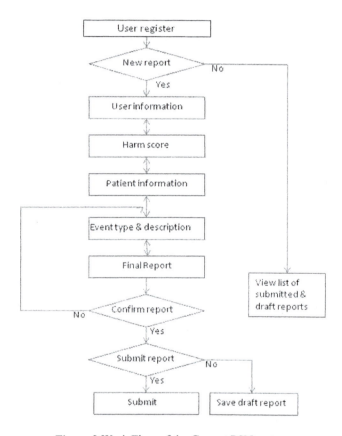

Figure 2 Work Flow of the Current PSN system

Task analysis resulted in a high level work flow chart which illustrates the reporting procedure that all individual reporters need to follow in the current PSN system (Figure 2). The GOMS analysis indicates the current system requires a high memory load from reporters which could be alleviated by re-engineering the interface. For example:

1. The current system requires users choose between "anonymous" and "un-anonymous" radio buttons. Since most reporters use "anonymous"

reporting (no release of their work ID), new designs should offer a check box for "un-anonymous" and thus make "anonymous" a default value.

2. Rather than asking reporters to recall an entire incident and type a long, time consuming free text description. A set of procedure-based questions, with conditional skips given previous answers, would guide reporter better through the entire recall process meanwhile an additional free text box will also be offered in case any information not included in those questions deemed valuable for reporting.

3. Auto-completion feature would increase the efficiency of pull-down menu selection, which is not offered in the current system. This is especially true when hundreds of names are listed in the menus.

4. Since most cases were reported with 24-48 hours, a convenient time stamp button should be very useful for reporters to click, rather than manually typing in the day and time which is laborious and prone to typos.

5. Providing a holistic view that contains all entries of an individual report on a single page, which would be easy for reporter to conduct final editing and confirmation, rather than forcing reporters flip pages back and forth to verify.

6. Providing a navigational bar as an indicator of progress towards the completion of a quality report. This will improve user transparency of the system and allow user to estimate if time slot is adequate to finish a report.

DISCUSSION

In a voluntary reporting system, human factors play a key role in reporting quality and overall system acceptance(Y Gong, 2009). We notice that designers of voluntary reporting systems should pay more attention to the balance between efficiency and expressiveness of data entry. Current voluntary reporting systems are mainly template based, which has an intent to increase data entry speed. Meanwhile, it may have the unintended effect of homogenizing incident descriptions with a loss of detail. We suggest an intelligent interface for voluntary reporting based on existing data repository that can predict term requirements and offer intelligent guesses during data entry.

It is generally agreed that cultivating a non-punitive reporting culture for voluntary reporting is essential for improving quality of reports. In addition, researcher argued that human factors associated with user interface greatly affect users' acceptance of the reporting system, thus reporting rate, completeness and accuracy of the incidents in the systems (C. Johnson, 2003). Those reporting systems with more considerations of human factors tend to have better user acceptance, therefore may generate quality reports(Harris et al., 2007).

To better understand the term requirements and reporters' cognitive characteristics, we continue working on human factors with human-centered approaches and reporting terms with ontological technologies. Cognitive human factors can greatly affect the decision making process when reporters interact with reporting interfaces(Y. Gong & J. Zhang, 2005). It can be a challenge to identify an ideal match between details of an incident and predefined categories. If predefined categories do not match contributors' mental model of incident type, it may result in underreporting, incomplete or inaccurate issues. In addition, reporters may wish to capture varying levels of details. Therefore, each high-level category (broad concepts) ideally should contain more details through low-level categories (narrow concepts). This strategy may guide reporters to identify an optimal granularity as well as build an interoperable system at the local, regional, and national level.

This paper reports our initial efforts in developing a human-centered prototype of voluntary medical incident reporting system. Further analyses that reveal improvements through human centered design will be obtained once the prototype becomes available. We plan to apply the keystroke level model which will allow us to uncover finer differences than those reported by GOMS. We will also further investigate the relationship between the harm scores and reporters' description at the semantic level, which may better serve in a human-centered voluntary reporting system. All efforts we make will contribute to a better reporting rate and quality in voluntary reporting for patient safety research.

REFERENCES

Arroyo, D. A. (2005). A nonpunitive, computerized system for improved reporting of medical occurrences. *Advances in patient safety: From research to implementation, 4.*

Barach, P., & Small, S. D. (2000). Reporting and preventing medical mishaps: lessons from non-medical near miss reporting systems. [Review]. *BMJ, 320*(7237), 759-763.

Bates, D. W., Cullen, D. J., Laird, N., Petersen, L. A., Small, S. D., Servi, D., et al. (1995). Incidence of adverse drug events and potential adverse drug events. Implications for prevention. ADE Prevention Study Group. *JAMA, 274*(1), 29-34.

Beasley, J. W., Escoto, K.H., & Karsh, B. . (2004). Design elements for a primary care medical error reporting system. . *Wisconsin Medical Journal, 103.*

Cohen, M. R. (2000). Why error reporting systems should be voluntary. *BMJ, 320*(7237), 728-729. doi: 10.1136/bmj.320.7237.728

Cullen, D. J., Bates, D. W., Small, S. D., Cooper, J. B., Nemeskal, A. R., & Leape, L. L. (1995). The incident reporting system does not detect adverse drug events: a problem for quality improvement. *Jt Comm J Qual Improv, 21*(10), 541-548.

Gong, Y. (2009). *Toward learning from a voluntary medical incident reporting system.* Paper presented at the HCI International 2009, San Diego, CA.

Gong, Y., Richardson, J., Zhijian, L., Alafaireet, P., & Yoo, I. (2008). Analyzing voluntary medical incident reports. *AMIA ... , Annual Symposium Proceedings/AMIA Symposium.*

Gong, Y., & Zhang, J. (2005). A distributed information analysis for information search tasks. *AMIA ... Annual Symposium Proceedings/AMIA Symposium*, 965.

Gong, Y., & Zhang, J. (2005). A human-centered design and evaluation framework for information search. *AMIA ... Annual Symposium Proceedings/AMIA Symposium*, 281-285.

Harris, C. B., Krauss, M. J., Coopersmith, C. M., Avidan, M., Nast, P. A., Kollef, M. H., et al. (2007). Patient safety event reporting in critical care: a study of three intensive care units. *Critical Care Medicine, 35*(4), 1068-1076.

Holden, R. J., & Karsh, B. . (2005). *Applying a theoretical framework to the research and design of medical error reporting systems*. Paper presented at the Proceedings of the International Conference on Healthcare Systems Ergonomics and Patient Safety.

Holden, R. J., & Karsh, B.-T. (2007). A Review of Medical Error Reporting System Design Considerations and a Proposed Cross-Level Systems Research Framework. *Human Factors: The Journal of the Human Factors and Ergonomics Society, 49*(2), 257-276. doi: 10.1518/001872007x312487

Johnson, C. (2003). *A handbook of accident and incident reporting*. Glasgow: Glasgow University Press.

Johnson, C. W. (2003). How will we get the data and what will we do with it then? Issues in the reporting of adverse healthcare events. *Qual Saf Health Care, 12*(90002), ii64-67. doi: 10.1136/qhc.12.suppl_2.ii64

Kaplan, H. S., & Rabin Fastman, B. (2003). Organization of event reporting data for sense making and system improvement. *Qual Saf Health Care, 12*(90002), ii68-72. doi: 10.1136/qhc.12.suppl_2.ii68

Karsh, B. T., Escoto, K. H., Beasley, J. W., & Holden, R. J. (2006). Toward a theoretical approach to medical error reporting system research and design. *Appl Ergon, 37*(3), 283-295. [pii]10.1016/j.apergo.2005.07.003

Kieras, D. (1994). A Guide to GOMS Task Analysis. Ann Arbor: University of Michigan.

Kingston, M. J., Evans, S. M., Smith, B. J., & Berry, J. G. (2004). Attitudes of doctors and nurses towards incident reporting: a qualitative analysis. *Med J Aust, 181*(1), 36-39. doi: kin10795_fm [pii]

Kohn, L. T., Corrigan, J. M., & Donaldson, M. S. (1999). *To err is human*. Washington, DC: National Academy Press.

Leape, L. L. (1994). Error in medicine. *JAMA, 272*(23), 1851-1857.

Leape, L. L. (2002). Reporting of Adverse Events. *N Engl J Med, 347*(20), 1633-1638. doi: 10.1056/NEJMNEJMhpr011493

Marx, D. A. (2001). Patient safety and the "just culture": A primer for health care executives.

Runciman, B., Merry, A., & Smith, A. M. (2001). Improving patients' safety by gathering information. *BMJ, 323*(7308), 298-. doi: 10.1136/bmj.323.7308.298

Unkown. (2006). Patient Safety Network Learning Center, 2009, from https://apps.muhealth.org/psn/psn_learning/faq.html

Wakefield, B. J., Uden-Holman, T., & Wakefield, D.S. (2005). Development and validation of the Medication Administration Error Reporting Survey. *Advances in patient safety: From research to implementation, 4.*

CHAPTER 66

Development and Comfort Analysis of a Chair for Microscopic Surgery

Kageyu Noro[1], Tetsuya Naruse[2], Nobuhisa Nao-I,[3] Maki Kozawa[3]

[1]Wasea University, ErgoSeating Co., Ltd.
Tokyo, Japan

[2]Gifu Pref. Research Institute for Human Life Technology
Takayama, Japan

[3]Department of Ophthalmology, Faculty of Medicine
Miyazaki University, Miyazaki, Japan

ABSTRACT

A special chair for microscopic surgery was developed from studies on Zen sitting to facilitate long hours of surgery operations on a comfortable seat. This paper describes design principle for the chair along with a proposal for its manufacture. The chair development places an emphasis on 3-dimensional contours formed on an interface between muscles, especially those on lumbar, buttock, and thigh parts, and a sitting tool. The design principle includes 1) maximization of contact surface area formed when a person sits on a seat and 2) adjustment of peak body pressures. The essential point of the chair manufacture is creation of a shell emulating a 3D shape of primary user's buttocks. Evaluation experiments were conducted in an operation room to compare the prototype chair with a conventional chair. The evaluated items were pelvis tilt angles, body pressure distributions, physical measurements, and subjective ratings. Of the results, the pelvis tilt angles measured on the prototype chair demonstrated the backward inclination of pelvis being prevented by the sacral support as seen in *zazen*. The transitional variations of statistical figures from areas identified by palpation on the regionally-differentiated pressure maps (RDPM) along with changes in pelvis tilt angles served to objectively gauge levels of seating comfort. Accordingly, the newly developed chair was proved to provide improved seating comfort and stability to facilitate easier long duration work as compared with a conventional chair. It could be said as a conclusion that this research has justified the effectiveness of a concave-shaped chair that was made up in accordance with individual's buttocks size following the method adopted in the Zen sitting.

Keywords: Surgery chair, Concave-shaped chair, Seating comfort, Pelvis, Body pressure, Zen sitting

INTRODUCTION

Microscopic surgery operations—characteristics and issues to be addressed

Schurr and Buess (2000), Wallace (1999), and Nao-i et al (2009) described the physical demands on surgeons during surgery, referring to their risk of musculoskeletal disorders to increase. Specifically, those demands include the following items:

1) Operation of the surgical equipment requires the surgeon to maintain an elevated foot.
2) The chair seat bites into the surgeon's thighs.
3) The surgeon lacks hand support while operating.
4) The surgeon is unable to recline, or sustain pelvic support.
5) The surgeon must sustain a flexed neck position while looking into a microscope.
6) The surgeon must maintain a fixed posture for a long period of time.

Such restrictive postural demands are causing most surgeons acuter pains than those pointed out by Schurr and Buess (2000) and Wallace (1999).

What is a better chair for microscopic surgery like? Schurr et al (2000) suggested an ideal position for a surgeon to take is a semi-standing position to get the necessary areas within reach, being supported by a horn-shaped seat to avoid forward-sliding as well as equipped with user-friendly foot-pedals. Wallace (1999) suggested a way for improving surgical ergonomics in ophthalmic surgery by getting the microscope tilted to a certain extent with the patient's head tilted 45°. Congleton (1985) reported that a surgeon could reduce his or her fatigue by taking the neutral posture during surgical procedures. A chair used for helping take this posture was called a neutral posture chair. This approach sharply reflects the trend of relevant research in the mid 1980s. Namely, it was strongly affected by the recommended sitting posture proposed by Mandal (1982, 1982a), intensively publicized in the early 1980s. This is because the observed surgeon's posture looks very much like the one suggested by Mandal. It was Mandal that put into practice in the form of a chair the theories or ideas proposed by Keegan (1953, 1964). The Congleton's prototype chair for surgeons is an inherited form of the Keegan's theory.

Eastern view on chairs as especially seen in Zen approach

Figure 1 shows a priest taking a *zazen* posture with a cross-directional contour formed between *zafu* and sacrum/buttocks areas.

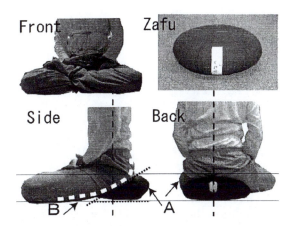

Figure 1. *Zazen* **posture by a priest sitting on a** *zafu* **A with cross-directional contour B formed along the interface between** *zafu* **and sacrum/buttocks areas.**

It was made clear after a 40-minute observation of a Zen priest A performing *zazen* that his cross-directional pelvic rotation angle was -5.04° on average with a standard deviation of 0.6 against the corresponding angle of 0° for standing position. A priest B, on the other hand, posed an average cross-directional pelvic rotation angle of -2.6° with a standard deviation of 0.27 during his 20-minute *zazen* (See Figure 2)(Noro, 2007). Since the cross-directional pelvic rotation angle from ordinary chair sitting reportedly is around -19° (Noro, 2009), these figures represent an extreme superiority.

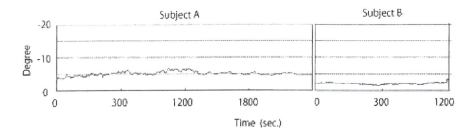

Figure 2. Pelvis tilt angles of priest A (40 min) and priest B (20 min)

The reason for improvement of pelvic rotation lies in a body being supported by a surface contact area; i.e., a large supporting area formed by both knees and buttocks along with a difference of elevation between buttocks and both knees, as seen in the side view of Figure 1, helps keep the pelvis rotation angle close to that of standing position, yet maintaining that position for a long period of time. These are the implications considered contributory to seating comfort and long hours of work. From the designing perspective, a chair is designed in line with the 3D shape of buttocks independent of bone framework and body size. This idea coincides with the fact that *zafus* required for Zen sitting are made so as to match individual's buttocks without consideration of general use.

AN EXPECTED SURGICAL CHAIR AND WORKING HYPOTHESIS

The key to designing the chair in question is to create a shell emulating the 3D shape of user's buttocks. It has commonly been seen that a seat and a back are dealt with in a separate manner. We consider the supporting surface of a chair with the concept of 'shell' (seat and back are treated in a unified manner). The scientific grounds for this approach lie in lessening pressures exerted on muscles relevant to maximization of surface contact area. Actually, it has been a long tradition since the 13th century in the field of Zen sitting that the floor sitting tools (*zafu* or *zabuton*) are optimally customized in line with physical features of each individual. The shell of the current surgery chair should be manufactured not only emulating the 3D shape of buttocks of a surgeon as an individual but also in accordance with the surgeon's working posture. This surgeon is to be called a primary user in this paper. This idea resulted in creation of a prototype model as shown in Figure 4 in comparison with a conventional chair with a convex seatpan.

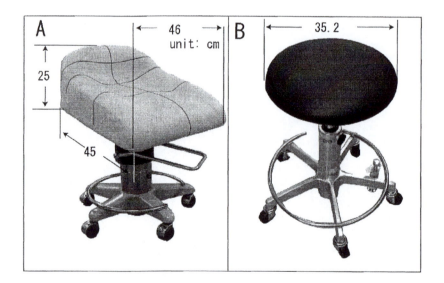

Figure 3. A concave-shaped prototype chair (A) and a current surgery chair with a convex seatpan (B) (The chair B is to be called 'the conventional chair' hereafter)

EXPERIMENTAL METHODS

A subjective survey and an objective measurement were conducted.

Subjective survey

Specifically designed questionnaire forms were used, with hearings performed before and after surgery in which a 5-point scale was adopted for rating. Prior to surgery, surgeons were requested to follow the paired comparison method making a comparison in comfortable or uncomfortable feelings—with or without any point hitting some body part—and senses of slippage between prototype and reference chairs. A 5-point scale was used for rating, while statistical methods including ANOVA were applied to analysis of ratings.

Objective measurement

Body pressure distributions. The sensor used was a polyester pressure imaging pad of a square type measuring 45 cm in side length embedded with 36 x 36 cells, sensing at a sampling rate of 10 frames/sec with a pressure range of 0-220 mmHg. Accuracy of measurement was approximately 10 % or 10 mmHg. Calibrations were conducted as appropriate.

Pelvis tilt angles. Pelvis tilt angles were measured using a patented gyroscopic device [1] that determines rotation angles against 3D axes with respective range between -60° and +60° for pitch angles and roll angles.

Chairs subject to experiments

Two chairs shown in Figure 4 were used for experiments.

Subjects

Three most experienced surgeons participated in the experiments as subjects of whom Sub A was the primary user. Their major attributes are given below:
    Sub A: male, 165 cm tall, 14.9 percentile;
    Sub B: female, 164 cm tall, 84.2 percentile;
    Sub C: female, 153 cm tall, 14.7 percentile.

---

[1] Japan Patent Office Certificate (2007): Title of invention, a pelvic-angle measuring device; Inventor, Kageyu Noro; Patent no. 3928103 2004.

Evaluation experiments in operation room

The current prototype chair was judged by the primary user as utilizable at clinical sites. Evaluation experiments were conducted in an operation room one year after it had got in use. The measurements described were performed during five incidences of surgery.

RESULTS

Subjective evaluation

Figure 4 gives compared ratings between before and after surgery of comfort sensed from body parts of sacrum, ischial tuberosities, and thighs through the use of the prototype chair.

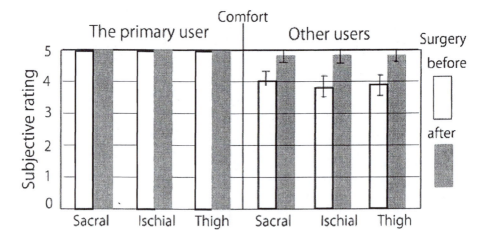

Figure 4. Ratings for the prototype chair of comfort from body parts of sacrum, ischial tuberosities, and thighs

The primary user Sub A gave a score of 5 point to all the ratings of comfort from the body parts concerned through the use of the prototype model as the survey was performed before and after surgery. On the contrary, the scores given by the other subjects increased to 5 point for every body part after the prototype chair was used, 1 point up from 4 point before its use. The reason is in the following. While, before the use of the chair, the evaluations were conducted with the positional relationship between surgeons and the surgical table remaining rough, the evaluations after its use reflected the fact that the positional relationship became exact in the course of surgery operations. This was the reason revealed by their subsequent comments.

There follow comparisons between the two chairs concerned. Figure 5 demonstrates an interval scale chart obtained from the paired comparison applied to them.

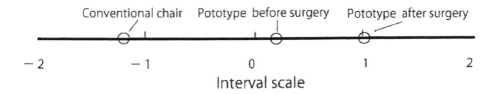

Figure 5. Comparison between prototype and conventional chairs

The paired comparison was conducted in terms of comfort and resistance to sliding before and after surgery taken from the body parts of sacrum, ischial tuberosities, and thighs with the aim of obtaining a psychological scale to identify the difference in subjects' evaluations between the two chairs concerned. On the scale of comfort shown in Figure 5, it was clarified that the prototype chair gained even a higher mark after it was actually used in surgery with regard to all the three body parts in question, compared with the one marked before surgery. The ratings for the conventional chair as the reference proved to be lowest in all the items in question. On the scale of resistance to sliding, however, the conventional chair as the reference gained the highest grade. This fact stems from a circular metal ring around the seat of the conventional chair acting to prevent a body from sliding, which at the same time served as a cause for a pain in thighs. Regarding resistance to sliding of the prototype chair, an improvement was seen after surgery compared with before surgery, as demonstrated by a difference in mark on the scale. This implication backs the comments from the subjects on the improvement of comfort as shown in Figure 6. The comparison between the prototype and conventional chairs seen in Figure 6 can be said to pose an extreme difference.

Analysis of objective data

Static measurement. The evaluation of the contact surface created between a surgeon and the prototype chair was conducted by measuring his or her pressure distribution during his or her work prior to a surgical operation. Figure 6 compares body pressure distributions between the prototype chair and the conventional chair.

The six areas are depicted by rectangles in Figure 6 with identification by palpation of locations of surgeon's buttocks, thighs, and sacrum. This idea is named the regional differentiation of pressure distributions, with its associated pressure maps being dubbed "regionally-differentiated pressure maps" (RDPM). A glance at Figure 6 reveals clear differences between two types of chairs in the size of pressure dispersion areas, peak pressure values and their associated locations. These distinctions can be said to stem from a difference between a concave-shaped prototype chair and a conventional chair with a convex seatpan.

598

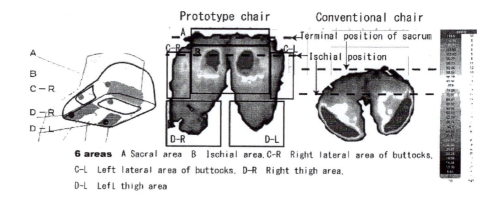

Prototype chair   Conventional chair

← Terminal position of sacrum

← Ischial position

6 areas A Sacral area B Ischial area. C-R Right lateral area of buttocks.
C-L Left lateral area of buttocks. D-R Right thigh area.
D-L Left thigh area

Figure 6. Compared body pressure distributions between prototype chair (Sub A) and conventional chair (Sub C)

Table 1 summarizes the measured results mentioned above taken from 5 instances of surgery performed during the experiment, comprising the items of surgeons, chairs used, time required, pelvic pitch angles, and mean pressures. The last two items are averaged over the duration of experiment.

Table 1. Summary of measured results

Type of surgery	Sub.	Chair	Time required	Pelvis angle degree (pitch)	Pressure (ave.)	Area cm^2 (ave.)
1) Vitreous surgery	A	Prototype	63min.	-6.2	67.0	1165
2) Vitreous surgery combined with phacoemulcification	C	Prototype	75min.	-8.4	55.8	963
3) Vitreous surgery combined with phacoemulcification	B	Prototype	100min.	-1.1	75.4	875
4) Vitreous surgery combined with phacoemulcification	B	Conventional	57min.	-7.4	61.6	702
5) Cataract surgery	C	Conventional	22min.	-10.2	90.4	558

The pelvis tilt angles of the prototype chair listed in Table 1 are considerably favorable as a whole. The larger pelvis tilt angles for the conventional chair are seen compared with the prototype chair. This fact presumably indicates greater degrees of pelvis backward rotation taking place for the conventional chair. The higher pressures are seen for the conventional chair. Besides, the size of the body supporting area of the prototype chair ranges between 1165 cm^2 for Sub A and 875 cm^2 for Sub B, nearly doubling from the conventional chair for which 702 cm^2 for Sub B and 558 cm^2 for Sub C are observed. Given this, the objective data are considered consistent with the subjective data; i.e., the prototype chair is superior to the conventional chair in terms of provision of seating comfort, yet a further increase in the comfort level is seen after surgery.

DISCUSSIONS

Has our hypothesis been verified that larger supporting area will contribute to improved seating comfort to assure long hours of work? The prototype chair has gained a favorable result compared with the conventional chair. As indicated in Table 1, the prototype chair has a larger supporting area than a conventional chair. In addition, it has gained returns from subjects expressing improvement in seating comfort during the working hours extending as long as around an hour. This result can be said to support validity of our hypothesis.

Schurr et al (2000), as cited at the beginning, referred to the requirement for an ideal chair that it should provide an environment for a surgeon to take a semi-standing position to get the necessary areas within reach, being equipped with user-friendly footpedals as well as with a horn-shaped seat to prevent a forward-slide. This research has employed the Zen sitting approach, but has not run counter to the theory proposed by Schurr et al. The horn-shaped seat cited in Schurr et al (2000) is dubbed a concave-shaped chair in this paper, compared with a conventional chair with a convex seatpan. It could be said as a conclusion that this paper has justified the effectiveness of a concave-shaped chair made up in accordance with individual's buttocks size following the method adopted in the Zen sitting. Topics of mass production of concave-shaped chairs would be dealt with in Noro (2009).

REFERENCES

Congleton, J.J., Ayoub, M.M., and Smith, J.L., (1985). The Design and Evaluation of the Neutral Posture Chair for Surgeons. Human Factors. 1985 27 5 589-600.

Keegan, J.J., (1953). Alterations of the lumbar curve. Journal of Bone and Joint Surgery. 35, 589-603.

Keegan, J.J., (1964). The Medical Problem of Lumbar Spine Flatting in Automobile Seats. In: Booklet 838A 1-12, Society of Automotive Engineering.

Mandal, A.C., (1982). The Seated Man. Private Edition 11.

Mandal, A.C., (1982a). Correct height of school furniture. Human Factors, 24 257-269.

Nao-i, N., Noro, K., Kawano, N., (2009). Analysis and improvement of a chair for microscopic surgeons. The 32nd Annual Meeting of the Japanese Society of Ophthalmic Surgeons 2009 (in Japanese).

Noro, K., (2007). Za saikou. Journal of Society of Biomechanisms(SOBIM) Japan. 31 (1) (in Japanese).

Noro, K., (2009). A development of chairs based on studies on the Asian traditional ways of seating. In : Schlick, C. (Ed.) Industrial Engineering and Ergonomics. Springer.

Schurr, M.O., Buess,G.F., Wieth, F., Saile, F-J.and Botsch, M., (1999). Ergonomic surgeon's chair for use during minimally invasive surgery. Suygical Laparoscopy Endoscopy & Percutaneous Techniques. 1999 9 (4) 244-7.

Schurr, M.O. and Buess, G.F., (2000). Systems technology in the operating theatre: a prerequisite for the use of advanced devices in surgery. Min Invas & Allied Technol. 2000 9 3/4 179-184.

Wallace III, R.B., (1999). The 45 degree tilt: Improvement in surgical ergonomics. J. Cataract refract surg. vol 25, February 1999.

CHAPTER 67

Importance of Human Factors Engineering in Healthcare and Implementation in Patient Safety

Ozlem Yildirim, Canan Cetin, Tulay Aytekin Aktas, Atif Akdas

Marmara University
Institute of Health Sciences,
Tıbbiye Avenue, Haydarpaşa 34668
Istanbul,Turkey

ABSTRACT

Objective of this study is to define importance of human factors engineering importance and relationship in improving patient safety at hospitals by communicating related patient safety processes. "Implementation of Human Factors Engineering in Patient Safety Scale" that consisted of "medication management", "facility management", "equipment management", "environment of care", "safety culture" factors was developed. Healthcare staff perceived "medication management" at highest and "safety culture" at lowest level as means of human factors engineering implementation in patient safety. As "safety culture" increases in the future, it will have an effect on implementation of human factors engineering in patient safety field.

Key Words: Environment of care, human error, human factors engineering, medication management, medical equipment, patient safety, safety culture.

INTRODUCTION

Human Factors Engineering (HFE); is a branch of science that focuses on human and the components that lead to human error, analyzing systems and relationship between human-machine systems by using appropriate quality techniques. HFE aims to implement proactive and reactive approaches to calculate and eliminate risk and helps to increase the quality of health care services and the productivity of clinical processes as well as reduces health care costs, increases patient and physician satisfaction. In recent years, as part of quality improvement activities and accreditation process considerable attention is now being given to the concept of patient safety in this context in Turkey. Systems are being defined in order to ensure safety of processes such as medication management, medical equipment management and management of healthcare environment. Objective of this paper is to contribute to the literature on Patient Safety and Human Factors Engineering and to provide a resource and a scale from which hospitals may benefit.

HUMAN FACTORS ENGINEERING AND ITS IMPORTANCE IN HEALTH CARE

There are many dictionary definitions for "error". "Error" can be defined as "something done wrong because of neglect", "violation of law or duty" or "moral defect". An error becomes visible when small factors come together and has negative impact on results (Reason J. 1990). HFE is a term which scientists use to define human role in any particular system. Main task of human factors engineers is to design work environments and equipments in a way to determine and prevent errors by assuming that people will do errors (Gosbee JW, Lin L. 2001). HFE is used in the United States and in some other countries. The same concept is mostly referred to in Europe as "ergonomics." Research in this area is only just beginning to be implemented in healthcare services. HFE has been implemented in various organizations that have encountered problems of design that reflect human limitations and competencies. Lack of information, excess of information, unrelated information with the task, ineffective presentation of information to the operator that is unclear in how decisions are to be made; unavailable work flows to the users, inappropriate default settings, difficulty in perceiving errors during data entry and difficulty or slowness in later correcting mistakes are example of some problems that HFE can be used in eliminating those problems (Engelke C, Olivier D. 2002,Gosbee JW. 2002, Gosbee JW. 2004). HFE also plays a significant role in improving patient safety processes. The benefits may be summarized as; reducing usability errors, facilitating easy use of medical equipments, increasing safety of manual actions, ensuring easy reading of control panels and indicators, establishing safe connections between instruments and more effective warning systems. To review a healthcare system in terms of HFE, certain steps may be defined and analyzed through a systematic approach and current systems may be improved. Steps in this plan may be cited as; development of a mission statement, determining key staff such as management team, staff, specialists, auxiliary users and users, determining processes, process leaders and participants, defining problems and/or current status through prioritizing, defining roles in choosing equipment, setting goals and comparison data, agreeing on measurement methods and available data, developing strategies to meet the set goals, agreeing on key assessment points and defining evaluation process of design changes (Bogner MS. 1994, Grout JR. 2006, Cook RI, Woods DD. 1994,Cook RI, Woods DD. 1994).

IMPORTANCE OF PATIENT SAFETY IN HEALTH CARE, RELATIONSHIP WITH HUMAN FACTORS ENGINEERING

Medicine is a branch of science that focuses primarily on patients' health and preventing human from diseases as much as possible. Patient safety can be defined as the complete set of measures taken by hospitals and hospital staff to prevent possible harm to patients. The ultimate goal of patient safety is to create a physical and psychological environment that will positively affect patients and their relatives as well as hospital staff (Joint Commission Resources. 2006). In the first report of Institute of Medicine published in 1999, it is estimated that number of people who loose their lives in hospitals annually due to medical errors is more then the number of deaths due to serious illnesses such as cancer (Kohn LT, Corrigan JM, Donaldson MS. 1999). It can be argued that patient safety related errors can be prevented through implementation of HFE and causes of errors can be determined. Institute of Medicine's second report, published in 2001 addresses the fact that responsibility of individuals, confidentiality of information, systems requirements and professional approaches have given way to a new approach based on systems responsibility, expertise knowledge, transparency, satisfaction of requirements and evidence-based decisions. This creates the fact that attention needs to be given to the recommendations about patient safety generated by hospital professionals, patients and their families and to improvement opportunities (Leape LL, Berwick DM, Bates DW. 2002, Piotrowski MM. Cohen M, Mercier J, Saint S, Steinbinder A, Thompson M. 2005, Richardson WC, Briere R. (Eds), 2001). Carelessness, fatigue, ineffective communication, badly designed equipment, negligence, noisy working environments are all among the factors that can lead the way to human error. Since this is the case, it is of great importance that systems are designed to prevent human error. People live within systems and if the systems they work in are open to error, it may not be possible to work according to principles governing safety. For systems to be developed that focus on patient safety, the culture, organizational structure, communication channels and management concept within the particular organization must all be formulated and made operative around the concept of "patient safety" (Mohr JJ, Barach P, Cravero JP, Blike GT, Godfrey MM, Batalden PB, Nelson EC. 2003).

METHOD

This research study was designed on the basis of a cross-hatching model to define the role of HFE in the improvement of patient safety and to develop a scale reflecting the perceptions of healthcare professionals regarding linkage between usage of HFE and its impact on improving patient safety at the hospitals.

UNIVERSE AND SAMPLING

Six separate sample groups were used in the study. Detailed information about these sample groups is given below. A total of five volunteer specialist faculty members in the university departments of Industrial Engineering, Management and Organization and Nursing Services were enlisted to evaluate the content validity of the "Implementation of Human Factors Engineering in Patient Safety Scale." A group of 300 healthcare professionals chosen from the universe of the study as a purposive sample participated voluntarily in the validity and reliability analyses of the scale. Universe of the study comprised professionals working at hospitals on the European side of Istanbul. Total of 310 professionals were chosen from the universe as a purposive sample to participate voluntarily in the study (Table 1).

Table 1: Frequency and Percentage Distribution of Volunteers Chosen as a Purposive Sample

Gender	Male	Female			Total
%	23.4 (n=71)	76.6 (n=233)			100 (n=304
Sector	Private	Public			
%	57.9 (n=173)	42.1(n=126)			100 (n=299
Job Title	Medical Manager	Administrative Manager	Physician	Nurse	
%	1.6 (n=5)	2.6 (n=8)	22.2 (n=68)	52.3 (n=160)	
	Medical Technician	Pharmacist	Engineer	Other	
%	6.2 (n=19)	0.7 (n=2)	1.3 (n=4)	13.1 (n=40)	100 (n=306
Education	Elementary	High School	Graduate	Postgraduate	
%	0.7 (n=2)	20.5 (n=60)	56.7 (n=166)	22.2 (n=65)	100 (n=29?
Age	18-30	31-40	41-50	51+	
%	165 (n=59.6)	87 (n=31.4)	24 (n=8.7)	0.4 (n=1)	100 (n=277
Seniority	0-10	11-15	16-20	21+	
%	81.9 (n=230)	11 (n=31)	5 (n=14)	2.1 (n=6)	299

DATA COLLECTION TOOL

"Implementation of Human Factors Engineering in Patient Safety Scale" and "Personal Data Form" was used as data collection tools. 119 items were drawn up according to the five factors determined following the review of related literature. As a result of "Principal Component Analysis" used in factor analysis and the parallel analysis carried out with the "Varimax Vertical Axis Rotation Technique" it was found that the scale items could be grouped into five sub-dimensions. Total variance for the scale in the five sub-dimensions was 53%. The Cronback Alpha internal consistency reliability coefficient of the scale in the sub-dimensions was between 0.793 - 0.962 and found to be 0.985 for the scale as a whole. "Personal Data Form" was developed to collect information on each professional's gender, age, seniority, type of hospital worked in, job title and education level.

ANALYSIS OF DATA

The "Implementation of Human Factors Engineering in Patient Safety Scale used in the research comprised 66 items divided into five factors which were "Medication Management" (18 items), "Facility Management" (8 items), "Medical Equipment" (22 items), "Safety Culture" (13 items) and "Environment of Care" (5 items). The 5-grade Likert scale was used as the scale of assessment from 1-5, divided into five equal parts with score intervals corresponding to each choice determined as well. Before starting statistical analysis in the research, demographic variables were first grouped and then the scale applied to the healthcare professionals were assigned scores. Later, frequency (η), mean (X) and standard deviation (ss) were calculated from the points derived from the scales. Within the groups, non-parametric techniques were used for groups that did not display normal distribution features ($n<30$) and parametric analysis techniques were used for those that did. In this context, the scores attained by the healthcare professionals in the sample group were examined with the t-test to determine whether differences appeared according to gender and type of hospital variables and the one-way Variance Analysis was performed to determine whether there were differences according to the level of education variable. When differences were found between groups as a result of the one-way Variance Analysis test, the source of these differences [between which groups the difference appeared] was examined with Tukey's Post Hoc Test. Scores of the healthcare professionals of the sample group on the scale were then tested with the non-parametric Kruskal Wallis-H test to determine whether there were differences according to the type of job variable. When differences were found between groups as a result of the non-parametric Kruskal Wallis-H test, the non-parametric Mann Whitney-U test was used to determine the source [between which groups they appeared] of the differences. The data thus obtained was then analyzed using the SPSS for Windows 13.0 (SPSS Inc. 2007) program to find the level of significance ($p<0.05$) indicated by the techniques mentioned above and the findings were compiled in tables appropriate to the aims of the research.

FINDINGS

The perception of the healthcare professionals who participated in the study about the implementation of HFE in the area of patient safety varied between 2.96 - 3.93. Healthcare professionals had the highest perception of the implementation of HFE in patient safety in the sub-dimension of "Medication Management" and the lowest perception in the sub-dimension of "Safety Culture." In addition, the perception of healthcare professionals about HFE in patient safety generally fell in the "Frequently" category [X=3.57].

The scores of the healthcare professionals participating in the study in the sub-dimensions of the Implementation of Human Factors Engineering in Patient Safety Scale also indicate that when the findings are analyzed according to differences in terms of the hospital type variable, a significant difference [p<0.05] can be seen in favor of A-Group private hospital professionals in the sub-dimensions "Medication Management", "Facility Management", "Medical Equipment", "Safety Culture" and "Environment of Care" (Table 2).

Table 2: Results of Independent Group t-Test Performed to Determine Whether Differences Exist in terms of Hospital Type Variable in Scores of Sub-Dimensions

MEDICATION MANAGEMENT						
					t-test	
Groups	η	X	SS	t	SD	p
Private hospital	161	4.32	0.63			
Public hospital	114	3.38	0.97	9.720	273	0.000
FACILITY MANAGEMENT						
					t-test	
Groups	η	X	SS	t	SD	p
Private hospital	170	4.55	0.55			
Public hospital	116	2.13	0.94	27.237	284	0.000
MEDICAL EQUIPMENT MANAGEMENT						
					t-test	
Groups	η	X	SS	t	SD	p
Private hospital	169	04.04	0.804			
Public hospital	124	3.204	0.864	8.542	291	0.000
SAFETY CULTURE						
					t-test	
Groups	η	X	SS	t	SD	p
Private hospital	170	3.42	0.87			
Public hospital	130	2.36	0.89	10.292	298	0.000

				t-test		
ENVIRONMENT OF CARE						
Groups	η	X	SS	t	SD	P
Private hospital	167	4.43	0.63			
Public hospital	112	3.17	0.87	13.945	277	0.000

In terms of the hospital classification variable, a significant statistical difference could be seen in favor of private hospital professionals [p<0.05] in their perception of the implementation of HFE in patient safety.

Results show no statistically significant difference [p>0.05] in terms of level of education of the professionals in the sub-scales "Medication Management", "Medical Equipment", "Safety Culture" in the perception of health professionals regarding the implementation of human factors engineering in patient safety. There was however a statistically significant difference [p>0.05] in the other dimensions. Post-hoc analysis was carried out to determine the groups responsible for significant difference. Following ANOVA test, Levene's test was first performed to decide upon the post-hoc multiple comparison technique to be used. Variances were found to be homogeneous [Facility Management: LF=0.631; p>0.05], [Environment of Care: LF=2.020; p>0.05]. The Tukey multiple comparison technique was preferred at this point since this is a technique that is usually used in the case of homogeneous variance. With regard to the sub-dimension of "Facility Management" in the context of the perception of healthcare professionals of the implementation of HFE in patient safety, there was a difference between professionals with high school degree and post-graduate degrees in favor of post-graduate group and a difference between professionals with graduate degrees and those with post-graduate degrees in favor of those with graduate degrees. With regard to the "Environment of Care" sub-dimension, there was a statistically significant difference between high school graduate professionals and professionals with post-graduate degrees in favor of high school graduates and a difference between professionals with graduate degrees and post-graduate degrees in favor of those with graduate degrees.When the results of the one-way variance analysis of the perception of healthcare professionals regarding the implementation of HFE in patient safety in terms of the variable of level of education is analyzed, a statistically significant difference [p>0.05] in terms of level of education was observed in how healthcare professionals perceived the implementation of HFE in patient safety. To decide which post-hoc multiple comparison technique would be used after the ANOVA, Levene's test was first used to test whether the variance of group distribution was homogeneous. It was established that the variances were homogeneous [LF=1.665; p>0.05]. With regard to the perception of healthcare professionals regarding the implementation of HFE in patient safety, it was found that there were significant differences between the group of high school graduate professionals and the professionals with graduate degrees in favor of the group with graduate degrees and between the group of professionals with graduate degrees and the professionals with post graduate degrees in favor of those with graduate degrees (Table 3).

Table 3: Results of Tukey's Multiple Comparison Analysis Performed to Determine Whether Differences Exist in Scores on the Implementation of HFE in Patient Safety Scale in terms of the Variable of Level of Education

(I) Level of Education	(J) Level of Education	XI-XJ	Shx	P
High School	Graduate	-0.33*	0.13	0.034
	Post Graduate	0.02	0.15	0.984
Graduate	High School	0.33*	0.13	0.034
	Post Graduate	0.36*	0.12	0.015
P. Graduate	High School	-0.02	0.15	0.984
	Graduate	-0.36*	0.12	0.015

In "Medication Management" sub-dimension, significant differences were found between the group of doctors and the group of other professionals in favor of the group of other professionals and between the group of doctors and the administrative managers in favor of the administrative manager group. In "Facility Management" sub-dimension, significant differences were found between the group of doctors and the group of other professionals in favor of the group of other professionals and between the group of doctors and the group of engineers in favor of the group of engineers. In "Medical Equipment" sub-dimension significant differences were found between the group of doctors and the group of other professionals in favor of the group of other professionals and between the group of doctors and the group of engineers in favor of the group of engineers. In "Safety Culture" sub-dimension significant differences were found between the group of doctors and the group of pharmacists in favor of the group of pharmacists and between the group of doctors and the group of engineers in favor of the group of engineers. In "Environment of Care" sub-dimension significant differences were found between the group of doctors and the group of administrative directors in favor of the group of administrative directors and between the group of doctors and the group of engineers in favor of the group of engineers.

Table 4 shows the results of non-parametric Kruskal Wallis-H analysis of the perception of healthcare professionals regarding the implementation of HFE in patient safety in terms of the variable of job title. With respect to how healthcare professionals perceived the implementation of human factors engineering in the area of patient safety, it was established that there were significant differences between the group of doctors and the group of other professionals in favor of the group of other professionals and between the group of doctors and the group of engineers in favor of the group of engineers.

Table 4: Results of Kruskal Wallis-H Analysis Performed to Determine Whether Differences Exist in the Implementation of Human Factors Engineering in Patient Safety Scale in terms of the Variable of Type of Job Title

Groups	η	X	SS	Kruskal Wallis-H Results			
				X	X2	SD	p
Medical Manager	5	3.74	1.08	170.60			
Administrative Manager	8	4.03	0.47	196.19			
Doctor	68	3.22	0.90	116.92			
Nurse	156	3.49	0.88	142.70	37.940	7	0.000
Medical Technician	19	3.86	0.89	183.16			
Pharmacist	2	3.97	0.32	184.50			
Engineer	4	4.36	0.49	235.75			
Other	40	4.12	0.57	208.18			

DISCUSSION AND RESULTS

Health care staff perceived "Medication Management" at highest level and "Safety Culture" at lowest as means of HFE implementation in patient safety. Statistically, perceptions of staff regarding implementation of HFE in patient safety do not show any significant difference based on gender and age of employees. Perceptions of staff regarding implementation of HFE in patient safety are statistically meaningful in the advantage of private hospital employees. Statistically, perceptions of staff regarding implementation of human factors engineering in patient safety do not show any significant difference based on education level of employees and professions. Perceptions of staff regarding implementation of HFE in patient safety show significant difference between high school and graduates in the advantage of graduates, and between graduates and post-graduates in the advantage of post-graduates. Perceptions of staff regarding implementation of HFE in patient safety show significant difference between "other" staff and "physicians" in the advantage of "other" staff, and between "physicians" and "engineers" in the advantage of "engineers". A negative relation is determined between total score of the scale and "Medication Management", "Facility Management", "Safety Culture" and "Environment of Care" sub-dimensions. A negative relation is determined between total scores and "Medication Management", "Facility Management", "Equipment Management", "Safety Culture" and "Environment of Care" sub-dimensions. The scale can be used by the health care organizations in measuring the perceptions of health care staff regarding implementation of HFE in patient safety. Scale developed and the topic had an added value for the literature in the way that it was the first Turkish resource. Perceptions of staff regarding implementation of HFE in patient safety are at level frequent in general. Perceptions of staff regarding the reflection of HFE approaches to hospital processes in the field of patient safety are higher in private health care staff. It can be said that as the studies in public health care industry speeds up and implementation becomes widespread, there might be differences in the results of future evaluations. Conducting the study by sampling more health care organizations from public and private industry and factor based analysis in between these groups can be a leading step for future researches. For further research studies, researches can focus on the factors on which the physician perceptions are low. Besides, a detailed research study can be conducted for different professions.

REFERENCES

Reason J. (1990). Human Error. 1st ed, Cambridge University Press, Cambridge.

Gosbee JW, Lin L. (2001). *The role of human factors engineering in medical device and medical system errors. In: Vincent C. Clinical Risk Management*, s.301–318.

Bogner MS. (1994). *Human Error in Medicine. 1st ed,* Hillsdale, New Jersey.

Gosbee JW. (2002). *Human factors engineering and patient safety. Quality and Safety in Health Care,* 11:352-354.

Gosbee JW. (2004). *Introduction to the human factors engineering series. Joint Commission Journal on Quality and Patient Safety,* 4(30):215-219.

Grout JR. (2006). *Mistake proofing: changing designs to reduce error. Quality and Safety in Health Care,* 15;44-49.

Kohn LT, Corrigan JM, Donaldson MS. (1999). To Err is Human: Building a Safer Health System. National Academy Press, Washington. p.32, 53, 57, 63, 144, 145.

Leape LL, Berwick DM, Bates DW. (2002). *What practices will most improve safety: Evidence-based medicine meets patient safety. Journal of the American Medical Association,* 288:501–507.

Piotrowski MM. Cohen M, Mercier J, Saint S, Steinbinder A, Thompson M. (2005). *Introducing the national patient safety goals department: Sharing programs of excellence from individual organizations. Joint Commission Journal on Quality and Patient Safety,* 31(1):43-46(4).

Richardson WC, Briere R. (Eds), (2001). *Crossing the Quality Chasm: A New Health System for the 21st Century. 1st ed, National Academies Press. Washington, DC.*

SPSS Inc. (2007*). SPSS for Windows.* Version 13.00, Chicago

Cook RI, Woods DD. (1994). *Operating at the sharp end: The complexity of human error. In: Human Error in Medicine. Bogner MS, (Ed),. Lawrence Erlbaum Associates Publishers, Hillsdale, New Jersey,* s. 255–310.

Engelke C, Olivier D. (2002). *Putting human factors engineering into practice. Medical Device and Diagnostic Industry,*60.

Joint Comission Resources. (2006). Patient Safety Essentials for Health Care. Joiınt Comission International, Oakbrook Terrace, Illinois. p.17-19.

Leape LL. (1997). A systems analysis approach to medical error. Journal of Evaluation in Clinical Practice, 3(3):213-222.

Mohr JJ, Barach P, Cravero JP, Blike GT, Godfrey MM, Batalden PB, Nelson EC. (2003). Microsystems in health care: Part 6. Designing patient safety into the microsystem. Joint Commission Journal on Quality and Patient Safety, 29(8):401-8.

Chapter 68

An Analytical Framework to Measure Effective Human Machine Interaction

Juan Wachs[1], Brad Duerstock[2]

[1]School of Industrial Engineering
Purdue University, West Lafayette, IN 47907

[2]Center for Paralysis Research
Purdue University, West Lafayette, IN 47907

ABSTRACT

Initially, human machine interface (HMI) was understood as the hardware and software through which human and machine could communicate. Gradually, is being recognized that many human factors such as usability, emotion, user's physical and cognitive characteristics, domain knowledge, contribute as much to the effectiveness and efficiency of HMIs as robust, reliable and sophisticated algorithms do. Clearly, both the human centered factors and the technical factors have direct or indirect relations with the effectiveness of the HMI. Nevertheless, the degree of influence of these factors on the effectiveness of human machine interaction is not well understood. Most of the work in the human machine interaction area is focused on creating and refining techniques and algorithms, application-driven efforts, or heuristic procedures, but there is a lack of basic or foundational work. In this work, we present a novel development of an accessible interface for the control of a robotic arm based on natural, effortless hand gestures designed for students with mobility impairments, and we provide a systematic framework to measure the effectiveness of this interface.

Keywords: Human machine interfaces, hand gestures recognition, assistive technologies, robotics, intelligent wheelchairs.

INTRODUCTION

Physical access to classrooms, laboratories and learning resources is crucial for students with disabilities. Active participation once present, however, is also vital, encompassing interaction with teachers, other students, and engagement with course materials and equipment (Salend, 1998). In order for students with mobility impairments to gain educational experiences comparable to those of able-bodied students they must perform comparable tasks (Warger, 1998). Actively exploring and interacting with scientific concepts and practices grants a more thorough educational experience as a whole. This need for active learning, however, creates serious hurdles for students with disabilities.

Robotic assembly tasks and navigation planning and control are the most common tasks in automation and production labs. Able-bodied students are expected to personally control robots in assembly tasks, and analyze and design robot manipulations, in both undergraduate academic courses and high school science classes. In order to perform independent graduate or postgraduate research, or to pursue an engineering career, such as automation engineer, manufacturing engineer, or controls engineer, students with disabilities must be able to independently operate a robot in real-time.

There has been extensive research on the use of sensors that allow people with disabilities to interact with machines. Sensors allow the control of devices by eye-blinking, gaze, breathing, EEG and EMG signals, posture and gestures, lip reading and tongue movements. There are two main problems with these interfaces: (a) they are non-adaptive. Most of these methods leverage the strength of a single limb or body part that functions relatively well (Kim et al, 2006). Different solutions are needed, however, in cases of progressive illness where limb control skills decay gradually with time, or when the user is rehabilitating and hence has improving motor skills. In fact, technology permitting a single modality of interaction is appropriate only when the user's condition is stable. As most paralyzed people experience a change in their condition throughout their lives, a new paradigm is needed; (b) their design does not follow an analytic methodology.

The term "effective interface" in the context of human-robot interaction is relatively new and so far there is no universally accepted definition for this term. Most of the existing definitions are unstructured and only focus on one aspect of effectiveness, for example Olsen and Goodrich (Olsen and Goodrich, 2003) only focus on effectiveness as a function of task effort. In this proposal, the PI identifies a set of factors that influence the effectiveness of the interface, and attempts to organize then in a comprehensive and coherent framework. In order to evaluate the effectiveness of a given interface for robotic control, first performance measures need to be defined.

METHODOLOGY

STUDY THE FACTORS THAT INFLUENCE THE EFFECTIVENESS OF INTERACTION MODALITIES FOR ROBOTIC CONTROL

This section addresses foundational problems of the human-machine interaction area – how to define the effectiveness of a modality (or interface) used for interaction, and how to measure it? In order to evaluate the effectiveness of hand gestures over standard interface techniques for robotic control, performance measures must be defined. Interface effectiveness can be defined as a function (1), which is optimum when the interface used is the best among the options available. Different users may prefer a different interface according to their physical abilities (joystick, keyboard, hand gestures, sip-n-puff, EEG and EMG based signals, tongue control, etc).

$$\underset{I \in \Gamma}{Max}\ e(I) = f(T,U,M,E,L) \tag{1}$$

where:
$e(I)$ = is the effectiveness for a given interface
f is some inverse function of e, including the following:
T = task completion time
U = user skills, expertise and knowledge domain
M = the number of discrete user expressions (physical or physiologic) required to complete a single operation.
E = number of user errors while completing the task.
L = learning rate (based on learning curve)
I = is an interface modality
Γ = the set of all feasible interface modalities (e.g. joystick, keyboard, hand gestures)

The function e defines the relationship between the interface and its effectiveness. For example, an interface that is easy to learn will improve task efficiency. The common measure of efficiency is the time to complete a task (T). Benchmark robotic tasks will be used in the user studies so that the results are comparable across different interfaces.

An interesting feature of this formulation is that it considers the user's knowledge and experience, since they affect the task performance, which in turn affects the interface's efficiency. The algorithm used to decode the signals into significant robot commands is not considered in this scheme, since it is extrinsic to the interface adopted. For example, there are several algorithms used for tongue movement recognition, and even that is possible to measure the superiority of some algorithms over the others, the success of these systems does not rely as much on the algorithm refinement, as it relies on the particular way the interface is used. The number of discrete movements required for the user to operate the robot is an indication of the cognitive load, the complexity, and the level of performance required in the task. The measurement of effectiveness involves the evaluation of (1), the analytical form of which is unknown. Therefore, a set of multiple objective performance measures are proposed to act collectively as proxies for (1): task completion time Z_1, number of movements Z_2, number of errors Z_3, and learning rate Z_4. These proxies do not include U, since use experience is directly related to the learning rate. The recognition accuracy of the interface is not included, since it impacts the task performance indirectly. The goal is to use a formulation that is independent of the sensing technology. Since f is some inverse function of e, bringing the different objectives to minima will lead maximum interface effectiveness.

$$Min\ Z_1(I), Min\ Z_2(I), Min\ Z_3(I), Min\ Z_4(I) \tag{2}$$
$$I \in \Gamma$$

This multiobjective optimization problem may have conflicting solutions when all the objectives are minimized simultaneously. As with most multiobjective problems, this difficulty is overcome by allowing the decision maker (the user) to select the best I according to his preferences. Another method of overcome the conflicting multiobjective values is to adopt a goal programming approach: map the four performance measures into a single measure using weights w_i to reflect the relative importance of each objective.

$$Min\ f(I)\ \underset{I \in \Gamma}{\propto}\ Max\ Z(I) = w_1 Z_1(I) + w_2 Z_2(I) + w_3 Z_4(I) + w_4 Z_4(I)$$
$$s.t. \tag{3}$$
$$\sum_i w_i > 1$$
$$w_i > 0\ \forall i$$

where:
w_i = the relative importance of factor Z_i.

The weights in (3) can be found empirically by letting the decision maker assign importance to each factor according to his/her needs and preferences. Alternatively, the weights can be varied, and for each unique weighting scheme the corresponding solution can be presented to the user for acceptance or rejection. The objectives Z_i will be calculated by running simulations of a task using a virtual model of a robot (like the one in Figure 1), which each of the interfaces considered, for example: (a) standard joystick, (b) voice and gestures, (c) EEG signals. Then, (3) can be computed and comparisons can be made among the interfaces.

A NEW METHOD FOR GESTURE BASED ROBOTIC CONTROL

The concepts described in Tele-gest project (Wachs et al, 2005) to achieve a real-time implementation of a teleoperated control using static hand gestures can be extended to a highly adaptable and robust recognition system for users with mobility impairments. The main components of such as system are described in the following sections and their implementation is left for future work.

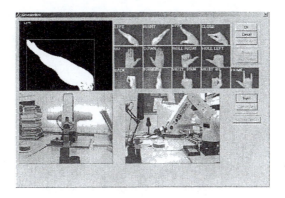

Figure 1. User interface for robot control

Mobility Capabilities - Hand Gesture and Face Movements:

The framework described here will allow the user to control a robot using a wide variety of user customized gestures. For instance, an able-bodied user would want to control the robot while moving his hand in the direction of the robot's intended movement. Moving his hand to the right causes the robotic arm to rotate to the right; moving the hand forward causes the robotic arm to move forward, and so on. While this mapping is perfectly natural to most users, some individuals with severe mobility impairments cannot move their hands in a straight line. Thus the logic used must be able to detect and recognize non-standard hand movements.

Hand and Facial Gesture Recognition

A software application is developed to enable the control of a robotic arm by hand

and facial movements. The main hardware components (presented in Figure 2) are: an electric power wheelchair (EPW), a netbook running the recognition system, a Ladybug2 © spherical digital video camera system, and a 6-axis robotic arm. This video camera system has six digital cameras arranged in such a way that it can collect video from more than 75% of its full perimeter. The first and second cameras are oriented towards the hand and face, respectively. The netbook processes the images, recognizing the actions for wheelchair control, and provides visual feedback. The LCD display shows two windows: one for the hand gesture recognition feedback and the other for the facial movement and gesture recognition feedback (see Figure 3).

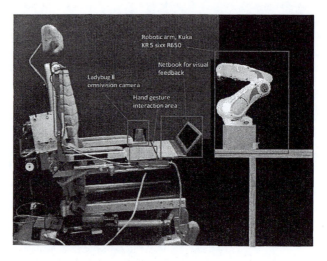

FIGURE 2. Prototype of the interface on the EPW

The data flow of the system is as follows: the user performs gestures with her hand. Sensors on the spherical digital camera system capture face and hand views. The images from the six sensors are 'stitched' together in one large image so the hand and face appear in a common system of coordinates.

The images from the sensors are sent to the netbook, where they are processed. The software searches the images for known movements, hand shapes, or facial expressions. If one of these patterns is recognized in the image, the type of movement is translated into a navigational command which was predetermined in an earlier stage of the system's operation. A netbook display shows a feedback window. The window shows the area where the hand gesture was detected. On the left corner of the window, a caption with the name of the command associated with the recognized gesture is presented. At the same time, the action given by the command is carried out by a robotic arm for a parts assembly task. This type of task was chosen since the feasibility of this approach can be evaluated easily using the measures expressed in (1).

FIGURE 3. Two-view screen: left side: the hand; right side: the face.

When the system is used for the first time, a calibration process sets up the hardware and allows the student to interactively teach the gestures that will be used for robotic control. The calibration routine takes only a few seconds, and can be evoked again at any point during the robotic control. In calibration mode the user determines the neutral area and the interaction area. The interaction area is determined by the distance from the hand to the camera, the sensor's focal length, and the extent to which the hand can be moved. The focal length is fixed, while the distance to the camera can be modified by moving the camera further or closer to the user–this process takes place before the calibration.

The neutral area is determined automatically by detecting the user's hand position. It is a rectangular area around the hand's position with a size equal to the product of the minimum bounding box around the hand by a constant factor (to compensate for non precise hand movements, such as tremors) (see Figure 4). This process takes a few seconds.

The user teaches the gestures to the system by showing the same gesture multiple times when prompted. A vocabulary of 12 commands is designed for robot world coordinates control. The 'forward' and 'back' commands control the X-axis, the 'right' and 'left' commands control the Y-axis, and the 'up' and 'down' commands control the Z-axis of the robotic arm. The 'roll right' and 'roll left' commands rotate the wrist joint, and the 'open grip' and 'close grip' commands control the robot gripper. The 'stop' command stops any action the robot performs. The 'home' command resets all robot joints in the home position. Each of the robotic navigation commands are displayed one after the other with a delay of 30 seconds. When the command is displayed, the user must move her hand in any trajectory, leaving the neutral area, and then bring the hand back to the neutral area. The trajectory, velocity and shape are registered by the system and stored in a database for further use.

Each of the navigational commands is presented five times, and the user is prompted to show a gesture each time. Once the system has been calibrated it is ready be used in operation mode.

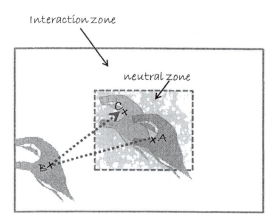

FIGURE 4. Performing a gesture

In operation mode, a graphical user interface (GUI) is presented on the netbook's LCD screen. The image displayed on the GUI is the camera system's view of the scene. On the image, two rectangles with different colors are plotted representing the interaction and the neutral area. Initially, the user's hand is placed inside the neutral area, so no action is carried out by the robot. If the user wants the robot to move, he moves his hand according to the type of action desired. Then the system tries to detect the hand and identify the gesture. If the gesture is recognized, the system displays a caption with the name of the recognized action printed on the screen, otherwise the 'try again' message is displayed.

The neutral area is dynamically updated according to the hand's preferred resting place. This is important because subjects with mobility impairments may find it difficult to move their hands back to the origin of the gesture after performing a movement. Suppose that the user performs a gesture which involves moving her hand from point A to point B. The user could then move her hand back to the origin of the gesture, point A, or move her hand to a point C (see Figure 4). In either case, the hand's final destination (point A or C) will be considered as the resting place (neutral area) for the hand until the next gesture is evoked. This requires only that points A and C be close to each other.

Continuous control is used to resemble joystick operation. With a joystick, the robotic arm continues moving in the indicated direction until the user tilts the handle back to the origin point or releases the handle. Analogously, in our system, as long as the user's hand is outside the neutral area, robot movement occurs. As in the previous example, if a gesture starts at point A (inside the neutral area) and ends at point B (outside the neutral area), then the action requested is continuously carried out until the subject returns to point A or any point inside the neutral area. Continuous control was selected because navigational actions are continuous; discrete operation is more appropriate when the actions required are discrete events in time, such as 'stop' or 'turn-on-engine' on a car.

When a person with mobility impairments is able to make different hand shapes (hand poses), this information can be used to discriminate between non-intended movements and navigational gestures. For example, the hand of a user holding an imaginary or virtual joystick will probably show the 'fist' shape. The user moves the fist left, right, upwards or downwards. Each of these movements can be translated to an action, as with standard joystick operation. The pointing pose could also be used. A subject pointing his index finger forward and tilting his hand in one of the four directions could indicate his desire to move the robot tip in that direction. Four different hand shapes are planned to be included that will be automatically recognized: fist, pointing gesture, palm up and palm down.

Facial gestures can be used as modifiers of the hand gestures. A 'modifier' alters how an action is interpreted by the recognition system. Certain facial expressions indicate that the preceding hand gestures will be interpreted as navigational actions, or that they should not be used (unintentional gestures). Facial expressions will change the operation mode from 'active' to 'sleep' or the other way around. For this switch only one facial expression is required, the open mouth.

EXPERIMENTS

A combination of qualitative and quantitative assessments and usability experiments will be used to measure the effectiveness of the interface with students with physical impairments. The repeated measures design approach, in which participants will serve as their own controls in rating the usability of the vision based interface, will be adopted. In the control condition, the subjects will perform an assembly task (Towers of Hanoi) with the robotic arm using a conventional isometric joystick to control the robot. The task will consist of moving the disks from one of the rods onto another rod, on top of other disks that may already be present on that rod. This task requires several manipulation tasks such as sliding, grasp, move, and release. These operations are very common in automation labs and thus this exercise provides a good case scenario for this system as a pedagogic tool. In the experimental condition, the students complete the assembly task using the interface. To control for extraneous variables such as practice effects across the two scenarios, the subjects will repeat the assembly task three times for each condition (standard joystick, hand gestures). The order of the control (conventional joystick) and experimental conditions will be counterbalanced for participants.

Performance measures: Six measures will be used to assess the user performance for the control and experimental conditions: four quantitative and two qualitative measures.

Quantitative measures:

1) Usability: User ratings of functionality, ease of use and additional human-centered measures for the vision interface and the conventional joystick will be collected using the Likert 5 point scale (1 = very hard, 5 = very easy). The subjects will rate several features for each of the two control cases, reflecting the level of suitability to the user. The questions will assess easiness of use and learn of the interface, memorability, comfort, intuitiveness, safety and precision.

2) Task completion time: The time required to complete the task from beginning to end will be recorded for each task.

3) Number of mistakes: The number of times that the robot gripper collided with an obstacle, released the object too early or imprecisely, or did not grasp the object at all.

4) Recognition accuracy: The number of gestures that were recognized correctly during the assembly task. For this ground truth annotations of the images used by the system are necessary.

Paired sample T-tests will be used to examine whether there is a significant difference in the user performance between the experimental condition and the conventional control (standard joystick).

Qualitative measures: After students complete the assembly task using the different control strategies, they will be interviewed individually to obtain feedback about their experiences using vision based interface compared to a conventional joystick. The interview questions will focus mostly on determining which features of the hand gesture interface were the most important and to identify which additional features would be particularly important. These responses will be tabulated and analyzed qualitatively to assess the perceived benefits and challenges of using each interface from the users' perspective. This information will be used to improve future versions.

CONCLUSIONS

The research presented in this paper is an attempt to offer a new methodology to assess effectiveness in interface design for human-machine interface and to implement a hand gesture based interface for robotic control. The main target population to use the propose system are students with physically and mobility impairments. The motivation for this choice is to develop an infrastructure to allow students with motor impairments to independently operate a robot, a fundamental piece of laboratory equipment in secondary and postsecondary automation classes.

Two fundamental problems are analyzed in this paper: – how to define effective human robot interaction and how to measure it? We offer an analytic framework based on maximization of multi-objectives to enhance the interface design. Finally, we suggest a procedure including quantitative and qualitative measures for the assessment and evaluation of the hand gesture based interface described. The implementation of the system is left for future work.

REFERENCES

Kim, K.H., Kim, H.K., Kim, J.S., Son, W., Lee, S.Y., (2006). A biosignal-based human interface controlling a powerwheelchair for people with motor disabilities. *ETRI Journal*, 28(1):111-114.

Olsen, D. R. Jr., and Goodrich, M. A. (2003) .Metrics for evaluating human-robot interactions. Proceedings *of Performance Metrics for Intelligent Systems (PERMIS), 2003.*

Salend, S.J. (1998) Using an activities-based approach to teach science to students with disabilities. *Intervention in school and clinic*, 34(2), 67-72.

Wachs, J., Stern, H., Edan, Y. (2005) Cluster Labeling and Parameter Estimation for the Automated Setup of a Hand-Gesture Recognition System. *IEEE Trans. on Systems, Man and Cybernetics. Part A. v*ol. 35, no. 6, Nov. 2005, pp. 932-944.

Warger, C. (1998) Integrating assistive technology into the standard curriculum. ERIC/OSEP Digest E568. ERIC Clearinghouse on Disabilities and Gifted Education, Office of Educational Research and Improvement, Washington, DC.

Chapter 69

A Review of Web-Based Dietary Interventions From the HF/E Perspective

Inkyoung Hur, Bum chul Kwon, Ji Soo Yi

School of Industrial Engineering
Purdue University
West Lafayette, IN 47907-2023, USA

ABSTRACT

Web-based dietary interventions have been widely used to change people's dietary behaviors and improve their health. Copious features and technologies have been developed and used: from a simple HTML page to interactive online games. Due to the variety of these features, a practitioner who plans to create or improve an intervention website may wonder which features should be used to maximize the impacts of a web-based intervention. Unfortunately, in spite of extensive literature review, we failed to find straightforward guidelines for practitioners. Thus, the aim of this study is to collect preliminary information to eventually create practical guidelines. We selected 47 papers regarding web-based dietary interventions and reviewed the papers to identify what kinds of technologies have been used and how these have impacted on measured outcomes. Through this review, we identified several interesting patterns, and their implications were discussed.

Keywords: Web-based dietary intervention, review, human-computer interaction

INTRODUCTION

Web-based interventions have been widely adopted by patients and family members to educate and manage their nutrition intakes (ChanLin, Huang, and K. C. Chan 2002). Many studies have been conducted to investigate the effectiveness of different approaches in web-based intervention, and several review studies (e.g., Lustria et al., 2009) summarized the outcomes of these studies. These studies generally proved the positive effects of web-based dietary interventions.

More specifically, the review studies generally emphasize the importance of the three general features, such as tailoring, monitoring, and social support. Rimer and Kreuter (2006) highlighted the importance of tailored information to individual users in health interventions. Self-monitoring to allow users to voluntarily assess the progress of their dietary interventions was also found effective on behavioral changes (Siu et al., 2007; Kubiak et al., 2006). The social support was realized in many ways such as motivational interviews and counseling, which led to improve self-efficacy (Thijs, 2007).

Though these review studies identified important general features (tailoring, monitoring, and social support), those guidelines may not be sufficiently detailed for practitioners who create or improve a web-based dietary intervention. For example, "provide more social support" could be helpful advice, but what kinds of detailed feature should be added to "provide more social support?" When one finds out that an online discussion board and an online chatting service with experts are potential social-support features, which one is more effective? Unfortunately, we could not answer these practical questions even after extensive literature review of existing review studies. Many of these review studies conducted from the perspective of healthcare (Rimer and Kreuter 2006; Thijs 2007), so the detailed descriptions that may be necessary for practitioners were not discussed.

Thus, the main goal of this study is to identify detailed features used in web-based dietary interventions and their effectiveness. The findings from this study will be a stepping-stone toward practical design guidelines for successful web-based dietary interventions.

METHODS

We conducted an extensive search for academic studies on web-based dietary intervention published in peer-reviewed journals, both in the healthcare and human-computer interaction (HCI) domains, from 2001 through 2009. The search was conducted in the following electronic journal archives: Medline, Science Direct, and Springer for the healthcare studies, and IEEE Xplore and ACM Digital Library for the HCI studies. Search keywords used in the selection process were every possible combination of terms from each block: (Block1: "Internet", "web", "online"), (Block2: "dietary", "nutrition"), and (Block3: "intervention" and "system").

Explicit inclusion and exclusion criteria were applied to the selection process.

First, titles of the papers were reviewed to judge their general eligibility for the review. Included studies involved dietary interventions for dietary changes or weight loss. We excluded review papers and studies irrelevant to healthcare interventions in this step. Second, we judged the foci of study interventions in the abstract. Studies including non-dietary interventions (e.g., physical exercise) were excluded. Third, we interpreted "web-based" broadly, so we also included other information technologies using the Internet, such as mobile technologies (Tanguay, 2007). Last, we iteratively evaluated once-selected papers whether or not to keep in the final list based on details of technical components described in the paper or website. We included studies that utilized simple techniques used in web-based health intervention systems only if the features were described in sufficient details. Eventually, total 47 papers[1] were chosen for the review.

After selection, two reviewers codified the 47 studies separately on several aspects, such as study characteristics, detailed features, and outcomes of studies. In addition, the purposes of interventions, theories used, demographic information of research participants, and periods of evaluation study were also codified. Then, the two reviewers combined the two separately codified data. When conflicts between the two codified data occurred, the two reviewers made an agreement on a single code based on evidence found in the corresponding papers.

RESULTS

GENERAL CHARACTERISTICS

Each paper was unique, but there were some general trends. User's diet control was the most popular theme (25 papers) among 47 papers, and we also found weight loss (6 papers) and nutrition education (6 papers) as other popular themes. The other papers (10 papers) dealt with other topics, such as the effectiveness of nutrition labels. Popular theories used in selected papers were the social cognitive theory (5 papers), transtheoretical model (TTM) (3 papers), and planned behavior (3 papers), but most of the papers (36 papers) did not specify any theories. Target population varied among studies. Studies were targeted to the general population (20 papers), senior adults (4 papers), children (6 papers), obese adults (6 papers), and patients (2 papers), and the other studies did not clearly specify target users (9 papers). The mean age of the participants in studies ranged from 13 years old (Williamson et al. 2006) to 63 years old (Verheijden et al. 2004). Six studies out of 47 studies dealt with only female participants. The durations of evaluation studies vary, but most of them were less than or equal to 6 months (33 papers). Relatively small number of

---

[1] Due to the space limitation, the 47 papers are listed in the following webpage: https://engineering.purdue.edu/HIVELab/wiki/pmwiki.php/Main/FoodForTheHeartFullReferencesForAHFE2010

evaluation studies took more than 6 months (6 papers). Among 47 papers we reviewed, 20 papers were from healthcare, and 18 from HCI.

In order to measure the effectiveness of web-based interventions, multiple measures were used, which can be categorized into five different phases: intervention usages, knowledge level, psychological influences, dietary behavior, and health status. Intervention usages usually shed a light on usability of web-based intervention, and they are often measured by usefulness and ease of use. The change of the knowledge level is measured by a set of multiple-choice questions regarding specific healthcare facts (Silk et al 2008). The gained knowledge may cause psychological influence, so the intention to change behaviors or self-efficacy was measured to understand this psychological aspect. The actual behaviors changes were measured by nutrition intakes or reported food-eating habits (using food frequency questionnaire (FFQ), eating disorder inventory, or binge eating behavior). Health statuses, which could be ultimate measures of dietary interventions, were measured by weight, body mass index (BMI), high-density lipoprotein cholesterol (HDL), and triglyceride. It should be noted that one study often used a combination of these measures. Among total 47 studies, 34 studies tested the effects of dietary interventions using measures that belong to more than one category.

DETAILED FEATURES

Table 1. Summary of features used in web dietary intervention and corresponding outcomes.

Features	E/T[1]	K[2]	BI[3]	BC[4]	HI[5]
Self-monitoring with graphs	10/20	+2/2	+4q/4	+5/5	(-2 +6)/8
Online discussion board	11/17	+1/1	+5/5	+7/7	(-1 +6)/7
Self-goal-setting	10/12	+1/1	+4/4	+8/8	+6/6
Newsletter	9/10	+1/1	+2/2	+7/7	(-2 +2)/4
Personal feedback from experts	7/10	+1/1	+3/3	+4/4	(-1 +3)/4
Link to resources	6/9	+2/2	+4q/4	+4/4	+2/2
Offline class / module	6/8	+2/2	+3q/3	+2/2	+2/2
Calculator	5/7	0/0	+1/1	+1/1	(-1 +3)/4
Chatting rooms with experts	4/7	0/0	+4/4	+3/3	+3/3
Online quiz	4/6	+2/2	+2/2	+4/4	+2/2
Offline brochure and handbook	4/4	+1/1	+2q/2	+3/3	+3/3
Audio and video documentary	2/4	+2/2	0/0	+2/2	0/0
Offline assignment	2/3	+1/1	0/0	+1/1	+1/1
Video game	1/2	+1/1	+1q/1	0/0	0/0
Self-monitoring using a mobile device	1/1	0/0	0/0	+1q/1	0/0

[1] E/T: This column shows the total number of evaluation studies and the total number of studies that include the feature. For example, "10/20 for self-monitoring using a graphic report" means that 20 studies include the feature, but 10 out of 20 studies actually include evaluation studies.
[2-5]K: Change of knowledge level, BI: Change of behavior intention, BC: Change of behavior, and HI: Change of health status; The cell represent the number of negative and positive outcomes over the total number of evaluation studies. For example, "(-2 +6)/8" means that there were total 8 evaluation studies tested a feature, 2 studies reported negative outcomes and 6 studies reported positive outcomes.
q One of the studies used qualitative method to measure the outcomes.

As shown in Table 1, multiple features have been used in web-based intervention.

Some of features could be included in the three general features (tailoring, monitoring, and social support). Newsletters sent to a group of participants (10 papers) and individual feedbacks with well-matched advices and tips (10 papers) are examples of tailored information. Online discussion boards (17 papers) are examples of social supports. Self-goal-setting (12 papers) and self-reports or food diary and progress reports with graphs to assess the health status along with feedback (20 studies) would be examples of monitoring. However, there are many techniques are out of these three categories. General information with offline classes or module (8 papers), guidelines through brochure or handbooks (4 papers), and links to other sources for diet-related and health-related information (9 papers) were reported in the papers we reviewed, but they are more like untailored information. Sometimes, one feature could be related with multiple general categories. For example, a chatting service with experts could provide both tailored information and social support.

Some features are more frequently used in the reviewed studies, but some are not. Self-monitoring (e.g., food diary and track current calorie), personalized messages including advices, tips, goal-setting, using graphs for monitoring progress, and online discussion boards are often used in our studies. In contrast, sensors, providing a food planner, and generating ideal characters were tried recently, but there have not been tested by multiple studies. Few studies tested the effects of using those brand-new technologies.

Dietary interventions produced either positive or negative effects on dietary or health changes. Table 1 shows the numbers of positive or negative results in four outcome variables (i.e., knowledge, behavior intention, behavior, and health status). In most cases, we did not find negative outcomes on change of knowledge level, behavior intention, and behavior. Only negative outcomes were noticed in health status. The only feature that reported only positive outcomes on health status is self-goal setting (six positive outcomes out of six evaluation studies) after having more than four evaluation studies.

Technologies for interactivity with users, quiz and assignments always showed the positive influences on knowledge gain and behavior changes in six studies. However, calculators were found effective in health changes in three studies, but one study resulted in negative result. As technologies for visuals, graphics in reports and images were found effective in dietary changes in six studies out of eight, but the rest reported negative effects on health improvement. However, multimedia components such as video and audio were not tested on health improvement. Sensors were not tested in any study.

COMBINATIONS OF FEATURES

Single feature in the web-based dietary intervention did not guarantee dietary changes. Many of our interventions incorporate with more than one feature. In a study of Brown et al. (2004), behavior and health status were changed positively with their dietary intervention, which incorporated documentary style audio and video, personalized feedback, self-quizzes, self-monitoring, goal-setting, weekly reading and writing assignments, and participation in the Internet discussion group.

Some features were identified together in a study producing positive effects in dietary and health changes. Self goal-setting and self-monitoring were used together in nine studies. Among the nine studies, five studies utilized graphical reports for the two functional features. Three out of the five studies targeted to overweight adults resulted in positive effects.

Online technologies and offline technologies were used together. Online discussion board and classes were found in five studies to deliver information. Self-monitoring and handbooks, and quiz and graphical reports were pairs frequently appeared together in three studies resulting in all positive results. Expert meeting and graphical reports existed in three studies with two positive results on behavior and health changes.

These combinations of features are summarized in Table 2.

Table 2. Summary of results of features combinations used in web dietary intervention

Features	E/T[1]	K[2]	BI[3]	BC[4]	HI[5]
Online discussion boards + Self-monitoring with graphs	4/6	0/0	+2/2	+1/1	(-1 +3)/4
Personal Feedback from experts + Online discussion boards	3/6	+1/1	+1/1	+1/1	(-1 +1)/2
Self-monitoring with graphs + Self-goal-setting	4/5	0/0	+2/2	+2/2	+4/4
Online discussion boards + Offline class / module	4/5	+1/1	+1/1	+2/2	+2/2
Personalized feedback (newsletter) + Self-monitoring with graphs	4/5	+1/1	0/0	+2/2	(-2 +1)/3
Offline brochure and handbook + Self-goal-setting	4/4	+1/1	+2[q]/2	+3/3	+3/3
Personal Feedback from experts + Self-goal-setting	4/4	+1/1	+2/2	+3/3	+2/2
Personal Feedback from experts + Self-monitoring with graphs	4/4	0/0	+2/2	+1/1	(-1 +3)/4
Personalized feedback (newsletter) + Calculator	3/4	0/0	0/0	+1/1	(-1 +1)/2
Offline class / module + Self-goal-setting	3/4	+1/1	0/0	+2/2	+1/1
Self-monitoring with graphs + Online quiz	3/3	+1/1	+2/2	+3/3	+2/2
Self-monitoring with graphs + Chatting rooms with Experts	2/3	0/0	+2/2	+1/1	+2/2

[1] E/T: This column shows the total number of evaluation studies and the total number of studies that include the feature. For example, "10/20 for self-monitoring using a graphic report" means that 20 studies include the feature, but 10 out of 20 studies actually include evaluation studies.

[2-5]K: Change of knowledge level, BI: Change of behavior intention, BC: Change of behavior, and HI: Change of health status; The cell represent the number of negative and positive outcomes over the total number of evaluation studies. For example, "(-2 +6)/8" means that there were total 8 evaluation studies tested a feature, 2 studies reported negative outcomes and 6 studies reported positive outcomes.

[q] One of the studies used qualitative method to measure the outcomes.

Different features were used depending on who is the target population. Interventions for children resulted in positive health status, adopted mentor support or mandatory assignments (Tate, Jackvony, and Wing 2006; Bruning Brown, Winzelberg, Abascal, and C. B. Taylor 2004). Dietary interventions for overweight adults utilized self goal-setting and self-monitoring with graphical reports (Svetkey et al. 2008). Personal feedbacks, online discussions, and calculators of calorie were used in the successful systems for obese users. For the old, tailored personal feedback including advices on recommended alternative meals was one of the commonly used components in web-based dietary interventions (ChanLin, Huang, and K. C. Chan 2002; Oenema, Tan, and Brug 2005). Dietary interventions for patients with certain diseases provided nutrition guidelines suggested by governments without many technological components such as interactivity, multimedia or reinforcement (Verheijden et al. 2004; Probst, Faraji, Batterham, Steel, and Tapsell 2008). Dietary interventions for normal adults utilized various features such as information (Papadaki and Scott 2008; Park et al. 2008), multimedia (Park et al. 2008; Silk et al. 2007), online discussion (G. Block et al. 2004; Carpenter, Finley, and Barlow 2004; Mamykina et al. 2008), but no dominant technologies used were found in the successful dietary interventions for the normal adults.

DISCUSSION AND CONCLUSIONS

As a preliminary review of existing literature regarding web-based dietary interventions, it was difficult to provide any definitive guidelines. However, we identified some interesting patterns. First, a few intervention feature were more commonly used in successful dietary interventions, such as personal feedback via email, a discussion board, and self goal-setting and self-monitoring using graphical reports. A practitioner might consider these features for his or her own web-based intervention technologies since there is evidence showing that these features led to successful intervention outcomes. Second, most of the successful web-based interventions used multiple features, not a single feature. This allude that web-based dietary interventions were multifaceted activities, and target users should be reached through multiple channels. Third, some of new features have been developed in the domain of HCI, such as interactive quizzes, multimedia, games, cell phones, and sensors. Though their effectiveness has not been fully tested, these new media and technologies could generate interesting innovation in the future. Fourth, different target populations appeared to be better served by different combinations of features. Thus, a practitioner should pay attention on whom the target user group is, and make a conscious decision in selecting technological components according to the target users.

However, this paper also has several limitations. First, we intentionally excluded studies having physical exercise components in order to focus on dietary changes and subsequent healthcare outcomes. However, we later found that this exclusion criterion might bias the review results because exercise is often essential component to improve health outcomes. Second, the discrepancy between two different domains, healthcare and HCI, impedes more in-depth analysis of different features and their outcomes. As shown in the Result section, healthcare studies tend to focus on rigorous evaluation, but their descriptions for used features are not detailed enough. In contrast, HCI studies tend to focus on developing innovative intervention technologies and describe them in details, but their evaluation tends to be not rigorous and detailed enough. The discrepancy should be resolved through collaborative effort between healthcare and HCI professionals in the future.

In future work, more comprehensive set of literature will be incorporated for our review in order to reveal trustworthy relationships between health outcomes and features in web-based dietary intervention systems. Meanwhile, this preliminary review of features can be utilized to improve and create web-based dietary interventions.

ACKNOWLEDGEMENTS

This work is partially supported by the seed grant from Regenstrief Center for Healthcare Engineering (2009-2010) at Purdue University.

REFERENCES

Block, G., Block, T., Patricia Wakimoto, R. D., Block, C. H. (2004). "ORIGINAL RESEARCH Demonstration of an E-mailed Worksite Nutrition Intervention Program." *Preventing Chronic Disease*, 04_0034.

Bruning Brown, J., Winzelberg, A. J., Abascal, L. B., Taylor, C. B. (2004). "An evaluation of an Internet-delivered eating disorder prevention program for adolescents and their parents." *Journal of Adolescent Health*, 35(4), 290–296.

Carpenter, R. A., Finley, C., Barlow, C. E. (2004). "Pilot test of a behavioral skill building intervention to improve overall diet quality." *Journal of nutrition education and behavior*, 36(1), 20–26.

ChanLin, L. J., Huang, R. F., Chan, K. C. (2002). "Applying web-based instruction to food nutrition course." In *International Conference on Computers in Education, Auckland, New Zealand, December 3rd–6*.

Kubiak, T., Hermanns, N., Schreckling, H. J., Kulzer, B., Haak, T. (2006). "Evaluation of a self-management-based patient education program for the treatment and prevention of hypoglycemia-related problems in type 1 diabetes." *Patient Education and Counseling*, 60(2), 228–234.

Lustria, M. L., Cortese, J., Noar, S. M., Glueckauf, R. L. (2009). "Computer-tailored health interventions delivered over the web: Review and analysis of key components." *Patient Education and Counseling*, 74(2), 156–173.

Mamykina, L., Mynatt, E., Davidson, P., Greenblatt, D. (2008). "MAHI: investigation of social scaffolding for reflective thinking in diabetes management." In *Proceeding of the twenty-sixth annual SIGCHI conference on Human factors in computing systems, April* (pp. 05–10).

Oenema, A., Tan, F., Brug, J. (2005). "Short-term efficacy of a web-based computer-tailored nutrition intervention: main effects and mediators." *Annals of Behavioral Medicine*, 29(1), 54–63.

Papadaki, A., Scott, J. A. (2008). "Follow-up of a web-based tailored intervention promoting the Mediterranean diet in Scotland." *Patient Education and*

630

Counseling, 73(2), 256–263.

Park, A., Nitzke, S., Kritsch, K., Kattelmann, K., White, A., Boeckner, L., Lohse, B., et al. (2008). "Internet-based interventions have potential to affect short-term mediators and indicators of dietary behavior of young adults." *Journal of Nutrition Education and Behavior, 40*(5), 288–297.

Probst, Y. C., Faraji, S., Batterham, M., Steel, D. G., Tapsell, L. C. (2008). "Computerized dietary assessments compare well with interviewer administered diet histories for patients with type 2 diabetes mellitus in the primary healthcare setting." *Patient Education and Counseling.*

Rimer, B., Kreuter, M. (2006). "Advancing tailored health communication: A persuasion and message effects perspective." *Journal of Communication, 56,* 184.

Silk, K. J., Sherry, J., Winn, B., Keesecker, N., Horodynski, M. A., Sayir, A. (2007). "Increasing nutrition literacy: Testing the effectiveness of print, web site, and game modalities." *Journal of Nutrition Education and Behavior.*

Siu, A. M., Chan, C. C., Poon, P. K., Chui, D. Y., Chan, S. C. (2007). "Evaluation of the chronic disease self-management program in a Chinese population." *Patient Education and Counseling, 65*(1), 42–50.

Svetkey, L. P., Stevens, V. J., Brantley, P. J., Appel, L. J., Hollis, J. F., Loria, C. M., Vollmer, W. M., et al. (2008). "Comparison of strategies for sustaining weight loss: the weight loss maintenance randomized controlled trial." *Jama, 299*(10), 1139.

Tanguay, S., Heywood, P. (n.d.). "MyFoodPhone: the start of a mobile health revolution." *Mobile Persuasion: 20 Perspective on the Future of Behavior Change,* 21–28.

Tate, D. F., Jackvony, E. H., Wing, R. R. (2006). "A randomized trial comparing human e-mail counseling, computer-automated tailored counseling, and no counseling in an Internet weight loss program." *Archives of Internal Medicine, 166*(15), 1620.

Taylor, C. B., Bryson, S., Luce, K. H., Cunning, D., Doyle, A. C., Abascal, L. B., Rockwell, R., et al. (2006). "Prevention of eating disorders in at-risk college-age women." *Archives of General Psychiatry, 63*(8), 881.

Thijs, G. A. (2007). "GP's consult & health behaviour change project Developing a programme to train GPs in communication skills to achieve lifestyle improvements." *Patient Education and Counseling, 67*(3), 267–271.

Verheijden, M., Bakx, J. C., Akkermans, R., van den Hoogen, H., Godwin, N. M., Rosser, W., van Staveren, W., et al. (2004). "Web-based targeted nutrition counselling and social support for patients at increased cardiovascular risk in general practice: randomized controlled trial." *Journal of Medical Internet Research, 6*(4).

Williamson, D. A., Walden, H. M., White, M. A., York-Crowe, E., Newton, R. L., Alfonso, A., Gordon, S., et al. (2006). "Two-Year Internet-Based Randomized Controlled Trial for Weight Loss in African-American Girls." *Obesity, 14*(7), 1231–1243.

Chapter 70

HFMEA™ of a Radiotherapy Information System – Challenges & Recommendations for Future Studies

Liam Chadwick, Enda F. Fallon

Centre for Occupational Health & Safety Engineering and Ergonomics
College of Engineering and Informatics
National University of Ireland Galway
Galway, Ireland

ABSTRACT

A risk assessment of the radiotherapy information systems in a large public hospital was completed. The primary focus of the assessment was to determine the risk to the patient as a result of the use of both an Electronic Patient Record (EPR) and paper-based chart. The former was only used by the radiotherapy department studied while the latter was used by the remainder of the departments in the hospital. Following the risk assessment, a prioritized subset of the hazardous processes was analyzed in-depth using Health Care Failure Mode and Effect Analysis (HFMEA™). Although the overall analysis was successful in identifying the potential failures related to each of the analyzed processes, a number of unexpected challenges were encountered related to its implementation. These challenges are presented and the methods used to overcome them are described. Recommendations are also made to enhance the HFMEA™ methodology to support its use as a training aid, by providing enhanced process descriptions, and to facilitate continuous improvement of health care processes through changes related

632

to the HFMEA™ flowchart and worksheet.

Keywords: Radiotherapy, HFMEA™, IDEFØ modeling, Risk Assessment, Clinical Information Systems

INTRODUCTION

Every year 10 million people worldwide are diagnosed with cancer. Of these, 40-50% will receive radiotherapy treatment (International Atomic Energy Agency, 2002). Radiotherapy is one of the major treatment modalities for cancer and the treatment is normally quite safe. However, if an error occurs it can be very serious for the patient with the potential to damage healthy tissue, and in the worst case can have fatal consequences (WHO, 2008). The quantification of errors in radiotherapy is difficult, but values of 5% for error rates has been reported (Yeung et al., 2005).

The prevention of errors is an important aspect of all procedures in the radiotherapy treatment process and advanced technology systems have been implemented to support and address this issue. For example, Record and Verify systems (R&V), in-vivo Dosimetry, Electronic Patient Records (EPR) and Picture Archiving and Communication Systems (PACS) have been introduced to mitigate potential errors in the treatment process. However, the processes by which these new technologies are integrated and utilized do not necessarily adhere to best engineering practice, in terms of safety and risk (Chadwick et al., 2009). The resulting absence of a systematic approach to the safety and risk of high technology systems in health care can introduce new error pathways, and consequently increase the potential for errors to propagate throughout the treatment process (Ash, J. S. et al., 2004; Han et al., 2005; Patton et al., 2003).

Unfortunately, from a review of the errors reported in radiotherapy, the interventions and recommendations introduced to radiation therapy treatment processes typically appear to occur after a serious event (Aspley, 1996; Patton et al., 2003; WHO, 2008; Williams, M. V., 2007; Yeung et al., 2005). Examples of interventions include, the requirement for formal analysis of new software after the Therac-25 incidents in the US and Canada in the 1980s and the introduction of external audits and formal QA including ISO accreditation in the United Kingdom after the radiotherapy incidents in Exeter in the 1980s and the North Staffordshire incident in the mid 1990s (Aspley, 1996; The Royal College of Radiologists et al., 2008; WHO, 2008). In other countries similar steps were taken after incidents there (c.f. Zaragosa, Spain (Nenot, 1998) and Epinal, France (Ash, D., 2007)).

The purpose of the study presented in this paper was to complete a risk assessment of the information systems in the radiotherapy department of a large public hospital. The use of a conventional paper record in the main hospital and an Electronic Patient Record (EPR) within the radiotherapy department resulted in concerns related to the updating and completeness of each record, and consequently the availability of patient treatment information outside of the radiotherapy department. The focus of the study was on the extent to which there was a risk to patient safety in the management of the patient medical record using both hardcopy (patient chart) and softcopy (EPR) mediums.

This work discusses the many challenges that were encountered during the final stage of assessment, which involved in-depth analysis of prioritized processes using HFMEA™, and recommendations are presented for consideration in future studies.

RADIOTHERAPY SYSTEM STUDIED

The study was completed is one of the four designated large public centers for cancer treatment in Ireland, under the National Cancer Control Programme. It was commissioned in 2004-2005 and at the time of the study employed 3 radiation oncology consultants, 24 radiotherapists, 10 medical physicists and 4.5 Whole-Time-Equivalent (WTE) nurses. The department consists of 3 Linear Accelerators, CT and conventional simulators, 3D- treatment planning, High Dose Rate (HDR), Low Dose Rate (LDR) and orthovoltage systems. On average the department treats 1,500 patients per year. It provides conventional conformal radiotherapy, Intensity-Modulated Radiation Therapy (IMRT) & more recently Image-Guided Radiation Therapy (IGRT).

The equipment specification for the department at the time of its commissioning was performed with technical specifications and costs considered in a standard "Value for Money" assessment, using a conventional Option Appraisal methodology (van der Putten et al., 2001). The emphasis of such an appraisal was on the value of the technical and service components of the system, and not on patient safety or risk exposure. The department utilizes the latest in radiotherapy treatment support technology including Record and Verify systems (R&V), in-vivo Dosimetry, Electronic Patient Records (EPR), Picture Archiving and Communication Systems (PACS) and 3D- treatment planning. Despite the sophistication of the technology, no attempt was made to apply a 'systems' engineering approach to its overall development. Neither was consideration given to the application of Human Factors Engineering methods, including human error or organizational issues (Fallon et al., 2009a). Furthermore, there was little systematic consideration of risk mitigation in its specification and development.

DESCRIPTION OF THE RISK ASSESSMENT

The assessment was completed in 5 stages, as follows:
1. Process Modeling – an IDEFØ model of the treatment process was developed to establish a clearly defined and accurate representation of the treatment process (IEEE, 1998). The full IDEFØ model consisted of 24 separate IDEFØ diagrams and included over 100 processes and sub-processes.
2. Hazard Analysis – a total of 20 individual sub-processes, which had

implications for or interactions with the electronic or hardcopy patient file (i.e. the EPR or patient chart) were identified as hazardous in the context of this study, from the completed IDEFØ model. They were subsequently verified by a group consisting of the radiotherapy services manager, lead medical physicist for the department, clinical nurse manager and the radiotherapy administration manager.

3. Risk Assessment – A detailed risk assessment was completed for each of the 20 identified processes, using the Irish Health Service Executive (HSE) Risk Assessment Tool, which is based on the AS/NZS 4360 Risk Management Standard (Australia Standards, 1999; Hughes, 2008). This analysis was completed by the same team who completed the hazard analysis, in (2) above.

4. Risk Prioritization – A risk score was calculated using only the High Risk Scores for each process, i.e. risk scores of 15 or higher. Processes were prioritized based on this Total High Risk Score, which focused the HFMEA™ analysis on the high risk processes.

5. HFMEA™ – HFMEA™ (Derosier et al., 2002) was used to perform an in-depth analysis of the prioritized processes. The participative nature of HFMEA™ was particularly suited to the team orientated operating environment of radiotherapy (Fallon et al., 2009b).

CHALLENGES AND SOLUTIONS FOR APPLYING HFMEA™

HFMEA™ was chosen to facilitate the in-depth analysis of the identified high risk processes for the following reasons:

- Established and explicitly designed for use as a health care analysis tool.
- Clearly defined methodology for use in health care *process* analysis.
- Synthesized from a number of analysis techniques (Failure Modes and Effects Analysis (FMEA), Root Cause Analysis (RCA), Hazard Analysis and Critical Control Point (HACCP)), of which the authors had extensive prior knowledge and experience.
- Successfully applied in several published studies (Esmail et al., 2005; Florence and Calil, 2006; Linkin et al., 2005; Ouellett-Piazzo et al., 2007; van Tilburg et al., 2006).
- Learning opportunity for staff who would gain experience using the technique, which could be used for future studies within the radiotherapy department and the wider hospital.

The HFMEA™ process consists of 5 steps, outlined below (Derosier et al., 2002):

1. Define the HFMEA™ Topic

2. Assemble the Team
3. Graphically Describe the Process
4. Conduct a Hazard Analysis
5. Actions and Outcome Measures

HFMEA™ TOPIC DEFINITION AND TEAM ASSEMBLY

In this study, the topic of analysis was defined by the assessment objective, i.e. the risk assessment of the radiotherapy information system, with the focus of the HFMEA™ study on the high risk processes that had implications for the patient record, as prioritized by the risk assessment, as described above.

The team was assembled from staff members within the radiotherapy department including those who took part in the IDEFØ model development and the hazard and risk analysis described above. The team also included additional staff who volunteered to participate in the analysis following the circulation of a department notification.

GRAPHICAL DESCRIPTION OF THE PROCESS

The IDEFØ model served as the graphical description of the treatment process. The model represented the actual treatment process as it occurred in the department at the time of the analysis, and not the ideal treatment process as observed in the existing quality oriented process maps. From this it was possible to identify discrepancies between the ideal and actual treatment process.

The IDEFØ model was particularly useful during the HFMEA™ analysis as it provided extensive supporting knowledge regarding the inputs to each of the analyzed processes, expected outputs, factors controlling their occurrence and resources required. This information was used to support the determination of potential failure modes. Furthermore, using the IDEFØ model it was possible to identify the preceding processes that must have been successfully completed in the treatment schedule for each subsequent process to occur. Also, it was possible to examine the knock-on effect of each potential process failure within the treatment schedule. This provided information regarding the tightly-coupled nature and inter-dependencies that existed within the treatment schedule.

HFMEA™ HAZARD ANALYSIS

Unfortunately, due to national economic factors there had been a moratorium on the hiring of health care staff within the Irish health service for some time prior to and during the analysis. This resulted in a staff shortage within the department and significantly affected the availability of staff to participate in the analysis. As a result arranging a suitable time for the analysis team to meet was extremely difficult. To overcome this issue, team members were gathered and a group analysis was begun to allow stakeholders gain experience with the technique, and also to identify potential failure modes for a number of the processes under analysis. The analysis team was then broken into smaller groups (sub-teams) which analyzed processes under the supervision of the authors (acting as facilitators). The sub-teams reviewed and cross-checked each other's work and provided additional information as necessary. This distributed team approach facilitated the completion of the analysis while ensuring multiple viewpoints of each process were considered. However, it was slightly more complicated to organize than the specified HFMEA™ group meetings. This approach is similar to aspects of the Delphi Methodology, which results in a general consensus of findings through a process of iterative review and refinement (Browne, 1968).

In keeping with the HFMEA™ methodology and to allow comparison with future studies, each process failure mode and potential cause was analyzed using the HFMEA™ hazard analysis scoring method. Unfortunately, the hazard scoring severity descriptions were not well suited to the requirements of the analysis. Participating stakeholders believed that the 4-level scoring system and descriptions recommended by the technique were too general, i.e. minor, moderate, major and catastrophic, stating that they did not allow for sufficient specification of the scoring. This issue regarding the HFMEA™ technique has been reported in other studies (Habraken et al., 2009; Wetterneck et al., 2006). Using a five level scoring mechanism would facilitate greater specificity in the hazard scoring, e.g. as used in the HSE Risk Assessment Tool (Hughes, 2008).

Another issue encountered was that no element of the HFMEA™ hazard descriptions relates to 'Data', i.e. necessary patient information, e.g. patient histology report. To overcome this problem, the analysis used the more generic severity descriptions for the hazard score provided for in the HFMEA™ documentation, e.g. "Moderate Event – Failure can be overcome with modifications to the process or product, but there is minor performance loss". These are closely related to conventional engineering severity descriptions. However, it is proposed that a severity categorization for 'Data' be considered for future studies where 'Data' is a key part of the analysis. An early draft of IEC 80001: Application of risk management to information technology (IT) networks incorporating medical devices - Part 1 - Roles, responsibilities and activities, contained a proposed severity of occurrence category for health care 'Network' failures. Using similar descriptions and categories for 'Data' failures would facilitate its inclusion as a guideline for consideration in future HFMEA™ studies, as illustrated in Table 1 below.

Table 1: Proposed Severity of Occurrence Descriptions for Data Failures

Description	Description
Catastrophic	A data failure could directly cause death.
Major	A data failure could result in serious injury, permanent impairment; irreversible; major intervention required.

Moderate	A data failure could result in minor injury, temporary impairment; self-limiting illness; reversible; minor intervention required.
Minor	A data failure could result in less than minor injury or transient injury – no intervention would be required.

Team participants also experienced difficulties using the HFMEA™ flowchart, regularly continuing to Step 5 – Actions and Outcome Measures, when they should have 'Stopped'. This problem was also reported in the work of Habraken et al. (2009). The reason for this, in the context of the current analysis, was explained by the analysts' desire to effectively describe the process, noting that controls may have existed or that the failure mode may have been detectable, but a description of these was not required. In this way, HFMEA™ did not meet the stakeholder's expectations as a technique that would sufficiently document the treatment process so that it could be used to support the training of new staff. Furthermore, the method does not support or easily facilitate recommendations to improve the existing process if an existing measure was in place, i.e. promote continuous improvement. This was another desirable objective of stakeholders performing the analysis, who identified that improvements to the existing controls or detection methods could be implemented to make the process safer, efficient or more reliable than it was. It is recommended that the HFMEA™ flowchart be re-worked to support greater levels of process description and documentation, and to facilitate continuous process improvement.

HFMEA™ ACTIONS AND OUTCOME MEASURES

The participating analysts expressed difficulty in relating the identified potential failure modes to suitable corrective actions. Habraken et al. (2009) identified the same deficiency in the HFMEA™ methodology and support information. This was particularly true for potential causes related to human errors, which analysts had little practical experience with. Corrective measures related to potential human errors were recommended by the authors during the analysis, to support this deficiency. To support future HFMEA™ work it is suggested that support information be provided to analysts in the form of appropriate potential human errors, in the form of External Error Modes (EEMs) and associated Performance-Shaping Factors (PSFs). These should be coupled with suitable remedial measures and Error Reduction Mechanisms (ERMs), which have been identified in the health care domain or generic recommendations from existing analysis techniques. This information could be derived from the literature dealing with existing Human Error Assessment (HEA) methods, e.g. Human Error Assessment and Reduction Technique (HEART: Williams, J. C., 1988), Systematic Human Error Reduction and Prediction Analysis (SHERPA: Kirwan, 1994), or through the formal analysis of health care disciplines.

DISCUSSION AND CONCLUSION

Through the completion of the HFMEA™ analysis a number of challenges came to light, which had not been considered prior to commencement of the work. Each challenge required modification to the recommended HFMEA™ methodology while ensuring that the objective of the assessment and technique was not undermined. Similarly, the approaches adopted and recommendations discussed above suggest modifications to the method as a whole to improve its value as an analysis tool, i.e. support detailed process descriptions to support stakeholder training needs and facilitate continuous process improvement.

The use of the IDEFØ model was particularly useful in linking the different stages of the treatment process and provided information regarding resource requirements and controlling factors. It also provided contextual information regarding the potential consequences and impact of individual process failures on the treatment process and schedule.

The distributed team approach utilised as part of the HFMEA™ analysis ensured that it was completed within a reasonable time frame and reduced the pressure of coordinating large group meetings in the understaffed radiotherapy department.

Changes to the hazard scoring are suggested to provide a greater level of specificity for scoring hazards. Also, the inclusion of 'Data' as a severity descriptor would prove beneficial for future studies, which are similar in focus to the study described. It is recommended that the HFMEA™ flowchart should be modified to facilitate continuous improvement and improved process descriptions.

Information to support the identification of potential causes (in particular detailed descriptions of human errors) and applicable Error Reduction Mechanisms is required. This information should be developed using a combination of literature review, existing HEA methods or formal analysis of health care disciplines.

In conclusion, the HFMEA™ method was applied to the analysis of a radiotherapy information system with some modification to the recommended methodology. These modifications combined with the described recommendations could produce a more functional methodology that would ultimately be of greater ease, value and support to organizations.

ACKNOWLEDGEMENTS

The authors would like to thank the radiotherapy staff for participating in this study. HFMEA™ is a trademark of CCD Health Systems.

REFERENCES

Ash, D. (2007), "Lessons from Epinal." *Clinical Oncology,* 19(8), 614-615.

Ash, J. S., Berg, M., and Coiera, E. (2004), "Some Unintended Consequences of Information Technology in Health Care: The Nature of Patient Care Information System-related Errors." *J Am Med Inform Assoc.*, 11(2), 104-112.

Aspley, S. J. (1996), "Implementation of ISO 9002 in cancer care." *International Journal of Health Care Quality Assurance,* 9(2), 28-30.

Australia Standards, (1999), *AS/NZS 4360:1995: Australian/New Zealand Standard of Risk Management.* Sydney, Australia, Standards Australia.

Browne, B. B. (1968), *Delphi Process: A Methodology Used for the Elicitation of Opinions of Experts.* Report P-3925, Rand Corporation, Santa Monica, CA.

Chadwick, L., Fallon, E. F., Kelly, J., and van der Putten, W. J. (2009), "Risk assessment of patient records in radiotherapy " 26th International Conference of the International Society for Quality in Health Care: Designing for Quality, Dublin, Ireland.

Derosier, J., Stalhandske, E., Bagian, J. P., and Nudell, T. (2002), "Using Health Care Failure Mode and Effect Analysis™: The VA National Center for Patient Safety's Prospective Risk Analysis System." *Journal of Quality Improvement,* 28(5), 248-267.

Esmail, R., Cummings, C., Dersch, D., Duchscherer, G., Glowa, J., Liggett, G., Hulme, T., and The Patient Safety Adverse Events Team Calgary Health Region Calgary Alberta Canada. (2005), "Using Healthcare Failure Mode and Effect Analysis Tool to Review the Process of Ordering and Administrating Potassium Chloride and Potassium Phosphate." *Healthcare Quarterly,* 8(Special Issue), 73-80.

Fallon, E. F., Chadwick, L., and van der Putten, W. J. (2009a), Learning from Risk Assessment in Radiotherapy. In: Duffy, V. G. (Ed.), *Digital Human Modeling, LNCS 5620.* pp. 502-511. Springer, Heidelberg, Germany.

Fallon, E. F., Chadwick, L., and van der Putten, W. J. (2009b), "Risk assessment in Radiotherapy. Lessons from systems engineering." WC 2009, IFMBE Proceedings 25/XII.

Florence, G., and Calil, S. J. (2006), "Health Failure Mode and Effect Analysis for Clinical Engineering Application on Cardiac Defibrillators." *Journal of Clinical Engineering,* 31(2), 108-113.

Habraken, M. M. P., Van der Schaaf, T. W., Leistikow, I. P., and Reijnders-Thijssen, P. M. J. (2009), "Prospective risk analysis of health care processes: A systematic evaluation of the use of HFMEA® in Dutch health care." *Ergonomics,* 52(7), 809 - 819.

Han, Y. Y., Carcillo, J. A., Venkataraman, S. T., Clark, R. S. B., Watson, R. S., Nguyen, T. C., Bayir, H., and Orr, R. A. (2005), "Unexpected Increased Mortality After Implementation of a Commercially Sold Computerized Physician Order Entry System." *Pediatrics,* 116(6), 1506-1512.

Hughes, S. (2008), *'Your Service, Your Say' The Policy and Procedures for the Management of Consumer Feedback to include Comments, Compliments and Complaints in the Health Service Executive (HSE).* Health Service Executive,, Dublin.

640

IEEE, (1998), *IEEE standard for functional modeling language - syntax and semantics for IDEF0. IEEE Std 1320.1-1998.* New York, IEEE.

International Atomic Energy Agency. (2002), International action for the protection of radiological patients. In. IAEA, Vienna.

Kirwan, B. (1994), *A Guide to Practical Human Reliability Assessment.* Taylor and Francis, London.

Linkin, Darren R., Sausman, C., Santos, L., Lyons, C., Fox, C., Aumiller, L., Esterhai, J., Pittman, B., and Lautenbach, E. (2005), "Applicability of Healthcare Failure Mode and Effects Analysis to Healthcare Epidemiology: Evaluation of the Sterilization and Use of Surgical Instruments." *Clinical Infectious Diseases,* 41(7), 1014-1019.

Nenot, J. C. (1998), "Radiation accidents: lessons learnt for future radiological protection." *International Journal of Radiation Biology,* 73(4), 435 - 442.

Ouellett-Piazzo, K., Asfaw, B., and Cowen, J. (2007), "CT Healthcare Failure Mode Effect Analysis (HFMEA®): The Misadministration of IV Contrast in Outpatients." *Radiology Management,* 29(1), 36-44.

Patton, G. A., Gaffney, D. K., and Moeller, J. H. (2003), "Facilitation of radiotherapeutic error by computerized record and verify systems." *International Journal of Radiation Oncology*Biology*Physics,* 56(1), 50-57.

The Royal College of Radiologists, Society and College of Radiographers, Institute of Physics and Engineering in Medicine, National Patient Safety Agency, and British Institute of Radiology. (2008), *Towards Safer Radiotherapy* The Royal College of Radiologists, London.

van der Putten, W. J., McLean, B., Hollywood, D., Davidson, Y., Clancy, K., Folan, J., and Higgins, W. (2001), "Any colour as long as it's black... Equipment purchasing for radiotherapy." European Congress of Medical Physics and Clinical Engineering, 12-15 September, Belfast, N.I.

van Tilburg, C. M., Leistikow, I. P., Rademaker, C. M. A., Bierings, M. B., and van Dijk, A. T. H. (2006), "Health care failure mode and effect analysis: a useful proactive risk analysis in a pediatric oncology ward." *Quality and Safety in Health Care,* 15(1), 58-63.

Wetterneck, T. B., Skibinski, K. A., Roberts, T. L., Kleppin, S. M., Schroeder, M. E., Enloe, M., Rough, S. S., Hundt, A. S., and Carayon, P. (2006), "Using failure mode and effects analysis to plan implementation of smart i.v. pump technology." *American Journal of Health-System Pharmacy,* 63(16), 1528-1538.

WHO (2008), *Radiotherapy Risk Profile.* World Health Organization, Geneva.

Williams, J. C. (1988), "A data-based method for assessing and reducing human error to improve operational performance." IEEE Fourth Conference on Human Factors and Power Plants, Monterey, CA.

Williams, M. V. (2007), "Improving patient safety in radiotherapy by learning from near misses, incidents and errors." *British Journal of Radiology,* 80(953), 297-301.

Yeung, T. K., Bortolotto, K., Cosby, S., Hoar, M., and Lederer, E. (2005), "Quality assurance in radiotherapy: evaluation of errors and incidents recorded over a 10 year period." *Radiotherapy and Oncology,* 74(3), 283-291.

CHAPTER 71

Enabling Pre-Hospital Documentation Via Spoken Language Understanding on the Modern Battlefield

*Lisa Anthony[1], Kenny Sharma[1], Susan Harkness Regli[1],
Kathleen Stibler[1], Patrice D. Tremoulet[1], Robert T. Gerhardt[2]*

[1]Lockheed Martin Advanced Technology Laboratories
3 Executive Campus, Cherry Hill, NJ, USA

[2]U.S. Army Institute of Surgical Research
3400 Rawley Chambers Avenue, Fort Sam Houston, TX, USA

ABSTRACT

Lockheed Martin Advanced Technology Laboratories (LM ATL) and the U.S. Army Institute of Surgical Research (USAISR) have been collaborating to improve methods of pre-hospital documentation. The collaboration is focused on creating a pre-hospital documentation grammar and spoken natural language understanding capability to support a hands-free and eyes-free interaction paradigm. This interaction paradigm will minimize the impact on the field medic during treatment of patients in stressful combat situations. The MediTRA-PH (Medical Treatment and Reporting Assistant: Pre-Hospital) prototype is a proof-of-concept of this approach. This paper presents the user-focused concept of operations for MediTRA-PH, the design and development cycles undertaken and preliminary evaluation results in terms of recognition accuracy and user satisfaction.

Keywords: Pre-hospital documentation, Military medical care, Trauma care, Field medics, User-centered design, Spoken language understanding, SLICE.

INTRODUCTION

Pre-hospital trauma care in the military theater is the first link in a chain of medical care as a casualty moves from point of injury to combat support hospitals and eventually to larger facilities. At each transfer point, there is a high risk of information breakdown. Beginning at the patient's point of injury, there is a significant lack of consistent documentation of medically-relevant information, including: mechanism of injury, demographics, medical interventions performed, and outcome. Field medics, lacking any other resource, often resort to writing the details of the patient's identity and treatments on bandages or medical tape on the patient's skin. Lockheed Martin Advanced Technology Laboratories (LM ATL) and the US Army Institute of Surgical Research (USAISR) have collaborated on a prototype system, aimed at improving pre-hospital documentation. This system, MediTRA-PH, or the Medical Treatment and Reporting Assistant: Pre-Hospital, uses a hands-free, eyes-free spoken-language interaction paradigm. This design choice was inspired by military medical subject matter experts (SMEs) input that, while early and frequent documentation does help medical providers later in the treatment chain to be more effective, field medics will only comply with documentation requirements that do not interfere with treatment and saving lives. MediTRA-PH uses LM ATL's Spoken-Language Interfaces for Computing Environment (SLICE) spoken-language understanding technology as the primary method of interaction with the system. This paper presents work to date on MediTRA-PH, including consultation with military medical SMEs on the spoken language interface to the system, and preliminary evaluation results: recognition accuracy and user satisfaction.

RELATED WORK

CURRENT MILITARY MEDICAL SYSTEMS

Since 2003, the US military has deployed a progressive series of components under the Medical Communications for Combat Casualty Care (MC4) system. Goals of the MC4 system include "enabling lifelong electronic medical records, streamlined medical logistics and enhanced situational awareness for Army tactical forces.[1]" MC4 consists of many different inter-operable systems and each focuses on a different aspect of the healthcare process. Handheld devices are carried by medical personnel, allowing them to enter information about initial treatment procedures so that it becomes part of the patient's electronic medical record. However, military medical personnel we consulted cited MC4's point-of-injury care support as

---

[1] https://www.mc4.army.mil/about.asp#how

644

cumbersome. Also, interacting with MC4 systems can interfere with patient interaction because doctors and other medical personnel are required to split their attention between the system and the patient. Both of these limitations are targets of the MediTRA-PH prototype system described in this paper. MediTRA-PH will be a hands-free, eyes-free, point-of-injury component of MC4 that integrates with the electronic medical records (EMR) pipeline created and enabled by MC4.

SPOKEN LANGUAGE UNDERSTANDING SYSTEMS

To build the MediTRA-PH prototype, we leverage our spoken language understanding capabilities, called Spoken Language Interaction for Computing Environments (SLICE), and our experience building multimodal systems for the Office of Naval Research (ONR) and DARPA. Based on nine years of spoken language understanding research begun under the DARPA Communicator program, SLICE (Daniels & Hastie, 2003) provides hands-free operation of computer systems, allowing warfighters in the field to use voice to interact with distributed information sources. Other spoken language systems have been developed for the medical domain, such as Q-Med (Johnson et al, 1992) and Dragon Medical (Nuance, 2008). However, these systems have focused on recognition rather than understanding and are used primarily for dictation and diagnostic interviews during office visits. The environments in which MediTRA-PH is used impose much more demanding challenges on the software, including ambient noise levels, faster pace, and potentially nonresponsive patients.

MEDITRA-PH PROTOTYPE

DESIGN PROCESS AND METHODOLOGY

We engaged in a hybrid user-centered design process called Interface Design for Engineering and Advanced Systems (IDEAS) (Regli & Tremoulet, 2007) to create a demonstration interface that shows the spoken-language understanding in action. IDEAS, an extension to traditional User-Centered Design (UCD), directly facilitates discussion and translation between engineering subject matter experts and domain subject matter experts. The LM ATL team worked with military medical personnel early and often while designing the MediTRA-PH prototype to ensure the eventual prototype system would meet target end-user goals. USAISR arranged a two-day workshop early in the effort to accomplish two primary goals: (1) to elicit the context in which the system would be used; and (2) to elicit specific grammar and vocabulary requirements. USAISR provided four field medics (two from the US Army, and two from the US Army Ranger Regiment), one US Army nurse, one US Army flight surgeon, and one US Army Reserve physician to consult on this panel.

Day one of the workshop was organized as a focus group, in which the LM ATL team asked directed questions designed to elicit domain context information and to engage the medical SMEs in the process of developing a vision for the MediTRA-PH system. Day two involved engaging the field medics in scenario role-playing activities to allow the system designers, as well as all stakeholders, to witness re-enactments of several trauma care incidents in as close to the natural context as possible. Two of the medics each performed two scenario re-enactments of actual events they experienced (minus any identifying details), in two situations: (1) care under fire and (2) tactical field care. These re-enactments revealed that a fielded MediTRA-PH system would have to be extremely robust, flexible, and ruggedized, as well as passive and non-intrusive, in order to successfully integrate with the field medic's environment and context, confirming the utility of a hands-free and eyes-free spoken-language interaction. To inform the design of this spoken-language interaction, we used both direct and indirect elicitation methods to define the vocabulary and grammar that medical personnel use in the field. In the structured interviews, we asked the field medics to directly report commands that they would like to give to the system. In the re-enactments, we recorded the language that they used in context. Both types of elicitation are important because users often cannot explicitly describe how they do a task once it becomes second nature (Beyer & Holtzblatt, 1998). After the workshop, the LM ATL team sent follow-up questions and intermediary design artifacts to the medical SMEs for review and comment, to keep them involved and engaged as the prototype was iteratively developed.

PROTOTYPE CONCEPT OF OPERATIONS

Based on the workshop, we developed the following concept of operations for the MediTRA-PH system. A warfighter sustains an injury during contact. As a field medic moves within treatment distance, the medic's ruggedized handheld platform buzzes to indicate connection with the patient's electronic information carrier (EIC), similar to a traditional identification tag, or "dog tag." The medic says "get PHR" (Personal Health Record), signaling his MediTRA-PH device to query the patient's digital dog tag for complete medical history and present a visual summary of the most important elements. The medic can also request that the information be read in his headset, or ask to be warned only if a proposed treatment is contraindicated based on prior medical history. As the medic treats the patient, he speaks aloud his treatment steps, and MediTRA-PH passively captures the information. It adds this information to the incident report as well as the patient's PHR. Throughout the treatment, the patient's digital dog tag receives updates from MediTRA-PH to ensure that documentation from this point of injury treatment care episode is carried forward when the patient is evacuated from the area. If the medic is working on several patients, as in for example a mass casualty event, the digital dog tag syncing solution ensures documentation is mapped to the right patient. When the medic returns to base, he docks his handheld platform to a MediTRA-PH-enabled machine, which can assist him in filling out his required paperwork, leveraging the data captured from the field. Using the desktop interface, field medics can update MediTRA-PH's vocabulary as new treatments and medications are deployed.

RECOMMENDATIONS FOR REPORTING SYSTEMS

Although the main product of this effort is a prototype system, another important result is the generation of design recommendations for medical domain field reporting systems. These design recommendations were developed as a result of our interactions with medical SMEs during this effort, and reflect the participatory, collaborative nature of our design process, which involves end users and technologists in developing a user-centered vision and, ultimately, system. We have divided the recommendations into three areas: (1) general interaction preferences; (2) spoken language reporting; (3) other desired features and functionality.

General Interaction Preferences

The medical personnel strongly recommended a system designed to seamlessly integrate into existing treatment episodes via a passive hands-free and eyes-free interaction paradigm. Medics use both their hands to treat patients. A trauma care reporting system should be usable without physical interaction. Furthermore, medical personnel want to maximize the quality of the time they spend with their patients and are unwilling to split attention between a computer and a patient, or ignore a patient completely for a minute or two as a report is filled out. Because of the speed and urgency of treatment episodes during pre-hospital trauma care, a system like MediTRA-PH must be configurable to provide feedback on-demand or not at all. Entering information into the PHR must also be extremely rapid so as not to interfere with treatment. Point-of-care documentation systems must also support fluid transition between multiple patients for one or more medics. In mass casualty events, a medic might be treating multiple patients interchangeably, jumping back and forth quickly between injuries. The system must be enable rapid transition, preferably automatically if possible, so that information about injuries and treatments is matched to the correct patient in the electronic medical record (EMR). Finally, point-of-care documentation systems must support interactions for both experienced and novice medics. According to our data from the workshop, 6500 new medics are trained every year at Ft. Sam Houston, leading to a large influx of deployed medical personnel who may be seeing battle injuries for the first time. Stress and inexperience sometimes lead medics to forget important elements of their training. A system like MediTRA-PH should support treatment guides and reminders or alerts when the medic requests it or the system detects him faltering. Rapid deployment is crucial to the military medical provider, so a system like MediTRA-PH should require as little end-user training as possible.

Spoken-Language Reporting

The key reporting modality that can support all interaction preferences and fit well into the context of trauma care is spoken-language reporting. As our military medical SMEs told us, many medics already speak aloud while treating patients, and most recognize that this is beneficial in order to keep focused on the moment and also to interact with the patient to keep him conscious. To best support the medic and require the least amount of user-training, a medical voice documentation system must focus on keywords people use in the field already, such as treatment devices and medication names, as a means of interaction with the system, rather than introducing new language. Because medics vary in experience, it is important for a system like MediTRA-PH to support multiple ways to report a similar thing, including using expert medical terminology (i.e., "anterior femur") and layman's terms (i.e., "right thigh").– Also, medical terminology changes frequently as new treatment devices and medications are deployed, and a system for documenting pre-hospital trauma care must be able to handle all current vocabulary. Furthermore, individual medics might use shorthand or other localized terminology that vary from user to user. An essential design recommendation is to provide an end-user focused interface for vocabulary editing and refining that can be used by medics to add, remove or change vocabulary terms and potentially even new reporting structures to the extent possible with current technological capability. Finally, the spoken-language understanding mechanism must be able to handle a wide variety of speakers, including people with accents, male or female voices, and voices of people in situations of stress, which could change their speaker characteristics. Furthermore, it must be able to listen passively and only capture language that is medically relevant, ignoring and not recording off-topic conversations with other medical staff or the patient, for reasons of privacy.

Other Desired Features and Functionality

Although the primary context of use for a system like MediTRA-PH is the point of injury, patient care documentation does not end when the medic leaves the battlefield. Medics need to be able to enter and edit reports when they have returned to the base, to ensure that all information was captured accurately and completely. Medics have paperwork obligations that a system like MediTRA-PH can make less tedious by feeding directly into the medic's incident report forms such as the SF-600 from the spoken-language report. A system like MediTRA-PH must integrate into the existing electronic medical records pipeline (i.e., MC4), supplying information about trauma care into a patient's long-term health record.

DEVELOPMENT

A working prototype was built for demonstration and evaluation of the MediTRA-PH approach, based on LM ATL's SLICE architecture. SLICE converts spoken audio into a representation of the context-dependent meaning, or user intent. The underlying language model representation provides a rule-based classification system that enables the system to process spoken language input to perform tasks.

Architecture and Language Model Approach

The MediTRA-PH SLICE implementation uses an XML-based language model that defines the medic domain in terms of vocabulary and contextual relationships between categories of concepts. Figure 1 shows a diagram of how the language model concepts are related to each other.

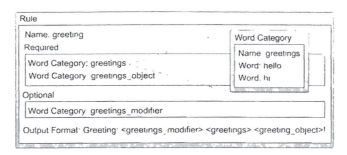

Figure 1. Language model used in the MediTRA-PH prototype system.

Words are the base unit of the language model. A word consists of a literal text representation and one or more pronunciations. Word Categories are logical groupings of particular types of words that can be used interchangeably. The words may have different meanings, but represent the same general concept in relation to each other. For example "fractured" and "broken" can both be categorized as breaks in bones. Rules define logical groupings of categories that must be spoken together to have meaning. Each rule provides both optional and required categories, allowing for flexibility in how much information is required to satisfy a particular rule. The order in which words in each category are spoken is unimportant, compensating for natural variation in how individuals describe the same event or object. Rule Mappings define how to route the data within the SLICE architecture. Particular rules can be handled differently than other rules; for example, an injury should be routed and handled differently than a system command.

Live Speech Recognition and Segmentation

For the initial prototype development, it is assumed that the field medic user will use a key phrase (several variations on "start / stop recorder") to begin and end a particular patient's record. Once the key phrase is uttered, MediTRA-PH does not require any further explicit segmentation of words, phrase or sentences. In order to support the kinds of language our military medical SMEs delivered, the speech recognizer is configured to interpret both continuous words and bursts of phrases, rather than requiring a complete phrase before trying to recognize. This configuration results in a constant streaming of text recognized from the user's raw audio, which is matched against one underlying language model. MediTRA-PH's language model is constructed to minimize the amount of information required for each phrase to be successfully understood, while also minimizing potential conflicts (in the form of redundancy and overlaps) between phrases. Minimizing the complexity of language model components also attempts to reduce the amount of information lost through misrecognition or other errors. To identify the correct break between phrases, disambiguation and conflict management are performed, allowing for delineation of phrases at the earliest possible and logical point in order to prevent error propagation and maximize language model matching efficiency.

Proof-of-concept Graphical User Interface (GUI)

The MediTRA-PH prototype includes a proof-of-concept graphical user interface (GUI) to demonstrate possible visualizations of treatment episodes. Figure 2 shows four possible representations of information classified as injuries, treatments, and mechanisms of injury. In Figure 2, Section 1 illustrates a categorized display of information, representing the basic information types captured in separate fields for ease of readability. Section 2 illustrates a moving timeline of categorized information, where the update interval is user-configurable, which illustrates what was done and how long ago. Section 3 shows a standard body injury diagram that can be used to quickly assess where injuries are located. Section 4 shows a time-tagged transcript of all available information, which is valuable during the post-treatment analysis period. These display visualizations are only preliminary and require iterative user-oriented refinement to ensure they meet user needs.

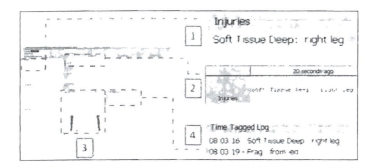

Figure 2. Proof-of-concept graphical user interface (GUI) for MediTRA-PH.

DEMONSTRATION AND QUALITATIVE FEEDBACK

The MediTRA-PH prototype was demonstrated during the 2009 Advanced Technology Applications for Combat Casualty Care (ATACCC) conference. Several formal sessions and informal discussions occurred with field medics of varying experience, including some of the SMEs consulted during initial concept development. Reaction to MediTRA-PH was positive. Specifically, the hands-free, eyes-free collection of speech was viewed as a viable method of nonintrusive information capture. We received positive confirmation that our supported language model was accurate and of high utility, and received constructive feedback on both additional vocabulary to support and information to display in the GUI. Overall, the SMEs consulted at ATACC indicated that use of the MediTRA-PH prototype was intuitive and met their conceptual expectations.

SPOKEN LANGUAGE UNDERSTANDING PERFORMANCE

To evaluate the performance of MediTRA-PH in terms of how well it can understand spoken medical language, we collected a corpus of in-domain spoken language from 15 participants with varying levels of medical domain familiarity. The corpus was comprised of individual treatment episodes with a total of 32 domain-relevant utterances. The audio files for each vignette and participant were fed into MediTRA-PH automatically and the results of the recognition step and the language model matching step were recorded. Finally, the results were compared to ground truth. Three possible outcomes for each utterance within a vignette were possible: (1) true positive, or the correct rule fired; (2) false positive, or an incorrect rule fired; and (3) false negative, or a failure to fire any rule. Because the corpus did not contain any out-of-domain utterances, there were no true negatives. The results of this analysis are as follows. Over all utterances in the corpus, 84.89% of them were correctly classified by MediTRA-PH, meaning the correct rule fired, enabling the system to automatically populate the patient's PHR from spoken language input. Of the remainder, only 0.68% were false positives, while most of the errors MediTRA-PH made were in the form of false negatives (14.43% overall), meaning that no rule was fired. This occurred because of the low spoken-language recognition accuracy achieved by MediTRA-PH's SLICE core in the medical domain; we computed the standard NIST metric, word error rate (WER) (Pallett et al, 1990), to be 73.93%. Speech recognition can be tuned to perform better in domain-specific applications if the recognizer models are trained to specific dictionaries. It is possible that by decreasing the SLICE word error rate, performance of MediTRA-PH at matching rules would significantly improve.

CONCLUSIONS AND FUTURE WORK

This paper has presented MediTRA-PH, a prototype enabling hands-free, eyes-free spoken language reporting of pre-hospital care. This prototype is the product of a multi-disciplinary collaboration between technologists, human factors specialists, and medical SMEs. If deployed, such a system would greatly alleviate the problem of loss of medical treatment information from point of injury contexts and improve patient treatment and hand-off. Moreover, the technology described here could be applied to other domains such as disaster relief, civilian health records, Homeland Security, and so on. Further work in this area could include the following: (1) refining the pre-hospital grammar through further work with medical SMEs; (2)

extending the SLICE speech recognition technology to perform better for domain-specific medical terminology, yielding an increase in recognition accuracy and therefore language model matching accuracy; (3) more formal evaluations of the refined MediTRA-PH prototype in lab and field exercises, including testing of out-of-domain vocabulary; (4) exploring user interface designs and visualizations that allow practical use of MediTRA-PH during conflict scenarios; and (5) expanding the scope of the work by consulting with other medical SMEs to validate our design recommendations with the broader military medical community.

ACKNOWLEDGEMENTS

The authors thank the military medical SMEs who participated in the workshops and demonstrations as part of this effort. Privacy concerns prevent us from listing them by name. This work was funded and carried out jointly by Lockheed Martin Advanced Technology Laboratories and the U.S. Army Institute of Surgical Research (USAISR). COL David G. Gilbertson, formerly of the USAISR, provided invaluable support to this research. The opinions or assertions contained herein are the private views of the authors and are not to be construed as official or as reflecting the views of the Department of the Army or the Department of Defense.

REFERENCES

Beyer, H. and Holtzblatt, K. (1998), *Contextual Design*. Morgan Kaufmann Publishers, New York.

Daniels, J.J. and Hastie, H.W. (2003), "The pragmatics of taking a spoken language system out of the laboratory." In *Proceedings of the HLT-NAACL 2003 Workshop on Research Directions in Dialogue Processing*, Volume 7, Human Language Technology Conference. Association for Computational Linguistics, Morristown, NJ, 16-18.

Johnson, K., Poon, A., Shiffman, S., Lin, R., and Fagan, L.M. (1992), *Q-Med: a spoken-language system to conduct medical interviews*. Knowledge Systems, AI Laboratory (KSL-92-09).

Nuance. (2008), *Optimizing clinical productivity: using speech recognition with medical features vs. a general-purpose solution.* Whitepaper: http://www.nuance.com/healthcare/pdf/wp_healthcare_MDComparative.pdf

Pallet, D.S., Fisher W.M., and Fiscus, J.G. (1990), "Tools for the Analysis of Benchmark Speech Recognition Tests." In *Proceedings IEEE International Conference on Acoustics, Speech, and Signal Processing*, Volume 1, 97-100.

Regli, S.H. and Tremoulet, P.D. (2007), "The IDEAS Process: Interaction Design and Engineering for Advanced Systems." *APA Division 21, Division 19, & HFES Potomac Chapter Annual Symposium on Applied Research*. George Mason University, Fairfax, VA, March 1-2.

CHAPTER 72

Mobile Based Personalized Learning for People with Learning Disabilities

Saeed Zare, Heidi Schelhowe, Michael Lund

Digital Media in Education (dimeb)
TZI, University of Bremen
Bremen, Germany

ABSTRACT

Learning based on mobile technology can be seen as a bridge between theory and practical experiences, which inherits the advantages of e-learning and covers the restrictions of time and place of learning. This paper describes the main ideas and the architecture of an ongoing study on a learning system called Intelligent Mobile Learning Interaction System (IMLIS) that provides a mobile learning environment for people with learning disabilities. IMLIS is based on personalized learning strategies and feeds the learners with personalized contents, interactions and presentation formats in order to engage them for learning activities with new motivations. The suggested methodology defines a comprehensive learning process for the target group to support them in formal and informal learning. We apply knowledge from the field of research and practice with learning disabled people as well as pedagogical and didactical aspects in the design.

Keywords: Mobile Learning, Personalized Learning, Interaction Design, People with Learning Disabilities, Inclusive Design, Accessibility

653

INTRODUCTION

Learning process considers both social and personal sides. To address today's educational needs, analyze of didactical and educational aspects and trying to apply personalization on learning process is required. We integrated the idea of mobile technology and personalized learning in a dynamic context to focus specifically on the factors that reduce complexities and limitations in a learning process for people with learning disabilities and improve their learning performance.

Personalized learning is currently discussed according to develop of educational sectors in relation to the usage of digital media. Our concept of personalization is embedded in a learning context of a school (classes for students with special needs) and connected to self-regulated learning. As guidance and conduction during the learning process are important for our target group, these factors should be so adaptive that can foster to build self-regulation skills for the learners. Convergence offers the opportunity of ubiquitous learning "anytime, anywhere", so the learners should not wait for a fixed time for learning. The scope of this study consists of three areas of mobile technology, disability and education.

PEOPLE WITH LEARNING DISABILITIES

The first impression of normal people from disabled people is a person sitting on a wheelchair. But the persons on wheelchair belong to just one category of disabled people. According to the World Health Organization (WHO), a disability is defined as "Any restriction or lack (resulting from any impairment) of ability to perform an activity in the manner or within the range considered normal for a human being".

People who are physically disabled are "differently-abled", which means that they can adapt themselves in a way with limitations as well as environment. But people with mental disabilities can hardly adapt themselves with the situations; so they should be supported to be able to adapt themselves with the environment. Many institutions classified and categorized disabilities (e.g. ICF, ICD-10 from WHO). In this study, the target group is people with mental disabilities with focus on learning disabilities.

Meta-cognition is defined as learner's ability to manage and control his/her cognitive process (Wilson et al. 1993) and allows the transfer of tacit knowledge (Polanyi 1983) to explicit knowledge. Tacit knowledge is embodied in the personality and in order to communicate this knowledge it must be explainable that it can be put in words. For this process of making knowledge explicit, meta-cognition is needed to enable a higher level of abstraction. The meta-cognition extents the cognitive dimension of learning and prepare the individual for further ongoing learning experiences. People with learning disabilities have serious problems with meta-cognition because the step to reflect own action and to put this on an abstract level is often part of their disabilities. They need mostly guidance during the process of adapting explicit knowledge as well as process of translating tacit knowledge into language. Mobile learning offers bridges that can help to reduce the gap between tacit knowledge and language. On one hand the system take the role of a guide that moderate between both worlds that foster with its structure meta-cognitive processes and on the other hand mobile technologies connects contexts so that the translation between the two worlds become less abstract.

The learning process for people with learning disabilities should be carefully designed as they can be easily deflected from learning. Instructors should analyze the learning activities and interactions cautiously before implementing them in system. The analysis should be based on learner's abilities, synthetic skills, learning environment and system technical supports. Enough exercises, feedbacks, repetitions and emphasizes should be provided during learning process. Pitsch points to use sensory perceptions at the same time like simulations or touch and move (if applicable) as well as using the activities with words and pictures (Pitsch 2003). The completing time of every activity should be calculated, as they need mostly more time than normal people. The learning sessions should be adjusted in small sessions due to stamina and capacity of learners.

PERSONALIZED MOBILE LEARNING

MOBILE LEARNING

Mobile Learning is a widely accepted term for describing a learning process on mobile technologies. The new generations of mobile technology is trying to cover the disadvantages of previous generations and realize the optimum expectations from mobile technology in new contexts. Many studies believe that mobile learning is situated in the future of learning (Keegan 2005), (Sharples et al. 2007). Winters says "Mobile learning applications are best viewed as mediating tools in the learning process" and Walker continues with: "Mobile learning is not something that people do; learning is what people do. With technology getting smaller, more personal, ubiquitous, and powerful, it better supports a mobile society (...) Mobile learning is not just about learning using portable devices, but learning across contexts" (Sharples et al. 2007).

Mobile learning is not only an extended version of e-learning, but also it has the own characteristics and didactical methods. The specific quality of this learning activity is that the learner is not fixed to a certain predetermined location. The mobile activity is embedded in a didactical framework; a leading aspect of this framework is that mobile learning is adapting to context. By this context the social interaction becomes meaningful in the sense of cognition. Thus, not only social contexts but also relations to objects became an important part of the context. Mobile learning develops also new concepts of learning in contexts that enables to respect the individual needs of the learner in real life. Physical, mental, emotional, cognitive and cultural aspects influence these needs. The power of mobile learning lays in participation of a whole cycle of work processes that can be learned also in real contexts or contexts that fits to the needs of the user.

Currently, the focus in learning solutions is moving from learning objects to learning activities. The processes of design in recent learning approaches are emphasizing on integrating of learning contents with learning activities and interactions (Down 2007). In this way, learners are responsible for their own learning activities under unified structures in a framework.

Corbeil explains the following aspects based on Naismith hypothesis (Naismith et al. 2004), which believes that mobile technology has a huge impact on learning (Corbeil et al. 2007).

- Learning provides conditions for learners to create their learning by meaningful connections to learning contents as well as other learners.
- Learning force learners to organize and publish their conclusions, experiences and observations in their own learning process.
- Collaborative learning is enhancing day by day and can be supported by mobile solutions.
- Learning process is now in center of learner's environment more than traditional classrooms.
- Learners can easier recall and reflect the daily-life learning material as they are facing and capturing the daily life events.

The mentioned points bring enough pedagogical implications to reinforce the usage of mobile learning and its advancement.

PERSONALIZED LEARNING

Personalization considered as a principal in design of this study and points to the specific presentation of information to fit the individual learners. Personalization is tailoring of learning contents and interactions according to learner needs. Successful trainers and tutors in traditional learning had applied personalization unintentionally by differentiations in learner's attitude and behaviors.

The report of the teaching and learning, review group (2020Vision 2006) points to personalization as a moral purpose and social justice and says: "Put simply, personalizing learning and teaching means taking a highly structured and responsive approach to each child's and young person's learning, in order that all are able to progress, achieve and participate. It means strengthening the link between learning and teaching by engaging pupils (…)".

Traxler indicates to diversities, differences and individualities that can be recognized by personalized learning and adapted to the learner. It provides a productive and meaningful process for the learners to enhance their abilities according to their own autonomy (Ally 2009). In this process the learner becomes responsible for his/her own learning. An optimum personalized learning system should focus on important influence of the learner autonomy, self-motivation and self-management. The learner's autonomy can be highly increased in a personalized learning process, as learner has a good feeling regarding the personalized learning environment.

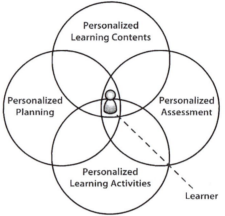

Figure 1. Learner's position in a personalized learning system

A personalized mobile learning system identifies the individualities and history of the learner and tries to provide the appropriate learning patterns, attributes and interactions based on learners abilities when and where the learner needs them. Personalized learning contains different key aspects that have direct affects of psychological factors like cognitive and social abilities of the learner. In non-personalized systems, learners may receive high leveled, redundant or irrelevant blocks of information, which causes a boring and unmotivated learning process.

METHODOLOGY

The methodological approach utilized to the review and analysis of both quantitative and qualitative aspects in this study. A huge part of approach is carried out on the empirical research based on experimental prospect through literature review on mobile technology, learning theories and previous research as well as interviews with experts and structured workshops with target group.

The approach was participatory design and different design stockholders are participated during the study in diverse times and supported us with feedback; in this way, the developmental process became an iterative process in design. As a learner-centered design method, we focused on learner's needs and abilities. In learner-centered design scenarios, the development of study growth with observations and information from the learner and enables a direct focus on learner's requirements during learning process. We also conducted different analytical workshops to observe our target group as well as their initial reactions and behaviors on working with mobile devices. With help of these workshops, we could analyze our basic requirements and highlight the impediments.

THE IMLIS SYSTEM

SYSTEM ARCHITECTURE

The IMLIS is a client-server based system and emphasizes on a dynamic structure for learning contents instead of a fixed-static structure. By adapting of learning contents and activities to the learners and their special abilities, the system should be able to prepare an appropriate profile or model for the learners, which can actuate the way for an personalized learning process in presenting and memorizing the learning contents. The system architecture consists of three parts, mobile clients, stationary server and teacher portal. The client can be any touch screen mobile device, which is connected via a wireless protocol to the server. The server is a standalone stationary device that manages, personalizes and feeds the mobile client with appropriate learning process. Teacher-portal is a gateway for uploading the learning contents and activities, as well as monitoring center for learner's behavior and personalization statistics.

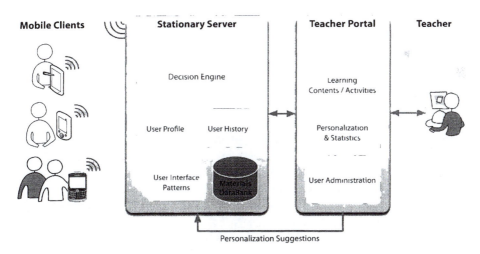

Figure 2. IMLIS system architecture

Figure 2 depicts the IMLIS architecture. The learning process in IMLIS starts with generating a personal profile based on strength and weaknesses of the learner. This profile can be generated with the help of tutors or parents with small analytical games, interactions and short questionnaire, which evaluate the level of personal development. The learner's profile is not fixed and can be updated during the learning process according to learner's behaviors and history. By gathering the analytical information from the server, the decision engine decides about the personalized appropriate learning contents and interactions based on the learner's profile criteria. The selected contents and interactions from learning material pool will be matched with the user interface patterns and will be sent to the mobile client. The mobile client devices can be different for the learners mostly according to their physical abilities, level of disability and preferences. As soon as receiving the learning packet by client, the learning process will start.

Teachers and instructors have the possibility via teacher-portal to upload, define, edit or administrate the predefined learning materials, lessons and interactions according to specific metadata. Each learning-asset or interaction is assigned to metadata and specifications according to their relations to learning abilities. Teacher-portal also provides the statistics from learner's behavior, history and progress via messages and learning curve graphs.

The user interface design parameters, which we included in our design considerations for mobile clients, are:

- Screens sizes of mobile devices
- The target group, people with learning disability
- Simplicity, to reduce learners irrelevant thinking load and fostering orientations
- Velocity of content loading, with some transforming methods

We have selected a very simple user interface, which is consisting of a simple self-expressed navigational "back" and "forward" buttons. In different interactions, we have used special buttons due to interaction's functionality.

PERSONALIZATION IN IMLIS

Personalization in IMLIS is based on a three-dimensional cubic according to level of knowledge, level of mental ability and the work tasks for the learners. The learning behavior of learners are recorded and analyzed during the process and informs the system to decide for appropriate learning process. During and at the end of every learning process, a series of feedback tests and interactions will be offered to the learner to verify the performance of the process. The result of these feedbacks will be synchronized at the same time with the server to update the profile criteria and to be applied for the next decisions on the same (type of) user. On the other hand, based on all learners' behaviors, system analyzes the learning assets and interactions in order to find weaknesses of the material for the appropriate group of learners and according to these analysis, system suggests the weaknesses and pitfalls to the teachers/experts whether the material and workflow should be adapted more to the specific needs of the learners.

The learning entity will be selected with respect to three dimensions mentioned above. Figure 3 depicts a comparison between other learning approaches and IMLIS approach for selecting an entity based on three-dimensional personalization cubic.

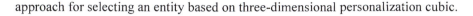

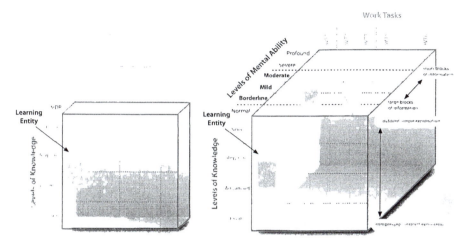

Figure 3. Two-dimensional learning approach in comparison with IMLIS three-dimensional learning strategy (adapted from Zare et al. 2009)

We considered the following parameters in our personalization approach to build a system that consist on technology, learner and teaching efforts.
- Adaptation of needs for the individual user and learning style
- The importance of personal well-being
- Motivating and encouraging design
- Orienting the learner through learning process
- Fostering competencies and self-confidence
- Supporting of self-control or self-awareness of own behavioral during the performance, by attention focusing and behavioral recording
- Modeling workflow according to the description of educational experts with respect to fortify perceived competence and self-directness of the learners

The quality and type of the personalized feedbacks can impact on learning quality. In some cases feedbacks can be combined with interactions to influence more on the learner. For our target group, feedbacks from system are extremely vital.

WORKSHOPS

During the study, we have conducted different workshops, which helped us to carry the practical side of our study beside theoretical side. The workshops have been done in three phases as pre-workshops, developmental workshops and post-workshops. The pre-workshops made the initial impressions of mobile learning clear. With the knowledge and experiences in pre-workshops, we could start implementation parallel with developmental workshops. The developmental workshops helped us not to be distracted from our goal and the post-workshops helped us to see the result of our approach in real context.

After finishing every workshop, we were able to evaluate our approach and identify the advantages and disadvantages of our solution. The overall results of outcomes from the questions made a clear overview of our system. In our workshops with target group, nearly always it became obvious that for most of them a mobile device represents to be connected to society. This symbolic representation has a strong effect on their motivation to learn with a mobile device. We recognized that nearly all knew something about mobile devices. Most of them have seen advertisements and knew pretty much about device brands. For them a mobile device symbolizes the world to get connected to. A lot of them had a private cell phone and they knew how to use them. In most cases they need a mobile to organize help if needed or to get useful information. Some could work with other functions of cell phones like SMS or using the camera. Encouragement and praise were the important issues to motivate them to stay in the process. We tried to give them immediate feedbacks after every favorable outcome to push them with positive reactions toward the goals. In many cases, we assumed a small effort as a success.

As our approach was participatory design, our participants were always informed about their role and impacts in design of the system. We validated continuously their comments and tried to implement them in our prototype.

CONCLUSIONS AND FUTURE WORKS

This research empowers the situation of people with learning disabilities and at the same time contributes the comprehension of the media specification of mobile technologies. The convergence from both aspects provides the concept of personalization. The situated and contextual learning offers possible use cases for guided learning in work places or for group works. Beside this core, system results gives contribution for a broader concept of barrier-free screen design and usability of mobile devices.

The strength of this system lays in functions of personalization. It adapts according the behavior of the learner. This potential can be extended in various aspects. The aspect to give the learner ownership or leadership in the process of personalization can be fostered and the system can support the work in teacher-portal by personalization. The profile of the learners and metadata of the material can be extended in multiple layers with other sub-criteria, that enable to decide when and under which conditions this criteria should be used. In teacher-portal, the teacher and administrator can monitor and analyze the learning activities of each individual learner. The teacher can also upload learning contents tagged with metadata according to criteria. The advanced learners can decide for taking leadership of their own learning process. According to personal profile, the system selects the appropriate material and supports the learner with selected learning process.

Currently, IMLIS prototype has been successfully implemented on different mobile devices to support personalization for promoting the learning abilities. Several workshops and experiments are conducted by our team and indicate that this system provides an efficient and effective personalized learning method for individual learners with learning disabilities.

REFERENCES

2020 Vision (2006). Report of the Teaching and Learning in 2020 Review Group. DfES Publications. ISBN: 978-1-84478-862-0.

Ally, Mohamed. (2009). Mobile Learning. Transforming the Delivery of Education and Training. Published by AU Press. Athabasca University. ISBN 978-1-897425-43-5

Corbeil, Joseph Rene. Valdes-Corbeil, Maria Elena. (2007). Are You Ready for Mobile Learning?. EDUCAUSE Quarterly: EQ Volume 30, Number 2.

Down, Kerry Ann. e-Learning Pedagogy programme. (2007). Bristol, UK. JISC Publication. http://www.jisc.ac.uk/whatwedo/programmes/elearningpedagogy.aspx

Keegan, Desmond. (2005). Mobile Learning: The Next Generation of Learning. Distance Educational International.

Naismith, Laura. Lonsdale, Peter. Vavoula, Giasemi. Sharples, Mike. (2004). Report 11: Literature Review in Mobile Technologies and Learning. FutureLab series. ISBN: 0-9548594-1-3. University of Birmingham.

Pitsch, Hans-Jürgen. (2003). How does a trainer working with the mentally disabled differ from any other teacher or trainer?: Agora XII. Training for mentally disabled people and their trainers. Permitting the mentally disabled a genuine and appropriate exercise of their rights. Cedefop Hrsg.: Thessaloniki, 6 July 2001. pp. 127-140. Luxembourg.

Polanyi, Michael. (1983). The Tacit Dimension. First published Doubleday & Co. 1966. Reprinted Mass.: Peter S.

Sharples, Mike. Walker, Kevin. Winters, Niall. et al. (2007). Big Issues in Mobile Learning: Report of a workshop by the Kaleidoscope Network of Excellence Mobile Learning Initiative. The Learning Sciences Research Institute. University of Nottingham.

Wilson, Brent. Jonassen, David H. Cole, Peggy. (1993). Cognitive Approaches to Instructional Design. New York: McGraw-Hill. http://carbon.ucdenver.edu/~bwilson

WHO - World Health Organization. (2010). http://www.who.int

Zare, Saeed. Krannich, Dennis. (2009). Concept and Design of a Mobile Learning Support System for Mentally Disabled People at Workplace. Paper presented at International Conference on E-Learning in the Workplace, ICELW. New York. USA.

CHAPTER 73

A Novel Input Method for Trepidant Users of Telemedical Services

Alexander Mertens, Nicole Jochems, Christopher M. Schlick [1]
Daniel Dünnebacke, Jan Henrik Dornberg [2]

[1]Chair and Institute of Industrial Engineering and Ergonomics
of RWTH Aachen University

[2]Research Institute for Operations Management (FIR)
at RWTH Aachen University

ABSTRACT

People with tremor find it often very difficult to use IT services and therefore attain very low levels of effectiveness, efficiency and satisfaction, when doing so. The currently available user interfaces do not guarantee sufficient precision for information input. Individuals suffering from intention tremor show a significant worsening when moving towards a button, precisely because target-oriented movements produce shaking in the affected body parts. For this target group a new method regarding the information input has been developed and evaluated. This technique enables the persons concerned to input information on a touch screen by using a wiping movement. Variations caused by the tremor are compensated with a continuous movement rather than a single direct movement towards a target field. Moreover, the screen surface causes a significant friction that helps to damp tremor forces. The user input can be identified by the computer with high accuracy by means of special heuristics which also assist barrier free access among the target group.

Keywords: HCI, Interaction Technique, Tremor, Touchscreen, AAL

INTRODUCTION

Electronic support systems and telemedical services are often the only measures handicapped people may take if they desire to maintain their self determining lifestyle (Eberspächer et al., 2006). Especially people affected by tremor are facing many difficulties using IT Systems, telemedical devices and other user interfaces (Mertens et al., 2009). If the interaction process requires information input by traditional input devices and standard graphical user interfaces (GUI), the average error rate is either very high or the input is not executable at all for the patient (Monesko et al., 2009). This leads to reduced efficiency, effectiveness and satisfaction of the user and adjunctive social isolation because of the hindered maintaining and establishing of contacts (Martínez-Martín, 1998).

IT-Systems within the eHealth sector are continuously increasing their usability and many user interfaces consider particular requirements of specific target groups (Korhonen et al., 2003; Sommerlatte, 2008). For the most part those considerations regard limitations of perception, cognition or the total musculoskeletal system (Akram et al., 2007; Kobsa et al., 2009; Rutgersson et al., 2007).

People suffering deficits in their fine motor skills generally do not have adequate support even though many theoretical concepts that address this problem exist. The few existing solutions require extra hardware and therefore reduce the overall mobility as well as they increase the cost factor.

A well known method of dealing with this problem is voice interaction, but this technology still requires a sustained time for calibration the system for each individual user (Keshet et al., 2009; Pfister et al., 2008). Also, the use of voice recognition is not feasible in many public situations, due to data security, privacy concerns or background noise.

New concepts as eye tracking (Duchowski, 2009; Gollücke, 2009) or brain-computer interfaces (BCI) (Dzaack et al., 2009) offer promising prospects. These still remain in their developing stage currently and most likely will not work without any additional hardware.

The target group for this concept includes primarily elderly and handicapped people, who only possess a limited degree of control for their upper extremities (including all forms of tremor), disregarding its cause, and different problems coordinate the hand-arm apparatus (Plumb et al., 2006). Due to the increasing likelihood with age for such a disease, the requirements of elderly people, operating electronic devices, are carefully considered (Raskin, 2005).

In addition to the integrative aspect for handicapped users, economic interests play an important role as well, especially in times of demographic change (AAL, 2006). The increasing number of elderly people that want an independent and self-determined life tends to demand capacities from the health care system that cannot be satisfied with common means of medical care. (Statistisches Bundesamt, 2008). The use within (tele)medical supply contexts (e.g. a telemedical platform monitoring vital functions) may significantly help to reduce the costs to the health care system. Especially in regions with a lower developed infrastructure, new potentials, regarding medical aftercare, occur for those who have previously been excluded from a self determined handling of IT-Solutions and telemedical services due to their tremor. The integrative appliance opens up new vistas to increase the overall quality of life (Wahl et al., 2004). In Germany, approximately 0.5-4% of the population below 65yrs and at least 5% of those above 65yrs are suffering from a tremor (Klaffke et al., 2009). Additional feasible scenarios for application are all interaction procedures with electronic systems via touch screen.

The touch screen may either act as a simple tool of information input and therefore help the user to prompt information with a high ratio of accuracy, or the touch screen may act as a information output as well and the user interface for input will be overlayed on top of the standard software only when interaction is necessary.

In addition to the prototypical implementation within a telemedical project, the authors plan the use e.g. with ATMs or ticket machines where a touch screen generally is available. No modifications to the hardware are required, only a software application, which analyses the user input impulses and transforms them into usable user interaction, needs to be installed. Additionally, a touch screen may be installed in a home scenario and used for communication, environmental control or general IT-supported work. For this also, standard technology along the hardware is sufficient.

METHODOLOGY

The deviation in accuracy entailed by restrictions of motor functions of people interacting with telemedical services when using a computer mouse is compensated with help of a common touch screen. The improvement of efficiency and satisfaction for the interaction of elderly people with computer-based systems via help of direct input devices was already proven (Schneider et al., 2007). The here presented concept adapts and evaluates the assumptions to decrease error probability and associated effectiveness for people suffering of tremor. Therefore the normal tangency of a touch screen is transformed to a swabbing interaction movement. Through this modification additional parameters like direction, speed, acceleration, starting point and drift can be detected which help to estimate the intended choice.

This is realized by not stopping the user's input-movement at the screen border but letting him swab beyond. The detection is solely on the touch screen-surface but because of the sustained movement the tremble does not augment while approximating the target and tracked parameters provide a basis for calculating higher accuracy. Additionally the frictional resistance during the whole interaction process damps the deviation even more and assists the user.

DESIGN PATTERN APPROACH

The interaction technique is formulated as a design pattern, whose concept has proven good outcomes in the domains of architecture, software engineering and human-computer interaction (Borchers, 2001). A special template for the application area of eHealth has been composed that respects the demands and requirements of the actors and scenarios (Mertens et al., 2009). The adoption of design patterns for the knowledge management helps to prevent the "reinvention" of solutions for specific problems that were already solved in this and in related domains.

DESIGN PATTERN: *TREMOR SWABING AWAY*

CONTEXT: People who are unable to perform exact movements, caused through kinetic tremor, often find it very unsatisfying to use electronic devices that are provided with "standard " user interfaces. This further increases the difficulty handling electronic devices when a lack of experience already exists and no satisfying input accuracy is achieved.

TARGET GROUP: This design pattern enables people who face problems coordinating exact movement, for information input via touchscreen. This design pattern is applicable for all characteristics of uncontrollable tremor and for people who have a general problem performing precise physical movements. The field of application is not conditioned by the age, but since the probability to be afflicted with tremor increases significantly with rising age, focus is on elderly people. For application of the interaction technique, it is not necessary to involve any medical staff.

USAGE SITE: This design pattern is usable at any location where the user is allowed to use electronic devices autonomously. This is particularly true in most daily life scenarios and many areas of prevention and rehab. For scenarios requiring a high security of input, the system is not usable as a user may not be controlled on the validity of his entries. Implementations in a clinic environment need to consider that the used systems must not emit any restricted radiation and offers a feasible security check when private inputs are demanded.

PROBLEM: Generally the problem of inaccurate inputs from the user may be compensated through an uncommonly large input box on the screen, in order to compensate any tremor movement. This method, however, is limited to medium occurrences of a tremor and will find problems when facing strong tremor symptoms, simply because every screen has a limited size. For a high number of options, this leaves the developer with the only choice of decreasing the box size and therefore unavoidably an increase in the number of wrong inputs.

SOLUTION: In order to make the preferred technical devices accessible to the described target group, the existing user surface is virtually enlarged to guarantee a big enough screen for the actual input. The principal behind this is based on Fitts' Law (Fitts, 1954; MacKenzie et al., 1992), as the virtual depth of the control elements is increased by several degrees. This is realized with a wiping technique which allows the user to move beyond the physical borders of the touch screen. The measure only takes place on the screen; however, with an adjustment towards a floating or wiping movement, the relevant parameters as orientation, speed and

666

starting point may be determined, so that the targeted control unit can be judged with a significantly higher accuracy. The desired selection on the screen is interpolated with a balancing curve. By doing so, the starting point may be chosen freely, as the determined direction and covered way indicate the intended target and are autonomous from other partitions that were crossed with the fingertip. A determination by integrating the touched areas or even pixels would force the user to start his movement in the center. Affirmation or emendations can be accomplished with the help of very simple gestures like swabbing with a clockwise or anti-clockwise direction. An additional increase in precision is given through the continuous contact of the finger with the screen, as the friction coefficient serves as a damping effect on the symptoms of a tremor.

LIMITATIONS: A use of this design pattern is only given if the patient has a high level of understanding of the standard interaction processes, symbols, numbers and letters used. As the described pattern is mainly an assistance for compensating physical disruptions, a use makes only sense among those people, who still need to have a certain degree of mobility in order to reach towards the screen. Unwanted effects with other third-party systems may be generally excluded since the interaction technique is only an add-on to the software layer. A disproportional physical strain when handling this technology is not expected if the user uses it according to its purpose.

ILLUSTRATION:

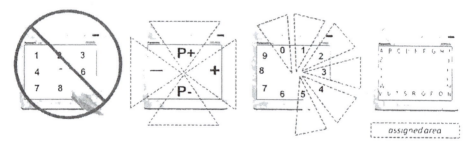

Figure 1. Interaction technique TRABING for a virtual numeric keypad [1]

DIRECTIVES: The guidelines and standards that need to be obeyed for this design pattern are mainly dependant on the purpose and classification of the used system. In accordance with the Medical Device Directive (MDD) in Germany the used software must be declared and evaluated as a medical product, a corresponding risk evaluation must be undertaken in order to verify its conformity among the CE classification. Principally the design pattern supports the requirements towards a barrier-free system design and the different standards for usability, e.g. ISO 13407, ISO 14915 or DIN EN 62366.

EVALUATION

The concept has been developed within a scientific project about establishing a telemedical platform for monitoring of elderly cardiovascular patients during homely rehab (Mertens at al., 2009). As the system offers access to aligned value added services, an adequate interaction technique for all users had to be assured, to allow self-reliant access. A first evaluation showed high acceptance from the subjects, a learning curve with swift practice effects and a significant improvement of the input procedure for all tremor patients. The testing was performed with 15 seniors, 9 female and 6 male between 56 and 91 years (average age: 72.8 years, SD=6.9) who were afflicted with diverse tremor characteristics. The causation and the origin of the affection were not included, as no ethics vote for a retrospective analysis was consented. In a related survey prior to the trial the subjective attitude towards computers for seven dimensions (Comfort, Efficacy, Gender Equality, Control, Dehumanization, Interest and Utility) was ascertained (see Figure 2). According to this, only two of the participants indicated a daily computer use and just four of them have a PC in their home. The appraisal was accomplished with help of 15 questions that had to be answered by dint of a Likert scaling (Gina et al., 1992).

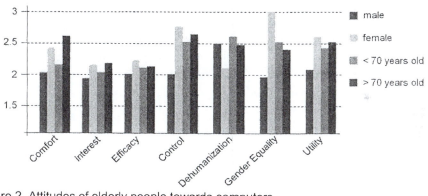

Figure 2. Attitudes of elderly people towards computers

The used apparatus for conducting the trials was a Panasonic Toughbook H1 Mobile Clinical Assistant (MCA) ® with a 10.4" resistive touch display and a resolution of 1024 x 768 pixels.

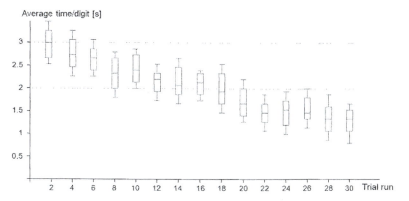

Figure 3. Training curve of 15 elderly users applying TRABING. The average time for entering one digit of a 20-digit number is plotted for the 30 trials/person

The manipulation was solely done by finger touch, no pen was used (see Figure 1). The implemented test cases showed ten numeric digits (0-9) that were arranged equidistant in a circle near to the boundaries of the device. Each trial a new randomized 20-digit number was shown in the middle of the screen and the subjects had to select the numeric digit in strict rotation. The aggregated time for entering a complete 20-digit number was measured, here wrong selections were ignored. The error ratio was regarded by the extended time that was necessary to process the complete number.

The measured execution times show a high stability after the 20th trial for most of the probands (see Figure 3). It turned out that after a short phase of orientation (trial 1-6) the learning process took place (trial 7-21) and after that stable phase regarding the time for entering a digit was attained. The respective average time mainly depends on the degree of muscle tremor, the computer literacy and associated attitude towards the use of computers (see Table 1).

Table 1: Correlation of the seven dimensions: Attitudes towards computers of older adults

	Comfort	Interest	Efficacy	Control	Dehumanization	Gender Equality	Utility
Comfort	1						
Interest	.777	1					
Efficacy	.798	.388	1				
Control	.866	.55	.931	1			
Dehumanization	-.075	-.259	-.121	-.151	1		
Gender Equality	-.021	.142	-.014	.169	-.777	1	
Utility	.532	.657	.575	.627	-.17	-.01	1

cursive: significant at α = 0.05; **bold**: significant at α = 0.01

FUTURE TOPICS

Actually additional parameters, like surface morphology, pen vs. finger, number of parallel symbols (character, numeric) as well as custom-built algorithms for analyzing the tracking data are evaluated to determine the impact on the dependent variables (time, error ration and user satisfaction).

The application of linguistic heuristics is expected to even increase the recall factor as the probability for inputs can be considered. The angle of the sector for each screen compartment may be adjusted flexibly to the input alphabet with help of formula (1).

$$A(s) = \left(\frac{180°}{|I|}\right) + (180° \cdot p(s))$$ (1)

In equation (1) $A()$ determines the recommended angle,

s is the symbol,

I is the sample space and

$p()$ is the probability for occurrence (see Table 2).

The formula (1) assures that for every symbol a lower bound for the partition size is maintained and therefore all symbols stay attainable.

An even better adjustment for a specific sample space can be achieved if sequences are considered and impossible choices are determined based on former input.

Table 2: Frequency of Occurrence for German Alphabet

A	06,51 %	H	04,76 %	O	02,51 %	V	00,67 %
B	01,89 %	I	07,55 %	P	00,79 %	W	01,89 %
C	03,06 %	J	00,27 %	Q	00,02 %	Y	00,03 %
D	05,08 %	K	01,21 %	R	07,00 %	Y	00,04 %
E	17,40 %	L	03,44 %	S	07,27 %	Z	01,13 %
F	01,66 %	M	02,53 %	T	06,15 %	ß	00,31%
G	03,01 %	N	09,78 %	U	04,35 %		

In case of any input situations where many options are required, these may be designated to the same input area, similar to the T9 method used with mobile phones. According to the actual tremors deviation an ideal compartment size may be identified, allowing a maximum efficiency in the relationship between the error rate and number of simultaneous options.

CONCLUSION

In summary, this paper reports on three contributions:

(1) Requirement analysis for people with tremor and development of a novel interaction input technique for relevant scenarios which currently are not provided with sufficient solutions. For this, the economic influences coming with demographic change is given special regard with new access to supply potentials of medical care.

(2) Specifications and refinement for the implementation of *TRABING*.

Identification of relevant parameters for an efficient use among the target group mentioned at (1) and following evaluation.

(3) Semi-formal specification with help of design patterns for easier perusal and comprehension. This concept is part of the groundwork for establishing a pattern language for the domain of eHealth. Herewith the formalizing, structuring, propagation and acquisition of expertise for novices as well as mavens can be facilitated.

REFERENCES

Akram, W. Tiberii, L. and Betke, M. (2007) A Customizable Camera-Based Human Computer Interaction System Allowing People with Disabilities Autonomous Hands-Free Navigation of Multiple Computing Tasks; Springer.

Ambient Assisted Living – European Overview Report: Europe Is Facing a Demographic Challenge: Ambient Assisted Living Offers Solutions, VDI/VDE 2006.

Borchers, J. (2001) A pattern approach to interaction design. Chichester: Wiley.

Duchowski, A. T. (2009) Eye Tracking Methodology: Theory and Practice, Springer, Berlin, 2nd ed.

Dzaack J. Zander T.O. Vilimek R. Trösterer S. and Rötting M. (2009). Brain activity and eye-movements: Multimodal interaction in human-machine systems. In A. Lichtenstein, C. Stößel & C. Clemens (Hrsg.), Der Mensch im Mittelpunkt technischer Systeme. 8. Berliner Werkstatt - Mensch-Maschine-Systeme, 7. - 9. Okt. 2009, S. 189-190 und 464-469 [CD], Düssedorf: VDI Verlag,

Eberspächer, J. and Reden, J. von, ed. (2006) Umhegt oder abhängig: Der Mensch in einer digitalen Umgebung. Springer, Heidelberg.

Fitts, Paul. M. (1954) The information capacity of the human motor system in controlling the amplitude of movement. Journal of Experimental Psychology, volume 47, number 6.

Gina, M.J. and Sherry, L. W. (1992) Influence of Direct Computer Experience on Older Adults′ Attitudes Toward Computers, In: Journal of Gerontology: Psychological Science, Vol. 47, no 4, pp. 250-257.

Gollücke, V. (2009) Eye-Tracking - Grundlagen, Technologien und Anwendungsgebiete, Grin Verlag.

Keshet, J. (2009) Bengio, S.: Automatic Speech and Speaker Recognition: Large Margin and Kernel Methods, John Wiley & Sons.

Klaffke, S. and Trottenberg, T. (2009) Essentieller Tremor, www.Charite-Berlin.de.

Kobsa, A. and Stephanidis, C. (2009) Adaptable and Adaptive Information Access for All Users, Including Disabled and Elderly People; Pittsburgh, USA.

Korhonen, P. and Paavilainen, A. (2003) Application of ubiquitous computing technologies for support of independent living of the elderly in real life settings; Ubicomp 2003 UbiHealth workshop.

MacKenzie, I.S. and Buxton, W. (1992) Extending Fitts′ Law to two-dimensional tasks. In: Proceedings of the SIGCHI Conference on Human Factors in Computing Systems 1992, P. Bauersfeld, J. Bennett, G. Lynch, Eds. ACM, New York.

Mertens, A. Dünnebacke, D. Kausch, B. Laing, P. and Schlick, C. M. (2009) Innovation of homely rehab with help of telemedical services. In: IFMBE Proceedings of World Congress on Medical Physics and Biomedical Engineering, Munich, Springer.

Monekosso, D. Remangnino, P. and Kuno, Y., ed. (2009): Intelligent Enviroments – Methods, algorithms and applications. London: Springer.

Martínez-Martín, P. (1998): An introduction to the concept of "quality of life in Parkinson's disease" In: Journal of Neurology, Springer, Berlin.

Pfister, B. Kaufmann, T. (2008) Sprachverarbeitung: Grundlagen und Methoden der Sprachsynthese und Spracherkennung, Springer, Berlin.

Plumb, M. Bain, P. (2006) Essential Tremor: The Facts, Oxford Univ. Pr.

Raskin, J. (2005) The Humane Interface, New Directions for Designing Interactive Systems.Addison –Wesley.

Rutgersson, S. Arvola, M. (2007) User Interfaces for Persons with Deafblindness, LNCS 4397, Springer.

Schneider, N. Schreiber, S. Wilkes, J. Grandt, M. and Schlick, C. (2007) Investigation of Adaptation Dimensions for Age-Differentiated Human-Computer Interfaces, In: Universal Access in HCI, Part I, HCII 2007, 12th International Conference on Human-Computer Interaction ,Beijing, China, Hrsg.: Stephanidis, C., Springer, Berlin.

Sommerlatte, T. (2008) Technikgestaltung aus Sicht des Nutzers, Digitale Visionen, Springer, Berlin Heidelberg.

Statistisches Bundesamt, ed. (2008) Demografischer Wandel in Deutschland - Heft 1: Bevölkerungs- und Haushaltsentwicklung im Bund und in den Ländern, Onlineveröffentlichung.

Wahl, H. and Naegele, G. (2004) Zukunft des Alters in einer alternden Gesellschaft: Szenarien jenseits von Ökonomie und Demografie. Sozialer Fortschritt, 53 (2004) 11-12.

CHAPTER 74

Using Goal Directed Task Analysis to Identify Situation Awareness Requirements of Advanced Paramedics

Harris Abd Hamid, Patrick Waterson

Department of Ergonomics (Human Sciences)
Loughborough University
Loughborough, LE11 3TU England

ABSTRACT

Advanced paramedics, known in the UK as emergency care practitioners (ECPs) are employed by ambulance services to meet the demand for urgent care in the community. The role has been evaluated in terms of patient outcomes. To further understand and evaluate the role, Goal Directed Task Analysis (GDTA) was conducted. Based on the analysis, situation awareness requirements among ECPs were identified. The results of the GDTA are presented and their implication discussed in terms of goal hierarchy structure, patient as source of information, and sociotechnical system. We discuss recommendations for the role with regard to the ECPs' SA. Further SA studies to evaluate and develop the ECP role are also outlined.

Keywords: Situation Awareness, Goal Directed Task Analysis, Decision Making, Pre-hospital Care, Socio-technical Systems

INTRODUCTION

In the UK, advanced paramedic, known as emergency care practitioner (ECP), is a work role designed and introduced to the Ambulance Services to lighten the burden on ambulance and hospital resources. Not all 999 calls require the use of ambulance to transport patients to hospital. The ECP role was introduced to meet "the urgent care needs of patients by providing the right skill at the right time in the right place" (Department of Health, 2004, p. 2). As practitioners providing pre-hospital care in the community, the ECPs have more autonomy and freedom to make decisions compared to other ambulance crews. They are able to treat patient on scene and refer patients to other care pathway including the hospital (emergency department, ward), other healthcare services (out-of-hour services, general practitioner, bed bureau), and other services (social, psychiatry).

By providing pre-hospital care in the community, ECPs can reduce the number of unnecessary patient transport to the hospital. The avoidance of hospital attendance can free up resources of the ambulance services as well as the hospitals. Nonetheless, ECPs are not exclusively assigned cases that can be treated on location. Because they are part of the pool of ambulance services' human resource, they are also being deployed to respond to emergency cases. Therefore, for any given patient, the need to treat patients in the community or somewhere else is not always apparent and decided at the start.

Available evidence shows that there are significant avoidance of patient transports and admissions to hospital (Mason et al., 2007). These patient outcomes are in line with the role's objectives. However, it is not clear what the factors that influence these outcomes are. Given the autonomy of the role, the decisions made by the ECPs are important to consider. The cognitive aspect of ECP, especially their decision-making, needs to be examined to gain a better and more comprehensive understanding of this role.

To support decision-making in domains as complex and safety-critical as healthcare, situation awareness (SA) has increasingly receive attention from researchers. SA is defined as "...the perception of the elements in the environment within a volume of time and space, the comprehension of their meaning, and the projection of their status in the near future" (Endsley, 1998, p.97). Based on this definition, Goal Directed Task Analysis (GDTA) was introduced as a method to identify SA requirements (Endsley et al, 2003). This method is used to identify the goals and the data of which the operators require to be aware. This method is particularly useful to create a user-centred system.

GDTA has been applied in various domains including healthcare (Wright et al., 2004), military (Jones et al., 2003), and power transmission and distribution industry (Connors et al., 2007). More recently, the GDTA was used to develop a protocol (Goal Directed Information Analysis) to identify information requirements of emergency first responders in the fire and rescue service (Prasanna et al., 2009). In their study, Prasanna et al developed the new protocol to overcome the constraints of GTDA. However, the protocol produced goal-decision-information hierarchy which is similar to the hierarchy produced using GDTA.

For this study, the GDTA which is based on Endsley's definition of SA is used. The GDTA is applied in a novel setting namely pre-hospital. By identifying the SA requirements of ECPs, the role can be further understood. This understanding can help to evaluate how the role is developing against its original aims and objectives. Additionally, in a recent development, electronic patient records are being introduced to an Ambulance Services in England. It is envisaged that an analysis of the ECPs' SA can provide input for the integration of this information technology into the existing work system.

An analysis of the existing literature reveals gaps in method applications and work practices. The GDTA as a method to analyse cognitive task has not been reported for pre-hospital care domain. On the practical side, the ECP role has not been investigated using a cognitive approach. It is recognised that the ECP role need to be further developed and designed especially in light of the introduction of new information technology. Towards fulfilling this need, we address the identified gaps in the literature. For this paper, the objective is to describe an application of GDTA in the domain of pre-hospital care. The second objective is to discuss the SA requirements of advanced paramedics within a sociotechnical system.

METHOD

The participants in this study were recruited at one Ambulance Services NHS (National Health Services) Trust in the England. Prior to data collection, ethics approval was obtained from the NHS. The participants were recruited through the Director of Operations of the Ambulance Services. Documentations (job description, organisational chart, work guidelines) about the ECP work role from the Ambulance Services were obtained to understand the context of the work role. In addition, observations of ECPs on the job were conducted through ride-out sessions with two ECPs during their shifts. The ride-out sessions involved meeting the ECPs at their base (ambulance station) and following them to treat patients at different locations. The ride out sessions were carried out in four sessions covering 40 hours. The documentations and observations form the basis of the interviews to discover the goal structures of the work role.

Fifteen ECPs (including one from the ride-out session) and three ambulance dispatch staff were interviewed using semi-structured interview schedules. The ambulance dispatch staff were considered as subject matter experts due to their knowledge of the different pre-hospital care personnel. Moreover, they provide a view of the ECP job from a control room point of view where the decisions on the assignments of 999 calls are made.

The ambulance dispatch staff and the ECPs work full-time with the Ambulance Services and had working experience ranging from 1 to 6.5 years. All except one interview were conducted during their regular working hours. The interviews were used to discover the physical tasks, and later, the goals associated with each tasks. Due to time restrictions imposed by an unpredictable work pace, each ECP was interviewed once. The tasks and goals structure was validated by each successive ECP to produce the hierarchy of goals, decisions, and information requirements.

RESULTS AND IMPLICATIONS

As with other types of cognitive task analysis, GDTA can generate a huge amount of detailed information. This section presents selected portions of the information obtained from the GDTA. These portions are selected to aid the presentation of implications and recommendations for the ECP role.

GOAL HIERARCHY

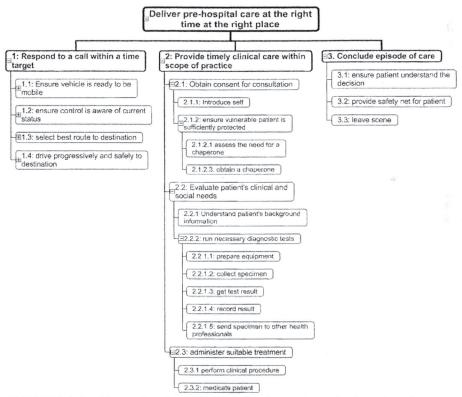

FIGURE 1.1 Goal hierarchy of an ECP showing the goals and sub-goals only.

The scope for the task analysis covers activities of responding to a call to discharging the patient (Figure 1.1). Both emergency (life-threatening) and urgent (non-life threatening) patient care were considered. The main goal for the ECPs is derived from the document published by the Department of Health (2004). The goal of "delivering pre-hospital care at the right time at the right place" runs in close

676

parallel with the mission statement of the Ambulance Services. However, this goal seem to be more directly applicable with ECPs than other ambulance crews (paramedics, ambulance technicians) whose main goals are more towards transporting patients to an appointed care environment.

PATIENTS AS SOURCE INFORMATION

The information that the ECP receive from the ambulance dispatcher room are usually brief. The information that patients provide to the call taker or triage nurse at the ambulance dispatcher room will be condensed before relayed to the ECPs. They would receive a code representing the general condition of the patient (based on a computerised triage system), age, sex, and address (postcode) on their mobile data terminal that is installed at their vehicle. Therefore, the patients themselves are important source of information to understand what kind of medical attention that they need.

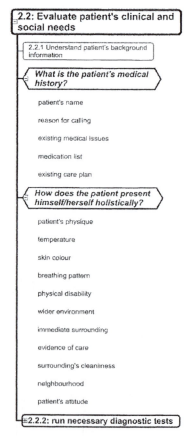

FIGURE 1.2 Information requirements for evaluating a patient's need.

The interaction with the patients would help to unpack the initial information that the ECPs received from the ambulance dispatchers. The information can be obtained through direct interaction with the patient or through observation. Evaluating a patient's clinical and social needs (Figure 1.2) involves both ways of gathering information. The ECPs need to get directly the information from the patient or an accompanying person regarding the patient's medical history. In contrast, the holistic evaluation of the patient involves observation that may or may not be supplemented by direct verbal report from the patient. The holistic evaluation is important to make a decision on the best care pathway for the patient.

The information that provide a picture of underlying physical conditions can be obtained by running diagnostic tests (sub-goal 2.2.2) that may involve tools like ECG machine (recording electrical activity of heart muscles) and oximeter (measuring oxygen saturation). These tools provide information via printed and electronic display. In this light, the patients and their immediate surrounding can be thought of as *displays* from which the ECP get vital information. The patients are a rich source of information including visual (skin colour), auditory (breathing sound), and thermal (skin temperature). Their display can change during the consultation time. Of great importance is sudden deterioration of the patient. This highly dynamic situation requires the ECPs to regularly update their SA. The diagnostic tools can be combined with the information obtained from the patient to create a better understanding and projection of the likely patient conditions.

Having patient as an integral part of *information display* poses challenges to the ECPs for building and maintaining accurate SA. Endsley et al. (2003) outlined major threats to SA such as attentional tunnelling, workload, data overload, and misplaced salience. These threats are applicable to the interaction with technology as well the interaction with humans. The field of social psychology can provide further insight into the threats to the SA of ECPs. Our understanding of the world via social interaction can be distorted by priming effect of the schema, confirmation bias, stereotyping, and other reasoning biases. More recently, Rahman (2009) explored the role of affect in naturalistic decision making.

SOCIOTECHNICAL SYSTEM

The GDTA generates information requirements that are technology and process independent. Useful information gained from this analysis can be considered for the integration of electronic patient record (EPR) into the work of the ECPs. EPR can support the goal of evaluating clinical and social needs (goal 2.2). Not only the EPR can offer a source of reliable information, the timeliness of the information can be improved. The EPR can potentially start feeding information to the ECP even before they arrive at the patient's location. This advantage can overcome the limitations of current technology that represents the micro-level influences to the ECP work.

The GDTA also highlight the clarity of the ECP role. The ECP are expected to provide pre-hospital care at the right time at the right place using the right skills.

Currently, the emphasis is on the right time: the nationally set response time target is used to measure their performance. This macro-level influence place emphasis on part of the goal hierarchy that may not be shared by the ECPs themselves. With a time-based performance assessment, the all important goal is goal 1 (respond to a call within a time target). In contrast, with a clinical-outcome performance assessment, the emphasis will be on goal 3 (conclude episode of care) where an appropriate safety netting are likely to meet the patient's clinical and social needs. There are calls to review the assessment of ECPs performance which is currently based on time-target to clinical outcome-based (Gray and Walker, 2008).

CONCLUSION AND FUTURE WORKS

This paper presented the application of GDTA to identify SA requirements of advanced paramedics. The GDTA is expected to be a valuable input for fully implementing the EPR in the Ambulance Services. More important, the GDTA is found to be useful to understand the cognitive aspect of the ECP job. Not only the micro level of the work system is identified as influencing SA, the macro level influence is also recognised. The information generated by the GDTA is valuable to address and consider the cognitive, technical, and social aspects of a work system.

The work reported here is based on a limited number of participants due to staffing constraints. The range of clinical cases that can be observed was also limited because of practical constraints. The occurrence of the more dynamic emergency cases (relative to the urgent cases) cannot be predicted. The range of cases observed during the ride out sessions was not as diverse as hoped for. Nonetheless, the observations had proved useful in producing a draft of the goal hierarchy.

Further works are proposed to follow up the present study. Case studies involving observations and ride-out sessions will be conducted to further validate the extant GDTA. The case studies are hoped to be able to overcome the limited range of cases observed in this present study. A measure of SA can be developed based on the SA requirements identified. These measures can be used to evaluate how much the EPR support, or impede, the ECP's SA. The third line of investigation that can be pursued is validating the source of information available to the ECPs. Not only the source should be identified, the timeliness and reliability should also be measured. This validation can reveal what additional advantages can be offered by the EPR system. The role of ECP has much to offer in delivering unscheduled care in the community. A comprehensive evaluation would assist in developing the role to maximise its potentials.

Acknowledgement

The study is supported by a scholarship from the International Islamic University Malaysia and Ministry of Higher Education Malaysia to the first author.

REFERENCES

Connors, E.S., Endsley, M.R., and Jones, L. (2007), Situation Awareness in the Power Transmission and Distribution Industry. *Human Factors and Ergonomics Society Annual Meeting Proceedings, Cognitive Engineering and Decision Making*, pp. 215-219(5)

Department of Health. (2004), *The Emergency Care Practitioner Report*. London: COI Communications for the Department of Health.

Endsley, M.R. (1988), Design and evaluating for situation awareness enhancement. *Proceedings of the Human Factors Society 32nd Annual Meeting*, Santa Monica, CA: Human Factors Society, 97-101.

Endsley, M.R., Bolte, B., and Jones, D.G. (2003), *Designing for Situation Awareness: An Approach to Human-Centred Design*. Taylor & Francis: London.

Gray J,, and Walker, G. (2008), AMPDS categories: are they an appropriate method to select cases for extended role ambulance practitioners? *Emergency Medicine Journal*, 25, 9, 601-603.

Jones, D.G., Bolstad, C.A., Riley, J.M, and Endsley, M.R., and Shattuck, L. (2003), *Situation Awareness Requirements For The Future Objective Force*. Paper presensented at the Collaborative Technology Alliances Symposium April 29th – May 1st 2003 College Park, MD.

Mason, S., O'Keeffe, C., Coleman., P., Edlin, R., Nicholl, J. (2007), Effectiveness of emergency care practitioners working within existing emergency service models of care. *Emergency Medicine Journal*, 24, 4, 239-243.

Prasanna, R., Yang, L., and King, M. (2009), GDIA: a Cognitive Task Analysis Protocol to Capture the Information Requirements of Emergency First Responders. In J. Landgren & S. Jul. *Proceedings of the 6th International ISCRAM Conference* – Gothenburg, Sweden, May 2009. Retrieved from http://www.iscram.org/ISCRAM2009/papers/ on 15 February 2010.

Rahman, M. (2009), Understanding naturalistic decision making under life threatening conditions. In B.L.W. Wong and N.A. Stanton. *Proceedings of NDM9, the 9th International Conference on Naturalistic Decision Making*, London, UK, June 2009: British Computer Society.

Riley, J.M., and Endsley, M.R. (2004), Situation Awareness in HRI with Collaborating Remotely Piloted Vehicles. In *Human Factors and Ergonomics Society Annual Meeting Proceedings, Cognitive Engineering and Decision Making*, pp. 407-411(5).

680

Wright, M.C., Taekman, J.M., and Endsley, M.R. (2004), Objective measures of situation awareness in a simulated medical environment, *Quality & Safety in Health Care*, 13:i65-i71.

Chapter 75

Integrating Ergonomics in an Environment of Care: Application of the NIOSH Model Ergonomic Program in a Healthcare Setting

Manny Halpern

Occupational & Industrial Orthopaedic Center
NYU Hospital for Joint Diseases
63 Downing Street, New York, NY 10014

ABSTRACT

An Environment of Care can be defined as the experience of a user with the healthcare delivery system. The Agency for Healthcare Research & Quality (AHRQ) states that quality in health care means doing the right thing at the right time in the right way for the right person and having the best results possible. To anchor ergonomics in this context we need to examine where the ergonomics program contributes to the performance of the healthcare delivery system. The National Institute for Occupational Safety & Health (NIOSH) published guidelines for setting up a good ergonomics program focused on prevention of the work-related musculoskeletal disorders in the manufacturing and service industries. This pilot study applied the NIOSH model in a hospital and attempted to examine its relevance to healthcare settings.

With few recordable musculoskeletal injuries in the organization, the program was set up in a pathology department where management was willing to collaborate rather than where a need was urgent. The program was therefore set as a standalone and adapted to address ergonomic issues related to computer terminals in the lab and offices. A task force was set up and assessed the design of the computer workstations using a standard checklist, musculoskeletal symptoms were surveyed and a series of focus groups and training sessions in Healthy Computing were conducted for labor and management. Products were selected to address the ergonomic deficiencies noted in the design of the workstations. The action steps that could be implemented during the project were compared against the guidelines of the NIOSH model ergonomics program. The pilot study revealed that of 41 items listed in NIOSH model, 30 (73%) were applicable to the hospital environment. However, the ergonomic concerns in healthcare go beyond musculoskeletal injuries of the care providers and the NIOSH guidelines do not address ergonomic concerns related to patient safety. Other criteria may have to serve for selecting an intervention site and assessing the impact of the program. Three conclusions can be drawn about integrating ergonomics in an Environment of Care: 1) the impact of the ergonomic program may not be manifested in OSHA logs but in medical insurance data; 2) to make the ergonomic program more effective, link it to patient safety and risk management; 3) more research is needed to link the ergonomic concerns of both employees and patients in a healthcare setting. Out of 14 dimensions of healthcare performance listed by the Organization for Economic Cooperation & Development (OECD), this pilot study showed that at least 4 may be impacted by an ergonomics program: environment & amenities, efficiency, responsiveness to the needs of the users, and safety.

Keywords: Ergonomics, Healthcare, Quality of Care, Safety, Performance

INTRODUCTION

With a growing interest in evidence-based design of healthcare facilities, the American Institute of Architects (AIA) defined the notion of Environment of Care as the experience of a user with the healthcare delivery system (AIA, 2010). This experience is a function of six factors: the *concepts* driving the delivery of care, the *people* using the facility, organizational, technological and building *systems*, *layout*, the *physical environment* such as light and way finding, and an *implementation* process that includes the stakeholders. These factors constitute the functional requirements of the healthcare environment. To integrate an ergonomic program in an Environment of Care, we need to consider these functional requirements.

Guidelines on how to design ergonomic programs have been available for some time in the manufacturing and service industries. However, the challenge is to adapt these guidelines to the healthcare services. A pilot project was initiated to apply the NIOSH model for setting up an ergonomic program for prevention of work-related musculoskeletal disorders (WMSD) and examine its relevance to a particular department at a specialty hospital. The aim was to identify who can participate, how, and what measures might be feasible. The program could serve as a model for other departments in the organization, with the results benefiting also other healthcare facilities.

The pilot project was conceived as a standalone, rather than a part of a hospital-wide health and safety program. With few recordable musculoskeletal injuries, the data of the hospital could not help in implementing the first step of the program: identify targets for intervention. Other criteria were needed upfront to select an intervention site: management willingness to host the program and a manageable staff size. Pathology & Laboratories met these criteria. The department had 28 employees, the majority (about 68%) consisted of technical staff (lab technicians and pathologists), and the rest administrative or clerical. Since all staff was increasingly using Video Display Terminals (VDT), the initial focus of the ergonomic program was on Healthy Computing. However, it is the process and the impact on research and practice that are the focus of this report as these may guide us in how to integrate the ergonomic program with the functional requirements of an Environment of Care.

METHOD

The NIOSH program consists of seven steps (NIOSH, 1997). A review of the content identified 41 action items. These served as the basis for the actions needed to be taken in the pilot project.

Instead of injury records, a symptoms survey (Goldsheyder et al, 2002) was distributed during three kick-off meetings, which introduced the project to the staff and functioned as focus groups for eliciting ergonomic issues encountered by the staff.

The office manager assigned two staff members – a supervisor and a lab technician - to an ergonomic task force that would assist the investigator. The office manager supervised and supported the activities undertaken.

A VDT checklist - *Performance Oriented Ergonomic Checklist for Computer (VDT) Workstations* (Cornel University, 1996) - served to assess the design of all 19 computer workstations. The investigator completed the checklist; a subset was assessed by one of the task force members and compared to the findings of the investigator. These results served as context-specific examples in four one-hour Healthy Computing training sessions. A walk-through conducted a month after the training sessions attempted to note layout changes undertaken by the staff.

Based on the checklist, the investigator and the task force developed a list of features that addressed the design deficiencies noted in the facility. Office products were identified, which met the requirement to alleviate risk factors related to musculoskeletal disorders. These products were then prioritized based on the impact on the number of workstations and the tasks that take place. The products were selected from catalogs of vendors doing business with the hospital and recommended as guidelines for Purchasing.

RESULTS

The action steps that could be implemented during the project were compared against the items in the NIOSH model program. The seven steps of the NIOSH program are listed in Table 1, indicating the actions that were implemented in each of the steps of the model program. The pilot study revealed that of 41 items listed in NIOSH model ergonomics program, 30 (73%) were applicable to the hospital environment.

Table 1. A summary of the application of the NIOSH ergonomic program at a department in a specialty hospital

NIOSH program	Pilot program
1. Looking for signs of potential musculoskeletal problems at work	None available (see Step 4) ➔OSHA logs at the hospital do not reflect prevalence or incidence of work-related musculoskeletal disorders. Management willingness was the criterion that served to select an intervention site.
2. Setting the stage for action ➔Used 10 out of 12 guideline elements	The pilot project is a stand-alone in one department. The hospital had no organization-wide program for handling specific employee health & safety issues.
3. Offering training ➔Used 9 out of 10 guideline elements	4 one-hour sessions attended by 23 of 28 employees. Limit training to Healthy Computing; the other ergonomic concerns were not broad enough or transferable to other operations.
4. Gathering data to identify problem areas ➔ Used 3 out 5 guideline elements	MSD symptom survey: 19 employees, 68% reporting symptoms in the last 12 months; 5 sought medical care; 3 lost work time (wrist/hand problems). VDT checklist: 19 design deficiencies common to more than 70% of the 20 stations result in trunk and head twisting while working with computers, and near-falls due to slipping stools. Focus group: awkward postures while drawing blood in the phlebotomy lab and computer data entry are attributed to facility design features that raise patient safety concerns.
5. Identifying effective controls and evaluating these approaches once they have been instituted ➔ Used 3 out of 5 guideline elements	Engineering controls - flat monitors, new chairs and stools, and desk accessories. Administrative control – consider transferring some ICD code assignation from the pathologist to the typist. Personal protective equipment – Not applicable. Vendors visited the facility to set up demo workstations. The task force decided to separate the lab-specific solutions from the ergonomic pilot project.
6. Establishing health	As the pilot study took place in hospital, staff has numerous

NIOSH program	Pilot program
care management → Used 5 out of 9 guideline elements	venues for addressing musculoskeletal problems.
7. Minimizing risk factors for MSD when planning new work processes and operations	11 process outcomes could serve to evaluate the pilot program as a whole 5 factors act as barriers for sustaining the program 3 factors act as facilitators
	→ The ergonomic concerns in healthcare go beyond MSD of the staff. These are beyond the scope of NIOSH program.

Similar to other professional service industries, the absence of recordable OSHA logs does not indicate the absence of musculoskeletal problems; these may be better gauged by symptoms surveys. In the target department, 68% of the respondents complained of musculoskeletal disorders and 15% lost work time because of the condition (wrist/hand problems). These data were not reflected in the OSHA logs since they were not considered by the professional lab staff as work-related injuries, which require reporting to Employee Health located in another part of the city. In this particular hospital, the impact should have translated to about 120 cases annually (assuming all are at a risk equal to that of the intervention site).

The VDT checklist revealed that 19 design deficiencies common to more than 70% of the 20 stations result in trunk and head twisting while working with computers, and near falls due to stools slipping over the floors because the casters were inappropriate for hard floors. The walk-through conducted after the Healthy Computing training sessions did not detect any changes in layout undertaken by the staff. This may reflect more on the limitations imposed by the current equipment, such as bulky CRT monitors occupying the limited surface area on the workbenches, rather than on the effectiveness of the educational component.

However, the focus groups revealed numerous risk factors for WMSD only partially related to the use of VDT:
- Repeated and prolonged stooping while drawing blood, mainly from patients on wheelchairs (Figure 1)
- Awkward posture while entering data into the computer (Figure 2)
- Awkward posture while writing labels by hand

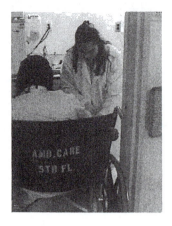
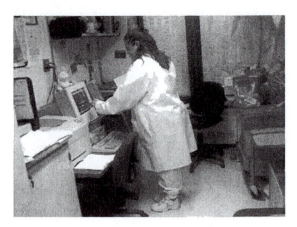

FIGURE 1. Drawing blood   FIGURE 2. Data entry in the lab.

These risk factors are attributable to the size, layout and lighting in the lab. For example:

- The doorway and the sink at the entrance limit wheelchair access.
- Lighting is insufficient for drawing blood at the doorway.
- The lab room is too narrow for having patients recline.
- The lab room has no curtains around the patient chairs.
- The floor space is limited when patients are in the lab so the technicians enter data into the computer standing; however the desk is designed for sedentary operations.
- The software and the printer are not capable of printing labels.

Additional shortcomings have also been noted:

- The patients and their companions are standing in the corridor; there is no seated waiting area.
- The blood drawing carts are difficult to maneuver in and out of the lab or around the patient beds.
- The carts do not enable data entry on the floor, so this has to be done in the crowded lab.

These ergonomic issues raised unexpected questions about compliance with the requirements of the Americans with Disabilities Act (ADA), the safety of patients in the lab, and the consequences of making errors while entering data into the computer in the lab. This raised questions regarding the responsibility for addressing patient comfort and safety separately from those of the staff.

Finally, the project identified several performance measures that serve to evaluate process outcomes, facilitators and barriers that affected the introduction of the ergonomic program. At the departmental level, the main facilitator proved to be management support in allocating time and funds through the operating budget, while the absence of an action plan secured by funding proved to be a barrier for expanding the program a year after the submission of the report.

CONCLUSIONS

In order to integrate ergonomics in an Environment of Care, this pilot study showed that we need to answer the following questions:

1. *How to assess the need and impact of an ergonomic program?*
Most of NIOSH guidelines can be transferred to healthcare services. However, OSHA logs in hospitals may not reflect prevalence or incidence of WMSD. Other criteria may have to serve for selecting an intervention site and assessing the impact of the program. As a lesson from research to practice, **the impact of the ergonomic program may not be manifested in OSHA log but rather in reduction of lost work time due to non-work-related health problems.**

2. *Who is responsible for ergonomic concerns related to patient safety?*
NIOSH guidelines do not address ergonomic concerns related to patient safety. Thus, the ergonomic concerns in health care go beyond musculoskeletal injuries. Addressing these concerns requires a systems approach – an examination of the processes within the department but also their interface with other units. The lesson from research to practice is: to be effective, **link the ergonomic program to patient safety and risk management.**

3. *Do we need a conceptual model that addresses employee and patient safety?*
While the NIOSH ergonomic program focuses on prevention of WMSD of the employees, it became evident that in healthcare facilities, ergonomic issues need to consider the relationship between employee and patient safety. In addition to collecting information on employee injuries and WMSD, we need to collect information on patient injuries. To study the relationship, we need access to two separate reporting systems, one that is public and the other is confidential. The lesson from practice to research: **relating patient injuries to employee WMSD requires a more complex investigation practices.** More research is needed in this area.

The Agency for Healthcare Research & Quality (AHRQ, 2003) states that quality in healthcare means doing the right thing at the right time in the right way for the right person and having the best results possible. *How can ergonomics contribute to these performance criteria?*

To better answer this question, we need to agree on dimensions of performance of healthcare delivery systems. The Organization of Economic Cooperation and Development (OECD), which encompasses most of the developed countries, agreed on 14 criteria (Arah et al, 2006): Acceptability, accessibility, appropriateness, environment & amenities, competence, continuity, cost, efficiency, equity, governance, responsiveness to the needs of the users, safety, sustainability and timeliness. This pilot study suggests that by integrating ergonomics within the functional requirements of an Environment of Care, we may impact quality of healthcare in at least four dimensions of performance: the *environment* and its amenities, *efficiency*, *responsiveness*, and *safety*.

REFERENCES

AHRQ (2003), *Annual Report on Research and Management, FY 2003*, Chapter 6. Agency for Healthcare Research & Quality, Rockville, MD. http://www.ahrq.gov/about/annrpt03/annrpt03f.htm [checked 3/11/10]

American Institute of Architects AIA (2010). *Guidelines for Design and Construction of Healthcare Facilities* (FGI Guidelines).

Arah OA, Westert GP, Hurst J, Klazinga NS. (2006) , "A conceptual framework for the OECD Health Care Quality Indicators Project." *International Journal of Quality in Health Care. Journal of Medicine,* Sep 18, Suppl 1:5-13.

Cornell University Ergonomics Web, 1996, http://ergo.human.cornell.edu/CUVDTChecklist.html [checked 3/11/10]

Goldsheyder, D., Nordin, M., Schecter Weiner, S., Hiebert, R. (2002). "Musculoskeletal Symptom Survey Among Mason Tenders." *American Journal of Industrial Medicine*, 42:384–396

NIOSH (1997), *Elements of Ergonomics Programs: A primer based on workplace evaluations of musculoskeletal disorders.* National Institute for Occupational Safety and Health, U.S. Department of Health and Human Services, Public Health Service, Centers for Disease Control and Prevention CDC, NIOSH Publication No. 97-117. http://www.cdc.gov/niosh/docs/97-117/default.html

Chapter 76

The Anthropometric Difference Between Curaçao and the Netherlands/USA with Regard to Table Height

Malcolm G. Lancelot , Glenn J. Smith

Occupational Hygienist and Safety
ARBO CONSULT, Occupational Health Department
Curaçao, Netherlands Antilles
mlancelot@arboconsult.com, gsmith@arboconsult.com

ABSTRACT

Incorrect work table (workstation) height with regard to users' stature, often leads to musculoskeletal health complaints of neck, shoulders, arms, wrists and back. In Curaçao these complaints frequently lead to absenteeism of employees.

Many companies in Curaçao request ARBO CONSULT, an Occupational Health Department, to do a workplace investigation for employees who suffer from the above mentioned complaints.

The objective of this paper was to analyze collected data of Curaçao and compare this data with international standards, to determine whether or not there is a significant difference between the two data groups with reference to work table height.

The collected mixed data (n = 200) with regard to gender represents workers with abovementioned health complaints and is a cross sectional representation of the companies in Curaçao. The data was analyzed to obtain the real work table height (non- adjustable work tables), the advised work table height and the difference between these two in relation to complaints of employees.

The data of the advised work table height was respectively compared with Netherlands' standards derived from an anthropometric mixed data study from the Netherlands and with an anthropometric mixed data study from the USA.

There was a significant difference between the real work table height and the advised work table height in Curaçao. The comparison between the Mean of the real work table height and the Mean of the advised work table height in Curaçao also showed a considerable difference, as did the difference between the advised Netherlands' standard and the advised work table height in Curaçao.

An interesting finding was that the data of the advised work table height of Curaçao showed a close relationship with the data of the American study

Keywords: Ergonomics, Work table Height, Musculoskeletal Health Complaints, Curaçao, ARBO CONSULT

INTRODUCTION

In this paper a report will be given of workplace investigations with regard to the results of the analysis of table height. The investigations were done by the occupational hygienists of the Occupational Hygiene Division of the ARBO CONSULT.

Background information on ARBO CONSULT

The Occupational Health Department was established in 1985 by the government of Curaçao and became a section of the Public Health Department.

Due to a high absenteeism of the public servants which was not always related to illnesses but more related to working conditions, the government found it necessary to establish an Occupational Health Department which was able to successfully reduce the number of absenteeism per year in the civil services from 10% to 6%.

During the years the Occupational Health Department expanded its activities and was able to become a private organization namely ARBO CONSULT in July 2002. The ARBO CONSULT team consists of several professionals such as occupational health physicians, occupational social workers, an occupational consultant, occupational hygienists/ safety professionals and doctor assistants. One of the occupational health physicians is also a Master of Law. Furthermore there is an agreement with a psychologist who whenever is needed will provide care for the clients of ARBO CONSULT.

Objective

The purpose for the workplace investigations was most often to determine the causes of complaints of shoulder, neck, arms and wrists of the workers of an establishment. In the majority of the investigations there was a discrepancy between the height of the table on location and the suitable height in accordance with users' stature.

Therefore the objective of this paper is to analyze the collected data of Curaçao concerning table height and compare this data with international standards, in order to determine whether or not there is a significance difference between the collected data group of Curaçao and standards derived from anthropometric data groups of the Netherlands and USA.
To accomplish an optimal ergonomic arrangement of the workplace, it is important that the table height is adapted to sitting height of the chair, the type of work and the stature of the worker. For every worker there is a different relation with reference to body height, the adjusted sitting height and the adjusted working height of the table. In most cases this was not taken in to consideration. For this reason there were many health complaints of the worker at the workplace.
Therefore the objective of ergonomists to minimize health complaints must be, **"fitting the job to the man"** instead of **"fitting the man to the job".**

METHOD

Two hundred (200) workplace investigations were randomly selected out of the workplace investigations done by ARBO CONSULT. The investigations were done due to health complaints of workers concerning shoulders, neck, arms, wrists and back.

The collected data is a cross sectional representation of companies in Curaçao and represents mixed data with regard to gender.

The collected data was analyzed to obtain the real work table height (non-adjustable work tables), the advised work table height and the difference between these two in relation to complaints of employees. This was done by means of descriptive statistical analysis. The real work table height is the height of the users' table and the advised table height refers to a suitable table height according to the users' specific stature. It should be mentioned that the advised table height is determined after the correct sitting height is established.

Per group the absolute frequency, cumulative absolute frequency, relative frequency and the cumulative frequency were determined. The mean, median and mode of the data groups were also defined.

692

The differences between real table height and advised table height were established. These differences were then compared with work table height standards of the Netherlands (75 cm for 50th percentile) (1-2) and with the mean (Gaussian distribution) of an anthropometric mixed data study from the USA (3).

The results of all the above were plotted in to graphs.

RESULTS

Graph I shows the absolute frequencies of advised table height from the collected mixed data of Curaçao (Female and Male).

Graph I

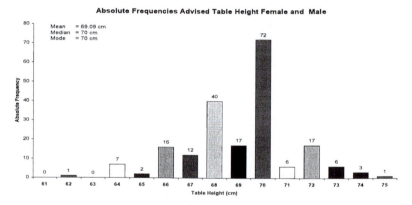

The difference between the real work table height and the advised work table height of the mixed data of Curaçao has **a range of 12**, a **minimum of 2 cm difference** and **a maximum of 14 cm difference** (Graph I and II).

Graph II

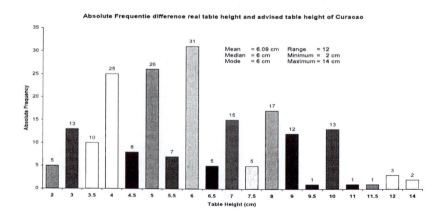

Graph III

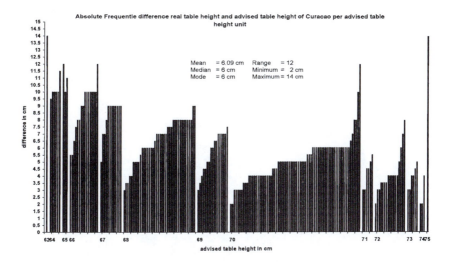

Absolute Frequentie difference real table height and advised table height of Curacao per advised table height unit

The comparison between the Mean of the real work table height of Curaçao **(75.2 cm)** and the Mean of the advised work table height of Curaçao **(69.1 cm)** is a **difference of 6.1 cm**.

Graph IV

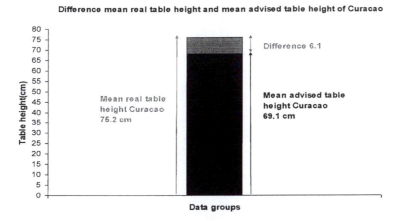

Difference mean real table height and mean advised table height of Curacao

694

The difference between the advised standard of the Netherlands (**75 cm at the 50th percentile**) and the advised work table height in Curaçao has a **range of 13** with a **minimum of 0 cm** and a **maximum of 13 cm**.

Graph V

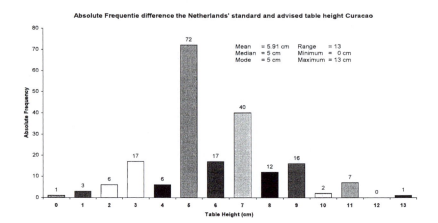

Absolute Frequentie difference the Netherlands' standard and advised table height Curacao

The difference between the Netherlands Standard (75 cm) and the Mean of the advised work table height of Curaçao (69.1 cm) is **5.9 cm**. *The Mean, Median and Mode are the same in a normal distribution.*

Graph VI

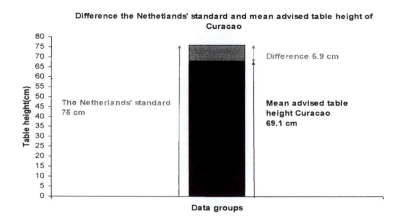

Difference the Nethetlands' standard and mean advised table height of Curacao

The comparison between the Mean of the American study (68.1 cm) and the Mean of the advised work table height of Curaçao (69.1 cm) is a difference of **1 cm**.

Graph VII

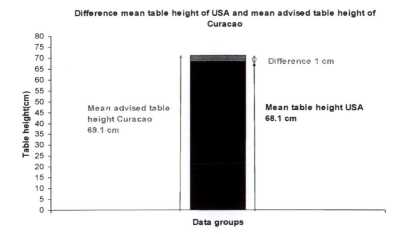

The difference between the Mean of the American study (68.1 cm) and the Median and Mode of the advised work table height of Curaçao (70 cm) is **1.9 cm**.

Graph VIII

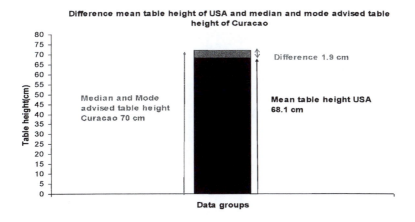

DISCUSSION, CONCLUSIONS AND RECOMMENDATIONS

Discussion

The comparison between the real table height and advised table height of the collected data in Curaçao shows a significant difference range from 2 cm to 14 cm and a mean difference of 6.1 cm. It can be expected that these differences will lead to musculoskeletal complaints involving the back, neck and upper extremities, which are in accordance with complaints registered.

The comparison of the standard of the Netherlands with the Curaçao data with regard to the advised table height and also the mean difference, illustrates more or less the same pattern as the abovementioned. A difference range from 0 cm to 13 cm and a mean difference of 5.9 cm

It is clear that there is not a significant difference between the two abovementioned comparisons. This can be explained by the fact that in the Netherlands Antilles which also includes Curaçao, most of the worktables (workstations) are imported from the Netherlands and some other European countries. A great part of these worktables are non-adjustable height tables.

An interesting finding was that the mean of the data of the anthropometric data study of USA data shows a much closer relationship with the mean of the advised table height of the sample study of Curaçao, namely a 1 cm difference. This is also true for the mean of the American study with the median and mode of the advised table height of the Curaçao study, namely 1.9 cm.

One could argue that because of these almost negligible differences, importing non-adjustable height worktables form the USA would lead to less musculoskeletal health complaints of back, neck and upper extremities.

To validate the abovementioned assertion, a more in depth data study needs to be done between Curaçao and the USA.

No ergonomic standards and guidelines are established in the Netherlands Antilles (Curaçao) and consequently there is not a standard with regard to table height and office chairs.

Conclusions and recommendations

Looking at the outcome of the descriptive statistical analysis and from an occupational hygiene/ergonomics point of view, the following conclusions and recommendations can be drawn.

- Because of the stature difference between the average worktable user of Curaçao and the average worktable user of the Netherlands, importing and using non-adjustable height worktables from the Netherlands will lead to the development of musculoskeletal health complaints of back, neck and upper extremities.
- A way to minimize these health complaints is to import adjustable worktables from the Netherlands whereas the table must be adjustable in a range of 62 cm to 82 cm in height. This could give coverage from the 0 percentile to the 100 percentile of the working class in Curaçao with regard to their stature.
- A wide-ranging (rural) anthropometric study needs to be done in the Netherlands Antilles with the objective to develop ergonomic standard and guidelines for the Netherlands Antilles.

But even with ergonomic standards and guidelines in place, the maxim **"fitting the job to the man"** should always be used for correcting the workplace arrangement ergonomically.

Correcting the workplace arrangement ergonomically, may include rearranging the workplace (work station), readjusting the height of the table, and (re)adjusting the office chair.

REFERENCES

Kantoor inrichten: informatie, organisatie en besluitvorming bij kantoorinrichtingsprojecten/ Arbeidsinspectie – Voorburg (1990), Directoraat-Generaal van de Arbeid van het Ministerie van Sociale Zaken en Werkgelegenheid. -III-, ISBN 90-5307-072-9, (1) 41

Creemers M.R., e.a. (1987), Polytechnisch zakboekje, Arnhem, P.B.N.A., (2) J3/12 - J3/13.

Donelson S. M., and Gordon C. C. (1996), 1995 matched anthropometric database of U.S. Marine Corps personnel: summary statistics, United States army soldier systems command Natick research, development and engineering center, Natick Massachusetts 01760-5020, (3) 78-79, 112-113

Chapter 77

Medicoergonomics – Industrial Ergonomics Adapted to Clinical Requirements

Wolfgang Friesdorf, Daniela Fuchs, Beatrice Podtschaske

Department Human Factors Engineering & Product Ergonomics
Berlin Institute of Technology
Fasanenstr. 1, 10623 Berlin, Germany

ABSTRACT

Increasing complexity and cost pressure characterize the health care systems. Ergonomics has knowledge and experience to optimize working systems. Due to patient's individuality an adaptation is necessary: "Medicoergonomics" has been coined (Donchin, 2007). The recursive and hierarchical task-process-task-model is designed to model medical treatment processes. It consists of 6 layers, on the highest system layer: a case, on the lowest layer: medical procedures. Decision elements on all layers enable modeling of standardized as well as non-standardized treatment processes.

Keywords: Medicoergonomics, Systems Engineering, Patient Treatment, Working Processes, Work Analysis

INTRODUCTION

Patient treatment has become a complex task. A wide variety of medical procedures is available due to new drugs and high sophisticated medical devices. Patients can be treated successfully even if their disease pattern seems to be hopeless. Today's Medicine is excellent. And the progress is still ongoing. Some decades ago the guiding principle in Intensive Care Medicine was to apply whatever medical procedure is available, no matter the cost. Today alternative procedures are available; decision making has become much more complicated. Patients with several diseases (multimorbidity) require parallel treatment with the risk of interferences and no clear relation between medical procedures and clinical findings. Side effects have to be considered and detected carefully. Altogether the demands on clinical experts have increased on the one side. And on the other side health care has to deal with limited resources. We see a distinct cost pressure, the need to safe money wherever possible, and – at least in Germany – a shortage of qualified personnel at the labor market: Treatment must be effective i.e. evidence based, which is a medical challenge. Treatment must also be efficient: Excellent medicine requires excellent processes with an optimal interaction between human, technology and organization to safe resources – a core competence of Ergonomics.

Precondition of an optimization is transparency of the working processes.

In Industry this transparency is gained with a hierarchical systems engineering (SE) concept (Daenzer & Huber, 1992, Bubb & Schmidtke, 1993). Starting on a high system layer a task (e.g. the production of products) is divided into sub tasks; the sub tasks into sub-sub tasks etc. On the lowest layer single procedure are focused – typically a human-machine-interaction. The whole production is a well organized composition of these single procedures. This method is widely evaluated in repetitive tasks which are fulfilled by standardized procedures. In contrast non-recurring and individual projects in complex environment (e.g. planning of a hospital building) are also structured hierarchically: project phases and work packages with milestones in between, but a project management has to plan, control, and adapt the activities according to the project progress (Daenzer & Huber, 1992).
The question is, can we learn from these concepts in clinical environment to structure a patient's treatment (see also Friesdorf et al., 1994)?

OBJECTIVES

In this paper we want to investigate whether the patient's treatment is

- a recurring task with the potential of standardization or
- an individual project with the requirements of a patient related project management.

Based on these results a clinical working process model shall be designed, which could serve for the analyses of these processes, their comparison (benchmarking), their optimization, and treatment related work flow control.

TREATMENT: RECURRING TASK OR PROJECT?

A patient course might help answering the question.

PATIENT FOR HIP IMPLANT

A healthy 50 years old lady, overweight, with severe hip arthrosis is planned for a total hip endoprothesis. The hospital is using a clinical pathway for this surgical intervention. According to this pathway the patient is admitted and prepared. Anesthesia and the operation are performed as planned, but in the recovery phase she develops a respiratory insufficiency due to her overweight. She has to be treated in an ICU (intensive care unit). As a consequence she is taken off from the pathway and her treatment (diagnostics, therapy and monitoring) is planned and applied individually according to the clinical situation. A huge "tool box" of medical and nursing measures is available to be composed according to the requirements. Guidelines (e.g. respiratory therapy) are used and protocols (e.g. application of antibiotics) are adapted to clinical findings. After three days in the ICU the lady has stabilized and she is transferred to a peripheral ward. Rehabilitation follows; the hip implantation was successful.

What do we see with industrial glasses when we follow such a clinical treatment?

- Hospitals try to standardize recurring treatments. The concept of Diagnosis Related Groups (DRGs) is supporting this effort. Almost 1000 cases are listed in the German DRG catalogue. Our experience shows, that up to now few hospitals are using pathways. They have defined some 100 cases, and only 2 to 3 % of all patients are treated following a pathway (mainly in obstetrics, which actually is no illness!).
- The pathways are defined by experts (physicians and nurses): measures (small working packages) on a time line. From the degree of granulation this is comparable to a very detailed project plan. The working processes are not considered (this would correspond to the lowest system layer in recurring industrial production).
- As soon as a patient's status differs from the expected course he/she is taken from the pathway and the following treatment is defined conventionally.
- As far as we saw the use of pathways in our clinical consulting projects, they were paper based with very little flexibility to adapt the pathway according to the individual progress.

- Guidelines and protocols (standard operating procedures, SOPs) are available. In general their use is not supported by information technology (IT). They are defined for quality certification purpose and filed in folders.

PATIENT'S TREATMENT: A PROJECT WITH STANDARDIZED MODULES

Three different types of treatment courses can be differentiated:
1. *Standard-Pathway*: The patient's progress is in accordance to a detailed treatment plan. The single measures are preset; parameters of these measures are adapted to patient's parameters (e.g. doses of a drug related to body weight).
2. *Modified-Pathway*: Patient treatment starts on a Standard-Pathway, but the individual progress enforces a modification (in the example given above a temporary ICU treatment is necessary due to respiratory insufficiency). The ICU treatment in itself follows guidelines and protocols.
3. *Off-Pathway*: Patient's state of health is so complex, that one DRG cannot describe the situation. E.g. multimorbid patients do not fit in any given pathway. Their treatment has to be composed individually using all medical skills and considering guidelines and SOPs.

In answering the question "recurring task or project?" we consider a patient treatment as a "project" with recurring sub-tasks on different sub-layers which can be described as standardized modules.

Even if a patient seems to be a standard case, the decision to put him/her on a pathway requires high medical experience. In any case a patient individual risk management must guarantee safe treatment.

Using an analogy to industrial production must fail.

REQUIREMENTS

A clinical working process model must cope with all three types of treatment (Standard-, Modified-, and Off-Pathway). It should be useable for the analyses of processes, their comparison (benchmarking), and their optimization. This model should serve as a core for treatment related work flow control.

CONCEPT

Starting point is a general recursive hierarchical task – process – task model (Carayon & Friesdorf, 2006, Marsolek & Friesdorf, 2006), which we also used in risk analyses (Friesdorf, Buss & Marsolek, 2007). It allows a decomposition of complex tasks. Working with this model we could adjust it in details. Figure 1 shows the current version. A little, but important adjustment was the inclusion of

702

decisions (see the rhombus in figure 1). Depending on a decision the treatment process continues accordingly with the corresponding sub-tasks.

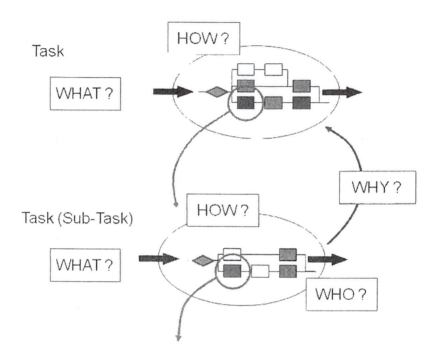

FIGURE 1. Recursive hierarchical task – process – task – model

SIX SYSTEM LAYERS

The general recursive approach does not define the numbers of layers. Our experience shows, that six layers are appropriate for our clinical analyses:

1. Layer: The case, including the whole patient treatment from the first symptom till the completion of treatment.
2. Layer: Phases of treatment, mainly defined by health care structures (e.g. general practitioner, emergency service, hospital, rehabilitation)
3. Layer: Steps within structures (e.g. emergency department, operation area, ICU)
4. Layer: Partial treatments (e.g. surgical intervention including

interdisciplinary cooperation between anesthetist and surgeon)
5. Layer: Disciplinary procedures (e.g. induction of anesthesia)
6. Layer: Single procedures (e.g. application of a drug)

Starting from layer 1 a given case can be decomposed step by step (layer by layer). The included decision elements define the specific sub-tasks like a sub-tree. This concept enables the modeling of a S*tandard-Pathway* as well as a *Modified-Pathway*. If this concept is used for modeling an *Off-Pathway* patient modules from different layers can be put together; the treatment is composed according to the clinical needs. Modules from higher system layers stand for more standardization; a composition of modules from lower layers shows a more individualized non-standardized treatment.

On the lowest layer we find all available medical procedures. A set of parameters describes these procedures (e.g. task which is fulfilled, quality parameters, environment requirements, responsibilities, qualification, costs, risks). In a bottom-up approach we can analyze in which modules, and on the highest system layer in which patients (cases), a special medical procedure is used. This will be interesting in risk analyses and when new procedures might substitute a previous.

CONCLUSION

Ergonomics has extensive experience mainly in industrial production. We can and should use this knowledge and experience, but we have to adapt it to medical needs. Accordingly Yoel Donchin has coined "Medicoergonomics" (Donchin, 2007, Friesdorf & Marsolek, 2009). Our model is meant to support his approach. Up to now the presented model is practically used and evaluated in several working process analysis. These projects were typically focused on parts of a whole patient treatment (e.g. patient in an emergency department). The goal is to fill the model with each project, step by step, and to file all tasks, sub-tasks etc. in a task library. This is a long way to go and only realizable if the model is widely accepted and used; so we are looking for cooperation. Digital technology must support the use, the filing, and the communication; we are working on software support.

REFERENCES

Bubb, H. & Schmidtke, H. (1993). Systemergonomie. In Schmidtke, H. (Ed.): Ergonomie (305-458*).* München, Wien: Hanser Verlag.
Carayon, P. & Friesdorf, W. (2006). Human Factors and Ergonomics in Medicine. In Salvendy, G. (Ed.): Handbook of the Human Factors and Ergonomics (1517–1537). New Jersey: John Wiley & Son.
Daenzer, W. F. & Huber, F. (1992). Systems Engineering. Zürich: Verl. Industrielle

704

Organisation.

Donchin, Y. (2007). Forecasting the Next Error is More Important than to Analyze the Previous One. Journal of Clinical Monitoring and Computing, 21 (3), 179-180.

Friesdorf, W., Konichezky, S., Gross-Alltag, F., Geva, D., Nathe, M. & Schraag, S. (1994). Decision Making in High Dependency Environments – Can We Learn from Modern Industrial Management Models? International Journal of Clinical Monitoring and Computing, 11, 11–17.

Friesdorf, W., Buss, B., Marsolek, I. (2007). Patient Safety by Treatment Standardization and Process Navigation – A Systems Ergonomics Management Concept. Theoretical Issues in Ergonomics Science, 8 (5), 469-479

Friesdorf, W. & Marsolek, I. (2009). Medicoergonomics – A Human Factors Engineering Approach for the Healthcare Sector. In Schlick, C. M. (Ed.): Industrial Engineering and Ergonomics. Visions, Concepts, Methods and Tools. Festschrift in Honor of Professor Holger Luczak (pp. 165-176). Berlin, Heidelberg: Springer Verlag.

Marsolek, I. & Friesdorf, W. (2006). Work Systems and Process Analysis in Health Care. In: Carayon, P. (Ed.): Handbook of Human Factors and Ergonomics in Healthcare and Patient Safety (649-662). Mahwah, New Jersey: Lawrence Erlbaum Associates.

Chapter 78

Methods of Clinical Process Analysis – Systematically Replacing a "Banal Rationing" by a "Balanced Rationalization"

Ingo Marsolek

University of Applied Sciences for
Technology and Economics Berlin
Treskowallee 8 in 10318 Berlin, Germany

ABSTRACT

Clinical work systems in industrialized nations all over the world are being characterized by a steadily increasing cost pressure, increasing quality as well as customer demands and a growing system complexity. In this context a sustainable work system improvement should not only focus on the release of a work system's existing optimization potential (= a "banal rationing" of resources), but on the establishment of a high treatment quality as well as a lasting staff motivation based on a sustainable reinvestment of the released resources into further system improvements (a so called "balanced rationalization"). Therefore a participatory analysis and optimization approach is needed for the improvement of clinical work processes systematically focusing on a sustainable staff participation and qualification within each of the following project steps: 1. Work System Preparation

& Staff Participation, 2. Process Visualization & Verification, 3. Hierarchical Structuring & Process Quantification, 4. Identification of Process Characteristic Strengths & Deficits, 5. Development of an Optimized Process, 6. Evaluation of Changes & Initiation of a Continuous Process Control, and 7. Initiation of a Continuous Process Flow Management. The optimization potential, which can be found by using such a systematic improvement approach, is often not only obvious but also significant. However, it can only be set free, if the achieved process optimizations are also carefully integrated into the clinical work system's existing organizational structure as well as already established improvement approaches and additional staff resources are made available for the actual project work.

Keywords: Clinical Work Processes, Analysis, Optimization, Control, Banal Rationing, Balanced Rationalization, Staff Participation, Staff Qualification, Process Visualization

SITUATION

Clinical work systems in industrialized nations all over the world are being characterized by three different developments (Marsolek and Friesdorf, 2006):
1. A steadily increasing cost pressure resulting from diminishing financial healthcare resources, dramatic demographic changes towards more and more elderly and chronically ill patients and a growing number of innovative and promising (but in most cases also cost intensive) diagnostic and therapeutic treatment possibilities.
2. Increasing quality and customer demands because of the patient's health/life being at risk, a growing number of media reports about medical progresses as well as malpractices, an increasing necessity/willingness within patients to pay for selected services on a private basis and a growing competition among healthcare providers.
3. A growing system complexity resulting from each patient's individual health status, unpredictable treatment dynamics, unavoidable ethical problems and an increasing fragmentation of the entire patient treatment process.

STRATEGY

The need for a systematic analysis and organizational (re)design of clinical work processes and systems is obvious. But a system ergonomic work system improvement should not only focus on the release of a work system's existing optimization potential (= a "banal rationing" of resources), but also needs to establish a high treatment quality as well as a lasting staff motivation based on a sustainable reinvestment of the released resources into further system improvements on all system layers (a so called "balanced rationalization" – see also Marsolek and

Friesdorf, 2007):

1. At the "management level" by finding the right management balance between the necessary medical, management, innovation and communication competence through an early initiation and careful monitoring of all necessary change projects.

2. At the "process level" by finding the right management balance between the definition of realistic goals for the release of the existing optimization potential (top-down) and an adequate staff enabling/qualification (bottom-up).

3. At the "staff level" by finding the right management balance between the release of the existing optimization potential (staff resources) and its re-investment for future system improvements (e.g. the development of new markets and innovative services such as integrated patient care, ambulatory surgery, home care etc.).

For the realization of a "balanced rationalization" the work system's management vision and main strategy have to be successively realized (top-down) in well-defined optimization projects, which need to be consequently controlled (bottom-up) with the help of according project benchmarks. While within the realization of each defined optimization project various common management techniques can be used for eliminating existing organizational weaknesses, it is of utmost importance that the same work system's characterizing organizational strengths are not being destroyed or neglected throughout the very same approach. Therefore all already existing (and well functioning) optimization approaches of a work system have to be systematically integrated into the "balanced rationalization" as well. (see also Marsolek and Friesdorf, 2007)

METHODS

Especially for clinical work systems a typical organizational weakness can be found within the "in-transparency" of the work system's underlying work processes caused by the complexity of the patient treatment. Therefore a continuous analysis, redesign and control of clinical work processes (based on a Continuous Process Flow Management Cycle as shown in Figure 1) has to be the fundamental basis for all other system improvements.

"Continuous Process Flow Management"

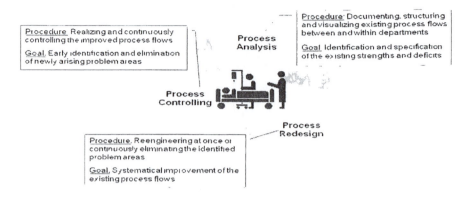

Figure 1. Continuous Process Flow Management Cycle

For this reason the already existing medical task orientation within clinical work systems (defining "WHAT has to be done with a patient") has to be systematically supported by building up an additional system ergonomic process management competence. Because the best task definition ("WHAT has to be done") can not lead to the desired result, if it is not realized with absolutely correct work processes ("HOW to do it"). With each incorrectness (in a decision, communication etc.) a loss of quality or efficiency becomes unavoidable. Therefore, besides the classical work organization ("top-down") the establishment of a sustainable process orientation is needed, which can only be achieved by a "bottom-up-oriented" staff participation and qualification within the entire continuous process flow management cycle (see also Marsolek and Friesdorf, 2005 as well as Marsolek, 2008). For this reason the following 7 change process steps are required:

1. Work System Preparation & Staff Participation:
Define the project's objective; inform all involved divisions and staff members; collect existing process information as well as process strengths and deficits.

2. Process Visualization & Verification:
Observe the existing processes; establish process transparency by visualizing all activities in a process flow diagram using specific process symbols easy to understand (see also Figure 2. Exemplary Process Flow Visualization as well as Fuermann and Dammasch, 1997); encourage the participation of all involved staff members; verify and correct the process flow diagram with each input until a final version is achieved, that realistically displays the existing processes and is understood from all involved persons.

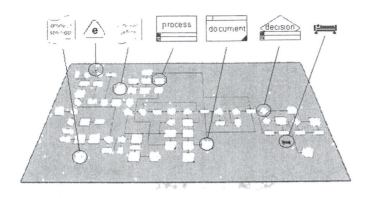

Figure 2. Exemplary Process Flow Visualization

3. Hierarchical Structuring & Process Quantification:

Define process modules by grouping all activities within the verified process flow diagram to superior process tasks/modules (see also Figure 3. Exemplary Hierarchical Process Structuring); observe or estimate the proportional distributions after each decision or distribution; elaborate a catalogue with all used documents for the verified process flow diagram; observe or estimate the needed staff, machinery, rooms, materials and times for each identified process activity and module.

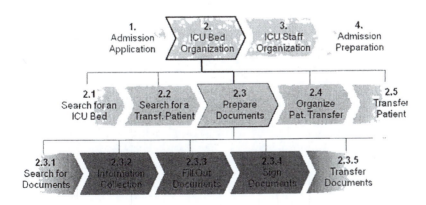

Figure 3. Exemplary Hierarchical Process Structuring

4. Identification of Process Characteristic Strengths & Deficits:

Detect all existing process strengths and deficits together with the involved clinical experts; use common management tools such as process data analysis, process benchmarking, information flow analysis, value assessment, output assessment, process check lists etc. (see also Camp, 1995; Chang, 1996; Fuermann and Dammasch 1996; Hammer and Champy, 1993; Vorley and Tickle, 2001); group all detected process characteristics according to their desired optimization: "unnecessary and redundant processes" for a later "Key Process Concentration", "process strengths and deficits" for a later "Process Deficit Elimination" and "stable and unstable processes" for a later "Process Flow Stabilization".

5. Development of an Optimized Process:

Reengineer and visualize an improved process flow; develop organizational, technical and architectural changes together with the involved staff; focus on the following strategies: 1. Key Process Concentration: Avoid all unnecessary and redundant process activities, that do not help to fulfill superior process tasks/modules, 2. Process Deficit Elimination: Eliminate all identified process deficits by developing an optimized process flow, that still includes already existing system strengths, 3. Process Flow Stabilization: Develop an optimized process flow, that stabilizes all identified unstable activities but also supports already stable ones; calculate or estimate the existing improvement potential based on the quantified processes; elaborate adequate realization strategies and define "WHO" has to do "WHAT" until "WHEN" together with the responsible staff.

6. Evaluation of Changes & Initiation of a Continuous Process Control:

Define characteristic and objective benchmarks to be collected before and after the actual process redesign (not only for an objective evaluation but also for forcing the actual realization and initiating a continuous process control): use timely benchmarks (e.g. waiting times, work times, usage times etc.), financial benchmarks (staff costs, room costs, treatment costs etc.), quality benchmarks (e.g. number of mistakes, missing information, complaints etc.) or costumer/staff satisfaction benchmarks (from patients, external physicians, internal staff members etc.); decide whether to assess process changes with the help of the involved staff or to measure them by hard facts; define an adequate collection strategy (e.g. interviews, questionnaires, self-documentation, external observation or automatic documentation with the help of (already existing) data management systems).

7. Initiation of a Continuous Process Flow Management:

Use the ongoing staff participation as the starting point for an additional qualification of selected staff members from "process segment experts" to "process flow partners" and "process flow managers" (see also Figure 4. Systematic Staff Qualification within the Continuous Process Flow Management Cycle); aim on the establishment of a self-learning organization.

„Continuous Process Flow Management"

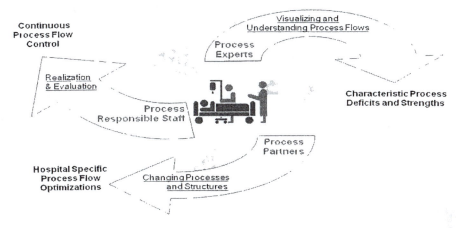

Figure 4. Systematic Staff Qualification within the Continuous Process Flow Management Cycle

In addition to that the entire change process needs to be adequately integrated into all already well functioning work system improvement approaches: e.g. using a well functioning critical incident reporting system for the identification of all sub-optimal work processes, using a well established (financial or non-financial) incentive system for motivating the desired staff participation, using a well defined staff training program to multiply the identified optimization potential and to qualify selected staff members to become "process flow managers".

RESULTS

The existing optimization potential, which can be found by using such a systematic improvement approach, is indeed significant (see also Marsolek and Friesdorf, 2004 as well as Marsolek and Friesdorf, 2006). Nevertheless, it can only be set free, if the achieved process optimizations are also carefully integrated into the clinical work system's existing organizational structure as well as all already established improvement approaches such as its incentive system, staff training programs etc. Furthermore, for the establishment of a continuous optimization process additional staff resources have to be made available to allow selected staff members to become "process flow managers" within their clinical work system.

DISCUSSION

Although the optimization potential, which can be found within clinical work systems by using such a systematic improvement approach, is often not only significant but also obvious, setting it free within the clinical everyday life is much harder to realize. Major reasons for this are not only the limited financial and personal resources of many clinical work systems, but also the widespread system ergonomic incapability of singular medical devices as well as information technology and the healthcare sector's characteristic treatment fragmentation causing divided responsibilities as well as conflicting interests. Therefore it becomes absolutely unavoidable to focus not only on a sustainable process improvement of entire clinical patient treatment chains (e.g. from the accident scene via first aid, ER OR, ICU and ward treatment until home care and rehab), but also to develop corresponding process oriented and system ergonomic medical work places as an alternative for today's chaotic conglomeration of singular treatment devices and information technology.

REMARKS

This research has been generously supported by the Japan Society for the Promotion of Science (JSPS), the German Alexander von Humboldt-Foundation (AvH), the German Academic Exchange Service (DAAD) as well as many clinical project partners in Germany, Austria, Israel, Japan and the US.

REFERENCES

Camp, R.C. (1995), *Business Process Benchmarking – Finding and Implementing Best Practices*. Milwaukee, WI: Quality Press.

Chang, R.Y. (1996), *Process Reengineering in Action*. London: Kogan Page Ltd.

Fuermann, T. and Dammasch, C. (1997), *Prozessmanagement – Anleitung zur staendigen Verbesserung aller Prozesse im Unternehmen*. Muenchen: Carl Hanser Verlag.

Hammer, M. and Champy, C. (1993), *Reengineering the Corporation: A Manifesto for Business Revolution*. New York: Harper Business.

Marsolek, I. (2008): *Clinical Process Optimization – Methodische Grundlagen und praktische Umsetzung*. Saarbrücken: VDM Verlag.

Marsolek, I. and Friesdorf, W. (2004), Optimising Clinical Process Flows: Experiences from the Expert Systems. In: *International Journal of Intensive Care* (Winter Issue), 172-179.

Marsolek, I. and Friesdorf, W. (2005), Process Reengineering in the Healthcare Sector – Reducing Complexity through Process Visualization and Staff

Participation. In: Carayon, P.; Robertson, M.; Kleiner, B. & Hoonakker, P.L.T. (Eds.): *Human Factors in Organizational Design and Management.* Santa Monica, CA: IEA Press, 299-304.

Marsolek, I. and Friesdorf, W. (2006), Work Systems and Process Analysis in Health Care. In: Carayon, P. (Ed.): *Handbook of Human Factors and Ergonomics in Healthcare and Patient Safety.* Mahwah, New Jersey: Lawrence Erlbaum Associates, 649-662.

Marsolek, I. and Friesdorf, W. (2007), Arbeitswissenschaft im Gesundheitswesen – Balancierte Rationalisierung statt banaler Rationierung. In: Gesundheitsstadt Berlin e.V. (Ed.): *Handbuch Gesundheitswirtschaft – Kompetenzen und Perspektiven der Hauptstadtregion.* Berlin, Medizinisch Wissenschaftliche Verlagsgesellschaft, 88-94.

Vorley, G. and Tickle, F. (2001), *Quality Management – Principles & Techniques.* Guildford Surrey, UK: Quality Management & Training Limited.

Chapter 79

The Relative Importance of Usability and Functionality Factors for E-Health Web Sites

Fethi Calisir,Ayse Elvan Bayraktaroglu,
Cigdem Altın Gumussoy,Hande Topaloglu

Industrial Engineering Department
Istanbul Technical University

ABSTRACT

With the growing usage of Internet, the demand for online healthcare information and advice, and as a reply to that, the number of health related web sites are increasing. In case of online health information and advice, user interface replaces the face to face communication. To ensure users' needs are met, it is critical to balance functionality and usability in the design of the web site. This study seeks to identify the complex interrelationships between various factors of usability and functionality in case of e-health web sites. Two Turkish e-health web sites are assessed for this evaluation. As not only functionality and usability of a web site, but also some of their other factors are related to each other, Analytical Network Process (ANP) is used to analyze the relative importance of the factors. Findings show that users of health information web sites give higher priority to functionality and the functionality factors with the highest relative importance are services/facilities and personalization/categorization of information.

Keywords: Usability, Functionality, E-health Web Sites, ANP

INTRODUCTION

With growing usage of the Internet, it has become a commonly used medium for health information and advice. Researches indicate that both the demand for online healthcare information and the number of health related web sites are ascending (Goetzinger et al., 2007; Fisher et al., 2008). While users demand trustworthiness (exact and relevant information) from these web sites, web sites demand loyalty from their users. There is a link between consumer trust and loyalty, and user interface design is one of the many factors that affect the perception of trust (Gummerus et al., 2004; Roy et al., 2001). In case of online health information and advice, user interface replaces the face to face communication. The success of the user's visit to the web site depends largely on user interface design, because "the design of a website is the on-line consumers' gateway to the organization" (Bontis and De Castro, 2000). As usability and functionality are quality characteristics that evaluate an interface design, balancing them in the design of the web site is significant to ensure users' needs are met (Nielsen, 2003).

Usability assesses the extent a web site facilitates users utilize the offered functions easily and appropriately. Functionality estimates the extent the web site operates in the way it is structured and is expected to perform as users desire (Bertot et al., 2006; Nielsen, 2003). It also may be that a functional web site is not usable or vice versa (Yeung and Law, 2004; Lu and Yeung, 1998; Nielsen, 2003).

The interface of a web site is a medium for users to interact with the system and the functionality of a web site becomes obvious to users through its interface. Although usability is generally seen as a feature related only to the interface and not to the logic of available functions, Seffah et al. (2008) indicated that attributes of a system can affect the usability of the whole system. Without understanding the functionality of the system, successful usage of a web site is not possible. Functionality can affect the usability of the system in a positive way, such as a "cut" function (Bass and John, 2003). At some point, the greater the functionality offered to the users the more skilled users must cope with the complex and time-consuming structure of the system (Kavadias et al., 2007). Therefore, excessive functionality can decrease the usability of the web site (Saarloos et al., 2008; Furnell, 2005).

As usability and functionality are interrelated in web design and only functionality-focused web site designs disregards usability needs (Kline and Seffah, 2005), in the design process of a web site, both usability and functionality should be taken into account (Weir et al., 2007). Therefore, this study aims to examine the relative importance of both usability and functionality factors for e-health web sites from the perspective of Turkish young users.

USABILITY AND FUNCTIONALITY FACTORS USED IN THIS RESEARCH

The factors related with usability and functionality of e-health web sites were selected and classified on the basis of literature review and the personal judgment of the expert authors. The literature reviewed on usability and functionality included studies on web sites envisaged for different purposes and software packages designed for organizations such as libraries. The list of usability and functionality factors is presented in Table 1.

Table 1 Usability and functionality factors used in the evaluations

Usability Factors	Functionality Factors
A1.Navigation	B1.Personalization/Categorization of Information
A2.Interaction	B2.Search Options
A3.Learnability	B3.Differentiation of the Information Types
A4.Ease of Use	B4.Services/Facilities
A5.Response Time	B5.Thesaurus
A6.Memorability	
A7.Efficiency	
A8.Satisfaction	

Usability factors determined for this study are navigation, interaction, learnability, ease of use, response time, memorability, efficiency, and satisfaction.

Navigation: It refers to finding one's way to the desired information through menus, graphical components, links and page sequence, and layout (Palmer, 2002) as well as, even while doing this, knowing where one is in the site (Roy et al., 2001). An appropriate navigation decreases the cognitive load on the user visiting a web site (Tarafdar and Zhang, 2005) and the navigability of the site affects the web site's success (Palmer, 2002).

Interaction: Responses to the user's actions are produced by the system (Palmer, 2002). Along with navigability, interaction with the site helps users to find easily the desired information in a web site (Chakraborty et al., 2005). Ghose and Dou (1998) suggest that interactivity is a component which has a fortifying effect on the attractiveness of an e-commerce web site.

Learnability: It is associated with the skill levels of web site users and thereby the level of effort needed to learn how to operate the system (Calero et al., 2005). For success of a web site, the time a user needs to learn how to operate the system should be very short (Nielsen, 2000a). Learnability is about the cognitive load imposed upon a user when he/she accesses a web site for the first time. Thus, as noted by Tarafdar and Zhang (2005) good navigation is critical to reduce the cognitive load, and it can be said that navigation and learnability of a web site are related.

Ease of use: It refers to being able to operate a web site without experiencing any difficulty and trouble. Ease of use is an important criterion for web users as it affects web sites effectiveness and user sentiment toward the web site (Chakraborty et al., 2005). Tarafdar and Zhang (2005) and Palmer (2002) specify navigation mechanism to be one of the main design elements for easy operation of the site.

Response time: It is the time needed by the system to respond to the activity of a user (Palmer, 2002). Short download times are essential to provide short response times (Nielsen, 2000b). Minimal download times are a critical consideration in preventing users getting bored and switching to another web site (Thelwall, 2000). The shorter a site's response time is the more the customers perceive it as high quality and the more gratifying they find the experience (Novak et al., 2000).

Memorability: It is the ease of recall of the main functions and their presentation on the web site when a user revisits the page (Nielsen, 2003). According to Ferreira and Pithan (2005), a web site, which is constructed in conformity with the model visualized by users, is expected to have higher memorability, because users tend to remember much better structures that appear logical to them. Seffah et al. (2008) point out that an inconsistent interface structure raises the memory load on users.

Efficiency: It is the ability of the web site to allow users to work quickly (Reilly et al., 2003; Nielsen, 2003) to attain their desired goal with the minimum number of clicks. In their research, Ferreira and Pithan (2005) point out that inefficiencies caused by technical errors or poor functionalities of the site can affect users' satisfaction negatively. According to De Marsico and Levialdi (2004), navigation by offering logical- and suitable links and paths through pages of a web site strongly influences the efficiency of the web site.

Satisfaction: It is the general pleasure a user feels making use of a web site. Satisfaction is primarily affected by the perceived efficiency and effectiveness, and emotions and thoughts arising from the usage of the web site (Ferreira and Pithan, 2005). Satisfaction may be a critical factor especially for non-compulsory usage of web sites, because it is generally expected that users tend to revisit a web site when they find it satisfactory on their first visit (Ha and Janda, 2008).

Functionality factors determined for e-health web sites are personalization/categorization of information, search options, differentiation of the information types, services/facilities and thesaurus.

Personalization/categorization of information: It is the flexibility offered by the web site to customize the quantity and content of the information accessed. Through personalization/categorization of information users can easily find the most pertinent information, such as children's health, pregnancy or age/gender related information, for their situation (Fisher et al., 2008). Liang et al. (2007) suggest that customization increases user satisfaction by limiting information overload on users with respect to their preferences (Pearson et al., 2007). Palmer (2002) points out that customization is an extension of the interaction provided by the web site.

Search options: Web sites offer both simple- and advanced search strategies and enable additional eliminations in retrieved results (Bertot et al., 2006; Kapoor and Goyal, 2007). The search function helps users quickly and precisely find what they are looking for. Nielsen (1999) suggests that web sites having over 200 pages should have a search component.

Differentiation of the information types: It includes both the variety of information offered under different categories such as well-being, medical research, patient stories etc. and the format of the information offered to different user groups such as layperson users or medical professionals (Fisher et al., 2008).

Services/facilities: They are purposive services/facilities offered to the customer to assist in achieving the related goal of the site (Kapoor and Goyal, 2007).

Thesaurus: It helps layperson users to fully understand the content by offering the meanings of the medical terms (Fisher et al., 2008).

To examine the relative importance of both usability and functionality factors for e-health web sites from the perspective of Turkish young users two Turkish health information web sites (Doktorumonline.net and Hekimce.com), which have the functionalities mentioned above, are evaluated. As seen above, not only functionality and usability of a system, but also some of their other factors are related to each other. Hence, these interactions create a complex model which consists of dependence and feedback among the factors. In evaluating web sites, such a model can be treated with the Analytic Network Process (ANP) proposed by Saaty (1996) in order to determine the relative importance of both usability and functionality factors.

ANALYTICAL NETWORK PROCESS

A complex structure of a decision making process consisting of interrelated factors, which means there are dependence and feedback among the factors, can only be modeled as a network. To deal with these interrelated factors in the decision making problem Analytical Network Process (ANP) is a convenient tool for decision makers as individuals or groups (Saaty, 1996).

ANP has three stages: structuring (design), assessment (comparison), and synthesis (computation).

At the structuring stage, after determining related factors and alternatives, if necessary, the relationships between each factor pairs are designated by experts. As a result, a network model, which consists of factors and relations among them, is constructed.

At the assessment stage, a nine-point scale suggested by Saaty (1996) is used by the decision makers to do pairwise comparisons of the factors in the network.

With Saaty's scale the question "of the dependent factors, which one influences the common factor more and how much more?" is answered. According to this scale, a value of 1 shows that both factors compared have equal influence levels on the affected factor, while a value of 9 shows that one factor has extremely more influence than that of the other on the affected factor. To obtain the aggregated group judgment, the geometric means of all individual paired-comparison judgments for each question are calculated. Using these aggregated group judgments, pairwise-comparison matrices are generated.

At the synthesis stage the relative importance of the factors is computed. Importance is viewed as the influence of the factors on a common goal. To synthesize aggregated judgments to compute the relative importance of the factors, the computation of the eigenvector for each pairwise-comparison matrix, the generation of a supermatrix and a weighted supermatrix (if necessary), and the computation of the convergence of the supermatrix (limit matrix) are requisite. The relative weights (desired priorities) of the factors in the decision network are the values of the limit matrix.

EVALUATION OF THE FACTORS AND THE SITES

First, the structuring stage was performed; the usability and functionality factors were determined. After the determination of usability and functionality factors, the group of experts whose working areas are usability engineering filled in a pairwise relationship matrix separately. To aggregate these matrices into a group matrix majority rule was used (Fig. 1). The asterisk (*) entered in this matrix indicated that there is a direct relationship of factor i to factor j: If factor i affects factor j, the cell aij was filled with an asterisk (*). Where there was no relationship, the cell was left blank. Then the ANP model representing the associations between factors are generated using the Super Decisions software.

| | USABILITY | | | | | | | | FUNCTIONALITY | | | | |
	A1	A2	A3	A4	A5	A6	A7	A8	B1	B2	B3	B4	B5
A1		*	*	*		*	*	*					
A2	*		*	*		*		*		*	*	*	
A3				*		*	*	*	*				
A4			*			*	*	*			*		
A5	*					*	*	*		*			
A6	*		*	*			*	*		*	*	*	*
A7			*	*		*		*					
A8			*	*									
B1	*	*	*	*		*	*	*		*	*	*	
B2	*	*	*	*	*	*	*	*					*
B3	*	*	*	*		*	*	*	*	*		*	
B4	*	*	*	*	*	*	*	*	*	*	*		*
B5	*	*	*	*		*	*	*		*	*	*	

(A1 navigation; A2 interaction; A3 learnability; A4 ease of use; A5 response time; A6 memorability; A7 efficiency; A8 satisfaction; B1 personalization/categorization of information; B2 search options; B3 differentiation of the information types; B4 services/facilities; B5 thesaurus)

Figure 1 Aggregated pairwise relationship matrix

In the judgment assessment stage, the second stage of the process, 40 bachelor students from Istanbul Technical University filled in the pairwise comparison questionnaire. An example question from the questionnaire can be seen in Fig. 2. In the questionnaire both the related usability and functionality factors and evaluated web sites were pairwise compared with respect to all usability and functionality factors. Before the respondents filled in the questionnaire they were separated into 2 groups. Each group was asked to consider themselves as sick and look up some

720

information about their disease in the evaluated websites, one group beginning with the site Doktorumonline.net and the other one beginning with the site Hekimce.com. Afterwards they filled in the questionnaire.

Of the factors given below which one influences "navigation" more and how much more?

1=Equally   3=Moderately more   5=Strongly more   7=Very strongly more   9=Extremely more

Interaction	9	8	7	6	5	4	3	2	1	2	3	4	5	6	7	8	9	Response Time
Response Time	9	8	7	6	5	4	3	2	1	2	3	4	5	6	7	8	9	Memorability
Memorability	9	8	7	6	5	4	3	2	1	2	3	4	5	6	7	8	9	Interaction

Figure 2 A part of pairwise comparison questionnaire

For all possible pairs the respondents evaluated the relative importance of the affecting factors on the affected factor. Then, to obtain the aggregated group judgment the geometric means of all paired comparison judgments for each question were calculated. Utilizing the Super Decisions software, the aggregated group judgment was formed into pairwise comparison matrices.

In the synthesis stage of the process, the relative importance of the factors and the global preferences for the e-health web sites were computed using the Super Decisions software for algebraic matrix computations. The output of the limit matrix can be converted to the descending priority order: the relative importance of the factors (Table 2). The global preferences for the evaluated web sites can be seen in Table 3.

Table 2 The importance of the factors

Sub-Factors	Priorities
B4.Services/Facilities	12.5554
B1.Personalization/Categorization of Information	12.3872
B2.Search Options	10.9038
B3.Differentiation of the Information Types	10.8426
A6.Memorability	9.9098
A2.Interaction	9.665
A1.Navigation	7.3712
A5.Response Time	7.2335
A4.Ease of Use	6.729
B5.Thesaurus	6.1019
A3.Learnability	4.282
A7.Efficiency	1.4681
A8.Satisfaction	0.5505

Table 3 The global preferences for the web sites

E-Health Site	Priorities (%)
Doktorumonline.net	51.8
Hekimce.com	48.2

CONCLUSION

As it can be seen in Table 2, the most important factors in terms of usability and functionality are "services/facilities" and "personalization/categorization of information" with a relative importance of 12% for each factor. "Search options" and "differentiation of information types" are the third and fourth factors with similar priorities, almost 10% for both. This shows that for young users of e-health web sites the functionality factors have higher importance than usability factors and that the users value personalized health information and additionally offered services, like opportunity to ask questions to a medical professional. The only functionality factor which has been evaluated to be less important than some usability factors is "thesaurus". For less educated users of the same web sites, "thesaurus" could have found a higher place in the priority list. In general functionality has been assessed to be slightly more important than usability for e-health web sites, with relative importance levels 53% for functionality and 47% for usability.

The most important usability factors for young users are "memorability" and "interaction" with almost the same importance level, 9%, followed by 2 other usability factors with importance levels of 7%, "navigation" and "response time". This shows that getting quick responses to their actions to find the relevant information with the least effort through menus, graphical components, sequences, and layout in a memorable way is important to young users of health web sites.

"Ease of use" and "learnability" are the usability factors which have lower importance levels in the evaluation. The reason for this result may be that the respondents are young and familiar with information technologies. This may have led the respondents to think that these factors are less important. The least important usability factors are "efficiency" and "satisfaction".

The global preferences of two e-health web sites show that Doktorumonline.net is considered slightly better than Hekimce.com in terms of usability and functionality factors.

The findings of this study show that for young users of e-health web sites, functionality is slightly more important than usability. They require highly personalized, easily and quickly found health information from memorable web sites which offers not only plain text information but also additional services.

REFERENCES

Bass, L., and John, B.E. (2003), "Linking usability to software architecture patterns through general scenarios." *The Journal of Systems and Software*, 66, 187-197.

Bertot, J.C., Snead, J.T., Jaeger, P.T., and McClure, C.R. (2006), "Functionality, usability and accessibility." *Performance Measurement and Metrics*, 7(1), 17-28.

Bontis, N., and De Castro, A. (2000), "The first world congress on the management of electronic commerce: review and commentary." *Internet Research: Electronic Networking Applications and Policy*, 10(5), 365-373.

Calero, C., Ruiz, J. and Piattini, M. (2005), "Classifying web metrics using the web quality model" *Online Information Review*, 29(3), 227-248.

Chakraborty, G., Srivastava, P. and Warren, D.L. (2005), "Understanding B2B corporate web sites' effectiveness from North American and European perspective" *Industrial Marketing Management*, 34, 420-429.

De Marsico, M. and Levialdi, S. (2004), "Evaluating web sites: exploiting user's expectations" *International Journal of Human-Computer Studies*, 60, 381-416.

Ferreira, S.M. and Pithan, D.N. (2005), "Usability of digital libraries" *OCLC Systems and Services*, 21(4), 311-323.

Fisher, J., Burstein, F., Lynch, K., and Lazarenko, K. (2008), ""Usability + usefulness = trust" : an exploratory study of Australian health web sites." *Internet Research*, 18(5), 477-498.

Furnell, S. (2005), "Why users can not use security." *Computers and Security*, 24, 274-279.

Ghose, S. and Dou, W. (1998), "Interactive functions and their impacts on the appeal of internet presence sites" *Journal of Advertising Research*, 38(2), 29-43.

Goetzinger, L., Park, J., Lee, Y.J., and Widdows, R. (2007), "Value-driven consumer e-health information search behavior." *Int. J. of Pharmaceutical and Healthcare Marketing*, 1(2), 128-142.

Gummerus, J., Liljander, V., Pura, M., and Van Riel, A. (2004), "Customer loyalty to content-based Web sites: the case of an online health-care service." *Journal of Service Marketing*, 18(3), 175-186.

Ha, H. and Janda, S. (2008), "An empirical test of a proposed customer satisfaction model in e-services" *Journal of Services Marketing*, 22(5), 399-408.

Kapoor, K. and Goyal, O.P. (2007), "Web-based OPACs in Indian academic libraries: a functional comparison" *Program: Electronic Library and Information Systems*, 41(3), 291-309.

Kavadias, C.D., Rupp, S., Tombros, S.L., and Vergados, D.D. (2007), "A P2P technology middleware architecture enabling user-centric services deployment on low-cost embedded networked devices." *Computer Communications*, 30, 527-537.

Klein, R.B., and Seffah, A. (2005), "Evaluation of integrated software development environments: Challenges and results from three empirical studies." *International Journal of Human-Computer Studies*, 63, 607-627.

Liang, T., Lai, H. and Ku, Y. (2007), "Personalized content recommendation and user satisfaction: theoretical synthesis and empirical findings" *Journal of Management Information Systems*, 23(3), 45-70.

Lu, M., and Yeung, W. (1998), "A framework for effective commercial web application development." *Internet Research: Electronic Networking Applications and Policy*, 8(2), 166-173.

Nielsen, J. (1999), "User interface directions for the web" *Communications of the ACM*, 42(1), 65-72

Nielsen, J. (2000a), "End of web design." *Useit.com Alertbox: Current Issues in Web Usability*, July, available at: www.useit.com/alertbox/20000723.html (accessed 7 September 2008).

Nielsen, J. (2000b), *Designing Web Usability*, New Riders Publishing, Indianapolis, IN.

Nielsen, J. (2003), "Usability 101: Introduction to usability." *Useit.com Alertbox: Current Issues in Web Usability*, August, available at: www.useit.com/alertbox/20030825.html (accessed on 7 September 2008).

Novak, T.P., Hoffman, D.L. and Yung, Y. (2000), "Measuring the customer experience in online environments: a structural modeling approach" *Marketing Science*, 19(1), 22-42.

Palmer, J.W. (2002), "Web site usability, design, and performance metrics" *Information Systems Research*, 13(2), 151-67.

Pearson, J.M., Pearson, A. and Green, D. (2007), "Determining the importance of key criteria in web usability" *Management Research News*, 30(11), 816-828.

- Reilly, E.F., Leibrandt, T.J., Zonno, A.J., Simpson, M.J. and Morris, J.B. (2003), "General surgery residency program websites: usefulness and usability for resident applicants", *Current Surgery*, Vol. 61 No. 2, pp. 236-240.

Roy, M.C., Dewit, O., and Aubert, B.A. (2001), "The impact of interface usability on trust in web retailers." *Internet Research: Electronic Networking Applications and Policy*, 11(5), 388-398.

Saarloos, D.J.M., Arentze, T.A., Borgers, A.W.J., and Timmermans, H.J.P. (2008), "A multi-agent paradigm as structuring principle for planning support systems." *Computers, Environment and Urban Systems*, 32, 29-40.

Saaty, T.L. (1996), *Decision-making with dependence and feedback: The Analytic Network Process*, RWS Publishing, Pittsburgh.

Seffah, A., Mohamed, T., Habieb-Mammar, H., and Abran, A. (2008), "Reconciling usability and interactive system architecture using patterns." *The Journal of Systems and Software*, 81(11), 1845-1852.

Tarafdar, M. and Zhang, J. (2005), "Analyzing the influence of web site design parameters on web site usability" *Information Resources Management Journal*, 18(4), 62-80.

Thelwall, M. (2000), "Effective web sites for small and medium-sized enterprises" *Journal of Small Business and Enterprise Development*, 7(2), 149-159.

Weir, C., McKay, I., and Jack, M. (2007), "Functionality and usability in design for eStatements in eBanking services." *Interacting with Computers*, 19, 241-256.

Yeung, T.A., and Law, R. (2004), "Extending the modified heuristic usability evaluation technique to chain and independent hotel websites." *International Journal of Hospitality Management*, 23, 307-313.

Chapter 80

Factors Affecting Acceptance of Internet as a Healthcare Information Source

Ceren B. Cakir, Tufan V. Koc

Industrial Engineering Department
Istanbul Technical University
Istanbul, Turkey

ABSTRACT

By the help of internet technology, people have opportunity to perform much kind of activities over the virtual world. Improving internet technology extends its usage area alternatives, and simultaneously its users. Mailing, information seeking and providing, librarianship and education are the widely used areas while money-containing issues like banking and shopping have limited use. As being one of the most widely used areas, information seeking and providing, also have some difficulty in the users perspective, as finding the most effective information among the poor or incorrect information pool. In healthcare dimension, that defective information may result much more serious problems. Growing evidence indicates that a significant proportion of internet health information consumers are engaging treatment strategies inconsistent with professional recommendations. To face with the misuse of the internet technology as in healthcare concept, it's essential to define and understand users' acceptance of internet as a healthcare information source.

In the literature, there is a huge variety of papers that study users' acceptance of internet technology in different usage areas and there are also research analyzing internet as being healthcare information source, that mostly focusing on its limitations. However there exists just little research on people's acceptance of internet as a healthcare information source.

This study aims to define factors affecting users' healthcare information seeking over internet. To this end, models proposed in studies will be thoroughly analyzed and a model will be generated.

Keywords: e-health, healthcare information, internet, acceptance

INTRODUCTION

By accelerating and simplifying daily life activities, Internet has become literally essential for almost every kind of people. In the reports of 2009 September, it was declared that 25,6% of world population uses Internet which increased about 380% in nine years (Internet World Stats, 2009). Similarly, in Turkey, overall Internet usage ratio increased from 33,4% (2007) to 40% (2009) in two years, and this recent usage ratio was around 62,2% for younger population and 88,5% for university and higher education (Statistics Institute of Turkey, 2009).

Although Internet enables many activities to be performed virtually, not the whole of these activities are considered as a necessity by the users. In Turkey, the most popular Internet usage activities of 2009 were sending and receiving e-mails (72,2%), reading or downloading online newspapers (70%), posting messages to chat sites, newsgroups or online discussion forums (57,8%), Playing or downloading games, images, music or film (56,3%), besides Internet banking was one of the least popular with 14% (Statistics Institute of Turkey, 2009). In general, information seeking and providing may be the main topic of the common activities.

Depending on the raise in the number of Internet users, the amount of Internet sites also increases. According to reports of 2010 February, although 207 million hostnames have been taken worldwide, approximately 84 million of them were active (Netcraft, 2010). Increasing Internet user number and easiness of taking a hostname triggers those high numbers. Indicated pollution in the virtual world results one of the most widely used areas of Internet, information seeking and providing, also having some difficulties in the users' perspective, as finding the most effective information among the poor or incorrect information pool.

In information seeking and providing perspective, information contains excessively much fields, meanwhile healthcare is the one of the severest. Importance of the issue comes from the risk of internalizing inaccurate information, continues with implementing without validating and possibility of resulting more serious situations. Because of there is no limitation about healthcare information providing, it is important to understand healthcare information seekers perspective.

Growing evidence indicates that a significant proportion of internet health information consumers are engaging treatment strategies inconsistent with professional recommendations (Weaver et al., 2009). To face with the misuse of the internet technology as in healthcare concept, it's essential to identify and understand users' acceptance of internet as a healthcare information source.

The study focuses on defining healthcare information seekers' perspective, understanding the factors affecting them, and their decision making mechanism. In this concept, literature has been reviewed and a model proposal consists of demographic, health-related and personal characteristic factors have been developed.

RESEARCH METHODOLOGY

Relevant literature is reviewed and models proposed in these studies are thoroughly analyzed in the research methodology step of the research. Following the research and the analysis, based on examined papers a model is designed.

LITERATURE REVIEW

Very limited studies have encountered about users' acceptance of internet as a healthcare information resource. Conversely, internet's being healthcare information source, and users' acceptance of internet technology has been studied separately widely in literature.

Acceptance of healthcare information provided on Internet was analyzed for specific user groups; like patients (Wilson and Lankton, 2004), physicians (Anonymous, 2000), elderly people (Botella et. al., 2009) and also more specific groups like women in menopause stage (Sillence et. al. 2004). Kwon and Kim (2009) conducted a survey to the people searching for cancer information, based on their preference in resource alternatives; library or internet. In a research published by Gauld (2010), a cross-natural study applied in Australia and New Zealand, investigated internet health information acceptance and reliability with respect to patients' having a e-mail contact with their doctor or not. One another study analyzed factors affecting trust and mistrust feelings about online health sites (Sillence et al. 2004). A review study have analyzed 52 articles in which overall 94 different factors have been tested, and 62 of them found to predict acceptance in at least one study (Or and Karsh 2010).

MODEL PROPOSAL

In the first step, it is important to understand users' acceptance of internet technology as a healthcare information resource. In this concept, previous experience will be questioned, to detect the rate of acceptance. Yearly and monthly frequencies of health information search are used in previous research (Gauld 2010, Freeman and Spyridakis 2009). Afterwards tendency to search in the future is supplemented with regard to technology acceptance models (Hu et al. 1999,), and tendency to implement information in daily life, with or without asking a professional is added with respect to recommendation of a previous e-health research (Or and Karsh 2009).

Demographic factors have been studied in most of the research. Age, gender and education level are the ones tested in most studies (Kwon and Kim 2009), while internet and computer experience are also included in some papers (Castaneda et al 2007).

Patient factors have been studied widely, as health status, experiencing a severe disease himself/herself or someone he/she cares (Rice 2006).

Not proposed in previous research, hypochondriasis is added as a personal characteristic that has an influence on health information seeking, which may affect the acceptance of online health information. Hypochondriacal disorder is defined as a chronic condition that is characterized by a preoccupation with fears of having a serious illness and also fears that are often based on a misinterpretation of bodily signs and symptoms (Katz and Zenger 1999).

RESEARCH MODEL

Developed research model is summarized in Figure 1.1. The acceptance of internet as a healthcare information source is the dependent variable as being the behavior tried to understand. Arguments have been told in the previous part become the independent variables, as demographic factors, health-related factors and personal characteristic.

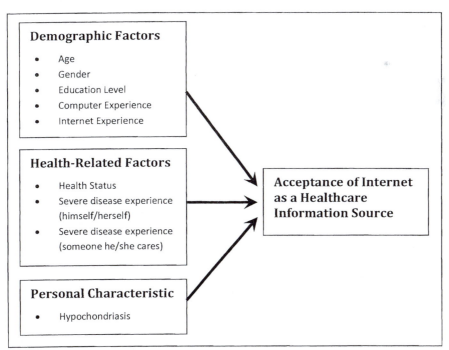

FIGURE 1.1 Research Model.

728

Acceptance of internet as a healthcare information source is the attitude of people against the health information on internet. It is concentrate on if they use it frequently, if they rely on the information, and if they apply or advice the information.

Demographic factors are age, gender, education level, computer experience and internet experience.

Health related factors are current health status of the user and severe disease background of that person or any people close to him/her that he cares.

Personal characteristic is a factor, to refer to different characteristics of people in the health manner. This factor expressed as hypochondrias, the people with high anxiety about getting sick.

CONCLUSION AND FURTHER STUDIES

As being the indispensable member of daily life, Internet gives an opportunity to humanity to reach any kind of information easily. In the healthcare dimension, it may become even dangerous, due to inaccurate and inadequate information. Therefore, it is essential to define the factors affecting people to accept these information, and their criteria. In this study factors affecting acceptance of internet as a healthcare information source have been determined by the help of previous research, and a model proposal have been developed.

In further studies, with the aim to analyze the model, a survey about healthcare information seeking will be conducted over the internet and the results will be discussed.

REFERENCES

Anonymous (2000), "e-health spanding fails to spur physician use, acceptance of internet.", *Direct Marketing*, 63(5), 60-61.

Botella, C., Etchemandy, E., Castilla, D., Banos, R.M., Garcia-Palacios, A., Quero, S., Alcaniz, M., and Lozano, J.A. (2009), "An e-health system for the elderly (butler project): a pilot study on acceptance and satisfaction." *Cyber Psychology & Behavior*, 12(3), 255-262.

Castaneda, J.A., Munoz-Leiva, F., and Luque, T. (2007), "Web acceptance model (WAM): moderating effects of user experience." *Information & Management*, 44, 384-396.

Freeman, K.S., and Spyridakis J.H. (2009). " Effect of contact information on the credibility of online health information." *IEEE Transactions on Professional Communication*, 52(2), 152-166.

Gauld, R. (2010), "Factors associated with e-mail and internet use for health information and communications among Australians and New Zealanders."

Social Science Computer Review, 0(0), 1-11.

Hu, P.J., Chau, P.Y.K., Sheng, O.R.L., and Tam, K.Y. (1999). "Examining the technology acceptance model using physician acceptance of telemedicine technology." *Journal of Management Information Systems*, 16(2), 91-112.

Internet World Stats (2009), www.internetworldstats.com

Katz, R.C., and Zenger, N. (1999) "Assessing hypochondriasis: findings on the survey of health concerns." *Journal of Psychopathology and Behavioral Assessment*, 21(3), 183-189.

Kwon, N., and Kim, K. (2009), "Who goes to a library for cancer information in the e-health era? A secondary data analysis of the health information national trends survey." Library & İnformation Science Research, 31, 192-200.

Malhotra, Y., and Galletta, D.F. (1999). "Extending the technology acceptance model to account for social influence: theoretical bases and empirical validation." *Proceedings of the 32nd Hawaii International Conference on System Sciences.*

Netcraft (2010), www.netcraft.com

Or, C.K.L., and Karsh, B.-T. (2009), "A systematic review of patient acceptance of consumer health information technology." *Journal of the American Medical Informatics Association*, 16(4), 550-560.

Rice, R.E. (2006), "Influences, usage, and outcomes of internet health information searching: multivariate results from the pew surveys." *International Journal of Medical Informatics*, 75, 8-28.

Sillence, E., Briggs, P., Fishwick, L., and Harris, P. (2004), "Trust and mistrust of online health sites." *Proceedings of the SIGCHI (Special Interest Group on Computer-Human Interaction) conference on Human factors in computing systems* 6(1), 663-670.

Statistics Institute of Turkey (2009), www.tuik.gov.tr

Weaver J.B., Thompson, N.J., Weaver, S.S., and Hopkins, G.L. (2009), "Healthcare non-adherence decisions and internet health information." Computers in Human Behavior, 25(6), 1373-1380.

Wilson, E. V., and Lankton, N. K. (2004), "Modeling patients' acceptance of provider-delivered e-health." *Journal of the American Medical Informatics Association (JAMIA)*, 11, 241-248.

Chapter 81

Ergonomic Considerations on the Implementation of Small-Caliber Trans-Nasal Gastroduodenoscopy

Kazuhiko Shinohara

School of Bionics
Tokyo University of Technology
Tokyo, 1920982, JAPAN

ABSTRACT

We identified problems regarding the implementation of small-caliber trans-nasal gastroduodenoscopy through work-analysis of physicians and nurses during endoscopic examination by trans-nasal gastroduodenoscopy with a 5.9 mm diameter endoscope compared with conventional trans-oral gastroduodenal endoscopy with a 10 mm diameter endoscope in this study. We found that trans-nasal gastroduodenoscopy entailed additional work and caused problems such as complicated preparations for the nasal cavity to avoid nasal bleeding and pain, difficulties in spatial orientation in the nasal and upper pharyngeal canal, and physical difficulties in maneuvering the small-caliber endoscope and accessory devices. These ergonomic problems can be resolved by the development of a packaged-preparation kit, appropriate training to address problems using a simulator, and adequate holding devices for the small-caliber endoscope through careful analysis of the workflow of the endoscopy staff.

Keywords: Trans-nasal gastroduodenoscopy, Ergonomic problem

INTRODUCTION

The use of small-caliber trans-nasal gastroduodenoscopy (NGS) for the diagnosis of upper gastrointestinal disease has become widespread in Japan owing to its advantage of causing minimal discomfort to the patient. However, there has been some resistance to its use by medical staff because of its impact on their workflow and workload. In this study, we analyzed the workflow of physicians and nurses during small-caliber NGS in order to identify the ergonomic problems of performing this examination.

MATERIALS AND METHODS

We compared the workflow of physicians and nurses during NGS and conventional trans-oral gastroduonenoscopy (OGS). The caliber of the trans-nasal endoscope was 5.9 mm and that of the conventional trans-oral endoscope was 10 mm. Endoscopic examination was performed under local anesthesia to the nasopharyngeal or oropharyngeal mucosa. The endoscope was inserted to the second portion of the duodenum, and the duodenum, stomach, esophagus and pharynx were closely observed. Both examinations (n=12, each) were performed by a same physician with 25years' clinical experience. (Fig 1, Fig 2)

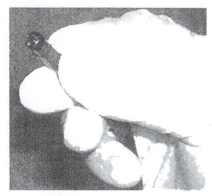 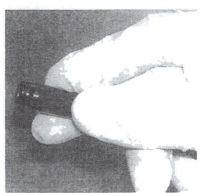

Endoscope for NGS(ϕ 5.9mm)   Endoscope for OGS(ϕ 10mm)

Figure 1. Endoscope for NGS and OGS

732

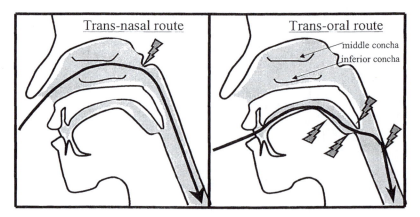

Figure 2. Insertion routes of trans-nasal endoscopy and trans-oral endoscopy

RESULTS

The additional workload associated with NGS arose for the nurses during the preparation of patients for the procedure. Nurses first must prepare and infuse the vasoconstrictors and local anesthesia to the vaporizer and syringe. They must prepare an elastic catheter with local anesthetic gel for dilation of the nasal canal. After breathability through each nasal canal is checked in order to decide which canal will be prepared for NGS, the vasocontrictor is vaporized and local anesthetic gel is infused into the nasal canal. The elastic catheter with local anesthetic gel is then inserted into the nasal canal. (Fig.3)

The additional workload and ergonomic problems associated with NGS arose for the physician during the NGS examination itself. At first, when they bend the tip of endoscope downward to insert it into the pharyngeal cavity from the nasal canal, the endoscopic image on the display moves upward. This is somewhat confusing for gastrointestinal endoscopists because they are not familiar with the spatial orientation of the nasal and upper pharyngeal cavities. (Fig.4) The shaft of the small-caliber endoscope for NGS is difficult to grasp and rotate. As the small-caliber endoscope is more flexible than the conventional endoscope, transmission of the pushing force to the tip of the endoscope is occasionally disturbed by bending of the endoscope's shaft. The lighting power for the distant view of the stomach is slightly dark, and the suction and insufflation powers are weak compared with the conventional trans-oral endoscope. Also, maneuverability of the biopsy forceps applicable for NGS is poor, and the amount of obtainable specimen is smaller due to their small caliber. The mean examination times without biopsy maneuvers were 234.0 s for NGS (SD 118, n=12) and 120.1 s for OGS (SD 31.6, n=12).

DISCUSSION

OGS with a flexible endoscope became widely performed in routine health check programs in Japan in the last half century. OGS examination has contributed to the diagnosis of early gastric cancer and has reduced deaths from gastric cancer, which is prevalent in eastern Asia. In addition, with OGS and accessory devices, early mucosal gastric cancers are resectable without laparotomy. However, OGS generates discomfort in the patient's throat which can cause nausea and vomiting. These discomforts result from compression of the tongue and dorsal pharyngeal wall by the endoscope, and cannot be completely removed by local anesthesia. Intravenous general anesthesia is more effective for reducing such discomfort during OGS, but there are problems with adapting intravenous general anesthesia for OGS in terms of cardiorespiratory suppression and the high economic costs incurred. The advantage of NGS is minimal discomfort to the patient's throat. This has led to widespread use of NGS with a small-caliber endoscope for the diagnosis of upper gastrointestinal disease in Japan. Also, since the nasal and upper pharyngeal cavities can be closely observed during NGS without the patient experiencing nausea and vomiting, it is expected that the possibilities for early diagnosis of nasal and upper pharyngeal cancers will be increased. (Fig.2)

The disadvantage and side-effects of NGS are as follows: potential nasal bleeding and osseous pain at the cranial portion of upper pharynx from the endoscope and contra-indication for patients with impaired coagulation function and/or with a small nasal orifice. Most of the additional workload created for nurses in preparations for NGS was found to concern the prevention of nasal bleeding and reduction of pain. Thus, these procedures cannot be omitted. However, the additional workload on nurses during preparation for the procedure can be reduced by standardization of the preparation procedure, with development of a packaged preparation kit that contains disposable preparation devices and premedication drugs. (Fig.3)

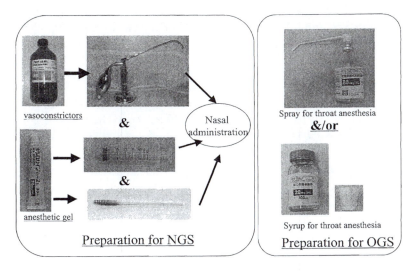

Figure 3. Preparation of NGS & OGS

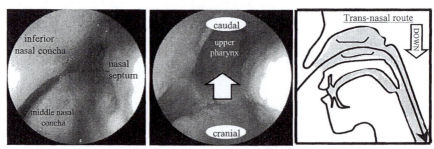

Figure 4. Endoscopic views are vertically inverted in the nasal meatus and upper pharynx

Additional workload and ergonomic problems for the physician are intrinsic to the method of trans-nasal insertion by small-caliber endoscopes. Discrepancies between the bending direction of the endoscope and endoscopic image on the display at the entrance of the pharynx can be easily handled through training using a suitable training problem and mock-up training for gastrointestinal physicians.(Fig.4) Impaired maneuverability of the small-caliber shaft of the endoscope could be helped by the development of adequate holding devices and/or gloves through careful analysis of the physicians' maneuver. Problems of impaired force transmission in the small-caliber endoscope, the weaker lighting for distant views, weaker suction and insufflation powers as well as the impaired maneuverability of the biopsy forceps are in a trade-off relationship to the caliber of the endoscope itself. As a tactical resolution to these problems, physicians should limit the purpose of NGS to the observation and diagnosis of upper gastrointestinal disease, and further examination and endoscopic treatment should be performed by conventional OGS.

The mean examination time for NGS was 234.0 s, 120.1 s longer than that for OGS. This difference stems from the gentle insertion of the endoscope through the nasal cavity and impaired suction and insufflation abilities, and an examination time of 234.0 seconds itself is acceptable for upper gastrointestinal examination. In addition, the diagnostic accuracy of NGS is reported to be equal to conventional OGS in Japan. Thus, the mechanical performance of small-caliber endoscopes for NGS can be evaluated as suitable for clinical use. However, further improvements and developments of accessory devices should be undertaken for the ergonomic improvement in the workload of medical staff and device usability.

CONCLUSION

The workflow and workload of medical staff and the ergonomic problems they faced while preparing and performing small-caliber NGS were investigated in this study. We found that NGS entailed additional work in preparation of the nasal cavity and caused maneuverability problems with the small-caliber endoscope and accessory devices. For the smooth and safe implementation of NGS as an examination for gastrointestinal disease, ergonomic problems should be resolved by developing a packaged preparation kit, providing appropriate training through simulation, and providing adequate holding and accessory devices for the small-caliber endoscope through further analysis of the workflow of the medical staff involved.

REFERENCES

Shinokara,K. (2008), Safety and ergonomic problems on the clinical practice of gastrointestinal endoscopy, Proceeding of 2nd International conference on Applied Ergonomics, (CDR)

Chapter 82

The Epidemiology of Work-Related Musculoskeletal Disorders Among Turkish Computer Users

Emel Ozcan, Sina Esmaeilzadeh, Halim Issever

Istanbul Faculty of Medicine
Istanbul University
Fatih, 34093, Istanbul, Turkey

ABSTRACT

Objectives: The aims of this study were to estimate the prevalence of work-related musculoskeletal disorders (WMSDs) among computer users in Turkish office workers and to investigate the association between WMSDs and both physical and psychosocial risk factors in workplaces.

Materials and Methods: The study base consisted of 400 computer users in Istanbul University, Istanbul Faculty of Medicine. The participants completed a modified version of the standardized "Nordic Musculoskeletal Questionnaire" and "Ergonomic Questionnaire" covering demographics data, musculoskeletal symptoms (MSS), physical and psychosocial risk factors during computing. Any participant who had self-reported MSS in his or her neck, upper back, shoulders, elbows, wrists, hands and/or low back regions during the last 12 months considered as a symptomatic subject. The intensity of MSS during the past 12-month was evaluated by Visual Analogue Scale (VAS). The case definition of WMSDs developed by the National Institute for Occupational Safety and Health's (NIOSH) was used to estimate the prevalence of these disorders. The relationships between ergonomic risk factors and WMSDs and also the association between intensity of MSS and psychosocial risk factors were estimated.

Results: The prevalence of MSS according to its frequency was 64.0% in neck,

63.7% in upper back, 59.8% in low back, 56.9% in shoulders, 38.9% in wrists and 21.9%, respectively. Total prevalence of WMSDs was 58.5% (n=182). While there was a positive association between VAS scores and duration of computer use per day (r=0.152, p=0.007), no relationships were found between VAS scores and age (r=0.083, p=0.146), and between VAS scores and body mass index (r=0.056, p=0.328). There was a positive relationship between wrists complaints and wrists radial/ulnar deviation (OR=1.741, 95% CI; 1.099-2.759, p=0.018), also between wrists complaints and wrists flexion/extension (OR=2.094, 95% CI; 1.299-3.377, p=0.002). Similar findings were observed between forward trunk inclination during sitting and both upper back complaints (OR=2.788, 95% CI; 1.658-4.689, p<0.001), and low back complaint (OR=2.788, 95% CI; 1.658-4.689, p=0.011). In subjects who had psychosocial risk factors, the mean of VAS scores was higher as compared to whom without any psychosocial risk factors and these differences were statistically significant (p<0.05).

Conclusions: These results suggest that WMSDs are common in Turkish computer users. Neck and upper back complaints are the most common MSS. Use of computer for a long period, existence of inappropriate body postures and presence of psychosocial risk factors in workplace are important risk factors in severity of MSS and in the development of WMSDs. This results support that intervention strategies that aim to minimize physical and psychosocial risk factors in workplace may be effective in prevention of the WMSDs among computer users.

Keywords: Work-related Musculoskeletal Disorders, Computer Users, Ergonomics, Epidemiology

INTRODUCTION

The presence of work-related musculoskeletal disorders (WMSDs) is a widespread problem among computer users (2000; Punnett and Bergqvist, 1997). Studies conducted by the National Institute for Occupational Safety and Health (NIOSH) demonstrated that more than 75% of workers who use a computer have reported occasional discomfort of the back, neck, and shoulders (Sauter et al., 1991). Cross-sectional studies of computer users have reported a prevalence of 10–62% of musculoskeletal symptoms (MSS) in the neck/shoulder region among computer users (Wahlström, 2005; Bergqvist et al., 1995).

WMSDs are believed to have a multi-factorial etiology. Non-neutral wrist, arm and neck postures, the work station design and the duration of computer work as well as psychosocial factors, such as time pressure are believed to interact in the development of these symptoms (Wahlström, 2005).

Despite above studies, there isn't any comprehensive research on Turkish computer users about prevalence and risk factors of WMSDs. The aims of this study are to estimate the prevalence of WMSDs among computer users in Turkish office workers and to investigate associations between WMSDs and workplace ergonomic as well as psychosocial risk factors.

MATERIAL and METHODS

STUDY DESIGN AND PARTICIPANTS

This cross sectional study was conducted among office workers in Istanbul University, Istanbul Faculty of Medicine who used computer in workplace. This study was approved by the local ethics committee of the Istanbul Faculty of Medicine. The study base consisted of 400 computer users. The participants completed a self administered questionnaire covering demographics data, MSS and workplace ergonomics and psychosocial risk factors. The case definition of WMSDs developed by the NIOSH was used to estimate the prevalence of WMSDs in computer users (Feuerstein et al., 2000; Hales et al., 1994). Any participant who had self-reported MSS symptoms in his/her neck, upper back, shoulders, elbows, wrists and/or back regions during the last 12 months considered as a symptomatic subject in our study. These regions were defined separately as shaded areas in a human drawing. The questionnaire for symptoms screening was a modified version of the standardized "'Nordic Musculoskeletal Questionnaire" (m-NMQ) (Kuorinka et al., 1987; Hildebrandt et al., 2001). The following individual characteristics were assessed: gender, age, height and weight (from which body mass index was subsequently calculated), marital status, formal education and employment. Labor characteristics of the study population were assessed by questions about the duration of computer use in workplace (hours per day and years at current job). Workplace ergonomic risk factors such as abnormalities in working postures and incorrect computer equipments positioning were evaluated by modified version Ergonomic Questionnaire (EQ). The questionnaire was based on the NIOSH Symptom Survey questionnaire, which has been shown to be reliable and valid for the purposes of workplace ergonomics studies. The questionnaire elicited physical workload during computing (body postures in neck, upper back, shoulders, elbows, wrists, hands and low back regions, also equipment poisoning such as monitor, keyboard and mouse positions) (Lewis et al., 2001; Delisle et al., 2006; Baron et al., 1996). The psychological demand and control items of the modified "Job Content Questionnaire" were used in assessment of psychosocial risk factors in workplace (Karasek, 1985; Ostry et al., 2001). The test-retest reliability of Turkish translation of these questionnaires has been demonstrated in beginning of the study.

STATISTICAL ANALYSIS

All statistical analyses were conducted using the Statistical Package for the Social Sciences (SPSS) version 16.0. The relationship between individual factors such as gender, marital status and education level and the presence WMSDs was evaluated by chi-square test. The correlation between severity of MSS and some of the continuous variables such as age and duration of computer use were estimated by Pearson's correlation test. The relationship between ergonomic risk factors at workplace and presence of regional MSS was assessed by chi-square test. The results were presented as odds ratios (OR), with 95% confidence intervals (CI) for each independent variable. The Mann-Whitney U test was used to compare the mean of VAS scores between subjects with and without psychosocial risk factors.

RESULTS

DEMOGRAPHIC CHARACTERISTICS OF THE SUBJECTS

A total of 400 questionnaires were distributed of which 311 responded [RR (response rate) = %77.75], of whom 215 (69.1%) were females and 96 (30.9%) were males. The mean age of subjects was 36.40 ± 8.06 years and the mean of BMI was 24.09 ± 4.14 years. 61.1% of subjects were married and 56.3% graduated from high school or below it. The mean of daily working time with computer was 5.99 ± 2.03 hours and the mean of working years with computer was 7.82 ± 4.90 years. Table 1 shows demographic and labor characteristics of the study population.

WORK RELATED MUSCULOSKELETAL DISORDERS AND SYMPTOMS

The 12-month self reported prevalence for WMSDs according to NIOSH criteria was 58.5%. In the study population, neck was the most common anatomical region for MSS in the past 12-month (64.0%), followed by the upper back (63.7%), low back (59.8%), shoulders (56.9%), wrists/hands (38.9%) and elbows (21.9%). In all anatomical regions the mean of VAS scores were higher in women than in men. These findings are summarized in Table 2.

Analyses of the relationship between intensity of MSS (mean of VAS scores in the past 12-month) and continuous variables such as age, BMI, daily duration of computer use in workplace and working years with computer in current position revealed that there was a positive correlation between VAS scores and duration of computer use per day (Pearson's r=0.152, p=0.007), also between VAS scores and duration of working years with computer (Pearson's r=0.171, p=0.003). There was not found any statistically significant correlations between VAS scores and age (Pearson's r=0.083, p=0.146), and between VAS scores and BMI (Pearson's r=0.056, p=0.328). These findings are summarized in Table 1.

Analyses of the association between WMSDs and individual characteristics of the subjects revealed that WMSDs were common in females (OR=1.654, 95% CI; 1.018-2.689, p=0.042) and the prevalence of WMSDs was higher in married participants (OR=1.824, 95% CI; 1.147-2.900, p=0.011). The association between demographic characteristics of the study population and presence of WMSDs are shown in Table 1.

740

Table 1: Demographic and labor characteristics of the study population and the correlation between some of the individual/labor factors and the intensity of MSS or the presence of WMSDs in the study population †

Continuous variables	Mean ± SD	r	p
Age (years)	36.40 ± 8.06	0.083	0.146
BMI (kg/m²)	24.09 ± 4.14	0.056	0.328
Daily computer use (hours)	5.99 ± 2.03	0.152	0.007**
Working years with computer (years)	7.82 ± 4.90	0.171	0.003**
VAS score	3.68 ± 2.57		

Categorical variables		n (%)	p	OR	95% CI
Gender	Male	96 (30.9)	0.042*	1.654	1.018–2.689
	Female	215 (69.1)			
Marital status	Single	121 (38.9)	0.011*	1.824	1.147–2.900
	Married	190 (61.1)			
Education level	≥ Bachelor's degree	136 (43.7)	0.743	0.927	0.588–1.461
	≤ High school	175 (56.3)			

† Total Number of Subjects = 311, **SD:** Standard Deviation, **BMI:** Body Mass Index, **WMSDs:** Work-related Musculoskeletal Disorders, **MSS:** Musculoskeletal Symptoms
* p <0.05, ** p <0.01.

Table 2: Prevalence of WMSDs and intensity of symptoms in the past 12 months and distribution of symptoms according to anatomical regions

	Female n (%)	Male n (%)	Total n (%)
Symptoms in			
Wrists/hands	89 (41.4)	32 (33.3)	121 (38.9)
Elbows	52 (24.2)	16 (16.7)	68 (21.9)
Shoulders	137 (63.7)	40 (41.7)	177 (56.9)
Neck	150 (69.8)	49 (51.0)	199 (64.0)
Upper back	152 (70.7)	46 (47.9)	198 (63.7)
Low back	133 (61.9)	53 (55.2)	186 (59.8)
Intensity of symptoms	3.93 ± 2.58	3.11 ± 2.57	3.68 ± 2.57
Total prevalence of WMSDs (according to NIOSH criteria)	134 (62.3)	48 (50.0)	182 (58.5)

WMSDs: Work-related Musculoskeletal Disorders, **NIOSH:** National Institute of Occupational Safety and Health.

ERGONOMIC RISK FACTORS

Analyses of the association between regional MSS and ergonomics risk factors, revealed that MSS in wrist/hand regions were significantly associated with wrist radial/ulnar deviation (OR=1.741, 95% CI; 1.099-2.759, p=0.018) and wrist flexion/extension (OR=2.094, 95% CI; 1.299-3.377, p=0.002). Similar findings were found between forward trunk inclination during sitting and presence of MSS in both upper back (OR=2.788, 95% CI; 1.658-4.689, p<0.001) and low back (OR=2.788, 95% CI; 1.658-4.689, p=0.011) regions (see Table 3).

Table 3: The association between ergonomic risk factors presence of MSS in different anatomical regions

WMSS	Ergonomic factors	OR	95% CI	p
Wrist/hand	Wrist radial or ulnar deviation	1.741	1.099 - 2.759	0,018*
	Wrist flexion or extension	2.094	1.299 - 3.377	0,002**
	Keyboard deviation to right or left	1.333	0.844 - 2.106	0,217
Elbow	Elbow flexion > or < 90°	1.326	0.758 - 2.319	0,321
Shoulder	Faraway (abnormal) mouse position	0.847	0.523 - 1.460	0,606
Neck	Neck Flexion or extension	1.027	0.643 - 1.640	0,912
	Monitor deviation to right or left	1.214	0.758 - 1.944	0,420
Upper back	Forward trunk inclination during sitting	2.788	1.658 - 4.689	<0,001***
Low back	Forward trunk inclination during sitting	1.866	1.150 - 3.027	0,011*

WMSS: Work-related Musculoskeletal Symptoms, * p <0.05, ** p<0.01, *** p<0.001.

WORK RELATED PSYCHOSOCIAL FACTORS

In subjects with psychosocial factors, the mean of VAS scores was higher than subjects without these factors and the statistical analysis revealed that the increasing of VAS scores were significantly (p<0.05) associated with the presence of psychosocial factors in workplaces (see Table 4).

Table 4: The association between intensity of MSS and psychosocial risk factors

Psychosocial risk factors		% (n)	Mean of VAS	Mann-Whitney U Z	P
The job requires learning new things†	Yes	17.7 (55)	5.35	-5.265	< 0.001***
	No	82.3 (256)	3.32		
The job involves a lot of repetitive work†	Yes	20.9 (65)	5.54	-6.524	< 0.001***
	No	79.1 (246)	3.18		
The job requires creativity†	Yes	5.8 (18)	5.78	-3.526	< 0.001***
	No	94.2 (293)	3.68		
The job has a variety of tasks†	Yes	17.0 (53)	4.60	-2.841	0.004**
	No	83.0 (258)	3.48		
The job requires working very fast††	Yes	29.6 (92)	4.98	-5.919	< 0.001***
	No	70.4 (219)	3.13		
The job requires working very hard††	Yes	10.9 (34)	4.65	-2.439	0.015*
	No	89.1 (277)	3.56		

MSS: Musculoskeletal Symptoms, † Control Items, †† Psychological Demand Items, * p <0.05, ** p<0.01, *** p<0.001.

DISCUSSION

In this cross-sectional study, the prevalence of WMSDs during the past 12 months was 58.5%. Studies among office workers have reported different results. From questionnaire studies, the prevalence of symptoms ranged from 17% to 75% (Halford and Cohen, 2003). The differences between studies could be due to different questionnaire tools used in researches, differences in populations studied or in the criteria used for defining MSS (Wahlström et al., 2004).

Similarly to many other studies (Punnett and Bergqvist, 1997; Tittiranond et al., 1999; de Zwart et al. 2001), the prevalence of WMSDs in our study was higher in women than in men. These gender differences in the prevalence of MSS might be explained by differences in exposures to work-related physical and psychosocial risk factors (Hooftman et al., 2005).

We found a positive significant correlation between duration of computer use and intensity of MSS in this research. Duration of computer use in workplace is a strong risk indicator for WMSDs among office workers (Punnett and Bergqvist, 1997; Jensen et al., 2001). In some researches no association was found between duration of computer use and WMSDs (Marcus et al., 2002; Korhonen et al., 2003). In this research there was a significant association between wrist ulnar/radial deviation and wrist MSS also between wrist flexion/extension and wrist MSS. There were few comparable studies on wrist posture and its relation to MSS in wrist/hand region. In a prospective study Marcus et al. were not find any relation between wrist postures and wrist/hand MSS (Marcus et al., 2002).

In this study, we did not find any association between neck posture and MSS in neck region. Sauter et al. reported similar findings to our study in neck region (Sauter et al., 1991). However many researchers reported an association between poor neck posture and a higher prevalence of WMSDs (Bergqvist et al., 1995; Cook et al., 2000; Hoekstra et al., 1994). We found a significant association between psychosocial factors and intensity of MSS among computer users. The association between psychosocial risk factors at work such as job demands and MSS has not been widely researched in office workers and conflicting evidence exists (Devereux et al., 1999). Exposure to psychosocial risk factors at workplace may contribute to the reporting of symptoms. Three studies conducted by the NIOSH in office workers who used computers have also shown an association between psychosocial factors and WMSDs in the upper extremity (Hurrel et al., 1996). One study revealed that these psychosocial factors were not associated with MSS of the upper limbs, but organizational and psychological factors were important factors (Marcus and Gerr, 1996).

CONCLUSIONS

According to the findings in this research, WMSDs is common among office workers specially in female computer users. Neck and upper back complaints are the most common MSS. Daily duration of computer use postural abnormality in wrist region, forward trunk inclination in sitting position and psychosocial factors in workplace are potential risk factors for the occurrence of the MSS. These results support that ergonomic intervention strategies that aim to minimize physical and psychosocial risk factors in workplace may be effective in prevention of the WMSDs among computer users.

REFERENCES

Baron, S., Hales, T., Hurrell, J. (1996), *"Evaluation of symptom surveys for occupational musculoskeletal disorders"*, Am J Ind Med. Volume 29 No. 6 pp. 609-617.

Bergqvist, U., Wolgast, E., Nilsson, B., Voss, M. (1995), *"Musculoskeletal disorders among visual display terminal workers: individual, ergonomic, and work organizational factors"*, Ergonomics. Volume 38 No. 4 pp. 763-776.

Brogmus, G.E., Sorock, G.S., Webster, B.S. (1996), *"Recent trends in work-related cumulative trauma disorders of the upper extremities in the United States: an evaluation of possible reasons"*, J Occup Environ Med. Volume 38 No. 4 pp. 401-411.

Burt, S., Hornung, R., Fine, L. (1990) *"Health Hazard Evaluation and Technical Assistance Report: Newsday, Inc. Melville, NY: National Institute of Occupational Safety and Health"*, HETA Report 89-250–2046.

Cook, C., Burgess-Limerick, R., Chang, S. (2000), *"The prevalence of neck and upper extremity musculoskeletal symptoms in computer mouse users"*, Int J Ind Ergonom. Volume 26 pp. 347–356.

de Zwart, B.C., Frings-Dresen, M.H., Kilbom, A. (2001), *"Gender differences in upper extremity musculoskeletal complaints in the working population"*, Int Arch Occup Environ Health. Volume 74 No. 1 pp. 21-30.

Delisle, A., Larivière, C., Plamondon, A., Imbeau, D. (2006), *"Comparison of three computer office workstations offering forearm support: impact on upper limb posture and muscle activation"*, Ergonomics. Volume 49 No. 2 pp. 139-160.

Devereux, J.J., Buckle, P.W., Vlachonikolis, I.G. (1999), *"Interactions between physical and psychosocial risk factors at work increase the risk of back disorders: an epidemiological approach"*, Occup Environ Med. Volume 56 No. 5 pp. 343-353.

Feuerstein, M., Huang, G.D., Haufler, A.J., Miller, J.K. (2000), *"Development of a screen for predicting clinical outcomes in patients with work-related upper extremity disorders"*, J Occup Environ Med. Volume 42 No. 7 pp. 749-761.

744

Hales, T.R., Sauter, S.L., Peterson, M.R., Fine, L.J., Putz-Anderson, V., Schleifer, L.R., Ochs, T.T., Bernard, B.P. (1994), *"Musculoskeletal disorders among visual display terminal users in a telecommunications company"*, Ergonomics. Volume 37 No. 10 pp. 1603-1621.

Halford, V., Cohen, H.H. (2003), *"Technology use and psychosocial factors in the self-reporting of musculoskeletal disorder symptoms in call center workers"*, J Safety Res. Volume 34 No. 2 pp. 167-173.

Hildebrandt, V.H., Bongers, P.M., van Dijk, F.J., Kemper, H.C., Dul, J. (2001), *"Dutch Musculoskeletal Questionnaire: description and basic qualities"*, Ergonomics. Volume 44 pp. 1038–1055.

Hoekstra, E., Hurrell, J., Swanso, N. (1994), *"Health Hazard Evaluation Report: Social Security Administration Teleservices Centers, Boston, MA: National Institute of Occupational Safety and Health"*, HETA Report 92-0382-2450.

Hooftman, W.E., van der Beek, A.J., Bongers, P.M., van Mechelen, W. (2005), *"Gender differences in self-reported physical and psychosocial exposures in jobs with both female and male workers"*, J Occup Environ Med. Volume 47 No. 3pp. 244-252.

Hurrell, J.J.J., Bernard B.P., Hales, T.R. (1996), *"Psychosocial factors and musculoskeletal disorders: summary and implications of three NIOSH health hazard evaluations of video display terminal work"* in: Beyond Biomechanics: Psychosocial Aspects of Musculoskeletal Disorders in Office Work, Moon, Sauter, (Eds.). London: Taylor and Francis. pp. 99–105.

Jensen, C., Ryholt, C.U., Christensen, C. (2001), *"Computer work-related factors and musculoskeletal symptoms"*, in: Prevention of Muscle Disorders in Computer Users: Scientific Basis and Recommendations, Sandsj, Kadefors, (Eds.). Goteborg, Sweden, pp. 18–22.

Karasek, R. (1985), *"Job Content Instrument Users Guide: revision 1.1"*, Los Angeles, CA: Department of Industrial and Systems Engineering, University of Southern California.

Karlqvist, L.K., Hagberg, M., Köster, M., Wenemark, M., nell, R. (1996), *"Musculoskeletal Symptoms among Computer-assisted Design (CAD) Operators and Evaluation of a Self-assessment Questionnaire"*, Int J Occup Environ Health. Volume 2 No. 3 pp. 185-194.

Korhonen, T., Ketola, R., Toivonen, R., Luukkonen, R., Häkkänen, M., Viikari-Juntura, E. (2003), *"Work related and individual predictors for incident neck pain among office employees working with video display units"*, Occup Environ Med. Volume 60 No. 7 pp. 475-482.

Kuorinka, I., Jonsson, B., Kilbom, A., Vinterberg, H., Biering-Sorensen, F., Andersson, G., Jørgensen, K. (1987), *"Standardised Nordic questionnaires for the analysis of musculoskeletal symptoms"*, Appl Ergon. Volume 18 No. 3pp. 233-237.

Lewis, R.J., Fogleman, M., Deeb, J., Crandall, E., Agopsowicz, D. (2001), *"Effectiveness of a VDT ergonomics training program"*, Int J Ind Ergonom. Volume 27 No. 2 pp.119-131.

Marcus, M., Gerr, F., Monteilh, C., Ortiz, D.J., Gentry, E., Cohen, S., Edwards, A., Ensor, C., Kleinbaum, D. (2002), *"A prospective study of computer users: II. Postural risk factors for musculoskeletal symptoms and disorders"*, Am J Ind Med. Volume 41 No. 4 pp. 236-249.

Marcus, M., Gerr, F. (1996), *"Upper extremity musculoskeletal symptoms among female office workers: associations with video display terminal use and occupational psychosocial stressors"*, Am J Ind Med Volume 29 pp. 161–170.

Ostry, A.S., Marion, S.A., Demers, P.A., Hershler, R., Kelly, S., Teschke, K., Hertzman, C. (2001), *"Measuring psychosocial job strain with the job content questionnaire using experienced job evaluators"*, Am J Ind Med. Volume 39 No. 4 pp. 397-401.

Punnett, L., Bergqvist, U. (1997), *"Visual display unit work and upper extremity musculoskeletal disorders. A review of epidemiological findings"*, Arbete och Halsa. Volume16 pp. 1–161.

Sauter, S.L., Schleifer, L.M., Knutson, S.J. (1991), *"Work posture, workstation design, and musculoskeletal discomfort in a VDT data entry task"*, Hum Facltors. Volume 33 No. 2 pp. 151-167.

Tittiranonda, P., Burastero, S., Rempel, D. (1999), *"Risk factors for musculoskeletal disorders among computer users"*, Occup Med. Volume 14 No. 1 pp. 17-38.

Wahlström, J., Hagberg, M., Toomingas, A., Wigaeus Tornqvist, E. (2004), *"Perceived muscular tension, job strain, physical exposure, and associations with neck pain among VDU users; a prospective cohort study"*, Occup Environ Med. Volume 61 No. 6 pp. 523-528.

Wahlström J. (2005), *"Ergonomics, musculoskeletal disorders and computer work"*, Occup Med. Volume 55 No. 3 pp. 168-176.

Chapter 83

Reduction of Physical Load at the Waist of a Care Worker Using Muscle Suit

*Masato Sakurai[1], Takashi Inoue[1],
Miwa Nakanishi[2], Sakae Yamamoto[1]*

[1]Tokyo University of Science
Japan

[2]Chiba University
Japan

ABSTRACT

To examine the physical load at the waist of a care worker when shifting the position of a patient in with and without a wearable robot (Muscle Suit), we measured the spine angle in forward bend during the task. Using Muscle Suit improves the posture of high forward bend for the observer with supporting the muscle at the waist. It is suggested that there is a possibility to reduce the risk of low back pain for care worker in daily task.

Keywords: Care worker, waist, low back pain, wearable robot, body mechanics

INTRODUCTION

According to world population prospects from the United Nations Population Division database revised in 2008, the world population over the age of 65 will reach approximately 523 million, 7.6% of the total in 2010. In Japan it will be about 29 million, 22.6% of the total Japanese population.[United Nation Population Division (2008)] As many countries progress to more aged societies and elderly people who need nursing care correspondingly increase year to year, the role of the care worker becomes more and more important.

Several daily tasks cause the care workers low back pain.[National Institute for Occupational Safety and Health (1997)] For example, a care worker changing the diaper of a patient in bed needs to bend over and maintain a posture of high forward bend (30 – 70 degrees between the spine and vertical axis of body). The muscle at the waist of care workers must bear a physical load that is one cause of low back pain.

A wearable robot (Muscle Suit), just developed at Hiroshi Kobayashi laboratory, Tokyo University of Science, that supports human movement has been studied from the application point of view.[Aida et al. (2009)] Fig. 1 shows Muscle Suit used in this paper. The Muscle Suit consists of McKibben artificial muscles to the physical load for the wearer when moving.

A previous study examined the physical load at neck and waist with or without a Muscle Suit when making a bed, and it is suggested that a Muscle Suit is effective in reducing the physical load at the waist of a care worker.[Nakanishi and Yamamoto (2009)] In order to shed some light on this suggestion, we studied the reduction of physical load and resulting deformation of posture by the reduction in physical load at the waist of a care worker when wearing a Muscle Suit for the waist.

Fig. 1 Muscle Suit.

PURPOSE

The purpose of this paper is to examine the physical load at the waist of a care worker when shifting the position of a patient with and without a wearable robot (Muscle Suit). Also it is to evaluate the effects of supporting the muscle at the waist as well as improving the posture of a care worker with the Muscle Suit.

METHODOLOGY

TASK

The task is to turn a patient on their back and to the right on a bed with and without a Muscle Suit since it is confirmed from a previous experiment that this task requires the care worker to bend forward at the waist about 70 deg resulting in a heavy physical load at the waist. In nursing facilities, the care worker has this task of four – seven times per a day for a patient in a case such as changing a diaper of a patient in bed. Fig. 2 shows each scene of task in this experiment. The worker is

748

observed to maintain a posture of high forward bend during this task. The observers were asked to hold a left arm of a patient with their left hand and a left side of back or left leg of a patient with their right hand in this task. It was used a model of human body (*Sakura*, KYOTOKAGAKU) as a patient.

Fig. 2 The task in this experiment. It is to turn a patient on their back and to the right on a bed.

CONDITIONS

Each observer performed this task five times for both conditions of with and without a Muscle Suit, respectively. The interval task to task was 15 sec. To remove the order effect in two conditions, observers were divided into two groups, one doing the task with Muscle Suit at first and the other doing the task without it at first. The inter-condition interval had enough time to refresh from the physical load in each condition. To learn how to use Muscle Suit, each observer had an adequate practice with it.

APPRATUS

Fig. 3 shows the apparatus in this experiment. In the experimental room, the model of human body as a patient, a manikin, layed in bed. The height of bed was 48cm and the weight of the manikin was 21kg. The observer stood on the left side of the bed and performed the task there. A camcorder (HDR-S12, SONY) was placed at a distance of 231cm from the bed to record the posture and movement of the observer during the task. The Muscle Suit's weight was 3.5kg, and it was connected to the air compressor (6-4, JUN-AIR) to

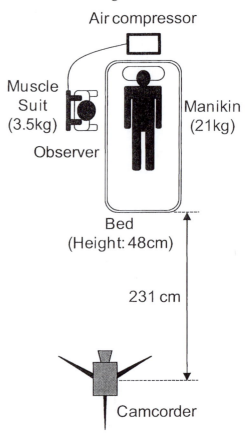

Fig. 3 Apparatus.

send air to the McKibben artificial muscles of the Muscle Suit to reduce the physical load for the wearer.

MEASUREMENT

We measured the maximum value of spine angle in forward bend during the task from the images in the record of the camcorder for each observer. Fig. 4 shows the spine angle of forward bend in this experiment. It was defined as the angle between the vertical axis and the line of the right shoulder to the right waist of the observer. It indicates a high physical load at the waist with increasing spine angle of the observer with a high risk for lower back pain. In this paper, the spine angle represents the index of physical load at the waist of the observer.

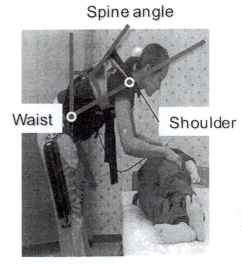

Fig. 4 Spine angle in this experiment.

After the experiment, each observer was asked to fill out the questionnaire regarding to the movement when wearing Muscle Suit and the capability in nursing facilities. This is also an important evaluation from the point of view of a care worker.

OBSERVERS

Six care workers (two males and four females) participated in this experiment. They work at the nursing facilities (Ginnofune Yokohama, Yokohama) for approximately five years in average. The average of age and height was 46 years old (S.D. 8.3) and 161cm (S.D. 7.4), respectively.

RESULTS AND DISCUSSIONS

MAXIMUM SPINE ANGLE

Fig. 5 shows the results of maximum spine angle for each observer with and without the Muscle Suit. The horizontal axis represents each observer and the vertical axis indicates the average values of the maximum spine angle. The values are the average of the maximum spine angle measured over five times for each observer, and the error bars represent the standard deviation in intra-observer variability for each condition.

As shown in Fig. 5, the values of the maximum spine angle with Muscle Suit are smaller than those of without it in all the observers. Four of the observers indicate significantly differences in these conditions. It is suggested that using the Muscle Suit improves the posture of high forward bend for observers whose muscles are supported at the waist.

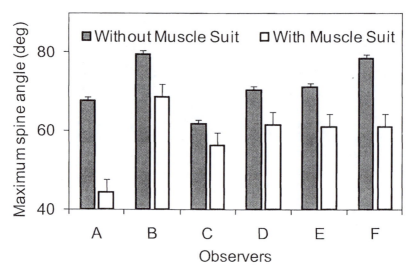

Fig. 5 The results of maximum spine angle for each observer with or without Muscle Suit conditions during the task.

RELATIONSHIP BETWEEN MUSCLE SUIT AND BODY MECHANICS

It is well-known that "Body Mechanics" is a technique for care worker to reduce load applying principals of physics and mechanics. It is recommended to use body mechanics when performing the task such as lifting in the work place.[Gassett et al. (1996), Karahan et al. (2009)] In a task such as shifting the position of a patient in this experiment, a low center of gravity for care worker is recommended with bending at the knees and ankles when performing the task. According to this, the waist of a care worker goes down, and it is not necessary to maintain a posture of high forward bend in the spine angle. To evaluate the relationship between the uses of Muscle Suit and body mechanics, the knee and ankle angle of each observer was measured with and without Muscle Suit in the task from the images recorded by the camcorder. Fig. 6 shows the definition of knee and ankle angles of the observer in this experiment. The knee angle is the angle between the right thigh and leg with the right knee in the vertex, the ankle angle is the angle between the right leg and vertical axis with the right ankle in the vertex for observer.

Fig. 7 (a) and (b) show the results of the knee and ankle angles for each observer based on the average of five repetitions. The vertical axes of Fig. 7 (a) and (b) indicate the angles of knee and ankle without the Muscle Suit, respectively. The horizontal axes of those graphs represent the angle of knee and ankle with the Muscle Suit, respectively.

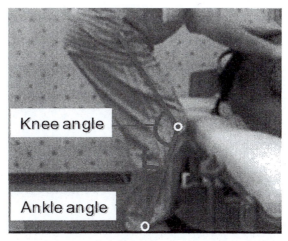

Fig. 6 Knee and ankle angles in this experiment.

In Fig. 7 (a), the right-bottom region refers to the use of body mechanics when wearing the Muscle Suit compared of the results without it since decreasing the knee angle of the observer lowers their center of gravity. In Fig. 7 (b), the left-top region refers to the use of body mechanics when wearing the Muscle Suit compared to the results without it since increasing the ankle angle of the observer lowers their center of gravity as well. As shown in Fig. 7 (a) and (b), more than four observers bend their knees and ankles with the Muscle Suit compared with the results without it, and it seemed to work well for observers in terms of the use of body mechanics with wearing the Muscle Suit. It is suggested that care workers can effectively use body mechanics wearing Muscle Suit compared with the results without it.

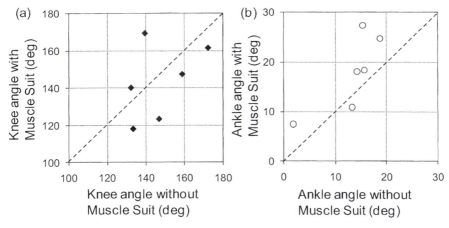

Fig. 7 Knee and ankle angles with and without Muscle Suit. (a): Knee angle. (b): Ankle angle.

752

SUBJECTIVE EVALUATION IN CARE WORKERS

Fig. 8 shows the results of the questionnaire regarding the effect at the waist and the ease of movement with the Muscle Suit based on 10 observers responses including the pilot experiment. In Fig. 8, 70% of all the observers indicated an effect at the waist with the Muscle Suit. On the other hand, the 30% of the females (three) did not feel it. The mean height of these females is 152cm, much lower than that of the average of 10 observers. Therefore, it is considered that the Muscle Suit used in this experiment does not work well for lower height observers because of the size.

As shown in Fig. 8, most of observers feel a limitation in the movement of the upper body with the Muscle Suit. The Muscle Suit used in this experiment is strapped to the shoulder by the observer in Fig. 1, consequently it is not easy to move their upper body. From the capability point of view in nursing facilities, the variable size of the Muscle Suit and the free movement of the upper body when wearing the Muscle Suit will be future work.

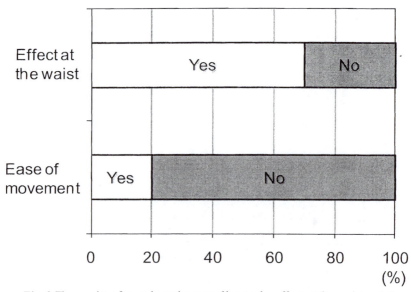

Fig. 8 The results of questionnaire regarding to the effect at the waist and the ease of movement with Muscle based on 10 observers responses.

CONCLUSIONS

Using Muscle Suit improves the posture of high forward bend for observers by supporting the muscles at the waist when shifting the position of a patient. It is suggested that there is a possibility to reduce the risk of low back pain for care worker in daily tasks. In addition, wearing the Muscle Suit lowers the observer's center of gravity. Therefore, it is considered that the care worker can effectively use body mechanics in tasks such as this experiment. From the subjective evaluation, most of observers feel the effect at the waist of observer with the Muscle Suit indicating the possibility for use of the Muscle Suit in nursing facilities

REFEFENCES

Aida, T., Nozaki, H., and Kobayashi, H. (2009), "Development of muscle suit and application to factory laborers." *Proceedings of the 2009 IEEE International Conference on Mechatronics and Automation* (Changchun, China, 2009), 1027-1032.

Gassett, R.S., Hearne, B., and Keelan, B. (1996), "Ergonomics and body mechanics in the work place." *Orthop. Clin. North Am.*, 27(4), 861-879.

Karahan, A., Kav, S., Abbasoglu, A., and Dogan, N. (2009), "Low back pain: prevalence and associated risk factors among hospital." *J. of Advanced Nursing*, 65(3), 516-524.

Nakanishi, M. and Yamamoto, S. (2009), "Effectiveness of using muscle suit in daily nursing care scenes." *J. of Ergonomics in Occupational Safety and Health*, 11 Supplement, 90-93 (in Japanese).

National Institute for Occupational Safety and Health (1997), "Musculoskeletal disorders and workplace factors." *NIOSH Publication* No.97-141.

United Nations Population Division (2008), "world population prospects: the 2008 revision population database." http://esa.un.org/unpp/.

Chapter 84

Practicing User-Centered Design to Reduce Call Time and Improve Accuracy in a Medical Triage System

Becky Reed, Patricia Hyle, Judy Dundas, Steven Schneider

Healthwise
Boise, ID 83702, USA

ABSTRACT

This work is a case study of the redesign and deployment of a medical triage system that supports nurse users performing symptom triage in a call center environment. The significance of this project comes from the user-centered design approach to achieving product design.

Keywords: Case Study, User-centered Design, Usability Testing, Nurse Call Center, Telephone Triage, Tele-Health

INTRODUCTION

Thousands of call center nurses use computer-based symptom triage systems to answer health-related questions, triage callers' symptoms, and provide care recommendations ranging from calling for emergency service to using home treatment. This work outlines the iterative user-centered design process utilized during redesign, testing, and launch of such a system.

Improving the quality of care advice and managing the costs associated with time spent per call are key measures of success for this system. With this in mind, the primary objectives included reducing overall call time, increasing treatment recommendation reliability, and improving the overall user experience of nurses.

Secondary objectives included reducing the training time required to train new nurses on the system and minimizing the amount of retraining required of nurses experienced in the existing system.

EXPLORING THE PROBLEM

The current triage system began as a hypertext system with aspects familiar to traditional, paper-based protocols. For each chief complaint (symptom), all possible triage questions were displayed, all possible dispositions (care recommendations) were available, and general health education information was incorporated between and within questions.

Utilization data and experiential feedback suggested that the depth of content provided and the hypertext model of the system might be placing greater system operation responsibility on the nurse-users than necessary. In addition, the "all-available," manual nature of the hypertext system meant that there were no systematic controls on the reliability or consistency of a treatment recommendation. An early goal of this project was to preserve the medical accuracy of the triage process while redesigning the system to be in step with nurse users' workflow and expectations.

Because such a large number of nurses were using the existing system as a primary application, the project had further constraints.

- Triage topics could not be updated incrementally. At rollout of the new system, all topics needed to be converted to the new process to ensure consistency across topics and to reduce confusion by the nurse end-users.
- Disruption to the call center staff had to be minimal. Call center management, technical support, and nurses had to be fully engaged throughout the project lifecycle, providing input and feedback to the requirements and design to ensure acceptance of the redesigned system.
- Resistance to change, even with improvements, was predictable and had the potential of being significant to nurse satisfaction and effectiveness. Such a significant change required early introduction of the changes. And allowing active participation of key stakeholders in the design reduced resistance to the changes.

AUDITING THE EXISTING APPLICATION

An audit of the existing system established preliminary requirements and pinpointed areas for further observation and analysis. The audit included heuristic evaluations, user feedback, direct observation, and task analysis.

A three-part heuristic evaluation was conducted of the existing system. Heuristic evaluation is often a component solely assigned to an interface or interaction designer. In this case, the medical team offered perspective on medical logic and known areas for improvement. The content team offered perspective on language of the protocol questions and supporting text. They were also tasked with assessing how plain language could be used to better fit the telephonic modality of the system. The design team focused on interface components and behaviors. These evaluations were combined and used as a single steering document.

Feedback about the existing system was also collected, clarified, and moved into the requirements stream for consideration. One obstacle we had with a significant portion of the feedback was that it was symptomatic of the existing system and didn't offer much insight into underlying needs. This type of feedback was noted for exploration during direct observation or user interviews.

Direct observation requests were initiated early in the process, and several members of the team were able to do live 1:1 observation with nurse users. While direct observation is frequently used during training to assess nurses' system competency, this observation was designed to assess system performance. The observations also included design and product team members to allow them first-hand experience with call center workflow and the tasks nurse users perform. The observers used the time between calls to conduct exploratory interviews to better understand the gaps in the existing system. Non-medical team members debriefed with medical team domain experts to ensure observation context was captured accurately.

Heuristic evaluation and observation findings evolved into four key product development goals:

1. Improve triage workflow to reduce nurse variability in the triage process. Because the existing system was so complex, new nurses often had trouble determining what they must discuss with the caller and what they could easily circumvent. Experienced nurses knew how to use the hypertext structure of the system. Faster calls were often a result of a nurse "knowing just where to go."

2. Improve the usability and availability of essential triage information. Nurses were observed scanning large pieces of information to find explanatory text or definitions to help callers understand questions or medical concepts. Other issues, such as compound questions, were difficult for nurses to communicate.

3. Improve the overall user experience of the triage nurse, and reduce the amount of information nurses were expected to retain in working memory while speaking to a caller.

4. Make the conclusion of the triage clear. In the hypertext structure, nurses were given all possible outcomes. They had to work through several triage possibilities to ensure that a more urgent triage outcome was no longer possible. Nurses knew they'd reached the conclusion of their triage

because they were trained to know the signs, not because the interface gave indictors.

Two common triage situations were used to conduct task analysis (KLM-GOMS model) for time benchmarks. Mouse travel, scroll distance, and clicks/keystrokes were also measured. Reducing scrolling and transitioning between screens represented a significant opportunity for the redesign to reduce call time or afford nurses more clinical time within a call. The task analysis showed us those system usage tasks that could be automated or redesigned to reduce the user's responsibility.

PROTOTYPING THE REDESIGN

While the technical and user experience teams were developing the new user interface, the medical and content teams were redesigning the triage logic and content. These two efforts came together in the triage prototype. Early in the process the prototype was developed interactively using Axure, a prototyping application. The interactive prototype gave the disparate teams the opportunity to model each other's perspective and allowed feedback to be quickly incorporated. Medical and product teams could run scenarios with a sample triage topic to get an immediate sense of the experience. Design teams could show areas of opportunity and concern in a far more compelling way than conjecture and opinion.

USER TESTING AND EVALUATION

Testing groups were recruited very early in the project. We chose to recruit local samples of call center nurses who were unfamiliar with the current system, as well as national samples of nurses who were system users. Usability testing sessions were conducted in a lab environment with local samples to evaluate product design. Heuristic evaluation and quality assurance testing were also conducted on protocols, medical logic, and product design with the national group.

Our interactive prototype afforded the opportunity to do usability testing and demonstrations long before development resources were sourced for the project. We completed two types of early prototype reviews. First, local samples of triage nurses were recruited to do lab-based usability testing of early prototype versions. This process allowed us to identify and adjust significant problem areas before moving into the field.

The second prototype review involved web-based prototype demonstrations to the clinical and operations management staff of existing call center clients. These demos included discussions about the medical and content development process. This early feedback allowed the medical and content teams to discuss the rationale for the product changes, provided an overview of the development process, and gave clients an early view into the scope of the planned changes. Clients gave feedback on both medical and product changes. The early view built client stakeholders' confidence in the process and allowed them to express their views on the revisions to the product. These early conversations also enabled the project team to request on-site review and feedback by call center staff.

Field studies began with the intent of observing the nurses' ability to navigate the new system without coaching, reach the expected treatment recommendation, and identify any obstruction points in the triage workflow. During field studies, several triage protocols were available so that nurses could triage scenarios common to them. Field studies included task-based usability testing as well as Q&A between nurses and a domain expert.

When the field studies were completed, we released representative prototypes to nursing managers and call center stakeholders. We encouraged them to evaluate the triage decision trees, and we incorporated their feedback into the iteration process.

ITERATIONS AND REVISIONS

Numerous prototype revisions were iterated based on nurse feedback. Highlighted are a few of the changes that emerged as a result of different evaluation methods.

Task Analysis - Auto Scroll Functionality

Initial task analysis, scrolling measurement, and direct observation identified a need to reduce overall scroll. While scrolling was reduced through revision to the information model, the interface design also specified an "auto scroll" feature. During development, this feature proved difficult to implement. But the task analysis and scrolling measurement provided our team with clear, quantitative evidence of the importance and benefit of succeeding with this feature.

Usability Testing - Triage Dashboard

Early prototypes included a dashboard feature. Very early usability testing sessions revealed that a comprehensive dashboard did not offer as much benefit as the screen real estate it occupied. Three iterations of usability testing and more workflow analysis (through direct observation) allowed this dashboard concept to evolve into a simpler and more concise form.

Observation - Warning Message Behavior

Certain pieces of medical information are critical for nurses to cover with patients. The design team initially translated this importance into modal containers of varying forms. But observation of live calls and input from domain experts showed that a modal dialogue solution would be an unmerited disruption.

FINAL DESIGN

BEHAVIOR AND INTERACTION

The final version of the system presents nurses with a structured triage system. Triage questions are ordered based on the severity of the caller's symptoms. Multiple iterations though similar questions of differing severity are no longer needed. Instead, questions about a single symptom are asked one time, and the system remembers the responses and uses them to generate the final care recommendation. In addition, questions are conditionally displayed depending on the answers to previous questions. Nurses no longer have to remember or combine responses to different questions to rule out possible concerns. The triage logic is built into the system itself.

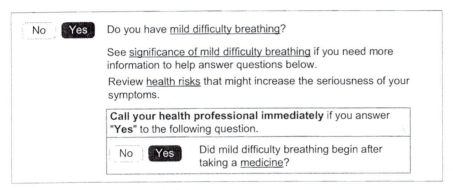

FIGURE 1.0: Triage Question (before).

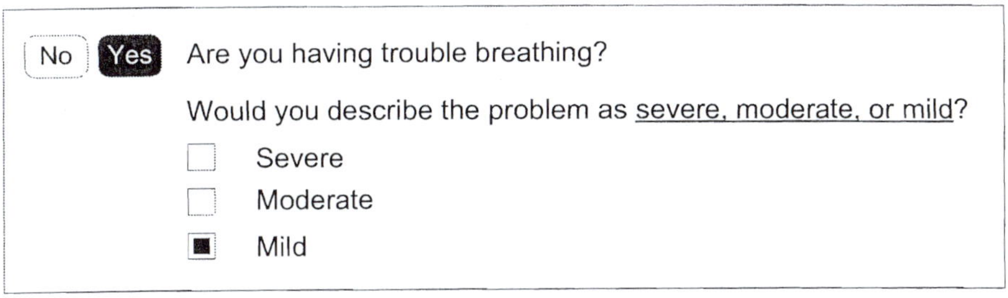

FIGURE 2.0: Triage Question (after).

INFORMATION MODEL

Information model changes began with reduction of the general health information provided during a triage. Content not critical during the triage experience was moved to the end of the triage. Information deemed critical or necessary but supplemental was rewritten to be concise and reflective of the question it was paired with. In the end, the information the nurse needs to present to the patient during the call is directly associated with the triage question. Nurses no longer rely on their knowledge of the system to find the content they need to help callers understand their symptoms or the nurse's recommendations.

Particular attention was paid to the telephonic context of the triage interaction. Triage questions and supplemental information are written using plain language that does not require a nurse to translate from medical to consumer-friendly language.

INTERFACE DESIGN

Interface design focused on reducing the system operations burden on the nurses so that they can focus on the callers' needs. All questions are now available within a single container that auto-scrolls as the triage is performed. Interface controls are unified and are in close proximity to reduce mouse travel. Supplemental information has been differentiated from primary information using distinct visual treatments and system behaviors. Current and remaining questions in the triage have been given design treatments to focus the user's attention on the current question. And to reduce change where possible, we have intentionally designed the primary interface controls in the new system with a very familiar feeling. Heavily used "yes" and "no" controls remain in the same position and order.

Finally, triage completion has been improved. A modal "disposition reached" behavior has been incorporated to clearly indicate to nurses when they have completed sufficient triage to confidently provide a care recommendation to the caller. We also provide access to the appropriate health education next steps that might accompany a care recommendation (from first aid to home care) in this location.

Yellow "Yes"

SEEK CARE TODAY

Based on your answers, **you may need care soon.** Symptoms probably will not get better without medical care.

- Call your doctor today to discuss the symptoms and arrange for care.
- If you cannot reach your doctor or don't have one, seek care today.
- If it is evening, watch the symptoms and seek care in the morning.
- If the symptoms get worse, seek care sooner.

Return to the Check Your Symptoms section and answer the questions. You may need to see a doctor sooner if you have other more serious symptoms.

FIGURE 3.0 Disposition/Care Recommendation (before).

Seek Care Today [X]

Based on your answers, **you may need care soon.** The problem probably will not get better without medical care.

- Call your doctor today to discuss the symptoms and arrange for care.
- If you cannot reach your doctor or don't have one, seek care today.
- If it is evening, watch the symptoms and seek care in the morning.
- If the symptoms get worse, seek care sooner.

Learn what you can do at home.
Get help with other symptoms.

Response Summary

Confirmed
Breathing problems are slowly worsening*
Mild difficulty breathing*
Difficulty breathing

Denied
Medicine may be causing symptoms

FIGURE 4.0: Disposition/Care Recommendation (after).

SYSTEM COMPARISON

TASK ANALYSIS

Task analysis (KLM-GOMS) was used to estimate the time being spent on system operations by users in the existing system. Task analysis was also conducted on prototypes to estimate possible efficacy of design changes. We included an estimated time for scrolling ("S") and standard estimates for mental preparation (M), pointing (P), and button press (B). The time estimates from this approach combined with examples of scrolling and mousing distances (measured by other means) helped in making the case for moving the triage system into a significantly different interaction model.

Triage protocols vary in length and duration depending on the severity of the caller's situation. Table 1.0 shows a representative comparison of the time needed for a "seek care today" disposition from the existing system and the redesign. System time was reduced by over 50% by the design changes.

Table 1.0 System Operation Time Compared

Task: "Seek Care Today" Disposition in Cough Symptom Topic		
System Version:	**Predictive Equation (KLM-GOMS)**	**Total Time**
Existing	$T_{execute} = 19T_M + 6_S + 18T_P + 25T_B$	63.1 sec
Redesign	$T_{execute} = 7T_M + 16T_P + 27T_B$	28.7 sec

Note: Illustrates mouse-only usage. Additional predictions were made for keyboard-only usage.

One limitation that we felt with using the KLM-GOMS model was adequate prediction of total call time. Triage calls are somewhat variable in duration, given the nurse's responsibilities of ensuring that the caller understands the advice being given and how to take appropriate action. Context-specific adaptations were devised (not shown) to model the time required to read, explain, and confirm questions aloud. Time estimates were made using lab data and are being confirmed in ongoing field studies.

Future work will include cognitive load assessment measures as particular changes reflected a significant reduction in non-system aspects such as organization of content and phrasing of questions for telephonic usage.

USABILITY TESTING

Usability testing offered feedback on the simplicity of the new system. Observation of the existing system had shown new users "going to the wrong spot" and "not quite knowing which to select." In the existing system, training was required to shape nurses to the "correct" behaviors, and artifacts such as "cheat sheets" and training supplements were frequently used to help reduce error. In the redesign, testing indicated that the key actions were accomplished with greater success and that key tasks could be completed with lower levels of priming.

DEPLOYMENT AND CHANGE MANAGEMENT

Preparation of nurse users was planned early in the process to help increase acceptance of the new system. In addition to involving nurse users and managers early in the project, we also provided regular updates about the evolution of the design. Protocol changes were of the utmost importance to the users, so we communicated regularly with them about the evolution of particular protocols, letting users know that user feedback had been utilized and preparing trainers and managers for changes to come.

Training that included best practices and lessons learned was developed in partnership with internal trainers and client trainers. The team developed a training plan that addressed both the substance of the training requirements and the typical training scenario for call centers. Call centers rarely have the ability to pull a shift of nurses off the telephone for a lengthy training. Instead, they develop materials that nurses can access on their own during slow or off-phone hours. Our materials were designed to meet these needs. We developed recorded training videos that provided insight and direction on interface and system changes and on standard and complex triage scenarios. In addition, we prepared "train the trainer" materials that included written manuals, self-tests, and change management best practice guidelines for nurse trainers to adapt and use while preparing their nurses for the change. Finally, a post-implementation feedback process was developed to address any technical or medical issues that required immediate response.

CONCLUSIONS

The use of user-centered design was key to the success of this initiative. Every aspect of the project was influenced by UCD techniques, which allowed the team to uncover and address unknown user needs.

Involvement of stakeholders at many levels of our client organizations was also critical to the success of the project. Input and buy-in from nurses and nurse management opened the doors to field studies, review sessions, and collaborative efforts.

The interactive prototype was a critical tool for both internal cross-functional teams and the external stakeholders. This tool provided visibility into the implications of design and development choices. The use of the prototype created consensus, facilitated iterative design, and allowed various workstreams (engineering, medical logic, and content development) to coalesce around a shared solution.

Building time into the project schedule for iterative user feedback and the resulting product changes is required to enable user-centered design. Without backing from project sponsors, these can be an easy target for project cuts.

Chapter 85

The Impact of Usability in Emergency Telemedical Services

*Shirley Beul[1], Sarah Mennicken[2], Martina Ziefle[2], Eva-Maria Jakobs[1],
Daniel Wielpütz[3], Max Skorning[3], Rolf Rossaint[3]*

[1]Textlinguistics and Technical Communication

[2]Communication Science, Human Technology Centre

[3]Clinic for Anesthesiology
RWTH Aachen University, Germany

ABSTRACT

The aim of this study is to identify key usability requirements in telemedical services; in particular, services specialized on emergencies. To generate an optimal user-
centered workflow, the communicative and organizational structure of an exemplary
system was analyzed. We focused on the impact of new technologies, which are
introduced to the medical workflow. The graying of societies implicates two major
consequences in this context: An increasing number of older people needing medical care and a shortage of medical staff in the next generation reveal a threatening
supply gap concerning the security of people's health. To bridge this gap, it is necessary to develop alternative systems, which can guarantee an area-wide supply of
emergency medical services, especially in rural regions with an insufficient infrastructure. Results show a high potential of telemedical emergency care, though the
benefit critically depends on the given system's usability and the appropriateness of
the workflow. On this basis, we introduce first guidelines, in both, communicative
and ergonomic criteria, which should be carefully followed by medical practitioners, and designers, likewise.

Keywords: Telemedical service, teledoctor, emergency, usability, workflow analysis

INTRODUCTION

When people call an ambulance in case of emergency, they depend on the medical aid provided by emergency medical services (EMS). Due to the demographic change in western countries, the area-wide supply of EMS cannot be maintained in the future in its current form and quality. Because of the graying of societies, the population of experts in the medical sector, especially in EMS, declines while the number of older people relying on medical support increases. In rural areas of Germany, the shortage of emergency physicians has already been noticeable.

Looking behind the scenes of today's EMS in Germany, several shortcomings are revealed. Up to now, these systems lack of a systematic and standardized quality control, for both, communicative and or organizational structures: No nationwide standard or rather standardized quality control is implemented in EMS, so far. This is especially harmful as a rescue process entails a highly complex time-critical organizational workflow, with many actors on-site and remotely (hospital, police, fire brigades) as well as a high probability of pitfalls within this process. As a consequence of the complex circumstances a rescue mission is executed in, the documentation of the process often is insufficient in order to guide the rescue team in chaotic and time-critical situations. Also, the legibility of hard copy materials suffers from illegible handwritings of doctors, and the general problem of noting data while riding fast in an ambulance over bumpy streets.

One possible solution to bridge this upcoming supply gap and to overcome these present shortcomings is the implementation of telemedical services in the rescue chain, allowing the emergency doctor (referred to as ED) or paramedics on-site to access telemedical support if necessary. To derive full advantage of the telemedical services, requirements of its users (i.e. paramedics, the ED on-site (if present) as well as of the medical staff in the central unit remotely) have to be studied. How should an emergency telemedical service (ETS) be designed to be efficient, usable, and, most importantly, perceived as a surplus regarding time-critical information supply? How can ICT be smoothly integrated in the workflow without burdening, but supporting the medical personnel?

The majority of studies concerning ICT usage in EMS context so far concentrated on the usage of technology as such. Few studies focused on organizational issues and the communicative needs within different emergency teams [Reddy 2009]. Yet, the focus of studies dealing with emergency cases is mainly patient-centered [Trout 2000], while the perspective of emergency doctors has been widely neglected, even though he/she is the main actor taking the full responsibility for the treatment, organization of the whole operation, and the team coordination [Chisholm 2000]. In German EMS, the ED is the most important person as he is legally responsible for the patient's safety and the medical treatment of the first aid.

THE USAGE OF ICT IN EMS

One of the key factors for effective emergency management is designing and implementing ICT within the EMS workflow. ICT has the potential to effectively support the coordination and cooperation between staff involved in the EMS workflow. To date, there is an upcoming number of studies reporting on the usage of ICT in the emergency-rescue-chain. In the United States, telemedical consultations between physicians in the hospitals and paramedics on-site are widely used and mostly accomplished via radio. By the use of ICT, the quality of primary care and intra-hospital procedures could be optimized respecting a faster and a more focused transfer of information [Wuerz et al. 1995, Adams et al. 2006, van Halteren et al. 2006]. As a reaction to recent crises, as 9/11, or hurricane Katrina in the United States, medical informatics researchers started to develop ICT as corrective measures to be used in disaster situations [Walderhaug et al. 2008, McCurdy et al. 2005]. Another innovate approach [Na et al. 2010] aims at the broad implementation of a telematic system in EMS, a so-called "teledoctor". We will describe this concept in more detail in the following chapter. Across studies, a major claim is a high quality of ICT data reliability, security and safety, when designing innovative interactive systems for emergency response in a major incident [Kyng et al. 2006, Li et al. 2006].

Nowadays, it is widely accepted that ICT plays an increasingly important role in EMS and the workflow in hospital emergency departments. Though, to date, only few studies concentrate on communicative, as well as organizational issues and the coordination and cooperation of the different persons involved in the rescue chain. In a very recent study [Reddy et al. 2009], the coordination between a hospital emergency department and EMS team in the United States was examined, uncovering the enormous importance of social and communication aspects in the EMS workflow. Sociotechnical aspects [Berg, 1999] in the EMS-context are especially multifaceted and highly complex, accompanied by a high time pressure and responsibility. [Reddy et al. 2009] claim that the ICT usage within the EMS process must be based on a thorough understanding of the workflow and should face the potential areas of breakdown in the coordination between emergency personnel. In addition, the interaction of communication and interaction of humans with technology is an extremely important success factor to be considered in technology-supported EMS settings [Aarts et al. 1998, Manoj & Hubenko 2007, Wears & Perry 2002]. Among other factors, the technical competence of EMS staff, but also their acceptance barriers towards technology usage should be carefully studied prior to implementing technology in such a sensitive area [Arning & Ziefle 2009a+b, Ziefle 2002].

THE TELEDOCTOR CONCEPT

In the following, we explain the course of events in an emergency situation, actors, and the involved technology (Figure 1) in the telemedical process.

Firstly, a person calls 911 [112 in Germany] to inform the staff of the primary control unit about the emergency and to request the aid of a rescue team. The control unit transmits some incident describing keywords to the rescue team sent on-site. Usually, one emergency vehicle with two paramedics is sent out, but depending on the severity of the case, an additional car with an ED is dispatched as well. In this telemedical concept we evaluate the utility of additional medical staff: an experienced emergency medical physician in a competence center, the so-called teledoctor. Assisted by advanced mobile data transmission, all vital parameters of a patient, as well as video and pictures of the scene, are transferred in real time to the competence center. The teledoctor and one paramedic on-site can communicate verbally via headsets. The other staff on-site is equipped with headsets, too in order to track the communication.

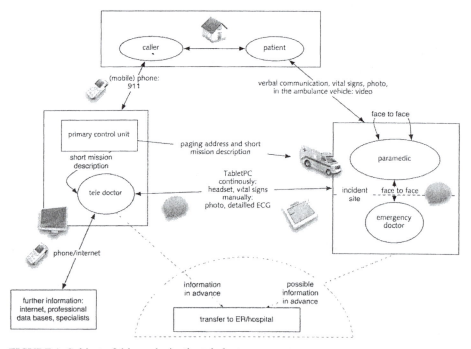

FIGURE 1: Subject of this analysis: the teledoctor concept

Besides verbal communication, the personnel on-site and the teledoctor can exchange information via a software installed on a tablet PC (see Figure 2). The teledoctor has the same software at his disposal and can enter the information with a multi-display workspace setup (see Figure 3).

FIGURE 2: ED using the tablet PC

FIGURE 3: Teledoctor in the multi-display workspace setup of the competence center

The software allows entering information as of a common emergency log. Besides this software, the teledoctor has access to information online as databases, e.g. with detailed information on pharmaceuticals. Via a protected connection he can track the local position of the emergency vehicle on an online maps application and he can look inside the car by using a controllable camera. With this information, he can help the staff on-site with organizational support e.g. to contact a hospital, transfer data about the incident to the doctor in charge, consult with other institutions (e.g., family members, cardiologists, poisoning centers) or assist the completion of the emergency log. The EMS personnel on-site can mainly focus on the patient's care and does not have to handle the various demands of an emergency situation.

ANALYSIS AND EVALUATION OF AN EMERGENCY TELEMEDICAL SERVICE

In the following we describe the methods we used for the analysis and evaluation and the results we derived from them.

METHODS

To evaluate the teledoctor concept a triangulation of methods was undertaken: semi-structured interviews, a participatory observation, and an ergonomic evaluation of the most important mobile technical device in the teledoctor concept, the tablet PC.

For the *semi-structured interviews*, an interview guide was designed. Medical professionals (n=11, $ED_{1...11}$) volunteered as participants. These doctors (anesthesiologists) were on duty on several departments of hospitals (e.g. operating room, intensive care unit etc.) Apart from this, all queried physicians are taking shifts in EMS. Partially, they were experienced in using telemedical support (n=5). The interviews were recorded, transcribed and qualitatively analyzed regarding content.

Also, a *participatory observation* of real emergency cases was run: Two observers took part in eight rescue missions to identify problematic issues within the workflow. Observers were taking shifts alternately in the ambulance and the tele-doctor office. Remarkable actions were recorded, as well as problems and individual coping strategies (measures taken by the actor to handle and solve the problem).

In order to evaluate *usability and ergonomic issues* of the tablet PC, we created a hardware dummy with the same measurements and shape as the original tablet PC, but made of wood with an adjustable weight. Each participant had to perform different writing tasks on a dummy with a predefined weight. The dummy was given to the people without explaining its handling to observe their intuitive way of grabbing and holding it. This is visualized in Figure 4.

30 participants (18-60 years, 50% of each gender (male: M=34 years, SD=14; female: M=39 years, SD=13) took part in our setting, which were allocated to three groups: Group 1 dealt with the dummy prepared with the original weight of the tablet PC (3.31 lbs/ 1500 g). A 2.43lbs/ 1100 g dummy was handed out to the second group and 1.54 lbs/ 700 g to group 3.

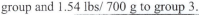

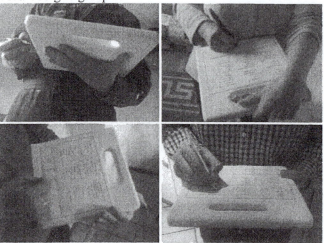

FIGURE 4: Snapshots of participants when handling the dummy and making notes

In the participatory observation on the emergency vehicle, we were able to watch the original tablet PC in use. Since it was never used no longer than five minutes without interruption when standing, we conducted our study not to last any longer than that. To create a realistic task we read out a story to users, while they answered some questions about it by marking the correct answer on a paper questionnaire placed on the dummy. After answering story related questions, users evaluated the appropriateness of the dummy's weight (using another questionnaire).

RESULTS

Integrating ICT in the traditional workflow changes the process of a rescue mission. In particular, technical devices and its usage become an additional issue the medical staff has to deal with. The system presents strengths and weaknesses, which will be discussed. We will provide guidelines to overcome the mentioned risks as well.

Communication, Information and Organization Flow

Due to the integrated ICT, the communication between the medical personnel changes. The actors on-site communicate with the physically absent teledoctor. Neither eye contact nor physical actions (e.g. gestures) can be used to attract the attention of the people on the other location or to realize a natural turn-taking within the conversation. In this case, actors have to ensure with verbal comments that the interlocutor is listening, and to bide the correct moment to speak. Therefore, communicative guidelines have to be defined regarding the identification of communicative paths, and the successful turn-taking between teledoctor and speaker on-site. „You must develop an own communication culture [within the system]. [...] You have to learn to let people finish speaking, and to announce before asking a question as each question may interrupt actions on-site and deflect [actors'] attention" (ED_{10}). In addition, the teledoctor must turn off her/his microphone when not needed in order to avoid the transmission of unnecessary noise or comments, which distract the rescue team on-site.

Furthermore, the communication on-site is reduced within the rescue team as well as between the medical staff and the patient because of the communicative integration of the teledoctor. The paramedics and the ED on-site are concerned with new tasks: They have to communicate with the teledoctor respectively among the team and interact with ICT (e.g. tablet pc) or medical devices (e.g. ECG etc.) simultaneously, which takes attention and cognitive resources. On the other site, the teledoctor's job demands high multitasking ability, too. Thus, the amount of information and the workload is to be reduced to the most necessary extent. For this purpose, informational requirements of the medical personnel have to be identified. Mainly, by investigating informational needs of querying EDs (and paramedics) and the actual usage of the system (e.g. input modes) should detect the most suitable way of information representation and the appropriate amount of information.

Additional factors, which determine the success of this ETS, are a consequent report of feedback and a clear division of labor. By analyzing the communication between the teledoctor and the speaker on-site, it becomes obvious that redundant conversations were conducted and one party has to inquire the other about the progress of current actions. Both must give feedback about her/his actions frequently to share the workload efficiently. Therefore, feedback rules are needed or an EMS –related feedback etiquette. Moreover, tasks within the rescue operation have to be divided clearly across the team. The teledoctor should execute actions appropriate to his workspace. To support the rescue team on-site, s/he should be responsible for the enrolment at the hospital, the writing of the emergency log, and the mining of patient data, for instance, retrieving data regarding the patient's anamnesis (physician's letters, medical history) from hospital's or surgery's databases. For this reason, it has to be ensured that the teledoctor can access important patient data.

However, completing these tasks depends on the quality of the information flow: "As the teledoctor, I often take care of the enrolment at the hospital and the transfer of the patient data to the doctor in charge there to support the ED on-site. Thus, I am not having all [required] information I can never answer questions [of the emergency room receptionist] regarding the patient. Then, I have to interrupt the talk. In this case, I do not know if data gets lost or is transmitted incorrectly, […] although it [my support] reduces the workload of the emergency physician on-site" (ED_{11}). As a consequence, necessary information must been transmitted (or requested) before beginning to process the task.

Apparently, the patient is not exclusively in the focus of attention in the emergency operation any more. In the interviews, EDs with experience of being telemedically-supported remark an alleged attention shift from the emergency situation and the patient to the advice or discussion with the teledoctor. Also, sometimes the patient is confused by the voice connection between the "invisible" teledoctor and the ED on-site (ED_8). For this, guidelines have to be established in order to control the communication between the teledoctor and the rescue team.

Usage of ICT within the Workflow

The implemented ICT should meet four general requirements concerning appropriateness, robustness, durability, and weight. The technical equipment should fit into the workflow according to the varying circumstances on the incident site.

First, the set of devices must be reduced to a minimum extent at least. If possible, multifunctional devices should be used. Second, regarding the results of our participatory observation, used hardware must be rugged and robust because it is placed everywhere possible on-site. Third, the battery of devices must last for a few hours. Presently, it often has to charge in the ambulance vehicle for which it has to be mounted and cannot be used meanwhile. Recharge and mounting units should be designed in a way that the device can be used while charging. Fourth, it must be lightweight because it is an add-on to the current medical equipment. In our ergonomic evaluation of this device, the majority of participants assessed the original weight (3.31 lbs/ 1500 g) as too heavy (Figure 5). Due to our results, we highly recommend the usage of a handheld device in ETS with a weight between 1.54 lbs/ 700 g and 2.43 lbs/ 1100g.

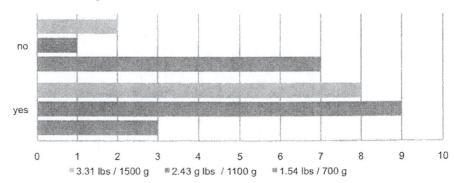

FIGURE 5: Should the tablet PC be more lightweight? For each weight class n = 10.

Concerning the tablet PC in particular: Software should deliver a real time feedback about the teledoctor's notes to the ED on-site to avoid taking the same notes twice. Thus, the teledoctor can write the log of the rescue mission by listening the conversation between ED on-site and the patient, which the ED on-site can see. Therefore, a feature of the system should be a collaborative, real time text editing for both users by employing high-quality audio equipment.

The design of the graphical user interface (GUI) is even more important than adding or improving functions of devices. The common principles of ergonomic design should be followed carefully as the quality of the GUI often determines the user's performance and her/his degree of satisfaction while using the device.

Finally, reliability of the technical equipment in general, and of data transfer in particular is another key requirement, which should be taken for granted in high-quality ETS. The system has to report feedback immediately to the teledoctor and the ED on-site because an emergency case people's health is in high-risk and time sensitive situation.

CONCLUSION

The evaluation and analysis of the teledoctor concept as a telemedical support for emergency situations basically showed positive results. All interviewed emergency doctor agreed that the teledoctor is a powerful concept, which can compensate for the lack of emergency physicians on-site, support less experienced colleagues (with a possibly) different specialization, and to context-adaptively advise paramedics as well as to authorize medical treatment. Also, respondents agreed on the increasing efficiency of the emergency process: with the help of a teledoctor, the diagnosis on-site can be accomplished by virtue of a more detailed and targeted medical information, which basically expedites the logistics and the organization of the rescue operation. Basically, any technical support in the high-demanding and time-critical rescue-chain is a relief to the emergency staff involved in the process.

However, the interviewed emergency doctors also reported shortcomings and barriers towards the usage of a teledoctor. One concern is the low trust in the reliability of the technology involved. The second one was the alleged attention shift from the emergency situation and the patient as such to the advice or discussion with the teledoctor. With respect to optimization of the emergency situation, the teledoctor concept seems to be a promising way of meeting the upcoming challenges in emergency medicine. Future studies will have to examine if the concerns raised by the EDs will vanish with increasing experience in both roles: as ED on-site or as supervising teledoctor in the primary control unit remotely.

REFERENCES

Aarts, J., Peel, V. and Wright, G. (1998), "Organizational Issues in health informatics: a model approach." *International Journal of Medical Informatics*, 52(1-3), 235–242.

Adams, G.L., Campbell, P.T., Adams, J.M. et al. (2006), "Effectiveness of prehospital wireless transmission to a cardiologist via handheld device for patients with acute myocardial infarction." *American Journal of Cardiology*, 98, 1160–1164.

Arning, K., and Ziefle, M. (2009a), "Different perspectives on technology acceptance." In A. Holzinger et al. (eds.). *HCI for eInclusion*. Berlin: Springer, 20–41.

Arning, K., and Ziefle, M. (2009b), "Cognitive and personal predictors for PDA navigation performance." *Behaviour and Information Technology*, 28(3), 251–263.

Berg, M. (1999) Patient care information systems and healthcare work: A sociotechnical approach. *International Journal of Medical Informatics*, 52(2), 87–101.

Chisholm, C.D., Collisson, E.K., Nelson, D.R., and Cordell, W.H. (2000), "Emergency Department Interruptions: Are Emergency Physicians "Interrupt-driven" and "Multi-tasking"?" *Academic Emergency Medicine*, 7(11(, 1239–1243.

van Halteren, A., Bults, R., Wac, K., Dokovsky, N., Koprinkov, G., et al. (2004), "Wireless body area networks for healthcare: the MobiHealth project." *Studies of Health Technolology Informatics*, 108, 181–193.

Kyng, M., Nielsen, E., and Kristensen, M. (2006). Challenges in designing interactive systems for emergency response. *6th Conference on Designing Interactive Systems*, 301–310.

Li, J., Wilson, L., Stapleton, S., and Cregan, P. (2006), "Design of an advanced telemedicine system for emergency care." *18th Australia conference on Computer-Human Interaction: Design: Activities, Artefacts and Environments*, 416.

Manoj, B.S., and Hubenko, A. (2007), "Communication challenges in emergency response." *Communication of the ACM*, 50(3), 51–53.

McCurdy, N.J., Griswold, W.G., and Lenert, L.A. (2005), "RealityFlythrough: enhancing situational awareness for medical response to disasters using ubiquitois video." *American Medical Informatics Association*, 510-514.

Na, I.-S., Skorning, M., May, A., Schneiders, M.-T., Protogerakis, M., Beckers, S., Fischermann, H., Brodziak, T. and Rossaint, R. (2010), "Med-on-@ix: Real-time Teleconsultation in Emergency Medical Services – Promising or Unnecessary?" In: Ziefle, M., and Röcker, C. (eds.). *Human-Centered Design of eHealth Technologies*. Hershey, P.A., IGI Global.

Reddy, M., Paul, S., Abraham, J., McNeese, M., DeFlitch, C., and Yen, J. (2009), "Challenges to effective crisis management: Using information and communication technologies to coordinate emergency medical services and emergency department teams." *International Journal of Medical Informatics*, 78 (4), 259–269.

Trout, A., Magnusson, A., and Hedges, J. (2000), "Patient Satisfaction Investigations and the Emergency Department: What Does the Literature Say?" *Academic Emergency Medicine*, 7 (6), 695-709).

Walderhaug, S., Meland, P., Mikalsen, M., Sagern, T., and Brevik, J. (2008), "Evacuation support system for improved medical documentation and information flow in the field." *International J of Medical Informatics*, 77 (2), 137-151.

Wears, R., and Perry, S. (2002), "Human factors and ergonomics in the emergency department." *Annals of Emergency Medicine*, 40 (2), 206-212.

Wuerz, R.C., Swope, G.E., Holliman C.J., and Vazquez-de Miguel, G. (1995), "On-line medical direction: a prospective study." *Prehospital Disaster Medicine*, 10, 174-177.

Ziefle, M. (2002), "The influence of user expertise and phone complexity on performance, ease of use and learnability of different mobile phones." *Behaviour and Information Technology*, 21(5), 303-311.

ACKNOWLEDGEMENTS

We owe gratitude to the emergency doctors, who volunteered to take part in this study. Thanks also to the EMS staff of the professional fire brigade Aachen, Germany to allow us to gain insights into a thrilling professional area. Also, we thank Oliver Sack, Simon Himmel, Stefan Ladwig, and Teresa Schmidt for their research support, as well as Jochen Dahlhausen for helping creating the dummy. This work was funded by the Excellence initiative of German State and Federal government.

Chapter 86

Reducing the Risk of Heat Stress Using Artificial Neural Networks Based Job-Combination Approach

Sanjay Srivastava, Yogesh K Anand

Department of Mechanical Engineering
Dayalbagh Educational Institute Agra 282110, India

ABSTRACT

We design and implement a system to reduce the risk of heat stress, a recognized occupational health hazard (OHH), in two labor intensive industries using a job-combination approach. A novel feature of the system is employing artificial neural networks (ANNs) as model free estimators to evaluate perceived discomforts (PDs) of workers for different job combinations proposed in the work.

Keywords: Occupational Health, Heat Stress, Artificial Neural Networks

INTRODUCTION

Current guidelines define working environment that cause an increase above 38°C (heat stress) as potentially hazardous (ACGIH, 2004). However, the effectiveness of these guidelines is limited by the individual variation among employee and variation in work practices (Gun and Budd, 1995). Hot conditions give rise to physiological heat strain (Candi *et al*, 2008), and cognitive decrements (Hancock and Vasmatzidis, 1998; Enander and Hygge, 1990). We present artificial neural networks (ANNs) based system to reduce the risk of heat stress of workers of two labor intensive industries of India namely glass/bangles manufacturing (GBM) units and brick manufacturing (BM) units. GBM consists of four jobs namely *tarwala* job, bangles rolling, making *gundhi* and finishing (*Tarwala*, and *gundhi* are local terms used in Glass/bangles manufacturing in India), whereas BM comprises firing work, molding, and three types of lifting (lifting-1, lifting-2 and lifting-3). *Tarwala* job, and firing work are found to be the most severe jobs in GBM and BM respectively as these involve undue exposure of workers to excessive heat. The risk of OHH, in general, is significantly influenced by factors such as working hours (WH), duration of rest breaks (RB), and number of rest breaks (NoRB). Workers of a GBM unit at Firozabad, India are trained to perform *tarwala* job along with each one of the remaining three jobs with predefined WH distribution – resulting in three job combinations. Similarly, four job combinations are generated in a BM unit at Hathras, India. The risk of OHH for a job combination is evaluated based on the perceived discomforts (PDs) of workers using interview method in the form of linguistic variables such as very low, moderate, extremely high etc., which in turn are converted to suitable numeric values for their use in ANNs. It is extremely difficult to evaluate PDs for every possible amalgamation of WH, RB, & NoRB for each job combination, therefore, we use artificial neural networks (ANNs) with backpropagation learning, also called backpropagation neural networks (BPNNs), as model free estimators. ANNs effectively evaluate perceived discomforts of workers for different values of WH, RB and NoRB for each job combination. Seven ANNs, three for GBM, and four for BM are trained, one for each job combination, with data set generates as above. The proposed system act as an advisor to a worker to choose a job combination and the corresponding values of WH, RB, & NoRB to decide his/her occupational risks and earnings suitably.

BACK PROPAGATION NEURAL NETWORK

Artificial Neural Networks (ANNs) are gaining wide popularity in the intelligent decision making systems. The data driven approach of the ANNs enables them to behave as model free estimators, i.e., they can capture and model complex input-output relationships even without the help of a mathematical model. In this work, an effort is made to utilize the function approximation capabilities of ANNs using back propagation neural networks in the evaluation of PDs of workers. The back propagation neural network is a multiple layer network with an input layer, output layer and some hidden layers between input and output layers (Fausett, 1994). Its learning procedure is based on gradient search with least sum squared optimality criterion. Calculation of the gradient is done by partial derivative of sum squared error with respect to weights. After the initial weights have been randomly specified and the input has been presented to the neural network, each neuron currently sum outputs from all neurons in the preceding layer. The sums and activation (output) values for each neuron in each layer are propagated forward through the entire network to compute an actual output and error of each neuron in the output layer.

The error for each neuron is computed as the difference between actual output and its corresponding target output, and then the partial derivative of sum-squared errors of all the neurons in the output layer is propagated back through the entire network and the weights are updated. In course of the back propagation learning, a gradient search procedure is used to find connection weights of the network, but it tends to trap itself into the local minima. The local minima may be avoided by adjusting value of the momentum. This algorithm can be expressed succinctly in the form of a pseudo-code as given below.

1. Pick a rate parameter R.
2. Until performance is satisfactory

For each sample input, compute the resulting output. Compute β for nodes in the output layer using

$$\beta_z = D_z - O_z$$

where D represents the desired output and O represents the actual output of the neuron. Compute β for all other nodes using

$$\beta_j = \Sigma k \, w_{j \to k} O_k \left(1 - O_k\right) \beta_k$$

Compute weight changes for all weights using

$$\Delta w_{i \to j} = r O_i O_j \left(1 - O_j\right) \beta_j$$

Add up the weight changes for all sample inputs and change the weights.

LEVENBERG - MARQUARDT APPROXIMATION

This algorithm uses Levenberg - Marquardt learning rule, which uses an approximation of the Newton's method to get better performance. This technique is relatively faster as demonstrated by (Raj et al., 1999, Srivastava et al., 2004) while modeling input/output relationships of complex processes using this algorithm. The Levenberg - Marquardt (LM) Approximation update rule is:

$$\Delta W = \left(J^T J + \mu I\right)^{-1} J^T e$$

where J is the Jacobean matrix of derivatives of each error to each weight, μ is a scalar and e is an error vector. If the scalar is very large, the above expression approximates the Gradient Descent method while it is small the above expression becomes the Gauss-Newton method. The Gauss-Newton method is faster and more accurate near error minima. Hence, the aim is to shift towards the Gauss-Newton as quickly as possible. Thus μ is decreased after each successful step and increased only when step increases the error. The architecture of ANN employed to evaluate PDs (the risks of OHH) is shown in Figure 1.

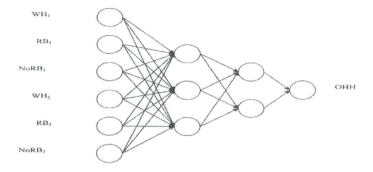

Figure 1. BPNN architecture to evaluate PDs (the risks of OHH)

METHODOLOGY

The procedural steps are as follows – (1) Visiting GBM and BM units to identify the working conditions of workers. (2) Taking feedback from selected workers using interview method to evaluate their perceived discomforts (present risk of OHHs). (3) Training these workers for each job-combination by varying (i) WH, (ii) RB, and (iii) NoRB. (4) Analyzing and recording the perceived discomforts of trained workers with respect to variation in (i) WH, (ii) RB, and (iii) NoRB for each job combination. The exhaustive data so collected act as training data for each of seven ANNs. (4) Training ANNs with available data set for each job combination, (5) Testing and validating the accuracy of ANN models with data set which has not been used in training. A sample training data set of an ANN for one job combination is shown in Table 1.

Table 1. Sample input data for training ANN for (tarwala job + bangles rolling)

WH1	RB1	NoRB1	WH2	RB2	NoRB2	Experimental perceived discomforts
12	0	0	0	0	0	100
11	10	1	1	0	0	93
10	15	1	2	0	0	88
9	20	1	3	0	0	82
8	30	1	4	0	0	78
7	0	0	5	10	1	88

ANN MODELING AND SIMULATION RESULTS

The design and implementation of the ANN models is far from an exact science. Several design issues need to be carefully finalized to obtain a functional model. The selection of the number of neurons in the two hidden layers is critical for the success of training the ANN model. The ANN attempts to create a function mapping by adjusting the weights in the inner layers. If the number of these neurons is too large, the ANN may be over-trained giving spurious values in the testing phase. If too few neurons are selected, the function mapping may not be accomplished due to under-training. The number of neurons has to be carefully selected with appropriate experimentation, as there are no standard procedures available for all kinds of ANN training problems. Similarly, the selection of the learning rate and μ affect the convergence of the network. In the BPNN with LM rule the learning rate and μ are continually modified based on the training results. Therefore, only the initial values have to be specified in the model.

As mentioned earlier, seven exhaustive training data, one for each ANN, are collected by interview method. Each ANN comprises a three layer network with six inputs i.e., WH1, RB1, and NoRB1 for job involving the risk of heat stress, and WH2, RB2, and NoRB2 for second job under consideration, and one output i.e. perceived discomfort (risk of OHH). After the training, the weights are frozen and the model is tested for validation. For this purpose, the input parameters to the network are sets of values that have not been used for training the network but are in the same range as those used for training. This enables us to test the network with regard to its capability for interpolation. The perceived discomfort is thus obtained for this set of parameters. Then a worker is asked to perform jobs with same sets of parameters to evaluate the perceived discomfort. The level of agreement between the perceived discomfort predicted by ANN and the actual one predicted by worker indicates the efficacy of the ANN model.

For training problem at hand the following parameters were found to give rapid convergence of the training network with good performance in the estimation. The first and second layers of the neurons are modeled with a log of the sigmoid function, and the third layer was a purely linear function. Neurons taken in the first and second hidden layers are three and two respectively. Maximum epochs are considered as 200, the error goal was set at 1e-005, and the learning rate for training the ANN is taken as 0.2.

The results of the validation procedure described above for GBM units as well BM units are given in Table 2 and Table 3 respectively. Perceived discomforts estimated by ANNs are in close proximity with actual experimental values.

Table 2. Comparison of results of perceived discomforts for GBM units
(tarwala job + bangles rolling)

						Perceived discomfort	
WH	RB				NoRB	Experimental	ANN
1	1	NoRB1	WH2	RB2	2	results	results
12	30	4	0	0	0	78	80
9	20	3	3	10	1	77	76
7	45	1	5	20	1	81	80
1	20	1	11	20	4	60	60

Table 3 Comparison of results of perceived discomforts for BM units
(firing work + molding)

						Perceived discomfort	
				RB		Experimental	ANN
WH1	RB1	NoRB1	WH2	2	NoRB2	results	results
12	30	5	0	0	0	60	59
7	45	1	5	20	1	46	44
5	30	1	7	30	1	46	45
0	0	0	12	30	3	22	23

DISCUSSIONS

The proposed system not only depicts the effects of extreme conditions on the health of the workers but also help in selecting a job combination which would reduce the risk of heat stress at the desired earnings. Further, the top management faces the problem of monopoly of workers of tarwala job in GBM units, and firing work in BM units, as both are very high skilled jobs. The system presented in this work will alleviate this problem as job combination approach will make other workers getting trained for tarwala job/firing work. In fact the system will help in 'work generalization' to take over 'work specialization'. Therefore, the feasibility of implementing this system is high as it is beneficial to both the parties – workers as well as owners. In view of these facts, the work presented here forms an important basis to effectively address the issues in health management of workers.

REFERENCES

ACGIH. 2004, *Threshold limit values for chemical substances and physical agent and biological exposure indices*, American Conference of Governmental Industrial Hygienists, Cincinnati, OH

Gun, R.T., and Budd, G.M. 1995, Effects of thermal, personal and behavioral factors on the physiological strain, thermal comforts and productivity of Australian shearers in hot weather, *Ergonomics*, **38**, 1368-1384

Candi D. A., Christina L. L., Skai S. S., Maeen Z. I., and Thomas E. B. 2008, Heat strain at the critical WBGT and the effects of gender, clothing and metabolic rate, *International Journal of Industrial Ergonomics*, **38**, 640-644

Hancock, P.A., and Vasmatzidis, I. 1998, Human occupational and performance limits under stress: the thermal environment as a prototypical example, *Ergonomics*, **41**, 1169-1191

Enander, A.E., and Hygge, S. 1990, Thermal stress and human performance, *Scandinavian Journal of Work, Environment and Health*, **16**, 44-50

Fausett L. 1994, *Fundamentals of Neural Networks*, (Prentice Hall, Englewood Cliffs, NJ)

Raj, K.H., Sharma R. S., Srivastava S., and Patvardhan C.1999, Modelling of manufacturing process with ANNs for intelligent manufacturing, *International Journal of Machine Tools & Manufacture*, **40**, 851-868

Srivastava S., Srivastava K., Sharma R. S. & Hansraj K. 2004, Modelling of hot closed die forging of an automotive piston with ANN for intelligent manufacturing, *Journal of Scientific and Industrial Research*, **63** 997-1005

CHAPTER 87

Exposure of Vibration to Bus and Train Passengers

[1]*Ahmad Rasdan Ismail,* [2]*Nur Farhana Kamaruddin,* [2]*Mohd Zaki Nuawi and* [2]*Mohd Jailani Mohd Nor*

[1]Faculty of Mechanical Engineering
Universiti Malaysia Pahang
26300 UMP, Kuantan, Pahang, Malaysia

[2]Department of Mechanical and Materials Engineering
Faculty of Engineering and Built Environment
Universiti Kebangsaan Malaysia, 43600 UKM Bangi, Malaysia

ABSTRACT

Trains and buses are the most important forms of public transportation worldwide. However, high magnitude whole-body vibration (WBV) that can be associated with train travel may lead to various diseases and health problems, such as lower back pain, in humans. This study gives an account of the Daily Exposure to Vibration A(8) and Vibration Dose Value (VDV) experienced by train passengers, with care taken to elucidate the effects of WBV on the human body. This study was conducted on a national train travelling from the east coast of Malaysia to the southern part of the country. The WBV exposure was measured for duration of 8 hours, which is equivalent to a typical travel time of a passenger or worker. Data was collected using an IEPE(ICP^{TM}) accelerometer sensor connected to a DT9837 device, capable of effectively measuring and analyzing the vibrations. The vibration results were displayed on a personal computer using a custom graphical user interface (GUI). Matlab software was used to interpret the results and determine the WBV exposure level. The values of Daily Exposure to Vibration A(8) and the Vibration Dose Value (VDV) during one stretch of train travel were measured as

0.3749 m/s^2 and 1.2513 m/s$^{1.75}$ respectively. The results here confirm that WBV absorbed by the human body increases with an increase in the duration of vibration exposure and the number of trips taken by a passenger, illustrated by the increase in the value of Daily Exposure to Vibration A(8) and the calculated Vibration Dose Value (VDV).

Keywords: Whole-body vibration, Vibration dose value, Low back pain

INTRODUCTION

Ergonomics is the application of scientific principles, methods and data drawn from a variety of disciplines to the development of engineering systems in which people play a significant role. Among the basic contributing disciplines are psychology, cognitive sciences, physiology, biomechanics, applied physical anthropometry and industrial systems engineering (Kroemer et. al., 2003). The importance of safety and ergonomics has grown significantly over the past years (Matilla et. al., 1996). The latest technology has allowed for the expanded used of ergonomics and additional safety features in products and equipment. At the same time, new technology has created new risks, and the management of these risks is more complicated. For this reason, it is important for a designer to use his knowledge of ergonomics during the design process of machines, equipment, products and systems. There is substantial epidemiologic evidence that links physical ergonomics exposures at the workplace, such as lifting, constrained postures, repetitive movements, fast work pace, handling of heavy material, forceful exertions and vibration, to the occurrence of upper extremity musculoskeletal disorders (Bernard et. al., 1997; Grieco et. al., 1998; Hagberg et. al., 1995; National Research Council and Institute of Medicine, 2001; van der Windt et. al., 2000). Ergonomics (also called human factors or human engineering in the United States) can be defined as the study of human characteristics for the appropriate design of the living and work environment. Its fundamental aim is that all human-made tools, devices, equipment, machines, and environments should advance, directly or indirectly, the safety, well-being, and performance of human beings (Kroemer et. al., 2003). Several ergonomic interventions, such as employee training, redesign of process tools or workstations and improvement of work conditions, have been suggested and implemented to tackle musculoskeletal problems related to industrial work (Wang et. al., 2003; Weestgard and Winkel et. al., 1997).

Depending on the source, WBV has been given a variety of definitions. From the Directive 2002/44/EC of the European Parliament and of the Council, the term 'whole-body vibration' refers to the mechanical vibration that, when transmitted to the whole body, entails risks to the health and safety of workers, in particular lower-back morbidity and trauma of the spine (Directive 2002/44/EC). WBV is defined as the vibration that occurs when the greater part of a person's weight is supported on a vibrating surface. WBV principally occurs in vehicles and wheeled working machines. In most cases exposure to WBV occurs when the person is in a sitting position and the vibration is thus primarily transmitted through the seat pan, with additional transmittance through the backrest. WBV may impair performance and comfort. It may also contribute to the development of various injuries and disorders. In many work situations, WBV is a prominent and troublesome occupational health problem (Griffin et. al., 1990).

Lower back pain (LBP) is among the most common and costly health problems (Garg et. al., 1992; Van Tulder et. al., 1995). Occupational, non-occupational, and individual risk factors play a role in the development, the duration, and the recurrence of LBP. Several critical reviews have discussed the occupational risk factors that result in back disorders (Wilder et. al., 1996; Burdorf et. al., 1997; Bovenzi et. al., 1999; Lings et. al., 2000; Waddell et. al., 2000). All of these reviews conclude that there is strong epidemiological evidence relating occupational WBV exposure to LBP. In five European countries (Belgium, Germany, Netherlands, France, Denmark), LBP and spinal disorders due to WBV are currently recognized as occupational diseases (Hulshof et. al., 2002). However, WBV remains a common occupational risk factor for LBP, with high exposures and the resulting injuries affecting 4% to 8% of the workforce in industrialized countries (Palmer et. al., 2000). Important high risk groups include drivers of off-road vehicles (such as those used for earth moving, forestry, and agriculture), drivers of forklift trucks, lorries, and buses, crane operators, and helicopter pilots.

MATERIALS AND METHODS

WBV measurements were conducted according to ISO 2631-1:1997. The triaxial accelerometer sensor was located between the passenger's contact points and the vibration source. During the test, a randomly chosen passenger sat on the accelerometer, shown in Figure 1.

Figure 1. Triaxial accelerometer sensor used for WBV measurement

The WBV was sampled 1000 times per second. For each experiment, the exposure time was set to 8 hours, which is equivalent to the duration of a typical occupational exposure time. The study was conducted at different locations. The measurement location for each experiments is listed in Table 1. After the accelerometer and DT9837 were connected, the data collection began. The total vibration along each axis (x, y, and z) felt by the passenger was displayed in a graph using Matlab.

Table 1: Location of measurement

Experiment	Location
1	From Kajang to Seremban
2	From Seremban to Gemas
3	From Segamat to Tampin

Excessive exposure of WBV typically occurs when the exposure duration is long and accompanied with a large vibration magnitude. Of the various modes of public transportation, the train most often produces high magnitude of vibration. The Malaysian national train traveling from the east coast to the southern part of the country was chosen for this study. WBV measurements using a randomly chosen passenger were performed three times at three different locations. Measurements were taken as the train passed from Kajang to Seremban, Seremban to Gemas, and Segamat to Tampin.

The two measurement devices used in the study were the IEPE(ICP^{TM}) accelerometer sensor, and a DT9837 instrument. The IEPE(ICP^{TM}) accelerometer sensor (also known as a triaxial seat accelerometer) was a DYTRAN Model 5313A. The sensor was used to assess the vibration level. The DT9837 instrument was a highly accurate five channel data acquisition module that is ideal for portable noise and vibration measurements.

In this study, Matlab software was used to analyze the vibration signal gathered by the DT9837 instrument via a USB port. The Matlab graphical user interface (GUI) scripts were examined using the *GUIDE* function for ease of measurement and assessment of WBV exposure. Using this Matlab script, three graphs, displaying the three different axes of vibration, were displayed for real time observation. The collected data were also saved in the computer for subsequent analysis. Thus, in this work, the total daily exposure to vibration towards humans was evaluated using an accelerometer sensor, a DT9837 instrument and Matlab software.

RESULTS

From the three experiments conducted, the Daily Exposure to Vibration A(8) value, Vibration Dose Value (VDV) and Exosure Points value were evaluated using Matlab features, according to formulas (1), (2), and (3) listed below. The results were displayed in the custom made graphical user interface (GUI).

The Daily Exposure to Vibration A(8) was calculated as follows;

$$A(8) = vibration\ value\ \left(\frac{m}{s^2}\right) \times \sqrt{\frac{exposure\ time\ (min)}{480\ (min)}} \tag{1}$$

The Vibration Dose Value (VDV) was calculated as follows;

$$VDV = \left(\int_0^T a^4(t)dt\right)^{0.25} \tag{2}$$

Where,
a(t) = frequency-weighted acceleration (m/s^2)
T   = the total period of the day during which vibration may occur (s)

The Exposure Point was calculated as follows;
$$exposure\ points = 2 \times (vibration\ value)^2 \qquad (3)$$

All the data obtained in the experiments are shown in Table 2. WBV graphs were demonstrated in Figure 2. The WBV data shown in Figure 2(a), were collected during travel from Kajang to Seremban, while the data in Figure 2(b) were collected from Seremban to Gemas and the data in Figure 2(c) were collected on route from Segamat to Tampin. The Vibration Dose Value (VDV) measurement GUI is shown in Figure 3.

Table 2: Whole-body vibration measurement data collected in train

Analysis Method	Experiment 1		Experiment 2		Experiment 3	
	From	To	From	To	From	To
	Kajang	Seremban	Seremban	Gemas	Segamat	Tampin
Daily exposure to vibration A(8)	0.3221 m/s^2		0.2884 m/s^2		0.3749 m/s^2	
Exposure points system	41.4867 point		33.2716 point		56.206 point	
Vibration dose value (VDV)	1.1014 m/s$^{1.75}$		1.0973 m/s$^{1.75}$		1.2513 m/s$^{1.75}$	
Daily exposure action value time (0.5 m/s^2)	9 hrs 50 min		12 hrs 16 min		7 hrs 16 min	
Daily exposure limit value time (1.15 m/s^2)	52 hr 3 min		64 hr 54 min		38 hr 25 min	
Points per hour	5.1858 point		4.159 point		7.0258 point	
Time achieving 1.75 m/s$^{1.75}$	1 hr 29 min		1 hr 30 min		53 min	

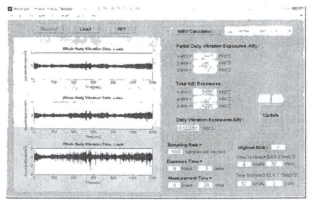

(a)

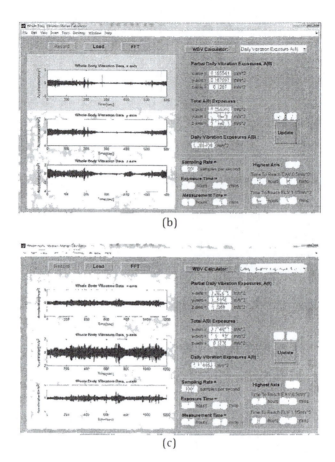

Figure 2. Whole-body vibration custom data acquisition system (a) from Kajang to Seremban, (b) from Seremban to Gemas, and (c) from Segamat to Tampin

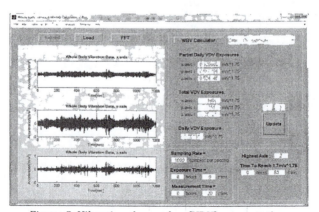

Figure 3. Vibration dose value (VDV) computation

DISCUSSION

This study finds that, the frequency-weighted acceleration value associated with train travel in Malaysia is close to the permissible exposure limit stated in ISO 2631-1:1997. Hence, the high magnitude of WBV experienced by a passenger of a moving train may cause musculoskeletal disorders.

The basic method discussed in ISO 2631-1 involves a frequency-weighted root-mean-square (r.m.s.) calculation that is primarily applicable to the assessment of health risks from stationary vibrations and does not account for severe single or multiply occurring shock events. Single shocks events can be analyzed with additional methodology that involves a running r.m.s. calculation as described in 2631-1, although no information on the health risk levels of single event shocks is provided in the literature. This additional method involves a (frequency-weighted fourth power vibration dose value (VDV)), and is more sensitive to shocks than the basic method. However, the VDV method will underestimate the health risks of vibration that contains severe shocks in comparison to the health risks of vibration that does not containing severe shocks. The EU Physical Agents Directive uses the basic method for the assessment of health risks, with VDV as an alternative. The two methods give different assessment results (Kjell et. al.).

The (r.m.s) vibration magnitude is expressed in terms of the frequency-weighted acceleration at the seat of a seated person or the feet of a standing person, in units of meters per second squared (m/s^2). The r.m.s vibration magnitude represents the average acceleration over a measurement period. The vibration exposure is assessed using the highest of the three orthogonal axes values ($1.4a_{wx}$, $1.4a_{wy}$ or a_{wz}). A frequency-weighted acceleration value less than $0.45m/s^2$ mean that no negative health effect should be expected. A frequency-weighted value between $0.45m/s^2$ and $0.90m/s^2$ implies the possibility of negative health effects. A frequency-weighted acceleration value greater than $0.90m/s^2$ suggests high risks of negative health problems.

Table 3 shows the r.m.s acceleration value limits for exposures up to 8 hours. Vibrations experienced by the train passengers must not exceed the given limits, or they risk the development of vibration-related health problems.

Table 3: Standard value of RMS acceleration

Exposure Limit	8 hrs	4 hrs	2.5 hrs	1 hr	30 min	5 min	1 min
RMS acceleration	2.8 m/s^2	4.0 m/s^2	5.6 m/s^2	11.2 m/s^2	16.8 m/s^2	27.4 m/s^2	61.3 m/s^2

In our third experiment, which was conducted on the tracks between Segamat and Tampin, the values of Daily Exposure to Vibration A(8) and Vibration Dose Value (VDV) were much higher than the other two experiments. The values of daily exposure to vibration A(8) and vibration dose value (VDV) were 0.3749 m/s^2 and 1.2513 $m/s^{1.75}$, respectively. These high values may arise because of the condition of the track, the train operation style, or the speed of the train. In addition, the daily exposure action value time took only 7 hours and 16 minutes to exceed the vibration exposure limit for an eight hour exposure. Surprisingly, this result was found to exceed the standard time of WBV assessment, which is stated in ISO 2631-1:1997 to be 8 hours. The reason for this is unclear, but the speed of the train may contribute to the higher vibration magnitude. The high WBV exposure may result in musculoskeletal disorders to the passengers. There are several examples in the literature that relate WBV exposure from occupational vehicles to musculoskeletal disorders.

The term musculoskeletal disorder refers to conditions that involve the nerves, tendons, muscles, and supporting structures of the body (Bernard et. al., 1998). Exposure to WBV is another occupational risk factor that may cause LBP in participants of occupational vehicles (Bovenzi et. al., 1999). In western countries, an estimated 4–7 percent of all employees are exposed to potentially harmful doses of WBV. Experimental studies have found that resonance frequencies of most of the organs or other parts of the body lie between 1 and 10 Hz, which are in the range of frequencies found in occupational machines and vehicles. Six million workers are exposed to WBV, typically while in a seated position. Workers at risk include delivery vehicles drivers, forklift operators, helicopters pilots, and construction equipment operators (Griffin et. al., 2006). Tractor drivers have reported a 61–94% prevalence of LBP and pathological changes in the spine, and heavy-equipment drivers report a 70% prevalence of LBP. WBV is recognized as an important risk factor for occupational LBP in a variety of occupational groups (Joubert, D.M. et. al., London, L. et al. 2007). At least four European countries have placed WBV injury on their official lists of occupational diseases (Hulshof et al., 2002). Among such physical exposures encountered in working conditions, WBV has repeatedly been identified as a risk factor for LBP (Santos, B.R. et. al. 2008). Several epidemiologic studies conducted in the past several years have found strong evidence of a correlation between WBV exposure and the onset of LBP (Noorloos, D. et. al., 2008). The National Research Council (2001) reported that there is evidence of a "clear relationship between back disorders and whole-body vibration". Joubert, D.M. and London, L. et. al. (2007) studied the association between back belt usage and back pain among forklift drivers that were frequently exposed to WBV. LBP has been identified as one of the most costly disorders among the working population worldwide, and sitting has been associated with the risk of developing LBP (Lis, A. M. et. al., 2007). It was shown that sustained truck sitting postures maintained by mining vehicle operators generates back muscle fatigue and postural balance issues (Santos, B.R. et. al., 2008).

CONCLUSIONS

In this work, it was found that the exposure of the human body to WBV increased with an increase in the vibration exposure duration and total number of train trips, as illustrated. The increase in the Daily value of Exposure to Vibration A(8) values and Vibration Dose Values (VDV) with increasing travel time and trip quantity. In this work, the frequency-weighted acceleration value recorded on the Malaysian national train was found to be close to the accepted exposure limit set in ISO 2631-1:1997. Hence, most of the passengers on these trains are being subjected to potentially dangerous levels of WBV during their travel. WBV exposure is known to cause health problems in humans. Empirical studies have shown that passengers that are exposed to WBV while in occupational vehicles often have musculoskeletal disorders. However, in Malaysia there is currently insufficient research dedicated to the problem. Because the general public is unaware of the seriousness of WBV, train passengers are often not given information concerning WBV exposure and the related health risks. In conclusion, more studies are needed to provide clear evidence of the association between WBV and musculoskeletal disorders, especially involving to Malaysian occupational vehicles. A future study should focus on a train driver's exposure to WBV and the consequent health problems.

REFERENCES

Angela Maria Lis, Katia M. Black, Hayley Korn, Margareta Nordin., (2007). Association between sitting and occupational LBP. Eur Spine J, 16, 283–298.

Bernard B.P., (1998). Musculoskeletal Disorders and Workplace Factors: A Critical Review of Epidemiologic Evidence for Work-Related Disorders of the Neck, Upper Extremities, and Low Back. NASA no. 19980001289.

Bernard, B.P., (1997). Musculoskeletal Disorders and Workplace Factors: A Critical Review of Epidemiologic Evidence for Work-related Musculoskeletal Disorders of the Neck, Upper Extremity, and Low Back. National Institute for Occupational Safety and Health, Cincinnati, OH.

Bovenzi, M., Hulshof, C.T.J., (1999). An updated review of epidemiologic studies on the relationship between exposure to whole-body vibration and low back pain. Int. Arch. Occup. Environ. Health, 72(6), 351–365.

Brenda R. Santos, Christian Larivie're, Alain Delisle, Andre' Plamondon, Paul-E'mile Boileau, Daniel Imbeau, (2008). A laboratory study to quantify the biomechanical responses to whole-body vibration: The influence on balance, reflex response, muscular activity and fatigue. International Journal of Industrial Ergonomics, 38, 626–639.

Burdorf A, Sorock G., (1997). Positive and negative evidence of risk factors for back disorders. Scand J Work Environ Health, 23, 243–56.

Danielle Noorloos, Linda Tersteeg, Ivo J.H. Tiemessen, Carel T.J. Hulshof, Monique H.W. Frings-Dresen., (2008). Does body mass index increase the risk of low back pain in a population exposed to whole body vibration? Applied Ergonomics, 39, 779–785.

Darren M. Jouberta, Leslie London., (2007). A cross-sectional study of back belt use and low back pain amongst forklift drivers. International Journal of Industrial Ergonomics, 37, 505–513.

Directive 2002/44/EC of the European Parliament and of the Council of 25 June 2002, Official Journal of the European Communities.

Garg A, Moore JS., (1992). Epidemiology of low-back pain in industry. Occupational Medicine, 7, 593–608.

Grieco, A., Molteni, G., De Vito, G., Sias, N., (1998). Epidemiology of musculoskeletal disorders due to biomechanical overload. Ergonomics, 41, 1253–1260.

Griffin MJ., 1990. Handbook of human vibration. Academic Press, London, 171-220.

792

Griffin, M.J., (2006). Health effects of vibration – the known and unknown. Conference on Human Vibration, Morgan Town, WV: DHHS/CDC/NIOSH, pp. 3–4.

Hagberg, M., Silverstein, B., Wells, R., Smith, M.J., Hendrick, H.W., Carayon, P., Perusse, M., (1995). In: Kuorinka, I., Forcier, L. (Eds.), Work related musculoskeletal disorders (WMSDs): a reference book for prevention. Taylor & Francis, London, Bristol.

Hulshof, C., Van Der Laan, G., Braam, I., Verbeek, J., (2002). The fate of Mrs. Robinson: criteria for recognition of whole-body vibration injury as an occupational disease. Journal of Sound and Vibration, 253, 185–194.

Kjell Spang, KS-Miltek. Project OSHA SME/2002/4668/DK. Ongoing standardization work of interest to the project.

Kroemer K., Kroemer H.K., and Elbert K.K., (2003). Ergonomics how to design for ease and efficiency. Second edition. Prentice Hall.

Lings S, Leboeuf-Yde C., (2000). Whole-body vibration and low back pain: a systematic, critical review of the epidemiological literature 1992–1999. Int Arch Occup Environ Health, 73, 290–297.

Matilla M., (1996). Computer-aided ergonomics and safety – A challenge for integrated ergonomics. International Journal of Industrial Ergonomics, 17, 309-314.

National Research Council, Institute of Medicine, (2001). Musculoskeletal Disorders and the Workplace: Low Back and Upper Extremities. National Academy Press, Washington, D.C.

Palmer KT, Griffin MJ, Bendall H., (2000). Prevalence and pattern of occupational exposure to whole body vibration in Great Britain: findings from a national survey. Occupational Environmental Medicine, 57, 229–236.

van der Windt, D.A., Thomas, E., Pope, D.P., de Winter, A.F., Macfarlane, G.J., Bouter, L.M., Silman, A.J.,(2000). Occupational risk factors for shoulder pain: a systematic review. Occup. Environ. Med., 57, 433–442.

Van Tulder MW, Koes BW, Bouter LM., (1995). A cost-of-illness study of back pain in The Netherlands. Pain, 62, 233–240.

Waddell G, Burton AK., (2001). Occupational health guidelines for the management of low back pain at work; evidence review. Occupational Medicine, 51, 124-135.

Wang, M.J.J., Chung, H.C., Wu, H.C., (2003). The evaluation of manual FOUP handling in 300 mm wafer fab. IEEE Transactions on Semiconductor Manufacturing, 16, 551–554.

Weestgard, R.H., Winkel, J., (1997). Ergonomic intervention research for improved musculoskeletal health: a critical review. International Journal of Industrial Ergonomics, 20, 463–500.

Chapter 88

Management of Work Site Health Promotion: A Success Story

Bernhard Zimolong, Gabriele Elke

Department of Work and Organizational Psychology
Ruhr-University Bochum
44780 Bochum, Germany

ABSTRACT

Based on the framework of a healthy work organization and of a holistic health management system, the aim of this project was the development, implementation, evaluation and transfer of a holistic health management system within the tax administration of North-Rhine Westphalia, Germany.

Results from a three years longitudinal study indicate moderate to small benefits in the development of organizational and individual health resources and health effects. To reach long term sustainability, health promotion structures and processes have been adapted and readjusted at all levels of the fiscal authorities.

Keywords: Health and safety management, workplace health promotion, human resource management, health promoting leadership, longitudinal study, health climate, musculoskeletal disorders

INTRODUCTION

Worksite Health Promotion (WHP) programs are initiatives directed at improving the health and well-being of workers and, in some cases, their relatives. They include programs designed to prevent the occurrence of disease or the progression of disease from its early unrecognized stage to one that's more severe. At their core, WHP support primary, secondary, and tertiary prevention efforts (Goetzel & Ozminkowski, 2008).

The main driving force behind employers' growing interest in providing WHP services to their workers is undoubtedly rapidly raising health care costs. Employers' health care costs, primarily focused on sickness care, are increasing exponentially with no immediate attenuation in sight. The most recent worksite health promotion survey in Germany in 2004 reports that 20% of the enterprises from a representative panel of 16,000 implemented some form of health promotion activities. Most frequently mentioned were analyses of status of employee's illness, and surveys on health and sickness status (9%), followed by health education (6%), health circles (6%) and other activities (5%) (www.iab.de). If structural, e.g., management and environmental activities as well as individually focused health activities are simultaneously considered, less than 10% of the enterprises make use of an holistic management approach.

Worksite health promotion programs range from single component to multicomponent programs, facing the multicausal causation of several disorders. Heaney and Goetzel (1998) examined 47 peer-reviewed studies over a 20-year period. They reported that WHP varied widely in terms of their comprehensiveness, intensity, and duration. Consequently, the measurable impact of these programs varied significantly because different intervention and evaluation methods were employed. Despite the variability in programs and study designs, the authors concluded that there was "indicative to acceptable" evidence supporting the effectiveness of multicomponent WHP in achieving long-term behavior change and risk reduction among workers.

A recent review of workplace-based health promotion and disease-prevention programs was reported by the Community Preventive Services Task Force in 2007 (Task Force Comm. Prev. Serv., 2007). Results from 50 studies which qualified for inclusion in the review, included a range of health behaviors, physiologic measurements, and productivity indicators linked to changes in health status. Most of the changes in these outcomes were small when measured at an individual level. For example, the review found strong evidence of WHP effectiveness in reducing tobacco use among participants (with a median reduction in prevalence rates of 1.5 percentage points), dietary fat consumption as measured by self-report (median reduction in risk prevalence of 5.4 percentage points), high blood pressure (median prevalence risk reduction of 4.5 percentage points), total serum cholesterol levels (median prevalence reduction of 6.6 percentage points), the number of days absent from work for the reason of illness or disability (median reduction of 1.2 days per year), and improvements in other general measures of worker productivity.

Aside from changes in health risks, the review reported additional benefits associated with work site programs. These include increasing worker awareness of health topics; increasing detection of certain diseases, or risk for disease at an earlier stage; referral to medical professionals for employees at high risk for disease; and creation of need-specific health promotion programs based on the analysis of aggregate results.

From reviews of benchmarking and best-practice studies the following system elements of holistic health programs are described repeatedly as effective WHP practices (Goetzel et al. 2007): (a) integrating WHP into the organization's central operations; (b) addressing individual, environmental, policy, and cultural factors; (c) development of a healthy organizational culture, (d) targeting several health issues simultaneously; (e) implementation of health-screenings, (f) tailoring programs to address specific needs of the population (e.g., provision of a menu-approach), (g) effectively communicating; (h) attaining high participation rates; (i) networking with local and regional healthcare providers and institutions; (j) evaluation and continuous improvement of the program.

From the research on best practices in Occupational Health and Safety (OHS) management and based on findings from literature, Zimolong, Elke & Trimpop (2006) proposed a holistic framework of OHS management. The domain-specific management system must be integrated into the processes of an organization to assure a sustainable effect. OHS activities have to be incorporated into the daily routines of managers, supervisors and employees, and OHS standards and processes into the life cycle of products, services and work systems. Best practices of Human Resource Management support long term commitment and involvement of employees to OHS. The system elements to be managed are risk control and health promotion systems. Key elements of the systems are human resources management, management of information and communication, (re)design of work and technology, and development of an OHS supporting culture.

OHS intervention strategy

The objective of the research project was the development, implementation, evaluation and transfer of a holistic health management framework within the German tax administration. The organizational change process followed the dual approach of the structural and psychosocial model drawing on top level commitment and from bottom up on strong participatory support of employees, members of employee committees, and health and safety representatives (Zimolong & Elke, 2009). Starting point was an update of previous and ongoing work site health activities and programs at the level of the local tax offices, an assessment of their outcomes, a health survey, and the participatory implementation of steering committees in each of the tax offices. Their responsibility was to plan, coordinate, evaluate and improve health promotion activities. Participants of the local steering committees were the senior and deputy manager, first-level managers, employees, members of the local employee committee, and health and safety representatives.

Taken together, 8 -12 members joined the local committee. The president of the tax administration Rhineland chaired the central steering committee of the 9 tax offices. This committee incorporated all senior managers of the tax offices, central health and safety representatives, members of the central employee committee and the scientific consultants.

Key elements of the management framework to be implemented incorporated processes (structures) and activities of HRM, e.g., leadership accountability for OHS objectives, linked with feedback systems from annual surveys, top- down and bottom-up health goal settings and negotiations, setting up of monitoring and feedback systems, promoting of OHS responsibility of self-managed teams and individuals, and training systems linked to managers' and subordinates' needs. Additionally, peer support with respect to health activities and positive health attitudes generated by work teams served as supplement for both managerial influence and internalized member commitment.

Managerial tasks in the information and communication domain included the establishment of top-down and bottom-up information and communication channels and platforms on health issues, installation of a web-based communication platform, and the start-off of internal communication processes promoted by incentives and personal communication ownerships. Management of job and work design emphasized the allocation of accountability to first line managers and teams. This was linked to an ongoing monitoring and improvement loop with the focus on physical (ergonomic) as well as psychosocial aspects of work place environment. Teams (health circles) developed work process improvements and were encouraged to adjust characteristics of their computer software.

The implementation and continuing improvement of the generic OHS management system was guided and redirected by annual health surveys. Specifically tailored health interventions such as a 12 month 'Healthy back' program (Schwennen & Zimolong, 2009), and ongoing menu offers of physical and psychosocial activities at the local tax offices completed the OHS system to be implemented. A notable part of the offers resulted from networking with local and regional healthcare providers and institutions.

METHOD

The tax administration of North Rhine-Westphalia operates 137 local tax offices counting approximately 30,000 employees. The project started in ten local tax offices of the tax administration Rhineland whose managers decided to voluntarily participate in the pilot project. One tax office dropped off in the beginning of 2006. Each tax office counts between 138 and 380 employees organized in 10-15 functional units summing up to 2,136 people. The implementation of the OHS management system started in 2005 (time1) and was guided and redirected by annual health surveys end of 2005 (time 1), 2006 (time 2), and 2007 time 3). Scales were drawn from a short version of the Occupational Health and Safety questionnaire (OHS, Gurt, Uhle & Schwennen, 2010). The questionnaire covers different aspects of work demands, organizational resources such as health promoting leadership, promotion of work autonomy, i.e. of decision latitude and job variety, health resources of employees such as coping strategies and personal health promotion activities. Health effects of employees are captured on various indicators of physical and psychological employee health.

The intervention period for this study was three years. Data were collected at three time points (end of intervention year one, two, and three.) Participation rates varied between 1.627 at time 1(n=10) to 1.127 at time 3 (n=9) equaling response rates of 72,3% to 52,7%. Questionnaires were administered online during a two-week period. Responses were completely anonymous; participation was voluntary, but encouraged via emails send out by the head of the tax office. Socio-demographic data from time 3 showed that 13.0% were younger than or equal to 30, 24.2% between 30 and 40 years, 33.4% between 41 and 50 years, and 29.3% older than 50 years. About 66.4% of participants were female and 67.3% employed full-time. These figures reflect the actual distribution of the socio-demographic profile in the tax administration.

We assessed participant perceptions of health promoting leadership, work design, information and communication, health culture, health resources of employees and health effects of employees. All measures used 5-point Likert-type scales ("strongly agree" to "strongly disagree") with a neutral midpoint. Health promoting leadership was assessed using three five-item scales and features the scales "Adoption of responsibility for health promotion", "Participation and involvement" and "Promotion of self-initiative". A sample item of the scale participation and involvement reads "My supervisor/ top management discusses health-related topics with us". Internal consistency (Cronbach α) was .83 for adoption of responsibility (n=2 items; see table 2). Work design was assessed using a 3-item scale. This scale was designed to capture the aspects of work autonomy. One item reads: „I am able to plan my work autonomously. " Information and communication captured one general and one health specific aspect of the transfer of information. We assessed health culture by adopting a 2-item scale on health climate and participation rates of the health surveys and the back screening. Climate for health assessed the perception of the individual employee to which degree "health" is already part of the organizational norms (2 Items, $\alpha = .72$). The scale features items such as: "Health initiatives in my organization are either insufficient or inadequate" (reverse coded).

Health resources of employees were measured with OHS scales and participation rates of health promotion activities. OHS scales contained the 4-item scale "Coping strategies" ("When I am under pressure, I can rely on well-proven strategies, which I have applied in comparable situations"), one 1-item scale "Empowerment" (degree of exerting influence on ongoing decision processes) and one 1-item scale "Personal health accountability". Attendance rate of health promotion activities included activities at the tax office during work and leisure time activities off the work respectively such as physical activities, healthy nutrition, and stress management. Health promotion activity was assessed at time 3 asking for percentages of activities at time 1 (prior to the intervention of HMS) and time 3 (end of intervention).

Health of employees was assessed by seven health scales which were grouped into three categories (1) demands, (2) short term and (3) long term health consequences. Demands were assessed by one 3-item scale "High Work Demands" („I have so much work to do that I can hardly handle it") and one 3-item scale "Emotional Demands" („In my job I have to suppress my emotions to appear "neutral" at the surface"). Short-term health consequences contain the 3-item scale "Strain Intensity" („I react harshly though I don't want to"). Persisting tiredness and irritability (2 items) features items such as „Do you feel tired and worn-out throughout the whole day? ". The scale "Musculoskeletal Pain" (2 items) contains items such as „Do you feel pain in you neck or shoulders?". Back pain intensity (1 item) reads: „How intense has your back pain been throughout the last 4 weeks on average". Disability due to back pain (1 item) features „To what degree has your back pain impaired you concerning routine tasks (e.g. work, school, housework)? ".

Results

ORGANIZATIONAL AND INDIVIDUAL HEALTH RESOURCES

Table 1 shows the mean scores and standard deviations of the scales, participation and attendance rates of time 3 (2007) together with the standardized effect sizes between ratings of 2005 and 2007 (d) if available. CI represents the 95% confidence interval of d. We computed multivariate MANOVA with SPSS 17 to test for significant differences between time 1 and time 3.

As a general result, between time standardized mean differences d show mostly the expected significant changes in time 2. Exceptions are promotion of self initiative, strain intensity and persisting tiredness. Significant between time effect sizes range from d=.20 (information and communication) to .36 (coping strategies). In the following section we refer to change in percentage for convenience.

With respect to organizational resources, health promoting leadership has been clearly improved by feedback from the surveys, goal agreements, and specific health promoting leadership trainings. As assessed by the subordinates adoption of responsibility for health promotion of subordinates increased by 9.2% and promotion of participation and involvement of subordinates into health promoting activities by 16.4%. Additionally, promotion of work autonomy, i.e. of decision latitude and job variety, was improved by 7.4% as well as transfer of information including health issues by 6.2%. Health culture and climate increased by 7.7%. Whereas the attendance rate of back health screening was 49.7%, the attendance rate of the health surveys ranged from 72,9% to 52,2%.

Health resources of employees indicated by coping strategies with stress (+ 7.9%), attendance rates of health promotion activities offered by the tax offices (+527%) and private health promotion activities such as physical activities, nutrition, and stress management (+ 49%) increased clearly. With respect to work demands and health effects of employees, it has to be noted that the present negative trend on health effects could be stopped and even reversed. Despite increased work demands by 5.5%, the health effect indicators remained constant or showed even slight decreases over the whole project phase, as far as strain intensity, persisting tiredness and cardiovascular diseases are concerned. In the second project phase after the implementation of the target specific "Healthy Back" program in 2006, the negative trend could be stopped and even reversed: The frequency of back pain dropped by

Table 1 Means, Standard Deviations (SD), and Standardized Effect Sizes d between time 1 (2005) and 3 (2007) for perceived organizational and individual resources, work demands and health effects

Key elements of OHS management system	Indicators: OHS questionnaire ; 5 point Likert type scale (1=strongly disagree; 5= strongly agree; (n=Items; Cronbachs α for 2007); participation rates in %; d=standardized effect size 2005-2007; 95%$_{CId}$= Confidence Interval of d					
Policy, Structure, Strategy	New set up of - position of a health manager at first level of ministry of finance - health promotion group at second level of administration - intranet platform on health topics - regular health surveys every two years - embedding health promotion as policy into the tax offices					
Organizational resources		Mean Time 3 2007	SD	d 2005-2007	p value <	95% CI of d
Health promoting leadership (assessment by subordinates)	Adoption of responsibility for health promotion (2; .83)	2.93	1.11	.22	.01	.079
	Participation and involvement (1)	2.57	1.10	.32	.01	.080
	Promotion of self-initiative (1)	2.66	.98	-.08	n.s.	.079
Work design	Work autonomy (3; .80)	3.96	.79	.34	.01	.081
Information and communication	Transfer of information (2; .60)	3.35	.97	.20	.01	.079
Health culture	Attendance rates health survey	52,7%	73 –52%	- 27,1 %		-

Attendance rates back screening	49,7%	40 – 66%	-		-
Health climate (2; .72)	3.73	.94	.28	.01	.080

Table 2 (continued)

Key elements of OHS management system	Indicators	Mean Time 3 2007	SD	d 2005 -2007	p value <	95% CI of d
Health resources of employees	Coping strategies (4; .54)	3.37	.64	.36	.01	.080
	Empowerment (1)	3.82	.86	-		
	Personal health accountability (1)	3.03	1.04	-		
	Attendance rates of health promotion activities offered by employer	29,8%		difference 527%		
	Attendance rates of private health promotion activities: physical activities, nutrition, stress management	73,6%		difference 49%		
Work demands	Work demands (3; .54);	2.91	.72	.21	.01	.079
	Emotional demands (3; .71)	2.92	.89	.26	.01	.079
Health effects	Strain Intensity (3; .75)	2.61	.88	.07	n.s.	.079
	Persisting tiredness, irritability (2; .66)	2.91	1.04	.14	n.s.	.079
	Weekly/daily musculoskeletal pain* (2; .81) 2006/07	41,8%		difference - 5.4%	.05	

Back pain intensity** (1) 2006/07	2.24	.99	-.13	.05	.081
Disability due to back pain** (1) 2006/07	1,95	.85	-.18	.01	.081

* 5 point frequency scale: once per year ;daily;
** 5 point intensity scale: very low; very strong

3.9%, the weekly/daily back pain by 5.5%, and impairment due to back pain by 8.7%. Consequently lost workdays due to severe back pain dropped by 12.3%. This percentage accounts for an estimated reduction of 540 lost workdays due to severe back pain for 2,000 employees of the nine tax offices.

POLICY, STRUCTURE, AND STRATEGY

To reach long term sustainability, new responsibilities and structures have been established at all levels of the fiscal authority. At the first level of the ministry of finance a new position of a health manager has been set up. At the second level - tax administration Rhineland - the health promotion group, responsible for promotion, coordination and evaluation of the health management system, has been set up as a permanent function. Additionally, at the third level, new health promotion structures and processes have been introduced into the tax offices such as a standing health promotion group at management level, regular health surveys every two years, reports on health promotion activities to the management, and health promotion as a regular topic in meetings.

CONCLUSIONS

This trial shows that a health management system is effective in managing health resources of organizations and individual resources of employees. Stabilization of strain effects despite an increase in work demands is consistent with psychological hypotheses that organizational and individual resources are important mediators between work demands and health effects. In combination with a specific target program on back health, intervention had a significant effect in reducing weekly/daily back pain, pain intensity and disability scores. There is evidence of a moderate to small effect with the health management system ranging from d= .20 to .36.

Success factors are affiliated with the introduction and continuous improvement of the holistic health management system: Integrating worksite health promotion into the organization's central operations; addressing individual, environmental, policy, and cultural factors; implementing health-screenings, tailoring programs to address specific needs of the employees and specific target groups, promoting participation and involvement of employees into health promoting activities, evaluating and continuously improving worksite health promotion. To reach long term sustainability, new responsibilities and structures have been established at all levels of the fiscal authority.

The benefits of the participative implementation of the health management system were broad ranging and maintained at 36 months, supporting amongst others health climate and culture at the workplace (Gurt & Elke, 2009). In the course of the project, several health issues were targeted simultaneously (e.g. implementation of health-screenings, offering specific mental health courses, participation and involvement, health promoting leadership). By tailoring the measures to the specific needs of the employees, the participation and attendance rates of health promoting activities have been increased significantly over the course of the project. Networking with healthcare providers and institutions such as the Unfallkasse Nordrhein-Westfalen and the University Bochum provided additional support as well as scientific expertise.

Acknowledgments. INOPE (Health promotion and prevention supported by Integrated Network, Organizational, and Personnel development) is a project funded by the German Federal Ministry of Education and Research (BMBF, www.inope.de)

REFERENCES

Goetzel, R. Z. & Ozminkowski, R. J. (2008), "The health and cost benefits of work site health promotion programs." *Annu. Rev. Public Health*, 29, 303–23.

Goetzel, R. Z., Shechter, D., Ozminkowski, R. J., Marmet, P. F. & Tabrizi, M. J. (2007), "Promising practices in employer's health and productivity management efforts: Findings from a benchmarking study." *J. Occup. Environ. Med.*, 49, 111–30.

Gurt, J. & Elke, G. (2009), Health Promoting Leadership: The Mediating Role of an Organizational Health Culture, in: Proceedings of HCI International 2009, 19-24 July, 2009, San Diego: Ergonomics and Health Aspects of Work with Computers, J. A. Jacko et al., (Eds.). pp. 29-38. Berlin/Heidelberg: Springer.

Gurt, J. Uhle, T. &Schwennen, C. (2010). „Fragebogen zum Arbeits- und Gesundheitsschutz-Betriebliche Gesundheitsförderung [Health and Safety Management Questionnaire–Occupational Health Promotion]", in: Handbuch wirtschaftlicher Testverfahren, W. Sarges & H. Wottawa (Eds.). pp. 45-54. Pabst: Lengerich.

Heaney C. & Goetzel, R. Z. (1998),. "A review of health-related outcomes of multi-component worksite health promotion programs." *Am. J. Health Promot.*, 11, 290–307.

Task Force Comm. Prev. Serv. (2007), Proceedings of the Task Force Meeting: Worksite Reviews.

Wilson, M. G., DeJoy, D. M., Vandenberg, R. J., Richardson, H. A. & McGrath, A. L. (2004), "Work characteristics and employee health and well-being: Test of a model of healthy work organization." *Journal of Occupational and Organizational Psychology*, 77, 565–588.

Zimolong, B., & Elke, G. (2006), "Occupational health and safety management." In G. Salvendy (Ed.), Handbook of Human Factors and Ergonomics (pp. 673-707). New York: Wiley.

Zimolong, B. & Elke, G. (2009), "Management of Work Site Health promotion Programs: A Review", in: Proceedings of HCI International 2009, 19-24 July, 2009, San Diego: Ergonomics and Health Aspects of Work with Computers, J. A. Jacko et al., (Eds.). pp. 131-140. Berlin/Heidelberg: Springer.

Zimolong, B., Elke, G., & Trimpop, R. (2006), Gesundheitsmanagement (Health management). In B. Zimolong & U. Konradt (Eds.), Enzyklopädie der Psychologie (Encyclopedia of Psychology): Band 2 Ingenieurpsychologie (Engineering Psychology) (pp. 633-668). Göttingen: Hogrefe.

Chapter 89

Instilling an Organizational Climate for Health: Does Top-Management Make the Difference?

Jochen Gurt, Christian Schwennen

Department of Work and Organizational Psychology
Ruhr-University Bochum
44780 Bochum, Germany

ABSTRACT

Shared norms, values and basic assumptions are at the core of organizational culture and facilitate employee behavior giving orientation and providing a higher meaning to their work. Therefore, organizational culture is regarded as an instrument to coordinate and integrate employee behavior towards shared long-term organizational goals. It is widely agreed among health promotion researchers that leader support and engagement for health promotion is essential for the emergence of a climate for health and a corresponding culture. Supervisors and top-management support the development process. While research is well advanced within other contexts, empirical studies on the leadership–climate link are rare for health. The present study examines the impact of leadership on the emergence of a climate for health, taking into account the hierarchical structure of the organization. Sampling took place in nine local tax offices in North Rhine-Westphalia, Germany, who participated in a three-year health promotion project. HLM-Results indicate that supervisors are the primary source of a climate for health, while most of the influence of top-management seems to trickle trough the hierarchical levels of the organization rather than directly influencing employee perceptions. For organizational health promotion practitioners it is therefore indispensible to secure supervisor commitment for health promotion initiatives.

Keywords: Leadership, organizational culture, occupational health promotion, hierarchical regression, multi-level analysis

INTRODUCTION

Recent predictions of the demographic development make it obvious that organizations will have to deal with an aging workforce in the future. The anticipated influx of younger qualified workers will not be sufficient to replace retiring employees (Musich, McDonald, & Chapman, 2009). The most likely consequence will be that employees will have to work longer before retirement and employee health will become an even more important issue for organizations. It is therefore not surprising that worksite health promotion has received increased attention in the past decade (Goetzel & Ozminkowski, 2008). Findings indicate that comprehensive health promotion programs and management-oriented approaches making employee health and well-being a long-term management issue are far more effective than individual training and small-scale interventions targeting single individual risk factors (Fjell, Alexanderson, Kalqvist, & Bildt, 2007; Goetzel & Ozminkowski, 2008). O'Donnell (2002) states that health promotion programs that are creating supportive environments are the most effective in leading to long-term healthy life-style changes. From an organizational perspective this incorporates ongoing efforts concerning design and organization of the workplace and work itself as to make it more health inducing and facilitate healthy behaviors through e.g. changing the physical work setting, adapting organizational structures, corporate policies and making employee health an strategic corporate objective. Such comprehensive approaches are also known as health management systems emphasizing their ongoing nature and consideration of health promotion as a management routine (Zimolong & Elke, 2006). For such approaches an organizational health culture is both an important targeted outcome and a pre-requisite for successful future health promotion (Goetzel & Ozminkowski, 2008).

WORPLACE HEALTH PROMOTION: THE IMPORTANCE OF CULTURE AND CLIMATE

The importance of organizational culture is highlighted by practitioners first of all because it is supposed to guide employee behavior even *when nobody is looking* (Gandossy et al., 2009). Empirical studies reveal relationships to core organizational objectives like service and customer orientation, innovation and business performance (e.g. Kotter & Heskett, 1992). At the core of organizational culture are shared basic assumptions. Schein's three layer model (2000) suggests that these are not entirely perceivable and may even be unconscious. Still, they form the basis of evaluation of what is right or wrong, i.e. the organizational norms and values. These become perceivable in the form of organizational artifacts, like rules and regulations (or their absence), which posses symbolic value. Symbolic value can be attributed to anything that allows for a "conscious or unconscious association with some wider concept or meaning" (Hatch, 2000, p. 249). Important for this study are two things: First, that social norms and values within an organization can either promote employee health, well-being and healthy lifestyles or have a detrimental effect (Allen, 2002). And second, the assumption that at least to some degree artifacts and therefore organizational culture can be intentionally shaped by management actions. While in general organizational culture is influenced by all organizational factors and processes that carry symbolic value, a lot of studies show that leaders and leadership behavior are among the most important factors in establishing, modifying or transforming a certain culture.

CULTURE & CLIMATE FOR HEALTH AS LEADERSHIP OUTCOMES

The assumption that leaders impact culture draws from social cognitive learning theory (Bandura, 1977). Employee behavior is (among other factors) a function of perceived behavior of their environment. Leaders serve as role-models – employees are influenced by observing and imitating his behavior. Hatch (2000) proposes a second mechanism: Leaders can also introduce new artifacts into the cultural field. The recognition of these artifacts as organizational symbols leads to cultural changes. According to Hatch (2000) leaders in particular facilitate this symbolizing activity for example through exemplary behavior because their (social) power relation magnifies the symbolic value oft their behavior and provides them the capacity to lead change. A couple of studies investigated climate instead of culture, which is considered as its surface-level manifestation (Schein, 2000). It is supposed to comprise various domain-specific dimensions for example service or safety, which in turn relate to the corresponding success variables like service quality or safety performance (Parker et al., 2003; Zohar, 2000). It seems reasonable that cultural change may be preceded by a climate change, making climate an appropriate research variable in this study (Parker et al., 2003). The most convincing evidence that leadership influences climate perceptions exists for transformational leadership (Bass & Avolio, 1993), which is aimed at changing employees' perceptions, expectations, aspirations and values through the establishment of vision and goals. Quite a number of studies reveal a significant leadership-climate-outcome link for example concerning safety (Hofman & Morgeson, 2004), but the findings for health are rare. While there are a lot of normative claims that leaders' engagement for employee health and well-being may well enhance climate for health (Allen, 2002), "both the volume of actual studies and the quality of the methodologies drops off sharply" as far as the domain of health promotion is concerned (Bennet, Cooke, & Pelletier 2003, p. 88). Rosen (1996) identified leadership as the most important factor to build a healthy culture in a qualitative interview study. Recently a study by Golaszewski, et al. (2008)

showed that supervisor modeling concerning healthy behavior significantly correlated with perceptions of a climate for health. Bass and Avolio (1993, p. 130) noted also that "cultural norms arise and change because of what leaders focus their attention on". Therefore, more recent studies point out the importance of domain-specific leadership (Barling, Loughlin, & Kelloway, 2002). The rational for this is that leaders who score high on transformational leadership might still neglect for example the issue of safety and direct the attention of the employees to productivity or customer service. This also should also hold for the topic of health and workplace health promotion. Health-specific (HS) leadership can therefore be regarded as the leaders' explicit and perceptible consideration of and engagement for employee health (Gurt & Elke, 2009). Exemplary leader behaviors are e.g. setting the agenda for health by bringing up the topic, modeling of healthy behavior, supporting organizational health-promotion activities. Such leader support for worksite health promotion is considered to be crucial for the success of organizational health promotion (Zimolong & Elke, 2006). A Swedish study found that the impact of workplace health promotion initiatives was dependent on various leadership behaviors, such as participation, the attitude towards the cause of sick leave as well as respect, trust and open discussion (Dellve, Skagert, & Vilhelmsson, 2007). Although not explicitly mentioned in this study the relatedness of such behaviors to values and norms self-evident.

LEADERSHIP - AN ISSUE ON MULTIPLE LEVELS

Leadership is an organizational phenomenon that takes place on all different levels of the hierarchy. The meta-analytic findings of Lowe, Kroeck, & Sivasubramaniam (1996) stress the importance to take the hierarchical structure into account. Their results revealed that higher level leaders exhibited significantly less transformational leadership than their lower level counterparts. Therefore, the hierarchical level not only seems to define power and responsibilities, but may also entail a different set of adequate leadership behaviors. This leads to two important considerations for this study. First, if we follow the reasoning of Hatch (2000), power is a determinant of the symbolic value members of the organization attach to organizational artifacts – the higher the hierarchical position, the stronger the symbolic value of the leaders' behavior. Formal influence mechanisms on the top-level comprise setting the corporate agenda, explicitly developing a vision, mission and the corporate policy of the organization. Indirectly top-management also signals the importance of competing organizational initiatives by providing the necessary resources. Another aspect is that social learning does not only take place as far as employees are concerned, but also in terms of the supervisors – in an even more immediate sense. For them members of top-management serve as role-models and imitation is even easier as they can apply the learned behaviors without alteration. Bass and Avolio (1993, p.115) state that "desired role models of leadership begin at the top and are encouraged at each successive level below". This leads us to the first hypothesis:

H1a: Top-management HS leadership behavior will have a beneficial influence on supervisor HS leadership.

H1b: Top-management HS leadership behavior will have a beneficial influence on supervisor perception of a climate for health.

A second important issue to notice is that the hierarchical level also determines the intensity of interactions with employees. While top-management executives meet employees only occasionally face-to-face, supervisors interact with them on a daily basis. Despite their relatively smaller power-base it is therefore widely agreed that supervisors are important promoters of culture and cultural change. This follows the logic that values are incorporated in the organizational reality through social validation, i.e. by the experience of employees that they actually help to reduce uncertainty and subsequently anxiety in the daily work (Klenke, 2005). This experience can only happen on the supervisor level. Therefore it is reasonable to assume that normative postulates of top-management executives will only be integrated in the set of organizational values if they are mirrored by supervisor behavior (e.g. Zohar, 2000). This leads to our second hypotheses:

H2: Supervisor HS leadership will have a beneficial influence on employee perception of a climate for health.

Taken together with the assumption that symbolic value is the highest if artifacts are consistent with other artifacts within the symbolic field (Hatch, 2000), consistent supervisor and top-management behavior should mutually enhance their impact on climate for health leading us to a third hypothesis.

H3: Employee perceptions of a climate for health should be highest if both supervisor and top-management HS leadership are high.

EMPIRICAL STUDY

SAMPLE, MEASURES AND ANALYSIS

The sample consisted of 1,969 employees from nine local tax offices of the German tax administration in North-Rhine Westphalia. Tax office size ranged from 131 to 345 employees. Employees and supervisors were provided an inventory comprising a set of different questionnaires. Questionnaires were administered online during a two-week period. Responses were completely anonymous; participation was voluntary, but encouraged via emails send out by the head of the tax office. 1,024 responses from employees and 89 responses from supervisors were gathered equaling a response rate of 52.1 %. Participation rates in the tax offices ranged from 33.3% to 56.5%. Socio-demographic data showed that 13.0% were younger than or equal to 30, 24.2% between 30 and 40 years, 33.4% between 41 and 50 years, and 29.3% older than 50 years. About 66.4% of participants were female and 67.3% employed full-time. These figures mirror the actual distribution of the socio-demographic profile in the tax administration. For the hierarchical multi-level analysis a total of 353 employees could be matched with 54 supervisors.

HS leadership and climate for health scales were drawn from a short version of the Organizational Health and Safety (OHS) questionnaire (FAGS, Gurt, Uhle & Schwennen, 2010). Climate for health captures the perception of the individual employee to which degree "health" is already part of the organizational norms (2 Items, α = .66/.60). A sample item reads "Health initiatives in my organization are either insufficient or inadequate" (reverse coded). HS leadership was measured in terms of the engagement of the leader for health and health promotion (7 items, α = .91/.92). A sample item reads "My supervisor/ top management discusses health-related topics with us". Gender and type of employment (full-time vs. part-time) were integrated as control variables. Linear regression and hierarchical multi-level analysis were applied using SPSS 17.0 and HLM 6.0. Means, standard deviations and correlations among the study variables for the full sample of employees and supervisors are printed in table 1. Concerning the HLM computations we used full-information likelihood estimation in order to be able to compare nested models using a χ^2 –test. To capture the explained variance Bryk und Raudenbush's R^2 was calculated. All continuous explanatory variables were entered into the equations grand mean centered; categorical variables were coded 0/1.

Table 1: Descriptive Statistics, Cronbach's Alphas and Correlations*

		FK*	MA	1	2	3	4
1	Gender	.30 (.46)	.67 (.47)	- / -	-.42	.10	-.01
2	Employment Type	.83 (.38)	.64 (.48)	-.48	- / -	-.10	-.04
3	Climate for Health	3.99 (.74)	3.83 (.91)	.21	-.03	.66/.60	.30
4	HS Leadership	3.58 (.86)	2.75 (1.04)	.17	-.08	.40	.91/.92

* standard deviations are in parenthesis, gender: 1=female, employment type: 1= full-time, Cronbach's alphas are on the diagonal, $n_{supervisors}$ =89 (below diagonal), $n_{employees}$ =1,024 (above), all correlations significant (p≤.01), except figures printed in italic.

ANALYSIS RESULTS

To test H1a, 2 and 3 we applied hierarchical linear regression as variables were measured on different organizational levels. We used a stepwise approach as recommended by van Dick et al. (2005). Results for H1a, i.e. the impact of top-managements leadership on supervisor leadership, are printed in table 2. The ICC reveals a significant 18.5% of level-2 variance. The personal characteristics model shows no significant relation of gender and employment type to the criterion and does not decrease the total deviance of the model (χ^2= 2.00; df=3, p=.37). The next model, which includes top-managements' HS leadership behavior as a level-2 variable (intercept as outcome) leads to a significant decrease in total deviance ($\Delta \chi^2$= 8.21; df=1, p=.001). Level-2 variance drops from .20 to .15, which indicates that 23.2% percent of the between supervisors variance is explained by top-management HS leadership. Therefore, hypothesis 1a is confirmed.

Table 2: HLM-Results for Supervisor's HS Leadership

	Intercept-Only			Personal Characteristics			Level 2 HS-Leadership		
	β	se	p	β	se	p	β	se	p
Level 1									
Intercept	2.76	.08	.00	2.96	.14	.00	2.95	.16	.00
Employment type (1=full time)				-.16	.11	.13	-.17	.11	.11
Gender (1=female)				-.14	.12	.25	-.14	.12	.25
L1 Variance	.88			.88			.87		
Level 2									
Top-Management's HS leadership							.26	.08	.00
L2 Variance	.20			.20			.15		
Total Deviance	1001.35			999.35			991.14		
Δ χ²				2.00 (df=3, p=.37)			8.21 (df=1, p=.001)		

*dependent variable: supervisor HS leadership, β = parameter, se= standard error, p= significance level

To test H1b we applied simple linear regression as both variables were measured on the supervisor level. The coefficient for HS leadership (β=.40; p=.001) became significant, indicating a medium to strong relationship. In total, top-management HS leadership explained 16.0% of the variance in climate for health among supervisors confirming H1b.

In order to test H2 and 3 we regressed climate for health on supervisor HS leadership behavior (level 1) and top-management HS leadership behavior (level 2) using the stepwise HLM-approach again. The intercept-only model showed that climate for health varied very little between supervisor workgroups. Only 2.64% of the total variance was attributable to supervisors (ICC=.026). Adding personal characteristics in the second step did not explain any variance (R^2=.03, n.s.). In contrast, adding HS leadership behavior in the level 1 regression equation in step three decreased total variance considerably ($\Delta \chi^2$ =42.13, df=5, p=.00), explaining approximately 20% of level 1 variance as calculated by Byrk and Raudenbusch's R^2. The coefficient for HS leadership became highly significant (β=.28, p=.001) revealing a solid positive relationship. H2 can be confirmed.

To test hypothesis 3 we finally conducted a fourth step adding top-management HS leadership to the level 2 regression equation. This time we modeled it as a slopes-as-outcome model to capture the hypothesized cross-level interaction. In such a model the slope of supervisor's HS leadership (i.e. its impact on climate for health) is a function of top-management's HS leadership. By integrating the level 2 predictor this way, the model was moderately, but still significantly improved as indicated by the decrease in total deviance ($\Delta \chi^2$ =5.21, d.f. =1, p=.02). The significant coefficient for top-management HS leadership (β=.14, p=.01) reveals the sought-after cross-level effect indicating that the impact of supervisor behavior is the strongest when top-management behavior is accordingly. Therefore, also hypothesis 3 is confirmed.

CONCLUSION: THE INFLUENCE OF LEADERSHIP ON CLIMATE FOR HEALTH

All three hypotheses in our study could be confirmed exhibiting a close relationship between leadership on various levels and a climate for health. Our results therefore extend the findings in other domains to the domain of health (Barling et al., 2002). First, they show that, in order to create a climate for health it is not only important *how* leadership takes place, but also *which* topics and issues are put on the agenda by the leader (Bass & Avolio, 1993). Second, this seems to be true for both supervisor as well as top-management leadership. We found HS leadership to be related to climate for health on both levels. Top-management behavior was influential on both supervisors' perception of climate for health as well as their own HS leadership behavior, which in turn related to employees' perception of a climate for health. This suggests that the impact of top-management behavior trickles through the organizational hierarchy. It also indicates that frequent social interaction seems to be as influential on perceptions as power. Employees' perception of the organizational reality is shaped by their daily experience rather than the seldom top-management speeches and interaction – supervisors seem to be the promoters of climate and culture (Zimolong & Elke, 2006). Still, also the importance of consistency is stressed by our study as indicated by the significant cross-level effect we could establish for supervisor and top-management HS behavior. A drawback of this study is that only data from an administrative environment is analyzed. Future studies are therefore needed to validate the findings in private organizations. Also, the cross-sectional design of our study leaves all statements on causality to future studies incorporating a longitudinal design. Still, it seems reasonable to draw some conclusions for health management practitioners: First, our results indicate that supervisors should always be considered first hand when organizational cultural initiatives are implemented. They need to be qualified, informed and motivated to adapt their leadership behavior and guide employee attention e.g. by putting health and health promotion on their agenda. Second, top-management needs to be aware that its direct influence on employee perceptions of organizational climate seems to be fairly limited. It seems to rather have an indirect effect via its impact on supervisor behavior. Therefore, it has to closely monitor that its initiatives are not blocked on any organizational level in order to achieve the trickle through effect. On the contrary, if both the daily and highly perceptible supervisor behavior and the less visible, but highly symbolic top-management behavior consistently incorporate engagement for employee health and health promotion, is there a great chance to change the organizational climate in the intended health-promoting direction.

812

REFERENCES

Allen, J. R. (2002). "Building supportive cultural environments", in: Healthy promotion in the workplace, M. O'Donnell (Ed.). pp. 202-217. Albany, NY: Delmar.

Bandura, A. (1977). "Social Learning Theory" New York: General Learning Press.

Barling, J. Loughlin, C. and Kelloway, E.K. (2002). "Development and test of a model linking safety-specific transformational leadership and occupational safety", Journal of Applied Psychology Volume 87 No. 3.

Bass, B.M. and Avolio, B.J. (1993). "Transformational leadership and organizational culture", Public Administration Quarterly Volume 17 No. 1.

Bennet, J.B. Cook, R.F. and Pelletier, K.R. (2003). "Toward an integrated framework for comprehensive organizational wellness: Concepts, practices, and research in workplace health promotion", in: Handbook of occupational health psychology, J.C. Quick and L. E. Tetrick (Eds.). pp. 69-95. Washington, DC: American Psychological Association.

Dellve, L. Skagert, K. and Vilhelmsson, R. (2007). „Leadership in workplace health promotion projects: 1- and 2-year effects on long-term work attendance", The European Journal of Public Health Volume 17 No. 5.

Fjell, Y. Alexanderson, K. Kalqvist, L. and Bildt, C. (2007). „Self-reported musculoskeletal pain and working conditions among employees in the Swedish public sector", Work Volume 28 No. 1.

Gandossy, R. Peshawaria, R. Perlow, L. Trompenaars, F. and Dowling, D.W. (2009). "Driving Performance Through Corporate Culture: Interviews with Four Experts", Journal of Applied Corporate Finance Volume 21 No. 2.

Goetzel, R. and Ozminkowski, R.J. (2008). "The health and cost benefits of work site health-promotion programs", Annual Review of Public Health Volume 29.

Golaszewski, T. Hoebbel, C. Crossley, J. Foley, G. and Dorn, J. (2008). "The reliability and validity of an organizational health culture audit", American Journal of Health Studies Volume 23 No. 3.

Gurt, J. and Elke, G. (2009). "Health Promoting Leadership: The Mediating Role of an Organizational Health Culture", in: Proceedings of HCI International 2009, 19-24 July, 2009, San Diego: Ergonomics and Health Aspects of Work with Computers, J. A. Jacko et al., (Eds.). pp. 29-38. Berlin/Heidelberg: Springer.

Gurt, J. Uhle, T. and Schwennen, C. (2010). „Fragebogen zum Arbeits- und Gesundheitsschutz-Betriebliche Gesundheitsförderung [Health and Safety Management Questionnaire–Organizational Health Promotion]", in: Handbuch wirtschaftlicher Testverfahren, W. Sarges & H. Wottawa (Eds.). pp. 45-54. Pabst: Lengerich.

Hatch, M.J. (2000). "The Cultural Dynamics of Organizing and Change", in: Handbook of Organizational Culture & Climate , N.M. Ashkanasy, C. Wilderom and M.F. Peterson (Eds.) pp. 245-260. Thousand Oaks: Sage.

Hofmann, D.A. and Morgeson, F.P. (2004). „The role of leadership in safety", in: The Psychology of Workplace Safety, J. Barling and M.R. Frone (Eds.). pp. 159-180. Washington DC: American Psychological Assoc.

Klenke, K. (2005). "Corporate values as multi-level, multi-domain antecedents of leader behaviors", International Journal of Manpower Volume 26 No. 1.

Kotter, J.P. and Heskett, J.L. (1992). "Corporate culture and performance." New York: Free Press.

Lowe, K.B. Kroeck, K.G. and Sivasubramaniam, N. (1996). "Effectiveness correlates of transformational and transactional leadership: A meta-analytic review of the MLQ literature", Leadership Quarterly Volume 7 No. 3.

Musich, S. McDonald, T. and Chapman, L.S. (2009). "Health promotion strategies for the 'Boomer' generation: Wellness for the mature worker", American Journal of Health Promotion Volume 23 No. 3.

O'Donnell, M. (2002). "Employer's financial perspective on workplace health promotion", in: Health promotion in the workplace, M. O'Donnell (Ed.). pp. 23-48. Albany: Delmar Thompson Learning.

Parker, C.P. Baltes, B.B. Young, S.A. Huff, J.W. Altmann, R.A. Lacost, H.A. et al. (2003). "Relationships between psychological climate perceptions and work outcomes: A meta analytic review", Journal of Organizational Behavior Volume 24 No. 4.

Rosen, R.H. (1996). "Leading people: Transforming business from the inside out." New York: Viking.

Schein, E. (2000). "Sense and nonsense about culture and climate", in: Handbook of Organizational Culture and Climate, N.M. Ashkanasy, C. Wilderon and M.F. Peterson (Eds.). pp. 23-30. Thousand Oaks, CA: Sage.

Van Dick, R. Wagner, U. Stellmacher, J. and Christ, O. (2005). „Mehrebenenanalysen in der Organisationspsychologie: Ein Plädoyer und ein Beispiel. [Multilevel analyses in organizational psychology]", Zeitschrift für Arbeits- und Organisationspsychologie Volume 49 No. 1.

Zimolong, B., and Elke, G. (2009). "Management of Work Site Health-Promotion Programs: A Review", in: Proceedings of HCI International 2009, 19-24 July, 2009, San Diego: Ergonomics and Health Aspects of Work with Computers, J. A. Jacko et al., (Eds.). pp. 131-140. Berlin/Heidelberg: Springer.

Zimolong, B. and Elke, G. (2006). "Occupational Health and Safety Management", in: Handbook of Human Factors and Ergonomics, G. Salvendy (Ed.). pp. 673-707. New York: Wiley.

Zohar, D. (2000). "A Group-level model of safety climate: Testing the effect of group climate on microaccidents in manufacturing jobs", Journal of Applied Psychology Volume 85 No. 4.

CHAPTER 90

Speech Intelligibility and Visual Performance While Wearing Powered Air-Purifying Respirators (PAPRs)

Sehchang Hah[1], Tanya Yuditsky[1], Kenneth A. Schulz[2],
Henry Dorsey[3], Atul R. Deshmukh[2], Jill Sharra[1]

[1]Federal Aviation Administration
William J. Hughes Technical Center
Atlantic City International Airport, NJ 08405, USA

[2]Hi-Tec Systems, Inc.
William J. Hughes Technical Center
Atlantic City International Airport, NJ 08405, USA

[3]Northrop Grumman Information Technology
William J. Hughes Technical Center
Atlantic City International Airport, NJ 08405, USA

ABSTRACT

We evaluated a loose-fitting head cover Powered Air-Purifying Respirator (PAPR) and two full hood PAPRs for communication-intensive workers such as air traffic controllers. We measured sound levels of respirator blowers and analyzed the frequency spectrum of the blowers. Nine volunteers participated in speech intelligibility and visual performance experiments for three days. Our results showed that wearing a PAPR negatively affected both face-to-face and headset communications, usability, and comfort level significantly. The characteristics of the respirator, especially the sound level and frequency spectrum of the noise, played a significant role.

Keywords: Powered Air-Purifying Respirators (PAPRs), speech intelligibility, visual performance, sound levels, communication

INTRODUCTION

Various organizations use respirators to protect their employees against harmful agents such as asbestos and particles carrying flu viruses. Communication-intensive workers such as telephone operators and air traffic controllers may not perform their tasks well while wearing respirators. Controllers monitor tactical radar and auxiliary displays visually and communicate verbally with pilots (via headset), other controllers (via headset or face-to-face), and their supervisors (face-to-face).

There are different types of respirators. One type of respirator supplies fresh air from a tank or from an uncontaminated area through a hose. Another is the Air-Purifying Respirator (APR), which passes contaminated, ambient air through a filter delivered under positive pressure by a blower (Powered Air-Purifying Respirators [PAPRs]) or breathed in by a wearer (a negative-pressure APR). PAPRs provide a high Assigned Protection Factor (APF) of 1000 with a full hood and APF 25 with a loose-fitting head cover. APF is the ratio by which the filter reduces the concentration of contaminant. There were a few research reports about the effect of wearing the negative-pressure APRs on cognitive performance and communication. They reported detrimental effects on communication, but no cognitive performance degradation (Johnson, et al., 2000).

Our goal was to assess the feasibility of PAPR use in verbal communication-intensive work such as air traffic control. We did not intend to perform a market research on respirators or a selection of the best respirator. Before the experimental evaluation, we first surveyed PAPR configurations with loose-fitting head covers, full hood head covers, and three blowers that were available to us. We selected three configurations based on predefined criteria. During this survey, we found that it was impossible to use binoculars while wearing any of the PAPRs because of the long distance between the eyes and the plastic face shield, about 2.5 in. The field-of-view was too small. It was difficult to use telephones while wearing the full hood PAPRs. We also found that there was no degradation in visual acuity. In the first experimental evaluation phase, we measured speech intelligibility during electronic communication using headsets and evaluated visual performance with the selected three PAPRs. In the second phase, we evaluated the effect of wearing them during face-to-face communication.

PHASE ONE

METHOD

Participants

Nine volunteers (3 females and 6 males) participated in this phase. Three participants wore glasses and two participants wore contact lenses. The ages of the participants varied: two of the participants were younger than 25 years, four between 25 and 35 years, one between 46 and 60 years, and two over 60 years of age.

Materials and Equipment

We used a 3M Air-Mate PAPR with a loose-fitting head cover weighing about 3 lbs (1.4 kg), a Bullard PA20 with a full hood cover weighing about 3 lbs (1.4 kg), and a North CA101 with a full hood cover weighing about 4 lbs (1.8 kg) (see Figure 1). The National Institute of Occupational Safety and Health (NIOSH) requires all PAPRs to have an airflow amount above 6 cubic feet per minute (cfm). According to a North representative, the North CA101's airflow is 10.2 cfm with a freshly charged battery and a High Efficiency filter. The Bullard literature lists the PA20 airflow as ranging from 7.3 cfm to 8.5 cfm. The 3M Air-Mate literature specifies that their blowers meet the NIOSH requirement. The three blowers are shown in Figure 2.

For electronic communication between the speaker (i.e., the experimenter) and the listener (i.e., the participant) during the speech intelligibility experiment, we used headsets that are used by air traffic controllers (see Figure 3). All but one participant used the on-ear headset. We used a Brüel & Kjær 2260 Observer sound level meter with a Brüel & Kjær 1/4" pressure-field microphone (Model Number: 4938) to measure sound levels generated by the blowers. For the speech intelligibility task, we used the Modified Rhyme Test certified by the American National Standards Institute (ANSI) (ANSI, 1989). The test consisted of 50 sets of six monosyllabic words. The six words of each set had either the same initial consonant (e.g., save, same, sale, sane, sake and safe) or the same final consonant (e.g., hold, cold, told, fold, sold, and gold). In the Visual Performance Evaluation task, the participants played a game on a central 20-in. (50.8 cm) cathode-ray display while detecting a target randomly appearing on either of two 20-in. liquid-crystal displays (see Figure 4). The target was a Landolt C in one of three sizes and at one of four orientations.

Figure 1. Examples of a PAPR with a loose-fitting head cover (left) and a PAPR with a full hood cover (right).

Figure 2. PAPR blowers we evaluated (3M, Bullard, and North from left to right).

Figure 3. On-ear (left) and in-ear (right) headsets.

Figure 4. Visual Performance task set up (left) and the Landolt C in four orientations (right).

Procedure

To comply with Occupational Safety and Health Administration (OSHA) Regulations (OSHA, 2004), all nine participants completed the requirements to enroll in the Respiratory Protection Program (RPP). The participants completed speech intelligibility and visual performance experiments with one of three PAPRs each day for three days. For the speech intelligibility experiment, the participants and an experimenter were in separate rooms and used headsets to communicate. The participants completed four speech intelligibility sessions of 75 trials each day. The four sessions represented four combinations of PAPR wearing conditions

---

[1] Photos obtained from the Bullard® website (www.bullard.com).

between the speaker (i.e., the experimenter) and the listener (i.e., the participant): (1) speaker with a PAPR - listener without a PAPR, (2) speaker without a PAPR - listener with a PAPR, (3) both with PAPRs, and (4) neither of them with a PAPR. For each trial, the participant had to select the spoken word among six options displayed on the monitor, click on it with the mouse, and read it back to the experimenter. The presentation order of the conditions was randomly selected for each participant.

For the visual performance experiment, the participants wore PAPRs, and their task was to detect the target (Landolt C) and press one of the four buttons on a pad corresponding to the position of the opening of the C. Each visual performance session lasted an hour. We also measured the sound levels produced by each PAPR blower by the wearer's ear in an anechoic chamber at the fast mode (120 ms) with A-weighting.

Throughout the days of PAPR evaluations, the participants completed the Health and Well-Being Questionnaire approximately every 2 hours. After they finished evaluating each PAPR, they completed a Post-Test Survey and provided comments about their experiences with the respirator of the day. After they finished the experiment, we asked them to compare the PAPRs. They were also debriefed.

RESULTS

SOUND LEVELS AND SPECTRA OF PAPRs

The sound levels showed large differences between PAPRs (see Table 1). The range of the sound levels was between 52 dB (A) for the 3M and 75 dB (A) for the North. For reference, 50 dB (A) is the approximate noise level in an office, 60 dB (A) is the noise level near a freeway, inside a large store, or of normal speech (Bragdon, 1971; Levine & Shefner, 1981), and 70 dB (A) is the noise level of a freight train about 100 ft (30 m) away or speech at one foot (30 cm) away (Peterson & Gross, 1978).

The Speech Interference Level (SIL) has been used to evaluate environmental acoustics. For instance, SIL above 60 dB would make telephone conversation difficult (Cowan, 1994). The SILs of PAPRs showed a large range from 40 dB for the 3M to 64 dB for the North PAPR (see Table 1). The SIL is the average of the sound levels measured at 500 Hz, 1000 Hz, 2000 Hz, and 4000 Hz of octave bands without a weighting filter. We used 1/3-octave bands. These frequencies are critical for speech perception.

Table 1. Sound Levels Measured at the Wearer's Ear

Measure	3M	Bullard	North
Equivalent continuous sound level with A-weighting	52 dB (A)	66 dB (A)	75 dB (A)
Speech Interference Level (SIL)	40 dB	51 dB	64 dB

The spectra of the PAPRs showed that the sound levels (equivalent continuous sound levels with A-weighting) of the 3M were most similar to the anechoic room noise levels and substantially lower at high frequencies than those of the Bullard and the North (see Figure 5). High frequency sounds are perceived as more annoying than low frequency sounds even if they have the same sound level. Consonants typically produce high frequencies between 1,000 Hz and 5,000 Hz (Schiffman, 1996) and are more critical in speech intelligibility than vowels (Bragdon, 1971).

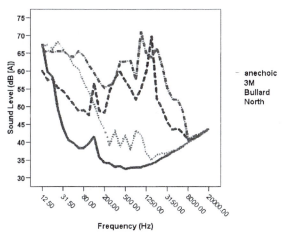

Figure 5. Spectra of 3M, Bullard, and North PAPRs measured at the wearer's ear.
Speech intelligibility

Participants' speech intelligibility errors ranged from 3% to 18% and varied widely depending on the PAPR worn and the condition they were in (see Table 2). The fewest errors occurred with the 3M and the most with the North. We examined the effect of wearing a respirator on speech intelligibility using the recognition error frequencies for each PAPR separately. We used non-parametric Friedman test for four conditions representing PAPR wearing conditions between the speaker and the listener. The Friedman test results were not significant with the 3M, but were significant with the Bullard, χ^2 (3, N= 9) =12.143, p =.007 and the North, χ^2 (3, N = 9) =18.341, p < .001. We performed follow-up tests using Multiple Matched-Pair Wilcoxon Tests at α=.0083 level to control the overall α level at .05. Only three pairs for the North were statistically significant. In general, there were fewer errors when the speaker did not wear the PAPR than when he wore the PAPR.: (Speaker On - Listener On vs. Speaker Off - Listener Off, z = -2.668 with p = .008), (Speaker On - Listener On vs. Speaker Off - Listener On, z = -2.692 with p = .007), (Speaker On - Listener Off vs. Speaker Off - Listener Off, z = -2.670 with p = .008). Although not statistically significant, there were more errors when the speaker wore the Bullard PAPR than he did not.

Table 2. Speech intelligibility Mean Error Percentages

		Speaker (i.e., Experimenter)					
		3M		Bullard		North	
		Respirator Off	Respirator On	Respirator Off	Respirator On	Respirator Off	Respirator On
Listener (i.e., Participant)	Respirator Off	3	4	4	12	4	18
	Respirator On	4	5	5	10	9	17

Visual Performance

Overall, the participants performed well in this task and missed only a small number of targets. There were no differences in the patterns of performance between conditions and between PAPRs, but two participants committed far more errors than others.

Well-Being Ratings, Surveys, and Comments

The participants reported a preference for the 3M PAPR over the others because it was the quietest and most comfortable. Also, they could hear surrounding sounds best with it because the 3M PAPR had a loose-fitting head cover exposing their ears. They reported that the North PAPR had a noticeably stronger air flow in the hood causing higher noise levels, problems with communications, and physical discomfort. The participants complained about the glare from the face shields of all the respirators. They also found that wearing the respirators in a seated position was very problematic because hoses were too stiff and short and blowers were bulky.

DISCUSSION

We conjecture that the full hood, noise levels, spectra patterns of blowers, and air flow intensity significantly contributed to high speech intelligibility errors with the Bullard and the North. The participants made most errors with the North, which was the noisiest and had the strongest air flow inside the hood. Also, it had high sound levels at high frequencies, which must have degraded speech perception (see Figure 5). The negative effect was more pronounced when the speaker spoke while wearing the respirator. In tasks where voice communication is critical such as air traffic control, a 25% error rate in the MRT is considered to be unacceptable (Ahlstrom & Longo, 2003). Thus, 17% and 18% are close to the unacceptable level. Ratings and comments were more favorable for the 3M PAPR which had a lighter, less restrictive, and loose-fitting head cover leaving the ears exposed. It was equipped with a soft headband suspension, which participants preferred over the hard plastic headgear inside the full hoods. In general, we did not observe significantly negative effects of PAPR use on visual performance. However, we observed large individual differences in the number of errors committed.

PHASE TWO

We measured speech intelligibility during face-to-face communication with and without PAPRs using all three PAPRs. We also measured the sound levels produced by the North PAPR in various face-to-face communication situations.

METHOD

Participants

Three members of our research team participated in the experiment.

Materials and Equipment

For the speech intelligibility experiment, we used all three PAPRs. For the sound level analysis, we used only the North that had the highest noise level. We used the same sound level meter with the same ¼-in. pressure-field microphone we had used in Phase One.

Procedure

During the speech intelligibility experiment, the speaker and the listener sat approximately 52 in. (1.32 m) apart in the same room. We used the same MRT stimuli we used in Phase One. In the baseline condition, they did not wear a PAPR. Both the speaker and the listener wore the same type of PAPR in the three experimental conditions. The presentation orders of the four conditions for the participants were counterbalanced. For the sound level measurements, the speaker and the listener sat about 52 in. (1.32 m) apart. The speaker spoke a word, and we measured sound levels at the ear of the listener.

RESULTS

Accuracy rates were 68% for the 3M PAPR, 48% for the Bullard PAPR, and 45% for the North PAPR compared to 94% for the baseline condition when neither the speaker nor the listener wore a PAPR. The sound level of the experimental room was 50 dB (A). The sound level at the listener's ear when words were spoken and no respirators were used (i.e., baseline condition) was 60 dB (A). As can be seen in Table 3, the noise levels increased substantially when the listener wore the North PAPR.

Table 3. Sound Levels of Face-to-Face Communication Conditions with the North

		Speaker	
		Without Respirator	With Respirator
Listener	Without Respirator	60 dB (A)	63 dB (A)
	With Respirator	75 dB (A)	74 dB (A)

DISCUSSION

Our results showed that face-to-face communication was very problematic with the use of PAPRs. The NIOSH requires that communication, as measured with the MRT, be at least 70% accurate when the speaker and listener are 3 m apart (NIOSH, 2008). Our participants sat at a distance of 1.3 m, and their accuracy rates with the Bullard PAPR and with the North PAPR are well below the NIOSH standard. The accuracy rate with the 3M PAPR was also slightly below the NIOSH standard. We believe these rates will get lower at the NIOSH distance of 3m.

The Federal Aviation Administration (FAA) Human Factors Design Standard (Ahlstrom & Longo, 2003) states that verbal signals for critical functions shall be at least 20 dB above the SIL measured at the operational position. As we reported in Phase One, the SIL of the North PAPR was 64 dB (see Table 1). For speech to be heard over this sound level, it should be at 84 dB (A), a much higher level than normal speech sound level, 60 dB (A).

GENERAL DISCUSSION

The participants experienced difficulties in both headset and face-to-face communication while wearing respirators. We found that the use of full hood PAPRs had a very large effect on speech intelligibility in electronic and face-to-face communication with accuracy rates far below those that are considered to be minimally acceptable by existing standards (Ahlstrom & Longo, 2003; NIOSH, 2008). Ahlstrom & Longo (2003) indicated that ambient noise in operational areas requiring phone use or speech communications should not exceed 55 dB (A). The two full hood PAPRS we tested generated higher noise levels. This would require a speaker to raise their voice significantly to be heard over the noise. The 3M that had loose-fitting head cover was just under the 55 dB (A) maximum. The participants rated the 3M loose-fitting head cover as more comfortable than the full hood PAPRs. However, the boom attached at the earpiece of the headset extended towards the mouth under the elastic around the jaw. This created a small gap between the head cover and the face, which might have affected the APF.

The participants complained about glare from the PAPR face-shield, but they completed visual performance tasks without much difficulty. We found that binoculars could not be used effectively with the PAPRs. This eliminates the possibility of using PAPRs in the Air Traffic Control Towers, where use of binoculars is required. The participants complained that all the respirators and hoses were too stiff and short to maintain a comfortable sitting posture. This is significant because most air traffic controllers work in a seated position.

Our evaluation used standardized tests that were quite different from the real tasks such as air traffic control tasks. Controllers use different types of words and speak very fast. We found controller speech rate during air traffic control simulations at about 300 words per minute. Therefore, many concerns need to be evaluated in a more operationally realistic environment before recommending any PAPRs for use by communication-intensive workers such as air traffic controllers.

If in the case of a flu pandemic the FAA decides to provide PAPRs to air traffic controllers, we recommend that they consider the findings of this study and conduct the following risk mitigation activities.

1. Identify the specific requirements that must be met to reduce the negative effects of PAPR use on air traffic control tasks.

Specific requirements of PAPRs can be derived from the findings reported here for a PAPR that optimizes controller performance and comfort. They include, but are not limited to, requirements for respirator noise level, air flow direction and intensity, hood material, size and weight of the blower unit, PAPR fit, flexibility and length of belts and tubes, and glare off the face shield.

2. Develop workarounds for tasks that cannot be performed with PAPRs.

Our results indicated that some tasks could not be accomplished effectively while wearing PAPRs. For example, controllers at air traffic control towers can not use binoculars while wearing PAPRs. Speech intelligibility for face-to-face communication while wearing PAPRs was too low to be acceptable. The electronic devices can be used for face-to-face communication even when the speaker and listener are next to each other.

3. Verify that the selected PAPR does not present problems that may compromise safety in an operational setting.

In the Speech Intelligibility task, the participants listened to one word at a time. Controllers use different types of words and speak very quickly. Although one of the PAPRs that we evaluated performed better than the others, its true impact on air traffic control tasks will need to be evaluated in a more realistic environment.

ACKOWLEDGEMENT

We express our deepest appreciation to the nine participants who volunteered for this study.

REFERENCES

Ahlstrom, V., & Longo, K. (Eds.). (2003). *Human factors design standard for acquisition of commercial-off-the-shelf subsystems, non-developmental items, and developmental systems* (DOT/FAA/CT-03/05/HF-STD-001). Atlantic City International Airport, NJ: FAA William J. Hughes Technical Center.

American National Standards Institute. (1989). *American National Standard method for measuring the intelligibility of speech over communication systems* (ANSI S3.2). DC: Author.

Bragdon, C. L. (1971*). Noise pollution: The unquiet crisis.* Philadelphia: University of Pennsylvania Press.

Cowan, J. (1994). Handbook of Environmental Acoustics. NY: John Wiley & Sons, Inc.

Johnson, A. T., Scott, W. H., Lausted, C. G., Coyne, K. M., Sahota, M. S., Johnson, M. M., et al. (2000). Communication using a telephone while wearing a respirator. *American Industrial Hygiene Association Journal, 61*(2), 264-267.

Levine, M. W., & Shefner, J. M. (1981). *Fundamentals of sensation and perception.* Reading, MA: Addison-Wesley Publishing Co.

National Institute of Occupational Safety and Health (NIOSH). (2008). *Statement of standard for full face-piece Air Purifying Respirators (APR).* Retrieved from http://www.cdc.gov/niosh/npptl/standardsdev/ cbrn/apr/standard/aprstd-a.html.

Occupational Safety and Health Administration. (2004). *Respiratory Protection. OSHA Regulations* (Standards - 29 CFR) 1910.134. DC: U. S. Department of Labor. Retrieved from http://www.osha.gov/pls/oshaweb/owadisp.show_document?p_table=STANDARDS&p_id=12716

Peterson, A. P. G., & Gross, E. E. Jr. (1978). *Handbook of noise measurement.* Concord, MA: GenRad Inc.

Schiffman, H. R. (1996). *Sensation and perception* (4th ed.). NY: John Wiley & Sons, Inc.

Chapter 91

Man-Machine-Interaction in Complex Intraoperative Orthopedic Work Systems

Wolfgang Lauer, Armin Janss, Bastian Ibach and Klaus Radermacher

Chair of Medical Engineering
Helmholtz-Institute for Biomedical Engineering of the
RWTH Aachen University, Aachen, Germany

ABSTRACT

Ergonomic optimization of surgical work systems and usability of medical products have become subjects of increasing interest during the past years as they have been proven to be a key factor in patient and user safety. Technical progress in the medical field has not only lead to new therapeutic possibilities but also to additional risks in terms of more complex man-machine-interaction and an increased number of devices in the OR. In the OrthoMIT project an approach for an integrated surgical workstation based on open standards was developed and implemented. This framework allows for flexible configuration of the surgical work system and safe interoperation between included devices according to specific interventions or workflows. As man-machine-interaction in such integrated work systems implies additional risks and questions (e.g. regarding situation and system awareness or versatile interaction modalities) diverse approaches were developed. A model based prospective risk assessment method for complex man-machine-interaction as well as a sterile multipurpose input device for planning and navigation tasks are presented as examples.

Keywords: Surgery, man-machine-interaction, operating room, ergonomics, usability, surgical integration, risk management, patient safety

INTRODUCTION

Technical progress in the medical area has lead to major changes concerning professional processes, tasks and tools. Besides a lot of therapeutic advantages the introduction of new, often computerized technological components with increased functionality has also lead to potential new risks for patients as well as for health care professionals e.g. in the interventional area.

Studies concerning human error in medical context showed that 77,2% of hazardous events are due to incorrect application although technical devices work properly (Leape, 1994). Preventable failures with technical systems in orthopedic surgery are in 72% due to human failure (Rau et al., 1996). According to (Weingart et al., 2000) the handling of technical equipment by itself constitutes a hazard for the patient, a problem which is even aggravated when new systems or complex tasks are introduced. Additionally, inadequate design of the work equipment and environment in relation to the task and situation, the erroneous interpretation of information provided by user interfaces and situational factors as well as external environmental factors and extrinsic stress are the main causes of "human error" in medicine (Bogner, 1994). Statistics on malpractice incidents and the handling of medical devices seem to prove this fact (Kohn et al., 2000).

Due to these findings ergonomics in surgery and usability of medical devices have become subjects of increasing interest during the past years (Stone & McCloy, 2004). Especially with the introduction of the technical standards IEC 60601-1-6 and IEC 62366 as well as with the actual revision of the Essential Requirements in the Medical Device Directive (2007/47/EG) the crucial importance of usability for patient and user safety has been emphasized. As an essential part of market approval manufacturers of medical devices have to install and document a usability engineering process which is closely linked to the product development process itself. Risk management for medical devices has to include ergonomic aspects potentially leading to use error with hazardous consequences.

Usability engineering of single medical components always includes the context of use and its implications on the specific man-machine-interaction (MMI). However, actual technological development towards integrated surgical work systems leads to more complex use cases and additional ergonomic challenges.

TECHNICAL INTEGRATION IN THE OR

In modern operating rooms (ORs) a lot of different technical assistance systems are used such as devices for endoscopy, surgical navigation or imaging. Almost all of these devices do have their own proprietary man-machine-interfaces, often consisting of visual displays and input devices (e.g. mice, keyboard, touchpad). Most of them operate independently or are incapable of communicating with other devices (Cleary & Kinsella, 2004). Besides technical bottlenecks (e.g. regarding interoperability or tele-surveillance) this situation also imposes ergonomic drawbacks:

- A lot of different interface structures and particularities have to be obeyed by the users. This may lead to potential risks under stressing conditions (e.g. in an emergency) due to time consuming change from rule-based behavior to a knowledge-based regulation level (Rasmussen, 1983). During low vigilance routine situations wrong mental rules may be applied especially during multitasking processes.
- As direct interaction between surgeon/assistants and non-sterile devices often is not possible indirect communication via other staff members has to be performed. In contrast to e.g. the aviation sector team communication in the OR is not standardized (except for instrument handling interaction with the scrub nurse). This often causes misunderstandings and may lead to latency and frustration or even hazards.
- Due to a lack of interoperability no comprehensive OR data logging can be performed. "Black boxes" in airplanes are very powerful tools for identification of reasons for adverse events - not only regarding technical failure. Also in case of critical incidents in surgery extensive device data concerning incident background and development is needed for technical as well as ergonomic analysis and optimization.

In order to reduce the number of man-machine-interfaces in the OR and to enhance the exchange and integrated use of information, one objective of actual activities in this domain is to provide the surgical team with integrated operating suites and optimized interfaces facilitating the interaction with different modules and components. The aim is to reduce the number of independent devices, to enhance flexibility and integration of information and to help overcome the classical separation between sterile and non-sterile areas within the OR regarding communication issues.

State of the art commercial integrated operating room solutions address some of the aspects mentioned above (Ibach et al., 2006). They offer a certain level of integration like central video routing and central control of some devices such as the operating table, OR lights, endoscopy camera and light e.g. by touch screen, voice- and/or remote-control. Almost all systems offer partial intra-system modularity. It is possible to add or remove certain proprietary modules of the same provider of the integrated operating room solution, such as endoscopy system, OR tables, HIS/PACS modules, allowing a customer specific configuration of the operating room. Most systems use proprietary communication protocols for inter-device communication.

The integration of third party devices (inter-provider modularity) is an aspect available systems do not at all or only partially cover yet. Integration of third party devices is only possible on the basis of a mutual agreement, i.e. in direct cooperation with the suite provider. Due to the lack of open standards and interfaces each device which shall be integrated into an existing OR solution must be reengineered and a new risk analysis must be carried out.

THE ORTHOMIT OPEN INTEGRATION CONCEPT

In the framework of the OrthoMIT project (funded by the German Ministry for Education and Research BMBF) an integrated work-system for smart hip-, knee- and spine-surgery is currently being developed (Ibach et al., 2008). This system comprises optimized surgical procedures, novel imaging techniques, smart instruments and implants as well as second generation surgical robot systems and innovative computer-assisted methods for combination, analysis and context-adapted representation of relevant information. The ergonomic quality of the devices as well as of the whole surgical work system is addressed by a special work package in the project.

One important aspect in OrthoMIT is the development of a novel network architecture for the OR including open interface standards. Thus, in contrast to existing solutions with vendor-specific, proprietary interfaces, medical products from different companies can be combined by the hospital in a flexible way without losing the operating approval. The aim is to provide a framework for flexible and modular integrated OR systems that can be easily and safely configured according to the specific needs of surgical disciplines, interventions and workflows. Hospitals shall be enabled to choose, integrate and exchange the appropriate devices rather due to their technical and ergonomic quality than due to the commercial relationship between their manufacturer and the provider of the surgical suite. Stand-alone as well as integrated use of the components shall be supported.

In this framework a service oriented architecture concept (SOA) has been developed (Ibach et al., 2006; Ibach et al., 2008) and this OrthoMIT approach has been accepted as a use case in the framework of the upcoming IEC 80001-1 (Application of risk management to information technology (IT) networks incorporating medical devices). The single components are connected through Ethernet as a shared transportation and communication medium. The basic component is the Service Manager, which is responsible for access control and information management. The different medical devices are either service provider, service consumer or both, providing or using services available within the network. Based on special services various device functions such as the movement of the OR table or the radiation control of the intraoperative x-ray C-arm can temporarily be used by other devices. They can also be accessed by the surgeon and the OR personnel through an Integrated Surgical Workstation (ISW) including context adapted man-machine-interfaces. In this context the OR table could e.g. offer services like upward or downward movement of the tabletop, actual position of certain table elements or geometry information of the table itself. Specific interaction modalities for sterile and non-sterile users can be integrated into the system as well as emergency input devices with override priority.

As an example for potential future applications automatic intraoperative x-ray imaging would need complex interaction between a surgical localization device, the OR table and a robotic x-ray C-arm. Currently in Orthopedics the sterile surgeon is not able to position the C-arm himself. He therefore has to direct an assistant moving the heavy and bulky device. This process can be very time consuming and erroneous due to communication   problems. In an integrated OR suite the surgeon could be enabled to use a navigated pointing device indicating the optimal x-ray position and orientation. The tracking server would then send this information to a specific control unit which also has access to the OR table and a robotic C-arm. Based on detailed information on the actual position and configuration of the OR table and the C-arm as well as on restricting boundary conditions (e.g. position of OR lamps and other devices, restrictions of table movement) the control unit could calculate an optimized imaging configuration. Avoiding possible collisions OR table and C-arm could be moved to these positions and the intended image could be provided to the surgeon on a dedicated interaction device in the sterile area. Although this process would be performed automatically using dedicated tracker-, C-arm and table-services the surgeon would still be able to stop the movement anytime by means of an override emergency or dead-man button. Due to safety reasons no other components would receive access to the OR table or the C-arm from the Service Manager during the imaging process. Afterwards the positioning services would again be offered to all appropriate devices and could e.g. be used by a conventional table remote control.

ERGONOMIC CHALLENGES AND APPROACHES

With this example it becomes obvious that these new functions and linking possibilities do not only provide new perspectives for therapeutic effectiveness and efficiency but may also lead to increased risk of use error and hazards e.g. related to the following aspects:

- Visual representation of different user interfaces in an Integrated Surgical Workstation - "real view" and device-specific interaction structure or standardized interfaces for all devices?
- Man-machine-interaction - which interaction concepts and devices are suitable for different situational requirements e.g. in the sterile and non-sterile area?
- Visual and acoustic alarms - how to handle different kinds of (simultaneous) alarms and how to represent them in prioritized and intuitive way?
- Mental modeling and situation awareness - how can the users be enabled to understand the current system configuration and functional state? How can they be supported in adapting their mental model to possible configuration changes during the intraoperative process?
- Education and Training - how can users be prepared for understanding

complex device structures in the OR? How can they be trained for stand-alone use as well as for routine and emergency interaction with integrated systems?

We are currently addressing these questions trying to assess potential risks for patients and OR personnel as well as to find ergonomic solutions.

As one approach for comprehensive risk management and usability engineering for complex man-machine-interaction a model based usability assessment method was developed within the framework of the INNORISK project (funded by the German Ministry for Economics BMWi). This method enables manufacturers of high risk products to perform prospective usability assessment and risk analysis including potential human-induced error already in early developmental phases (Janß et al., 2008). Based on this the mAIXuse method and a supporting software tool was developed and implemented. This method can be used during the whole product development process - in a prospective way or in terms of assessment and validation of existing prototypes.

Adapted from two model-based methods, the ConcurTaskTree and the CPM-GOMS (Cognitive Perceptual Motor – Goals Operators Methods Selection Rules) approach, the method uses formal, normative models to predict possible user and system behavior in order to support assessing the usability of a new or re-designed system. The software tool is intended to support the design engineer with building these models and analyzing them semi-automatically on the basis of different failure taxonomies concerning human error. It shall enable the design engineer to model not only the high- and low-level tasks of the system, the user and the interactions but also the performance shaping factors, human-human-interaction and additionally the different levels of cognitive regulations when interacting with a medical device. On the basis of these investigations the design engineer can then derive potential use errors and design countermeasures for the user interface.

The developed method has already been evaluated in comparison to classic risk assessment (Failure Mode and Effects Analysis FMEA) and has been successfully evaluated and implemented in different research projects together with our medical industrial partners.

One example for an approach towards novel devices for multipurpose man-machine-interaction out of the sterile area is the OrthoMIT Remote Pointer (RP, patent pending) (Janß et al., 2009).

This versatile device combines three different input modalities which are necessary for planning and navigation during surgical interventions. In the first instance, the RP is intended to replace the conventional tracking probes, which are used during the registration process for the palpation of anatomical landmarks. Furthermore, it can be used in order to control dialog-based applications with defined gestures whereas common dialog-based CAS-applications additionally need footswitches for back and forth commands. The third mode is a mouse emulation which enables the user to control common graphical user interfaces implementing the RP as a normal computer mouse. Actually, touchscreens and conventional computer mice are used in the OR but this implicates several disadvantages. For the use of both input devices by the surgeon and the OR-personnel sterile requirements have to be fulfilled. The mouse has to be sterilized before use and the touchscreen has to be covered by a sterile film. Furthermore, for the use of a mouse a horizontal plane is needed that has to be permanently reachable by the surgeon.

The Remote-Pointer basically is a tracked tool or probe respectively, common in conventional navigation systems, combined with an additional reflective marker, which is usually covered The interaction process is triggered by altering the number of visible markers of the rigid body. Especially, if the additional marker is uncovered (in a predefined position), this is interpreted by the system as a user command depending on the actual Remote-Pointer mode. The cursor position is then calculated as the intersection point between the pointer direction and an arbitrary virtual plane (typically the plane of the display system). The RP can be used from any position inside the OR using optical tracking. It doesn't need any source of energy and is fully sterilizable.

In first usability tests the performance character according to the "mouse emulation" mode of the RP (concerning accuracy, movement time, error rate, throughput etc.) was analyzed in comparison to a conventional computer mouse used on a table and the commercial Logitech AIR-Mouse which is controlled by means of an integrated motion sensor. The results showed that the RP performance characteristics concerning the mouse emulation were significantly better than with the Logitech Air mouse and that, as expected, the conventional computer mouse (used as gold standard for general man-computer-interaction) performed best. Additional tests will be conducted evaluating the other RP modes as well as the overall performance in combination with a commercial intraoperative support system for orthopedic surgery.

CONCLUSION

The importance of ergonomic optimization as a substantial part of risk management for patient and user safety is receiving increasing acceptance by legislative and normative stakeholders. Actual standards for medical products as well as the latest revision of the Medical Device Directive demand the implementation and documentation of a comprehensive usability engineering process as a substantial part of market approval.

Besides risk management for single medical products used in the OR additional questions regarding technical and ergonomic design of integrated surgical work systems are currently arising. Representation of user interfaces, situation and system awareness in complex and changing integrated setups, versatile and context-optimized man-machine-interaction modalities of the sterile area - all these aspects have to be considered in order to provide surgeons and supporting OR staff with usable, reliable and safe tools for their daily work.

In the framework of the OrthoMIT project a SOA-based integration concept for the OR was developed. This approach includes open standards and allows for flexible and adaptive configuration of the surgical work system. The concept was successfully introduced into the international standardization process. It was implemented together with scientific and industrial partners and will be further evaluated in the OrthoMIT demonstration OR.

To assure ergonomic quality and clinical usability of the OrthoMIT integrated surgical work system different approaches were developed. The software supported model based mAIXuse method enables developers and manufacturers of medical products to prospectively assess the ergonomic quality of complex man-machine-interaction. It serves as a tool in risk management regarding use-related human error and can be applied throughout the whole product development process including summative validation. As an approach for an ergonomic multi-purpose input device for MMI the OrthoMIT Remote Pointer was developed. It combines different functions needed for planning and navigation during surgical interventions and is fully compliant with sterile requirements.

Both, mAIXuse and the Remote Pointer, were positively evaluated together with medical and industrial partners and will be further developed with regard to the realization of a safe and ergonomic integrated OR.

ACKNOWLEDGEMENTS

The OrthoMIT project is funded by the German Federal Ministry of Education and Research (BMBF 01EQ0402).

The AiF/FQS project INNORISK (AiF 14879) was funded by the German Ministry of Economics and Technology (BMWi).

REFERENCES

Bogner, M.S. (1994), *Human Error in Medicine*. Lawrence Erlbaum Associates.
Cleary, K., Kinsella, A. (2004), *OR2020 – The operating room of the future*, Workshop Report, http://www.or2020.org (26.02.2010)
Ibach, B., Zimolong, A., Portheine, F., Niethard, F., and Radermacher, K. (2006), "Integrated medical workstations for Computer Integrated Smart Surgery (CISS) - state of the art, bottlenecks and approaches." *Proceedings CARS (Suppl. 1), International Journal of Computer Assisted Radiology and Surgery*, 1, 449-451.
Ibach, B., Kanert, A., and Radermacher, K. (2008), "Concept of a service-oriented integration architecture for the orthopaedic operating theatre." *International Journal of Computer Assisted Radiology and Surgery*, 3, 446-447.
Janß, A., Lauer, W., and Radermacher, K. (2008), "Using Cognitive Task Analysis for UI Design in Surgical Work Systems." in: *Proceedings of the 15th*

European Conference on Cognitive Ergonomics ECCE 2008, ACM International Conference Proceeding Series 250 ACM 2008, 66-69.

Janß, A., Lauer, W., and Radermacher, K. (2009), "Usability Evaluation Of A Non-Active Remote Pointing Device For Computer-assisted Orthopaedic Surgery." In: Davies, B., Joskowicz, L., and Murphy, S (editors) *Computer Assisted Orthopaedic Surgery – Proc. CAOS International 2009*, 326-328.

Kohn, L.T., Corrigan, J., and Donaldson, M.S. (2000), *To Err Is Human: Building a Safer Health System*. Washington, DC: National Academy Press.

Leape, L.L. (1994), "The Preventability of Medical Injury." In: Bogner, M.S. (editor), *Human Error in Medicine*. Hillsdale NJ, Erlbaum Publishers, 13-26.

Rasmussen, J. (1983), "Skills, Rules, and Knowledge; Signals, Signs, and Symbols, and Other Distinctions in Human Performance Models." *IEEE Transactions on Systems, Man and Cybernetics*, 13(3), May/June, 257-266.

Rau, G., Radermacher, K., Thull, B. and Pichler, C. v. (1996), "Aspects of an Ergonomic System Design of a Medical Worksystem." in: Taylor, R., Lavallée, St., Burdea, G., and Moesges, R.: *Computer Integrated Surgery*, MIT-Press, Cambridge, MA, 203-221.

Stone, R, and McCloy, R (2004), "Ergonomics in medicine and surgery." *BMJ*, 328, 1115-1118.

Weingart, S.N., Wilson, R.M., Gibberd, R.W., and Harrison, B. (2000), "Epidemiology of medical error." *BMJ*, 18, 774-777.

Zimolong, A., Radermacher, K., Stockheim, M., Zimolong, B., and Rau, G. (2003), "Reliability Analysis and Design in Computer-Assisted Surgery." in Stephanides, C. et al. (eds.), *Universal Access in HCI*. Lawrence Erlbaum Ass. Publ. 524-528.

CHAPTER 92

Methodology for a Combined Evaluation of Cognitive and Physical Ergonomics Aspects of Medical Equipment

Lars-Ola Bligård, Anna-Lisa Osvalder

Division Design & Human Factors
Chalmers University of Technology
SE-412 96 Gothenburg, Sweden

ABSTRACT

To achieve efficient use and high patient safety in work with medical equipment, it is important to evaluate both physical and cognitive ergonomics aspects of the devices. The aim of this paper is to describe an analytical methodology of a joint systematic search for potential deficiencies in the human-machine interaction; such as high physical and mental workload, use errors, usability problems, and physical ergonomic errors. The methodology is task-based, which makes it possible to use both with focus on the device design, as in development projects; as well as with focus on the procedure, as in a hospital setting. The methodology has successfully been tested in the development process of a dialysis machine.

Keywords: Medical Equipment, Cognitive Ergonomics, Physical Ergonomics, Usability, Human-Machine Interaction, Use Error, Ergonomic Evaluation

INTRODUCTION

In health care the use of medical technology plays a central part in the diagnostics and treatment of patients. To achieve efficient use and high patient safety, an important approach is to incorporate knowledge about the user and the use in the design process, and then utilise methods for evaluation of ergonomics and human factors aspects of the devices. The evaluation methods can be used in product development processes to detect and mitigate potential problems, but also in hospitals during equipment purchasing as well as a base for staff education.

Evaluation methods of today often focus on cognitive ergonomics (such as mental workload or usability) or physical ergonomics (such as physical workload or body posture). But, a well-known problem in the area of physical ergonomics is that working tasks are not always performed as intended, despite good ergonomic design of the working tools. A plausible reason in these cases could be lack of usability of the working tool; inadequate cognitive design as the origin of insufficient physical ergonomic design. Thus, to evaluate the true ergonomics of a device, both the mental and physical aspects of the use must be assessed together.

The aim of this paper is to describe an analytical methodology for a combined evaluation of cognitive and physical ergonomics useful in the area of medical technology. The idea with the methodology is to not separate cognitive and physical ergonomic aspects, but instead perform a joint systematic search for potential ergonomic deficiencies in the human-machine interaction; such as high workload (physical and mental), use errors, usability problems, and physical ergonomic errors. The purpose with the joint search is to achieve a more holistic evaluation approach and make the evaluation more cost effective than when using separate evaluation methods for cognitive and physical ergonomic aspects. When all ergonomic deficiencies are identified, it is possible to take counteracting measures in the design.

DESCRIPTION OF METHODOLODY

The goal of the methodology is to map the interaction between the human and the machine in a structured and systematic way and to predict and identify presumptive mismatches in the interaction, both physical and cognitive use errors as well as usability problems. Prediction means investigating when, where, and how mismatches may occur, and identification means determining the type and properties of the predicted mismatches.

The methodology is conducted by a person or a group that may consist of designers, software developers, marketing staff, and people with knowledge in human factors. Most important, however, is that knowledge about the users and knowledge about the usage of the equipment is present among those who conduct the assessment. The methodology consists of four phases (1) Definition of evaluation, (2) Human-machine system description, (3) Workload analysis, and (4)

Interaction of analysis. The exact content of each step may vary depending on the purpose of the evolution.

DEFINITION OF EVALUATION

The first activity of the methodology is to set the frame for the coming work. The goal of the evaluation activity is to answer the following five questions: (1) What is the purpose of the evaluation? (2) What machine shall be analysed? (3) Which usage will be analysed? (4) Who is the indented user? (5) What is the context for the use?

HUMAN-MACHINE SYSTEM DESCRIPTION

The system description is the base of the method, and illustrates how the human-machine system works. The system description is very important for the interaction analysis, because if the information is deficient, incomplete or wrong, the results of the analysis will suffer. The system description consists of four activities: User profiling, Task analysis, Context description, and Interaction description.

User Profiling

An important part of the methodology is the human acting in the system. The user profiling describes the user's abilities, limitations and characteristics (as knowledge and experience) that are relevant for the evaluation. As the goal of the evaluation is to find mismatches between the human and the machine, the user must be correctly and carefully described to get the evaluation accurate. Examples of user profiles are described by Janhager (2005) and IEC (2004).

Task Analysis

The next step in the system description is to choose which tasks are to be evaluated. The task selection depends on the goal of the study. Often tasks that are carried out on a regular basis or safety-critical are chosen. The selection of tasks must be based on the intended use, not on the design or function of the equipment. Above all, it is important that the tasks are realistic. Hierarchical task analysis, HTA (Stanton 2006), is used to describe how the overall goal of the working task can be achieved through sub-tasks and plans. HTA breaks down a task into elements or sub-tasks, called operations.

Context description

In the context description the physical, organisational, and psychosocial environment during use are specified and described. The purpose is to map the external performing shaping factors that affect the user (e.g. lighting, noise) and context characteristics (e.g. culture, organisation, room) that affect the interaction.

Interaction description

The last step in the description is the interaction between the human and the machine. Given the correct way of how a task is performed, as described in the HTA diagram, it should then be specified how the interface looks for the different user actions. In this way it becomes possible to evaluate the appearance of the user interface of the machine throughout the use. The interaction description can be compiled in a simple way by combining the appearance of the machine (for example screen dumps) with the HTA diagram. The interaction could also be described by using the User-Technical Process presented by Janhager (2005) or by Link Analysis (Stanton and Young, 1999).

WORKLOAD ANALYSIS

The workload analysis is based on the method Generic Task Specification framework, GTS, (Bligård and Osvalder, 2008). The framework uses different aspects derived from theories and methods such as the Skill-Rule-Knowledge (S-R-K) based decision making model (Rasmussen, 1983), the Human Information Process (Wickens and Hollands, 1999), Subjective Work Load Assessment (Wilson and Corlett, 1995), NASA Task Load Index (Hart and Staveland, 1988), and anthropometry (Pheasant and Haslegrave, 2005). The aspects are divided into three parts: Task demands, Mental workload, and Physical workload.

Task Demands

Task Demands means the demands that the task puts on the human for each operation. The demands from a task/operation are independent of the operator who performs the task/operation. The task demands are dependent on the task, but also on the interacting machine and the environment. If multiple machines or environments are studied for the same task, they require separate sets of task demands. The task demands are described by five aspects: (1) Task Type, (2) Task Category, (3) Performance/Accuracy, (4) Time Pressure, and (5) Performance Shaping Factors. Data for the aspects can be gathered by interviews with users, by observations of use, and by use of more advanced methods such as Applied Cognitive Task Analysis, ACTA (Militello and Hutton, 1998).

Mental workload

The mental workload describes the load that the human information process is exposed to when performing the tasks. The classification of mental workload is made for each operation and is dependent on the human who performs the task/operation, which is described by the user profile. This is in contrast to task demands, where the classification is independent of the human who performs the task/operation. The mental workload is described by six aspects: (1) Mental Task Type, (2) Attention Resources, (3) Memory Resources, (4) Processing Resources, (5) Frustration/Stress, and (6) Superimposed Mental Activates. Data for the aspects can be gathered by interviews with users and observations of use, and by use of more advanced methods such as Subjective Work Load Assessment (Wilson and Corlett, 1995) and NASA Task Load Index (Hart and Staveland, 1988).

Physical workload

The physical workload describes the load that the human body is exposed to when performing the tasks. The classification of physical work load is made for each operation and is dependent on the operator that performs the task/operation, who is described by the user profile. This is in contrast to task demands, where the classification is independent of the operator who performs the task/operation. The physical workload is described with five aspects: (1) Force Resources, (2) Fine Motor Resources, (3) Speed Resources, (4) Load of body parts, and (5) Body Contact. Data for the aspects can be gathered by interviews with users and observations of use, and by use of methods for load assessment such as Rapid Upper Limb Assessment, RULA, (McAtamney and Corlett, 1993), Rapid Entire Body Assessment, REBA, (Hignett and McAtamney, 2000). Also biomechanical calculations can be made or measurement of muscular strength with electromyography.

Result compilation

The final part in the work load analysis is the creation of matrices. The output data from GTS is semi-quantitative, and the results can therefore be displayed in different types of matrices. By this, the workload and demands can be compared to each other or assessed separately. By comparing the mental and physical workload with the task demands and the user profile, presumptive mismatches between the human and the machine can be detected, e.g. too high short term memory load or too high physical static loads on the shoulders.

INTERACTION ANALYSIS

The final step in the methodology is the interaction analysis, where the interplay between the human and machine is evaluated. The interaction analysis is based on the correct handling sequences described in the HTA diagram. In the analysis, the

HTA is divided into two parts; operations and functions. The operations are the lowest level in the HTA and the functions are the subtasks in the HTA. For each function and operation, questions are asked to identify presumptive mismatches. The question process tries to simulate how the user interacts with the device, where the analysts in the evaluation team play the role of the user. The interaction analysis consists of three parts (1) usability problem analysis, (2) use error analysis, and (3) physical ergonomic error analysis. The interaction analysis results in a list with identified human-machine mismatches. The mismatches will also be graded from low importance to high importance. A grade makes it easier to determine what is most important to rectify in the subsequent reworking of the user interface of the machine.

Usability problem analysis

The usability problem analysis is performed by the method Enhanced Cognitive Walkthrough, ECW, (Bligård and Osvalder, 2006b). ECW is an analytical method which looks into potential usability problems by investigating what hinders the user from performing correct actions and why that happens. ECW is applied both on the operational level and on the functional level. The questions used are listed in Table 1.

Table 1 Analysis questions for ECW (Bligård and Osvalder, 2006b)

Level 1: Analysis of Functions	Level 2: Analysis of Operations
Will the user know that the evaluated function is available?	Will the user try to achieve the right effect?
Will the user interface give clues that show that the function is available?	Will the user be able to notice that the correct action is available?
Will the user associate the right clue with the desired function?	Will the user associate the correct action with the desired effect?
Will the user get sufficient feedback to understand that the desired function has been chosen	If the correct action is performed, will the user see that progress being made towards the solution of the task?
Will the user get sufficient feedback to understand that the desired function has been performed?	-

The analyst answers the questions in Table 1 with a motivation and gives the answer a grade; a number between 1 and 5. Grade 5 is equal to 'always yes', and 1 means 'always no'. Grade 5 means that there is no usability problem. If the grade is set between 1 and 4 there exist a usability problem, and the next step is then to describe this problem. The problem is the cause which restrains the user from performing the correct action. Each problem is further categorised by a *problem type*. Different problem types can be used depending on the machine and the task that the user should solve.

A specific version of ECW can be used when evaluate alarm signals. This version is called alarm-ECW (Bligård and Thunberg, 207). In this case the questions used on the functional level are: (1) Will the operator be able to detect the there is an alarm condition? (2) Will the operator understand the seriousness of the alarm? (3) Will the operator be able to identify the alarm? (4) Will the operator be able to interpret the alarm? (5) Will the operator associate the correct measure/action with the alarm? The questions on the operational level are the same as shown for level 2 in Table 1.

Use error analysis

The use error analysis is performed by the method Predictive Use Error Analysis, PUEA, (Bligård and Osvalder, 2007). PUEA is an analytical method that searches for potential use errors by investigating what the user can do incorrect, why it happens, and what the consequences will be. PUEA is applied both on the operational level and on the functional level. The questions used are listed in Table 2.

Table 2 Analysis questions for PUEA (Bligård and Osvalder, 2007)

Level 1: Analysis of Functions	Level 2: Analysis of Operations
What happens if the user performs an incomplete operation or omits an operation?	What can the user do wrongly in this operation?
What happens if the user performs an error in the sequence of operations?	What happens if the user performs the operation at the wrong time?
What happens if the user performs functions/tasks correctly at the wrong time?	

The analysts, guided by the questions, try to predict as many use errors as possible that can arise in the human-machine interaction. Each predicted use error is noted in a list. During this process, they also eliminate errors that are considered too unlikely to occur. This elimination is done in relation to how the simulated user is expected to make decisions and perform, in view of the artefact, the social, the organisational and the physical contexts. However, it is important to be careful about dismissing improbable errors that would have serious consequences without further investigation, as these can also constitute a risk. If there are no use errors corresponding to the answers to the questions, this also should be noted.

For each predicted use error, an investigation is made according to eight items (Table 3). The first two concern the error itself, the next two its potential consequences, and the last four items concern mitigations of the errors and consequences. Four of the items also contain a categorisation, a judgment of probability, or a judgment of severity. This is done to facilitate a compilation and assessment of the investigation.

Table 3 Items of investigation for PUEA (Bligård and Osvalder, 2007)

Item	Explanation
Type	What is the type of use error? (categorisation)
Cause	Why does the use error occur? (description and categorisation)
Primary consequence	What is the direct effect of the use error? (description)
Secondary consequences	What effects can the use error have that lead to a hazardous situation for the user or other people, or to risk of machine damage or economic loss? (description and judgment of severity by a grade)
Detection	Can the user detect a use error before it has any secondary consequences?(description and judgment of probability by grade)
Recovery	Can the user recover from the error before any severe consequences arise? (description)
Protection from consequences	Which measures does the technical system employ to protect the user and the environment from the secondary consequences? (description)
Prevention of error	Which measures does the technical system employ to prevent occurrence of use errors? (description)

Physical ergonomic error analysis

The physical ergonomics errors are analysed with the method Predictive Ergonomic Error Analysis, PEEA, (Bligård and Osvalder, 2006a). PEEA is a method, which investigates if a work task is performed in an ergonomic correct way, and if the task can be performed incorrectly. PEEA is applied on the operation level of the HTA and works with questions in two steps (Table 4). First, the method examines if the actions will be performed in a correct ergonomic way, and then if the actions also can be performed in an incorrect ergonomic way. For each identified non-ergonomic action (ergonomic error) a further examination could be made. The analysis of the consequences of incorrect body postures can be made by e.g. RULA, or REBA analysis, as well as from knowledge and heuristics about human anatomy, physiology, anthropometry and biomechanics.

Table 4 Analysis questions for PEEA (Bligård and Osvalder, 2006a)

Error identification	Error investigation
Does the product give any information (cues) about how the action can be performed in an ergonomic correct way?	Which are the possible causes to that the action is performed in an incorrect ergonomic way?
Does the user know how the action can be performed in an ergonomic correct way?	Which are the short term consequences for the user?
Will the user try to perform the action in an ergonomic correct way?	Which are the long term consequences for the user?
Can the action be performed in a non-ergonomic way? How?	Is the machine designed to prevent the ergonomic error?

Result compilation

The result of the interaction analysis is primary a list in tabular form of the presumptive mismatches between the human and user interface of the machine in the form of usability problems, (cognitive) use errors and physical ergonomics errors. Matrixes are then employed to present the semi-quantitative results from the interaction analysis. The collected answers from the prediction and identification of errors and problems are ordered in different ways within the matrixes, to emphasise different aspects of the analysis. The matrixes can be combined in various ways, and the numbers in the matrix cells show the number of detected mismatches distributed according to the two types of data that are compared.

DISCUSSION AND CONCLUSION

The results from the workload analysis and the interaction analysis are combined to get an overview of the mismatches between the human and the machine. But since the mismatches are presumptive mismatches and that many steps in the methodology depends on subjective assessment of the analysts, the mismatches errors may not occur during a real use situation. The found mismatches need to be reviewed and confirmed in interaction with real users in usability tests, before changes are made in the analysed user interface.

The methodology including GTS, ECW, PUEA and PEEA are developed by the authors in research projects regarding evaluation and design of medical equipment with a high degree of usability and safety. The methodology has successfully been used in a real development process of a new dialysis machine. The methodology detected deficiencies that the design engineers were unaware of and provided them with a base for important design changes. The methodology can be employed for three different purposes in the development process.

- Evaluation of existing machines on the market, in order to find existing usability problems and potential use errors. This information could be used as input data to the design of new equipment.
- Evaluation during different stages in the development process, so problems and errors can be detected and mitigated
- Evaluation of the complete machine before launch, to confirm that the usage does not results in potential use errors or usability problems that can cause unacceptable risks for the patients.

REFERENCES

Bligård, L.-O., Osvalder, A.-L. (2006a), Predictive ergonomic error analysis – a method to detect incorrect ergonomic actions. *The 38th Annual Congress of the Nordic Ergonomics Society Conference*, Hämeenlinna, Finland.
Bligård, L.-O., Osvalder, A.-L. (2006b), Using enhanced cognitive walkthrough as a usability evaluation method for medical equipmented. *IEA*, Maastricht,

Netherlands

Bligård, L.-O., Osvalder, A.-L. (2007), An analytical approach for predicting and identifying use error and usability problem. *Lecture Notes in Computer Science,* 4799 427–440.

Bligård, L.-O., Osvalder, A.-L. (2008) Generic task specification – a framework for describing task demands and mental/physical work loads in a human-machine system. *2nd International Applied Human Factors and Ergonomics 2008.* Las Vegas.

Bligård, L., -O., Thunberg, A. (2007), An analytical usability method for alarm message evaluation: Alarm-ECW. *39th Nordic Ergonomics Society Conference* Lysekil, Sweden,.

Hart, S.G., Staveland, L.E. (1988), Development of nasa-tlx (task load index): Results of empirical and theoretical research. *In* Hancock, P.A. & Meshkati, N. eds. *Human mental workload.* Amsterdam: Elsevier, 139-183.

Hignett, S., Mcatamney, L. (2000). Rapid entire body assessment (REBA). *Applied Ergonomics,* 31 (2), 201-205

IEC. (2004), *IEC 60601-1-6:2004 medical electrical equipment - part 1-6: General requirements for safety - collateral standard: Usability* Geneva: IEC.

Janhager, J. (2005), *User consideration in early stages of product development – theories and methods.* The Royal Institute of Technology.

Mcatamney, L., Corlett, E.N. (1993), RULA: A survey method for the investigation of work-related upper limb disorders. *Applied Ergonomics,* 24 (2)

Militello, L.G., Hutton, R.J.B. (1998), Applied cognitive task analysis (acta: A practtitiober's toolkit for understanding cognitive task demands. *Ergonomics,* 41, 1618-1641.

Pheasant, S., Haslegrave, C.M. (2005), *Bodyspace: Anthropometry, ergonomics and design of work* 3ed. London Taylor & Francis.

Rasmussen, J. (1983), Skills, rules and knowledge; signals, signs and symbols, and other distinctions in human performance models. *IEEE Transactions on Systems, Man and Cybernetics,* SMC-13 (3), 257-66.

Stanton, N.A. (2006), Hierarchical task analysis: Developments, applications, and extensions. *Applied Ergonomics,* 37 (1 SPEC. ISS.), 55-79.

Stanton, N.A., Young, M.S. (1999), *A guide to methodology in ergonomics* London: Taylor & Francis.

Wickens, C.D., Hollands, J.G. (1999), *Engineering psychology and human performance,* 3rd ed. Upper Saddle River: Prentice Hall.

Wilson, J., Corlett, N. eds. (1995), *Evaluation of human work : Practical ergonomics methodology,* London Taylor & Francis.

Chapter 93

A Human Factors Evaluation of Flexible Endoscope Reprocessing

Qiawen Wu[1], Russell J. Branaghan[1],
Emily Hildebrand[2], Dana Epstein[2]

[1]Arizona State University
Mesa, AZ, 85212, USA

[2]Phoenix VA Health Care System
Phoenix, AZ, 85012, USA

ABSTRACT

Proper reprocessing of reusable medical devices is central to patient safety. Improperly reprocessed medical device can transmit viruses such as Hepatitis B, Hepatitis C, and HIV, as well as bacterial agents. Thus, errors may carry risks of preventable mortality, morbidity, and cost. Often, errors occur because of a failure to properly follow the manufacturer's instructions or Standard Operating Procedures. Unfortunately, this places much of the blame on the human operator rather than on the product design or other human factors. This paper reports findings of a heuristic evaluation of reprocessing procedures for flexible endoscopes, and provides design recommendations that target prominent heuristic violations.

Keywords: Reprocessing, Error, Human Factors, Heuristic Evaluation, Endoscope

INTRODUCTION

This paper examines human factors issues in reprocessing flexible endoscopes, and provides design recommendations for reducing potential hazards and reprocessing errors. The Olympus GIF 180, one of the most commonly used endoscopes, was chosen as a reusable medical device for the research. We identify human factors issues by conducting a heuristic evaluation of the reprocessing procedure. Then, we discuss design implications, and recommend future research endeavors. The research was motivated by the failure of reprocessing practices in numerous health systems throughout the United States.

In the United States, more than 46 million surgical procedures are conducted annually, and there are even more non-surgical invasive procedures (Rutala & Weber, 2004). Within that measure, there are approximately 10 million gastrointestinal endoscopies. The endoscope has become a valuable tool to diagnose and treat numerous medical disorders. While the incidence of infection associated with endoscopes is low, there are more healthcare-associated outbreaks linked to contaminated endoscopes than any other medical device (Rutala & Weber, 2004).

Following standard operating procedures (SOP) based on manufacturers' instructions is crucial to endoscope reprocessing. Research has shown that the transmission of infectious viruses, including Hepatitis B, and Hepatitis C, has occurred through the procedural re-use of endoscopes (Weber & Rutala, 2001; Mehta et al., 2006). One study found that almost 24% of the bacterial cultures from the internal channels of 71 gastrointestinal endoscopes grew substantial colonies of bacteria after all disinfection and sterilization (Rutala et al., 2008). In another study, more than one-third of facilities inspected were using a disinfecting agent that had been removed from the marketplace, was not FDA approved, or was not using a disinfecting agent at all (Kaczmarek et al., 1992). Despite public health threats, these errors often are unrecognized or unreported, and the science of measuring these errors and their effects is underdeveloped. Furthermore, errors are often blamed on the human operator, rather than on product design, manufacturer's instructions, training curriculum, operating procedures or other factors. This is problematic because design deficiencies are often disguised as human error.

The link between reprocessing errors and infection has received little public or scientific attention. For example, the Institute of Medicine's report 'To Err Is Human' (Kohn, Corrigan, and Donaldson, 2000), mentioned the term "Infection" only eight times whereas "medication" was mentioned 234 times. In contrast with seventy mentions of medication errors, reprocessing error was not even mentioned once in the IOM report. Furthermore, reprocessing practice is not specified in either the twenty evidence-based Patient Safety Indicators established by the Agency for Healthcare Research and Quality (AHRQ, 2003) or the thirty safe practices recommended by the National Quality Forum.

Although many studies have focused on human factors principles in healthcare settings (Reason, 1995, 2000; AHRQ, 2001; Barach, et al., 2008; Rich, 2008), there has been little study of the impact of human factors in reprocessing activities. Even though the critical nature of proper sterilization and reprocessing has been addressed in some literature (Wendt & Kampf, 2008; Centers for Disease Control, 2008; Rutala &Weber, 2001; Banerjee et al., 2007), the human factors in reprocessing errors have not yet been acknowledged, identified, or investigated. Furthermore, there is a gap between knowledge and practice. Currently, the standards practiced for reprocessing medical items are based on infection control principles and manufacturer instructions. However, these do not incorporate the role of human factors in actual practice. As a result, practical solutions have lagged behind in the arena of patient safety. Rather than focusing on the failure of medical personnel to comply with standards (so-called user error), we study the user-product interaction. That is our focus is not on user error, but on use-error.

ENDOSCOPE REPROCESSING

Endoscopes (see Figure 1 for a schematic) are used to examine the interior of the hollow viscera (ASTM, 2000). In this document, endoscope refers to flexible gastrointestinal endoscopes. These can be difficult to disinfect and easy to damage because of intricate design and delicate materials (Rutala & Weber, 2004). Also, air and water channels are difficult to clean manually and bioburden often remains even after cleaning (Ishino, Ido, & Koiwai, 2001).

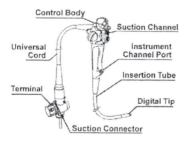

Figure 1. Schematic view of flexible endoscope, showing ports and channels.

Endoscope reprocessing is a multi-stepped procedure that makes a contaminated endoscope safe for reuse (Muscarella, 2000). It is a complex task composed of sequential subtasks. The steps below summarize what is typically a 75-page manufacturer's instruction manual or a 30-page SOP (Hildebrand, et al., 2010).

1. Pre-clean: Suction detergent channel and flush water through air/water channel. Remove valves and parts and soak in detergent solution.
2. Leak test: Connect the scope to an air source and submerge it in clean water to check for escaping bubbles, which indicate damage.

3. Clean: Mechanically clean internal and external surfaces. Brush internal channels and flush channels with water and detergent or enzymatic cleaner.
4. Disinfect: Immerse the endoscope in high level disinfectant and perfuse disinfectant into all channels and expose for a recommended time.
5. Rinse: Rinse the endoscope and all channels with sterile or filtered water.
6. Dry: Rinse the insertion tube and inner channels with alcohol and dry with forced air.
7. Store: Hang endoscope vertically in closed container.

It is apparent that this procedure is time consuming, physically engaging, and cognitively demanding. Humans have limitations, and are prone to mistakes. Furthermore, the frequency of mistakes is likely to rise with mental workload, number of steps, sub-optimal working environments, and stress. Therefore, it is essential to human factors to the endoscope reprocessing procedure.

RESEARCH METHOD

Heuristic evaluation is a quick and low cost method for identifying usability problems in a product (Nielsen & Mack, 1994). This method was employed to identify potential problems that may result in use-related hazards. This serves as a key step in our ongoing research, as it provides a preliminary glimpse into the human factors issues of reprocessing, and has informed our research direction. For these evaluations, we reviewed the Olympus GIF 180 Series model endoscope reprocessing procedure as it was described in an informational video used to train endoscope-reprocessing technicians. When confusion occurred, we referenced the instructional manual. The evaluation used the fourteen Nielsen-Shneiderman Heuristics (Zhang, Johnson, Patel, Paige, Kubose, 2003) modified based on Nielsen's (1994) and Shneiderman's (1998) heuristics specifically for evaluating medical devices (see Table 1). A severity scale was developed to rate the usability problems within this process. The scale was defined as: 1 = negligible likelihood of contamination; 2 = low likelihood for contamination; 3 = medium likelihood for contamination; 4 = high likelihood for contamination. A team of five completed the evaluation independently, applying the heuristics to the reprocessing procedure and the interface of the endoscope. All violations were assigned severity ratings.

Table 1: Nielsen-Shneiderman Heuristics (Zhang, et al, 2003)

Heuristic	Definition
Consistency	Users should not have to wonder whether different words, situations or actions mean the same thing.
Visibility	Users should always be informed what is going on with the system through appropriate display of information.
Match	The image of the system perceived by users should match their mental model.
Minimalism	Any extraneous information is a distraction and a slowdown.
Memory	Users shouldn't be required to memorize a lot of information. Memory load reduces user's capacity to carry out the tasks.
Feedback	Users should be given prompt and informative feedback about their actions.
Flexibility	Users always learn and are always different. Give users flexibility of customization and shortcuts to up performance.
Message	Messages should be informative enough so that users understand the nature of errors, learn from errors, and recover from errors.
Error	It's better to design interfaces that prevent errors from happening in the first place.
Closure	Every task has a beginning and an end. Users should be clearly notified about the completion of a task.
Undo	Users should be allowed to recover from errors. Reversible actions also encourage exploratory learning.
Language	The language should be always presented in a form understandable by the intended users.
Control	Don't give users the impression that they are controlled by the systems.
Document	Always provide help when needed.

RESULTS

Figure 2 shows the frequency and severities of heuristic violations for the reprocessing procedure. There were 277 heuristics violated. The most frequently violated heuristic (123 times) was Error, followed by Feedback and Memory (52 and 35 times respectively). These three heuristics accounted for 76% of violations.

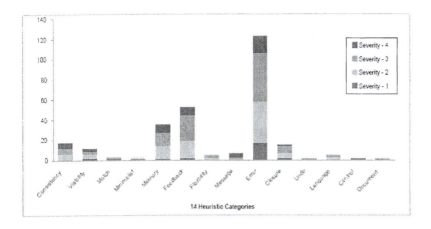

Figure 2. Numbers and severity scores of heuristic violated

Table 2 shows the mean severity score of each heuristic violated. The three most frequently violated heuristics (error, feedback, and memory) were not the most severe. Instead, message and control, had the highest severity, 3.33 and 4 respectively. Table 3 lists some of the common idntified heuristic violation. Space constraints preclude listing all of them.

Table 2: Mean Severity Ratings

Heuristic Violated	Mean Severity Rating	Heuristic Violated	Mean Severity Rating
Match	1.00	Feedback	2.75
Minimalist	1.50	Memory	2.83
Visibility	2.27	Consistency	3.00
Language	2.40	Undo	3.00
Closure	2.47	Document	3.00
Error	2.53	Message	3.33
Flexibility	2.60	Control	4.00

Table 3: Heuristics Violated and Usability Problems Related

Heuristic Violated	Problem Description	Design Recommendations
Error	Two channel entries are difficult to discriminate. They are the same size, shape and texture, but because of their dimensions, they require different brushes to clean.	• Design size and shape of the entries with a lockout, so only the proper size brush can be accommodated. • Provide a visual difference in channel size • Color-code channel entries and brushes.
Feedback	Users must use short strokes when cleaning internal channels, but there is no indication of how short these strokes should be.	• Provide visual marks (spaced one inch apart) on the plastic brush wire to provide visual feedback for short strokes.
Closure	It is not clear when a task is completed. For example, users must feed cleaning brushes through an insertion tube and clean the bristles in the detergent with their fingertips until visible debris is removed.	• Coat the brush ends with a colored chemical, which changes to white after appropriate cleaning.
Feedback Visibility Undo	When leak testing, excessive air pressure may damage scope.	• Place a governor on the air pump. • Display maximum pressure on gauge. • Provide audible alarm when air pressure is exceeded.
Feedback Visibility Undo	The screen on the sterilization machine is too small and does not display effective status information.	• Increase display size. Provide adequate contrast. Provide information regarding time remaining, and adequacy of sterilization solution.

Feedback	The keys on the sterilization machine are too close together, especially gloved use.	• Spread out keys enough for both female and male users to operate with gloves on.
Consistency	Keys on control panel are perceived as flat, and do not afford pressing, as buttons.	• Provide visual depth for pressing, like buttons. Provide audio/visual feedback such as a beep or color change.
	There is no 'START' button but a flat manufacturer's logo.	• Replace logo with a 'START' button. Illuminate the key after machine is in operation to indicate its status.

Below we provide examples of concepts we have been exploring to address some of the problems identified in Table 3. For example, technicians must clean multiple channels with a two-ended cleaning brush. They are to brush the suction channel using one brush, and then brush the suction cylinder with the Valve/Control head brush end. Unfortunately, as illustrated in Figure 2, it is difficult to visually discriminate the different sized brushes on each end.

Another problem is that two channels in the endoscope share the same opening (see Figure 3). To clean one channel, one inserts the brush straight into the opening. To clean the other, one enters the opening and bends the brush to 45° angle. It is easy to fail to do this, so that one channel does not get cleaned properly.

Third, technicians must insert the brush until it emerges from the distal end of the scope, and then clean the bristles with his fingertips before removing by gently pulling back the brush using short strokes. Yet, it is difficult to determine if all debris has been adequately removed, and how short the brush strokes should be.

Figure 3. A single-use brush being used in endoscope reprocessing

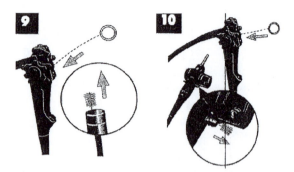

Figure 4. Illustrations for Brushing

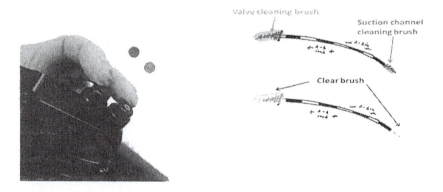

FIGURE 5. Design Concept to Solve Usability Problem cited in Table 2

CONCLUSION

The findings indicate that the most common problems were associatied with Error, Feedback, and Memory heuristics, whereas the most severe violations were associated with Message and Control heuristics. Design should be dedicated toward remedying these types of problems first. Our initial foray into design concepts suggests that remedying many of these problems may be fairly straightforward. Future research will involve observing how users interact with the endoscopes in the actual reprocessing facilities. Other areas to be investigated include the selection and training of reprocessing personnel, shift management, performance evaluation, quality control, device design, and reprocessing workspace design.

REFERENCES

Agency for healthcare Research and Quality (AHRQ) (2001), "Making Healthcare Care Safer: A Critical Analysis of Patient Safety Practices. Evidence Report/Technology Assessment "Publication No.43. AHRQ 01-E058. July 20, 2001, 662. Available at: http://www.ahrq.gov , accessed September 2003.

Agency for Healthcare Research and Quality (AHRQ) (2003), AHRQ Quality Indicators– Guide to Patient Safety Indicators. Rockville, MD; 2003. AHRQ Pub.03-R203.

American Society for Testing and Materials (ASTM). (2000). *ASTM standard practice for cleaning and disinfection of flexible fiberoptic and video endoscopes used in the examination of the hollow viscera* (F1518-00).West Conshohocken, PA. Author.

Banerjee, S., Nelson, D.B., Dominitz, J.A., Ikenberry, S.O., Anderson, M.A.,

Brooks, D.C., et al. (2007). Reprocessing failure. *Gastrointestinal Endoscopy*, 66, 869-871.

Barach, P., Johnson, J.K., Ahmad, A., Galvan, C., Bognar, A., Duncan, R. Starr, J.P. et al. (2008). A prospective observational study of human factors, adverse events, and patient outcomes in surgery for pediatric cardiac disease. *Journal of Thoracic Cardiovascular Surgery, 136*, 1422-1428.

Hildebrand, E,. Branaghan, R.J., Epstein, D.R., Wu, Q., Jolly, J., & Taggart, M. (2010). Exploring Human Factors in Endoscope Reprocessing (in preparation).

Rutala, W.A, Weber, D. J., and the Healthcare Infection Control Practices Advisory Committee (HICPAC),2008, Guideline for Disinfection and Sterilization in Healthcare Facilities, 2008, Centers for Disease Control

Ishino, Y., Ido, K., Koiwai, H., Sugano, K. (2001). Pitfalls in endoscope reprocessing: brushing of air and water channels is mandatory for high-level disinfection. Gatrointestinal Endoscopy, 53(2), 165-168.

Kaczmarek, R. G., Moore, R. M., Mccrohan, J., Goldmann, D. A., Reynolds, C., Caquelin, C., et al. (1992). Multi-state investigation of the actual disinfection/sterilization of endoscopes in health care facilities. *The American Journal of Medicine, 92*(3), 257-261.

Kohn, L. T., Corrigan, J. M., and Donaldson, M. S., Eds. (2000). *To Err is Human: Building a Safer* Kohn, Corrigan, and Donaldson, 2000

Mehta, A. C., Prakash, U. B. S., Garland, R., Haponik, E., Moses, L., Schaffner, W., et al. (2006). Prevention of flexible bronchoscopy-associated infection. *Chest, 128*, 1742.

Muscarella, L. (2000). Automatic flexible endoscope reprocessors. *Gastrointestinal Endoscopy Clinics of North America, 10*(2), 245-257.

Nielsen J, & Mack R, (ed.) (1994). Usability inspection methods. New York: John Wiley & Sons, Inc.

Nielsen J. (1993). Usability engineering. Boston: Academic Press

Reason, J. (1995). Understanding adverse events: human factors. *Quality in Healthcare, 4*,80-89.

854

Reason, J. (2000). Human error: Models and management. *British Medical Journal*, *320*, 768-770.

Rich, S. (2008). How human factors lead to medical device adverse events. *Nursing 2008*, *38*, 62-63.

Rutala, W.A. & Weber, D.J. (2001). Creutzfeldt-Jacob Disease: Recommendations for disinfection and sterilization. *Clinical Infectious Diseases*, *32*, 1348-1356.

Rutala, W. A., & Weber, D. J. (2004). Disinfection and sterilization in health care facilities: What clinicians need to know. *Clinical Infectious Diseases, 39*(5), 702-709.

Shneiderman B. Designing the user interface. 3rd ed. Reading, MA: Addison-Wesley; 1998.

Veterans Administration (2009).Use and Reprocessing of Flexible Fiberoptic Endoscopes at VA Medical Facilities. Report 09-01784-146, June 16,

Weber, D. J., & Rutala, W. A. (2001). Lessons from outbreaks associated with bronchoscopy. *Infection Control and Hospital Epidemiology, 22*(7), 403-408.

Wendt, C. & Kampf, B. (2008). Evidence-based spectrum of antimicrobial activity for disinfection of bronchoscopes, *Journal of Hospital Infection*, *70*, 60-69.

Zhang, J., Johnson, T.R., Patel, V.L., Paige, D.L., Kubose, T. (2003). Using usability heuristics to evaluate patient safety of medical devices. Journal of Biomedical Informatics. 36, 23-30.

Chapter 94

The Measure of Success: Quantifying the Acute Effects of Whole Body Vibration on Hamstring Flexibility

Selim Nurudeen, Dosun Shin, Donald Herring, Kristinn Heinrichs

School of Design Innovation,
Arizona State University
Temp, AZ 85287-2105, USA

ABSTRACT

Hamstring (HS) muscle strains are perhaps the most common muscle injuries in activities where sprinting is involved. An especially high rate of re-injury makes recovery a slow and frustrating process. Poor hamstring flexibility and strength imbalances are predisposing factors for hamstring muscle strains. Being flexible not only prevents injuries, but improves overall athletic performance.

The purpose of this investigation was two-fold: To determine the acute effects of whole body vibration (WBV) on hamstring flexibility using the Active Knee Extension (AKE) Test as a measure of HS flexibility or Knee Range of Motion

(ROM), and to design an apparatus to improve the accuracy of AKE measurement which would serve as a user-efficient device.

Thirty female recreationally active athletes were randomly divided into 3 groups (n=10 WBV + stretch (WBV-S), n=10 WBV only (WBV), n=10 control (C). One-way Analysis of Variance (ANOVA) demonstrated notable increased in HS flexibility (9.83% WBV-S, 3.50% WBV, and 1.56% C). Tukey Post-Hoc analysis indicated a significant increase in HS flexibility for the WBV-S group which lasted 6 minutes.

Based on the limitations from the AKE test employed in the study, design changes were made to the device to eliminate sources of error from body positioning and material flexibility to improve accuracy. The resulting device may have a profound influence on self-monitoring HS flexibility routines and test-retest accuracy.

Keywords: Whole Body Vibration, Hamstring Flexibility, Knee Range of Motion, Active Knee Extension Test, Measurement Device

INTRODUCTION

"Hamstring muscle injury is one of the most common muscle injuries sustained by athletes[4]". In sports such as track and field, where success is dependent upon competing at optimum performance, setbacks like this can be detrimental to ones performance and even athletic career. "The circumstances under which these injuries occur usually take place during periods where the mechanical limits of the muscles have been surpassed[2]". Even when the injury has been rehabilitated and the hamstring muscle is strong enough to resume regular activity, there is still a need for caution. It has been stated that the chance for re-injury is around 34%[7].

The hamstring muscle group is formed by the biceps femoris (short and long head), semitendonosus, and the semimembranosus. Causative factors for injury range from a lack of flexibility, to muscle imbalance, to age. WBV is a relatively new intervention which may provide beneficial gains to maximal force and flexibility by applying vibration. Studies have suggested that motor unit activity, blood flow, temperature, pain threshold, and muscle relaxation are also enhanced by this exercise[6,8].

While WBV is most commonly used to augment muscle force production, it is seldom used to increase flexibility. It is suggested that a lack of flexibility increases chances of a hamstring strain[3]. Stretching before activity prepares the muscle for the force loads which it must endure. This study sought to determine HS flexibility gains from this intervention. To accomplish this, the torso of the body is stabilized and the hip of the leg to be measured is flexed to 90 degrees. Any change in HS length is determined by the ROM of the knee. This form of measurement is known as the AKE test. The study utilized this test to measure flexibility under different conditions to try to form a correlation.

METHOD

SUBJECTS

The subjects for this study were 30 recreationally active females (age, 23.35 ± 6.78 years; height, 163 ± 15.33 cm; weight, 63.06 ± 18.06 kg). The subject gender was kept the same in order to rule out any errors created by using same sex subjects.

The subjects were randomly assigned to 3 groups of 10: Vibration (WBV), Vibration+Stretch (WBV-S), and control group (C). Subjects were tested for musculoskeletal injuries and were included if they showed visible evidence of HS tightness, defined as a limitation of 20° or more from full extension as determined by the AKE Test, which has been known to be a reliable and valid measurement [3,9].

MATERIALS

The test consisted of a polyvinylchloride (PVC) pipe crossbar (diameter = 1.6 cm, crossbar length = 66cm height = 50cm, base length = 64 cm), a magnetic protractor (Figure 1), a 16" goniometer to measure the angle of hip and knee flexion before the AKE test, a stopwatch to measure time intervals, and a (Power Plate pro5) to administer WBV.

The PVC served to maintain hip position and minimize measurement error due to passive insufficiency of the HS. Once the subject was properly positioned, hamstring flexibility was read by strapping the magnetic protractor to the proximal lower leg and determining the changes in knee ROM. All subjects read and signed an informed consent form approved by the Arizona State University Institutional Review Board (IRB) for the Protection of Human subjects.

Figure 1. The Active Knee Extension subject Positioning.

858

PROCEDURE

Participants completed three sessions: familiarization session one week prior to testing, pre-test, and post-test for each of the assigned conditions (WBV, WBV-S, and C).

FAMILIARIZATION SESSION

Three AKE measurements were taken and averaged for each subject. Five positions were held while standing on the vibration platform. While completing these positions, vibration was turned off. All subjects assumed each of these five positions to familiarize them to the protocol.

(Position 1): Upright standing with knees semi-locked
(Position 2): Isometric squat to 45° hip and knee flexion with feet pointed straight ahead and hips in neutral rotation
(Position 3): Same as position two but with feet slightly externally rotated
(Position 4): Lunge position with right leg on platform and left leg on the floor.
(Position 5): Lunge position with left leg on platform and right leg on the floor.

Position 1.   Position 2.   Position 3.   Position 4.   Position 5.

Figure 2. Test Positions.

PRETEST

All subjects performed a total of six AKE tests with a 60-second rest period between repetitions. The first five AKE tests served as a warm-up to decrease any effect that may have occurred with repeated measures performed from a cold start. The sixth AKE measurement was recorded as the pre-stretch measurement. When the subject could not extend the leg any further without the thigh moving away from the crossbar, the position was held while the measurement was taken.

VIBRATION PROTOCOL

The Bosco (1998) WBV protocol was utilized for this study[1]. WBV was applied using a vibration wave frequency of 26 Hz and 10mm peak to peak amplitude. The five positions were held for 30 seconds with a 40 second rest interval in between. The vibration platform was activated for all positions in the same order as the familiarization session.

WBV Group: Vibration protocol only.
WBV-S Group: Vibration protocol followed by the investigator passively stretching the hamstring muscle group until the subject first reported a mild stretching sensation and then held that position for seven seconds. The subject contracted the hamstrings isometrically for seven seconds by attempting to push the leg back into hip extension while maintaining knee extension toward the table against the resistance of the investigator. After the contraction, the subject relaxed for five seconds. Sequence was repeated five times on each subject in this group. All stretching was performed on the lower right extremity.
C Group: Performed vibration protocol with vibration turned off. No stretching.

POST TEST

The Post Test measurement was the same as the Pre-test except no warm-up contractions were performed and each of the AKEs were recorded. One AKE measurement was taken at 0 minutes (immediately) following the treatment protocol. Subsequent AKE measures were taken at time intervals of 2, 4, 6, 8, 16, and 32 minutes after the treatment ended.

RESULTS AND FINDINGS

A one-way between subjects Analysis of Variance (ANOVA) was used to compare the knee ROM means between groups followed by a Tukey Post-Hoc Analysis to determine which means were significantly different from the others[5].

Figure 3 plots the delta scores in degrees between the pre-test condition and subsequent measures. Zero on the x-axis indicates the measure taken immediately after the intervention. The negative values on the y-axis represent decreases in knee extension ROM, indicating a loss of flexibility. From the data it was determined that the vibration group (3.50% increase) and the vibration + stretch group (9.83% increase) produced higher averages in HS flexibility gains than the control group (1.56% increase).

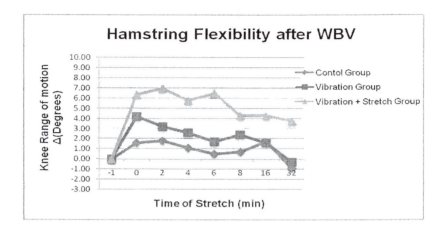

Figure 3. The change in hamstring flexibility for each group over time.

In order to determine how long WBV affected HS flexibility over time in each group, a Tukey Post-Hoc Analysis was used to compare group means. Significance was determined by $p<.05$. The results determined that only the WBV-S group possessed significantly greater gains in HS flexibility in comparison to the control group. This significance lasted six minutes following exercise. The data concluded that although acute WBV alone does have a positive effect on HS flexibility, the effect is not significant. However, when coupled with stretching, acute WBV significantly increases HS flexibility for up to six minutes.

CONCLUSIONS

The major finding in this study was that acute WBV alone was not enough to significantly increase HS flexibility. However, when coupled with stretching, WBV increased HS flexibility for up to 6 minutes. This estimate was in agreement with the six min decrease in HS flexibility from previous studies[3] which affirmed that a temporary creep effect occurred in which the viscoelastic component of the HS remained in the elastic range. This creep effect is characterized by the lengthening of the muscle tissue due to an applied load[10]. Regardless of the fact that Figure 3 showed a definitive decrease in flexibility over time by the vibration group, the effects of acute WBV could not be determined statistically because there was no significant increase in HS flexibility for that group during the entire test.

Current findings have stated, WBV combined with a contract-release stretching protocol significantly increases HS flexibility beyond simply using WBV[11]. While earlier studies[11] used a slightly different method of measuring HS flexibility, the premise was the same: the subject lying supine while an investigator used a goniometer to measure knee ROM[11]. While the present study was a repeated measures design conducted over a single session, previous studies collected data over a five day period. Each group, control included, underwent a hold-release protocol. The results affirmed that both groups, WBV and C, experienced a significant increase in knee ROM, 30% and 14% respectively.

Since the WBV treatment was brief (a single session), the response of the WBV-S was short term but may have been due to an increased pain threshold, blood flow, and muscle elasticity[6]. When WBV was coupled with stretching there was an adverse effect to the viscoelastic properties, thixotropic properties, and neural properties of the hamstring muscle group. This combination was successful in temporarily and significantly increasing HS flexibility.

FUTURE STUDY IMPLICATIONS

Throughout this study, attention has been paid to how the AKE apparatus could be improved and applied to future studies. This was done so by noting the difficulties encountered during the testing process as well as making alterations to the protocol.

The first implementation which could be added is the changing of the WBV intervention from short term to long term. Given the results of previous studies[11], it should be considered extremely useful for future experimentation to examine this route.

Another implication would be to diversify the stretching criteria. Though this study used a contract-release stretching protocol which has proven to be successful in temporarily increasing flexibility[11], it has not been determined which method of stretch is the most efficient.

FUTURE DESIGN IMPLICATIONS

Upon completion of this study, enhancements were made to the AKE testing apparatus in order to eliminate common sources of error (e.g. hip, pelvis, femoral position, and inaccuracies created by material flexibility) in test-retest measures observed in previous studies (Figure 4 and 5). These changes were made to the PVC device which served to stabilize the subject during measurement, and the magnetic protractor, which measured the gains in knee ROM.

Figure 4 and 5. AKE apparatus redesign.

862

Given that the measuring device was composed of flexible PVC which tended to "give" during the stabilizing of the subject, this material could be changed to more resistant yet lightweight material like aluminum.

Another problem was the positioning of the subject. As can be viewed in figure 1, the subject is positioned under the crossbar. In order to line up their flexed hip with the crossbar, the whole torso must be situated. This made the initial process tedious. To accommodate this problem, the crossbar was redesigned so that it was composed of two parts that could be separated to allow the subject to lie down. Once they were positioned, it could be fastened together (Figure 6).

Figure 6 and 7. Fastening crossbar and Adjustable crossbar.

Addressing the issue of positioning the torso for accurate measurement, the device was altered to where the crossbar itself was adjustable in reference to the hip. This concept presented "handle bars" which would allow the subject to easily adjust the crossbar by lifting the handles, pushing them forward along a rail to the desired position, and releasing it, locking it in place with a spring mechanism (Figure 7).

Targeting commercial usage, it was decided that the device be collapsible for convenience when not in use. Figures 8 and 9 present the frame in its collapsed state. The split crossbar pivots on both sides and comes together to conveniently act as a carrying point. Figures 10 and 11 add a thin mat to the apparatus in order to cushion the back of the subject as well as level them out for accurate measurement. When the frame is collapsed, the mat wraps around it, acting as a carrying device.

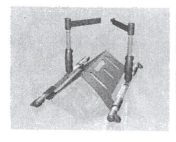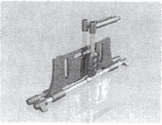

Figures 8 and 9. Collapsed Frame.

Figures 10 and 11. Added mat.

The second aspect of the device that required re-designing was the magnetic protractor which measured knee ROM. The device used in the actual study consisted of a magnetic protractor connected to a hook and loop strap that was attached to the proximal lower leg (See Figure 1). During the study, in order to read the results an investigator had to be present. The investigator had to ensure that the subject's hip was at 90 degrees measure accurate changes in knee ROM. The goal for the redesign of this product was to allow the subject to not only position their own hip accurately but also interpret their own knee ROM without the aid of an investigator.

Like its original, this device attaches to the subject's leg. Installed in its center is a pendulum which works like a balance to keep the blue plastic tabs on a constant horizontal plane (Figure 12). By cutting an opening in the grey plastic, the blue tabs are only be visible when the hip is at exactly 90 degrees. Since these alterations are based on the vantage point of the subject, the opening in the plastic will be positioned slightly under the horizon. The device still serves its primary purpose of measuring HS flexibility. By utilizing visible markers, the user will be able to observe the figures without the aid of an investigator (Figure 13). The Markers will correspond to both the upper and lower leg, stationary markers representing the upper leg while moving markers representing the lower leg (Figure 12).

Figures 12 and 13. Magnetic protractor.

Though these changes to the AKE design apparatus have not yet been realized in actual testing, they address key issues concerning the devices' current usage. Future implementations of these changes may not only have a positive effect on similar studies pertaining to hamstring measurement, but may also influence the area of self rehabilitation and device commercial usage.

REFERENCES

1. Bosco, C., Cardinale, M., Colli, R., Tihanyi, J., von Duvillard, S.P., and Viru, A. (1998). *The influence of whole body vibration on jumping performance.* Biol. Sport. 15, 157-164.
2. Croisier, J. (2004). Hamstring Injuries. Sports Med. 34(10), 681-695.
3. DePino, G. M., Webright, W.G., and Arnold, B.L. (2000). *Duration of Maintained Hamstring Flexibility after Cessation of an Acute Static Stretching* Protocol. Journal of Athletic Training. 35(1), 56-59.
4. Garrett, W.E., Rich, F.R., Nikolaou, P.K., Vogler, J.B. III. (1989). *Computed Tomography of hamstring muscle strains.* Med Sci Sports Exerc. 21, 506-514.
5. Hockey, R.D. (2008). SPSS *Demystified: A Step-by-Step Guide to Successful Data Analysis.* Pearson Education, Inc. Upper Saddle River, New Jersey.
6. Issurin, V.B., Liberman, D.G., and Tenenbaum, G. 1994. *Effect of vibratory stimulation training on maximal force and flexibility.* J Sports Sci. 12, 561-566.
7. Orchard, J. and Seward, H. (2002). *Epidemiology of injuries in the Australian passive muscle stiffness in symptoms of exercise-induced Football League, seasons 1997-2000.* Br J Sports Med. 36, 39-45
8. Shinohara, M. (2005). *Effects of Prolonged Vibration on Motor Unit Activity and Motor Performance.* Med. Sci. Sports Exerc. 37(12), 2120-2125.
9. Spernoga, S.G., Uhlt, T.L., Arnold, B.L., and Gansneder, B.M. (2001). *Duration of Maintained Hamstring Flexibility after a One-Time, Modified Hold-Relax Stretching Protocol.* Journal of Athletic Training. 36(1), 44-48.
10. Taylor, D.C., Dalton, J.D. Jr., Seaber, A.V., Garrett, W.E. Jr. (1990). *Viscoelastic properties of muscle-tendon units: the biochemical effects of stretching.* American Journal of Sports Medicine. 18, 300-309.
11. Tillaar, R.V. (2006). *Will Whole-Body Vibration Training Help Increase the Range of Motion of The Hamstrings?* Journal of Strength and Conditioning Research. 20(1), 192-196.

Chapter 95

Health 2.0: User Centering, Patient Centering and Human Centering

Steven V. Deal,[1] A. Patrick Jonas[2]

[1]Deal Corp
Yellow Springs, OH 45387, USA

[2]Center for Innovation in Family and Community Health
Beavercreek, OH 45431, USA

ABSTRACT

Health 2.0 introduced the defining features of Web 2.0. However, emerging technologies and processes support only the mechanical processes of the medical care; they fall short of capitalizing on the humanizing functionality that Web 2.0 affords. We briefly characterize the needs of the patient-physician dyad. We explore the ability of two popular healthcare repair strategies to satisfy those needs. We introduce a formative model for a dyad-centered which is intended to support improved outcomes, better products and more suitable and responsive compensation models.

Keywords: User-Centered Design, Medical Care, Health Care, Personalized Medicine, Human-Centered Health Home, Compensation Models

INTRODUCTION

The current American health system is bankrupting America, resulting in innumerable repair strategies. Among these are the 1) Patient-Centered Medical Home (PCMH) and 2) the institutionalization of Electronic Health Records (EHRs). These strategies represent two exclusive foci about which a medical care system can be organized -- client centering and technology centering. Other dominant care delivery influences, such as healthcare providers, payers, medical schools, hospitals, the government healthcare systems, insurance companies, research, or pharmaceutical companies, could also be adopted as care system centers. In practice, an effective health system must balance these points of view.

The selected focus influences how a system is conceptualized. Derivative priorities skew product and process requirements statements and alter constraint definitions. A poor centering selection can lead to local optimizations that subvert global purposes; it leads to unintended consequences. This has been the result of legislation and practices in the U.S. that currently define today's non-system which centers on the payer and provider.

We assume that the purpose of a care system is to help people achieve health and wellness goals. To serve that purpose, stable system functions must exist. An unstable system may exhaust its resources or shake itself apart. The current U.S. health care system is doing both.

We hold that an improved care system will center on the patient-physician dyad. In this paper, we examine the nature of that dyad and how its characteristics can be used to guide the development and selection of technology products. We describe how the PCMH and EHRs fail to support the dyad. We offer a formative description of the dyad-centered Human-Centered Health Home (HCHH) model. We conclude by suggesting changes to compensation models that might better support ecological health and wellness.

THE PATIENT-PHYSICIAN DYAD

Healthcare repair strategies treat the doctor-patient relationship mechanically. The doctor is a problem solver, databank, and billing machine. The patient is a data source.

The relationship and the actual work of the patient-physician dyad are largely ignored. The true work stems from the human-to-human relationship which focuses the dyad into more meaningful, cost-effective encounters.

Proposed processes and technologies support only the mechanics of medical care and detract from the humanity of the dyad. Database systems address the collection and storage of information. Measurement system interfaces are understandable only by members of the medical care establishment – if we're fortunate. Networks support secure data transmissions.

These systems don't address the inherent emotional, human-to-human nature of the encounter. They fail to incorporate the biological, psychological and social states which both patient and provider bring to the encounter. For example, the patient may be dehumanized by intense pain. He may have an intense fear of doctor's offices. The doctor may be dehumanized by schedule pressures or an emotionally intense encounter with a previous patient and elect to bring only her cognitive self to the patient.

In these examples, their wholeness has been violated. Breaches can lead to poor assessments of the patient's state and treatment plans that are misaligned with patient preferences. They induce provider burnout, a situation which, when broadly perceived, causes a decline in the number of students who will choose a career in primary care, ultimately producing a shortage.

'But you can't develop technology to support that squishy stuff.' Can't you? Human decision making and communication skills used in relationships improve as one is "coherent" with balance between the sympathetic and parasympathetic nervous systems (McCraty et al, 1998). Biofeedback measures of heart rate variability (HRV) can help to balance these two systems. Products based on HRV could be used in the clinical setting to allow the two humans to clarify how balanced their biology is.

'No one would pay for that!' May be. Therein is the problem with centering a care delivery on the payer-provider system. Providers may be wary of such a non-traditional clinical support strategy and reject it. Payer systems may dismiss the technology as unworthy of consideration due to unproven cost savings. However, the dyad may find that it leads to improved health, wellness and satisfaction of dyad members.

Technologies introduced to serve the dyad must support goal achievement and accommodate the states of the two participants. Tools for humanizing interactions such as those developed by and as a result of Carkhuff's (1969) problem solving models have been used for decades to train helpers, such as teachers and health professionals. The human dyad assessment model, modified from Carkhuff inspired models with the addition of a spiritual dimension, can be used by physicians to assess the impact of patient states on major activity performance (table 1). The model can also be used by physicians as a readiness self-assessment and planning tool.

Table 1 Human Dyad Assessment Model

	Physical		Intellectual		Emotional		Spiritual	
Living	Physician	1-5	Physician	1-5	Physician	1-5	Physician	1-5
	Patient:	1-5	Patient:	1-5	Patient:	1-5	Patient:	1-5
Learning	Physician	1-5	Physician	1-5	Physician	1-5	Physician	1-5
	Patient:	1-5	Patient:	1-5	Patient:	1-5	Patient:	1-5
Working	Physician	1-5	Physician	1-5	Physician	1-5	Physician	1-5
	Patient:	1-5	Patient:	1-5	Patient:	1-5	Patient:	1-5

Clients and physicians are participants in an intimate and supremely human encounter. As part of the encounter, patients and providers may need to be re-humanized. Patient re-humanization occurs when trust, respect, empathy and appreciation are established. Physician re-humanization occurs when the physical, intellectual, emotional and spiritual realities of a day of care giving are accounted for. Compensation models provide no payment for re-humanization. Only when the humanity and the mechanics of the encounter are simultaneously supported by process and technology can a stable care-delivery system emerge.

THE PATIENT CENTERED MEDICAL HOME

The PCMH is one popular strategy to fix the primary care shortage. The PCMH model is supported conceptually by major employers, physician groups and the government. The intent of the PCMH is to generate more money and less "administrivia" for primary care physicians, and therefore more interest in primary care specialties by medical students. The desperate need to redirect medical student specialty selection and reimbursement for family physicians, general pediatricians, and general internists initially leaves patients as a lower priority, in a "holding pattern."

The principles of PCMH (Patient-Centered Primary Care Collaborative (2007) are:

- A personal physician
- A physician-directed medical practice
- Whole person orientation (a physician responsibility)
- Care that is coordinated and/or integrated
- Quality and safety
- Enhanced [schedule] access
- Payment (compensation model definition)

From our description and the principles, it is easy to see that the model is centered on the problems of payers and providers. True patient centering is missing from the PCMH model. Like most suggested healthcare fixes, it is a forced, top-down approach. It fails to acknowledge the collaborative work and the complex relationships that exist in the patient-physician dyad. PCMH would benefit from analyses of cognitive work that define true patient-centering principles.

ELECTRONIC HEALTH RECORDS

Another repair strategy by the United States Department of Health and Human Services (HHS) is The Health Information Technology for Economic and Clinical Health (HITECH) Act (HITECH, 2009). This Act seeks to improve the medical outcomes and care delivery by means of web-enabled information technology (IT) products that embrace the elements of Web 2.0. Health 2.0, like Web 2.0, is intended to feature interactive information sharing, interoperability, user-centered design, and collaboration. HITECH enforces "meaningful use" of EHRs. Meaningful use standards were defined by two advisory committees, the Health Information Technology Policy Committee and the Health Information Technology Standards Committee. Table 1 summarizes and categorizes the list of 25 physician-related criteria taken from the Act (HHS, 2009).

Only the bolded items in the table are routinely used by physicians. Only those under patient access directly address patient needs, and these support patient health and wellness goals only in the context of the current healthcare system; they provide no help for achieving and maintaining a state of wellness. Meaningful use criteria fail to support the work of the dyad. Take, for example, vital sign tracking. Doctors are trained to expect that over the course of a person's life, blood pressure will trend higher. Traces of this vital sign over months and years are not useful; useful measures are those of the current visit and one or two previous visits. In other instances, we have listened to anecdotal stories about the need for web sites that allow doctors

Table 2 25 Meaningful Use Criteria for Eligible Providers (for Physicians)

Dr. Process Supports	Data for Clinical Use	Drug Management
Computerized order entry	**Diagnosis lists**	Drug interaction checks
Clinical decision rules	**Structured lab results**	Electronic prescriptions
Follow-up reminders	Vital sign tracking*	**Patient medications lists**
Insurance eligibility		**Medication allergies**
Electronic claims		Medication reconciliation
Electronic data exchange		
Transition care summaries		

Data for Payer Tracking	IT Constraints	Patient Access
Demographic records	Information security	Electronic records upon request
Smoking status		Timely patient record access
Syndromic surveillance data		Patient clinical summaries
Patient lists by conditions		
Care quality measures		
Immunization record submittals		

to share their EHR workarounds, and tales of clinical certifications that were achieved by modifying medical practice to conform to processes their EHRs require.

Another way in which HITECH fails to reach its promise is in interoperability. HITECH focuses on the EHR, a tool that will deliver data to the government, insurance companies and hospitals. There is no meaningful use standard that requires the EHR to interface with the physician's Electronic Medical Record (EMR) system nor with a patient's Personal Health Record (PCR). This serves neither patient nor provider needs. Full interoperability is necessary if the promise of electronic medicine – reduction in episodic and chronic illness and cost efficiency – is to be realized.

HUMAN-CENTERED HEALTH HOME: A MODEL FOR HEALTH SYSTEM IMPROVEMENT

We introduce a model (figure 1) that centers on the dyad.

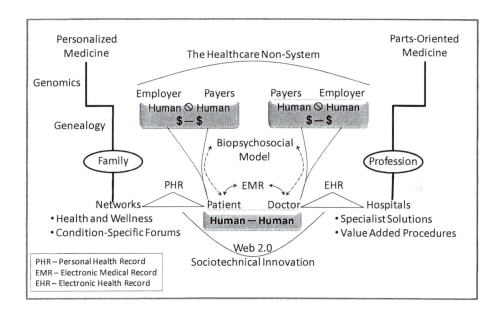

FIGURE 1 Family Influences and Social Networking Can Shield Patients from De-humanizing Influences of the Healthcare Non-System

The model starts with the patient side and connects to the physician side of the care system. The sociotechnical systems, networks and technologies on the left side of the figure act to shield patients and doctors from the dehumanizing aspects of the healthcare system.

The Human-Centered Health Home (HCHH) model emphasizes the humanness of the participants through the biopsychosocial model (BPSM). The BPSM is often used by medical professionals as a diagnostic or therapeutic aid. It is a hierarchical list developed by Engel (1977) that examines the human state from biosphere to subatomic particles. The BPSM, applied to both participants, insists upon mutual respect within the dyad, protection of the dyad from external forces, human-to-human connection of the dyad elements, followed by the detection of misalignment with good health and wellbeing.

The HCHH includes use of personal technology to enhance health and save medical dollars. The internet and the PCH afford people the freedom to learn and plan and record their progress. People may also use them to connect to individualized biological information to better clarify their genomic strengths and risks. Connecting to genealogy sites and the Surgeon General's Family Medical History site may enable people to filter unique strengths and risks for health planning purposes. They may wish to connect through condition-specific biological sites to other humans, send some of the information through social sites to family members or use their PHR to share information with their provider's EMR and the EHR of the medical care system. The personally chosen sociotechnical connections allow their humanness to play an important role in their decision processes about health and wellness.

We hold that the humanity of the patient serves as an anchor that allows an "equal person" connection point with the physician. The human centering promoted by this model is less driven by medical financial imperatives (physician salary, hospital budget, etc.) and more by patient humanity in the context of the patient's life. Importantly, the patient's humanity is relating to the physician's humanity in the context of the physician's life. This person-to-person connection becomes a force that may protect the dyad from succumbing to financially motivated medical- or insurance reimbursement- centering.

ALIGNING PAYMENT MODELS TO HUMAN-CENTERED WELLNESS CARE

As the human gets direct access to disruptive technologies, the sociotechnical elements filter them from becoming a patient prematurely and filter the physician from excessive centering on medical technologies. Cost implications of this human empowerment lean toward savings, while human implications lean toward increased human potential for the physician who is "pulled" toward human solutions.

Christensen (2009) expounded on a "disruptive solution for health care." He notes that there are three types of business activities now lumped into one reimbursement strategy: solution shops for undifferentiated patient problems that should be paid on a fee for service basis; value-added processes such as surgical procedures for already diagnosed problems that should be shopped around for cost and quality; and facilitated networks in which patients and professionals help each other by information exchange that should be funded as a membership fee. The physician or hospital that seeks to do everything for everyone can't survive with the current reimbursement model.

Sociotechnical disruptions that free patients to hide as humans in a Web 2.0 information world may serve to force the painful changes needed in the healthcare non-system. Technology that organically supports the humanity and sanity of the dyad will help. Physicians and their clients need to learn to discern, appreciate and favor approaches that are humanizing.

We contend that if the goals and needs of the dyad are respected and supported, a compensation system that supports those goals will necessarily emerge and be more cost effective. The result may be disruptive, but isn't that what is needed?

REFERENCES

Carkhuff, R. R. (1969). *Helping and Human Relations, Vols. I and II.* Holt, Rinehart and Winston, New York.

Christensen, C. M., Grossman, J. H., and Hwang, J. (2009). *The Innovator's Prescription, A Disruptive Solution for Health Care.* McGraw-Hill, New York.

Engel, G. L. (1977). "The need for a new medical model: A challenge for biomedicine." *Science* 196:129–136.

Health and Human Services (2009). CMS Proposes Definition of Meaningful Use of Certified Electronic Health Records (EHR) Technology. Centers for Medicare & Medicaid Services (CMS), Office of Public Affairs, Washington, DC. (http://www.cms.hhs.gov/apps/media/press/factsheet.asp?Counter=3564)

Health and Human Services (2010). The Health Information Technology for Economic and Clinical Health) Act. 42 CFR Parts 412, et al. Medicare and Medicaid Programs; Electronic Health Record Incentive Program; Proposed Rule, U.S. Department of Health and Human Services, Washington, DC.

HITECH (2009). Medicare and Medicaid Health Information Technology: Title IV of the American Recovery and Reinvestment Act. Centers for Medicare & Medicaid Services (CMS), Office of Public Affairs, Washington, DC. https://www.cms.hhs.gov/apps/media/press/factsheet.asp?Counter=3466&int NumPerPage=10&checkDate=&checkKey=&srchType=1&numDays=3500& srchOpt=0&srchData=&keywordType=All&chkNewsType=6&intPage=&sho wAll=&pYear=&year=&desc=&cboOrder=date.

McCraty, R., Atkinson, M., Tomasino, D., & Tiller, W. A. (1998). "The electricity of touch: Detection and measurement of cardiac energy exchange between people." In K. H. Pribram (Ed.), *Brain and values: Is a biological science of values possible?* (pp. 359-379). Mahwah, NJ: Lawrence Erlbaum Associates.

874

Patient-Centered Primary Care Collaborative (2007). Joint Principles of the Patient Centered Medical Home. American Academy of Family Physicians, American Academy of Pediatrics, American College of Physicians, American Osteopathic Association. http://www.pcpcc.net/content/joint-principles-patient-centered-medical-home.

Chapter 96

Sociotechnical Model of Inpatient Nursing Work System for Understanding Healthcare IT Innovation Diffusion

Renran Tian[1], Byung Cheol Lee[1], Jiyoung Park[1,3], Vincent G. Duffy[1,2]

1 School of Industrial Engineering
Purdue University
West Lafayette, IN 47906, USA

2 School of Agricultural and Biological Engineering
Purdue University
West Lafayette, IN 47906, USA

3 U-SCM Research Center
Dongguk University
82-1 Pil-dong, Jung-gu, Seoul, Korea

ABSTRACT

The conflicting implementation results of healthcare IT systems emphasizes the importance of further study about the change of work system during the implementation process, especially for nursing IT systems which are less studied. This study investigates the inpatient nursing job as one sociotechnical model to help understanding important individual and organizational factors that affect healthcare IT diffusion and continuance. One reciprocal relationship between nursing work system and IT diffusion will be proposed to demonstrate the potential connections and the process of confirmation when nurses are facing the implementation process. Initial data collection and analysis will be reported to facilitate any future works.

Keywords: Nursing Work System, Healthcare IT, Innovation Diffusion, Sociotechnical System

INTRODUCTION

Patient safety is the key issue for healthcare providers. Adverse drug events (ADE), including preventable medication errors, are a leading cause of patient harm. According to the Institute of Medicine report (Kohn and Corrigan, 2000) and Quality Interagency Coordination Task Force (QUIC) report (2000), IT systems are believed to help reduce existing medical problems and increase work efficiency as well as patient safety. QUIC report (2000) clearly demonstrated that "information technology has tremendous potential to reduce errors in health care by providing information when it is needed, providing clinical feedback, and alerting providers to potential problems".

Although much evidence and research has been proposed about the improvements of patient safety and working efficiency from healthcare IT systems implementation (Bates, 2000) (Bates, et al. 2001) (Johnson, et al., 2002) (Mekhjian, et al., 2002) (Kaushal, et al., 2001) (Kaushal, et al., 2003) (Paoletti, et al., 2007), problems and new medical errors related to these innovative IT systems have also been revealed and discussed (Patterson, et al., 2002) (Ash, et. al, 2004) (McDonald, 2006) (Mills, et al., 2006) (Halbesleben, et al., 2008). This conflicting result emphasizes the importance of understanding the change of working environment during healthcare IT innovation diffusion and the effects of these changes on IT system continuance. As described by Chaudhry et al. (2006), "more information is needed regarding the organizational change, workflow redesign, human factors, and project management issues involved with realizing benefits from healthcare information technology."

Nursing work system, as one sub-system of the overall medical environment, is more related to medication administration and patient care. Nurses play a unique role on medical process and work based on close cooperation and communication. In one hand, nursing job has many common features as other healthcare professionals; however, it also works based on its own principles. Comparing to physician support systems (for example, computerized order entry system) or healthcare management systems, nursing IT system as well as nursing work system are much less studied (Karsh and Holden, 2007).

In this study, we will investigate the features of nursing work system from the sociotechnical point of view to study how they will affect the nursing IT system diffusion and continuance. The reciprocal relationship between changes in nursing work system and IT innovation diffusion will be proposed. For future work, pool of potential influential features will be firstly constructed from interviews of nurses and pharmacists designed to fit aspects of sociotechnical models. Then during the implementation process of one computerized medication administration supporting system (BCMA, bar-coded medication administration), questionnaire-based analysis will be performed to distinguish the most influential features.

BACKGROUND AND MOTIVATION

SOCIOTECHNICAL SYSTEM AND ITS CHARACTERISTICS

Moray and Huey (1988) suggested that a sociotechnical system consists of several hierarchies. System or task is considered as base component and user or people who operate the system directly affect its performance. Organization and management infrastructure and environmental context provides the background of system or task operation. Especially, organization infrastructure with attitudinal and behavioral factors plays a significant role in the performance of the system and remarkably impacts its failure mode such as violation of a safety rule (Lawton and Parker, 1998). Such violation behavior also leads to inefficient work practice and error prone environment.

In spite of such structural classification, clear and decisive definition of a complex sociotechnical system hardly can be obtained. However, several characteristics are proposed: (a) large problem space, (b) social, (c) heterogeneous perspectives, (d) distributed, (e) dynamic, (f) potentially high hazards, (g) many coupled subsystems, (h) automated, (i) uncertain data, (j) mediated interaction via computers, and (k) disturbances (Vicente, 1999).

According to Carayon (2006), different sociotechnical models have been proposed to analyze those systems with high complexity and multiple boundaries, where health care is a typical sample. Along most of the dimensions of a work system, health care shows a higher level of complexity compared to traditional industry. Additionally, it contains diverse quality and service problems, its processes are complex and tangled with each other. Even simple treatment includes different and varying tasks, and various kinds of disturbance generate critical risk. Thus, the perspective of a sociotechnical approach is important to understand and solve the current healthcare system problems, and also it is crucial to recognize how the changes related to IT implementation will be reflected in various aspects.

MEDICATION PROCESS AND MEDICATION ADMINISTRATION OF NURSING JOB

As one essential part of overall medical process, improving medication process to reduce errors plays an important role in patient safety. According to Kaushal, Shaojania and Bates (2001), medication errors are defined as errors in drug ordering, transcribing, dispensing, administering, or monitoring. Bates (2000) summarized the computerized medication process following the sequence of prescribing (computerized physician order entry), transcription (electronic order transcription), dispensing (bar coding and automated dispensing), administration (bar coding), medication administration record (computerized medical record), and monitoring (computerized monitoring of ADE).

Along the chain steps of medication process, administration is very important. In one hand, it is the final chance to prevent the actual harm to patients from happening, and administration errors will be very difficult to intercept. On the other hand, there is a large proportion of medication errors happening in this step. According to Shane (2009), 38 percent of preventable medication errors occurred at the administration step, the frequency of administration errors ranges from 2.4% to 47.5% based on drug distribution system used; and in UK, 56.5 percent of reported errors associated with severe hard or death occurred at the administration step.

Because of its importance in improving patient safety, we choose medication administration process related to inpatient nursing job as our targeted healthcare environment to study. In US and most countries in the world, inpatient medication administration is primarily the responsibility of nurses. Although nurses' jobs are not limited to this, medication administration is the most important part. Also according to Shane (2009), nurses spend up to 40 percent of their time administering medications. Thus, the scope of this study is limited to the reciprocal relationship between work system changes and IT innovation diffusion within the inpatient nursing medication administration process.

CONCEPTUAL MODEL

INPATIENT MEDICATION ADMINISTRATION AS A SOCIOTECHNICAL MODEL

Healthcare is a complex sociotechnological system (STS). Among the various models for STS, the work system model developed by Carayon and Friesdorf (2006) has been proved to fit the healthcare environment very well. This model includes five elements and their complicated interactions among person, technology/tools, organization, tasks, and environment. Within the scope of this study, personnel involved are nurses and healthcare providers of nurses, physicians, pharmacists, nursing practitioners, and so on; also, medication administration process is composed of the tasks like order entry and modification, delivery, scheduling, monitoring, recording, and others. In such a complex system, communication and cooperation among these persons towards different tasks are essential.

In a typical healthcare system, the inpatient medication administration task has the features of large problem spaces, social system characteristics, heterogeneous perspective, and other systematic features including homeostatic activity, equifinality, dynamic and hazardous process, unstoppability, etc. (Carayon, 2006; Wears and Perry, 2007). The patient-staff-machine system of Marsolek and Friesdorf (2007) explains the uncertainty and highly dynamic features of the medication administration process. Instead of traditional human-machine interaction, this model also includes the interaction between patient and staff, and between patient and machine. Patient's efforts to come back to normal status and the provider's interventions based on patient status monitoring and recorded patient medical history together build up this non-linear process.

According to Wakefield and associates (2007), the patient safety during medication administration process is ensured and measured using a set of factors including the correctness of the treatment and its dose, route, time. Also, ensuring correct patient at correct location to be administered by correct person with awareness of enough information is important. Shane (2002) summarized the principles for safe medication administration as 5 rights: right patient, right drug, right dose, right route, and right time. The changes of work system factors correlated to IT innovation diffusion will be reflected in these measures.

NURSING IT ENVIRONMENT AND BAR-CODED MEDICATION ADMINISTRATION SYSTEM

Nursing informatics is related to the use of information management science and computer technology to enhance the use of nursing data. Based on the summary of healthcare IT functions provided by Menachemi (2008), the following figure 1 shows the inpatient nursing IT system functions and other related systems. Electronic Medical Record (EMR) is the basis for all IT systems working as the common database; thus, all data will be retrieved from EMR, and any changes made by clients will be updated in EMR. The two main functions for inpatient nursing IT are (1) bar-coded medical management and (2) nursing charting and care planning, and these two are main nursing tasks related to medication administration. Other important systems connecting to nursing systems are (1) CPOE to provide order information; (2) pharmacy systems to provide medication information and schedule; (3) laboratory information system to provide supplementary patient and medication information.

880

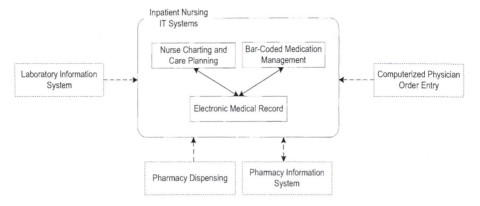

Figure 1. Inpatient nursing IT systems and related other IT systems

Bar-coded medication administration (BCMA) system is designed to aid the interaction between nurses and patients for medication administration process. This integrated system within the overall healthcare IT environment serves to assist two major parts of nursing jobs. Bar-coded medication management directly helps the delivery, storage, and administration of medications to improve patient safety and work efficiency. The basic idea is that bar coding can rapidly ensure the drug at hand is the one needed, when medical information related to patient bar code is compared with those of the medication bar code. More than this, current commercial BCMA system also includes functionalities to help with all other nursing jobs including charting, schedule planning, communication supporting, etc.

RECIPROCAL RELATIONSHIP OF WORK SYSTEM FACTORS AND IT INNOVATION DIFFUSION

Based on the work system model proposed by Carayon and Friesdorf (2006), inpatient nursing work system contains the elements of nurse, nursing tasks, medical tools and technologies, organization and environment, shown in figure 2. Here, organization refers to other nurses as well as other healthcare providers including pharmacists, physicians, nurse practitioners, etc. Environmental features may include interruptions, noise, crowded/messy environment, etc. Interactions among all these elements are essential for nursing work.

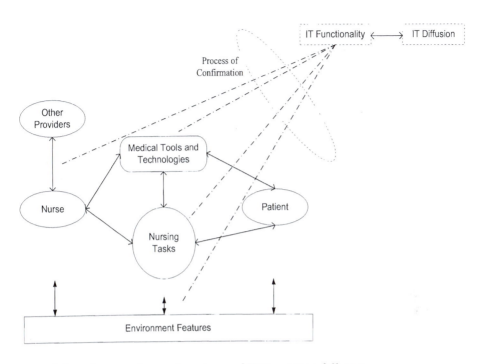

Figure 2. Inpatient nursing work system and IT innovation diffusion

Different from the original work system model, an inpatient nursing work system relies on the core interaction between nurses and patients. As described above, patients do not only act as the target of different tasks and medical devices, but also determine the medical process by their own status and historical data. Thus, figure 2 shows the element of patient and its interaction with nurses as well as tasks and tools in a nursing work system.

During the implementation of nursing IT innovation, some elements and interactions will be affected based on its functionalities. For example, the commercial BCMA system mentioned in previous subsection may change the way nurses cooperate with other providers, the tools and technologies need to be used, typical nursing tasks as well as many environmental features, shown as dotted lines in figure 2. However, not all these changes can be welcomed and are positive in improving nursing work system. Thus, it is important for nurses to compare these changes with the original work process as well as their expectation for confirmation, and the confirmation results will affect their attitudes on the new innovation. This can be summarized as the reciprocal relationship between nursing work system and healthcare IT innovation diffusion as: changes of inpatient work system caused by IT innovation diffusion will influence the IT diffusion process.

CONCLUSION AND FUTURE WORK

The proposed model is different from the traditional IT acceptance models mainly considering the features of IT system, features of users or individual/social cognitive processes. Because of the complexity of nursing work environment, work system features all around the nurses are proposed to be affected by and will affect the IT implementation process.

As one on-going study, the data collection and analysis has not been completed. Results will be presented at the conference. For future work, one urban hospital planning to adopt commercial BCMA system is selected for data collection. Interview policy based on the nursing work system model will be constructed, and collected interview data from nurses and pharmacists within different departments in the hospital will be analyzed to construct the large pool of potential work system features. After the IT implementation, these features will be tested based on online survey tools to find the most influential factors for IT acceptance.

ACKNOWLEDGEMENT

Authors would like to give special thanks to Steve R. Abel, Kyle Hultgren, and Carol Birk from School of Pharmacy and Pharmaceutical Sciences for their help in developing this study.

REFERENCES

Ash, J. S., Berg, M., and Enrico Coiera, (2004). Some Unintended Consequences of Information Technology in Health Care: the Nature of Patient Care Information System-related Errors, *Journal of the American Medical Informatics Association*, 11 (2).

Bates, D. W. (2000). Using information technology to reduce rates of medication errors in hospitals, British Medical Journal, 320 (7237).

Bates, D. W, Cohen, M., Leape, L. L., Overhage, J. M., Shabot, M. M., Sheridan, T. (2001). Reducing the Frequency of Errors in Medicine Using Information Technology, *Journal of the American Medical Informatics Association* 8 (4).

Carayon, P., and Smith, M. J., (2000). Work Organization and Ergonomics, *Applied Ergonomics*, 31, 649 - 662.

Carayon, P., (2006). Human Factors of Complex Sociotechnical Systems. *Applied Ergonomics*, 37 , 525 – 535.

Carayon, P., and Friesdorf, W., (2006). Human Factors and Ergonomics in Medicine. In G. Salvendy (Ed.), *Handbook of Human Factors and Ergonomics*, ,1517 – 1537.

Halbesleben, J. R. B., Wakefield, D. S., and Wakefield, B. J., (2008). Work-arounds in Health Care Settings: Literature Review and Research Agenda. *Health Care*

Management Review, 33 (1).

Johnson, C. L., Carlson, R. A., Tucker, C. L., and Willette, C., (2002). Using BCMA software to Improve Patient Safety in Veterans Administration Medical Centers. *Journal of Healthcare Information Management,* 16 (1).

Karsh, B., and Holden, R. J. (2007). "New Technology Implementation in Health Care", in P. Carayon (Ed.) Handbook of Human Factors and Ergonomics in Health Care and Patient Safety, pp. 393 – 410.

Kaushal, R., Barker, K. N., and Bates, D. W. (2001). How Can Information Technology Improve Patient Safety and Reduce Medication Errors in Children's Health Care?. *Archives of Pediatrics & Adolescent Medicine,* 155 (9).

Kaushal, R., Shaojania, K. G., and Bates, D. W. (2003). Effects of Computerized Physician Order Entry and Clinical Decision Support Systems on Medication Safety – a Systematic Review. *Archives of Internal Medicine*, 163 (12).

Kohn, L. T. and Corrigan, J. M., (2000). To Err is Human: Building a Safer Health System. From Committee on Quality of Health Care in America, Institute of Medicine. (Edited by Molla S. Donaldson). National Academy Press, Washington, D.C.

Lawton, R., Parker, D. (1998). Individual Difference in Accident Liability: A Review and Integrative Approach. *Human Factors*, 40.

Marsolek, I., and Friesdorf, W., (2007). "Work Systems and Process Analysis in Health Care", in P. Carayon, Handbook of Human Factors and Ergonomics in Health Care and Patient Safety, pp. 649 – 662.

McDonald, C. J. (2006). Computerization Can Create Safety Hazards: A Bar-Coding Near Miss. *Annals of Internal Medicine*, 144, 510 – 516.

Mekhjian, H. S., Kumar, R. R., Kuehn, L., Bentley, T. D., Teater, P., Thomas, A., Payne, B., and Ahmad, A. (2002). Immediate Benefits Realized Following Implementation of Physician Order Entry at an Academic Medical Center. *Journal of the American Medical Informatics Association*, 9 (5).

Menachemi, N., Chukmaitov, A. Saunders, C., and Trooks, R. G., (2008). Hospital Quality of Care: Does Information Technology Matter? The Relationship between Information Technology Adoption and Quality of Care. *Health Care Management Review*, 33 (1).

Mills, P. D., Neily, J., Mims, E., Burkhardt, M. E., and Bagian, J. (2006). Improving the Bar-Coded Medication Administration System at the Department of Veterans Affairs. *American Journal of Health-System Pharmacy,* 63, 1442 – 1447.

Moray, N. P., Huey, B. M. (1988). Human Factors Research and Nuclear Safety. National Academy Press, Washington, D.C.

Paoletti, R. D., Suess, T. M., Lesko, M. G., Feroli, A. A., Kennel, J. A., Mahler, J. M., and Sauders, T., (2007). Using Bar-code technology and medication observation methodology for safer medication administration. *American Journal of Health-System Pharmacy*, 64 (5).

Patterson, E. S., Cook, R., and Render, M. L. (2002). Improving Patient Safety by Identifying Side Effects from Introducing Bar Coding in Medication Administration. *Journal of the American Medical Informatics Association*, 9 (5).

Quality Interagency Coordination Task Force (QuIC), (2000). Doing What Counts for Patient Safety: Federal Actions to Reduce Medical Errors and Their Impact.

Shane, R., (2009). Current Status of Administration of Medicines. *American Journal of Health-System Pharmacy,* 66 (3).

Vicente, K. J. (1999). Cognitive work Analysis: Toward Safe, Productive, and Healthy Computer-Based work. Lawrence Erlbaum Associates, Mahwah, NJ.

Wakefield, D. S., Halbesleben, J.R.B., Ward, M. M., Qiu. Q., Brokel, J., and Crandall, D., (2007). Development of a Measure of Clinical Information Systems Expectations and Experiences. *Medical Care,* 45 (9).

Wears, R. L., and Perry, S., J., (2007). Human Factors and Ergonomics in the Emergency Department. In P. Carayon (Ed.). Handbook of Human Factors and Ergonomics in Health Care and Patient Safety, pp. 851 – 864.